Essentials of Mathematical Methods in Science and Engineering

Essentials of Mathematical Methods in Science and Engineering

SECOND EDITION

Selçuk Ş. Bayın

Institute of Applied Mathematics
Middle East Technical University
Ankara, Turkey

Registered Office
John Wiley & Sons, Inc., 111 River Street, Hoboken, NJ 07030, USA

Editorial Office
111 River Street, Hoboken, NJ 07030, USA

For details of our global editorial offices, customer services, and more information about Wiley products visit us at www.wiley.com.

Wiley also publishes its books in a variety of electronic formats and by print-on-demand. Some content that appears in standard print versions of this book may not be available in other formats.

Library of Congress Cataloging-in-Publication Data

Names: Bayın, Ş. Selçuk, 1951- author.
Title: Essentials of mathematical methods in science and engineering / Ş. Selçuk Bayın.
Description: Second edition. | Hoboken, NJ : Wiley, 2020. | Includes bibliographical references and index.
Identifiers: LCCN 2019027661 (print) | LCCN 2019027662 (ebook) | ISBN 9781119580249 (hardback) | ISBN 9781119580232 (adobe pdf) | ISBN 9781119580287 (epub)
Subjects: LCSH: Science–Mathematics. | Science–Methodology. | Engineering mathematics.
Classification: LCC Q158.5 .B39 2020 (print) | LCC Q158.5 (ebook) | DDC 501/.51–dc23
LC record available at https://lccn.loc.gov/2019027661
LC ebook record available at https://lccn.loc.gov/2019027662

Cover Design: Wiley
Cover Image: Courtesy of Selçuk Ş. Bayın

Set in 10/12pt CMR10 by SPi Global, Chennai, India

Printed in the United States of America

V10015769_112119

To my father,
Ömer Bayın

CONTENTS IN BRIEF

CONTENTS

Preface

After a year of freshman calculus, the basic mathematics training in science and engineering is usually completed during the second and the third years of the undergraduate curriculum. Students are usually required to take a sequence of three courses on the subjects of advanced calculus, differential equations, complex calculus, and introductory mathematical physics. Today, majority of the science and engineering departments are finding it convenient to use a single book that assures uniform formalism and a topical coverage in tune with their needs. The objective of *Essentials of Mathematical Methods in Science and Engineering* is to equip students with the basic mathematical skills required by majority of the science and engineering undergraduate programs.

The book gives a coherent treatment of the selected topics with a style that makes the essential mathematical skills easily accessible to a multidisciplinary audience. Since the book is written in modular format, each chapter covers its subject thoroughly and thus can be read independently. This also makes the book very useful for self-study and as reference or refresher for scientists. It is assumed that the reader has been exposed to two semesters of freshman calculus or has acquired an equivalent level of mathematical maturity.

The entire book contains a sufficient amount of material for a three-semester course meeting three to four hours a week. Respecting the disparity of the mathematics courses offered throughout the world, the topical coverage and the modular structure of the book make it versatile enough to be adopted for a number of different mathematics courses and allows instructors the flexibility to individualize their own teaching while maintaining the integrity and the uniformity of the discussions for their students.

About the Second Edition

The main aim of this book is to meet the demands of the majority of the modern undergraduate physics and engineering programs. It also aims to prepare students for a solid graduate program and establishes the groundwork of my graduate textbook *Mathematical Methods in Science and Engineering*, Wiley, second edition, 2018. The second edition, while maintaining all the successful features of the first edition, includes two new and extensive chapters (Chapters 6 and 7) entitled *Practical Linear Algebra* and *Applications of Linear Algebra*, respectively, and a computer file that includes Matlab codes.

The new chapters were developed and used as I taught linear algebra (3 hrs/week) and mathematical methods courses (3 + 1 h/wk) to engineering students. The file including the Matlab codes is self explanatory but assumes familiarity with the text in the book. These codes were used for the lab section of the mathematical methods course I taught to students with no prior Matlab experience. These codes will be available as open source in https://www.wiley.com, or in http://users.metu.edu.tr/bayin/.

In addition to these, numerous changes have been made to assure easy reading and smooth flow of the complex mathematical arguments. Derivations are given with sufficient detail so that the reader will not be distracted by searching for results in other parts of the book or by needing to write down equations. We have shown carefully selected keywords in boldface and framed key results so that the needed information can be located easily as the reader scans through the pages.

Chapter references are given at the end of each chapter with their full titles. Additional resources for the interested reader is listed at the back with respect to their subject matter. Our suggested references is by all means not meant to be complete. Nowadays, readers can locate additional references by a simple internet search. In particular, readers can use the websites: http://en.wikipedia.org and http://scienceworld.wolfram.com/. Of course, https://arxiv.org is an indispensable tool for researchers on any subject.

This book concentrates on analytic techniques. Computer programs like Mathematica® and Maple™ are capable of performing symbolic as well as numerical calculations. Even though they are extremely useful to scientists, one cannot stress enough the importance of a full grasp of the basic mathematical techniques with their intricacies and interdisciplinary connections. Only then the underlying unity and the beauty of the universe begins to appear. There are books specifically written for mathematical methods with these programs, some of which are included in our list for further reading at the back.

With their exclusive chapters and uniform level of formalism, this book connects with my graduate textbook *Mathematical Methods in Science and*

Engineering, Wiley, second edition, 2018, thus forming a complete set spanning a wide range of fundamental mathematical techniques for students, instructors, and researchers.

Summary of the Book

Chapter 1. Functional Analysis: This chapter aims to fill the gap between the introductory calculus and the advanced mathematical analysis courses. It introduces the basic techniques that are used throughout mathematics. Limits, derivatives, integrals, extremum of functions, implicit function theorem, inverse functions, and improper integrals are among the topics discussed.

Chapter 2. Vector Analysis: Since most of the classical theories can be introduced in terms of vectors, we present a rather detailed treatment of vectors and their techniques. Vector algebra, vector differentiation, gradient, divergence and curl operators, vector integration, Green's theorem, integral theorems, and the essential elements of the potential theory are among the topics covered.

Chapter 3. Generalized Coordinates and Tensors: Starting with the Cartesian coordinates, we discuss generalized coordinate systems and their transformations. Basis vectors, transformation matrix, line element, reciprocal basis vectors, covariant and contravariant components, differential operators in generalized coordinates, and introduction to Cartesian and general tensors are among the other essential topics of mathematical methods.

Chapter 4. Determinants and Matrices: A systematic treatment of the basic properties and methods of determinants and matrices that are much needed in science and engineering applications are presented here with examples.

Chapter 5. Linear Algebra: This chapter starts with a discussion of abstract linear spaces, also called vector spaces, and continues with systems of linear equations, inner product spaces, eigenvalue problems, quadratic forms, Hermitian matrices, and Dirac's bra and ket vectors.

Chapter 6. Practical Linear Algebra: In the previous chapter, we concentrate on the abstract properties of linear algebra. In this chapter, we introduce linear algebra from the practitioners point of view. In the first part, we start with systems of linear equations and discuss Gauss-Jordan reduction, row-echelon forms, elementary matrices, row space, column space and null space, rank and nullity, etc. In the second part, we introduce the numerical methods of linear algebra. Partial pivoting, LU-factorization, iteration method, interpolation, power method for eigenvalues, numerical integration, etc. are among the interesting topics discussed. In our accompanying website, we also have Matlab codes that the readers can experiment with the methods introduced in this chapter.

Chapter 7. Applications of Linear Algebra: This chapter introduces some of the important applications of linear algebra from different branches of science and engineering. We give examples from chemical engineering, linear programming, economics, geometry, elimination theory, coding theory, cryptography, and graph theory.

Chapter 8. Sequences and Series: This chapter starts with sequences and series of numbers and then introduces absolute convergence and tests for convergence. We then extend our discussion to series of functions and introduce the concept of uniform convergence. Power series and Taylor series are discussed in detail with applications.

Chapter 9. Complex Numbers and Functions: After the complex number system is introduced and their algebra is discussed, complex functions, complex differentiation, Cauchy–Riemann conditions and analytic functions are the main topics of this chapter.

Chapter 10. Complex Analysis: We introduce the complex integral theorems and discuss residues, Taylor series, and Laurent series along with their convergence properties.

Chapter 11. Ordinary Differential Equations: We start with the general properties of differential equations, their solutions, and boundary conditions. The most commonly encountered differential equations in applications are either first- or second-order ordinary differential equations. Hence, we discuss these two cases separately in detail and introduce methods of finding their analytic solutions. We also study linear equations of higher order. We finally conclude with the Frobenius method applied to first- and second-order differential equations with interesting and carefully selected examples.

Chapter 12. Second-Order Differential Equations and Special Functions: In this chapter, we discuss three of the most frequently encountered second-order differential equations of physics and engineering, that is, Legendre, Hermite, and Laguerre equations. We study these equations in detail from the viewpoint of the Frobenius method. By using the boundary conditions, we then show how the corresponding orthogonal polynomial sets are constructed. We also discuss how and under what conditions these polynomial sets can be used to represent a general solution.

Chapter 13. Bessel's Equation and Bessel Functions: Bessel functions are among the most frequently used special functions of mathematical physics. Since their orthogonality is with respect to their roots and not with respect to a parameter in the differential equation, they are discussed here separately in detail.

Chapter 14. Partial Differential Equations and Separation of Variables: Most of the second-order ordinary differential equations of physics and engineering are obtained from partial differential equations via the method of separation of variables. We introduce the most commonly encountered partial differential equations of physics and engineering and show how the method of separation of variables is used in Cartesian, spherical, and cylindrical coordinates. Interesting examples help the reader connect with the knowledge gained in the previous three chapters.

Chapter 15. Fourier Series: We first introduce orthogonal systems of functions and then concentrate on trigonometric Fourier series. We discuss their convergence and uniqueness properties along with specific examples.

Chapter 16. Fourier and Laplace Transforms: After a basic introduction to signal analysis and correlation functions, we introduce Fourier transforms and their inverses. We also introduce Laplace transforms and their applications to differential equations. We discuss methods of finding inverse Laplace transforms and their applications to transfer functions and signal processors.

Chapter 17. Calculus of Variations: We introduce basic variational analysis for different types of boundary conditions. Applications to Hamilton's principle and to Lagrangian mechanics is investigated in detail. The presence of constraints in dynamical systems along with the inverse problem is discussed with examples.

Chapter 18. Probability Theory and Distributions: Some of the interesting topics covered in this chapter include the basic theory of probability, permutations, and combinations, applications to statistical mechanics, and the connection with thermodynamics. We also discuss Bayes' theorem, random variables, distributions, distribution functions and probability, fundamental theorem of averages, moments, Chebyshev's theorem, and the law of large numbers.

Chapter 19. Information Theory: The first part of this chapter is devoted to classical information theory, where we discuss topics from Shannon's theory, decision theory, game theory, Nash equilibrium, and traveler's dilemma. The definition of Cbit and operations with them are also introduced. The second part of this chapter is on quantum information theory. After a general survey of quantum mechanics, we discuss Mach-Zehnder interferometer, Qbits, entanglement, and Bell states. Along with the no-cloning theorem, quantum cryptology, quantum dense coding, and quantum teleportation are among the other interesting topics discussed in this chapter. This chapter is written with a style that makes these interesting topics accessible to a wide range of audiences with minimum prior exposure to quantum mechanics.

Course Suggestions

Chapters 1–17 consist of the contents of the three, usually sequentially taught, core mathematical methods courses that most science and engineering departments require. These chapters also consist of the basic mathematical skills required by majority of the modern undergraduate science and engineering programs. Chapters 1–10 can be taught during the second year as a two-semester course. During the first or the second semester of the third year, a course composed of Chapters 11–17 can complete the sequence. Chapters 11–14 can also be used in a separate one-semester course on differential equations and special functions. Along with the Matlab codes, Chapters 4–7 can be used to design a separate course on linear algebra with a lab section. Instructors can also use these codes for in-class demonstrations with Matlab.

The two extensive chapters on probability theory and information theory (Chapters 18 and 19) are among the special chapters of the book. Even though most of the mathematical methods textbooks have chapters on probability, we have treated the subject with a style and level that prepares the reader for the following chapter on information theory. We have also included sections on applications to statistical mechanics and thermodynamics.

The chapter on information theory is unusual for the mathematical methods textbooks at both the undergraduate and the graduate levels. By selecting certain sections, Chapters 18 and 19 can be incorporated into the advanced undergraduate curriculum. In their entirety, they are more suitable to be used in a graduate course. Since we review the basic quantum mechanics needed, we require no prior exposure to quantum mechanics. In this regard, Chapter 19 is also designed to be useful to beginning researchers from a wide range of disciplines in science and engineering.

Examples and exercises are always an integral part of any learning process; hence, the topics are introduced with an ample number of examples. To maintain the continuity of the discussions, we have collected problems at the end of each chapter, where they are predominantly listed in the same order that they are discussed within the text. Therefore, we recommend that the entire problem sections be read quickly before their solutions are attempted. For communications about the book, we will use the website http://users.metu.edu.tr/bayin/.

Selçuk Ş. Bayın
METU/IAM
Ankara, Turkey
April 2019

Acknowledgments

For the first edition, I would like to thank Prof. J.P. Krisch of the University of Michigan for always being there whenever I needed advice and for sharing my excitement at all phases of the project. My special thanks go to Prof. J.C. Lauffenburger and Assoc. Prof. K.D. Scherkoske at Canisius College. I am grateful to Prof. R.P. Langlands of the Institute for Advanced Study at Princeton for his support and for his cordial and enduring contributions to METU culture. I am indebted to Prof. P.G.L. Leach for his insightful comments and for meticulously reading two of the chapters. I am grateful to Wiley for a grant to prepare the camera-ready copy, and I would like to thank my editor Susanne Steitz-Filler for sharing my excitement. My work on the two books *Mathematical Methods in Science and Engineering* and *Essentials of Mathematical Methods in Science and Engineering* has spanned an uninterrupted period of six years. With the time spent on my two books in Turkish published in the years 2000 and 2004, which were basically the forerunners of my first book, this project has dominated my life for almost a decade. In this regard, I cannot express enough gratitude to my darling young scientist daughter Sumru and beloved wife Adalet, for always being there for me during this long and strenuous journey, which also involved many sacrifices from them.

For the second edition, I am grateful to Prof. I. Tosun of the Middle East Technical University for bringing to my attention applications of linear algebra to chemistry and chemical engineering and for numerous discussions and for kindly reading the relevant sections of my book. I also thank Prof. A.S. Umar of the Vanderbilt University for comments and for using my book in his course. I thank Prof J. Krisch of the University of Michigan for continued support and encouragement. I thank our Chairman Prof. O. Ugur at the Institute of Applied Mathematics at METU for support. I also thank my editors Mindy Okura-Marszycki and Kathleen Santoloci and the publication team at Wiley who were most congenial and pleasant to work with. Last but not the least, I am grateful to my treasured wife Adalet and precious daughter Sumru, who is now a full fledged scientist in the field of stem cell research, for their everlasting love and care.

Selçuk Ş. Bayın

CHAPTER 1

FUNCTIONAL ANALYSIS

Functional analysis is the branch of mathematics that deals with spaces of functions and the transformation properties of functions between function spaces in terms of operators. Since these operators could be differential or integral, it makes functional analysis extremely useful in the study of differential and integral equations. Since the function and space concepts could be used to represent many different things, functional analysis has found a wide range of applications in science and engineering. It is also at the very foundation of numerical simulation. The most rudimentary concept of functional analysis is the definition of a function, which is basically a rule or a mapping that relates the members of one set of objects to the members of another set. In this chapter, we discuss the basic properties of functions like continuity, limit, convergence, inverse, differentiation, integration, etc.

1.1 CONCEPT OF FUNCTION

We start with a quick review of the basic concepts of **set theory**. Let S be a set of objects of any kind: points, numbers, functions, vectors, etc. When s

Essentials of Mathematical Methods in Science and Engineering, Second Edition. Selçuk Ş. Bayın.
© 2020 John Wiley & Sons, Inc. Published 2020 by John Wiley & Sons, Inc.

is an element of the set S, we show it as $s \in S$. For finite sets, we may define S by listing its elements as $S = \{s_1, s_2, \ldots, s_n\}$. For infinite sets, S is usually defined by a phrase describing the condition to be a member of the set, for example, $S = \{$All points on the sphere of radius $R\}$. When there is no room for confusion, we may also write an infinite set like the set of all odd numbers as $S = \{1, 3, 5, \ldots\}$. When each member of a set A is also a member of set B, we say that A is a **subset** of B and write $A \subset B$. The phrase B covers or contains A is also used. The **union** of two sets, $A \cup B$, consists of the elements of both A and B. The **intersection** of two sets, A and B, is defined as $A \cap B = \{$All elements common to A and $B\}$. When two sets have no common element, their intersection is called the **null set** or the **empty set**, which is usually shown by ϕ. The **neighborhood** of a point (x_1, y_1) in the xy-plane is the set of all points (x, y) inside a circle centered at (x_1, y_1) with the radius δ: $(x - x_1)^2 + (y - y_1)^2 < \delta^2$. An **open set** is defined as the set of points with neighborhoods entirely within the set. The interior of a circle defined by $x^2 + y^2 < 1$ is an open set. A **boundary point** is a point whose every neighborhood contains at least one point in the set and at least one point that does not belong to the set. The boundary of $x^2 + y^2 < 1$ is the set of points on the circumference, that is, $x^2 + y^2 = 1$. An open set plus its boundary is a **closed set**.

A **function** f is in general a rule or a relation that uniquely associates members of one set A with the members of another set B. The concept of function is essentially the same as that of **mapping**, which in general is so broad that it allows mathematicians to work with them without any resemblance to the simple class of functions with numerical values. The set A that f acts upon is called the **domain**, and the set B composed of the elements that f can produce is called the **range**. For **single-valued** functions, the common notation used is

$$f : x \rightarrow f(x).$$

Here, f stands for the function or the mapping that acts upon a single number x, which is an element of the domain, and produces $f(x)$, which is an element of the range. In general, f refers to the function itself, and $f(x)$ refers to the value it returns. However, in practice, $f(x)$ is also used to refer to the function itself. In this chapter, we basically concern ourselves with functions that take numerical values as $f(x)$, where the **argument** x is called the **independent variable**. We usually define a new variable y as $y = f(x)$, which is called the **dependent variable**.

Functions with multiple variables, that is, **multivariate** functions, can also be defined. For example, for each point (x, y) in some region of the xy-plane, we may assign a unique real number $f(x, y)$ according to the rule $f : (x, y) \rightarrow f(x, y)$. We now say that $f(x, y)$ is a function of two independent variables as x and y. In applications, $f(x, y)$ may represent physical properties like the temperature or the density distribution of a flat disc with negligible thickness. Definition of a function can be extended to cases with several independent variables as $f(x_1, \ldots, x_n)$, where n stands for the number of independent variables.

The term function is also used for the objects that associate more than one element in the domain to a single element in the range. Such objects are called **multiple-to-one** relations. For example,

$$
\begin{aligned}
f(x,y) &= 2xy + x^2 : &&\text{single-valued or one-to-one,}\\
f(x) &= \sin x : &&\text{many-to-one,}\\
f(x,y) &= x + x^2 : &&\text{single-valued,}\\
f(x) &= x^2,\ x \neq 0 : &&\text{two-to-one,}\\
f(x,y) &= \sin xy : &&\text{many-to-one.}
\end{aligned}
$$

Sometimes the term "function" is also used for relations that map a single point in its domain to multiple points in its range. As we shall discuss in Chapters 9 and 10, such functions are called **multivalued functions**, which are predominantly encountered in **complex analysis**.

1.2   CONTINUITY AND LIMITS

Similar to its usage in everyday language, the word continuity in mathematics also implies the absence of abrupt changes. In astrophysics, pressure and density distributions inside a solid neutron star are represented by continuous functions of the radial position as $P(r)$ and $\rho(r)$, respectively. This means that small changes in the radial position inside the star also result in small changes in the pressure and density. At the surface R, where the star meets the outside vacuum, pressure has to be continuous. Otherwise, there will be a net force on the surface layer, which will violate the static equilibrium condition. In this regard, in static neutron star models, pressure has to be a monotonic decreasing function of r, which smoothly drops to zero at the surface, that is, $P(R) = 0$. On the other hand, the density at the surface can change abruptly from a finite value to zero. This is also in line with our everyday experiences, where solid objects have sharp contours marked by density discontinuities. For gaseous stars, both pressure and density have to vanish continuously at the surface. In constructing physical models, deciding on which parameters are going to be taken as continuous at the boundaries requires physical reasoning and some insight. Usually, a collection of rules that have to be obeyed at the boundaries is called the **junction conditions** or the **boundary conditions**.

We are now ready to give a formal definition of continuity as follows:

Continuity: A numerically valued function $f(x)$ defined in some domain D is said to be continuous at the point $x_0 \in D$, if for any positive number $\varepsilon > 0$, there is a neighborhood N about x_0 such that $|f(x) - f(x_0)| < \varepsilon$ for every point common to both N and D, that is, $N \cap D$. If the function $f(x)$ is continuous at every point of D, we say it is continuous in D.

We finally quote the following theorems without proof [1, 2]:

Theorem 1.1. Let $f(x)$ be a continuous function at x, and let $\{x_n\}$ be a sequence of points in the domain of $f(x)$ with the limit $x_n \to x$ as $n \to \infty$, then the following is true:

$$\lim_{n\to\infty} f(x_n) \to f(x). \tag{1.1}$$

Theorem 1.2. For a function $f(x)$ defined in D, if the limit $f(x_n) \to f(x)$ as $n \to \infty$ exists whenever $x_n \in D$ and

$$\lim_{n\to\infty} x_n \to x \in D, \tag{1.2}$$

then the function $f(x)$ is continuous at x. For the limit $f(x_n) \to f(x)$ as $n \to \infty$ to exist, it is sufficient to show that the right and the left limits agree, that is,

$$\lim_{\varepsilon\to 0} f(x - \varepsilon) = \lim_{\varepsilon\to 0} f(x + \varepsilon), \tag{1.3}$$

$$f(x^-) = f(x^+) = f(x). \tag{1.4}$$

In practice, the second theorem is more useful in showing that a given function is continuous. If a function is discontinuous at a finite number of points in its interval of definition $[x_a, x_b]$, it is called **piecewise continuous**.

Generalization of these theorems to multivariate functions is easily accomplished by taking x to represent a point in a space with n independent variables as $x = (x_1, x_2, \ldots, x_n)$. However, with more than one independent variable, one has to be careful. Consider the simple function

$$F(x) = \frac{x^2 - y^2}{x^2 + y^2}, \tag{1.5}$$

which is finite at the origin. However, depending on the direction of approach to the origin, $f(x, y)$ takes different values:

$$\lim_{(x,y)\to(0,0)} f(x, y) \to 0 \text{ if we approach along the } y = x \text{ line,}$$

$$\lim_{(x,y)\to(0,0)} f(x, y) \to 0 \text{ if we approach along the } x \text{ axis,}$$

$$\lim_{(x,y)\to(0,0)} f(x, y) \to -1 \text{ if we approach along the } y \text{ axis.}$$

Hence, the limit $\lim_{(x,y)\to(0,0)} f(x, y)$ does not exist, and the function $f(x, y)$ is not continuous at the origin.

Basic properties of limits, which we give for functions with two variables, also hold for a general multivariate function: Let $u = f(x, y)$ and $v = g(x, y)$ be two functions defined in the domain D of the xy-plane.

Limit: If the limits

$$\lim_{(x,y)\to(x_0,y_0)} f(x, y) = f_0 \quad \text{and} \quad \lim_{(x,y)\to(x_0,y_0)} g(x, y) = g_0 \tag{1.6}$$

exist, then we can write

$$\lim_{(x,y)\to(x_0,y_0)} [f(x,y) + g(x,y)] = f_0 + g_0, \tag{1.7}$$

$$\lim_{(x,y)\to(x_0,y_0)} [f(x,y) \cdot g(x,y)] = f_0 \cdot g_0, \tag{1.8}$$

$$\lim_{(x,y)\to(x_0,y_0)} \left[\frac{f(x,y)}{g(x,y)}\right] = \frac{f_0}{g_0}, \quad g_0 \neq 0. \tag{1.9}$$

If the functions $f(x,y)$ and $g(x,y)$ are continuous at (x_0, y_0), then the functions

$$f(x,y) + g(x,y), \ f(x,y)g(x,y), \ \text{and} \ \frac{f(x,y)}{g(x,y)} \tag{1.10}$$

are also continuous at (x_0, y_0), provided that in the last case, $g(x,y)$ is different from zero at (x_0, y_0).

Let $F(u, v)$ be a continuous function defined in some domain D_0 of the uv-plane, and let $F(f(x,y), g(x,y))$ be defined for (x,y) in D. Then, if (f_0, g_0) is in D_0, we can write

$$\lim_{(x,y)\to(x_0,y_0)} F(f(x,y), g(x,y)) = F(f_0, g_0). \tag{1.11}$$

If $f(x,y)$ and $g(x,y)$ are continuous at (x_0, y_0), then so is $F(f(x,y), g(x,y))$. In evaluating limits of functions that can be expressed as ratios, L'Hôpital's rule is very useful.

L'Hôpital's rule: Let f and g be differentiable functions in the interval $a \leq x < b$ with $g'(x) \neq 0$, where the upper limit b could be finite or infinite. If f and g have the limits

$$\lim_{x\to b} f(x) = 0 \quad \text{and} \quad \lim_{x\to b} g(x) = 0, \tag{1.12}$$

or

$$\lim_{x\to b} f(x) = \infty \quad \text{and} \quad \lim_{x\to b} g(x) = \infty, \tag{1.13}$$

and if the limit

$$\boxed{\lim_{x\to b} \frac{f'(x)}{g'(x)} = L} \tag{1.14}$$

exists, where L could be zero or infinity, then

$$\boxed{\lim_{x\to b} \frac{f(x)}{g(x)} = L.} \tag{1.15}$$

1.3 PARTIAL DIFFERENTIATION

A necessary and sufficient condition for the derivative of $f(x)$ to exist at x_0 is that the left, $f'_-(x_0)$, and the right, $f'_+(x_0)$, derivatives exist and be equal (Figure 1.1), that is,

$$f'_+(x_0) = f'_-(x_0), \tag{1.16}$$

where

$$f'_+(x_0) = \lim_{\Delta x \to 0} \frac{f(x_0 + \Delta x) - f(x_0)}{\Delta x}, \tag{1.17}$$

$$f'_-(x_0) = \lim_{\Delta x \to 0} \frac{f(x_0) - f(x_0 - \Delta x)}{\Delta x}. \tag{1.18}$$

When the derivative exists, we always mean a finite derivative. If $f(x)$ has derivative at x_0, it means that it is continuous at that point. When the derivative of $f(x)$ exists at every point in the interval (a, b), we say that $f(x)$ is differentiable in (a, b) and write its derivative as

$$\frac{df(x)}{dx} \text{ or } f'(x). \tag{1.19}$$

Geometrically, derivative at a point is the **slope** of the tangent line at that point:

$$\tan \theta = \frac{df}{dx} = \lim_{\Delta x \to 0} \frac{\Delta f}{\Delta x}. \tag{1.20}$$

When a function depends on two variables, $z = f(x, y)$, the partial derivative with respect to x at (x_0, y_0) is defined as the limit

$$\lim_{\Delta x \to 0} \frac{f(x_0 + \Delta x, y_0) - f(x_0, y_0)}{\Delta x} \tag{1.21}$$

and we show it as in one of the following forms:

$$\frac{\partial f}{\partial x}(x_0, y_0), \ f_x(x_0, y_0), \ \frac{\partial z}{\partial x}(x_0, y_0), \text{ or } \left(\frac{\partial f}{\partial x}\right)_0. \tag{1.22}$$

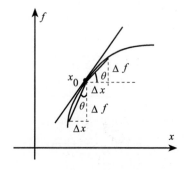

Figure 1.1 Derivative is the slope of the tangent line.

Similarly, the partial derivative with respect to y at (x_0, y_0) is defined as

$$\frac{\partial f}{\partial y}(x_0, y_0) = \lim_{\Delta y \to 0} \frac{f(x_0, y_0 + \Delta y) - f(x_0, y_0)}{\Delta y}. \tag{1.23}$$

A geometric interpretation of the partial derivative is that the section of the surface $z = f(x, y)$ with the plane $y = y_0$ is the curve $z = f(x, y_0)$; hence, the partial derivative $f_x(x_0, y_0)$ is the slope of the tangent line (Figure 1.2) to $z = f(x, y_0)$ at (x_0, y_0). Similarly, the partial derivative $f_y(x_0, y_0)$ is the slope of the tangent line to the curve $z = f(x_0, y)$ at (x_0, y_0). For a multivariate function, the partial derivative with respect to the ith independent variable is defined as

$$\frac{\partial f(x_1, \ldots, x_i, \ldots, x_n)}{\partial x_i}$$

$$= \lim_{\Delta x_i \to 0} \frac{f(x_1, \ldots, x_i + \Delta x_i, \ldots, x_n) - f(x_1, \ldots, x_i, \ldots, x_n)}{\Delta x_i}. \tag{1.24}$$

For a given $f(x, y)$, the partial derivatives f_x and f_y are functions of x and y, and they also have partial derivatives which are written as

$$f_{xx} = \frac{\partial^2 f}{\partial x^2} = \frac{\partial}{\partial x}\left[\frac{\partial f}{\partial x}\right], \quad f_{yy} = \frac{\partial^2 f}{\partial y^2} = \frac{\partial}{\partial y}\left[\frac{\partial f}{\partial y}\right], \tag{1.25}$$

$$f_{xy} = \frac{\partial^2 f}{\partial x \partial y} = \frac{\partial}{\partial x}\left[\frac{\partial f}{\partial y}\right], \quad f_{yx} = \frac{\partial^2 f}{\partial y \partial x} = \frac{\partial}{\partial y}\left[\frac{\partial f}{\partial x}\right]. \tag{1.26}$$

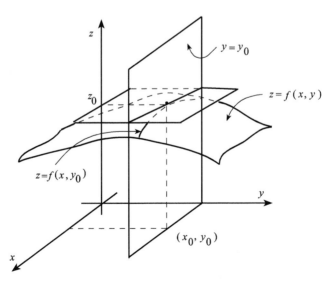

Figure 1.2   Partial derivative f_x is the slope of the tangent line to $z = f(x, y_0)$.

When f_{xy} and f_{yx} are continuous at (x_0, y_0), then the relation $f_{xy} = f_{yx}$ holds at (x_0, y_0). Under similar conditions, this result can be extended to cases with more than two independent variables and to higher order mixed partial derivatives.

1.4 TOTAL DIFFERENTIAL

When a function depends on two or more variables, we have seen that the limit at a point may depend on the direction of approach. Hence, it is important that we introduce a nondirectional derivative for functions with several variables. Given the function

$$f(x, y, z) = xz - y^2, \tag{1.27}$$

for a displacement of $\Delta r = (\Delta x, \Delta y, \Delta z)$, we can write its new value as

$$f(r + \Delta r) = (x + \Delta x)(z + \Delta z) - (y + \Delta y)^2 \tag{1.28}$$

$$= xz + x\Delta z + z\Delta x + \Delta x\Delta z - y^2 - 2y\Delta y - (\Delta y)^2 \tag{1.29}$$

$$= (xz - y^2) + (z\Delta x - 2y\Delta y + x\Delta z) + \Delta x\Delta z - (\Delta y)^2, \tag{1.30}$$

where r stands for the point (x, y, z), and Δr is the displacement $(\Delta x, \Delta y, \Delta z)$. For small Δr, the change in $f(x, y, z)$ to first order can be written as

$$\Delta f \simeq f(r + \Delta r) - f(r) = (xz - y^2) + (z\Delta x - 2y\Delta y + x\Delta z) - (xz - y^2), \tag{1.31}$$

$$\Delta f \simeq z\Delta x - 2y\Delta y + x\Delta z. \tag{1.32}$$

Considering that the first-order partial derivatives of f are given as

$$\frac{\partial f}{\partial x} = z, \quad \frac{\partial f}{\partial y} = -2y, \quad \frac{\partial f}{\partial z} = x. \tag{1.33}$$

Equation (1.32) is nothing but

$$\Delta f \simeq \frac{\partial f}{\partial x}\Delta x + \frac{\partial f}{\partial y}\Delta y + \frac{\partial f}{\partial z}\Delta z. \tag{1.34}$$

In general, if a function $f(x, y, z)$ is differentiable at (x, y, z) in some domain D with the partial derivatives

$$\frac{\partial f}{\partial x}, \frac{\partial f}{\partial y}, \frac{\partial f}{\partial z}, \tag{1.35}$$

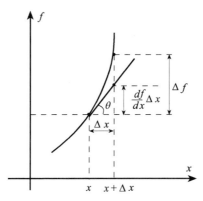

Figure 1.3 Total differential gives a local approximation to the change in a function.

then the change in $f(x, y, z)$ in D to first order in $(\Delta x, \Delta y, \Delta z)$ can be written as

$$\Delta f \simeq \frac{\partial f}{\partial x} \Delta x + \frac{\partial f}{\partial y} \Delta y + \frac{\partial f}{\partial z} \Delta z. \tag{1.36}$$

In the limit as $\Delta r \to 0$, we can write Eq. (1.36) as

$$df = \frac{\partial f}{\partial x} dx + \frac{\partial f}{\partial y} dy + \frac{\partial f}{\partial z} dz, \tag{1.37}$$

which is called the **total differential** of $f(x, y, z)$. In the case of a function with one variable, $f(x)$, the differential reduces to

$$\Delta f \simeq \frac{df}{dx} \Delta x, \tag{1.38}$$

which gives the local approximation to the change in the function at the point x via the value of the tangent line (Figure 1.3) at that point. The smaller the value of Δx, the better the approximation. In cases with several independent variables, Δf is naturally approximated by using the tangent plane at that point.

1.5 TAYLOR SERIES

The Taylor series of a function about x_0, when it exists, is given as

$$f(x) = \sum_{n=0}^{\infty} \frac{a_n}{n!} (x - x_0)^n = a_0 + a_1(x - x_0) + \frac{a_2}{2!} (x - x_0)^2 + \cdots. \tag{1.39}$$

To evaluate the coefficients, we differentiate repeatedly and set $x = x_0$ to find

$$f(x_0) = a_0,$$
$$f'(x_0) = a_1,$$
$$f''(x_0) = a_2, \tag{1.40}$$
$$\vdots$$
$$f^{(n)}(x_0) = a_n,$$

where

$$f^{(n)}(x_0) = \left(\frac{d^n f}{dx^n}\right)_{x_0} \tag{1.41}$$

and the zeroth derivative is defined as the function itself, that is,

$$f^{(0)}(x) = f(x). \tag{1.42}$$

Hence, the **Taylor series** of a function with a **single variable** is written as

$$\boxed{f(x) = \sum_{n=0}^{\infty} \frac{1}{n!} \left(\frac{d^n f}{dx^n}\right)_{x_0} (x - x_0)^n.} \tag{1.43}$$

This formula assumes that $f(x)$ is infinitely differentiable in an open domain including x_0. Functions that are equal to their **Taylor series** in the neighborhood of any point x_0 in their domain are called **analytic functions**. Taylor series about $x_0 = 0$ are called **Maclaurin series**.

Using the Taylor series, we can approximate a given differentiable function in the neighborhood of x_0 to orders beyond the linear term in Eq. (1.38). For example, to second order we obtain

$$f(x_0 + \Delta x) \simeq f(x_0) + \left(\frac{df}{dx}\right)_{x_0} \Delta x + \frac{1}{2}\left(\frac{d^2 f}{dx^2}\right)_{x_0} (\Delta x)^2, \tag{1.44}$$

$$f(x_0 + \Delta x) - f(x_0) \simeq \left(\frac{df}{dx}\right)_{x_0} \Delta x + \frac{1}{2}\left(\frac{d^2 f}{dx^2}\right)_{x_0} (\Delta x)^2, \tag{1.45}$$

$$\Delta^{(2)} f(x_0) = \left(\frac{df}{dx}\right)_{x_0} \Delta x + \frac{1}{2}\left(\frac{d^2 f}{dx^2}\right)_{x_0} (\Delta x)^2. \tag{1.46}$$

Since x_0 is any point in the open domain that the Taylor series exists, we can drop the subscript in x_0 and write

$$\Delta^{(2)} f(x) = \frac{df}{dx}\Delta x + \frac{1}{2}\frac{d^2 f}{dx^2}(\Delta x)^2, \tag{1.47}$$

where $\Delta^{(2)}f$ denotes the differential of f to the second order. Higher order differentials are obtained similarly.

The **Taylor series** of a function depending on several independent variables is also possible under similar conditions, and in the case of **two independent variables**, it is given as

$$f(x,y) = \sum_{n=0}^{\infty} \frac{1}{n!} \left[(x-x_0)\frac{\partial}{dx} + (y-y_0)\frac{\partial}{dy} \right]^n f(x_0,y_0), \qquad (1.48)$$

where the derivatives are to be evaluated at (x_0, y_0). For functions with two independent variables and to second order in the neighborhood of (x, y), Eq. (1.48) gives

$$f(x+\Delta x, y+\Delta y) \simeq f(x,y) + \left[\frac{\partial f}{\partial x}\Delta x + \frac{\partial f}{\partial y}\Delta y \right]$$
$$+ \frac{1}{2}\left[\Delta x\frac{\partial}{dx} + \Delta x\frac{\partial}{dy} \right]^2 f(x,y), \qquad (1.49)$$

which yields the differential, $\Delta^{(2)}f(x,y) = f(x+\Delta x, y+\Delta y) - f(x,y)$, as

$$\Delta^{(2)}f(x,y) = \left[\frac{\partial f}{\partial x}\Delta x + \frac{\partial f}{\partial y}\Delta y \right]$$
$$+ \frac{1}{2}\frac{\partial^2 f}{\partial x^2}(\Delta x)^2 + \frac{\partial^2 f}{\partial x \partial y}\Delta x \Delta y + \frac{1}{2}\frac{\partial^2 f}{\partial y^2}(\Delta y)^2 \qquad (1.50)$$

$$= \Delta^{(1)}f(x,y) + \left[\frac{1}{2}\frac{\partial^2 f}{\partial x^2}(\Delta x)^2 + \frac{\partial^2 f}{\partial x \partial y}\Delta x \Delta y + \frac{1}{2}\frac{\partial^2 f}{\partial y^2}(\Delta y)^2 \right]. \qquad (1.51)$$

For the higher order terms, note how the powers in Eq. (1.48) are expanded. Generalization to n independent variables is obvious.

Example 1.1. *Partial derivatives:* Consider the function

$$z(x,y) = xy^2 + e^x. \qquad (1.52)$$

Partial derivatives are written as

$$\frac{\partial z}{\partial x} = y^2 + e^x, \qquad (1.53)$$

$$\frac{\partial z}{\partial y} = 2xy, \qquad (1.54)$$

$$\frac{\partial}{\partial x}\left(\frac{\partial z}{\partial x}\right) = \frac{\partial^2 z}{\partial x^2} = e^x, \tag{1.55}$$

$$\frac{\partial}{\partial y}\left(\frac{\partial z}{\partial y}\right) = \frac{\partial^2 z}{\partial y^2} = 2x, \tag{1.56}$$

$$\frac{\partial}{\partial y}\left(\frac{\partial z}{\partial x}\right) = \frac{\partial^2 z}{\partial y \partial x} = 2y, \tag{1.57}$$

$$\frac{\partial}{\partial x}\left(\frac{\partial z}{\partial y}\right) = \frac{\partial^2 z}{\partial x \partial y} = 2y. \tag{1.58}$$

Example 1.2. Taylor series: Using the partial derivatives obtained in the previous example, we can write the first two terms of the Taylor series [Eq. (1.48)] of $z = xy^2 + e^x$ about the point $(0,1)$. First, the required derivatives at $(0,1)$ are evaluated as

$$z(0,1) = 1, \tag{1.59}$$

$$\left(\frac{\partial z}{\partial x}\right)_{(0,1)} = 2, \tag{1.60}$$

$$\left(\frac{\partial z}{\partial y}\right)_{(0,1)} = 0, \tag{1.61}$$

$$\left[\frac{\partial}{\partial x}\left(\frac{\partial z}{\partial x}\right)\right]_{(0,1)} = \left(\frac{\partial^2 z}{\partial x^2}\right)_{(0,1)} = 1, \tag{1.62}$$

$$\left[\frac{\partial}{\partial y}\left(\frac{\partial z}{\partial y}\right)\right]_{(0,1)} = \left(\frac{\partial^2 z}{\partial y^2}\right)_{(0,1)} = 0, \tag{1.63}$$

$$\left[\frac{\partial}{\partial y}\left(\frac{\partial z}{\partial x}\right)\right]_{(0,1)} = \left(\frac{\partial^2 z}{\partial y \partial x}\right)_{(0,1)} = 2, \tag{1.64}$$

$$\left[\frac{\partial}{\partial x}\left(\frac{\partial z}{\partial y}\right)\right]_{(0,1)} = \left(\frac{\partial^2 z}{\partial x \partial y}\right)_{(0,1)} = 2. \tag{1.65}$$

Using these derivatives, we can write the first two terms of the Taylor series about the point $(0,1)$ as

$$z(x,y) = z(0,1) + \left(\frac{\partial z}{\partial x}\right)_0 x + \left(\frac{\partial z}{\partial y}\right)_0 (y-1)$$

$$+ \frac{1}{2}\left(\frac{\partial^2 z}{\partial x^2}\right)_0 x^2 + \left(\frac{\partial^2 z}{\partial x \partial y}\right)_0 x(y-1) + \frac{1}{2}\left(\frac{\partial^2 z}{\partial y^2}\right)_0 (y-1)^2 + \cdots \tag{1.66}$$

$$= 1 + 2x + 0(y-1) + \frac{1}{2}x^2 + 2x(y-1) + \frac{1}{2}0(y-1)^2 + \cdots \tag{1.67}$$

$$= 1 + 2x + \frac{1}{2}x^2 + 2x(y - 1) + \cdots , \tag{1.68}$$

where the subscript 0 indicates that the derivatives are to be evaluated at the point $(0, 1)$. To find $\Delta^{(2)}z(0, 1)$, which is good to the second order, we first write

$$\Delta^{(1)}z(0, 1) = \left(\frac{\partial z}{\partial x}\right)_0 \Delta x + \left(\frac{\partial z}{\partial y}\right)_0 \Delta y = 2\Delta x \tag{1.69}$$

and then obtain

$$\Delta^{(2)}z(0, 1) = \Delta^{(1)}z(0, 1)$$

$$+ \frac{1}{2}\left(\frac{\partial^2 z}{\partial x^2}\right)_0 (\Delta x)^2 + \left(\frac{\partial^2 z}{\partial x \partial y}\right)_0 \Delta x \, \Delta y + \frac{1}{2}\left(\frac{\partial^2 z}{\partial y^2}\right)_0 (\Delta y)^2 \tag{1.70}$$

$$= 2\Delta x + \frac{1}{2}(\Delta x)^2 + 2\Delta x \, \Delta y. \tag{1.71}$$

1.6   MAXIMA AND MINIMA OF FUNCTIONS

We are frequently interested in the maximum or the minimum values that a function $f(x)$ attains in a closed domain $[a, b]$. The **absolute maximum** M_1 is the value of the function at some point x_0, if the inequality $M_1 = f(x_0) \geq f(x)$ holds for all x in $[a, b]$. An **absolute minimum** is also defined similarly. In general, we can quote the following theorem (Figure 1.4):

Theorem 1.3. If a function $f(x)$ is continuous in the closed interval $[a, b]$, then it possesses an absolute maximum M_1 and an absolute minimum M_2 in that interval.

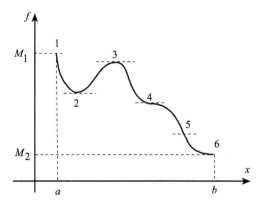

Figure 1.4   Maximum and minimum points of a function.

Proof of this theorem requires a rather detailed analysis of the real number system. On the other hand, we are usually interested in the **extremum** values, that is, the **local maximum** or the **minimum** values of a function. Operationally, we can determine whether a given point x_0 corresponds to an extremum or not by looking at the change or the variation in the function in the neighborhood of x_0. The total differential introduced in the previous sections is just the tool needed for this. We have seen that in one dimension, we can write the first, $\Delta f^{(1)}$, the second, $\Delta^{(2)} f$, and the third, $\Delta^{(3)} f$, differentials of a function with single independent variable as

$$\Delta^{(1)} f(x_0) = \left(\frac{df}{dx}\right)_{x_0} \Delta x, \tag{1.72}$$

$$\Delta^{(2)} f(x_0) = \left(\frac{df}{dx}\right)_{x_0} \Delta x + \frac{1}{2}\left(\frac{d^2 f}{dx^2}\right)_{x_0} (\Delta x)^2, \tag{1.73}$$

$$\Delta^{(3)} f(x_0) = \left(\frac{df}{dx}\right)_{x_0} \Delta x + \frac{1}{2}\left(\frac{d^2 f}{dx^2}\right)_{x_0} (\Delta x)^2 + \frac{1}{3!}\left(\frac{d^3 f}{dx^3}\right)_{x_0} (\Delta x)^3. \tag{1.74}$$

Extremum points are defined as the points where the first differential vanishes, which means

$$\left(\frac{df}{dx}\right)_{x_0} = 0. \tag{1.75}$$

In other words, the tangent line at an extremum point is horizontal (Figure 1.5a,b). In order to decide whether an extremum point corresponds to a local maximum or minimum, we look at the second differential:

$$\Delta^{(2)} f(x_0) = \frac{1}{2}\left(\frac{d^2 f}{dx^2}\right)_{x_0} (\Delta x)^2. \tag{1.76}$$

For a local maximum, the function decreases for small displacements about the extremum point (Figure 1.5a), which implies $\Delta^{(2)} f(x_0) < 0$. For a local minimum, a similar argument yields $\Delta^{(2)} f(x_0) > 0$. Thus, we obtain the following criteria:

$$\boxed{\left(\frac{df}{dx}\right)_{x_0} = 0 \text{ and } \left(\frac{d^2 f}{dx^2}\right)_{x_0} < 0 \quad \text{for a local maximum}} \tag{1.77}$$

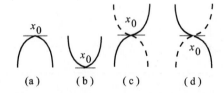

(a) (b) (c) (d)

Figure 1.5 Analysis of critical points.

and

$$\left(\frac{df}{dx}\right)_{x_0} = 0 \text{ and } \left(\frac{d^2f}{dx^2}\right)_{x_0} > 0 \quad \text{for a local minimum.} \tag{1.78}$$

In cases where the second derivative also vanishes, we look at the third differential $\Delta^{(3)}f(x_0)$. We now say that we have an **inflection point**; and depending on the sign of the third differential, we have either the third or the fourth shape in Figure 1.5.

Consider the function $f(x) = x^3$, where the first derivative, $f'(x) = 3x^2$, vanishes at $x_0 = 0$. However, the second derivative, $f''(x) = 6x$, also vanishes there, thus making $x_0 = 0$ a point of inflection. From the third differential:

$$\Delta^{(3)}f(x_0) = \frac{1}{3!}\left(\frac{d^3f}{dx^3}\right)_{x_0}(\Delta x)^3 = \frac{1}{3!}6(\Delta x)^3, \tag{1.79}$$

we see that $\Delta^{(3)}f(x_0) > 0$ for $\Delta x > 0$ and $\Delta^{(3)}f(x_0) < 0$ for $\Delta x < 0$. Thus, we choose the third shape in Figure 1.5 and plot $f(x) = x^3$ as in Figure 1.6. Points where the first derivative of a function vanishes are called the **critical points**.

Usually the **potential** in one-dimensional conservative systems can be represented by a (scalar) function $V(x)$. Negative of the derivative of the potential gives the x component of the force on the system:

$$F_x(x) = -\frac{dV}{dx}. \tag{1.80}$$

Thus, the critical points of a potential function $V(x)$ correspond to the points where the net force on the system is zero. In other words, the critical points are the points where the system is in **equilibrium**. Whether an equilibrium is stable or unstable depends on whether the critical point is a minimum or a maximum, respectively.

Analysis of the extrema of functions depending on more than one variable follows the same line of reasoning. However, since we can now approach the critical point from different directions, one has to be careful. Consider a continuous

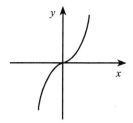

Figure 1.6   Plot of $y(x) = x^3$.

function $z = f(x, y)$ defined in some domain D. We say this function has a local maximum at (x_0, y_0) if the inequality

$$f(x, y) \leq f(x_0, y_0) \tag{1.81}$$

is satisfied for all points in some neighborhood of (x_0, y_0) and to have a local minimum if the inequality

$$f(x, y) \geq f(x_0, y_0) \tag{1.82}$$

is satisfied. In the following argument, we assume that all the necessary partial derivatives exist. Critical points are now defined as the points where the **first differential** $\Delta^{(1)} f(x, y)$ vanishes:

$$\Delta^{(1)} f(x, y) = \left[\frac{\partial f}{\partial x} \Delta x + \frac{\partial f}{\partial y} \Delta y \right] = 0. \tag{1.83}$$

Since the displacements Δx and Δy are arbitrary, the only way to satisfy this equation is to have both partial derivatives, f_x and f_y, vanish. Hence at the critical point (x_0, y_0), shown with the subscript 0, one has

$$\boxed{\left(\frac{\partial f}{\partial x} \right)_0 = 0, \quad \left(\frac{\partial f}{\partial y} \right)_0 = 0.} \tag{1.84}$$

To study the nature of these critical points, we again look at the **second differential** $\Delta^{(2)} f(x_0, y_0)$, which is now given as

$$\Delta^{(2)} f(x_0, y_0) = \frac{1}{2} \left(\frac{\partial^2 f}{\partial x^2} \right)_0 (\Delta x)^2 + \left(\frac{\partial^2 f}{\partial x \partial y} \right)_0 \Delta x \, \Delta y + \frac{1}{2} \left(\frac{\partial^2 f}{\partial y^2} \right)_0 (\Delta y)^2. \tag{1.85}$$

For a local maximum, the second differential $\Delta^{(2)} f(x_0, y_0)$ has to be negative and for a local minimum positive. Since we can approach the point (x_0, y_0) from different directions, we substitute (Figure 1.7)

$$\Delta x = \Delta s \cos \theta \quad \text{and} \quad \Delta y = \Delta s \sin \theta \tag{1.86}$$

to write Eq. (1.85) as

$$\Delta^{(2)} f(x_0, y_0) = \frac{1}{2} [A \cos^2 \theta + 2B \cos \theta \, \sin \theta + C \sin^2 \theta](\Delta s)^2, \tag{1.87}$$

where we have defined

$$A = \left(\frac{\partial^2 f}{\partial x^2} \right)_0, \quad B = \left(\frac{\partial^2 f}{\partial x \partial y} \right)_0, \quad C = \left(\frac{\partial^2 f}{\partial y^2} \right)_0. \tag{1.88}$$

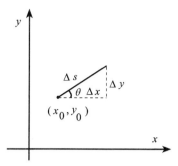

Figure 1.7 Definition of Δs.

Now the analysis of the nature of the critical points reduces to investigating the sign of $\Delta^{(2)} f(x_0, y_0)$ [Eq. (1.87)]. We present the final result as a theorem [2].

Theorem 1.4. Let $z = f(x, y)$ and its first and second partial derivatives be continuous in a domain D, and let (x_0, y_0) be a point in D, where the partial derivatives $\left(\frac{\partial z}{\partial x}\right)_0$ and $\left(\frac{\partial z}{\partial y}\right)_0$ vanish. Then, we have the following cases:

    I. For $B^2 - AC < 0$ and $A + C < 0$, we have a local maximum at (x_0, y_0).

    II. For $B^2 - AC < 0$ and $A + C > 0$, we have a local minimum at (x_0, y_0).

    III. For $B^2 - AC > 0$, we have a saddle point at (x_0, y_0).

    IV. For $B^2 - AC = 0$, the nature of the critical point is undetermined.

When $B^2 - AC > 0$ at (x_0, y_0), we have what is called a **saddle point**, where for some directions $\Delta^{(2)} f(x_0, y_0)$ is positive and negative for the others. When $B^2 - AC = 0$, for some directions $\Delta^{(2)} f(x_0, y_0)$ will be zero, hence one must look at higher order derivatives to study the nature of the critical point. When A, B, and C are all zero, then $\Delta^{(2)} f(x_0, y_0)$ also vanishes. Hence, we need to investigate the sign of $\Delta^{(3)} f(x_0, y_0)$.

1.7 EXTREMA OF FUNCTIONS WITH CONDITIONS

A problem of significance is finding the critical points of functions while satisfying one or more conditions. Consider finding the extremums of

$$w = f(x, y, z) \tag{1.89}$$

while satisfying the conditions

$$g_1(x, y, z) = 0 \tag{1.90}$$

and

$$g_2(x, y, z) = 0. \tag{1.91}$$

In principle, the two conditions define two surfaces, the intersection of which can be expressed as

$$x = x, \quad y = y(x), \quad z = z(x), \tag{1.92}$$

where we have used the variable x as a parameter. We can now substitute this parametric equation into $w = f(x, y, z)$ and write it entirely in terms of x as

$$w(x) = f(x, y(x), z(x)). \tag{1.93}$$

The extremum points can now be found by the technique discussed in the previous section. Geometrically, this problem corresponds to finding the extremum points of $w = f(x, y, z)$ on the curve defined by the intersection of $g_1(x, y, z) = 0$ and $g_2(x, y, z) = 0$. Unfortunately, this method rarely works to yield a solution analytically. Instead, we introduce the following method: At a critical point, we have seen that the change in w to first order in the differentials Δx, Δy, and Δz is zero:

$$\Delta w = \frac{\partial f}{\partial x} \Delta x + \frac{\partial f}{\partial y} \Delta y + \frac{\partial f}{\partial z} \Delta z = 0. \tag{1.94}$$

We also write the differentials of $g_1(x, y, z)$ and $g_2(x, y, z)$ as

$$\frac{\partial g_1}{\partial x} \Delta x + \frac{\partial g_1}{\partial y} \Delta y + \frac{\partial g_1}{\partial z} \Delta z = 0, \tag{1.95}$$

$$\frac{\partial g_2}{\partial x} \Delta x + \frac{\partial g_2}{\partial y} \Delta y + \frac{\partial g_2}{\partial z} \Delta z = 0. \tag{1.96}$$

We now multiply Eq. (1.95) with λ_1 and Eq. (1.96) with λ_2 and add to Eq. (1.94) to write

$$\left(\frac{\partial f}{\partial x} + \lambda_1 \frac{\partial g_1}{\partial x} + \lambda_2 \frac{\partial g_2}{\partial x} \right) \Delta x + \left(\frac{\partial f}{\partial y} + \lambda_1 \frac{\partial g_1}{\partial y} + \lambda_2 \frac{\partial g_2}{\partial y} \right) \Delta y$$

$$+ \left(\frac{\partial f}{\partial z} + \lambda_1 \frac{\partial g_1}{\partial z} + \lambda_2 \frac{\partial g_2}{\partial z} \right) \Delta z = 0. \tag{1.97}$$

Because of the given conditions in Eqs. (1.90) and (1.91), Δx, Δy, and Δz are not independent. Hence, their coefficients in Eq. (1.94) cannot be set to zero directly. However, the values of λ_1 and λ_2, which are called the **Lagrange undetermined multipliers**, can be chosen so that the coefficients of Δx, Δy, and Δz are all zero in Eq. (1.97):

$$\left(\frac{\partial f}{\partial x} + \lambda_1 \frac{\partial g_1}{\partial x} + \lambda_2 \frac{\partial g_2}{\partial x} \right) = 0, \tag{1.98}$$

$$\left(\frac{\partial f}{\partial y} + \lambda_1 \frac{\partial g_1}{\partial y} + \lambda_2 \frac{\partial g_2}{\partial y} \right) = 0, \tag{1.99}$$

$$\left(\frac{\partial f}{\partial z} + \lambda_1 \frac{\partial g_1}{\partial z} + \lambda_2 \frac{\partial g_2}{\partial z}\right) = 0. \tag{1.100}$$

Along with the two conditions, $g_1(x, y, z) = 0$ and $g_2(x, y, z) = 0$, these three equations are to be solved for the five unknowns, x, y, z, λ_1, and λ_2. The values that λ_1 and λ_2 assume are used to obtain the x, y, and z values needed, which correspond to the locations of the critical points. Analysis of the critical points now proceeds as before. Note that this method is quite general, and as long as the required derivatives exist and the conditions are compatible, it can be used with any number of conditions.

Example 1.3. ***Extremum problems:*** We now find the dimensions of a rectangular swimming pool with fixed volume V_0 and minimal area of its base and sides. If we denote the dimensions of its base with x and y and its height with z, the fixed volume is $V_0 = xyz$, and the total area of the base and the sides is

$$a = xy + 2xz + 2yz. \tag{1.101}$$

Using the condition of fixed volume, we write a as a function of x and y as

$$a = xy + \frac{2V_0}{y} + \frac{2V_0}{x}. \tag{1.102}$$

Now, the critical points of a are determined from the equations

$$\frac{\partial a}{\partial x} = 0, \quad \frac{\partial a}{\partial y} = 0, \tag{1.103}$$

which give the following two equations:

$$y - \frac{2V_0}{x^2} = 0, \tag{1.104}$$

$$x - \frac{2V_0}{y^2} = 0 \tag{1.105}$$

or

$$yx^2 - 2V_0 = 0, \tag{1.106}$$

$$xy^2 - 2V_0 = 0. \tag{1.107}$$

If we subtract Eq. (1.107) from Eq. (1.106), we obtain $y = x$, which when substituted back into Eq. (1.106) gives the critical dimensions

$$x = (2V_0)^{1/3}, \tag{1.108}$$

$$y = (2V_0)^{1/3}, \tag{1.109}$$

$$z = \left(\frac{V_0}{4}\right)^{1/3}, \tag{1.110}$$

where the final dimension is obtained from $V_0 = xyz$. To assure ourselves that this corresponds to a minimum, we evaluate the second-order derivatives at the critical point,

$$A = \frac{\partial^2 a}{\partial x^2} = \frac{4V_0}{x^3} = \frac{4V_0}{2V_0} = 2, \tag{1.111}$$

$$B = \frac{\partial^2 a}{\partial x \partial y} = 1, \tag{1.112}$$

$$C = \frac{\partial^2 a}{\partial y^2} = \frac{4V_0}{y^3} = \frac{4V_0}{2V_0} = 2, \tag{1.113}$$

and find

$$B^2 - AC = 1 - 4 = -3 < 0 \quad \text{and} \quad A + C = 2 + 2 = 4 > 0. \tag{1.114}$$

Thus, the critical dimensions we have obtained [Eqs. (1.108)–(1.110)] are indeed for a minimum by Theorem 1.4.

Example 1.4. *Lagrange undetermined multipliers:* We now solve the aforementioned problem by using the method of Lagrange undetermined multipliers. The equation to be minimized is now

$$f(x, y, z) = xy + 2xz + 2yz \tag{1.115}$$

with the condition

$$g(x, y, z) = V_0 - xyz = 0. \tag{1.116}$$

The equations to be solved are obtained from Eqs. (1.98)–(1.100) as

$$y + 2z - yz\lambda = 0, \tag{1.117}$$
$$x + 2z - xz\lambda = 0, \tag{1.118}$$
$$2x + 2y - \lambda xy = 0. \tag{1.119}$$

Along with $V_0 = xyz$, these give four equations to be solved for the critical dimensions x, y, z, and λ. Multiplying the first equation by x and the second one by y and then subtracting gives

$$x = y. \tag{1.120}$$

Substituting this into the third equation [Eq. (1.119)] gives the value of the Lagrange undetermined multiplier as $\lambda = 4/x$, which when substituted into Eqs. (1.117)–(1.119) gives

$$xy + 2xz - 4yz = 0, \tag{1.121}$$

$$x + 2z - 4z = 0, \tag{1.122}$$

$$2x + 2y - 4y = 0. \tag{1.123}$$

Using the condition $V_0 = xyz$ and Eq. (1.120), these three equations [Eqs. (1.121)–(1.123)] can be solved easily to yield the critical dimensions in terms of V_0 as

$$x = (2V_0)^{1/3}, \tag{1.124}$$

$$y = (2V_0)^{1/3}, \tag{1.125}$$

$$z = \left(\frac{V_0}{4}\right)^{1/3}. \tag{1.126}$$

Analysis of the critical point is done as in the previous example by using Theorem 1.4.

1.8 DERIVATIVES AND DIFFERENTIALS OF COMPOSITE FUNCTIONS

In what follows, we assume that the functions are defined in their appropriate domains and have continuous first partial derivatives.

Chain rule: If $z = f(x, y)$ and $x = x(t)$, $y = y(t)$, then

$$\boxed{\frac{dz}{dt} = \frac{\partial z}{\partial x}\frac{dx}{dt} + \frac{\partial z}{\partial y}\frac{dy}{dt}.} \tag{1.127}$$

Similarly, if $z = f(x, y)$, $x = g(u, v)$, and $y = h(u, v)$, then

$$\frac{\partial z}{\partial u} = \frac{\partial z}{\partial x}\frac{\partial x}{\partial u} + \frac{\partial z}{\partial y}\frac{\partial y}{\partial u}, \tag{1.128}$$

$$\frac{\partial z}{\partial v} = \frac{\partial z}{\partial x}\frac{\partial x}{\partial v} + \frac{\partial z}{\partial y}\frac{\partial y}{\partial v}. \tag{1.129}$$

A better notation to use is

$$\left(\frac{\partial z}{\partial u}\right)_v = \left(\frac{\partial z}{\partial x}\right)_y \left(\frac{\partial x}{\partial u}\right)_v + \left(\frac{\partial z}{\partial y}\right)_x \left(\frac{\partial y}{\partial u}\right)_v, \tag{1.130}$$

$$\left(\frac{\partial z}{\partial v}\right)_u = \left(\frac{\partial z}{\partial x}\right)_y \left(\frac{\partial x}{\partial v}\right)_u + \left(\frac{\partial z}{\partial y}\right)_x \left(\frac{\partial y}{\partial v}\right)_u. \tag{1.131}$$

This notation is particularly useful in thermodynamics, where z may also be expressed with another choice of variables, such as

$$z = f(x, y) \tag{1.132}$$

$$= g(x, w) \tag{1.133}$$

$$= h(u, y). \tag{1.134}$$

Hence, when we write the derivative $\frac{\partial z}{\partial x}$, we have to clarify whether we are in the (x, y) or the (x, w) space by writing

$$\left(\frac{\partial z}{\partial x} \right)_y \quad \text{or} \quad \left(\frac{\partial z}{\partial x} \right)_w . \tag{1.135}$$

These formulas can be extended to any number of variables. Using Eq. (1.127), we can write the differential dz as

$$dz = \left(\frac{\partial z}{\partial x} \frac{\partial x}{\partial t} + \frac{\partial z}{\partial y} \frac{\partial y}{\partial t} \right) dt = \frac{\partial z}{\partial x} dx + \frac{\partial z}{\partial y} dy. \tag{1.136}$$

We now treat x, y, and z as functions of (u, v) and write the differential dz as

$$dz = \frac{\partial z}{\partial u} du + \frac{\partial z}{\partial v} dv \tag{1.137}$$

$$= \left(\frac{\partial z}{\partial x} \frac{\partial x}{\partial u} + \frac{\partial z}{\partial y} \frac{\partial y}{\partial u} \right) du + \left(\frac{\partial z}{\partial x} \frac{\partial x}{\partial v} + \frac{\partial z}{\partial y} \frac{\partial y}{\partial v} \right) dv \tag{1.138}$$

$$= \left(\frac{\partial z}{\partial x} \right) \left(\frac{\partial x}{\partial u} du + \frac{\partial x}{\partial v} dv \right) + \left(\frac{\partial z}{\partial y} \right) \left(\frac{\partial y}{\partial u} du + \frac{\partial y}{\partial v} dv \right). \tag{1.139}$$

Since x and y are also functions of u and v, we have the differentials

$$dx = \frac{\partial x}{\partial u} du + \frac{\partial x}{\partial v} dv \tag{1.140}$$

and

$$dy = \frac{\partial y}{\partial u} du + \frac{\partial y}{\partial v} dv, \tag{1.141}$$

which allow us to write Eq. (1.139) as

$$dz = \frac{\partial z}{\partial x} dx + \frac{\partial z}{\partial y} dy. \tag{1.142}$$

This result can be extended to any number of variables. In other words, any equation in differentials that is true in one set of independent variables is also true for another choice of variables.

1.9 IMPLICIT FUNCTION THEOREM

A function given as

$$F(x, y, z) = 0 \tag{1.143}$$

can be used to describe several functions of the following forms:

$$z = f(x, y), \ y = g(x, z), \text{etc.} \tag{1.144}$$

For example,

$$x^2 + y^2 + z^2 - 9 = 0 \tag{1.145}$$

can be used to define the function

$$z = \sqrt{9 - x^2 - y^2}, \tag{1.146}$$

or

$$z = -\sqrt{9 - x^2 - y^2}, \tag{1.147}$$

both of which are defined in the domain $x^2 + y^2 + z^2 \le 9$. We say these functions are **implicitly** defined by Eq. (1.145). In order to be able to define a differentiable function, $z = f(x, y)$, by the implicit function $F(x, y, z) = 0$, the partial derivatives $\partial f / \partial x$ and $\partial f / \partial y$ should exist in some domain so that we can write the differential

$$dz = \frac{\partial f}{\partial x} \, dx + \frac{\partial f}{\partial y} \, dy. \tag{1.148}$$

Using the implicit function $F(x, y, z) = 0$, we write

$$F_x \, dx + F_y \, dy + F_z \, dz = 0 \tag{1.149}$$

and

$$dz = -\frac{F_x}{F_z} \, dx - \frac{F_y}{F_z} \, dy, \tag{1.150}$$

where

$$F_x = \frac{\partial F}{\partial x}, \quad F_y = \frac{\partial F}{\partial y}, \quad F_z = \frac{\partial F}{\partial z}. \tag{1.151}$$

Comparing the two differentials [Eqs. (1.148) and (1.150)], we obtain the partial derivatives

$$\frac{\partial f}{\partial x} = -\frac{F_x}{F_z}, \quad \frac{\partial f}{\partial y} = -\frac{F_y}{F_z}. \tag{1.152}$$

Hence, granted that $F_z \ne 0$, we can use the implicit function $F(x, y, z) = 0$ to define a function of the form $z = f(x, y)$.

We now consider a more complicated case, in which we have two implicit functions:

$$F(x, y, z, w) = 0, \tag{1.153}$$

$$G(x, y, z, w) = 0. \tag{1.154}$$

Using these two equations in terms of four variables, we can solve, in principle, for two of the variables in terms of the remaining two as

$$z = f(x, y), \tag{1.155}$$

$$w = g(x, y). \tag{1.156}$$

For $f(x, y)$ and $g(x, y)$ to be differentiable, certain conditions must be met by $F(x, y, z, w)$ and $G(x, y, z, w)$. First, we write the differentials

$$F_x \, dx + F_y \, dy + F_z \, dz + F_w \, dw = 0, \tag{1.157}$$

$$G_x \, dx + G_y \, dy + G_z \, dz + G_w \, dw = 0 \tag{1.158}$$

and rearrange them as

$$F_z \, dz + F_w \, dw = -F_x \, dx - F_y \, dy, \tag{1.159}$$

$$G_z \, dz + G_w \, dw = -G_x \, dx - G_y \, dy. \tag{1.160}$$

We now have a system of two linear equations for the differentials dz and dw to be solved simultaneously. We can either solve by elimination or use determinants and the Cramer's rule to write

$$dz = \frac{\begin{vmatrix} -F_x \, dx - F_y \, dy & F_w \\ -G_x \, dx - G_y \, dy & G_w \end{vmatrix}}{\begin{vmatrix} F_z & F_w \\ G_z & G_w \end{vmatrix}}, \tag{1.161}$$

$$dw = \frac{\begin{vmatrix} F_z & -F_x \, dx - F_y \, dy \\ G_z & -G_x \, dx - G_y \, dy \end{vmatrix}}{\begin{vmatrix} F_z & F_w \\ G_z & G_w \end{vmatrix}}. \tag{1.162}$$

Using the properties of determinants, we can write these as

$$dz = -\frac{\begin{vmatrix} F_x & F_w \\ G_x & G_w \end{vmatrix}}{\begin{vmatrix} F_z & F_w \\ G_z & G_w \end{vmatrix}} dx - \frac{\begin{vmatrix} F_y & F_w \\ G_y & G_w \end{vmatrix}}{\begin{vmatrix} F_z & F_w \\ G_z & G_w \end{vmatrix}} dy, \tag{1.163}$$

$$dw = -\frac{\begin{vmatrix} F_z & F_x \\ G_z & G_x \end{vmatrix}}{\begin{vmatrix} F_z & F_w \\ G_z & G_w \end{vmatrix}} dx - \frac{\begin{vmatrix} F_z & F_y \\ G_z & G_y \end{vmatrix}}{\begin{vmatrix} F_z & F_w \\ G_z & G_w \end{vmatrix}} dy. \tag{1.164}$$

For differentiable functions, $z = f(x, y)$ and $w = g(x, y)$, with existing first-order partial derivatives, we can write

$$dz = \frac{\partial f}{\partial x} dx + \frac{\partial f}{\partial y} dy, \tag{1.165}$$

$$dw = \frac{\partial g}{\partial x} dx + \frac{\partial g}{\partial y} dy. \tag{1.166}$$

Comparing with Eqs. (1.163) and (1.164), we obtain the partial derivatives:

$$\frac{\partial f}{\partial x} = -\frac{\frac{\partial(F,G)}{\partial(x,w)}}{\frac{\partial(F,G)}{\partial(z,w)}}, \quad \frac{\partial f}{\partial y} = -\frac{\frac{\partial(F,G)}{\partial(y,w)}}{\frac{\partial(F,G)}{\partial(z,w)}}, \tag{1.167}$$

$$\frac{\partial g}{\partial x} = -\frac{\frac{\partial(F,G)}{\partial(z,x)}}{\frac{\partial(F,G)}{\partial(z,w)}}, \quad \frac{\partial g}{\partial y} = -\frac{\frac{\partial(F,G)}{\partial(z,y)}}{\frac{\partial(F,G)}{\partial(z,w)}}, \tag{1.168}$$

where the determinants:

$$\frac{\partial(F,G)}{\partial(x,w)} = \begin{vmatrix} F_x & F_w \\ G_x & G_w \end{vmatrix}, \quad \frac{\partial(F,G)}{\partial(z,w)} = \begin{vmatrix} F_z & F_w \\ G_z & G_w \end{vmatrix}, \dots \tag{1.169}$$

are called the **Jacobi determinants**. In summary, given two implicit equations:

$$F(x, y, z, w) = 0 \quad \text{and} \quad G(x, y, z, w) = 0, \tag{1.170}$$

we can define two differentiable functions

$$z = f(x, y) \quad \text{and} \quad w = g(x, y) \tag{1.171}$$

with the partial derivatives given as in Eqs. (1.167) and (1.168), provided that the **Jacobian:**

$$\frac{\partial(F,G)}{\partial(z,w)} = \begin{vmatrix} F_z & F_w \\ G_z & G_w \end{vmatrix}, \tag{1.172}$$

is different from zero in the domain of definition.

This useful technique can be generalized to a set of m equations in $n + m$ number of unknowns:

$$F_1(y_1, \ldots, y_m, x_1, \ldots, x_n) = 0,$$

$$\vdots \tag{1.173}$$

$$F_m(y_1, \ldots, y_m, x_1, \ldots, x_n) = 0.$$

We look for m differentiable functions in terms of n variables as

$$y_1 = y_1(x_1, \ldots, x_n),$$

$$\vdots \tag{1.174}$$

$$y_m = y_m(x_1, \ldots, x_n).$$

We write the following differentials:

$$F_{1y_1}\, dy_1 + \cdots + F_{1y_m}\, dy_m = -F_{1x_1}\, dx_1 - \cdots - F_{1x_n}\, dx_n,$$

$$\vdots \tag{1.175}$$

$$F_{my_1}\, dy_1 + \cdots + F_{my_m}\, dy_m = -F_{mx_1}\, dx_1 - \cdots - F_{mx_n}\, dx_n$$

and obtain a set of m linear equations to be solved for the m differentials dy_i, $i = 1, \ldots, m$, of the dependent variables. Using Cramer's rule, we can solve for dy_i if and only if the determinant of the coefficients is different from zero:

$$\begin{vmatrix} F_{1y_1} & \cdots & F_{1y_m} \\ \vdots & \ddots & \vdots \\ F_{my_1} & \cdots & F_{my_m} \end{vmatrix} = \frac{\partial(F_1, \ldots, F_m)}{\partial(y_1, \ldots, y_m)} \neq 0. \tag{1.176}$$

To obtain closed expressions for the partial derivatives $\partial y_1 / \partial x_j$, we take partial derivatives of Eq. (1.173) to write

$$F_{1y_1} \frac{\partial y_1}{\partial x_j} + \cdots + F_{1y_m} \frac{\partial y_m}{\partial x_j} = -F_{1x_j},$$

$$\vdots \tag{1.177}$$

$$F_{my_1} \frac{\partial y_1}{\partial x_j} + \cdots + F_{my_m} \frac{\partial y_m}{\partial x_j} = -F_{mx_j},$$

which gives the solution for $\partial y_1 / \partial x_j$ as

$$\frac{\partial y_1}{\partial x_j} = -\frac{\begin{vmatrix} F_{1x_j} & F_{1y_2} & \cdots & F_{1y_m} \\ \vdots & \vdots & \ddots & \vdots \\ F_{mx_j} & F_{my_2} & \cdots & F_{my_m} \\ F_{1y_1} & F_{1y_2} & \cdots & F_{1y_m} \\ \vdots & \vdots & \ddots & \vdots \\ F_{my_1} & F_{my_2} & \cdots & F_{my_m} \end{vmatrix}}{} = -\frac{\dfrac{\partial(F_1, F_2, \ldots, F_m)}{\partial(x_j, y_2, \ldots, y_m)}}{\dfrac{\partial(F_1, F_2, \ldots, F_m)}{\partial(y_1, y_2, \ldots, y_m)}} \tag{1.178}$$

and similar expressions for the other partial derivatives can be obtained. In general, granted that the Jacobi determinant does not vanish:

$$\frac{\partial(F_1, F_2, \ldots, F_m)}{\partial(y_1, y_2, \ldots, y_m)} \neq 0,$$

we can obtain the partial derivatives $\partial y_i / \partial x_j$ as

$$\frac{\partial y_i}{\partial x_j} = -\frac{\dfrac{\partial(F_1, F_2, \ldots, F_m)}{\partial(y_1, \ldots, y_{i-1}, x_j, y_{i+1}, \ldots, y_m)}}{\dfrac{\partial(F_1, \ldots, F_m)}{\partial(y_1, \ldots, y_m)}}, \tag{1.179}$$

where $i = 1, \ldots, m$ and $j = 1, \ldots, n$. We conclude this section by stating the implicit function theorem [2]:

Implicit function theorem: Let the functions

$$F_i(y_1, \ldots, y_m, x_1, \ldots, x_n) = 0, \quad i = 1, \ldots, m, \tag{1.180}$$

be defined in the neighborhood of the point

$$P_0 = (y_{01}, \ldots, y_{0m}, x_{01}, \ldots, x_{0n}) \tag{1.181}$$

with continuous first-order partial derivatives existing in this neighborhood. If

$$\frac{\partial(F_1, \ldots, F_m)}{\partial(y_1, \ldots, y_m)} \neq 0 \quad \text{at } P_0, \tag{1.182}$$

then in an appropriate neighborhood of P_0, there is a unique set of continuous functions

$$y_i = f_i(x_1, \ldots, x_n), \quad i = 1, \ldots, m, \tag{1.183}$$

with continuous partial derivatives,

$$\frac{\partial y_i}{\partial x_j} = -\frac{\dfrac{\partial(F_1, F_2, \ldots, F_m)}{\partial(y_1, \ldots, y_{i-1}, x_j, y_{i+1}, \ldots, y_m)}}{\dfrac{\partial(F_1, \ldots, F_m)}{\partial(y_1, \ldots, y_m)}}, \tag{1.184}$$

where $i = 1, \ldots, m$ and $j = 1, \ldots n$, such that

$$y_{0i} = f_i(x_{01}, \ldots, x_{0n}), \quad i = 1, \ldots, m, \tag{1.185}$$

and

$$F_i(f_1(x_1, \ldots, x_n), \ldots, f_m(x_1, \ldots, x_n), x_1, \ldots, x_n) = 0, \quad i = 1, \ldots, m, \tag{1.186}$$

in the neighborhood of P_0. Note that if the Jacobi determinant [Eq. (1.182)] is zero at the point of interest, then we search for a different set of dependent variables to avoid the difficulty.

1.10   INVERSE FUNCTIONS

A pair of functions,

$$x = f(u, v), \tag{1.187}$$

$$y = g(u, v), \tag{1.188}$$

can be considered as a **mapping** from the xy space to the uv space. Under certain conditions, this maps a certain domain D_{xy} in the xy space to a certain domain D_{uv} in the uv space on a one-to-one basis. Under such conditions, an inverse mapping should also exist. However, analytically it may not always be possible to find the **inverse mapping** or the functions

$$u = u(x, y), \tag{1.189}$$

$$v = v(x, y). \tag{1.190}$$

In such cases, we may consider Eqs. (1.187) and (1.188) as implicit functions and write them as

$$F_1(x, y, u, v) = f(u, v) - x = 0, \tag{1.191}$$

$$F_2(x, y, u, v) = g(u, v) - y = 0. \tag{1.192}$$

We can now use Eq. (1.178) with $y_1 = u$, $y_2 = v$ and $x_1 = x$, $x_2 = y$ to write the partial derivatives of the inverse functions as

$$\frac{\partial u}{\partial x} = \frac{\dfrac{\partial(F_1, F_2)}{\partial(x, v)}}{\dfrac{\partial(F_1, F_2)}{\partial(u, v)}} = -\frac{\begin{vmatrix} -1 & \partial f/\partial v \\ 0 & \partial g/\partial v \end{vmatrix}}{\begin{vmatrix} \partial f/\partial u & \partial f/\partial v \\ \partial g/\partial u & \partial g/\partial v \end{vmatrix}} \tag{1.193}$$

$$= \frac{\partial g}{\partial v} \bigg/ \left[\frac{\partial f}{\partial u} \frac{\partial g}{\partial v} - \frac{\partial g}{\partial u} \frac{\partial f}{\partial v} \right]. \tag{1.194}$$

Similarly, the other partial derivatives can be obtained. As seen, the inverse function or the inverse mapping is well defined only when the **Jacobi determinant** J is different from zero:

$$J = \frac{\partial(f,g)}{\partial(u,v)} = \left[\frac{\partial f}{\partial u}\frac{\partial g}{\partial v} - \frac{\partial g}{\partial u}\frac{\partial f}{\partial v}\right] \neq 0, \tag{1.195}$$

where J is also called the **Jacobian of the mapping**. We will return to this point when we discuss coordinate transformations in Chapter 3. Note that the Jacobian of the inverse mapping is $1/J$. In other words,

$$\frac{\partial(f,g)}{\partial(u,v)}\frac{\partial(u,v)}{\partial(f,g)} = \begin{vmatrix} 1 & 0 \\ 0 & 1 \end{vmatrix} = 1. \tag{1.196}$$

Example 1.5. *Change of independent variable:* We now transform the Laplace equation:

$$\frac{\partial^2 z(x,y)}{\partial x^2} + \frac{\partial^2 z(x,y)}{\partial y^2} = 0, \tag{1.197}$$

into polar coordinates, that is, to a new set of independent variables defined by the equations

$$x = r\cos\phi, \quad y = r\sin\phi, \tag{1.198}$$

where $r \in (0,\infty)$ and $\phi \in [0,2\pi]$. We first write the partial derivatives of $z = z(x,y)$:

$$\frac{\partial z}{\partial r} = \frac{\partial z}{\partial x}\frac{\partial x}{\partial r} + \frac{\partial z}{\partial y}\frac{\partial y}{\partial r}, \tag{1.199}$$

$$\frac{\partial z}{\partial \phi} = \frac{\partial z}{\partial x}\frac{\partial x}{\partial \phi} + \frac{\partial z}{\partial y}\frac{\partial y}{\partial \phi}, \tag{1.200}$$

which lead to

$$\frac{\partial z}{\partial r} = \frac{\partial z}{\partial x}\cos\phi + \frac{\partial z}{\partial y}\sin\phi, \tag{1.201}$$

$$\frac{\partial z}{\partial \phi} = \frac{\partial z}{\partial x}(-r\sin\phi) + \frac{\partial z}{\partial y}(r\cos\phi). \tag{1.202}$$

Solving for $\partial z/\partial x$ and $\partial z/\partial y$, we obtain

$$\frac{\partial z}{\partial x} = \frac{\partial z}{\partial r}\cos\phi - \frac{\partial z}{\partial \phi}\frac{1}{r}\sin\phi, \tag{1.203}$$

$$\frac{\partial z}{\partial y} = \frac{\partial z}{\partial r}\sin\phi + \frac{\partial z}{\partial \phi}\frac{1}{r}\cos\phi. \tag{1.204}$$

We now repeat this process with $\partial z/\partial x$ to obtain the second derivative $\partial^2 z/\partial x^2$ as

$$\frac{\partial}{\partial x}\left[\frac{\partial z}{\partial x}\right] = \frac{\partial^2 z}{\partial x^2} = \frac{\partial\left[\frac{\partial z}{\partial x}\right]}{\partial r}\cos\phi - \frac{\partial\left[\frac{\partial z}{\partial x}\right]}{\partial\phi}\frac{1}{r}\sin\phi \tag{1.205}$$

$$= \cos\phi\frac{\partial}{\partial r}\left[\frac{\partial z}{\partial r}\cos\phi - \frac{\partial z}{\partial\phi}\frac{1}{r}\sin\phi\right]$$

$$- \frac{\sin\phi}{r}\frac{\partial}{\partial\phi}\left[\frac{\partial z}{\partial r}\cos\phi - \frac{\partial z}{\partial\phi}\frac{1}{r}\sin\phi\right] \tag{1.206}$$

$$= \frac{\partial^2 z}{\partial r^2}\cos^2\phi - \frac{\partial^2 z}{\partial r\partial\phi}\frac{2}{r}\cos\phi\sin\phi + \frac{\partial^2 z}{\partial\phi^2}\frac{1}{r^2}\sin^2\phi$$

$$+ \frac{1}{r}\frac{\partial z}{\partial r}\sin^2\phi + \frac{\partial z}{\partial\phi}\frac{2}{r^2}\sin\phi\,\cos\phi. \tag{1.207}$$

A similar procedure for $\partial z/\partial y$ yields $\partial^2 z/\partial y^2$:

$$\frac{\partial^2 z}{\partial y^2} = \frac{\partial^2 z}{\partial r^2}\sin^2\phi + \frac{\partial^2 z}{\partial r\partial\phi}\frac{2}{r}\sin\phi\,\cos\phi + \frac{\partial^2 z}{\partial\phi^2}\frac{1}{r^2}\cos^2\phi$$

$$+ \frac{1}{r}\frac{\partial z}{\partial r}\cos^2\phi - \frac{\partial z}{\partial\phi}\frac{2}{r^2}\sin\phi\,\cos\phi. \tag{1.208}$$

Adding Eqs. (1.207) and (1.208), we obtain the transformed equation:

$$\frac{\partial^2 z(r,\theta)}{\partial r^2} + \frac{1}{r}\frac{\partial z(r,\theta)}{\partial r} + \frac{1}{r^2}\frac{\partial^2 z(r,\theta)}{\partial\phi^2} = 0. \tag{1.209}$$

Since the Jacobian of the mapping is different from zero, that is,

$$J = \frac{\partial(x,y)}{\partial(r,\theta)} = \begin{vmatrix} \cos\phi & \sin\phi \\ -r\sin\phi & r\cos\phi \end{vmatrix} = r, \quad r \neq 0, \tag{1.210}$$

the inverse mapping exists:

$$r = \sqrt{x^2 + y^2}, \quad \phi = \tan^{-1}\frac{y}{x}. \tag{1.211}$$

1.11   INTEGRAL CALCULUS AND THE DEFINITE INTEGRAL

Let $f(x)$ be a continuous function in the interval $[x_a, x_b]$. By choosing $(n-1)$ points in this interval, $x_1, x_2, \ldots, x_{n-1}$, we can subdivide it into n subintervals, $\Delta x_1, \Delta x_2, \ldots, \Delta x_n$, which are not necessarily all equal in length. From Theorem 1.3, we know that $f(x)$ assumes a maximum, M, and a minimum, m, in $[x_a, x_b]$. Let M_i represent the maximum and m_i the minimum values that $f(x)$ assumes

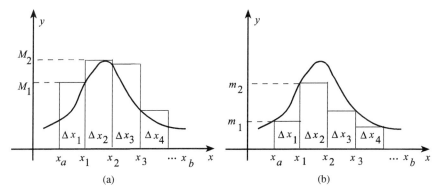

Figure 1.8   Upper (a) and lower (b) Darboux sums.

in Δx_i. We now denote a particular subdivision by d and write the sum of the rectangles shown in Figure 1.8a as

$$S(d) = \sum_{i=1}^{n} M_i \Delta x_i \tag{1.212}$$

and in Figure 1.8b as

$$s(d) = \sum_{i=1}^{n} m_i \Delta x_i. \tag{1.213}$$

The sums $S(d)$ and $s(d)$ are called the upper and the lower **Darboux sums**, respectively. Naturally, their values depend on the subdivision d. We pick the smallest of all $S(d)$ and call it the upper integral of $f(x)$ in $[x_a, x_b]$:

$$\overline{\int_{x_a}^{x_b}} f(x)dx. \tag{1.214}$$

Similarly, the largest of all $s(d)$ is called the lower integral of $f(x)$ in $[x_a, x_b]$:

$$\underline{\int_{x_a}^{x_b}} f(x)dx. \tag{1.215}$$

When these two integrals are equal, we say the definite integral of $f(x)$ in the interval $[x_a, x_b]$ exists and we write

$$\boxed{\overline{\int_{x_a}^{x_b}} f(x)dx = \underline{\int_{x_a}^{x_b}} f(x)dx = \int_{x_a}^{x_b} f(x)dx.} \tag{1.216}$$

This definition of integral is also called the **Riemann integral**, and the function $f(x)$ is called the **integrand**.

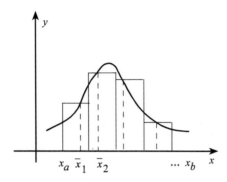

Figure 1.9   Riemann integral.

Darboux sums are not very practical to work with. Instead, for a particular subdivision, we write the sum

$$\sigma(d) = \sum_{k=1}^{n} f(\bar{x}_k)\Delta x_k, \tag{1.217}$$

where $\bar{x}_k$ is an arbitrary point in Δx_k (Figure 1.9). It is clear that the inequality

$$s(d) \leq \sigma(d) \leq S(d) \tag{1.218}$$

is satisfied. For a given subdivision the largest value of Δx_i is called the **norm** of d, which we will denote as $n(d)$.

1.12   RIEMANN INTEGRAL

We now give the basic definition of the Riemann integral as follows:
   Definition 1.1. Given a sequence of subdivisions $d_1, d_2, \ldots$ of the interval $[x_a, x_b]$ such that the sequence of norms $n(d_1), n(d_2), \ldots$ has the limit

$$\lim_{k \to \infty} n(d_k) \to 0 \tag{1.219}$$

and if $f(x)$ is integrable in $[x_a, x_b]$, then the Riemann integral is defined as

$$\int_{x_a}^{x_b} f(x)dx = \lim_{k \to \infty} \sigma(d_k), \tag{1.220}$$

where

$$\lim_{k \to \infty} S(d_k) = \lim_{k \to \infty} s(d_k) = \lim_{k \to \infty} \sigma(d_k). \tag{1.221}$$

Theorem 1.5. For the existence of the **Riemann integral**:

$$\boxed{\int_{x_a}^{x_b} f(x)dx,}$$

where x_a and x_b are finite numbers, it is sufficient to satisfy one of the following conditions:

(i) $f(x)$ is continuous in $[x_a, x_b]$.

(ii) $f(x)$ is bounded and piecewise continuous in $[x_a, x_b]$.

From these definitions, we can deduce the following properties of the Riemann integral [1]:

I. If $f_1(x)$ and $f_2(x)$ are integrable in $[x_a, x_b]$, then their sum is also integrable, and we can write

$$\int_{x_a}^{x_b} [f_1(x) + f_2(x)]dx = \int_{x_a}^{x_b} f_1(x)dx + \int_{x_a}^{x_b} f_2(x)dx. \tag{1.222}$$

II. If $f(x)$ is integrable in $[x_a, x_b]$, then the following are true:

$$\int_{x_a}^{x_b} \alpha f(x)dx = \alpha \int_{x_a}^{x_b} f(x)dx, \quad \alpha \text{ is a constant}, \tag{1.223}$$

$$\left| \int_{x_a}^{x_b} f(x)dx \right| \leq \int_{x_a}^{x_b} |f(x)|dx, \tag{1.224}$$

$$\left| \int_{x_a}^{x_b} f(x)dx \right| \leq M(x_b - x_a) \quad \text{if } |f(x)| \leq M \text{ in } [x_a, x_b], \tag{1.225}$$

$$\int_{x_a}^{x_b} f(x)dx = -\int_{x_b}^{x_a} f(x)dx. \tag{1.226}$$

III. If $f(x)$ is continuous and $f(x) \geq 0$ in $[x_a, x_b]$, then

$$\int_{x_a}^{x_b} f(x)dx = 0 \tag{1.227}$$

means $f(x) \equiv 0$.

IV. The **average** or the **mean**, $\langle f \rangle$, of $f(x)$ in the interval $[x_a, x_b]$ is defined as

$$\boxed{\langle f \rangle = \frac{1}{x_b - x_a} \int_{x_a}^{x_b} f(x)dx.} \tag{1.228}$$

If $f(x)$ is continuous, then there exist at least one point $x^* \in [x_a, x_b]$ such that

$$\boxed{\int_{x_a}^{x_b} f(x)dx = f(x^*)(x_b - x_a).} \tag{1.229}$$

This is also called the **mean value theorem** or **Rolle's theorem**.

V. If $f(x)$ is integrable in $[x_a, x_b]$ and if $x_a < x_c < x_b$, then

$$\int_{x_a}^{x_b} f(x)dx = \int_{x_a}^{x_c} f(x)dx + \int_{x_c}^{x_b} f(x)dx. \qquad (1.230)$$

VI. If $f(x) \geq g(x)$ in $[x_a, x_b]$, then

$$\int_{x_a}^{x_b} f(x)dx \geq \int_{x_a}^{x_b} g(x)dx. \qquad (1.231)$$

VII. Fundamental theorem of calculus: If $f(x)$ is continuous in $[x_a, x_b]$, then the function

$$F(x) = \int_{x_a}^{x} f(t)dt \qquad (1.232)$$

is also a continuous function of x in $[x_a, x_b]$. The function $F(x)$ is differentiable for every point in $[x_a, x_b]$ and its derivative at x is $f(x)$:

$$\boxed{\frac{dF}{dx} = \frac{d}{dx}\int_{x_a}^{x} f(t)dt = f(x).} \qquad (1.233)$$

$F(x)$ is called the **primitive** or the **antiderivative** of $f(x)$. Given a primitive, $F(x)$, then

$$F(x) + C_0,$$

where C_0 is a constant, is also a primitive. If a primitive is known for $f(x)$ in $[x_a, x_b]$, we can write

$$\int_{x_a}^{x_b} f(x)dx = \int_{x_a}^{x_b} \frac{dF}{dx} dx \qquad (1.234)$$

$$= F(x)|_{x_a}^{x_b} \qquad (1.235)$$

$$= F(x_b) - F(x_a). \qquad (1.236)$$

When the region of integration is not specified, we write the **indefinite integral**:

$$\int f(x)dx = F(x) + C, \qquad (1.237)$$

where C is an arbitrary constant and $F(x)$ is any function the derivative of which is $f(x)$.

VIII. If $f(x)$ is continuous and $f(x) \geq 0$ in $[x_a, x_b]$, then **geometrically** the integral

$$\int_{x_a}^{x_b} f(x)dx \qquad (1.238)$$

is the **area** under $f(x)$ between x_a and x_b.

IX. A very useful inequality in deciding whether a given integral is convergent or not is the **Schwarz inequality**:

$$\left[\int_{x_a}^{x_b} f(x)g(x)dx \right]^2 \leq \int_{x_a}^{x_b} f^2(x)dx \int_{x_a}^{x_b} g^2(x)dx. \tag{1.239}$$

X. One of the most commonly used techniques in integral calculus is the **integration by parts**:

$$\int_{x_a}^{x_b} vu' \, dx = [uv]_{x_a}^{x_b} - \int_{x_a}^{x_b} uv' \, dx \tag{1.240}$$

or

$$\int_{x_a}^{x_b} v \, du = [uv]_{x_a}^{x_b} - \int_{x_a}^{x_b} u \, dv, \tag{1.241}$$

where the derivatives u' and v' and u and v are continuous in $[x_a, x_b]$.

XI. In general, the following inequality holds:

$$\int_{x_a}^{x_b} f(x)dx \leq \int_{x_a}^{x_b} |f(x)|dx, \tag{1.242}$$

that is, if the integral $\int_{x_a}^{x_b} |f(x)|dx$ converges, then the integral $\int_{x_a}^{x_b} f(x) \, dx$ also converges. A convergent integral, $\int_{x_a}^{x_b} f(x) \, dx$, is said to be **absolutely convergent**, if $\int_{x_a}^{x_b} |f(x)|dx$ also converges. Integrals that converge but do not converge absolutely are called **conditionally convergent**.

1.13   IMPROPER INTEGRALS

We introduced Riemann integrals for bounded functions with finite intervals. Improper integrals are basically their extension to cases with infinite range and to functions that are not necessarily bounded.

Definition 1.2. Consider the integral

$$\int_a^c f(x)dx, \tag{1.243}$$

which exists in the Riemann sense in the interval $[a, c]$, where $a < c < b$. If the limit

$$\lim_{c \to b^-} \int_a^c f(x)dx \to A \tag{1.244}$$

exists, where the function $f(x)$ could be unbounded in the left neighborhood of b, then we say the integral $\int_a^b f(x)\,dx$ exists, or converges, and write

$$\int_a^b f(x)dx = A. \tag{1.245}$$

Example 1.6. *Improper integrals:* Consider the improper integral

$$I_1 = \int_0^1 \frac{x\,dx}{(1-x)^{1/2}}, \tag{1.246}$$

where the integrand, $x/(1-x)^{1/2}$, is unbounded at the end point $x = 1$. We write I_1 as the limit

$$I_1 = \lim_{c\to 1^-} \int_0^c \frac{x\,dx}{(1-x)^{1/2}} = \lim_{c\to 1^-} \left[\frac{2(1-x)^{3/2}}{3} - 2(1-x)^{1/2} \right]_0^c \tag{1.247}$$

$$= \lim_{c\to 1^-} \left[\frac{2}{3}(1-c)^{3/2} - 2(1-c)^{1/2} - \frac{2}{3} + 2 \right] = \frac{4}{3}, \tag{1.248}$$

thereby obtaining the value of I_1 as $4/3$. We now consider the integral

$$I_2 = \int_0^1 \frac{dx}{(1-x)}, \tag{1.249}$$

which does not exist since

$$I_2 = \lim_{c\to 1^-} \int_0^c \frac{dx}{(1-x)} = \lim_{c\to 1^-} \left[-\ln(1-x)\right]_0^c \tag{1.250}$$

$$= \lim_{c\to 1^-} \left[-\ln(1-c)\right] \to \infty. \tag{1.251}$$

In this case, we say the integral does not exist, or it is divergent, and for its value, we write $+\infty$.

A parallel argument is given if the integral $\int_c^b f(x)\,dx$ exists in the interval $[c, b]$, where $a < c < b$. We now write the limit

$$I = \lim_{c\to a^+} \int_c^b f(x)dx, \tag{1.252}$$

where $f(x)$ could be unbounded in the right neighborhood of a. If the limit

$$\lim_{c\to a^+} \int_c^b f(x)dx \to B \tag{1.253}$$

exists, we write

$$\int_a^b f(x)dx = B. \tag{1.254}$$

We now present another useful result from integral calculus:

Theorem 1.6. Let c be a point in the interval (a, b), and let $f(x)$ be integrable in the intervals $[a, a']$ and $[b', b]$, where $a < a' < c < b' < b$. Furthermore, $f(x)$ could be unbounded in the neighborhood of c. Then, the integral $I = \int_a^b f(x)\, dx$ exists if the integrals

$$I_1 = \int_a^c f(x)dx \tag{1.255}$$

and

$$I_2 = \int_c^b f(x)dx \tag{1.256}$$

both exist and when they exist, their sum is equal to I:

$$I = I_1 + I_2. \tag{1.257}$$

If either I_1 or I_2 diverges, then I also diverges.

Example 1.7. *Improper integrals:* Consider $I = \int_{-1}^3 dx/x$:

$$I = \int_{-1}^0 \frac{dx}{x} + \int_0^3 \frac{dx}{x}, \tag{1.258}$$

which converges provided that the integrals on the right-hand side converge. However, they both diverge:

$$\int_{-1}^0 \frac{dx}{x} = \lim_{c\to 0^-} \int_{-1}^c \frac{dx}{x} = \lim_{c\to 0^-} [\ln|x|]_{-1}^c = \lim_{c\to 0^-} \ln|c| \to -\infty \tag{1.259}$$

and similarly,

$$\int_0^3 \frac{dx}{x} \to \lim_{c\to 0^+} \int_c^3 \frac{dx}{x} = \lim_{c\to 0^+} [\ln|x|]_c^3 \to \ln 3 - \lim_{c\to 0^+} \ln|c| \to +\infty, \tag{1.260}$$

hence their sum also diverges.

When the range of the integral is infinite, we use the following results: If $f(x)$ is integrable in $[a, b]$ and the limit

$$\lim_{b\to\infty} \int_a^b f(x)dx \to A \tag{1.261}$$

exists, we can write

$$\int_a^\infty f(x)dx = A. \tag{1.262}$$

Similarly, we define the integral

$$\int_{-\infty}^b f(x)dx = B. \tag{1.263}$$

If the integrals

$$I_1 = \int_a^\infty f(x)dx \tag{1.264}$$

and

$$I_2 = \int_{-\infty}^a f(x)dx \tag{1.265}$$

both exist, then we can write

$$\int_{-\infty}^\infty f(x)dx = I_1 + I_2. \tag{1.266}$$

1.14 CAUCHY PRINCIPAL VALUE INTEGRALS

Since the integrals in Example 1.7, $I_1 = \int_{-1}^0 dx/x$ and $I_2 = \int_0^3 dx/x$, both diverge, we used Theorem 1.6 to conclude that their sum, $I = I_1 + I_2$, is also divergent. However, notice that I_1 diverges as $\lim_{c \to 0^-} \ln|c| \to -\infty$, while I_2 diverges as $\lim_{c \to 0^+}(-\ln|c|) \to +\infty$. In other words, if we consider the two integrals together, the two divergences offset each other, thus yielding a finite result for the value of the integral as

$$I = \int_{-1}^3 \frac{dx}{x} = \lim_{c \to 0^-} [\ln|x|]_{-1}^c + \lim_{c \to 0^+} [\ln|x|]_c^3 \tag{1.267}$$

$$= \lim_{c \to 0^-} \ln|c| - \ln 1 + \ln 3 - \lim_{c \to 0^+} \ln|c| \to \ln 3 \tag{1.268}$$

$$= \ln 3. \tag{1.269}$$

The problem with $\int_{-1}^3 dx/x$ is that the integrand, $1/x$, diverges at the origin. However, at all the other points in the range $[-1, 3]$, it is finite. In Riemann integrals (Theorem 1.6), divergence of either I_1 or I_2 is sufficient to conclude that the integral I does not exist. However, as in the aforementioned case, sometimes by considering the two integrals, I_1 and I_2, together, one may obtain a finite result. This is called taking the **Cauchy principal value** of the integral. Since it corresponds to a **modification** of the Riemann definition of integral, it has to be mentioned explicitly that we are taking the Cauchy principal value as

$$PV \int_{-1}^{3} \frac{dx}{x} = \ln 3. \tag{1.270}$$

Another example is the integral

$$I = \int_{-\infty}^{\infty} x^3 \, dx, \tag{1.271}$$

which is divergent in the ordinary sense, since

$$\int_{0}^{\infty} x^3 \, dx = \lim_{a \to \infty} \int_{0}^{a} x^3 \, dx = \lim_{a \to \infty} \frac{a^4}{4} \to \infty. \tag{1.272}$$

However, if we take its Cauchy principal value, we obtain

$$PV \int_{-\infty}^{\infty} x^3 \, dx = \lim_{a \to \infty} \left[\int_{-a}^{0} x^3 \, dx + \int_{0}^{a} x^3 \, dx \right] \tag{1.273}$$

$$= \lim_{a \to \infty} \left[\frac{a^4}{4} - \frac{a^4}{4} \right] = 0. \tag{1.274}$$

Example 1.8. *Cauchy principal value:* Considering the following integral:

$$I = \int_{-\infty}^{\infty} \frac{(1+x)dx}{1+x^2}, \tag{1.275}$$

which we write as

$$I = \lim_{c \to \infty} \left[\int_{-c}^{0} \frac{(1+x)dx}{1+x^2} + \int_{0}^{c} \frac{(1+x)dx}{1+x^2} \right]. \tag{1.276}$$

For finite c, we evaluate the second integral:

$$\int_{0}^{c} \frac{(1+x)dx}{1+x^2} = \tan^{-1}x + \frac{1}{2}\log(1+x^2) \Big|_{0}^{c} \tag{1.277}$$

$$= \tan^{-1}c + \frac{1}{2}\log(1+c^2), \tag{1.278}$$

which in the limit as $c \to \infty$ diverges as $\left[\tan^{-1}c + \frac{1}{2}\log(1+c^2)\right] \to \infty$. Hence, the integral I also diverges in the Riemann sense by Theorem 1.6. However, since the first integral also diverges, but this time as

$$\lim_{c \to \infty} \int_{-c}^{0} \frac{(1+x)dx}{1+x^2} \to \lim_{c \to \infty} \left[-\tan^{-1}(-c) - \frac{1}{2}\log(1+c^2) \right], \tag{1.279}$$

we consider the two integrals [Eq. (1.276)] together to obtain the Cauchy principal value of I as

$$PV \int_{-\infty}^{\infty} \frac{(1+x)dx}{1+x^2} = \pi. \tag{1.280}$$

1.15   INTEGRALS INVOLVING A PARAMETER

Integrals given in terms of a parameter play an important role in applications. In particular, integrals involving a parameter and with infinite range are of considerable significance. In this regard, we quote three useful theorems:

Theorem 1.7. If there exists a positive function $Q(x)$ satisfying the inequality $|f(\alpha, x)| \leq Q(x)$ for all $\alpha \in [\alpha_1, \alpha_2]$, and if $\int_a^\infty Q(x)dx$ is convergent, then the integral

$$g(\alpha) = \int_a^\infty f(\alpha, x)dx \tag{1.281}$$

is **uniformly convergent** in the interval $[\alpha_1, \alpha_2]$. This is also called the **Weierstrass** M-**test** for uniform convergence. If an integral, $\int_a^\infty f(\alpha, x)dx$, is uniformly convergent in $[\alpha_1, \alpha_2]$, then for any given $\varepsilon > 0$, there exists a number c_0 depending on ε but independent of α such that $\left| \int_c^\infty f(\alpha, x)dx \right| < \varepsilon$ for all $c > c_0 > a$.

Example 1.9.   *Uniform convergence:* Consider the integral

$$I = \int_0^\infty e^{-\alpha x} \sin x \, dx, \tag{1.282}$$

which is uniformly convergent for $\alpha \in [\varepsilon, \infty)$ for every $\varepsilon > 0$. To show this, we choose $Q(x)$ as $e^{-\varepsilon x}$ so that

$$\left| e^{-\alpha x} \sin x \right| \leq e^{-\varepsilon x} \tag{1.283}$$

is true for all $\alpha \geq \varepsilon$. Uniform convergence of I follows, since the integral

$$\int_0^\infty e^{-\varepsilon x} \, dx \tag{1.284}$$

is convergent. Note that by using integration by parts twice, we can evaluate the integral as

$$\int_0^\infty e^{-\alpha x} \sin x \, dx = \frac{1}{1+\alpha^2}, \quad \alpha > 0. \tag{1.285}$$

The case where $\alpha = 0$ may be excluded, since the integral $\int_0^\infty \sin x \, dx$ does not converge at all.

Theorem 1.8. Let $f(\alpha, x)$ and $\partial f(\alpha, x)/\partial \alpha$ be continuous for all $\alpha \in [\alpha_1, \alpha_2]$ and $x \in [a, \infty)$. If the integral

$$g(\alpha) = \int_a^\infty f(\alpha, x)dx \tag{1.286}$$

exists for all $\alpha \in [\alpha_1, \alpha_2]$ and if the integral

$$\int_a^\infty \frac{\partial f(\alpha, x)}{\partial \alpha}\, dx \tag{1.287}$$

is uniformly convergent for all $\alpha \in [\alpha_1, \alpha_2]$, then $g(\alpha)$ is differentiable in $[\alpha_1, \alpha_2]$ (at α_1 from the right and at α_2 from the left) with the derivative

$$\frac{dg}{d\alpha} = \int_a^b \frac{\partial f(\alpha, x)}{\partial \alpha}\, dx. \tag{1.288}$$

In other words, we can interchange the order of differentiation with respect to α and integration with respect to x as

$$\boxed{\frac{d}{d\alpha} \int_a^b f(\alpha, x)dx = \int_a^b \frac{\partial f(\alpha, x)}{\partial \alpha}\, dx.} \tag{1.289}$$

This is also called the **Leibnitz's rule** [2].

Theorem 1.9. Let $f(\alpha, x)$ be continuous for all $\alpha \in [\alpha_1, \alpha_2]$ and $x \in [a, \infty)$. Also let the integral

$$g(\alpha) = \int_a^\infty f(\alpha, x)dx \tag{1.290}$$

be uniformly convergent for all $\alpha \in [\alpha_1, \alpha_2]$. Then,

(a) $g(\alpha)$ is continuous in $[\alpha_1, \alpha_2]$ (at α_1 from the right and at α_2 from the left).

(b) The relation

$$\int_{\alpha_1}^\alpha g(\alpha')d\alpha' = \int_a^\infty \left[\int_{\alpha_1}^\alpha f(x, \alpha')\, d\alpha' \right] dx, \tag{1.291}$$

that is,

$$\boxed{\int_{\alpha_1}^\alpha \left[\int_a^\infty f(\alpha', x)dx \right] d\alpha' = \int_a^\infty \left[\int_{\alpha_1}^\alpha f(x, \alpha')d\alpha' \right] dx,} \tag{1.292}$$

is true for all $\alpha \in [\alpha_1, \alpha_2]$. In other words, the order of the integrals with respect to x and α' can be interchanged. Note that in case (a), the interval for α does not have to be finite.

Remark: In the aforementioned theorems, if the limits of integration are finite but the function $f(\alpha, x)$ or its partial derivative $\partial f(\alpha, x)/\partial \alpha$ is not bounded in the neighborhood of the segment defined by $x = b$ and $\alpha \in [\alpha_1, \alpha_2]$, we say that the integral

$$g(\alpha) = \int_a^b f(\alpha, x) dx \tag{1.293}$$

is uniformly convergent for all $\alpha \in [\alpha_1, \alpha_2]$, if for every $\epsilon > 0$, we can find a $\delta_0 > 0$ independent of α such that the inequality

$$\left| \int_{b-\delta}^b f(\alpha, x) dx \right| < \epsilon \tag{1.294}$$

is true for all $\delta \in [0, \delta_0]$. We can now apply the aforementioned theorems with the upper limit ∞ in the integrals replaced by b and the domain $x \in [a, \infty)$ by $x \in [a, b]$.

Example 1.10. *Integrals depending on a parameter:* Given the integral

$$g(\alpha) = \int_0^\infty \frac{\sin \alpha x}{x} dx = \frac{\pi}{2}, \quad \alpha \neq 0, \tag{1.295}$$

we differentiate with respect to α to write

$$\frac{\partial}{\partial \alpha} \int_0^\infty \frac{\sin \alpha x}{x} dx = \frac{\partial}{\partial \alpha} \left(\frac{\pi}{2} \right) = 0. \tag{1.296}$$

However, this is not correct. The integral on the right-hand side of

$$\int_0^\infty \frac{\partial}{\partial \alpha} \left[\frac{\sin \alpha x}{x} \right] dx = \int_0^\infty \cos \alpha x \, dx \tag{1.297}$$

does not exist, since the limit

$$\lim_{b \to \infty} \int_0^b \cos \alpha x \, dx = \lim_{b \to \infty} \frac{\sin \alpha x}{\alpha} \bigg|_0^b = \lim_{b \to \infty} \frac{1}{\alpha} \sin \alpha b \tag{1.298}$$

does not exist. Hence, the differentiation $dg/d\alpha$ is not justified (Theorem 1.8). On the other hand, given the integral

$$\int_0^{\pi/2} \frac{dx}{\alpha^2 \cos^2 x + \sin^2 x} = \frac{\pi}{2\alpha}, \quad \alpha > 0, \tag{1.299}$$

we can write

$$\frac{\partial}{\partial \alpha} \int_0^{\pi/2} \frac{dx}{\alpha^2 \cos^2 x + \sin^2 x} = \frac{d}{d\alpha} \left(\frac{\pi}{2\alpha} \right) \tag{1.300}$$

to obtain the integral

$$-\int_0^{\pi/2} \frac{2\alpha\cos^2 x\, dx}{(\alpha^2\cos^2 x + \sin^2 x)^2} = -\frac{\pi}{2\alpha^2}. \tag{1.301}$$

Example 1.11. *Integrals depending on a parameter:* Consider

$$f(\alpha, x) = \begin{cases} \dfrac{e^{-\alpha x}\sin x}{x}, & x \neq 0, \\[2mm] 1, & x = 0, \end{cases} \tag{1.302}$$

which is continuous for all x and α. Since

$$\frac{\partial f(\alpha, x)}{\partial \alpha} = -e^{-\alpha x}\sin x, \tag{1.303}$$

which is also continuous for all x and α, and the integral

$$\int_0^\infty \frac{\partial f(\alpha, x)}{\partial \alpha}\, dx = \int_0^\infty e^{-\alpha x}\sin x\, dx \tag{1.304}$$

converges uniformly for all $\alpha > 0$ (Example 1.9), using Theorem 1.8, we conclude that

$$g(\alpha) = \int_0^\infty e^{-\alpha x}\frac{\sin x}{x}\, dx, \quad \alpha > 0, \tag{1.305}$$

exists and can be differentiated to write

$$g'(\alpha) = \frac{\partial}{\partial \alpha}\int_0^\infty e^{-\alpha x}\frac{\sin x}{x}\, dx = \int_0^\infty \frac{\partial}{\partial \alpha}\left[e^{-\alpha x}\frac{\sin x}{x}\right] dx \tag{1.306}$$

$$= -\int_0^\infty e^{-\alpha x}\sin x\, dx = -\frac{1}{1+\alpha^2}, \tag{1.307}$$

where we have used the result in Eq. (1.285). We now use Theorem 1.9 to integrate $g'(\alpha)$ [Eqs. (1.306) and (1.307)], which is continuous for all $\alpha > 0$ to obtain

$$\int_0^\infty g'(\alpha)d\alpha = -\int_0^\infty \frac{1}{1+\alpha^2}\, d\alpha = -\tan^{-1}|_0^\infty = -\frac{\pi}{2}. \tag{1.308}$$

However, we can also write

$$\int_0^\infty g'(\alpha)d\alpha = -\int_0^\infty \left[\int_0^\infty e^{-\alpha x}\sin x\, dx\right] d\alpha = -\int_0^\infty \left[\int_0^\infty e^{-\alpha x}\sin x\, d\alpha\right] dx \tag{1.309}$$

$$= -\int_0^\infty \left[-\frac{e^{-\alpha x}\sin x}{x}\right]_0^\infty dx = -\int_0^\infty \frac{\sin x}{x}\, dx, \quad \alpha > 0, \tag{1.310}$$

which along with Eq. (1.308) yields the definite integral

$$\boxed{\int_0^\infty \frac{\sin x}{x}\,dx = \pi/2.}\tag{1.311}$$

1.16   LIMITS OF INTEGRATION DEPENDING ON A PARAMETER

Let $A(x)$ and $B(x)$ be two continuous functions with continuous derivatives in $[x_1, x_2]$, with $B(x) > A(x)$. Also let $f(t, x)$ and $\partial f(t, x)/\partial x$ be continuous in the region defined by $[x_1, x_2]$ and $[x_1 = A(x), x_2 = B(x)]$. We can now write the integral

$$\left[\int_{A(x)}^{B(x)} f(t, x)dt\right] = F(B(x), A(x), x)\tag{1.312}$$

and its partial derivative with respect to x as

$$\frac{d}{dx}\left[\int_{A(x)}^{B(x)} f(t, x)dt\right] = \frac{\partial F}{\partial B}\frac{dB}{dx} + \frac{\partial F}{\partial A}\frac{dA}{dx} + \frac{\partial F}{\partial x}.\tag{1.313}$$

Using the relations [Eq. (1.233)]:

$$\frac{\partial}{\partial v}\int_u^v f(t)dt = f(v),\tag{1.314}$$

$$\frac{\partial}{\partial u}\int_u^v f(t)dt = -f(u),\tag{1.315}$$

we can write

$$\frac{\partial F}{\partial B} = f(B, x), \quad \frac{\partial F}{\partial A} = -f(A, x).\tag{1.316}$$

We can also write

$$\frac{\partial}{\partial x}\left[\int_{A(x)}^{B(x)} f(t, x)dt\right] = \int_{A(x)}^{B(x)} \frac{\partial f(t, x)}{\partial x}\,dt,\tag{1.317}$$

thus obtaining the useful formula

$$\boxed{\frac{d}{dx}\left[\int_{A(x)}^{B(x)} f(t, x)dt\right] = \int_{A(x)}^{B(x)} \frac{\partial f(t, x)}{\partial x}\,dt + \frac{dB}{dx}f(B(x), x) - \frac{dA}{dx}f(A(x), x).}$$

$$\tag{1.318}$$

1.17   DOUBLE INTEGRALS

Consider a continuous and bounded function $f(x, y)$ defined in a closed region R of the xy-plane. It is important that R be bounded, that is, we can enclose it with a circle of sufficiently large radius. We subdivide R into rectangles by drawing parallels to the x and the y axes (Figure 1.10). We choose only the rectangles in R and numerate them from 1 to n. Area of the ith rectangle is shown as ΔA_i, and the largest of the diagonals, h, is called the **norm** of the mesh. We now form the sum

$$\sum_{i=1}^{n} f(x_i^*, y_i^*) \Delta A_i, \tag{1.319}$$

where, as in the one-dimensional integrals, (x_i^*, y_i^*) is a point arbitrarily chosen in the ith rectangle. If the sum converges to a limit as $h \to 0$, we define the double integral as the limit

$$\lim_{h \to 0} \sum_{i=1}^{n} f(x_i^*, y_i^*) \Delta A_i \to \int \int_R f(x, y) dx\, dy. \tag{1.320}$$

When the region R can be described by the inequalities

$$y_1(x) \leq y \leq y_2(x), \quad x_1 \leq x \leq x_2, \tag{1.321}$$

or

$$x_1(y) \leq x \leq x_2(y), \quad y_1 \leq y \leq y_2, \tag{1.322}$$

where $y_1(x)$, $y_2(x)$ and $x_1(y)$, $x_2(y)$ are continuous functions (Figure 1.11), we can write the double integral for the first case as the iterated integral

$$I = \int_{x_1}^{x_2} \left[\int_{y_1(x)}^{y_2(x)} f(x, y) dy \right] dx. \tag{1.323}$$

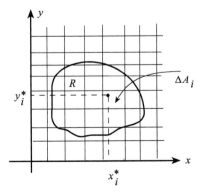

Figure 1.10   The double integral.

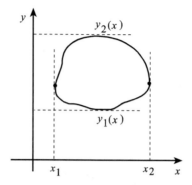

Figure 1.11   Ranges in the iterated integrals.

The definite integral inside the square brackets will yield a function $F(x)$, which reduces I to a one-dimensional definite integral:

$$\int_{x_1}^{x_2} F(x)dx. \tag{1.324}$$

A similar argument can be given for the second case [Eq. (1.322)]. We now present these results in terms of a theorem:

Theorem 1.10. If $f(x,y)$ is continuous and bounded in a closed interval described by the region $[1, 2]$

$$y_1(x) \le y \le y_2(x), \quad x_1 \le x \le x_2, \tag{1.325}$$

then

$$\int_{y_1(x)}^{y_2(x)} f(x,y)dy \tag{1.326}$$

is a continuous function of x and

$$\iint_R f(x,y)dx\,dy = \int_{x_1}^{x_2} \left[\int_{y_1(x)}^{y_2(x)} f(x,y)dy \right] dx. \tag{1.327}$$

Similarly, if R is described by

$$x_1(y) \le x \le x_2(y), \quad y_1 \le y \le y_2, \tag{1.328}$$

then we can write

$$\iint_R f(x,y)dx\,dy = \int_{y_1}^{y_2} \left[\int_{x_1(y)}^{x_2(y)} f(x,y)dx \right] dy. \tag{1.329}$$

1.18   PROPERTIES OF DOUBLE INTEGRALS

We can summarize the basic properties of double integrals, which are essentially same as the definite integrals of functions with single variable as follows:

I.

$$\int\int_R [f(x,y) + g(x,y)]dx\ dy = \int\int_R f(x,y)dx\ dy + \int\int_R g(x,y)dx\ dy,$$

$$(1.330)$$

$$\int\int cf(x,y)dx\ dy = c\int\int_R f(x,y)dx\ dy,\ c\ \text{is a constant},\quad (1.331)$$

$$\int\int_R f(x,y)dx\ dy = \int\int_{R_1} f(x,y)dx\ dy + \int\int_{R_2} f(x,y)dx\ dy,$$

$$(1.332)$$

where R is composed of R_1 and R_2, which overlap only at the boundary.

II. There exists a point (x_1, y_1) in R such that

$$\int\int_R f(x,y)dx\ dy = A_R f(x_1, y_1),\qquad (1.333)$$

where A_R is the area of R. The value $f(x_1, y_1)$ is also the **mean value** $\langle f \rangle_R$ of the function in the region R:

$$\boxed{\langle f \rangle_R = \frac{1}{A_R}\int\int_R f(x,y)dx\ dy.}\qquad (1.334)$$

III.

$$\left|\int\int_R f(x,y)dx\ dy\right| \le M_R \cdot A_R,\qquad (1.335)$$

where M_R is the absolute maximum in R:

$$|f(x,y)| \le M_R \qquad (1.336)$$

and A_R is the area of R.

IV. Uses of double integrals: If we set $f(x,y) = 1$ in $\int\int_R f(x,y)\ dx\ dy$, the double integral corresponds to the area A_R of the region R:

$$\boxed{\int\int_R dx\ dy = A_R.}\qquad (1.337)$$

For $f(x,y) \ge 0$, we can interpret the double integral as the volume between the surface $z = f(x,y)$ and the region R in the xy-plane. If we interpret $f(x,y)$ as the mass density of a flat object lying on the xy-plane covering the region R, the double integral

$$M = \int \int_R f(x, y) dx \, dy \tag{1.338}$$

gives its total mass M.

1.19   TRIPLE AND MULTIPLE INTEGRALS

The methods and the results developed for the double integrals can easily be extended to the triple and multiple integrals:

$$\int \int \int_R f(x, y, z) dx \, dy \, dz, \quad \int \int \int \int_R f(x, y, z, w) dx \, dy \, dz \, dw, \text{etc.} \tag{1.339}$$

Following the arguments given for the single and the double integrals, for a continuous and bounded function $f(x, y, z)$ in a bounded region R defined by

$$z_1(x, y) \leq z \leq z_2(x, y), \quad y_1(x) \leq y \leq y_2(x), \quad x_1 \leq x \leq x_2, \tag{1.340}$$

we can define the triple integral

$$\int \int \int_R f(x, y, z) dx \, dy \, dz = \int_{x_1}^{x_2} \left[\int_{y_1(x)}^{y_2(x)} \left[\int_{z_1(x,y)}^{z_2(x,y)} f(x, y, z) dz \right] dy \right] dx. \tag{1.341}$$

An obvious application of the triple integral is when $f(x, y, z) = 1$, which gives the volume V_R of the region R:

$$\int \int \int_R dx \, dy \, dz = V_R. \tag{1.342}$$

In physical applications, total amount of mass, charge, etc., with the density $\rho(x, y, z)$ are given as the triple integral

$$\int \int \int_R \rho(x, y, z) dx \, dy \, dz. \tag{1.343}$$

The **average** or the **mean** value of a function $f(x, y, z)$ in the region R with the volume V_R is defined as

$$\langle f \rangle_R = \frac{1}{V_R} \int \int \int_R f(x, y, z) dx \, dy \, dz. \tag{1.344}$$

Example 1.12. *Volume between two surfaces:* To find the volume between the cone $z = \sqrt{x^2 + y^2}$ and the paraboloid $z = x^2 + y^2$, we first write the triple integral

$$V = \int_0^1 \int_0^1 \left[\int_{x^2+y^2}^{\sqrt{x^2+y^2}} dz \right] dx\, dy = \int_0^1 \int_0^1 \left[\sqrt{x^2 + y^2} - x^2 - y^2 \right] dx\, dy.$$

$$(1.345)$$

We now use plane polar coordinates to write this as

$$V = \int_0^1 \int_0^{2\pi} (\rho - \rho^2)\rho\, d\phi\, d\rho = 2\pi \int_0^1 (\rho - \rho^2)\rho\, d\rho = 2\pi \left[\frac{\rho^3}{3} - \frac{\rho^4}{4} \right]_0^1 = \frac{\pi}{6}.$$

$$(1.346)$$

REFERENCES

1. Apostol, T.M. (1971). *Mathematical Analysis*. Reading, MA: Addison-Wesley, fourth printing.

2. Kaplan, W. (1984). *Advanced Calculus*, 3e. Reading, MA: Addison-Wesley.

PROBLEMS

1. Determine the critical points as well as the absolute maximum and minimum of the functions

    (i) $y = \ln x,\ 0 < x \le 2$,

    (ii) $y = x/(1 + 2x^2)$,

    (iii) $y = x^3 + 2x^2 + 1, -2 < x < 1$.

2. Determine the critical points of the functions

    (i) $z = x^3 - 6xy^2 + y^3$,

    (ii) $z = 1 + x^2 + y^2$,

    (iii) $z = x^2 - 4xy - y^2$.

and test for maximum or minimum.

3. Find the maximum and minimum points of $z = x^2 + 24xy + 8y^2$ subject to the condition $x^2 + y^2 = 25$.

4. Find the critical points of $w = x + y$ subject to $x^2 + y^2 + z^2 = 1$ and identify whether they are maximum or minimum.

5. Express the partial differential equation

$$\frac{\partial^2 \Psi(\overrightarrow{r})}{\partial x^2} + \frac{\partial^2 \Psi(\overrightarrow{r})}{\partial y^2} + \frac{\partial^2 \Psi(\overrightarrow{r})}{\partial z^2} = 0,$$

in spherical coordinates (r, θ, ϕ) defined by the equations

$$x = r \sin\theta \cos\phi, \quad y = r\sin\theta\sin\phi, \quad z = r\cos\theta,$$

where $r \in [0, \infty)$, $\theta \in [0, \pi]$, $\phi \in [0, 2\pi]$. Next, first show that the inverse transformation exists and then find it.

6. Given the mapping

$$x = u^2 - v^2,$$

$$y = 2uv.$$

(i) Write the Jacobian.

(ii) Evaluate the derivatives $\left(\dfrac{\partial u}{\partial y}\right)_x$ and $\left(\dfrac{\partial v}{\partial y}\right)_x$.

7. Find $\left(\dfrac{\partial u}{\partial x}\right)_y$ and $\left(\dfrac{\partial u}{\partial y}\right)_x$ for

$$e^u + xu - yv - 1 = 0,$$

$$e^v - xv + yu - 2 = 0.$$

8. Given the transformation functions

$$x = x(u, v),$$

$$y = y(u, v),$$

show that the inverse transformations

$$u = u(x, y),$$

$$v = v(x, y)$$

satisfy

$$\frac{\partial u}{\partial x} = \frac{1}{J}\frac{\partial y}{\partial v}, \quad \frac{\partial u}{\partial y} = -\frac{1}{J}\frac{\partial x}{\partial v}, \quad \frac{\partial v}{\partial x} = -\frac{1}{J}\frac{\partial y}{\partial u}, \quad \frac{\partial v}{\partial y} = \frac{1}{J}\frac{\partial x}{\partial u},$$

where $J = \dfrac{\partial(x, y)}{\partial(u, v)}$. Apply your result to Problem 6.

9. Given the transformation functions

$$x = x(u, v, w), \quad y = y(u, v, w), \quad z = z(u, v, w)$$

with the Jacobian $J = \dfrac{\partial(x, y, z)}{\partial(u, v, w)}$, show that the inverse transformation functions have the derivatives

$$\frac{\partial u}{\partial x} = \frac{1}{J}\frac{\partial(y, z)}{\partial(v, w)}, \quad \frac{\partial u}{\partial y} = \frac{1}{J}\frac{\partial(z, x)}{\partial(v, w)}, \quad \frac{\partial u}{\partial z} = \frac{1}{J}\frac{\partial(x, y)}{\partial(v, w)},$$

$$\frac{\partial v}{\partial x} = \frac{1}{J}\frac{\partial(y, z)}{\partial(w, u)}, \quad \frac{\partial v}{\partial y} = \frac{1}{J}\frac{\partial(z, x)}{\partial(w, u)}, \quad \frac{\partial v}{\partial z} = \frac{1}{J}\frac{\partial(x, y)}{\partial(w, u)},$$

$$\frac{\partial w}{\partial x} = \frac{1}{J}\frac{\partial(y, z)}{\partial(u, v)}, \quad \frac{\partial w}{\partial y} = \frac{1}{J}\frac{\partial(z, x)}{\partial(u, v)}, \quad \frac{\partial w}{\partial z} = \frac{1}{J}\frac{\partial(x, y)}{\partial(u, v)}.$$

Verify your result in Problem 5.

10. In a one-dimensional conservative system, potential energy can be represented by a (scalar) function $V(x)$, where the negative of the derivative of the potential gives the x component of the force: $F_x(x) = -dV/dx$. With the aid of a sketch, analyze the forces on a conservative system when it is displaced away from equilibrium by a small amount.

11. In one-dimensional potential problems, show that near equilibrium potential can be approximated by the harmonic oscillator potential

$$V(x) = \frac{1}{2}k(x - x_0)^2,$$

where k is a constant and x_0 is the equilibrium point. What is k?

12. Expand $z(x, y) = x^3 \sin y + y^2 \cos x$ in Taylor series up to third order about the origin.

13. If $x = x(u, v)$ and $y = y(u, v)$, then show the following:

(i) $\left(\dfrac{\partial x}{\partial u}\right)_v \left(\dfrac{\partial u}{\partial x}\right)_y = \left(\dfrac{\partial y}{\partial v}\right)_u \left(\dfrac{\partial v}{\partial y}\right)_x,$

(ii) $\left(\dfrac{\partial x}{\partial v}\right)_u \left(\dfrac{\partial v}{\partial x}\right)_y = \left(\dfrac{\partial u}{\partial y}\right)_x \left(\dfrac{\partial y}{\partial u}\right)_v.$

14. Show the integrals

(i) $\displaystyle\int_0^\infty \frac{\sin x \cos x}{x}\, dx = \frac{\pi}{4},$

$$\text{(ii)} \quad \int_0^\infty \frac{\sin^2 x}{x^2}\, dx = \frac{\pi}{2}.$$

Hint: Use $\displaystyle\int_0^\infty \frac{\sin x}{x}\, dx = \frac{\pi}{2}.$

15. Evaluate the improper integrals:

$$\text{(i)} \quad \int_0^1 \frac{dx}{\sqrt{1 - x^2}},$$

$$\text{(ii)} \quad \int_0^{1/2} \frac{dx}{\sqrt{x}(1 - 2x)}.$$

16. First show the following:

$$\text{(i)} \quad \int_1^\infty \frac{dx}{x^p} \text{ converges if and only if } p > 1,$$

$$\text{(ii)} \quad \int_0^1 \frac{dx}{x^p} \text{ converges if and only if } p < 1,$$

$$\text{(iii)} \quad \int_0^c \frac{dx}{|c - x|^p} \text{ converges if and only if } p < 1$$

and then check the convergence of

$$\text{(i)} \quad \int_0^\infty \frac{dx}{\sqrt{2x + x^3}},$$

$$\text{(ii)} \quad \int_0^\infty \frac{dx}{\sqrt{x + 2x^2}}.$$

17. Check the convergence of the integral

$$\int_0^1 \frac{x^2\, dx}{(1 - x^2)^{1/2}(2x^3 + 1)}.$$

18. Show that the following integral is convergent by using integration by parts:

$$\int_1^\infty \frac{\sin x}{x}\, dx.$$

19. Using the integral

$$I(a, b) = \int_0^{\pi/2} \frac{dx}{a^2 \cos^2 x + b^2 \sin^2 x} = \frac{\pi}{2ab}, \quad a > 0,\ b > 0,$$

where a and b are two parameters, show the integral

$$\int_0^{\pi/2} \frac{dx}{(a^2\cos^2 x + b^2\sin^2 x)^2} = \frac{\pi}{4ab}\left(\frac{1}{a^2} + \frac{1}{b^2}\right).$$

20. Determine the α values for which the following integrals are uniformly convergent:

(i) $\displaystyle\int_0^\infty \frac{\cos x\alpha}{1+x^2}\, dx,$

(ii) $\displaystyle\int_0^\infty \frac{1}{x^2+\alpha^2}\, dx.$

21. Can the order of integration be interchanged in the following integral (explain):

$$I = \int_0^1 \left[\int_0^1 \frac{x-\alpha}{(x+\alpha)^3}\, dx\right] d\alpha.$$

22. Use the result

$$g(\alpha) = \int_0^\infty \frac{\sin x\alpha}{x(x^2+1)}\, dx = \frac{\pi}{2}(1-e^{-\alpha}), \quad \alpha > 0,$$

to deduce the following integrals:

(i) $\displaystyle\int_0^\infty \frac{\sin x\alpha}{x(x^2+c^2)}\, dx = \frac{\pi}{2c^2}(1-e^{-c\alpha}), \quad c > 0,$

(ii) $\displaystyle\int_0^\infty \frac{\cos x\alpha}{(x^2+c^2)}\, dx = \frac{\pi e^{-c\alpha}}{2c}, \quad c > 0.$

23. Evaluate the following double integral over the triangle with the vertices $(-1,0)$, $(0,1)$, and $(2,0)$:

$$I = \int\int 2y\, dx\, dy.$$

24. Evaluate I over the triangle with the vertices $(0,0)$, $(1,1)$, and $(1,3)$, where

$$I = \int\int xy\, dx\, dy.$$

25. Evaluate the integral

$$\int_{x=0}^2 \int_{y=0}^{x^2} xy\, dy\, dx.$$

26. First evaluate the integral

$$\int_{x=0}^{2} \int_{y=0}^{x^2} (x^2 + 2y^2)xy \, dy \, dx$$

and then repeat the integration over the same region but with the x integral taken first.

27. Test the following integral for convergence:

$$\iiint_{x^2+y^2+z^2 \leq 1} \ln(x^2 + y^2 + z^2) dx \, dy \, dz.$$

28. Evaluate the integrals

(i) $$\int_{z=0}^{2} \int_{x=z}^{2} \int_{y=6x}^{z} dy \, dx \, dz,$$

(ii) $$\int_{y=-2}^{2} \int_{z=1}^{2} \int_{x=y+z}^{2y+z} y \, dx \, dz \, dy.$$

CHAPTER 2

VECTOR ANALYSIS

Certain properties in nature such as mass, charge, temperature, are scalars, which can be defined at a point by just giving a single number. In other words, they have only magnitude. Properties such as velocity and acceleration are on the other hand, vector quantities, which have both magnitude and direction. Most of the Newtonian mechanics and Maxwell's electrodynamics are formulated in terms of the language of vector analysis. In this chapter, we introduce the basic properties of scalars, vectors, and their fields.

2.1 VECTOR ALGEBRA: GEOMETRIC METHOD

Abstract vectors are defined as directed line segments. The length of the line segment describes the **magnitude** of the physical property and the arrow indicates its **direction**. As long as we preserve their magnitude and direction, we can move abstract vectors freely in space. In this regard, $\vec{A}$ and $\vec{A'}$ in Figure 2.1 are equivalent vectors:

$$\vec{A} \equiv \vec{A'}. \tag{2.1}$$

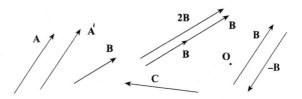

Figure 2.1   Abstract vectors.

We can use Latin letters with an arrow to show vector quantities:

$$\vec{A}, \vec{B}, \vec{a}, \vec{c}, \ldots \tag{2.2}$$

As we use in the figures, it is also customary to use boldface letters:

$$\mathbf{A}, \mathbf{B}, \mathbf{a}, \mathbf{c}, \ldots \tag{2.3}$$

The **magnitude** or the **norm** of a vector is a positive number shown as

$$|\vec{A}|, |\vec{B}|, |\vec{a}| \ldots, \text{ or simply as } A, B, a, \ldots \tag{2.4}$$

Multiplication of a vector with a positive number, $\alpha > 0$, multiplies the magnitude by the same number while leaving the direction untouched:

$$|\alpha \vec{A}| = \alpha |A| = \alpha A. \tag{2.5}$$

Multiplication of a vector with a negative number, $\beta < 0$, reverses the direction while changing the magnitude as

$$|\beta \vec{A}| = |\beta||\vec{A}| = |\beta|A. \tag{2.6}$$

Addition of vectors can be done by using the **parallelogram method** (Figure 2.2). A convenient way to add vectors is to draw them head to tail. This allows us to define the **null vector** $\vec{0}$ as

$$\boxed{\vec{A} + (-1)\vec{A} = \vec{0}.} \tag{2.7}$$

Using the cosine and the sine theorems, we can find the magnitude r and the angle ϕ of the vector $\vec{r} = \vec{A} + \vec{B}$ as (Figure 2.2)

$$r = (A^2 + B^2 + 2AB\cos\theta)^{1/2}, \tag{2.8}$$

$$\phi = \arcsin\left(\frac{A}{r}\sin\theta\right). \tag{2.9}$$

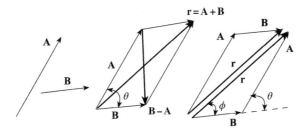

Figure 2.2 Addition of vectors, $\mathbf{r} = \mathbf{A} + \mathbf{B}$.

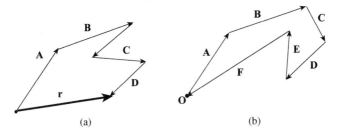

Figure 2.3 Addition of vectors by drawing them head to tail.

With respect to **addition**, vectors **commute**:

$$\boxed{\vec{A} + \vec{B} = \vec{B} + \vec{A}} \tag{2.10}$$

and **associate**:

$$\boxed{(\vec{A} + \vec{B}) + \vec{C} = \vec{A} + (\vec{B} + \vec{C}).} \tag{2.11}$$

A set of vectors $\{\vec{A}, \vec{B}, \vec{C}, \ldots\}$ can be added by drawing them head to tail, where in Figure 2.3a $\vec{r} = \vec{A} + \vec{B} + \vec{C} + \vec{D}$ is the **resultant**. If the resultant is a null vector, then the head of the last vector added and the tail of the first vector added meet (Figure 2.3b): $\vec{A} + \vec{B} + \vec{C} + \vec{D} + \vec{E} + \vec{F} = \vec{0}$.

Example 2.1. *Point of application:* In physics the point of application of a vector is important. Hence, we have to be careful when we move them around to find their resultant. In some equilibrium problems, forces act at the center of mass; hence, the net force is zero (Figure 2.4left). In Figure 2.4right, where there is a net force, the resultant also acts at point O.

2.1.1 Multiplication of Vectors

For the product of two vectors, there are two types of multiplication. The **scalar product**, which is also known as the **dot product** or the **inner product**, is

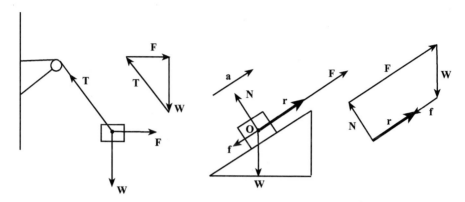

Figure 2.4   Force problems.

defined as

$$\boxed{\vec{A} \cdot \vec{B} = \vec{B} \cdot \vec{A} = AB \cos \theta,}$$ (2.12)

where θ is the angle between the two vectors. The dot product is also shown as $(\vec{A}, \vec{B})$. If we write $\vec{A} \cdot \vec{B}$ as

$$\vec{A} \cdot \frac{\vec{B}}{B} = A \cos \theta,$$ (2.13)

$$A_B = \vec{A} \cdot \hat{e}_B = A \cos \theta,$$ (2.14)

where $\hat{e}_B$ is a unit vector along the direction of $\vec{B}$, the dot product becomes a convenient way to find the **projection** of a vector along another vector, that is, the component A_B of $\vec{A}$ along $\hat{e}_B$.

In physics, **work** is a **scalar** quantity defined as the force times the displacement along the direction of the force. In other words, it is the dot product of the force with the displacement. For a particle in motion, the infinitesimal work is written as the dot product of the force $\vec{F}$ with the infinitesimal displacement vector $d\vec{s}$ along the trajectory of the particle (Figure 2.5):

$$\boxed{\delta W = \vec{F} \cdot d\vec{s}.}$$ (2.15)

We have chosen to write δW instead of dW to emphasize the fact that in general, work is path-dependent. To find the total work done between two points A and B, we have to integrate over a specific path C connecting the two points:

$$W_{A \to B}(\text{over } C) = \int_{\substack{A \\ C}}^{B} \vec{F} \cdot d\vec{s}.$$ (2.16)

A different path connecting A to B in general yields a different value for the work done.

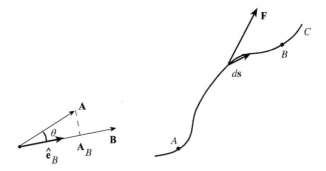

Figure 2.5   Dot or scalar product.

Another type of vector multiplication is called the **vector product** or the **cross product**, which is defined as the binary operation

$$\boxed{\overrightarrow{C} = \overrightarrow{A} \times \overrightarrow{B}.}$$ (2.17)

The result is a new vector $\overrightarrow{C}$, which is perpendicular to the plane defined by $\overrightarrow{A}$ and $\overrightarrow{B}$ with the magnitude C (Figure 2.6):

$$\boxed{C = AB\sin\theta.}$$ (2.18)

The **direction** is found by the **right-hand rule**, that is, when we curl the fingers of our right hand from the first vector to the second, the direction of our thumb gives the direction of $\overrightarrow{C}$. Note that when the order of the vectors multiplied is reversed, the direction of $\overrightarrow{C}$ also reverses. **Angular momentum**, $\overrightarrow{L} = \overrightarrow{r} \times \overrightarrow{p}$, and **torque**, $\overrightarrow{\tau} = \overrightarrow{r} \times \overrightarrow{F}$, are two important physical properties that are defined in terms of the vector product (Figure 2.6). In these expressions, $\overrightarrow{r}$ is the position vector defined with respect to an origin O and $\overrightarrow{p}$ and $\overrightarrow{F}$ are the momentum and the force vectors, respectively. In celestial mechanics, we usually choose the origin as the center of attraction. The gravitational force $\overrightarrow{F}_g$ on a planet m, is always directed toward the center of attraction M; hence, the torque is zero. Since the rate of change of the angular momentum is equal to the torque, in central force problems angular momentum is conserved,

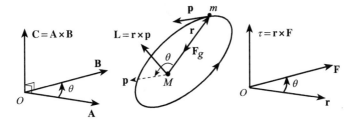

Figure 2.6   Cross or vector product.

that is, its magnitude and direction remains fixed. This means that the orbit of a planet always remains in the plane defined by the two vectors $\vec{r}$ and $\vec{p}$. This allows us to use plane polar coordinates in orbit calculations thereby simplifying the algebra significantly.

2.2   VECTOR ALGEBRA: COORDINATE REPRESENTATION

A convenient way to approach vector algebra came with Descartes through the introduction of **Cartesian coordinates**. We define a Cartesian coordinate system by choosing three mutually orthogonal straight-lines, which we identify as the x_1, x_2, x_3-axes, respectively. We also draw three **unit basis vectors,** $\widehat{e}_1$, $\widehat{e}_2$, $\widehat{e}_3$, along these axes (Figure 2.7). A point P in space can now be represented by the **position vector** $\vec{r}$, which can be written as the sum of three vectors, $x_1\widehat{e}_1$, $x_2\widehat{e}_2$, $x_3\widehat{e}_3$, along their respective axes as

$$\vec{r} = x_1\widehat{e}_1 + x_2\widehat{e}_2 + x_3\widehat{e}_3,$$ (2.19)

where x_1, x_2, x_3 are called the **coordinates** of the point P or the **components** of $\vec{r}$. We also use $\vec{x}$ for the position vector. In general, a **vector** $\vec{A}$, can be written as the sum of three vectors:

$$\vec{A} = A_1\widehat{e}_1 + A_2\widehat{e}_2 + A_3\widehat{e}_3,$$ (2.20)

where A_1, A_2, A_3 are called the **components** of $\vec{A}$. We can also write a vector as the set of ordered numbers:

$$\vec{A} = (A_1, A_2, A_3).$$ (2.21)

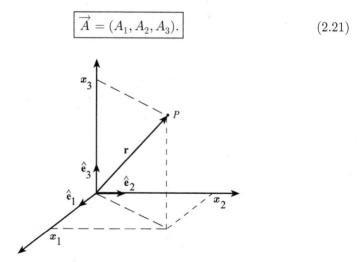

Figure 2.7   Cartesian coordinates.

Since the unit basis vectors are **mutually orthogonal**, they satisfy the relations

$$\widehat{e}_1 \cdot \widehat{e}_1 = 1, \quad \widehat{e}_1 \cdot \widehat{e}_2 = 0, \quad \widehat{e}_1 \cdot \widehat{e}_3 = 0, \tag{2.22}$$

$$\widehat{e}_2 \cdot \widehat{e}_1 = 0, \quad \widehat{e}_2 \cdot \widehat{e}_2 = 1, \quad \widehat{e}_2 \cdot \widehat{e}_3 = 0, \tag{2.23}$$

$$\widehat{e}_3 \cdot \widehat{e}_1 = 0, \quad \widehat{e}_3 \cdot \widehat{e}_2 = 0, \quad \widehat{e}_3 \cdot \widehat{e}_3 = 1, \tag{2.24}$$

which can be summarized as

$$\boxed{\widehat{e}_i \cdot \widehat{e}_j = \delta_{ij}, \quad i,j = 1,2,3.} \tag{2.25}$$

The right-hand side, δ_{ij}, is called the **Kronecker delta**, which is equal to one when the two indices are equal and zero when the two indices are different:

$$\delta_{ij} = \begin{cases} 0, & i \neq j, \\ 1, & i = j. \end{cases} \tag{2.26}$$

Using Eq. (2.25), we can write the square of the magnitude of a vector $\overrightarrow{A}$ in the following equivalent ways:

$$A^2 = \overrightarrow{A} \cdot \overrightarrow{A} = (\overrightarrow{A}, \overrightarrow{A}) = (A_1, A_2, A_3) \cdot (A_1, A_2, A_3) \tag{2.27}$$

$$= (A_1\widehat{e}_1 + A_2\widehat{e}_2 + A_3\widehat{e}_3) \cdot (A_1\widehat{e}_1 + A_2\widehat{e}_2 + A_3\widehat{e}_3) \tag{2.28}$$

$$= \sum_{i=1}^{3}\sum_{j=1}^{3} A_i A_j (\widehat{e}_i \cdot \widehat{e}_j) = \sum_{i=1}^{3}\sum_{j=1}^{3} A_i A_j \delta_{ij} = A_1^2 + A_2^2 + A_3^2. \tag{2.29}$$

The **components**, A_i, of $\overrightarrow{A}$ are obtained from the scalar products of $\overrightarrow{A}$ with the unit basis vectors:

$$\boxed{A_i = \overrightarrow{A} \cdot \widehat{e}_i, \quad i = 1,2,3.} \tag{2.30}$$

In component notation, two vectors are added by adding their respective components:

$$\overrightarrow{A} + \overrightarrow{B} = (A_1, A_2, A_3) + (B_1, B_2, B_3) = (A_1 + B_1, A_2 + B_2, A_3 + B_3) \tag{2.31}$$

$$= (A_1 + B_1)\widehat{e}_1 + (A_2 + B_2)\widehat{e}_2 + (A_3 + B_3)\widehat{e}_3. \tag{2.32}$$

Multiplication of a vector with a scalar α is accomplished by multiplying each component with that scalar:

$$\boxed{\alpha\overrightarrow{A} = (\alpha A_1, \alpha A_2, \alpha A_3).} \tag{2.33}$$

Dot product of two vectors can be written as

$$\vec{A} \cdot \vec{B} = (\vec{A}, \vec{B}) = A_1 B_1 + A_2 B_2 + A_3 B_3 = \sum_{i=1}^{3} A_i B_i. \qquad (2.34)$$

Using component notation one can prove the following properties of the dot product:

Properties of the dot product

$$(\vec{A}, \vec{B}) = (\vec{B}, \vec{A}), \qquad (2.35)$$

$$(\vec{A}, \vec{B} + \vec{C}) = (\vec{A}, \vec{B}) + (\vec{A}, \vec{C}), \qquad (2.36)$$

$$(\alpha \vec{A}, \vec{B}) = \alpha (\vec{A}, \vec{B}), \quad \alpha \text{ is a scalar}, \qquad (2.37)$$

$$\text{If } \vec{A} = 0, \quad \text{then } (\vec{A}, \vec{A}) = 0, \qquad (2.38)$$

$$\text{If } \vec{A} \neq 0, \quad \text{then } (\vec{A}, \vec{A}) > 0, \qquad (2.39)$$

$$(\vec{A}, \vec{B})^2 \leq (\vec{A}, \vec{A})(\vec{B}, \vec{B}), \qquad (2.40)$$

$$|\vec{A} - \vec{B}| \leq |\vec{A} + \vec{B}| \leq |\vec{A}| + |\vec{B}|. \qquad (2.41)$$

Equation (2.40) is known as the **Schwarz inequality**. Equation (2.41) is the **triangle inequality**, which says that the sum of the lengths of the two sides of a triangle is always greater than or equal to the length of the third side.

Before we write the cross product of two vectors in component notation, we write the following relations for the basis vectors:

$$\begin{array}{lll} \hat{e}_1 \times \hat{e}_1 = 0, & \hat{e}_1 \times \hat{e}_2 = \hat{e}_3, & \hat{e}_1 \times \hat{e}_3 = -\hat{e}_2, \\ \hat{e}_2 \times \hat{e}_1 = -\hat{e}_3, & \hat{e}_2 \times \hat{e}_2 = 0, & \hat{e}_2 \times \hat{e}_3 = \hat{e}_1, \\ \hat{e}_3 \times \hat{e}_1 = \hat{e}_2, & \hat{e}_3 \times \hat{e}_2 = -\hat{e}_1, & \hat{e}_3 \times \hat{e}_3 = 0. \end{array} \qquad (2.42)$$

The **cross product** of two vectors can now be written as

$$\vec{A} \times \vec{B} = (A_1 \hat{e}_1 + A_2 \hat{e}_2 + A_3 \hat{e}_3) \times (B_1 \hat{e}_1 + B_2 \hat{e}_2 + B_3 \hat{e}_3) \qquad (2.43)$$

$$= (A_2 B_3 - A_3 B_2)\hat{e}_1 + (A_3 B_1 - A_1 B_3)\hat{e}_2 + (A_1 B_2 - A_2 B_1)\hat{e}_3. \qquad (2.44)$$

We now introduce the **permutation symbol** ε_{ijk}:

$$\varepsilon_{ijk} = \begin{cases} 0 & \text{When any two indices are equal,} \\ 1 & \text{For even (cyclic) permutations: } 123, \, 231, \, 312, \\ -1 & \text{For odd (anticyclic) permutations: } 213, \, 321, \, 132. \end{cases} \qquad (2.45)$$

An important identity that the permutation symbol satisfies is

$$\sum_{i=1}^{3} \varepsilon_{ijk}\varepsilon_{ilm} = \delta_{jl}\delta_{km} - \delta_{jm}\delta_{kl}. \tag{2.46}$$

Using the permutation symbol, we can write the ith component of a cross product as

$$(\overrightarrow{A} \times \overrightarrow{B})_i = \sum_{j=1}^{3}\sum_{k=1}^{3} \varepsilon_{ijk}A_j B_k. \tag{2.47}$$

Using determinants, we can also write a cross product as

$$\overrightarrow{A} \times \overrightarrow{B} = \det \begin{pmatrix} \widehat{e}_1 & \widehat{e}_2 & \widehat{e}_3 \\ A_1 & A_2 & A_3 \\ B_1 & B_2 & B_3 \end{pmatrix} \tag{2.48}$$

$$= (A_2 B_3 - A_3 B_2)\widehat{e}_1 + (A_3 B_1 - A_1 B_3)\widehat{e}_2 + (A_1 B_2 - A_2 B_1)\widehat{e}_3. \tag{2.49}$$

Note that we prefer to use the index notation (x_1, x_2, x_3) over labeling of the axes as (x, y, z) and show the unit basis vectors as $(\widehat{e}_1, \widehat{e}_2, \widehat{e}_3)$ instead of $(\widehat{i}, \widehat{j}, \widehat{k})$. The advantages will become clear when we introduce generalized coordinates and tensors in n dimensions.

Example 2.2. *Triple product:* In applications, we frequently encounter the scalar triple product:

$$\overrightarrow{A} \cdot (\overrightarrow{B} \times \overrightarrow{C}) = A_1(B_2 C_3 - B_3 C_2) + A_2(B_3 C_1 - B_1 C_3) + A_3(B_1 C_2 - B_2 C_1), \tag{2.50}$$

which is geometrically equal to the volume of a parallelepiped defined by the vectors $\overrightarrow{A}, \overrightarrow{B}$, and $\overrightarrow{C}$ (Figure 2.8). Note that

$$\sigma_{BC} = |\overrightarrow{B} \times \overrightarrow{C}| = Bh_1 = B(C \sin \phi) \tag{2.51}$$

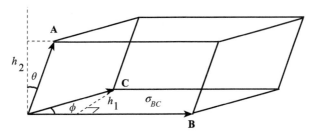

Figure 2.8 Tripple product is the volume of a parallelepiped.

is the area of the base and $h_2 = A \cos \theta$ is the perpendicular height to the base thereby giving the **volume** as

$$V = h_2 \cdot \sigma_{BC} = (A \cos \theta) BC \sin \phi = \vec{A} \cdot (\vec{B} \times \vec{C}). \tag{2.52}$$

Using index notation, one can easily show that

$$V = \vec{A} \cdot (\vec{B} \times \vec{C}) = \vec{B} \cdot (\vec{C} \times \vec{A}) = \vec{C} \cdot (\vec{A} \times \vec{B}). \tag{2.53}$$

The triple product can also be expressed as the determinant

$$\vec{A} \cdot (\vec{B} \times \vec{C}) = \det \begin{pmatrix} A_1 & A_2 & A_3 \\ B_1 & B_2 & B_3 \\ C_1 & C_2 & C_3 \end{pmatrix}. \tag{2.54}$$

Properties of the cross product can be summarized as follows:

Properties of the cross product

$$\vec{A} \times \vec{B} = -\vec{B} \times \vec{A}, \tag{2.55}$$

$$\vec{A} \times (\vec{B} + \vec{C}) = \vec{A} \times \vec{B} + \vec{A} \times \vec{C}, \tag{2.56}$$

$$(\alpha \vec{A}) \times \vec{B} = \alpha (\vec{A} \times \vec{B}), \quad \alpha \text{ is a scalar}, \tag{2.57}$$

$$\vec{A} \times (\vec{B} \times \vec{C}) = \vec{B}(\vec{A} \cdot \vec{C}) - \vec{C}(\vec{A} \cdot \vec{B}), \tag{2.58}$$

$$(\vec{A} \times \vec{B}) \cdot (\vec{A} \times \vec{B}) = A^2 B^2 - (\vec{A} \cdot \vec{B})^2, \tag{2.59}$$

$$\vec{A} \times (\vec{B} \times \vec{C}) + \vec{B} \times (\vec{C} \times \vec{A}) + \vec{C} \times (\vec{A} \times \vec{B}) = 0. \tag{2.60}$$

Using the index notation, we can prove Eq. (2.58) as

$$[\vec{A} \times (\vec{B} \times \vec{C})]_i = \sum_{j=1}^{3} \sum_{k=1}^{3} \varepsilon_{ijk} A_j \sum_{l=1}^{3} \sum_{m=1}^{3} \varepsilon_{klm} B_l C_m \tag{2.61}$$

$$= \sum_{j=1}^{3} \sum_{l=1}^{3} \sum_{m=1}^{3} \left[\sum_{k=1}^{3} \varepsilon_{kij} \varepsilon_{klm} \right] A_j B_l C_m \tag{2.62}$$

$$= \sum_{m=1}^{3} \sum_{j=1}^{3} \sum_{l=1}^{3} [\delta_{il} \delta_{jm} - \delta_{im} \delta_{jl}] A_j B_l C_m \tag{2.63}$$

$$= B_i \left[\sum_{m=1}^{3} A_m C_m \right] - \left[\sum_{l=1}^{3} A_l B_l \right] C_i \tag{2.64}$$

$$= B_i (\vec{A} \cdot \vec{C}) - (\vec{A} \cdot \vec{B}) C_i. \tag{2.65}$$

2.3   LINES AND PLANES

We define the parametric **equation of a line** passing through a point $\overrightarrow{P} = (x_{01}, x_{02}, x_{02})$ and in the direction of the vector $\overrightarrow{A}$ as

$$\boxed{\overrightarrow{x} = \overrightarrow{P} + t\overrightarrow{A},} \tag{2.66}$$

where $\overrightarrow{x} = (x_1, x_2, x_3)$ is a point on the line and t is a parameter (Figure 2.9). If the components of $\overrightarrow{A}$ are (a_1, a_2, a_3), we obtain the **parametric** equation of a line in space:

$$x_1(t) = x_{01} + ta_1, \tag{2.67}$$

$$x_2(t) = x_{02} + ta_2, \tag{2.68}$$

$$x_3(t) = x_{03} + ta_3. \tag{2.69}$$

In two dimensions, say on the x_1x_2-plane, the third equation above is absent. Thus, by eliminating t among the remaining two equations, we can express the equation of a line in one of the following forms:

$$x_2 = x_{02} + \left[\frac{x_1 - x_{01}}{a_1}\right]a_2, \tag{2.70}$$

$$x_2 = \frac{a_2}{a_1}x_1 + x_{02} - \frac{x_{01}}{a_1}a_2, \tag{2.71}$$

$$a_1x_2 - a_2x_1 = (x_{02}a_1 - x_{01}a_2). \tag{2.72}$$

Consider a plane that contains the point P with the coordinates (x_{01}, x_{02}, x_{03}). Let $\overrightarrow{N}$ be any nonzero vector normal to the plane at P and let $\overrightarrow{x}$ be any point

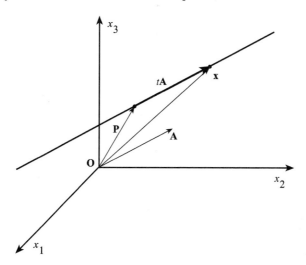

Figure 2.9   Equation of a line.

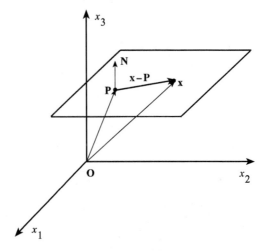

Figure 2.10   Equation of a plane.

on the plane (Figure 2.10). Since $(\vec{x} - \vec{P})$ is a vector on the plane whose dot product with $\vec{N}$ is zero, we can write

$$(\vec{x} - \vec{P}) \cdot \vec{N} = 0. \tag{2.73}$$

Since any $\vec{N}$ perpendicular to the plane satisfies this equation, we can also write this equation as

$$\boxed{(\vec{x} - \vec{P}) \cdot t\hat{n} = 0,} \tag{2.74}$$

where t is a parameter and $\hat{n}$ is the unit normal in the direction of $\vec{N}$. If we write $\hat{n}$ as

$$\hat{n} = (n_1, n_2, n_3), \quad n_1^2 + n_2^2 + n_3^2 = 1, \tag{2.75}$$

we can write the **equation of a plane**, that includes the point (x_{01}, x_{02}, x_{03}) and with its normal pointing in the direction $\hat{n}$ as

$$\boxed{n_1 x_1 + n_2 x_2 + n_3 x_3 = [x_{01} n_1 + x_{02} n_2 + x_{03} n_3].} \tag{2.76}$$

Example 2.3. *Lines and planes:* The parametric equation of the line passing through the point $\vec{P} = (3, 1, 1)$ and in the direction of $\vec{A} = (1, 5, 2)$ is

$$x_1(t) = 3 + t, \tag{2.77}$$

$$x_2(t) = 1 + 5t, \tag{2.78}$$

$$x_3(t) = 1 + 2t. \tag{2.79}$$

For a line in the $x_1 x_2$-plane passing through $\overrightarrow{P} = (2, 1, 0)$ and in the direction of $\overrightarrow{A} = (1, 5, 0)$ we write the parametric equation as

$$x_1(t) = 2 + t, \tag{2.80}$$

$$x_2(t) = 1 + 5t. \tag{2.81}$$

We can now eliminate t to write the equation of the line as

$$x_2 = 5x_1 - 9. \tag{2.82}$$

For a plane including the point $\overrightarrow{P} = (2, 1, -2)$ and with the normal $\overrightarrow{N} = (-1, 1, 1)$, the equation is written as [Eq. (2.76)]

$$-x_1 + x_2 + x_3 = -3. \tag{2.83}$$

In general, a line in the $x_1 x_2$-plane is given as

$$ax_1 + bx_2 = c. \tag{2.84}$$

Comparing with Eq. (2.72), we can now interpret the vector (a, b) as a vector orthogonal to the line, that is,

$$(a, b) \cdot (a_1, a_2) = (-a_2, a_1) \cdot (a_1, a_2) = 0. \tag{2.85}$$

To find the angle between two planes:

$$2x_1 + x_2 + x_3 = 2, \tag{2.86}$$

$$-x_1 + x_2 + 2x_3 = 1, \tag{2.87}$$

we find the angle between their normals:

$$\overrightarrow{N}_1 = (2, 1, 1), \tag{2.88}$$

$$\overrightarrow{N}_2 = (-1, 1, 2), \tag{2.89}$$

as

$$\overrightarrow{N}_1 \cdot \overrightarrow{N}_2 = N_1 N_2 \cos\theta, \tag{2.90}$$

$$\theta = \cos^{-1}\left[\frac{-2 + 1 + 2}{\sqrt{6}\sqrt{6}}\right] = \cos^{-1}\frac{1}{6}. \tag{2.91}$$

2.4 VECTOR DIFFERENTIAL CALCULUS

2.4.1 Scalar Fields and Vector Fields

We have mentioned that **tempeiature** is a **scalar**; hence, a single number is sufficient to define temperature at a given point. In general, the temperature inside a system varies with position. In order to define temperature in a system completely, we have to give the temperature at each point of the system. This is equivalent to giving temperature as a function of position $T = T(x_1, x_2, x_3)$. This is an example of what we call a **scalar field**. In general, a scalar field is a single-valued differentiable function $f(x_1, x_2, x_3)$, representing a physical property defined in some domain of space. In short, we also write $f(\overrightarrow{r})$ or $f(\overrightarrow{x})$. In thermodynamics, temperature is a well-defined property only for systems in **thermal equilibrium**, that is, when the entire system has reached the same temperature. However, granted that the temperature is changing sufficiently slowly within a system, we can treat a small part of the system as in thermal equilibrium with the rest and define a meaningful temperature distribution as a differentiable scalar field. This is called the local thermodynamic equilibrium assumption, and it is one of the main assumptions of the theory of stellar structure. Another example for a scalar field is the **gravitational potential $\Phi(\overrightarrow{r})$** in Newton's theory. For a point mass M located at the origin, the gravitational potential is written as

$$\Phi(\overrightarrow{r}) = -G\frac{M}{r}, \tag{2.92}$$

where G is the gravitational constant. For a **massive scalar field**, the potential is given as

$$\Phi(\overrightarrow{r}) = k\frac{e^{-\mu r}}{r}, \tag{2.93}$$

where μ^{-1} is the mass of the field quanta and k is a coupling constant.

We now consider **compressible flow** in some domain of space. Assume that the flow is smooth so that the fluid elements, which are small compared to the body of the fluid but still large enough to contain many molecules, are following well-defined paths called the **streamlines**. Such flows are called **irrotational** or **streamline** flows. At each point of the streamline, we can associate a vector tangent to the streamline corresponding to the velocity of the fluid element at that point. In order to define the velocity of the fluid, we have to give the velocity vector of the fluid elements at each point of the fluid as (Figure 2.11)

$$\overrightarrow{v}(\overrightarrow{r}) = v_1(\overrightarrow{r})\widehat{e}_1 + v_2(\overrightarrow{r})\widehat{e}_2 + v_3(\overrightarrow{r})\widehat{e}_3. \tag{2.94}$$

This is an example of a **vector field**. In general, we can define a vector field by assigning a vector to every point of a domain in space.

Figure 2.11   Velocity field in flow problems.

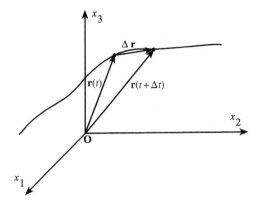

Figure 2.12 Vector differentiation.

2.4.2 Vector Differentiation

Trajectory of a particle can be defined in terms of the position vector $\overrightarrow{r}(t)$, where t is a parameter, which is usually taken as the time. The velocity $\overrightarrow{v}(t)$ and the acceleration $\overrightarrow{a}(t)$ are now defined as the derivatives (Figure 2.12)

$$\overrightarrow{v}(t) = \lim_{\Delta t \to 0} \frac{\overrightarrow{r}(t + \Delta t) - \overrightarrow{r}(t)}{\Delta t} = \frac{d\overrightarrow{r}(t)}{dt} \tag{2.95}$$

and

$$\overrightarrow{a}(t) = \lim_{\Delta t \to 0} \frac{\overrightarrow{v}(t + \Delta t) - \overrightarrow{v}(t)}{\Delta t} = \frac{d^2\overrightarrow{r}(t)}{dt^2}. \tag{2.96}$$

In general, for a differentiable vector field given in terms of a single parameter t:

$$\overrightarrow{A}(t) = A_1(t)\widehat{e}_1 + A_2(t)\widehat{e}_2 + A_3(t)\widehat{e}_3, \tag{2.97}$$

we can differentiate componentwise as

$$\boxed{\frac{d\overrightarrow{A}(t)}{dt} = \frac{dA_1(t)}{dt}\widehat{e}_1 + \frac{dA_2(t)}{dt}\widehat{e}_2 + \frac{dA_3(t)}{dt}\widehat{e}_3.} \tag{2.98}$$

Higher-order derivatives are found similarly according to the rules of calculus.

Basic properties of vector differentiation

$$\frac{d}{dt}(\overrightarrow{A} + \overrightarrow{B}) = \frac{d\overrightarrow{A}}{dt} + \frac{d\overrightarrow{B}}{dt}, \tag{2.99}$$

$$\frac{d[f(t)\overrightarrow{A}]}{dt} = \frac{df}{dt}\overrightarrow{A} + f\frac{d\overrightarrow{A}}{dt}, \tag{2.100}$$

$$\frac{d}{dt}(\overrightarrow{A} \cdot \overrightarrow{B}) = \frac{d\overrightarrow{A}}{dt} \cdot \overrightarrow{B} + \overrightarrow{A} \cdot \frac{d\overrightarrow{B}}{dt}, \tag{2.101}$$

$$\frac{d}{dt}(\vec{A} \times \vec{B}) = \vec{A} \times \frac{d\vec{B}}{dt} + \frac{d\vec{A}}{dt} \times \vec{B}. \tag{2.102}$$

Vector fields depending on more than one parameter can be differentiated partially. Given the vector field

$$\vec{A}(\vec{r}) = A_1(\vec{r})\hat{e}_1 + A_2(\vec{r})\hat{e}_2 + A_3(\vec{r})\hat{e}_3, \tag{2.103}$$

since each component is a differentiable function of the coordinates (x_1, x_2, x_3), we can differentiate it as

$$\frac{\partial \vec{A}(\vec{r})}{\partial x_i} = \frac{\partial A_1(\vec{r})}{\partial x_i}\hat{e}_1 + \frac{\partial A_2(\vec{r})}{\partial x_i}\hat{e}_2 + \frac{\partial A_3(\vec{r})}{\partial x_i}\hat{e}_3 \tag{2.104}$$

$$= \left(\frac{\partial A_1}{\partial x_i}, \frac{\partial A_2}{\partial x_i}, \frac{\partial A_3}{\partial x_i}\right), \quad i = 1, 2, 3. \tag{2.105}$$

2.5 GRADIENT OPERATOR

Given a scalar field $\Phi(\vec{r})$ defined in some domain of space described by the Cartesian coordinates (x_1, x_2, x_3), we can write the change in $\Phi(\vec{r})$ for an infinitesimal change in the position vector as

$$\Phi(\vec{r} + d\vec{r}) - \Phi(\vec{r}) = d\Phi(\vec{r}) = \frac{\partial \Phi}{\partial x_1} dx_1 + \frac{\partial \Phi}{\partial x_2} dx_2 + \frac{\partial \Phi}{\partial x_3} dx_3. \tag{2.106}$$

If we define two vectors:

$$\vec{\nabla}\Phi = \left(\frac{\partial \Phi}{\partial x_1}, \frac{\partial \Phi}{\partial x_2}, \frac{\partial \Phi}{\partial x_3}\right), \tag{2.107}$$

$$d\vec{r} = (dx_1, dx_2, dx_3), \tag{2.108}$$

we can write $d\Phi$ as

$$d\Phi(\vec{r}) = \vec{\nabla}\Phi \cdot d\vec{r}. \tag{2.109}$$

Note that even though Φ is a scalar field, $\vec{\nabla}\Phi$ is a vector field. We now introduce the **differential operator**

$$\boxed{\vec{\nabla} = \left(\frac{\partial}{\partial x_1}, \frac{\partial}{\partial x_2}, \frac{\partial}{\partial x_3}\right),} \tag{2.110}$$

which is called the **gradient** or the **del** operator. On its own, the del operator is meaningless. However, as we shall see shortly, it is a very useful operator.

2.5.1 Meaning of the Gradient

In applications, we associate a scalar field $\Phi(\overrightarrow{r})$ with physical properties such as the temperature, gravitational, or the electrostatic potential. Usually, we are interested in surfaces on which a scalar quantity takes a single value. In thermodynamics, surfaces on which temperature takes a single value are called the **isotherms**. In potential theory, **equipotentials** are surfaces on which the potential is a constant:

$$\Phi(x_1, x_2, x_3) = C. \tag{2.111}$$

If we treat C as a parameter, we obtain a family of surfaces as shown in Figure 2.13. Since $\Phi(\overrightarrow{r})$ is a single-valued function, none of these surfaces intersect each other. For two infinitesimally close points, $\overrightarrow{r}_1$ and $\overrightarrow{r}_2$, on one of the surfaces, $\Phi(x_1, x_2, x_3) = C$, the difference $d\overrightarrow{r} = \overrightarrow{r}_2 - \overrightarrow{r}_1$, is a vector on the surface (Figure 2.14). Thus, the equation

$$\boxed{\overrightarrow{\nabla}\Phi \cdot d\overrightarrow{r} = 0} \tag{2.112}$$

indicates that $\overrightarrow{\nabla}\Phi$ is a vector perpendicular to the surface $\Phi = C$ (Figure 2.14). This is evident in the special case of a family of planes, $\Phi(\overrightarrow{r}) = n_1 x_1 + n_2 x_2 + n_3 x_3 = C$, where the gradient $\overrightarrow{\nabla}\Phi = (n_1, n_2, n_3)$, is clearly normal to the plane. For a general family of surfaces, naturally the normal vector depends on the position in the surface.

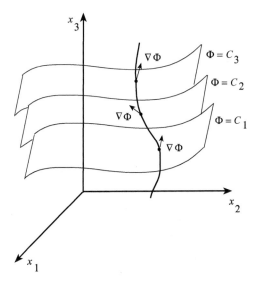

Figure 2.13 Equipotential surfaces.

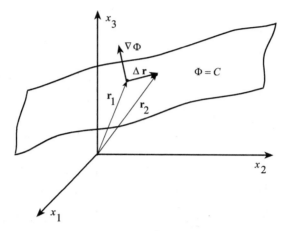

Figure 2.14   Gradient.

Example 2.4. *Equation of the tangent plane to a surface:* Since the normal to a surface $F(x_1, x_2, x_3) = C$ and the normal to the tangent plane at a given point $\vec{P} = (x_{01}, x_{02}, x_{03})$ coincide, we can write the equation of the tangent plane at P as $(\vec{x} - \vec{P}) \cdot \vec{\nabla} F = 0$, where $\vec{x}$ is a point on the tangent plane. In the limit as $\vec{x} \to \vec{P}$, we can write $(\vec{x} - \vec{P}) = d\vec{x}$. Hence, the aforementioned equation becomes $\vec{\nabla} F \cdot d\vec{x} = 0$. In other words, in the neighborhood of a point $\vec{P}$, the tangent plane approximately coincides with the surface. To be precise, this approximation is good to first order in $(\vec{x} - \vec{P})$.

2.5.2   Directional Derivative

We now consider a case where $\vec{r}_1$ is on the surface $\Phi(\vec{r}) = C_1$ and $\vec{r}_2$ is on the neighboring surface $\Phi(\vec{r}) = C_2$. In this case, the scalar product $\vec{\nabla} \Phi \cdot d\vec{r}$ is different from zero (Figure 2.15). Defining a unit vector in the direction of $d\vec{r}$ as $\hat{u} = d\vec{r}/|d\vec{r}|$, we write

$$\boxed{\frac{d\Phi}{du} = \vec{\nabla} \Phi \cdot \hat{u},} \qquad (2.113)$$

which is called the directional derivative of Φ in the direction of $\hat{u}$. If we move along a path A that intersects the family of surfaces $\Phi = C_i$, it is apparent from Figure 2.15 that the **directional derivative**:

$$\frac{d\Phi}{du} = |\vec{\nabla} \Phi| \cos \alpha, \qquad (2.114)$$

is zero when $\alpha = \pi/2$, that is, when we stay on the same surface. It is a maximum when we are moving through the surfaces in the direction of the gradient.

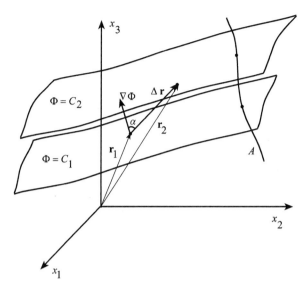

Figure 2.15   Directional derivative.

In other words, the gradient indicates the direction of maximum change in Φ as we move through the surfaces (Figure 2.15).

The gradient of a scalar field is very important in many applications and usually defines the direction of certain processes. In thermodynamics, heat flows from regions of high temperatures to low temperatures. Hence, the **heat current density** $\overrightarrow{J}$ is defined as proportional to the temperature gradient as

$$\overrightarrow{J}(\overrightarrow{r}) = -k\overrightarrow{\nabla}T(\overrightarrow{r}), \tag{2.115}$$

where k is the thermal conductivity. In transport problems, mass flows from regions of high concentration to low. Hence, the current density of the flowing material is taken as proportional to the gradient of the concentration $\overrightarrow{\nabla}C(\overrightarrow{r})$ as

$$\overrightarrow{J}(\overrightarrow{r}) = -\kappa\overrightarrow{\nabla}C(\overrightarrow{r}), \tag{2.116}$$

where κ is the diffusion constant.

2.6   DIVERGENCE AND CURL OPERATORS

The **del operator**:

$$\boxed{\overrightarrow{\nabla} = \widehat{e}_1\frac{\partial}{\partial x_1} + \widehat{e}_2\frac{\partial}{\partial x_2} + \widehat{e}_3\frac{\partial}{\partial x_3},} \tag{2.117}$$

can also be used to operate on a given vector field $\vec{A}$ either as

$$\boxed{\vec{\nabla} \cdot \vec{A},} \tag{2.118}$$

or as

$$\boxed{\vec{\nabla} \times \vec{A}.} \tag{2.119}$$

The first operation gives a **scalar field**:

$$\boxed{\vec{\nabla} \cdot \vec{A} = \frac{\partial A_1}{\partial x_1} + \frac{\partial A_2}{\partial x_2} + \frac{\partial A_3}{\partial x_3},} \tag{2.120}$$

called the **divergence** of the vector field $\vec{A}$, and the operator $\vec{\nabla}\cdot$ is called the **div operator**. The second operation gives another vector field, called the **curl** of $\vec{A}$, components of which are given as

$$\boxed{(\vec{\nabla} \times \vec{A})_i = \sum_{i=1}^{3} \varepsilon_{ijk} \partial_j A_k} \tag{2.121}$$

or as

$$\vec{\nabla} \times \vec{A} = \det \begin{pmatrix} \widehat{e}_1 & \widehat{e}_2 & \widehat{e}_3 \\ \partial_1 & \partial_2 & \partial_3 \\ A_1 & A_2 & A_3 \end{pmatrix} \tag{2.122}$$

$$= \widehat{e}_1 \left(\frac{\partial A_3}{\partial x_2} - \frac{\partial A_2}{\partial x_3} \right) + \widehat{e}_2 \left(\frac{\partial A_1}{\partial x_3} - \frac{\partial A_3}{\partial x_1} \right) + \widehat{e}_3 \left(\frac{\partial A_2}{\partial x_1} - \frac{\partial A_1}{\partial x_2} \right), \tag{2.123}$$

where $\vec{\nabla}\times$ is called the **curl operator** and ∂_i stands for $\partial/\partial x_i$.

Properties of gradient, divergence, and curl

$$\vec{\nabla}(\phi\psi) = \psi\vec{\nabla}\phi + \phi\vec{\nabla}\psi, \tag{2.124}$$

$$\vec{\nabla} \cdot (\vec{A} + \vec{B}) = \vec{\nabla} \cdot \vec{A} + \vec{\nabla} \cdot \vec{B}, \tag{2.125}$$

$$\vec{\nabla} \cdot (\phi\vec{A}) = \phi\vec{\nabla} \cdot \vec{A} + \vec{\nabla}\phi \cdot \vec{A}, \tag{2.126}$$

$$\vec{\nabla} \times (\vec{A} + \vec{B}) = \vec{\nabla} \times \vec{A} + \vec{\nabla} \times \vec{B}, \tag{2.127}$$

$$\vec{\nabla} \times (\phi\vec{A}) = \phi\vec{\nabla} \times \vec{A} + \vec{\nabla}\phi \times \vec{A}. \tag{2.128}$$

2.6.1 Meaning of Divergence and the Divergence Theorem

For a physical understanding of the divergence operator, we consider a tangible case like the flow of a fluid. The **density** $\rho(\overrightarrow{r}, t)$ of the fluid is a scalar field and gives the amount of fluid per unit volume as a function of position and time. The **current density** $\overrightarrow{J}(\overrightarrow{r}, t)$ is a vector field that gives the amount of fluid flowing per unit area per unit time. Another critical parameter related to the current density is the **flux** of the flowing material through an area element $\Delta\overrightarrow{\sigma}$. Naturally, flux depends on the relative orientation of $\overrightarrow{J}$ and $\Delta\overrightarrow{\sigma}$. For an infinitesimal area $d\overrightarrow{\sigma}$, flux $d\phi$ is defined as

$$d\phi = \overrightarrow{J}(\overrightarrow{r}, t) \cdot d\overrightarrow{\sigma} = \overrightarrow{J}(\overrightarrow{r}, t) \cdot \hat{n}\, d\sigma = J(\overrightarrow{r}, t) \cos\theta\, d\sigma, \qquad (2.129)$$

which gives the amount of matter that flows through the infinitesimal area element $d\sigma$ per unit time in the direction of the unit normal $\hat{n}$ to the surface (Figure 2.16). Notice that when the normal to the area element is perpendicular to the flow, $\theta = \pi/2$, flux is zero.

We now consider a compressible flow such as a gas flowing in some domain of space, which is described by the current density $\overrightarrow{J} = (J_1, J_2, J_3)$ and the matter density $\rho(\overrightarrow{r}, t)$. Take a small rectangular volume element

$$\Delta\tau = \Delta x_1 \Delta x_2 \Delta x_3 \qquad (2.130)$$

centered at $\overrightarrow{r} = (\frac{\Delta x_1}{2}, \frac{\Delta x_2}{2}, \frac{\Delta x_3}{2})$ as shown in Figure 2.17. The net amount of matter flowing per unit time in the x_2 direction into this volume element, that is, the net flux ϕ_2 in the x_2 direction, is equal to the sum of the fluxes from the surfaces 1 and 2:

$$\Delta\phi_2 = [\overrightarrow{J}(x_1, 0 + \Delta x_2, x_3, t) + \overrightarrow{J}(x_1, 0, x_3, t)] \cdot \hat{n}\, \Delta x_1\, \Delta x_3 \qquad (2.131)$$

$$= \left[\left(J_2|_{x_2=0} + \frac{\partial J_2}{\partial x_2}\Big|_{x_2=0} \Delta x_2 + \cdots \right) - J_2|_{x_2=0} \right] \Delta x_1\, \Delta x_3 \qquad (2.132)$$

$$= \frac{\partial J_2}{\partial x_2}\Big|_{x_2=0} \Delta x_2\, \Delta x_1\, \Delta x_3, \qquad (2.133)$$

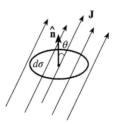

Figure 2.16 Flux through a surface.

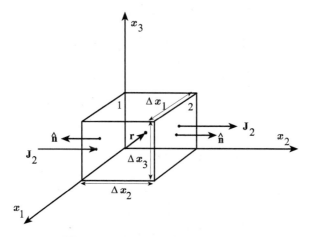

Figure 2.17   Flux through a cube.

where for the flux through the second surface we have used the Maclaurin series expansion of $\overrightarrow{J}(\overrightarrow{r}, t)$ for x_2 and kept only the first-order terms for a sufficiently small volume element. Note that the flux through the first surface is negative, since J_2 and the normal $\hat{n}$ to the surface are opposite in direction. Similar terms are obtained for the other two pairs of surfaces. Thus, their sum gives us the net amount of material flowing into the volume element:

$$\Delta\phi = \Delta\phi_1 + \Delta\phi_2 + \Delta\phi_3 \tag{2.134}$$

$$= \left[\frac{\partial J_1}{\partial x_1}\Big|_{x_1=0} + \frac{\partial J_2}{\partial x_2}\Big|_{x_2=0} + \frac{\partial J_3}{\partial x_3}\Big|_{x_3=0} \right] \Delta x_1 \, \Delta x_2 \, \Delta x_3. \tag{2.135}$$

Since the choice for the location of our rectangular volume element is arbitrary, for an arbitrary point in our domain, we can write

$$\Delta\phi = \left[\frac{\partial J_1}{\partial x_1} + \frac{\partial J_2}{\partial x_2} + \frac{\partial J_3}{\partial x_3} \right] \Delta\tau, \tag{2.136}$$

which is nothing but

$$\Delta\phi = \overrightarrow{\nabla} \cdot \overrightarrow{J}(\overrightarrow{r}, t)\Delta\tau. \tag{2.137}$$

Notice that when the net flux $\Delta\phi$ is positive, it corresponds to a net loss of matter within the volume element $\Delta\tau$. Hence, we equate it to $-\frac{d\rho}{dt}\Delta\tau$. Since the position of the volume element is fixed, that is, $\frac{d\overrightarrow{r}}{dt} = 0$, we can write

$$-\frac{d\rho}{dt}\Delta\tau = -\left[\frac{\partial\rho}{\partial x_1}\frac{dx_1}{dt} + \frac{\partial\rho}{\partial x_2}\frac{dx_2}{dt} + \frac{\partial\rho}{\partial x_3}\frac{dx_3}{dt} + \frac{\partial\rho}{\partial t} \right] \Delta\tau$$

$$= -\left[\vec{\nabla}\rho \cdot \frac{d\vec{r}}{dt} + \frac{\partial\rho}{\partial t}\right]\Delta\tau$$

$$= -\frac{\partial\rho}{\partial t}\Delta\tau \tag{2.138}$$

to obtain

$$\left[\vec{\nabla}\cdot\vec{J}(\vec{r},t) + \frac{\partial\rho(\vec{r},t)}{\partial t}\right]\Delta\tau = 0. \tag{2.139}$$

Since the **volume** element $\Delta\tau$ is in general different from zero, we can also write

$$\boxed{\vec{\nabla}\cdot\vec{J}(\vec{r},t) + \frac{\partial\rho(\vec{r},t)}{\partial t} = 0.} \tag{2.140}$$

For **compressible** flows, the current density can be related to the velocity field as

$$\boxed{\vec{J}(\vec{r},t) = \rho(\vec{r},t)\vec{v}(\vec{r},t),} \tag{2.141}$$

where $\vec{v}(\vec{r},t)$ is the velocity of the fluid element at $\vec{r}$ and t. Equation (2.140) is called the **equation of continuity**, and it is one of the most frequently encountered equations of science and engineering. It is a general expression for **conserved** quantities. In the fluid flow case, it represents conservation of **mass**. In electromagnetic theory, ρ stands for the electric charge density, and $\vec{J}$ is the electric current density. Now, the continuity equation becomes an expression of the conservation of **charge**. In quantum mechanics, the continuity equation is an expression for the conservation of **probability**, where $\rho = \Psi\Psi^*$ is the probability density, while $\vec{J}$ is the probability current density.

For a finite rectangular region R with the surface area S, we can use a network of n small rectangular volume elements, each of which satisfies

$$\left[\vec{\nabla}\cdot\vec{J}(\vec{r}_i,t) + \frac{\partial\rho(\vec{r}_i,t)}{\partial t}\right]\Delta\tau_i = 0, \tag{2.142}$$

where the subscript i denotes the ith volume element at $\vec{r}_i$. When we take the sum over all such cells and consider the limit as $n \to \infty$, fluxes through the adjacent sides will cancel each other, thus giving the integral version of the continuity equation as

$$\lim_{n\to\infty}\sum_{i=0}^{n}\left[\vec{\nabla}\cdot\vec{J}(\vec{r}_i,t) + \frac{\partial\rho(\vec{r}_i,t)}{\partial t}\right]\Delta\tau_i \to \int_V\left[\vec{\nabla}\cdot\vec{J}(\vec{r},t) + \frac{\partial\rho(\vec{r},t)}{\partial t}\right]d\tau$$

$$= 0, \tag{2.143}$$

$$\boxed{\int_V\vec{\nabla}\cdot\vec{J}(\vec{r},t)d\tau = -\int_V\frac{\partial\rho(\vec{r},t)}{\partial t}\,d\tau.} \tag{2.144}$$

Since the integral on the right-hand side is convergent, we can interchange the order of the derivative and the integral to write

$$\int_V \vec{\nabla} \cdot \vec{J}(\vec{r}, t) d\tau = -\frac{\partial}{\partial t}\left[\int_V \rho(\vec{r}, t) d\tau,\right] = -\frac{dm}{dt}, \tag{2.145}$$

where in the last step, we have used total derivative since $m = \int_V \rho(\vec{r}, t) d\tau$ is only a function of time. The right-hand side is the rate of change of the total amount of matter m within the volume V. When m is conserved, dm/dt is zero. Therefore, unless there is a net gain or loss of matter from the region, the divergence is zero. If there is net gain or loss of matter in a region, it implies the presence of **sources** or **sinks** within that region. That is, a nonzero divergence is an indication of the presence of sources or sinks in that region. It is important to note that if the divergence of a field is zero in a region, it does not necessarily mean that the field there is also zero, it just means that the sources are elsewhere.

Divergence theorem:

Another way to write the left-hand side of Eq. (2.145) is by using the definition of the total flux $\oint_S \vec{J} \cdot \hat{n}\, d\sigma$, that is, the net amount of material flowing in or out of the surface S per unit time, where S encloses the region R with the volume V. Equating the left-hand side of Eq. (2.145) with the **total flux** gives

$$\boxed{\int_V \vec{\nabla} \cdot \vec{J}(\vec{r}, t) d\tau = \oint_S \vec{J} \cdot \hat{n}\, d\sigma,} \tag{2.146}$$

where $\hat{n}$ is the outward unit normal to the surface S bounding the volume V and $d\sigma$ is the area element of the surface (Figure 2.18). Equation (2.146), which is valid for any piecewise smooth surface S with the volume V and the outward normal $\hat{n}$, is called **Gauss's theorem** or the **divergence theorem**, which can be used for any differentiable and integrable vector field $\vec{J}$. Gauss's theorem should not be confused with Gauss's law in electrodynamics, which is a physical law. A formal proof of the divergence theorem for any piecewise smooth surface that forms a closed boundary with an outward unit normal $\hat{n}$ can be found in [1].

Using the divergence theorem for an infinitesimal region, we can write an **integral** or an **operational definition** for the divergence of a vector field $\vec{J}$ as

$$\boxed{\vec{\nabla} \cdot \vec{J} = \lim_{\Delta\tau \to 0} \frac{1}{\Delta\tau} \oint_S \vec{J} \cdot \hat{n}\, d\sigma,} \tag{2.147}$$

where S is a closed surface enclosing the volume V. In summary, the divergence is a measure of the net in or out flux of a vector field over the closed surface S enclosing the volume V. It is for this reason that a vector field with zero divergence is called **solenoidal** in that region.

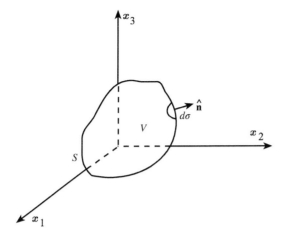

Figure 2.18   Area element on S.

Derivation of the divergence theorem has been motivated on the physical model of a fluid flow. However, the result is a mathematical identity valid for a general differentiable vector field. Even though $\vec{J} \cdot \hat{n}\, d\sigma$ represents the flux of $\vec{J}$ through $d\vec{\sigma}$, $\vec{J}$ may not represent any physical flow. As a mathematical identity, divergence theorem allows us to convert a volume integral to an integral over a closed surface, which then can be evaluated by using whichever is easier.

2.7   VECTOR INTEGRAL CALCULUS IN TWO DIMENSIONS

2.7.1   Arc Length and Line Integrals

A familiar line integral is the integral that gives the length of a curve:

$$l = \int_C ds = \int_C \sqrt{dx_1^2 + dx_2^2}, \qquad (2.148)$$

where C denotes the curve the length of which is to be measured and s is the arc length. If the curve is parameterized as

$$x_1 = x_1(t), \quad x_2 = x_2(t), \qquad (2.149)$$

we can write l as

$$l = \int_C \sqrt{\left(\frac{dx_1}{dt}\right)^2 + \left(\frac{dx_2}{dt}\right)^2}\, dt. \qquad (2.150)$$

We can also use either x_1 or x_2 as a parameter and write

$$l = \int_C \sqrt{1 + \left(\frac{dx_1}{dx_2}\right)^2}\, dx_2 = \int_C \sqrt{1 + \left(\frac{dx_2}{dx_1}\right)^2}\, dx_1. \qquad (2.151)$$

Line integrals are frequently encountered in applications with linear densities. For example, for a wire with linear mass density $\eta(s)$, we can write the total mass M as the line integral $M = \int_C \eta(s)ds$, or in parametric form as

$$M = \int_C \eta[x_1(t), x_2(t)] \sqrt{\left(\frac{dx_1}{dt}\right)^2 + \left(\frac{dx_2}{dt}\right)^2}\, dt. \qquad (2.152)$$

Extension of these formulas to n dimensions is obvious. In particular, for a curve parameterized in three dimensions as $(x_1(t), x_2(t), x_3(t))$, the **arc length** can be written as

$$l = \int_C \sqrt{\left(\frac{dx_1}{dt}\right)^2 + \left(\frac{dx_2}{dt}\right)^2 + \left(\frac{dx_3}{dt}\right)^2}\, dt. \qquad (2.153)$$

If the coordinate x_1 is used as the parameter, then the arc length becomes

$$l = \int_C \sqrt{1 + \left(\frac{dx_2}{dx_1}\right)^2 + \left(\frac{dx_3}{dx_1}\right)^2}\, dx_1. \qquad (2.154)$$

Example 2.5. *Work done on a particle:* An important application of the line integral is the expression for the work done on a particle moving along a trajectory under the influence of a force $\vec{F}$ as $W = \int_C F_T\, ds$. Here, F_T is the tangential component of the force, that is, the component along the displacement ds. We can also write W as $W = \int_C \vec{F} \cdot d\vec{r}$, where $\vec{F} = F_1 \hat{e}_1 + F_2 \hat{e}_2$ and $d\vec{r} = dx_1\, \hat{e}_1 + dx_2\, \hat{e}_2$, which gives

$$W = \int_C F_1(x_1, x_2)dx_1 + \int_C F_2(x_1, x_2)dx_2. \qquad (2.155)$$

Using the relations (Figure 2.19):

$$F_1 = F_T \cos\alpha + F_N \cos\left(\frac{\pi}{2} - \alpha\right) = F_T \cos\alpha + F_N \sin\alpha, \qquad (2.156)$$

$$F_2 = F_T \sin\alpha - F_N \sin\left(\frac{\pi}{2} - \alpha\right) = F_T \sin\alpha - F_N \cos\alpha, \qquad (2.157)$$

and

$$dx_1 = ds \cos\alpha, \quad dx_2 = ds \sin\alpha, \qquad (2.158)$$

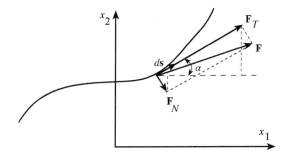

Figure 2.19 Normal and tangential components of force.

we can easily show the following:

$$W = \int_C [F_T \cos\alpha + F_N \sin\alpha] \cos\alpha \; ds + \int_C [F_T \sin\alpha - F_N \cos\alpha] \sin\alpha \; ds$$

(2.159)

$$= \int_C F_T \; ds.$$

(2.160)

In most applications, line integrals appear in combinations as

$$\boxed{\int_C P(x_1, x_2)dx_1 + \int_C Q(x_1, x_2)dx_2,}$$

(2.161)

which we also write as

$$\int_C P(x_1, x_2)dx_1 + Q(x_1, x_2)dx_2.$$

(2.162)

We can consider P and Q as the components of a vector field $\vec{w}$ as

$$\boxed{\vec{w} = P(x_1, x_2)\widehat{e}_1 + Q(x_1, x_2)\widehat{e}_2.}$$

(2.163)

Now the line integral [Eq. (2.161)] can be written as

$$\boxed{\int_C P(x_1, x_2)dx_1 + Q(x_1, x_2)dx_2 = \int_C w_T \; ds,}$$

(2.164)

where w_T denotes the tangential component of $\vec{w}$ in the direction of the unit tangent vector $\widehat{t}$ (Figure 2.20):

$$\widehat{t} = \frac{dx_1}{ds}\widehat{e}_1 + \frac{dx_2}{ds}\widehat{e}_2 = (\cos\alpha)\widehat{e}_1 + (\sin\alpha)\widehat{e}_2.$$

(2.165)

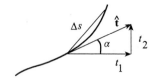

Figure 2.20 Unit tangent vector.

Using Eq. (2.165) we write

$$w_T = \overrightarrow{w} \cdot \hat{t} = P \cos \alpha + Q \sin \alpha. \tag{2.166}$$

Hence, proving

$$\int_C w_T \, ds = \int_C (P \cos \alpha + Q \sin \alpha) ds = \int_C P \, dx_1 + Q \, dx_2. \tag{2.167}$$

If we represent the path in terms of a parameter t, we can write

$$\int_C \overrightarrow{w} \cdot d\overrightarrow{r} = \int_C P \, dx_1 + Q \, dx_2 = \int_C \left(P\frac{dx_1}{dt} + Q\frac{dx_2}{dt} \right) dt \tag{2.168}$$

$$= \int_{t_1}^{t_2} \left(\overrightarrow{w} \cdot \frac{d\overrightarrow{r}}{dt} \right) dt. \tag{2.169}$$

Example 2.6. *Change in kinetic energy:* If we take $\overrightarrow{r}$ as the position of a particle of mass m moving under the influence of a force $\overrightarrow{F}$, the work done on the particle is written as

$$W = \int_C \overrightarrow{F} \cdot d\overrightarrow{r} = \int_C \overrightarrow{F} \cdot \frac{d\overrightarrow{r}}{dt} \, dt. \tag{2.170}$$

Substituting the second law of Newton, $\overrightarrow{F} = m\dfrac{d\overrightarrow{v}}{dt}$, where $\dfrac{d\overrightarrow{v}}{dt}$ is the acceleration of the particle, we can write W as

$$W = \int_C m\frac{d\overrightarrow{v}}{dt} \cdot \frac{d\overrightarrow{r}}{dt} dt = \int_C m\frac{d\overrightarrow{v}}{dt} \cdot \overrightarrow{v} \, dt \tag{2.171}$$

$$= \int_{t_1}^{t_2} \frac{d}{dt} \left[\frac{1}{2} m\overrightarrow{v} \cdot \overrightarrow{v} \right] dt = \int_{t_1}^{t_2} \frac{dT}{dt} \, dt \tag{2.172}$$

$$= T_2 - T_1. \tag{2.173}$$

The quantity we have defined as $T = \frac{1}{2}mv^2$, is nothing but the kinetic energy of the particle. Hence, the work done on the particle is equal to the change in kinetic energy.

2.7.2 Surface Area and Surface Integrals

We have given the expressions for the arc length of a curve in space. Our main aim is now to find the corresponding expressions for the area of a surface in space, which could either be given as $x_3 = x_3(x_1, x_2)$ or in parametric form as

$$x_1 = x_1(u, v), \tag{2.174}$$

$$x_2 = x_2(u, v), \tag{2.175}$$

$$x_3 = x_3(u, v). \tag{2.176}$$

Generalizations of the formulas [Eqs. (2.154) and (2.153)] to the area of a given surface are now written as

$$S = \int \int_{R_{x_1 x_2}} \sqrt{1 + \left(\frac{\partial x_3}{\partial x_1}\right)^2 + \left(\frac{\partial x_3}{\partial x_2}\right)^2} \, dx_1 \, dx_2, \tag{2.177}$$

or as

$$S = \int \int_{R_{uv}} \sqrt{EG - F^2} \, du \, dv, \tag{2.178}$$

where

$$E = \left(\frac{\partial x_1}{\partial u}\right)^2 + \left(\frac{\partial x_2}{\partial u}\right)^2 + \left(\frac{\partial x_3}{\partial u}\right)^2, \tag{2.179}$$

$$F = \frac{\partial x_1}{\partial u}\frac{\partial x_1}{\partial v} + \frac{\partial x_2}{\partial u}\frac{\partial x_2}{\partial v} + \frac{\partial x_3}{\partial u}\frac{\partial x_3}{\partial v}, \tag{2.180}$$

$$G = \left(\frac{\partial x_1}{\partial v}\right)^2 + \left(\frac{\partial x_2}{\partial v}\right)^2 + \left(\frac{\partial x_3}{\partial v}\right)^2. \tag{2.181}$$

A proper treatment of the derivation of this result is far too technical for our purposes. However, it can be found in pages 371–378 of *Treatise on Advanced Calculus* by Franklin [2]. An intuitive geometric derivation can be found in *Advanced Calculus* by Kaplan [1]. We give a rigorous derivation when we introduce the generalized coordinates and tensors in Chapter 3.

For the surface analog of the line integral

$$\int_C \overrightarrow{V} \cdot d\overrightarrow{r} = \int V_1 \, dx_1 + V_2 \, dx_2 + V_3 \, dx_3, \tag{2.182}$$

we write

$$I = \int \int_S \overrightarrow{V} \cdot d\overrightarrow{\sigma}. \tag{2.183}$$

Consider a smooth surface S with the outer unit normal defined as

$$\hat{n} = \cos\alpha\,\hat{e}_1 + \cos\beta\,\hat{e}_2 + \cos\gamma\,\hat{e}_3 \tag{2.184}$$

and take $\vec{V} = (V_1, V_2, V_3)$ to be a continuous vector field defined on S. We can now write the surface integral I as

$$\int\int_S (\vec{V} \cdot \hat{n})d\sigma = \int\int_S (V_1\cos\alpha + V_2\cos\beta + V_3\cos\gamma)d\sigma, \tag{2.185}$$

$$\boxed{I = \int\int_S V_1\,dx_2\,dx_3 + V_2\,dx_3\,dx_1 + V_3\,dx_1\,dx_2,} \tag{2.186}$$

where we used the fact that projections of the surface area element $d\vec{\sigma} = \hat{n}\,d\sigma$ onto the coordinate planes are given as $\cos\alpha\,d\sigma = dx_2\,dx_3$, $\cos\beta\,d\sigma = dx_3\,dx_1$, and $\cos\gamma\,d\sigma = dx_1\,dx_2$.

Similar to line integrals, a practical application of surface integrals is with surface densities. For example, the mass of a sheet described by the equation $x_3 = x_3(x_1, x_2)$ with the surface density $\sigma(x_1, x_2, x_3)$ is found by the surface integral

$$M = \int\int_{R_{x_1 x_2}} \sigma(x_1, x_2, x_3(x_1, x_2))\sqrt{1 + \left(\frac{dx_3}{dx_1}\right)^2 + \left(\frac{dx_3}{dx_2}\right)^2}\,dx_1\,dx_2, \tag{2.187}$$

or in terms of the parameters u and v as

$$M = \int\int_{R_{uv}} \sigma(u, v)\sqrt{EG - F^2}\,du\,dv. \tag{2.188}$$

2.7.3 An Alternate Way to Write Line Integrals

We can also write the line integral

$$\int_C P\,dx_1 + Q\,dx_2 \tag{2.189}$$

as

$$\boxed{\int_C P\,dx_1 + Q\,dx_2 = \int_C \vec{v} \cdot \hat{n}\,ds = \int_C v_n\,ds,} \tag{2.190}$$

where the vector field $\vec{v}$ is defined as

$$\boxed{\vec{v} = Q\hat{e}_1 - P\hat{e}_2} \tag{2.191}$$

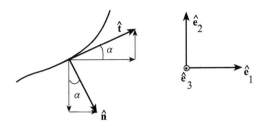

Figure 2.21 Normal and the tangential components.

and $\widehat{n}$ is the unit normal to the curve, that is, the perpendicular to $\widehat{t}$ (Figure 2.21):

$$\widehat{n} = \widehat{t} \times \widehat{e}_3 = \left[\frac{dx_1}{ds}\widehat{e}_1 + \frac{dx_2}{ds}\widehat{e}_2 \right] \times \widehat{e}_3 \tag{2.192}$$

$$= \frac{dx_2}{ds}\widehat{e}_1 - \frac{dx_1}{ds}\widehat{e}_2. \tag{2.193}$$

The normal component of $\overrightarrow{v}$ now becomes

$$v_n = \overrightarrow{v} \cdot \widehat{n} = (Q\widehat{e}_1 - P\widehat{e}_2) \cdot \left(\frac{dx_2}{ds}\widehat{e}_1 - \frac{dx_1}{ds}\widehat{e}_2 \right) \tag{2.194}$$

$$= Q\frac{dx_2}{ds} + P\frac{dx_1}{ds}, \tag{2.195}$$

which gives

$$\int_C v_n \, ds = \int_C \left(Q\frac{dx_2}{ds} + P\frac{dx_1}{ds} \right) ds \tag{2.196}$$

$$= \int_C P \, dx_1 + Q \, dx_2. \tag{2.197}$$

If we take $\overrightarrow{v}$ as

$$\boxed{\overrightarrow{v} = P\widehat{e}_1 + Q\widehat{e}_2,} \tag{2.198}$$

we get

$$\int_C v_n \, ds = \int_C \left(P\frac{dx_2}{ds} - Q\frac{dx_1}{ds} \right) ds, \tag{2.199}$$

$$\boxed{\int_C v_n \, ds = \int_C -Q \, dx_1 + P \, dx_2.} \tag{2.200}$$

Example 2.7. *Line integrals:* Let C be the arc $y = x^3$ from $(0,0)$ to $(-1,1)$. The line integral

$$I = \int_C xy^2 \, dx + x^3 y \, dy \tag{2.201}$$

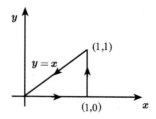

Figure 2.22   Closed path in Example 2.8.

can be evaluated as

$$I = \int_0^{-1} (x^7 + 3x^8)dx = \frac{x^8}{8} + \frac{x^9}{3} \Big|_0^{-1} = \frac{1}{8} - \frac{1}{3} = -\frac{5}{24}. \tag{2.202}$$

Example 2.8. *Closed paths:* Consider the line integral

$$I = \oint y^3 \, dx + x^2 \, dy \tag{2.203}$$

over the closed path in Figure 2.22. The first integral from $(0,0)$ to $(1,0)$ is zero, since $y = 0$ along this path. In the second part of the path from $(1,0)$ to $(1,1)$, we use y as our parameter and find $\int_0^1 1 \, dy = 1$. We use x as a parameter to obtain the integral over $y = x$ as

$$\int_1^0 x^3 \, dx + x^2 \, dx = \frac{x^4}{4} + \frac{x^3}{3} \Big|_1^0 = -\frac{7}{12}. \tag{2.204}$$

Finally, adding all these we obtain $I = \frac{5}{12}$.

2.7.4   Green's Theorem

Theorem 2.1. Let D be a simply connected domain of the $x_1 x_2$-plane and let C be a simple (does not intersect itself) smooth closed curve in D with its interior also in D. If $P(x_1, x_2)$ and $Q(x_1, x_2)$ are continuous functions with continuous first partial derivatives in D, then

$$\boxed{\oint_C P \, dx_1 + Q \, dx_2 = \int \int_R \left[\frac{\partial Q}{\partial x_1} - \frac{\partial P}{\partial x_2} \right] dx_1 \, dx_2,} \tag{2.205}$$

where R is the closed region enclosed by C.

    Proof: We first represent R by two curves,

$$a \leq x_1 \leq b, \quad f_1(x_1) \leq x_2 \leq f_2(x_1), \tag{2.206}$$

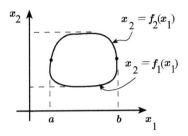

Figure 2.23 Green's theorem.

as shown in Figure 2.23 and write the second double integral in Eq. (2.205) as

$$-\int\int_R \frac{\partial P}{\partial x_2} \, dx_1 \, dx_2 = -\int_a^b \left[\int_{f_1(x_1)}^{f_2(x_1)} \frac{\partial P}{\partial x_2} \, dx_2 \right] dx_1. \tag{2.207}$$

The integral over x_2 can be taken immediately to yield

$$-\int\int_R \frac{\partial P}{\partial x_2} \, dx_1 \, dx_2 = -\int_a^b [\, P(x_1, f_2(x_1)) - P(x_1, f_1(x_1))] dx_1 \tag{2.208}$$

$$= \int_b^a P(x_1, f_2(x_1)) dx_1 + \int_a^b P(x_1, f_1(x_1)) dx_1 \tag{2.209}$$

$$= \oint_C P(x_1, x_2) dx_1. \tag{2.210}$$

Similarly, we can write the other double integral in Eq. (2.205) as

$$\int\int_R \frac{\partial Q}{\partial x_1} \, dx_1 \, dx_2 = \oint_C Q(x_1, x_2) dx_2, \tag{2.211}$$

thus proving Green's theorem.

Example 2.9. *Green's theorem:* Using Green's theorem [Eq. (2.205)], we can evaluate

$$I = \oint_C 16xy^3 \, dx + 24x^2 y^2 \, dy, \tag{2.212}$$

where C is the unit circle, $x^2 + y^2 = 1$, and $P = 16xy^3$ and $Q = 24x^2 y^2$ as

$$I = \int\int_R (48xy^2 - 48xy^2) dx \, dy = 0. \tag{2.213}$$

Example 2.10. *Green's theorem:* For the integral

$$I = \oint_C \frac{2y}{x^2 + y^2} \, dx + \frac{x}{x^2 + y^2} \, dy, \tag{2.214}$$

where C is the circle $x^2 + y^2 = 2$, we cannot apply Green's theorem since P and Q are not continuous at the origin.

Example 2.11. *Green's theorem:* For the integral

$$I = \oint_C (3x - y)dx + (x + 5y)dy \tag{2.215}$$

we find

$$I = \int\int_R (1 + 1)dx\, dy = 2A, \tag{2.216}$$

where A is the area enclosed by the closed path C.

2.7.5   Interpretations of Green's Theorem

I. If we take

$$\vec{w} = P(x_1, x_2)\widehat{e}_1 + Q(x_1, x_2)\widehat{e}_2 \tag{2.217}$$

in [Eq. (2.164)]:

$$\int_C P\, dx_1 + Q\, dx_2 = \int_C w_T\, ds, \tag{2.218}$$

where w_T is the tangential component of $\vec{w}$, that is, $w_T = \vec{w} \cdot \widehat{t}$ and notice that the right-hand side of Green's theorem [Eq. (2.205)] is the x_3 component of the curl of $\vec{w}$, that is,

$$(\vec{\nabla} \times \vec{w})_3 = \frac{\partial Q}{\partial x_1} - \frac{\partial P}{\partial x_2}, \tag{2.219}$$

we can write Green's theorem as

$$\boxed{\oint_C w_T\, ds = \int\int_R (\vec{\nabla} \times \vec{w})_3 dx_1\, dx_2.} \tag{2.220}$$

This is a special case of Stokes's theorem that will be discussed in Section 2.8.

II. We have seen that if we take $\vec{v}$ as

$$\vec{v} = Q\widehat{e}_1 - P\widehat{e}_2, \tag{2.221}$$

we can write the integral

$$I = \int_C P\, dx_1 + Q\, dx_2 \tag{2.222}$$

as [Eq. (2.190)]

$$I = \int_C \vec{v} \cdot \widehat{n}\, ds = \int_C v_n\, ds. \tag{2.223}$$

Now, the Green's theorem for $\vec{v}$ can be written as

$$\oint_C v_n \, ds = \int\int_R \left(\frac{\partial Q}{\partial x_1} - \frac{\partial P}{\partial x_2} \right) dx_1 \, dx_2, \tag{2.224}$$

$$\boxed{\oint_C v_n \, ds = \int\int_R \vec{\nabla} \cdot \vec{v} \, dx_1 \, dx_2.} \tag{2.225}$$

This is the two-dimensional version of the divergence theorem [Eq. (2.146)].

III. Area inside a closed curve: If we take $P = x_2$ in Eq. (2.210) or $Q = x_1$ in Eq. (2.211), we obtain the area of a closed curve as

$$\int\int_R dx_1 \, dx_2 = -\oint_C x_2 \, dx_1 = \oint_C x_1 \, dx_2. \tag{2.226}$$

Taking the arithmetic mean of these two equal expressions for the area of a region R enclosed by the closed curve C, we obtain another expression for the area A as

$$\boxed{A = \int\int_R dx_1 \, dx_2 = \frac{1}{2}\oint_C (x_1 \, dx_2 - x_2 \, dx_1),} \tag{2.227}$$

which the reader can check with Green's theorem [Eq. (2.205)].

2.7.6 Extension to Multiply Connected Domains

When the closed path C in Green's theorem encloses points at which one or both of the derivatives $\dfrac{\partial P}{\partial x_2}$ and $\dfrac{\partial Q}{\partial x_1}$ do not exist, Green's theorem is not applicable. However, by a simple modification of the path, we can still use Green's theorem to evaluate the integral

$$I = \oint_C P \, dx_1 + Q \, dx_2. \tag{2.228}$$

Consider the doubly connected domain D shown in Figure 2.24a defined by the boundaries a and b, where the closed path C_1 encloses the hole in the domain. As it is, Green's theorem is not applicable. However, if we modify our path as shown in Figure 2.24b, so that the closed path is inside the region, where the functions P, Q, and their first derivatives exist, we can apply Green's theorem to write

$$\oint_C P \, dx_1 + Q \, dx_2 = \oint_{\circlearrowleft C_1} + \int_{L_1} + \int_{L_2} + \oint_{\circlearrowleft C_2} [P \, dx_1 + Q \, dx_2] \tag{2.229}$$

$$= \int\int_R \left[\frac{\partial Q}{\partial x_1} - \frac{\partial P}{\partial x_2} \right] dx_1 \, dx_2, \tag{2.230}$$

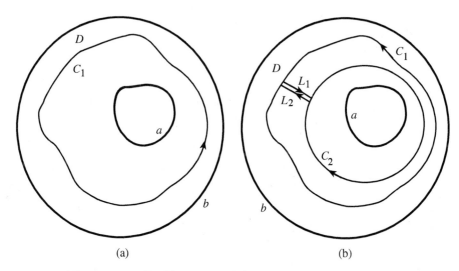

Figure 2.24 Doubly connected domain and the modified path.

where R is now the simply connected region bounded by the closed curve $C = C_1 + L_1 + L_2 + C_2$. We can choose the two paths L_1 and L_2 as close as possible. Since they are traversed in opposite directions, their contributions cancel each other, thereby yielding

$$\oint_{\circlearrowleft C_1} + \oint_{\circlearrowleft C_2} [P\,dx_1 + Q\,dx_2] = \int\int_R \left[\frac{\partial Q}{\partial x_1} - \frac{\partial P}{\partial x_2}\right] dx_1\,dx_2. \qquad (2.231)$$

In particular, when

$$\frac{\partial Q}{\partial x_1} = \frac{\partial P}{\partial x_2}, \qquad (2.232)$$

we obtain

$$\oint_{\circlearrowleft C_1} [P\,dx_1 + Q\,dx_2] = -\oint_{\circlearrowleft C_2} [P\,dx_1 + Q\,dx_2] \qquad (2.233)$$

$$= \oint_{\circlearrowright C_2} [P\,dx_1 + Q\,dx_2]. \qquad (2.234)$$

The advantage of this result is that by choosing a suitable path C_2, such as a circle, we can evaluate the desired integral I, where the first path, C_1, may be awkward in shape (Figure 2.24a).

Example 2.12. *Multiply-connected domains:* Consider the line integral

$$I = \oint_C \left(\frac{-x_2}{x_1^2 + x_2^2}\right) dx_1 + \left(\frac{x_1}{x_1^2 + x_2^2}\right) dx_2, \qquad (2.235)$$

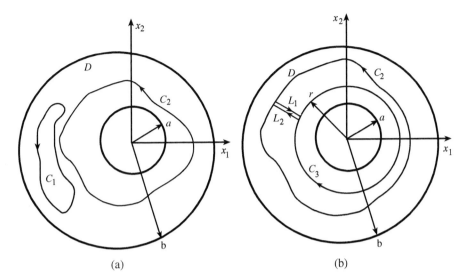

(a)   (b)

Figure 2.25   Paths in Example 2.12.

where C_1 and C_2 are two closed paths inside the domain D defined by two concentric circles $x_1^2 + x_2^2 = a^2$ and $x_1^2 + x_2^2 = b^2$ as shown in Figure 2.25a. For the path C_1, P, Q, and their first derivatives exist inside C_1. Furthermore, since $\dfrac{\partial P}{\partial x_2}$ and $\dfrac{\partial Q}{\partial x_1}$ are equal:

$$\frac{\partial P}{\partial x_2} = \frac{\partial Q}{\partial x_1} = \frac{x_2^2 - x_1^2}{(x_1^2 + x_2^2)^2}, \tag{2.236}$$

I is zero. For the second path C_2, which encloses the hole at the center, we modify it as shown in Figure 2.25b, where C_3 is chosen as a circle with radius $r > a$, so that the integral on the right-hand side of

$$\oint_{\circlearrowleft C_2} [P\, dx_1 + Q\, dx_2] = \oint_{\circlearrowleft C_3} [P\, dx_1 + Q\, dx_2] \tag{2.237}$$

can be evaluated analytically. The value of the integral over C_3, $x_1^2 + x_2^2 = r^2$, can be found easily by introducing a new variable θ :

$$x_1 = r\cos\theta, \quad x_2 = r\sin\theta, \quad \theta \in [0, 2\pi], \tag{2.238}$$

as

$$I = \oint_{\circlearrowleft C_3} [P\, dx_1 + Q\, dx_2] = \int_0^{2\pi} d\theta = 2\pi r. \tag{2.239}$$

2.8   CURL OPERATOR AND STOKES'S THEOREM

2.8.1   On the Plane

Consider a vector field on the plane and its line integral over the closed rectangular path shown in Figure 2.26 as

$$\oint_C \overrightarrow{V}(x_1, x_2) \cdot d\overrightarrow{r} = \left(\oint_{C_1} + \oint_{C_2} + \oint_{C_3} + \oint_{C_4} \right) \overrightarrow{V}(x_1, x_2) \cdot d\overrightarrow{r}. \qquad (2.240)$$

We first consider the integral over C_1, where $x_2 = x_{02}$:

$$\oint_{C_1} \overrightarrow{V}(x_1, x_2) \cdot d\overrightarrow{r} = \int_{x_{01}}^{x_{01}+\Delta x_1} dx_1 \, V_1(x_1, x_{02}). \qquad (2.241)$$

Expanding $\overrightarrow{V}(x_1, x_2)$ in Taylor series about (x_{01}, x_{02}) and keeping only the linear terms, we write

$$V_1(x_1, x_2) \simeq V_1(x_{01}, x_{02}) + \left. \frac{\partial V_1(x_1, x_2)}{\partial x_1} \right|_{(x_{01}, x_{02})} (x_1 - x_{01})$$

$$+ \left. \frac{\partial V_1(x_1, x_2)}{\partial x_2} \right|_{(x_{01}, x_{02})} (x_2 - x_{02}), \qquad (2.242)$$

which, when substituted into Eq. (2.241), gives

$$\oint_{C_1} \overrightarrow{V} \cdot d\overrightarrow{r} = \int_{x_{01}}^{x_{01}+\Delta x_1} dx_1 \left[V_1(x_{01}, x_{02}) + \left. \frac{\partial V_1(x_1, x_2)}{\partial x_1} \right|_{(x_{01}, x_{02})} (x_1 - x_{01}) \right]$$

$$(2.243)$$

$$= V_1(x_{01}, x_{02}) \Delta x_1 + \frac{1}{2} \left. \frac{\partial V_1(x_1, x_2)}{\partial x_1} \right|_{(x_{01}, x_{02})} \Delta x_1^2. \qquad (2.244)$$

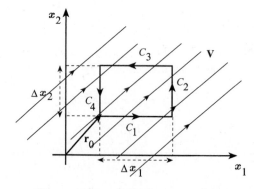

Figure 2.26   Curl of a vector field.

We now take the integral over C_3, where $x_2 = x_{02} + \Delta x_2$ and x_1 varies from $x_{01} + \Delta x_1$ to x_{01}:

$$\oint_{C_3} \overrightarrow{V}(x_1, x_2) \cdot d\overrightarrow{r} = \int_{x_{01}+\Delta x_1}^{x_{01}} dx_1 \, V_1(x_1, x_{02} + \Delta x_2). \tag{2.245}$$

Substituting the Taylor series expansion [Eq. (2.242)] of $V_1(x_1, x_2)$ about (x_{01}, x_{02}) evaluated at $x_2 = x_{02} + \Delta x_2$:

$$V_1(x_1, x_{02} + \Delta x_2) \simeq V_1(x_{01}, x_{02}) + \left. \frac{\partial V_1(x_1, x_2)}{\partial x_1} \right|_{(x_{01}, x_{02})} (x_1 - x_{01})$$

$$+ \left. \frac{\partial V_1(x_1, x_2)}{\partial x_2} \right|_{(x_{01}, x_{02})} \Delta x_2, \tag{2.246}$$

into Eq. (2.245) and integrating gives

$$\oint_{C_3} \overrightarrow{V}(x_1, x_2) \cdot d\overrightarrow{r} \simeq -V_1(x_{01}, x_{02})\Delta x_1 - \frac{1}{2} \left. \frac{\partial V_1(x_1, x_2)}{\partial x_1} \right|_{(x_{01}, x_{02})} \Delta x_1^2$$

$$- \left. \frac{\partial V_1(x_1, x_2)}{\partial x_2} \right|_{(x_{01}, x_{02})} \Delta x_2 \, \Delta x_1. \tag{2.247}$$

Note the **minus sign** coming from the dot product in the integral. Combining these results [Eqs. (2.244) and (2.247)] we obtain

$$\oint_{C_1} \overrightarrow{V}(x_1, x_2) \cdot d\overrightarrow{r} + \oint_{C_3} \overrightarrow{V}(x_1, x_2) \cdot d\overrightarrow{r} = - \left. \frac{\partial V_1(x_1, x_2)}{\partial x_2} \right|_{(x_{01}, x_{02})} \Delta x_2 \, \Delta x_1. \tag{2.248}$$

A similar procedure for the paths C_2 and C_4 yields

$$\oint_{C_2} \overrightarrow{V}(x_1, x_2) \cdot d\overrightarrow{r} + \oint_{C_4} \overrightarrow{V}(x_1, x_2) \cdot d\overrightarrow{r} = \left. \frac{\partial V_2(x_1, x_2)}{\partial x_1} \right|_{(x_{01}, x_{02})} \Delta x_1 \, \Delta x_2, \tag{2.249}$$

which, after combining with Eq. (2.248), gives

$$\oint_C \overrightarrow{V}(x_1, x_2) \cdot d\overrightarrow{r} = \left[\left. \frac{\partial V_2}{\partial x_1} \right|_{(x_{01}, x_{02})} - \left. \frac{\partial V_1}{\partial x_2} \right|_{(x_{01}, x_{02})} \right] \Delta x_1 \, \Delta x_2. \tag{2.250}$$

Since the location of $\overrightarrow{r}_0$ is arbitrary on the $x_1 x_2$-plane, we can drop the subscript 0. If we also notice that the quantity inside the square brackets is the x_3 component of $\overrightarrow{\nabla} \times \overrightarrow{V}$, we can write

$$\oint_C \overrightarrow{V}(x_1, x_2) \cdot d\overrightarrow{r} = (\overrightarrow{\nabla} \times \overrightarrow{V})_3 \Delta x_1 \, \Delta x_2. \tag{2.251}$$

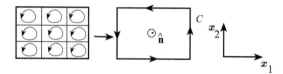

Figure 2.27 Infinitesimal rectangular path.

The approximation we have made by ignoring the higher-order terms in the Taylor series expansion is justified in the limits as $\Delta x_1 \to 0$ and $\Delta x_2 \to 0$. For an infinitesimal rectangle, we can replace Δx_1 with dx_1 and Δx_2 with dx_2. Similarly, $\Delta\sigma = \Delta x_1 \, \Delta x_2$ can be replaced with the infinitesimal area element $d\sigma_{12}$.

For a finite rectangular path C, we can sum over the infinitesimal paths as shown in Figure 2.27. Integrals over the adjacent sides cancel, thereby leaving only the integral over the boundary C as

$$\oint_C \vec{V}(x_1, x_2) \cdot d\vec{r} = \int\int_S (\vec{\nabla} \times \vec{V}) \cdot \hat{n} \, dx_1 \, dx_2 = \int\int_S (\vec{\nabla} \times \vec{V}) \cdot \hat{n} \, d\sigma_{12}, \tag{2.252}$$

where C is now a finite rectangular path. The right-hand side is a surface integral to be taken over the surface bounded by C, which in this case is the region on the $x_1 x_2$-plane bounded by the rectangle C with its outward normal $\hat{n}$ defined by the right-hand rule as the $\hat{e}_3$ direction.

Using Green's theorem [Eq. (2.220)] in Eq. (2.251), we see that this result is also valid for an arbitrary closed simple path C on the $x_1 x_2$-plane as

$$\oint_C \vec{V} \cdot d\vec{r} = \oint_C V_T \, ds = \int\int_R (\vec{\nabla} \times \vec{V}) \cdot \hat{n} \, dx_1 \, dx_2, \tag{2.253}$$

where V_T is the tangential component of $\vec{V}$ along C and R is the region bounded by C. The integral on the right-hand side is basically a surface integral over a surface bounded by the curve C, which we have taken as S_1 lying on the $x_1 x_2$-plane, with its normal $\hat{n}$ as defined by the right-hand rule (Figure 2.28a).

We now ask the question, What if we use a surface S_2 in three dimensions (Figure 2.28b), which also has the same boundary as the planar surface in Figure 2.28a? Does the value of the surface integral on the right-hand side in Eq. (2.253) change? Since the surface integral in Eq. (2.253) is equal to the line integral

$$\oint_C \vec{V} \cdot d\vec{r}, \tag{2.254}$$

which depends only on $\vec{V}$ and C, its value should not depend on which surface we use. In fact, it does not, provided that the surface is oriented. An oriented surface has two sides: an inside and an outside. The outside is defined by the right-hand rule. As in the first two surfaces in Figure 2.29, in an oriented surface

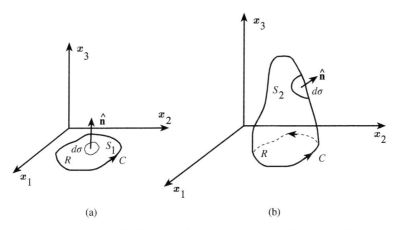

Figure 2.28 Different surfaces with the same boundary C.

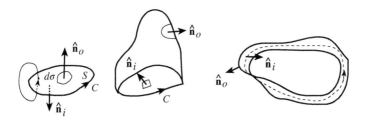

Figure 2.29 Orientable surfaces vs. the Möbius strip.

one cannot go from one side to the other without crossing over the boundary. In the last surface in Figure 2.29, we have a **Möbius strip**, which has only one side. Following a closed path, one can go from one "side" with the normal $\hat{n}_i$ to the other side with the normal $\hat{n}_0$, which points exactly in the opposite direction, without ever crossing a boundary.

Consider two orientable surfaces S_1 and S_2 with the same boundary C and cover them both with a network of simple closed paths C_i with small areas $\Delta\sigma_i$, each (Figure 2.30). In the limit as $\Delta\sigma_i \to 0$, each area element naturally coincides with the tangent plane to the surface of which it belongs at that point. Depending on their location, normals point in different directions. For each surface element, we can write

$$\oint_{C_i} \vec{V} \cdot d\vec{r} = (\vec{\nabla} \times \vec{V}) \cdot \hat{n}_i \, \Delta\sigma_i, \qquad (2.255)$$

where i denotes the ith surface element on either S_1 or S_2, with the boundary C_i. For the entire surface, we have to sum these as

$$\sum_i^l \oint_{C_i} \vec{V} \cdot d\vec{r} = \sum_i^l (\vec{\nabla} \times \vec{V}) \cdot \hat{n}_i \, \Delta\sigma_i \qquad (2.256)$$

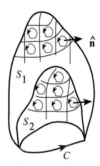

Figure 2.30   Two orientable surfaces.

for S_1 and

$$\sum_j^m \oint_{C_j} \vec{V} \cdot d\vec{r} = \sum_j^m (\vec{\nabla} \times \vec{V}) \cdot \hat{n}_j \, \Delta\sigma_j \qquad (2.257)$$

for S_2. Since the surfaces have different surface areas, l and m are different in general. In the limit as l and m go to infinity, contributions coming from adjacent sides will cancel. Thus, the sums on the left-hand sides of Eqs. (2.256) and (2.257) reduce to the same line integral over their common boundary C:

$$\oint_C \vec{V} \cdot d\vec{r}. \qquad (2.258)$$

On the other hand, the sums on the right-hand sides become surface integrals over their respective surfaces S_1 and S_2. Hence, in general, we can write

$$\oint_C \vec{V} \cdot d\vec{r} = \int\int_S (\vec{\nabla} \times \vec{V}) \cdot \hat{n} \, d\sigma, \qquad (2.259)$$

where S is any oriented surface with the boundary C.

2.8.2   In Space

In Figure 2.28, even though we took S as a surface in three-space, its boundary C is still on the $x_1 x_2$-plane. We now generalize this result by taking the closed simple path C also in space.

Stokes's Theorem: Consider a smooth oriented surface S in space with a smooth simple curve C as its boundary. For a given continuous and differentiable vector field $\vec{V}$:

$$\vec{V}(x_1, x_2, x_3) = V_1 \hat{e}_1 + V_2 \hat{e}_2 + V_3 \hat{e}_3, \qquad (2.260)$$

in some domain D of space, which includes S, we can write

$$\oint_C \overrightarrow{V} \cdot d\overrightarrow{r} = \oint_C V_T \, ds = \int \int_S (\overrightarrow{\nabla} \times \overrightarrow{V}) \cdot \hat{n} \, d\sigma, \qquad (2.261)$$

where $\hat{n}$ is the outward normal to S.

Proof: We first write Eq. (2.261) as

$$\oint_C \overrightarrow{V} \cdot d\overrightarrow{r} = \oint_C V_1 \, dx_1 + V_2 \, dx_2 + V_3 \, dx_3 \qquad (2.262)$$

$$= \int \int_S \left[\frac{\partial V_3}{\partial x_2} - \frac{\partial V_2}{\partial x_3} \right] dx_2 \, dx_3 + \left[\frac{\partial V_1}{\partial x_3} - \frac{\partial V_3}{\partial x_1} \right] dx_3 \, dx_1 \qquad (2.263)$$

$$+ \left[\frac{\partial V_2}{\partial x_1} - \frac{\partial V_1}{\partial x_2} \right] dx_1 \, dx_2,$$

which can be proven by proving three separate equations:

$$\oint_C V_1 \, dx_1 = \int \int_S \frac{\partial V_1}{\partial x_3} \, dx_3 \, dx_1 - \frac{\partial V_1}{\partial x_2} \, dx_1 dx_2, \qquad (2.264)$$

$$\oint_C V_2 \, dx_2 = \int \int_S \frac{\partial V_2}{\partial x_1} \, dx_1 \, dx_2 - \frac{\partial V_2}{\partial x_3} \, dx_2 \, dx_3, \qquad (2.265)$$

$$\oint_C V_3 \, dx_3 = \int \int_S \frac{\partial V_3}{\partial x_2} \, dx_2 \, dx_3 - \frac{\partial V_3}{\partial x_1} \, dx_3 \, dx_1. \qquad (2.266)$$

We also assume that the surface S can be written in the form $x_3 = f(x_1, x_2)$ and as shown in Figure 2.31, C_{12} is the projection of C onto the $x_1 x_2$-plane. Hence, when (x_1, x_2, x_3) goes around C a full loop, the corresponding point

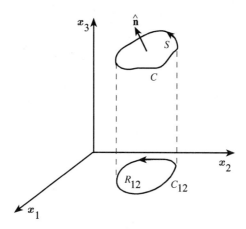

Figure 2.31 Stokes's theorem in space.

$(x_1, x_2, 0)$ also completes a full loop in C_{12} in the same direction. We choose the direction of $\hat{n}$ with the right-hand rule as the outward direction. Using Green's theorem [Eq. (2.205)] with $Q = 0$, we can write

$$\oint_C V_1(x_1, x_2, x_3)dx_1 = \oint_C V_1(x_1, x_2, f(x_1, x_2))dx_1 = \oint_{C_{12}} V_1(x_1, x_2, f(x_1, x_2))dx_1$$

$$(2.267)$$

$$= -\int\int_{R_{12}} \left[\frac{\partial V_1}{\partial x_2} + \frac{\partial V_1}{\partial x_3}\frac{\partial f}{\partial x_2}\right] dx_1\, dx_2. \qquad (2.268)$$

We now use Eq. (2.186) with $\overrightarrow{V}$ taken as

$$\overrightarrow{V} = \left(0, \frac{\partial V_1}{\partial x_3}, -\frac{\partial V_1}{\partial x_2}\right). \qquad (2.269)$$

Note that $\overrightarrow{V}$ in Eq. (2.186) is an arbitrary vector field. Since the normal $\overrightarrow{n}$ to $d\sigma$ is just the gradient to the surface $x_3 - f(x_1, x_2) = 0$, that is,

$$\overrightarrow{n} = \left(-\frac{\partial f}{\partial x_1}, -\frac{\partial f}{\partial x_2}, 1\right), \qquad (2.270)$$

we write Eq. (2.186) as

$$\int\int_S (\overrightarrow{V} \cdot \overrightarrow{n})\frac{d\sigma}{|\overrightarrow{n}|} = \int\int_S \frac{\partial V_1}{\partial x_3}\, dx_3\, dx_1 - \frac{\partial V_1}{\partial x_2}\, dx_1\, dx_2, \qquad (2.271)$$

$$-\int\int_S \left[\frac{\partial V_1}{\partial x_3}\frac{\partial f}{\partial x_2} + \frac{\partial V_1}{\partial x_2}\right]\frac{d\sigma}{|\overrightarrow{n}|} = \int\int_S \frac{\partial V_1}{\partial x_3}\, dx_3\, dx_1 - \frac{\partial V_1}{\partial x_2}\, dx_1\, dx_2. \quad (2.272)$$

Since

$$d\overrightarrow{\sigma} \cdot \hat{e}_3 = d\sigma\, \cos\gamma = d\sigma\, \hat{n} \cdot \hat{e}_3 = d\sigma\, \frac{\overrightarrow{n}}{|\overrightarrow{n}|} \cdot \hat{e}_3 = \frac{d\sigma}{|\overrightarrow{n}|}, \qquad (2.273)$$

where γ is the angle between $\hat{e}_3$ and $\overrightarrow{\sigma}$, $d\sigma \cos\gamma$ is the projection of the area element $d\overrightarrow{\sigma}$ onto the $x_1 x_2$-plane, that is, $d\sigma/|\overrightarrow{n}| = dx_1\, dx_2$. We can now rewrite the left-hand side of Eq. (2.272) as an integral over R_{12} to obtain the relation

$$-\int\int_{R_{12}} \left[\frac{\partial V_1}{\partial x_3}\frac{\partial f}{\partial x_2} + \frac{\partial V_1}{\partial x_2}\right] dx_1\, dx_2 = \int\int_S \frac{\partial V_1}{\partial x_3}\, dx_3\, dx_1 - \frac{\partial V_1}{\partial x_2}\, dx_1\, dx_2.$$

$$(2.274)$$

Substituting Eq. (2.274) into (2.268), we obtain

$$\oint_C V_1(x_1, x_2, x_3)dx_1 = \int\int_S \frac{\partial V_1}{\partial x_3}\, dx_3\, dx_1 - \frac{\partial V_1}{\partial x_2}\, dx_1\, dx_2, \qquad (2.275)$$

which is Eq. (2.264). In a similar fashion we also show Eqs. (2.265) and (2.266). Finally, by adding Eqs. (2.264)–(2.266) we establish Stokes's theorem.

Figure 2.32 Unit normal to circular path.

2.8.3 Geometric Interpretation of Curl

We have seen that the divergence of a vector field $\vec{V}$ is equal to the ratio of the flux through a closed surface S to the volume enclosed by S in the limit as the surface area of S goes to zero [Eq. (2.147)]:

$$\boxed{\vec{\nabla} \cdot \vec{V} = \lim_{S,V \to 0} \frac{1}{V} \oint_S \vec{V} \cdot \hat{n} \, d\sigma.}$$ (2.276)

Similarly, we can give an integral definition for the value of the curl of a vector field $\vec{V}$ in the direction $\hat{n}$ as

$$\boxed{(\vec{\nabla} \times \vec{V}) \cdot \hat{n} = \lim_{r \to 0} \frac{1}{A_r} \oint_{C_r} V_T \, ds = \lim_{r \to 0} \frac{1}{A_r} \oint_{C_r} \vec{V} \cdot d\vec{r},}$$ (2.277)

where C_r is a circular path with radius r and area A_r, and $\hat{n}$ is the unit normal to A_r determined by the right-hand rule (Figure 2.32). In the limit as the size of the path shrinks to zero, the surface enclosed by the circular path can be replaced by a more general surface with the normal $\hat{n}$. Note that this is also an operational definition that can be used to construct a "curl-meter," that is, an instrument that can be used to measure the value of the curl of a vector field in the direction $\hat{n}$ of the axis of the instrument.

2.9 MIXED OPERATIONS WITH THE DEL OPERATOR

By paying attention to the vector nature of the $\vec{\nabla}$ operator and also by keeping in mind that it is meaningless on its own, we can construct several other useful operators and identities. For a scalar field $\Phi(\vec{r})$, a very useful operator can be constructed by taking the divergence of a gradient as

$$\boxed{\vec{\nabla} \cdot \vec{\nabla} \Phi(\vec{r}) = \vec{\nabla}^2 \Phi(\vec{r}).}$$ (2.278)

The $\vec{\nabla}^2$ operator is called the **Laplacian** or the **Laplace operator**, which is one of the most commonly encountered operators in science. Two very important **vector identities** used in potential theory are

$$\boxed{\vec{\nabla} \cdot (\vec{\nabla} \times \vec{A}) = 0}$$ (2.279)

and

$$\boxed{\vec{\nabla} \times \vec{\nabla}\Phi = 0,}$$

$$(2.280)$$

where $\vec{A}$ and Φ are differentiable vector and scalar fields, respectively. In other words, the divergence of a curl and the curl of a gradient are zero. Using the definition of the $\vec{\nabla}$ operator, proofs can be written immediately. For Eq. (2.279) we write

$$\vec{\nabla} \cdot (\vec{\nabla} \times \vec{A}) = \det \begin{pmatrix} \partial_1 & \partial_2 & \partial_3 \\ \partial_1 & \partial_2 & \partial_3 \\ A_1 & A_2 & A_3 \end{pmatrix}, \qquad (2.281)$$

and obtain

$$\vec{\nabla} \cdot (\vec{\nabla} \times \vec{A}) = \frac{\partial}{\partial x_1}\left(\frac{\partial A_3}{\partial x_2} - \frac{\partial A_2}{\partial x_3}\right) + \frac{\partial}{\partial x_2}\left(\frac{\partial A_1}{\partial x_3} - \frac{\partial A_3}{\partial x_1}\right)$$

$$+ \frac{\partial}{\partial x_3}\left(\frac{\partial A_2}{\partial x_1} - \frac{\partial A_1}{\partial x_2}\right) \qquad (2.282)$$

$$= \frac{\partial^2 A_3}{\partial x_1 \partial x_2} - \frac{\partial^2 A_2}{\partial x_1 \partial x_3} + \frac{\partial^2 A_1}{\partial x_2 \partial x_3} - \frac{\partial^2 A_3}{\partial x_2 \partial x_1}$$

$$+ \frac{\partial^2 A_2}{\partial x_3 \partial x_1} - \frac{\partial^2 A_1}{\partial x_3 \partial x_2} \qquad (2.283)$$

$$= \left[\frac{\partial^2 A_1}{\partial x_2 \partial x_3} - \frac{\partial^2 A_1}{\partial x_3 \partial x_2}\right] + \left[\frac{\partial^2 A_2}{\partial x_3 \partial x_1} - \frac{\partial^2 A_2}{\partial x_1 \partial x_3}\right]$$

$$+ \left[\frac{\partial^2 A_3}{\partial x_1 \partial x_2} - \frac{\partial^2 A_3}{\partial x_2 \partial x_1}\right]. \qquad (2.284)$$

Since the vector field is differentiable, the order of differentiation in Eq. (2.284) is unimportant and so the divergence of a curl is zero.

For the second identity [Eq. (2.280)], we write

$$\vec{\nabla} \times \vec{\nabla}\Phi = \vec{\nabla} \times \left(\frac{\partial \Phi}{\partial x_1}, \frac{\partial \Phi}{\partial x_2}, \frac{\partial \Phi}{\partial x_3}\right) \qquad (2.285)$$

$$= \widehat{e}_1\left(\frac{\partial^2 \Phi}{\partial x_2 \partial x_3} - \frac{\partial^2 \Phi}{\partial x_3 \partial x_2}\right) + \widehat{e}_2\left(\frac{\partial^2 \Phi}{\partial x_3 \partial x_1} - \frac{\partial^2 \Phi}{\partial x_1 \partial x_3}\right)$$

$$+ \widehat{e}_3\left(\frac{\partial^2 \Phi}{\partial x_1 \partial x_2} - \frac{\partial^2 \Phi}{\partial x_2 \partial x_1}\right) \qquad (2.286)$$

$$= 0. \qquad (2.287)$$

Since for a differentiable Φ the mixed derivatives are equal, we obtain zero in the last step, thereby proving the identity.

Another useful identity is

$$\boxed{\vec{\nabla} \cdot (u\vec{\nabla}v) = \vec{\nabla}u \cdot \vec{\nabla}v + u\vec{\nabla}^2 v,} \tag{2.288}$$

where u and v are two differentiable scalar fields. We leave the proof of this identity as an exercise, but by using it we prove two very useful relations. We first switch u and v in Eq. (2.288):

$$\vec{\nabla} \cdot (v\vec{\nabla}u) = \vec{\nabla}v \cdot \vec{\nabla}u + v\vec{\nabla}^2 u, \tag{2.289}$$

and subtract from the original equation [Eq. (2.288)] to write

$$\vec{\nabla} \cdot (u\vec{\nabla}v) - \vec{\nabla} \cdot (v\vec{\nabla}u) = u\vec{\nabla}^2 v - v\vec{\nabla}^2 u. \tag{2.290}$$

Integrating both sides over a volume V bounded by the surface S and using the divergence theorem:

$$\int_V d\tau \, \vec{\nabla} \cdot \vec{A} = \oint_S d\vec{\sigma} \cdot \vec{A}, \tag{2.291}$$

we obtain **Green's second identity**:

$$\boxed{\oint_S d\vec{\sigma} \cdot [u\vec{\nabla}v - v\vec{\nabla}u] = \int_V d\tau [u\vec{\nabla}^2 v - v\vec{\nabla}^2 u].} \tag{2.292}$$

Applying the similar process directly to Eq. (2.288), we obtain **Green's first identity**:

$$\boxed{\oint_S d\vec{\sigma} \cdot u\vec{\nabla}v = \int_V d\tau [\vec{\nabla}u \cdot \vec{\nabla}v + u\vec{\nabla}^2 v].} \tag{2.293}$$

Useful vector identities $\qquad$ (2.294)

$$\vec{\nabla} \times \vec{\nabla}f = 0, \tag{2.295}$$

$$\vec{\nabla} \cdot (\vec{\nabla} \times \vec{A}) = 0, \tag{2.296}$$

$$\vec{\nabla} \cdot \vec{\nabla}f = \vec{\nabla}^2 f = \frac{\partial^2 f}{\partial x_1^2} + \frac{\partial^2 f}{\partial x_2^2} + \frac{\partial^2 f}{\partial x_3^2}, \tag{2.297}$$

$$\vec{\nabla} \times \vec{\nabla} \times \vec{A} = \vec{\nabla}(\vec{\nabla} \cdot \vec{A}) - \vec{\nabla}^2 \vec{A}, \tag{2.298}$$

$$\vec{\nabla} \cdot (\vec{A} \times \vec{B}) = \vec{B} \cdot (\vec{\nabla} \times \vec{A}) - \vec{A} \cdot (\vec{\nabla} \times \vec{B}), \tag{2.299}$$

$$\vec{\nabla}(\vec{A} \cdot \vec{B}) = \vec{A} \times (\vec{\nabla} \times \vec{B}) + \vec{B} \times (\vec{\nabla} \times \vec{A}) + (\vec{A} \cdot \vec{\nabla})\vec{B} + (\vec{B} \cdot \vec{\nabla})\vec{A}, \tag{2.300}$$

$$\vec{\nabla} \times (\vec{A} \times \vec{B}) = \vec{A}(\vec{\nabla} \cdot \vec{B}) - \vec{B}(\vec{\nabla} \cdot \vec{A}) + (\vec{B} \cdot \vec{\nabla})\vec{A} - (\vec{A} \cdot \vec{\nabla})\vec{B}. \tag{2.301}$$

2.10   POTENTIAL THEORY

The **gravitational force** that a point mass M located at the origin exerts on another point mass m at $\overrightarrow{r}$ is given by Newton's law (Figure 2.33) as

$$\boxed{\overrightarrow{F} = -G\frac{Mm}{r^2}\widehat{e}_r,} \tag{2.302}$$

where G is the gravitational constant and $\widehat{e}_r$ is a unit vector along the radial direction. Since mass is always positive, the minus sign in Eq. (2.302) indicates that the gravitational force is attractive. Newton's law also indicates that the gravitational force is central, that is, the force is directed along the line joining the two masses. We now introduce the **gravitational field** $\overrightarrow{g}$ due to the mass M as

$$\boxed{\overrightarrow{g} = -G\frac{M}{r^2}\widehat{e}_r,} \tag{2.303}$$

which assigns a vector to each point in space with a magnitude that decreases with the inverse square of the distance and always points toward the central mass M (Figure 2.33). Gravitational force that M exerts on another mass m can now be written as

$$\boxed{\overrightarrow{F} = m\overrightarrow{g}.} \tag{2.304}$$

In other words, M attracts m through its gravitational field, which eliminates the need for **action at a distance**. Field concept is a very significant step in understanding interactions in nature. Its advantages become even more clear with the introduction of the Lagrangian formulation of continuum mechanics and then the relativistic theories, where the speed of light is the maximum speed with which any effect in nature can propagate. Of course, in Newton's theory, the speed of light is infinite and the changes in a gravitational field at a given point are felt everywhere in the universe instantaneously. Today, the field concept is an indispensable part of physics, at both the classical and the quantum level.

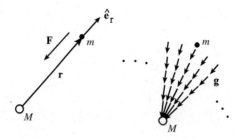

Figure 2.33   Gravitational force and gravitational field.

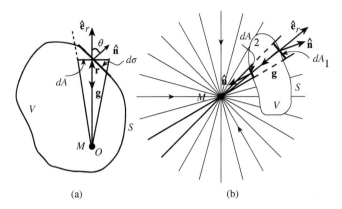

Figure 2.34   Flux of the gravitational field.

We now write the **flux** ϕ of the gravitational field of a point mass over a closed surface S enclosing the mass M (Figure 2.34a) as

$$\phi = \oint_S \vec{g} \cdot d\vec{\sigma} = -G \oint_S \frac{M}{r^2} \hat{e}_r \cdot d\vec{\sigma}. \tag{2.305}$$

Since the **solid angle** $d\Omega$ subtended by the area element $d\sigma$ is

$$d\Omega = \frac{\hat{e}_r \cdot d\vec{\sigma}}{r^2} = \frac{\hat{e}_r \cdot \hat{n} \, d\sigma}{r^2} = \frac{\cos\theta \, d\sigma}{r^2} = \frac{dA}{r^2}, \tag{2.306}$$

we can write the flux as

$$\phi = -GM \oint_S \frac{dA}{r^2} = -GM \oint_S d\Omega, \tag{2.307}$$

where dA is the area element in the direction of $\hat{e}_r$. Integration over the entire surface gives

$$\phi = -4\pi GM, \tag{2.308}$$

where the solid angle subtended by the entire surface is 4π. We now use the divergence theorem [Eq. (2.146)] to write the flux of the gravitational field as

$$\phi = \oint_S \vec{g} \cdot d\vec{\sigma} = \int_V \vec{\nabla} \cdot \vec{g} \, d\tau, \tag{2.309}$$

which gives

$$\boxed{\int_V \vec{\nabla} \cdot \vec{g} \, d\tau = -4\pi GM,} \tag{2.310}$$

where V is the volume enclosed by the closed surface S.

An important property of classical gravity is linearity; that is, when there are more than one particle interacting with m, the net force that m feels is the sum of the forces that each particle exerts on m as if it were alone. Naturally, for a continuous distribution of matter with density $\rho(\vec{r})$ interacting with a point mass m, the mass M in Eq. (2.310) is replaced by an integral:

$$M \rightarrow \int_V \rho(\vec{r})d\tau. \tag{2.311}$$

We now write Eq. (2.310) as

$$\int_V \vec{\nabla} \cdot \vec{g}\, d\tau = -4\pi G \int_V \rho(\vec{r})d\tau, \tag{2.312}$$

$$\int_V (\vec{\nabla} \cdot \vec{g} + 4\pi G\rho(\vec{r}))d\tau = 0. \tag{2.313}$$

For an arbitrary but finite volume element, the only way to satisfy this equality is to have the integrand vanish, that is,

$$\vec{\nabla} \cdot \vec{g} + 4\pi G\rho(\vec{r}) = 0, \tag{2.314}$$

which is usually written as

$$\boxed{\vec{\nabla} \cdot \vec{g} = -4\pi G\rho(\vec{r}).} \tag{2.315}$$

This is the **classical** gravitational field equation to which Einstein's theory of gravitation reduces in the limit of weak fields and small velocities. Given the mass distribution $\rho(\vec{r})$, it gives a partial differential equation to be solved for the gravitational field $\vec{g}$.

If we choose a closed surface that does not include the mass M, then the net flux over the entire surface is zero. If we concentrate on a pair of area elements, dA_1 and dA_2, in the figure on the right (Figure 2.34), we write the total flux as

$$d\phi_{12} = d\phi_1 + d\phi_2 \tag{2.316}$$

$$= -GM\, d\Omega_1 + GM\, d\Omega_2. \tag{2.317}$$

Since the solid angles, $d\Omega_1$ and $d\Omega_2$, subtended at the center by dA_1 and A_2, respectively, are equal, $d\phi_1$ and $d\phi_2$ cancel each other. Since the total flux is the sum of such pairs, the total flux is also zero. The gravitational field equation to be solved for a region that does not include any mass is given as

$$\boxed{\vec{\nabla} \cdot \vec{g} = 0.} \tag{2.318}$$

As we have mentioned before, this does not mean the gravitational field is zero in that region, but it means that the sources are outside the region of interest.

2.10.1 Gravitational Field of a Star

For a spherically symmetric star with density $\rho(r)$, the gravitational field depends only on the radial distance from the origin. Hence, we can write

$$\vec{g}(\vec{r}) = g(r)\hat{e}_r, \qquad (2.319)$$

where $\hat{e}_r$ is a unit vector pointing radially outwards. To find $g(r)$, we choose a spherical **Gaussian surface** $S(r)$ with radius r. Since the outward normal to a sphere is also in the $\hat{e}_r$ direction, we utilize the divergence theorem to convert the volume integral in Eq. (2.312) to a surface integral:

$$\int_{V(r)} \vec{\nabla} \cdot \vec{g} \, d\tau = \oint_{S(r)} \vec{g} \cdot d\vec{\sigma}, \qquad (2.320)$$

and write

$$\oint_{S(r)} g(r)(\hat{e}_r \cdot \hat{e}_r) d\sigma = -4\pi G \int_{V(r)} \rho(r) \, d\tau, \qquad (2.321)$$

$$\oint g(r)r^2 \, d\Omega = -4\pi G \int_{V(r)} \rho(r)r^2 \, dr \, d\Omega, \qquad (2.322)$$

$$g(r)r^2 \oint d\Omega = -4\pi G \int \rho(r)r^2 \, dr \oint d\Omega, \qquad (2.323)$$

where $d\sigma = r^2 \, d\Omega = r^2 \sin\theta \, d\theta \, d\phi$ is the infinitesimal surface area element of the sphere. Since $\oint d\Omega = 4\pi$, we obtain the magnitude of the gravitational field as

$$g(r) = -G\frac{\int_0^r \rho(r')4\pi r'^2 \, dr'}{r^2} = -G\frac{\int_0^{m(r)} dm_r}{r^2} \qquad (2.324)$$

$$= -G\frac{m(r)}{r^2}. \qquad (2.325)$$

An important feature of this result is that part of the mass lying outside the Gaussian surface, which is a sphere of radius r, does not contribute to the field at r and the mass inside the Gaussian surface acts as if it is concentrated at the center. Note that dm_r is the mass of an infinitesimal shell at r with thickness dr.

Similarly, if we find the gravitational field of a spherical shell of radius R, we find that for points outside, $r \geq R$, the shell behaves as if its entire mass is concentrated at the center. For points inside the shell, the gravitational field is zero. These interesting features of Newton's theory of gravity also remain intact in Einstein's theory, where they are summarized in terms of Birkhoff's theorem.

2.10.2 Work Done by Gravitational Force

We now approach the problem from a different direction. Consider a test particle of mass m moving along a closed path C in the gravitational field of another point particle of mass M (Figure 2.35). The work done by the gravitational field on the test particle is

$$W = \oint_C \overrightarrow{F} \cdot d\overrightarrow{r} = m \oint_C \overrightarrow{g} \cdot d\overrightarrow{r} \tag{2.326}$$

$$= m \oint_C g_T(s)ds, \tag{2.327}$$

where g_T is the tangential component of the gravitational field of M along the path. Using Stokes's theorem [Eq. (2.261)], we can also write this as

$$W = m \oint_C \overrightarrow{g} \cdot d\overrightarrow{r} \tag{2.328}$$

$$= m \int_S (\overrightarrow{\nabla} \times \overrightarrow{g}) \cdot d\overrightarrow{\sigma}. \tag{2.329}$$

If we calculate $\overrightarrow{\nabla} \times \overrightarrow{g}$ for the gravitational field of the point mass M located at the origin, we find

$$\overrightarrow{\nabla} \times \overrightarrow{g} = \overrightarrow{\nabla} \times \left(-G\frac{M}{r^2}\widehat{e}_r \right) \tag{2.330}$$

$$= -GM\overrightarrow{\nabla} \times \left[\frac{x_1\widehat{e}_1 + x_2\widehat{e}_2 + x_3\widehat{e}_3}{(x_1^2 + x_2^2 + x_3^2)^{3/2}} \right] = 0. \tag{2.331}$$

Substituting $\overrightarrow{\nabla} \times \overrightarrow{g} = 0$ into Eq. (2.329), we obtain the work done by the gravitational field on a point particle m moving on a closed path as zero.

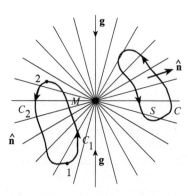

Figure 2.35   Work done by the gravitational field.

If we split a closed path into two parts as C_1 and C_2, as shown in Figure 2.35, we can write

$$m \oint_C \overrightarrow{g} \cdot d\overrightarrow{r} = m \int_{1 \atop C_1}^{2} \overrightarrow{g} \cdot d\overrightarrow{r} + m \int_{2 \atop C_2}^{1} \overrightarrow{g} \cdot d\overrightarrow{r} = 0. \qquad (2.332)$$

Interchanging the order of integration, we obtain

$$W_{1 \to 2} = m \int_{1 \atop C_1}^{2} \overrightarrow{g} \cdot d\overrightarrow{r} = m \int_{1 \atop C_2}^{2} \overrightarrow{g} \cdot d\overrightarrow{r}. \qquad (2.333)$$

Since C_1 and C_2 are two arbitrary paths connecting points 1 and 2, this means that the work done by the gravitational field is **path-independent**.

As the test particle moves under the influence of the gravitational field, it also satisfies the second law of Newton, that is,

$$\overrightarrow{F} = m \frac{d^2 \overrightarrow{r}}{dt^2}, \qquad (2.334)$$

$$m\overrightarrow{g} = m \frac{d^2 \overrightarrow{r}}{dt^2} = m \frac{d\overrightarrow{v}}{dt}. \qquad (2.335)$$

Using this in Eq. (2.333), we can write the work done by gravity as

$$W_{1 \to 2} = m \int_{1 \atop C}^{2} \overrightarrow{g} \cdot d\overrightarrow{r} = m \int_{1 \atop C_1}^{2} \frac{d\overrightarrow{v}}{dt} \cdot d\overrightarrow{r} \qquad (2.336)$$

$$= m \int_{1 \atop C_1}^{2} \frac{d\overrightarrow{v}}{dt} \cdot \frac{d\overrightarrow{r}}{dt} dt = m \int_{1 \atop C_1}^{2} \frac{d\overrightarrow{v}}{dt} \cdot \overrightarrow{v} \, dt \qquad (2.337)$$

$$= \int_{1 \atop C_1}^{2} \frac{d}{dt} \left[\frac{1}{2} m \overrightarrow{v} \cdot \overrightarrow{v} \right] dt = \int_{1 \atop C_1}^{2} d \left[\frac{1}{2} m \overrightarrow{v} \cdot \overrightarrow{v} \right] \qquad (2.338)$$

$$= \left[\frac{1}{2} m v^2 \right]_2 - \left[\frac{1}{2} m v^2 \right]_1. \qquad (2.339)$$

In other words, the work done by gravity is equal to the change in the kinetic energy:

$$T = \frac{1}{2} m v^2, \qquad (2.340)$$

of the particle as it moves from point 1 to 2. This result, $\overrightarrow{\nabla} \times \overrightarrow{g} = 0$, obtained for the gravitational field of a point mass M has several important consequences. First of all, since the gravitational interaction is linear, the gravitational field of an arbitrary mass distribution can be constructed from the gravitational fields of point masses by linear superposition. Hence,

$$\boxed{\vec{\nabla} \times \vec{g} = 0} \tag{2.341}$$

is a general property of Newtonian gravity, independent of the source and the coordinates used.

2.10.3   Path Independence and Exact Differentials

We have seen that for an arbitrary vector field $\vec{v}$, if the curl is identically zero:

$$\vec{\nabla} \times \vec{v} = 0, \tag{2.342}$$

then we can always find a scalar field, $\Phi(\vec{r})$, such that

$$\vec{v} = \vec{\nabla}\Phi. \tag{2.343}$$

The existence of a differentiable Φ is guaranteed by the vanishing of the curl of $\vec{v}$:

$$\frac{\partial v_1}{\partial x_2} - \frac{\partial v_2}{\partial x_1} = 0, \tag{2.344}$$

$$\frac{\partial v_2}{\partial x_3} - \frac{\partial v_3}{\partial x_2} = 0, \tag{2.345}$$

$$\frac{\partial v_3}{\partial x_1} - \frac{\partial v_1}{\partial x_3} = 0. \tag{2.346}$$

We now write the line integral

$$\int_1^2 \vec{v} \cdot d\vec{r} = \int_1^2 v_1 \, dx_1 + v_2 \, dx_2 + v_3 \, dx_3. \tag{2.347}$$

If we can find a scalar function, $\Phi(x_1, x_2, x_3)$, such that its partial derivatives are

$$\frac{\partial \Phi}{\partial x_1} = v_1, \quad \frac{\partial \Phi}{\partial x_2} = v_2, \quad \frac{\partial \Phi}{\partial x_3} = v_3, \tag{2.348}$$

then the line integral [Eq. (2.347)] can be evaluated as

$$\int_1^2 \vec{v} \cdot d\vec{r} = \int_1^2 \vec{\nabla}\Phi \cdot d\vec{r} = \int_1^2 \frac{\partial \Phi}{\partial x_1} \, dx_1 + \frac{\partial \Phi}{\partial x_2} \, dx_2 + \frac{\partial \Phi}{\partial x_3} \, dx_3 \tag{2.349}$$

$$= \int_1^2 d\Phi = \Phi(2) - \Phi(1). \tag{2.350}$$

In other words, when such a Φ can be found, the value of the line integral $\int_1^2 \vec{v} \cdot d\vec{r}$ depends only on the values that Φ takes at the end points, that is, it is **path-independent**. When such a function exists, $v_1 \, dx_1 + v_2 \, dx_2 + v_3 \, dx_3$ is called an **exact differential** and can be written as

$$v_1 \, dx_1 + v_2 \, dx_2 + v_3 \, dx_3 = d\Phi. \tag{2.351}$$

The existence of Φ is guaranteed by the following **sufficient** and **necessary** differentiability conditions:

$$\frac{\partial^2 \Phi}{\partial x_1 \partial x_2} = \frac{\partial^2 \Phi}{\partial x_2 \partial x_1}, \tag{2.352}$$

$$\frac{\partial^2 \Phi}{\partial x_2 \partial x_3} = \frac{\partial^2 \Phi}{\partial x_3 \partial x_2}, \tag{2.353}$$

$$\frac{\partial^2 \Phi}{\partial x_3 \partial x_1} = \frac{\partial^2 \Phi}{\partial x_1 \partial x_3}. \tag{2.354}$$

Using Eq. (2.348), we can write these as

$$\frac{\partial v_2}{\partial x_1} = \frac{\partial v_1}{\partial x_2}, \tag{2.355}$$

$$\frac{\partial v_3}{\partial x_2} = \frac{\partial v_2}{\partial x_3}, \tag{2.356}$$

$$\frac{\partial v_1}{\partial x_3} = \frac{\partial v_3}{\partial x_1}, \tag{2.357}$$

which are nothing but the conditions [Eqs. (2.344)–(2.346)] for $\vec{\nabla} \times \vec{v} = 0$.

2.10.4 Gravity and Conservative Forces

We are now ready to apply all this to gravitation. Since $\vec{\nabla} \times \vec{g} = 0$, we introduce a scalar function Φ such that

$$\vec{g} = -\vec{\nabla}\Phi, \tag{2.358}$$

where $\Phi(\vec{r})$ is called the **gravitational potential**, which for a point mass M is given as

$$\Phi(\vec{r}) = -G\frac{M}{r}. \tag{2.359}$$

The minus sign is introduced to assure that the force is attractive, that is, it is always toward the central mass M. We can now write Eq. (2.333) as

$$W_{1\rightarrow 2} = m \int_{1 \atop C}^{2} \vec{g} \cdot d\vec{r} = -m \int_{1 \atop C}^{2} \vec{\nabla}\Phi \cdot d\vec{r} \tag{2.360}$$

$$= -m[\Phi(2) - \Phi(1)]. \tag{2.361}$$

Using this with Eq. (2.339), we can write

$$-m[\Phi(2) - \Phi(1)] = \left[\frac{1}{2}mv^2\right]_2 - \left[\frac{1}{2}mv^2\right]_1.$$

(2.362)

If we rewrite this as

$$\left[\frac{1}{2}mv^2\right]_1 + m\Phi(1) = \left[\frac{1}{2}mv^2\right]_2 + m\Phi(2),$$

(2.363)

we see that the quantity

$$\boxed{\frac{1}{2}mv^2 + m\Phi(\vec{r}) = E}$$

(2.364)

is a constant throughout the motion of the particle. This constant, E, is nothing, but the conserved **total energy** of the particle. The first term, $\frac{1}{2}mv^2$, is the familiar **kinetic energy**. Hence, we interpret $m\Phi(\vec{r})$ as the **gravitational potential energy**, Ω, of the particle m:

$$\Omega(\vec{r}) = m\Phi(\vec{r}),$$

(2.365)

and write

$$\frac{1}{2}mv^2 + \Omega = E.$$

(2.366)

To justify our interpretation of Ω, consider m at a height of h from the surface of the Earth (Figure 2.36) and write

$$\Omega = m\Phi(\vec{r}) = -m\frac{GM}{(R+h)} = -\frac{GMm}{R\left(1 + \frac{h}{R}\right)},$$

(2.367)

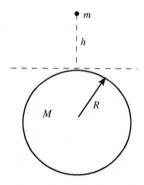

Figure 2.36   Gravitational potential energy.

where R is the radius of the Earth. For $h \ll R$, we can use the binomial expansion to write

$$\Omega \simeq -\frac{GMm}{R}\left(1 - \frac{h}{R}\right), \tag{2.368}$$

$$\Omega \simeq -\frac{GMm}{R} + m\left(\frac{GM}{R^2}\right)h. \tag{2.369}$$

If we identify GM/R^2 as the gravitational acceleration g, the average numerical value of which is 9.8 m/s^2, the second term becomes the gravitational potential energy, mgh, familiar from elementary physics. The first term on the right is a constant. Since from $\overrightarrow{g} = -\overrightarrow{\nabla}\Phi$, adding or subtracting a constant to Φ does not effect the fields and hence the forces, which are the directly accessible quantities to measurement, we can always choose the zero level of the gravitational potential energy. Thus, when $R \gg h$, we can take the gravitational potential energy as

$$\Omega = mgh. \tag{2.370}$$

From the definition of the **gravitational field** of a point particle:

$$\overrightarrow{g} = -G\frac{M}{r^2}\hat{e}_r, \tag{2.371}$$

it is seen that operationally the gravitational field at a point is basically the force on a unit test mass. Mathematically, the gravitational field of a mass distribution given by the density $\rho(\overrightarrow{r})$ is determined by the **field equation**

$$\boxed{\overrightarrow{\nabla} \cdot \overrightarrow{g} = -4\pi G\rho(\overrightarrow{r}),} \tag{2.372}$$

which is also known as **Gauss's law** for gravitation. Interactions with a vanishing curl are called **conservative forces**. Frictional forces and in general velocity-dependent forces are nonconservative, since the work done by them depends upon the path that the particles follow.

2.10.5 Gravitational Potential

We consider Eqs. (2.360) and (2.361) again and cancel m on both sides to write

$$\Phi(2) - \Phi(1) = -\int_{1}^{2}{}_{C}\, \overrightarrow{g} \cdot d\overrightarrow{r}, \tag{2.373}$$

$$\Phi(2) = \Phi(1) - \int_{1}^{2}{}_{C}\, \overrightarrow{g} \cdot d\overrightarrow{r}. \tag{2.374}$$

If we choose the initial point 1 at infinity and define the potential there as zero and the final point 2 as the point where we want to find the potential, we

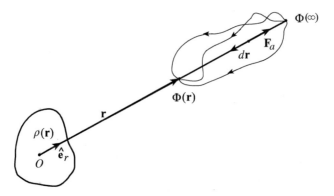

Figure 2.37 Gravitational potential.

obtain the **gravitational potential** as

$$\Phi(\overrightarrow{r}) = -\int_{\infty}^{\overrightarrow{r}} \overrightarrow{g} \cdot d\overrightarrow{r}. \tag{2.375}$$

From Figure 2.37 it is seen that the integral $-\int_{\infty}^{\overrightarrow{r}} \overrightarrow{g} \cdot d\overrightarrow{r}$ is equal to the work that one has to do to bring a unit test mass infinitesimally slowly from infinity to $\overrightarrow{r}$:

$$\Phi(\overrightarrow{r}) = -\int_{\infty}^{\overrightarrow{r}} \overrightarrow{g} \cdot d\overrightarrow{r} = \int_{\infty}^{\overrightarrow{r}} \overrightarrow{F}_{\text{applied}} \cdot d\overrightarrow{r} \tag{2.376}$$

$$= W_{\text{unit mass}}(\infty \rightarrow \overrightarrow{r}). \tag{2.377}$$

Note that for the test mass to move infinitesimally slowly, we have to apply a force by the amount

$$\overrightarrow{F}_{\text{applied}} = -\overrightarrow{g}, \tag{2.378}$$

so that the test particle does not accelerate toward the source of the gravitational potential. For a point mass M, this gives the gravitational potential as

$$\Phi(\overrightarrow{r}) = -G\frac{M}{r}. \tag{2.379}$$

What makes this definition meaningful is that gravity is a conservative field. Hence, Φ is path independent (Figure 2.37). We can now use

$$\overrightarrow{g} = -\overrightarrow{\nabla}\Phi \tag{2.380}$$

to write the gravitational field equation [Eq. (2.372)] as

$$\overrightarrow{\nabla} \cdot \overrightarrow{\nabla}\Phi = 4\pi G\rho, \tag{2.381}$$

or as

$$\boxed{\overrightarrow{\nabla}^2 \Phi = 4\pi G \rho,}$$

(2.382)

which is **Poisson's equation**. In a region where there is no mass, the equation to be solved is the **Laplace Equation**:

$$\boxed{\overrightarrow{\nabla}^2 \Phi = 0,}$$

(2.383)

where the operator $\overrightarrow{\nabla}^2$ is called the **Laplacian**.

The advantage of working with the gravitational potential is that it is a scalar; hence, it is easier to work with. Since gravity is a linear interaction, we can write the potential of N particles by linear superposition of the potentials of the individual particles that make up the system as

$$\boxed{\Phi(\overrightarrow{r}) = -G \sum_{i=1}^{N} \frac{m_i}{|\overrightarrow{r} - \overrightarrow{r}_i|},}$$

(2.384)

where $\overrightarrow{r}_i$ is the position of the ith particle and $\overrightarrow{r}$ is called the field point. In the case of a continuous mass distribution, we write the potential as an integral:

$$\boxed{\Phi(\overrightarrow{r}) = -G \int_V \frac{\rho(\overrightarrow{r}')d^3\overrightarrow{r}'}{|\overrightarrow{r} - \overrightarrow{r}'|},}$$

(2.385)

where the volume integral is over the source points $\overrightarrow{r}'$. After $\Phi(\overrightarrow{r})$ is found, one can construct the gravitational field easily by taking its gradient, which involves only differentiation.

2.10.6   Gravitational Potential Energy of a System

For a pair of particles, the gravitational **potential energy** is written as [Eqs. (2.359) and (2.365)]

$$\Omega = -G \frac{Mm}{r},$$

(2.386)

where r is the separation between the particles. For a system of N **discrete** particles, we can consider the system as made up of pairs and write the gravitational potential energy in the following equivalent ways:

$$\Omega = -G \sum_{\text{All pairs, } i \neq j} \frac{m_i m_j}{r_{ij}}, \quad \overrightarrow{r}_{ij} = \overrightarrow{r}_j - \overrightarrow{r}_i$$

(2.387)

$$= -G \sum_{i>j}^{N} \sum_{j=1}^{N} \frac{m_i m_j}{r_{ij}}$$

(2.388)

$$= -G\frac{1}{2}\sum_{i=1}^{N}\sum_{j=1}^{N}\frac{m_i m_j}{r_{ij}}, \quad i \neq j. \tag{2.389}$$

First of all, we do not include the cases with $i = j$, which are not even pairs. These terms basically correspond to the *self energies* of the particles that make up the system. We leave them out since they contribute as a constant that does not change with the changing configuration of the system. The factor of $1/2$ is inserted in the last expression to avoid double counting of the pairs. Note that Ω can also be written as

$$\Omega = \frac{1}{2}(m_1\Phi_1 + m_2\Phi_2 + \cdots + m_n\Phi_n) \tag{2.390}$$

$$= \frac{1}{2}\sum_{i=1}^{N} m_i\Phi_i, \quad \Phi_i = -G\sum_{j=1}^{N}\frac{m_j}{r_{ji}}, \quad i \neq j, \tag{2.391}$$

where Φ_i is the gravitational potential at the location of the particle m_i due to all other particles. If the particles form a continuum with the density ρ, we then write

$$\Omega = \frac{1}{2}\int_0^M \Phi \, dm \tag{2.392}$$

$$= \frac{1}{2}\int_V \Phi(\overrightarrow{r'})\rho(\overrightarrow{r'})d^3\overrightarrow{r'}, \tag{2.393}$$

where Φ is the potential of the part of the system with the mass $M - dm$ acting on $dm = \rho \, d^3\overrightarrow{r}$.

Example 2.13. *Gravitational potential energy of a uniform sphere:* For a spherically symmetric mass distribution with density $\rho(r)$ and radius R, we can write the gravitational potential energy as

$$\Omega(R) = \frac{1}{2}\int_V \Phi \, dm \tag{2.394}$$

$$= -G\int_0^M \frac{m(r)dm}{r}, \tag{2.395}$$

where $m(r)$ is the mass inside the radius r, and dm is the mass of the shell with radius r and thickness dr:

$$dm = 4\pi r^2 \rho(r)dr. \tag{2.396}$$

For uniform density ρ_0 we write Ω as

$$\Omega(R) = -G\int_0^R \frac{(4\pi\rho_0 r^3/3)4\pi r^2\rho_0 \, dr}{r}, \tag{2.397}$$

which gives

$$\boxed{\Omega(R) = -\frac{3GM^2}{5R}.}$$ (2.398)

Because of the minus sign, this is the amount of work that one has to do to disassemble this object by taking its particles to infinity.

2.10.7   Helmholtz Theorem

We now introduce an important theorem due to Helmholtz, which is an important part of the potential theory.

Theorem 2.2. A vector field, if it exists, is uniquely determined in a region R surrounded by the closed surface S by giving its divergence and curl in R and its normal component on S.

Proof: Assume that there are two fields, $\vec{v}_1$ and $\vec{v}_2$, that satisfy the required conditions, that is, they have the same divergence, curl, and normal component. We now need to show that if this be the case, then these two fields must be identical. Since divergence, curl, and dot product are all linear operators, we define a new field $\vec{w}$ as

$$\vec{w} = \vec{v}_1 - \vec{v}_2,$$ (2.399)

which satisfies

$$\vec{\nabla} \times \vec{w} = 0 \text{ in } R,$$ (2.400)

$$\vec{\nabla} \cdot \vec{w} = 0 \text{ in } R,$$ (2.401)

$$\hat{n} \cdot \vec{w} = 0 \text{ on } S.$$ (2.402)

Since $\vec{\nabla} \times \vec{w} = 0$, we can introduce a scalar potential Φ as

$$\vec{w} = -\vec{\nabla}\Phi.$$ (2.403)

Using Green's first identity [Eq. (2.293)]:

$$\oint_S d\vec{\sigma} \cdot (u\vec{\nabla}v) = \int_V d\tau [\vec{\nabla}u \cdot \vec{\nabla}v + u\vec{\nabla} \cdot \vec{\nabla}v]$$ (2.404)

with the substitution $u = v = \Phi$, we write

$$\oint_S d\vec{\sigma} \cdot (\Phi\vec{\nabla}\Phi) = \int_V d\tau [\vec{\nabla}\Phi \cdot \vec{\nabla}\Phi + \Phi\vec{\nabla} \cdot \vec{\nabla}\Phi].$$ (2.405)

When Eq. (2.403) is substituted, this becomes

$$\oint_S d\vec{\sigma} \cdot (\Phi\vec{w}) = \int_V d\tau [\vec{w} \cdot \vec{w} + \Phi\vec{\nabla} \cdot \vec{w}].$$ (2.406)

Since the first integral:

$$\oint_S d\vec{\sigma} \cdot \Phi\vec{w} = \oint_S d\sigma \, \Phi(\hat{n} \cdot \vec{w}),$$

(2.407)

is zero because of Eq. (2.402) and the integral

$$\int_V d\tau \, \Phi\vec{\nabla} \cdot \vec{w}$$

(2.408)

is zero because of Eq. (2.401), Eq. (2.406) reduces to

$$\int_V d\tau \, \vec{w} \cdot \vec{w} = \int_V d\tau |\vec{w}|^2 = 0.$$

(2.409)

Since $|\vec{w}|^2$ is always a positive quantity, the only way to satisfy this equation for a finite volume is to have

$$\vec{w} \equiv 0,$$

(2.410)

that is,

$$\vec{v}_1 \equiv \vec{v}_2,$$

(2.411)

thereby proving the theorem.

2.10.8   Applications of the Helmholtz Theorem

Helmholtz theorem says that a vector field is completely and uniquely specified by giving its divergence, curl, and the normal component on the bounding surface. When we are interested in the entire space, the bounding surface is usually taken as a sphere in the limit as its radius goes to infinity. Given a vector field $\vec{v}$, let us write its divergence and curl as

$$\vec{\nabla} \cdot \vec{v} = k_1 \rho(\vec{r}),$$

(2.412)

$$\vec{\nabla} \times \vec{v} = k_2 \vec{J}(\vec{r}),$$

(2.413)

where k_1 and k_2 are constants. The terms on the right-hand side, $\rho(\vec{r})$ and $\vec{J}(\vec{r})$, are known functions of position and in general represent **sources** and **current densities**, respectively.

There are three cases that we analyze separately:

(I) In cases for which there are no currents, the field satisfies

$$\boxed{\vec{\nabla} \cdot \vec{v} = k_1 \rho(\vec{r}),}$$

(2.414)

$$\boxed{\vec{\nabla} \times \vec{v} = 0.}$$

(2.415)

We have already shown that when the curl of a vector field is zero, we can always find a scalar potential $\Phi(\vec{r})$ such that

$$\boxed{\vec{v} = -\vec{\nabla}\Phi.}$$
(2.416)

Now the second equation [Eq. (2.415)] is satisfied automatically and the first equation can be written as **Poisson's equation**:

$$\boxed{\vec{\nabla}^2\Phi = -k_1\rho,}$$
(2.417)

the solution of which can be written as

$$\boxed{\Phi(\vec{r}) = \frac{k_1}{4\pi}\int_V \frac{\rho(\vec{r}')d\tau'}{|\vec{r} - \vec{r}'|},}$$
(2.418)

where the volume integral is over the source variable $\vec{r}'$ and $\vec{r}$ is the field point. Notice that the definition of scalar potential [Eq. (2.416)] is arbitrary up to an additive constant, which means that we are free to choose the zero level of the potential.

   (II) In cases where $\rho(\vec{r}) = 0$, the field equations become

$$\boxed{\vec{\nabla}\cdot\vec{v} = 0,}$$
(2.419)

$$\boxed{\vec{\nabla}\times\vec{v} = k_2\vec{J}(\vec{r}).}$$
(2.420)

We now use the fact that the divergence of a curl is zero and introduce a vector potential $\vec{A}(\vec{r})$ such that

$$\boxed{\vec{v} = \vec{\nabla}\times\vec{A}.}$$
(2.421)

We have already proven that the divergence of a curl vanishes identically. We now prove the converse, that is, if the divergence of a vector field $\vec{v}$ vanishes identically, then we can always find a vector potential $\vec{A}$ such that its curl gives $\vec{v}$. Since we want $\vec{A}$ to satisfy Eq. (2.421), we can write

$$\frac{\partial A_3}{\partial x_2} - \frac{\partial A_2}{\partial x_3} = v_1,$$
(2.422)

$$\frac{\partial A_1}{\partial x_3} - \frac{\partial A_3}{\partial x_1} = v_2,$$
(2.423)

$$\frac{\partial A_2}{\partial x_1} - \frac{\partial A_1}{\partial x_2} = v_3.$$
(2.424)

Remembering that the curl of a gradient is zero [Eq. (2.287)], we can always add or subtract the gradient of a scalar function to the vector potential $\vec{A}$,

$$\vec{A} \rightarrow \vec{A} + \vec{\nabla}\Psi, \tag{2.425}$$

without affecting the field $\vec{v}$. This gives us the freedom to set one of the components of $\vec{v}$ to zero. Hence, we set $A_3 = 0$, which simplifies Eqs. (2.422)–(2.424) to

$$-\frac{\partial A_2}{\partial x_3} = v_1, \tag{2.426}$$

$$\frac{\partial A_1}{\partial x_3} = v_2, \tag{2.427}$$

$$\frac{\partial A_2}{\partial x_1} - \frac{\partial A_1}{\partial x_2} = v_3. \tag{2.428}$$

The first two equations can be integrated immediately to yield

$$A_1 = \int_{x_{03}}^{x_3} v_2 \, dx_3, \tag{2.429}$$

$$A_2 = -\int_{x_{03}}^{x_3} v_1 \, dx_3 + f_2(x_1, x_2), \tag{2.430}$$

where $f_2(x_1, x_2)$ is arbitrary at this point. Substituting these into the third equation [Eq. (2.428)], we obtain

$$\frac{\partial f_2}{\partial x_1} - \int_{x_{03}}^{x_3} \left(\frac{\partial v_1}{\partial x_1} + \frac{\partial v_2}{\partial x_2} \right) dx_3 = v_3. \tag{2.431}$$

Using the fact that the divergence of $\vec{v}$ is zero:

$$\frac{\partial v_1}{\partial x_1} + \frac{\partial v_2}{\partial x_2} + \frac{\partial v_3}{\partial x_3} = 0, \tag{2.432}$$

we can write Eq. (2.431) as

$$\frac{\partial f_2}{\partial x_1} + \int_{x_{03}}^{x_3} \frac{\partial v_3}{\partial x_3} \, dx_3 = v_3. \tag{2.433}$$

The integral in Eq. (2.433) can be evaluated immediately to give

$$\frac{\partial f_2(x_1, x_2)}{\partial x_1} + v_3(x_1, x_2, x_3) - v_3(x_1, x_2, x_{03}) = v_3(x_1, x_2, x_3), \tag{2.434}$$

$$\frac{\partial f_2(x_1, x_2)}{\partial x_1} = v_3(x_1, x_2, x_{03}), \tag{2.435}$$

which yields $f_2(x_1, x_2)$ as the quadrature

$$f_2(x_1, x_2) = \int_{x_{01}}^{x_1} v_3(x_1, x_2, x_{03}) dx_1. \tag{2.436}$$

Substituting f_2 into Eq. (2.430), we obtain the vector potential as

$$A_1 = \int_{x_{03}}^{x_3} v_2(x_1, x_2, x_3) dx_3, \tag{2.437}$$

$$A_2 = \int_{x_{01}}^{x_1} v_3(x_1, x_2, x_{03}) dx_1 - \int_{x_{03}}^{x_3} v_1(x_1, x_2, x_3) dx_3, \tag{2.438}$$

$$A_3 = 0. \tag{2.439}$$

In conclusion, given a vector field $\vec{v}$ satisfying Eqs. (2.419) and (2.420), we can always find a vector potential $\vec{A}$ such that $\vec{v} = \vec{\nabla} \times \vec{A}$, where $\vec{A}$ is arbitrary up to the gradient of a scalar function.

Using a vector potential $\vec{A}$, [Eq. (2.421)], we can now write Eq. (2.420) as

$$\vec{\nabla} \times \vec{v} = k_2 \vec{J}(\vec{r}), \tag{2.440}$$

$$\vec{\nabla} \times \vec{\nabla} \times \vec{A} = k_2 \vec{J}(\vec{r}), \tag{2.441}$$

$$\vec{\nabla}(\vec{\nabla} \cdot \vec{A}) - \vec{\nabla}^2 \vec{A} = k_2 \vec{J}(\vec{r}). \tag{2.442}$$

Using the freedom in the choice of $\vec{A}$ we can set

$$\boxed{\vec{\nabla} \cdot \vec{A} = 0,} \tag{2.443}$$

which is called the **Coulomb gauge** in electrodynamics. The equation to be solved for the vector potential is now obtained as

$$\boxed{\vec{\nabla}^2 \vec{A} = -k_2 \vec{J}(\vec{r}).} \tag{2.444}$$

Since the Laplace operator is linear, each component of the vector potential satisfies Poisson's equation:

$$\vec{\nabla}^2 A_i = -k_2 J_i(\vec{r}), \quad i = 1, 2, 3, \tag{2.445}$$

Hence, the solution can be written as

$$\boxed{\vec{A}(\vec{r}) = \frac{k_2}{4\pi} \int_V \frac{\vec{J}(\vec{r}') d\tau'}{|\vec{r} - \vec{r}'|}.} \tag{2.446}$$

(III) In the general case, where the field equations are given as

$$\boxed{\vec{\nabla} \cdot \vec{v} = k_1 \rho(\vec{r}),} \tag{2.447}$$

$$\boxed{\vec{\nabla} \times \vec{v} = k_2 \vec{J}(\vec{r}),} \tag{2.448}$$

we can write the field in terms of the potentials Φ and $\vec{A}$ as

$$\boxed{\vec{v} = -\vec{\nabla}\Phi + \vec{\nabla} \times \vec{A}.} \tag{2.449}$$

Substituting this into the first equation [Eq. (2.447)] and using the fact that the divergence of a curl is zero, we obtain

$$-\vec{\nabla} \cdot \vec{\nabla}\Phi + \vec{\nabla} \cdot (\vec{\nabla} \times \vec{A}) = k_1\rho, \tag{2.450}$$

$$\vec{\nabla}^2 \Phi = -k_1\rho. \tag{2.451}$$

Similarly, substituting Eq. (2.449) into the second equation [Eq. (2.448)], we get

$$\vec{\nabla} \times (-\vec{\nabla}\Phi + \vec{\nabla} \times \vec{A}) = k_2 \vec{J}(\vec{r}), \tag{2.452}$$

$$-\vec{\nabla} \times \vec{\nabla}\Phi + \vec{\nabla} \times \vec{\nabla} \times \vec{A} = k_2 \vec{J}(\vec{r}), \tag{2.453}$$

$$\vec{\nabla}(\vec{\nabla} \cdot \vec{A}) - \vec{\nabla}^2 \vec{A} = k_2 \vec{J}(\vec{r}), \tag{2.454}$$

where we used the fact that the curl of a gradient is zero. Using the Coulomb gauge, $\vec{\nabla} \cdot \vec{A} = 0$, and Eq. (2.451), we obtain the two equations to be solved for the potentials as

$$\boxed{\vec{\nabla}^2 \Phi = -k_1\rho,} \tag{2.455}$$

$$\boxed{\vec{\nabla}^2 \vec{A} = -k_2 \vec{J}(\vec{r}).} \tag{2.456}$$

2.10.9   Examples from Physics

Gravitation

We have already discussed this case in detail. The field equations are given as

$$\boxed{\vec{\nabla} \cdot \vec{g} = -4\pi G\rho(\vec{r}),} \tag{2.457}$$

$$\boxed{\vec{\nabla} \times \vec{g} = 0,} \tag{2.458}$$

where $\rho(\vec{r})$ is the source of the gravitational field, that is, the mass density. Instead of these two equations, we can solve **Poisson's equation**:

$$\boxed{\vec{\nabla}^2 \Phi = 4\pi G \rho(\vec{r}),} \tag{2.459}$$

for the scalar potential Φ, which then can be used to find the gravitational field as

$$\vec{g} = -\vec{\nabla}\Phi. \tag{2.460}$$

Electrostatics

In electrostatics, the field equations for the electric field are given as

$$\boxed{\vec{\nabla} \cdot \vec{E} = 4\pi \rho(\vec{r}),} \tag{2.461}$$

$$\boxed{\vec{\nabla} \times \vec{E} = 0.} \tag{2.462}$$

Now, $\rho(\vec{r})$ stands for the charge density and the plus sign in Eq. (2.461) means like charges repel and opposite charges attract. **Poisson's equation** for the electrostatic potential is

$$\boxed{\vec{\nabla}^2 \Phi = -4\pi \rho(\vec{r}),} \tag{2.463}$$

where

$$\vec{E} = -\vec{\nabla}\Phi. \tag{2.464}$$

Magnetostatics

Now the field equations for the magnetic field are given as

$$\boxed{\vec{\nabla} \cdot \vec{B} = 0,} \tag{2.465}$$

$$\boxed{\vec{\nabla} \times \vec{B} = \frac{4\pi}{c}\vec{J},} \tag{2.466}$$

where c is the speed of light and $\vec{J}$ is the **current density**. The fact that the divergence of $\vec{B}$ is zero is a direct consequence of the fact that magnetic monopoles do not exist in nature. Introducing a vector potential $\vec{A}$ with the **Coulomb gauge** $\vec{\nabla} \cdot \vec{A} = 0$, we can solve

$$\boxed{\vec{\nabla}^2 \vec{A} = -\frac{4\pi}{c}\vec{J}} \tag{2.467}$$

and obtain the magnetic field via

$$\boxed{\vec{B} = \vec{\nabla} \times \vec{A}.} \tag{2.468}$$

Maxwell's Equations

The time-dependent Maxwell's equations are given as

$$
\begin{aligned}
&\vec{\nabla} \cdot \vec{E} = 4\pi\rho, \\
&\vec{\nabla} \times \vec{E} + \frac{1}{c}\frac{\partial \vec{B}}{\partial t} = 0, \\
&\vec{\nabla} \cdot \vec{B} = 0, \\
&\vec{\nabla} \times \vec{B} - \frac{1}{c}\frac{\partial \vec{E}}{\partial t} = \frac{4\pi}{c}\vec{J}.
\end{aligned}
\tag{2.469}
$$

These equations are coupled and have to be considered simultaneously. We now introduce the **potentials** Φ and $\vec{A}$:

$$
\begin{aligned}
&\vec{E} = -\frac{1}{c}\frac{\partial \vec{A}}{\partial t} - \vec{\nabla}\Phi, \\
&\vec{B} = \vec{\nabla} \times \vec{A},
\end{aligned}
\tag{2.470}
$$

and use the **Lorenz gauge**:

$$
\frac{1}{c}\frac{\partial \Phi}{\partial t} + \vec{\nabla} \cdot \vec{A} = 0,
\tag{2.471}
$$

to write the **Maxwell's equations** as

$$
\begin{aligned}
&\left[\frac{1}{c^2}\frac{\partial^2}{\partial t^2} - \vec{\nabla}^2\right]\Phi = 4\pi\rho, \\
&\left[\frac{1}{c^2}\frac{\partial^2}{\partial t^2} - \vec{\nabla}^2\right]\vec{A} = \frac{4\pi}{c}\vec{J}.
\end{aligned}
\tag{2.472}
$$

Irrotational Flow of Incompressible Fluids

For flow problems, the **continuity equation** is given as

$$
\vec{\nabla} \cdot \vec{J} + \frac{\partial \rho}{\partial t} = 0,
\tag{2.473}
$$

where $\vec{J}$ is the current density and ρ is the density of the flowing material. In general, the **current density** can be written as

$$
\vec{J} = \rho\vec{v},
\tag{2.474}
$$

where $\vec{v}$ is the velocity field of the fluid. Now, the continuity equation becomes

$$\boxed{\vec{\nabla} \cdot (\rho\vec{v}) + \frac{\partial \rho}{\partial t} = 0.}$$

(2.475)

For **stationary** flows $\partial\rho/\partial t = 0$. If we assume **incompressible** fluids, where $\rho = $ constant, the continuity equation reduces to

$$\vec{\nabla} \cdot \vec{v} = 0.$$

(2.476)

From the Helmholtz theorem, we know that this is not sufficient to determine the velocity field $\vec{v}$, thus if we also assume **irrotational flow**:

$$\vec{\nabla} \times \vec{v} = 0,$$

(2.477)

we can introduce a **velocity potential** Φ:

$$\vec{v} = \vec{\nabla}\Phi,$$

(2.478)

which along with Eq. (2.476), satisfies the **Laplace equation**:

$$\boxed{\vec{\nabla}^2 \Phi = 0.}$$

(2.479)

REFERENCES

1. Kaplan, W. (1984). *Advanced Calculus*, 3e. Reading, MA: Addison-Wesley.
2. Franklin, P.A. (1940). *A Treatise on Advanced Calculus*. New York: Wiley.

PROBLEMS

1. Using coordinate representation, show that

$$(\vec{A} \times \vec{B}) \cdot (\vec{C} \times \vec{D}) = (\vec{A} \cdot \vec{C})(\vec{B} \cdot \vec{D}) - (\vec{A} \cdot \vec{D})(\vec{B} \cdot \vec{C}).$$

2. Using the permutation symbol, show that the ith component of the cross product of two vectors, $\vec{A}$ and $\vec{B}$, can be written as

$$(\vec{A} \times \vec{B})_i = \sum_{j=1}^{3} \sum_{k=1}^{3} \varepsilon_{ijk} A_j B_k.$$

3. Prove the triangle inequality:

$$|\vec{A} - \vec{B}| \le |\vec{A} + \vec{B}| \le |\vec{A}| + |\vec{B}|.$$

4. Prove the following vector identity, which is also known as the **Jacobi identity**:

$$\vec{A} \times (\vec{B} \times \vec{C}) + \vec{B} \times (\vec{C} \times \vec{A}) + \vec{C} \times (\vec{A} \times \vec{B}) = 0.$$

5. Show that

$$(\vec{A} \times \vec{B}) \times (\vec{C} \times \vec{D}) = (\vec{A} \cdot \vec{B} \times \vec{D})\vec{C} - (\vec{A} \cdot \vec{B} \times \vec{C})\vec{D}.$$

6. Show that for three vectors, $\vec{A}$, $\vec{B}$, $\vec{C}$, to be noncoplanar the necessary and sufficient condition is

$$\vec{A} \cdot (\vec{B} \times \vec{C}) \neq 0.$$

7. Find a parametric equation for the line passing through the points
   (a) $(2, 2, -2)$ and $(-3, 1, 4)$,
   (b) $(-1, 4, 3)$ and $(4, -3, 1)$.

8. Find the equation of the line orthogonal to $\vec{A}$ and passing through $\vec{P}$:
   (a) $\vec{A} = (1, -1)$, $\vec{P} = (-5, 2)$,
   (b) $\vec{A} = (2, -1)$, $\vec{P} = (4, 2)$.

9. Show that the lines $2x_1 - 3x_2 = 1$ and $4x_1 + 3x_2 = 2$ are not orthogonal. What is the angle between them?

10. Find the equation of the plane including the point $\vec{P}$ and with the normal $\vec{N}$:
    (a) $\vec{P} = (2, 1, -1)$, $\vec{N} = (1, 1, 2)$,
    (b) $\vec{P} = (2, 3, 5)$, $\vec{N} = (-1, 1, 2)$.

11. Find the equation of the plane passing through the following three points:
    (a) $(2, 1, 1)$, $(4, 1, -1)$, $(1, 2, 2)$,
    (b) $(-5, -1, 2)$, $(2, 1, -1)$, $(3, -1, 2)$.

12. (a) Find a vector parallel to the line of intersection of $4x_1 - 2x_2 + 2x_3 = 2$ and $6x_1 + 2x_2 + 2x_3 = 4$.
    (b) Find a parametric equation for the line of intersection of the above planes.

13. Find the angle between the planes

    (a) $x_1 + x_2 + x_3 = 2$ and $x_1 - x_2 - x_3 = 3$,

    (b) $x_1 + 2x_2 + x_3 = 2$ and $-x_1 - 3x_2 + x_3 = 5$.

14. Find the distance between the point $\vec{P} = (1, 1, 2)$ and the plane $3x_1 + x_2 - 3x_3 = 2$.

15. Let P and Q be two points in n-space. Find the general expression for the midpoint of the line segment joining the two points.

16. If $\overrightarrow{x}(t)$ and $\overrightarrow{y}(t)$ are two differentiable vectors, then show that

$$\text{(a)} \quad \frac{d}{dt}[\overrightarrow{x}(t) \times \overrightarrow{y}(t)] = \overrightarrow{x}(t) \times \frac{d\overrightarrow{y}(t)}{dt} + \frac{d\overrightarrow{x}(t)}{dt} \times \overrightarrow{y}(t),$$

$$\text{(b)} \quad \frac{d}{dt}[\overrightarrow{x}(t) \times \overrightarrow{\dot{x}}(t)] = \overrightarrow{x}(t) \times \overrightarrow{\ddot{x}}(t).$$

17. Given the following parametric equation of a space curve:

$$x_1 = \cos t, \quad x_2 = \sin t, \quad x_3 = 2\sin^2 t,$$

(a) sketch the curve,
(b) find the equation of the tangent line at the point P with $t = \pi/3$,
(c) find the equation of a plane orthogonal to the curve at P,
(d) show that the curve lies in the surface $x_1^2 - x_2^2 + x_3 = 1$.

18. For the following surfaces, find the tangent planes and the normal lines at the points indicated:

(a) $x_1^2 + x_2^2 + x_3^2 = 6$ at $(1, 1, 2)$,

(b) $x_1^3 + x_1 x_2^2 + x_2 x_3^2 - 2x_3^3 = 2$ at $(1, 1, 1)$,

(c) $x_3 = 2x_1/x_2$ at $(1, 2, 1)$.

19. Find the directional derivative of $F(x_1, x_2, x_3) = 2x_1^2 + x_2^2 - x_3^2$ in the direction of the line from $(1, 2, 3)$ to $(3, 5, 1)$ at the point $(1, 2, 3)$.

20. For a general point, evaluate $\partial F/\partial n$ for $F = xyz$, where n is the outer normal to the surface $x_1^2 + 2x_2^2 + 4x_3^2 = 4$.

21. Determine the points and the directions for which the change in $f = 2x_1^2 + x_2^2 + x_3$ is greatest if the point is restricted to lie on $x_1^2 + x_2^2 = 2$.

22. Prove the following relations:

$$\text{(a)} \quad \overrightarrow{\nabla} r = \frac{\overrightarrow{r}}{r} = \hat{e}_r,$$

$$\text{(b)} \quad \overrightarrow{\nabla} r^n = nr^{n-2}\overrightarrow{r},$$

$$\text{(c)} \quad \overrightarrow{\nabla} \frac{1}{r} = -\frac{\overrightarrow{r}}{r^3},$$

$$\text{(d)} \quad \overrightarrow{\nabla} r^{-n} = -nr^{-n-2}\overrightarrow{r}.$$

23. Prove the following properties of the divergence and the curl operators:

(a) $\vec{\nabla} \cdot (\vec{A} + \vec{B}) = \vec{\nabla} \cdot \vec{A} + \vec{\nabla} \cdot \vec{B}$,

(b) $\vec{\nabla} \cdot (\phi\vec{A}) = \phi\vec{\nabla} \cdot \vec{A} + \vec{\nabla}\phi \cdot \vec{A}$,

(c) $\vec{\nabla} \times (\vec{A} + \vec{B}) = \vec{\nabla} \times \vec{A} + \vec{\nabla} \times \vec{B}$,

(d) $\vec{\nabla} \times (\phi\vec{A}) = \phi\vec{\nabla} \times \vec{A} + \vec{\nabla}\phi \times \vec{A}$.

24. Show that the following vector fields have zero curl and find a scalar function Φ such that $\vec{v} = \vec{\nabla}\Phi$:

(a) $\vec{v} = y^2z\widehat{e}_x + 2yxz\widehat{e}_y + y^2x\widehat{e}_z$,

(b) $\vec{v} = (3x^2y + z^2y)\widehat{e}_x + (x^3 + z^2x)\widehat{e}_y + 2zxy\widehat{e}_z$.

25. Using the vector field

$$\vec{v} = x^2yz\widehat{e}_x - 2x^3y^3\widehat{e}_y + xy^2z\widehat{e}_z,$$

show that $\vec{\nabla} \cdot \vec{\nabla} \times \vec{v} = 0$.

26. If $\vec{r}$ is the position vector, show the following:

(a) $\vec{\nabla} \cdot \vec{r} = 3$,

(b) $\vec{\nabla} \times \vec{r} = 0$,

(c) $(\vec{a} \cdot \vec{\nabla})\vec{r} = \vec{a}$.

27. Using the following scalar functions, show that $\vec{\nabla} \times \vec{\nabla}\Phi = 0$:

(a) $\Phi = e^{xy} \cos z$,

(b) $\Phi = \dfrac{1}{(x^2 + y^2 + z^2)^{1/2}}$.

28. An important property of the permutation symbol is given as

$$\sum_{i=1}^{3} \varepsilon_{ijk}\varepsilon_{ilm} = \delta_{jl}\delta_{km} - \delta_{jm}\delta_{kl}.$$

A general proof is difficult, but check the identity for the following specific values:

$$j = k = 1,$$

$$j = l = 1,$$
$$k = m = 2.$$

29. Prove the following vector identities:

(a) $\vec{\nabla} \times \vec{\nabla} \times \vec{A} = \vec{\nabla}(\vec{\nabla} \cdot \vec{A}) - \vec{\nabla}^2 \vec{A},$

(b) $\vec{\nabla} \cdot (\vec{A} \times \vec{B}) = \vec{B} \cdot (\vec{\nabla} \times \vec{A}) - \vec{A} \cdot (\vec{\nabla} \times \vec{B}),$

(c) $\vec{\nabla}(\vec{A} \cdot \vec{B}) = \vec{A} \times (\vec{\nabla} \times \vec{B}) + \vec{B} \times (\vec{\nabla} \times \vec{A}) + (\vec{A} \cdot \vec{\nabla})\vec{B}$
$\quad + (\vec{B} \cdot \vec{\nabla})\vec{A},$

(d) $\vec{\nabla} \times (\vec{A} \times \vec{B}) = \vec{A}(\vec{\nabla} \cdot \vec{B}) - \vec{B}(\vec{\nabla} \cdot \vec{A}) + (\vec{B} \cdot \vec{\nabla})\vec{A} - (\vec{A} \cdot \vec{\nabla})\vec{B}.$

30. Write the gradient and the Laplacian for the following scalar fields:

(a) $\quad \Phi = \ln(x^2 + y^2 + z^2),$

(b) $\quad \Phi = \frac{1}{(x^2+y^2+z^2)^{1/2}},$

(c) $\quad \Phi = \sqrt{x^2 - y^2}.$

31. Evaluate the following line integrals, where the paths are straight-lines connecting the end points:

(a) $\quad \int_{(0,0)}^{(1,1)} y^2 \, dx,$

(b) $\quad \int_{(1,2)}^{(2,1)} y \, dx + x \, dy.$

32. Evaluate the line integral

$$I = \int_C y^2 \, dx + x^2 \, dy$$

over the semicircle centered at the origin and with the unit radius in the upper half-plane.

33. Evaluate

$$I = \int_{(0,0)}^{(1,1)} y \, dx + x^2 \, dy$$

over the parabola $y = x^2$.

34. Evaluate

$$\oint (2xy - y^4 + 3)dx + (x^2 - 4xy^3)dy$$

over a circle of radius 2 and centered at the origin.

35. Evaluate the line integral

$$\int_{(0,1)}^{(1,2)} (x^2 + y^2)dx + 2xy \, dy$$

over the curve

$$y = e^x - ex^5 + 2x.$$

36. Evaluate $\int \int_S \vec{A} \cdot \hat{n} \, d\sigma$, where

$$\vec{A} = xy\hat{e}_x - x^2\hat{e}_y + (x + z)\hat{e}_z$$

and S is the portion of the plane $2x + 2y + z = 6$ included in the first octant and $\hat{n}$ is the unit normal to S.

37. Evaluate

$$\int_{(2,6)}^{(3,13)} (2x \ln xy + x)dx + \frac{x^2}{y} \, dy$$

over $y = x^2 + 2x - 2$.

38. Evaluate

$$I = \oint_C y^2 \, dx + xy \, dy,$$

where C is the square with the vertices $(1,1), (-1,1), (-1,-1), (1,-1)$.

39. Evaluate the line integral

$$I = \oint y^2 \, dx + x^2 \, dy$$

over the full circle $x^2 + y^2 = 1$.

40. Evaluate over the indicated paths by using the Green's theorem:

(a) $I = \oint_C y^2 \, dx + xy \, dy, x^2 + y^2 = 1,$

(b) $I = \oint_C (2x^3 - y^3)dx + (x^3 + 2y^3)dy, x^2 + y^2 = 1,$

(c) $I = \oint_C f(x)dx + g(y) \, dy,$ any closed path.

41. Evaluate

$$I = \int_C v_T \; ds,$$

where

$$\vec{v} = (x^2 + y^2)\widehat{e}_x + 2xy\widehat{e}_y$$

over $y = x^3$ from $(0,0)$ to $(1,1)$.

42. Use Green's theorem to evaluate

$$I = \oint_C v_n \; ds,$$

where

$$\vec{v} = (x^2 + y^2)\widehat{e}_x + 2xy\widehat{e}_y$$

and C is the circle $x^2 + y^2 = 2$.

43. Given the vector field $\vec{u} = -3y\widehat{e}_x + 2x\widehat{e}_y + \widehat{e}_z$, evaluate the following line integral by using the Stokes's theorem:

$$\oint_C u_T \; ds,$$

where C is the circle $x^2 + y^2 = 1$, $z = 1$.

44. Using the Stokes's theorem, evaluate

$$\oint_C [y^2 \; dx + z^2 \; dy + x^2 \; dz],$$

where C is the triangle with vertices at $(0,0,0)$, $(0,a,0)$, $(0,0,a)$.

45. Use Stokes's theorem to evaluate

$$I = \oint 8xy^2z \; dx + 8x^2yz \; dy + (4x^2y^2 - 2z)dz$$

around the path $x = \cos t$, $y = \sin t$, $z = \sin t$, where $t \in [0, 2\pi]$.

46. Evaluate the integral

$$\oint_S d\vec{\sigma} \cdot \vec{v}$$

for the surface of a sphere with radius R and centered at the origin in two different ways. Take $\vec{v}$ as

(a) $\vec{v} = x\widehat{e}_x + y\widehat{e}_y + z\widehat{e}_z$,

(b) $\vec{v} = x^n\widehat{e}_x + y^n\widehat{e}_y + z^n\widehat{e}_z$.

47. Given the temperature distribution

$$T(x_1, x_2, x_3) = x_1^2 + 2x_1x_2 + x_2^2 x_3,$$

   (a) determine the direction of heat flow at $(1, 2, 1)$,
   (b) find the rate of change of temperature at $(1, 2, 2)$ in the direction of $\widehat{e}_2 + \widehat{e}_3$.

48. Evaluate

$$\oint (2x - y + 4)dx + (5y + 3x - 6)dy$$

   around a triangle in the xy-plane with the vertices at $(0, 0), (3, 0), (3, 2)$ traversed in the counterclockwise direction.

49. Use Stokes's theorem to evaluate

$$\int\int_S (\vec{\nabla} \times \vec{V}) \cdot \widehat{n}\, d\sigma,$$

   where

$$\vec{V} = 2x_1x_2\widehat{e}_1 + (x_1^2 - 2x_2)\widehat{e}_2 + x_1^2 x_2^2 \widehat{e}_3$$

   over

$$x_3 = 9 - x_1^2 - x_2^2 \geq 0.$$

50. Obtain **Green's first identity**:

$$\oint_S d\vec{\sigma} \cdot u\vec{\nabla}v = \int_V d\tau [\vec{\nabla}u \cdot \vec{\nabla}v + u\vec{\nabla}^2 v].$$

51. Evaluate the following integrals, where S is the surface of the sphere $x^2 + y^2 + z^2 = a^2$:

$$\oint_S [x^3 \cos\theta_{x,n} + y^3 \cos\theta_{y,n} + z^3 \cos\theta_{z,n}] d\sigma$$

   and

$$\oint_S (x^2 + y^2 + z^2)^{1/2} [\cos\theta_{x,n} + \cos\theta_{y,n} + \cos\theta_{z,n}] d\sigma.$$

   For the vector field in the second part plot on the xy-plane and interpret your result.

52. Verify the divergence theorem for

$$\vec{A} = (2xy + z)\widehat{e}_x + y^2\widehat{e}_y - (x + 3y)\widehat{e}_z$$

taken over the region bounded by the planes

$$2x + 2y + z = 6, \quad x = 0, \ y = 0, \ z = 0.$$

53. Prove that
$$\vec{F} = (2xy + 3)\widehat{e}_x + (x^2 - 4z)\widehat{e}_y - 4y\widehat{e}_z$$

    is a conservative field and find the work done between $(3, -2, 2)$ and $(2, 1, -1)$.

54. Without using the divergence theorem, show that the gravitational force on a test particle inside a spherical shell of radius R is zero. Discuss your answer using the divergence theorem.

55. Without using the divergence theorem, find the gravitational field outside a uniform spherical mass of radius R. Repeat the same calculation with the gravitational potential and verify your answer obtained in the first part. Interpret your results using the divergence theorem.

56. Without using the divergence theorem, find the gravitational field for an internal point of a uniform spherical mass of radius R. Repeat the same calculation for the gravitational potential and verify your answer obtained in the first part. Discuss your results in terms of the divergence theorem.

57. Assume that gravitation is still represented by Gauss's law in a universe with four spatial dimensions. What would be Newton's law in this universe? Would circular orbits be stable?
    Note: You may ignore this problem. It is an advanced but fun problem that does not require a lot of calculation. The surface area of a sphere in four dimensions is $2\pi^2 R^3$. In three dimensions it is $4\pi R^2$.

CHAPTER 3

GENERALIZED COORDINATES AND TENSORS

Scalars are quantities that can be defined by just giving a single number. In other words, they have only magnitude. Vectors on the other hand, are defined as directed line segments, which have both magnitude and direction. By assigning a vector to each point in space, we define a vector field. Similarly, a scalar field can be defined as a function of position. Field is one of the most fundamental concepts of theoretical physics. In working with scalars and vectors, it is important that we first choose a suitable coordinate system. A proper choice of coordinates that reflects the symmetries of the physical system, simplifies the algebra and the interpretation of the solution significantly. In this chapter, we start with the Cartesian coordinates and their transformation properties. We then show how a generalized coordinate system can be constructed from the basic principles and discuss general coordinate transformations. Definition of scalars and vectors with respect to their transformation properties brings new depths into their discussion and allows us to introduce more sophisticated objects called tensors. We finally conclude with a detailed discussion of cylindrical and spherical coordinate systems, which are among the most frequently encountered coordinate systems in applications.

Essentials of Mathematical Methods in Science and Engineering, Second Edition. Selçuk Ş. Bayın. **133**
© 2020 John Wiley & Sons, Inc. Published 2020 by John Wiley & Sons, Inc.

3.1   TRANSFORMATIONS BETWEEN CARTESIAN COORDINATES

Transformations between Cartesian coordinates that exclude scale changes:

$$x' = kx, \quad k = \text{constant}, \tag{3.1}$$

are called **orthogonal transformations**. They preserve distances and magnitudes of vectors. There are basically three classes of orthogonal transformations. The first class involves **translations**, the second class is **rotations**, and the third class consists of **reflections** (Figure 3.1). Translations and rotations can be generated continuously from an initial frame; hence, they are called **proper transformations**. Since reflection cannot be accomplished continuously, they are called **improper transformations**. A general transformation is usually a combination of all three types.

3.1.1   Basis Vectors and Direction Cosines

We now consider orthogonal transformations with a common origin (Figure 3.2). To find the transformation equations, we write the position vector in two frames in terms of their respective unit basis vectors as

$$\vec{r} = x_1 \hat{e}_1 + x_2 \hat{e}_2 + x_3 \hat{e}_3, \tag{3.2}$$

$$\vec{r} = x_1' \hat{e}_1' + x_2' \hat{e}_2' + x_3' \hat{e}_3'. \tag{3.3}$$

Note that the point P that the position vector $\vec{r}$ represents exists independent of the definition of our coordinate system. Hence, in Eq. (3.3) we have written $\vec{r}$ instead of $\vec{r}'$. However, when we need to emphasize the coordinate system used

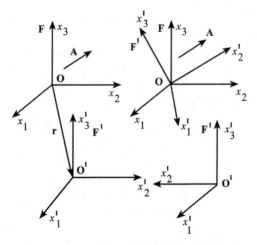

Figure 3.1   Orthogonal transformations.

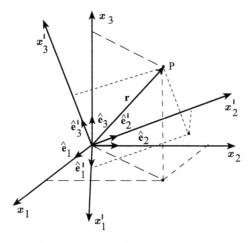

Figure 3.2   Orthogonal transformations with one point fixed.

explicitly, we can also write $\overrightarrow{r}'$. In other words, Eqs. (3.2) and (3.3) are just two different representations of the same vector. Obviously, there are infinitely many choices for the orientation of the Cartesian axes that one can use. To find a mathematical dictionary between them, that is, the transformation equations, we write the components of $\overrightarrow{r}$ in terms of $\widehat{e}'_i$ as

$$x'_1 = (\widehat{e}'_1 \cdot \overrightarrow{r}), \tag{3.4}$$

$$x'_2 = (\widehat{e}'_2 \cdot \overrightarrow{r}), \tag{3.5}$$

$$x'_3 = (\widehat{e}'_3 \cdot \overrightarrow{r}), \tag{3.6}$$

which, after using Eq. (3.2), gives

$$x'_1 = (\widehat{e}'_1 \cdot \widehat{e}_1)x_1 + (\widehat{e}'_1 \cdot \widehat{e}_2)x_2 + (\widehat{e}'_1 \cdot \widehat{e}_3)x_3, \tag{3.7}$$

$$x'_2 = (\widehat{e}'_2 \cdot \widehat{e}_1)x_1 + (\widehat{e}'_2 \cdot \widehat{e}_2)x_2 + (\widehat{e}'_2 \cdot \widehat{e}_3)x_3, \tag{3.8}$$

$$x'_3 = (\widehat{e}'_3 \cdot \widehat{e}_1)x_1 + (\widehat{e}'_3 \cdot \widehat{e}_2)x_2 + (\widehat{e}'_3 \cdot \widehat{e}_3)x_3. \tag{3.9}$$

These are the transformation equations that allow us to obtain the coordinates in terms of the primed system given the coordinates in the unprimed system. These equations can be conveniently written as

$$\boxed{x'_i = \sum_{j=1}^{3}(\widehat{e}'_i \cdot \widehat{e}_j)x_j, \quad i = 1, 2, 3.} \tag{3.10}$$

The coefficients $(\widehat{e}'_i \cdot \widehat{e}_j)$ are called the **direction cosines** and can be written as

$$\boxed{a_{ij} = (\widehat{e}'_i \cdot \widehat{e}_j) = \cos\theta_{ij}, \quad i = 1, 2, 3,} \tag{3.11}$$

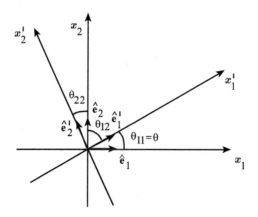

Figure 3.3   Direction cosines for rotations about the x_3-axis.

where θ_{ij} is the angle between the ith basis vector of the primed system and the jth basis vector of the unprimed system. For rotations about the x_3-axis (Figure 3.3), we can write a_{ij}, $i = 1, 2, 3$, as the array

$$S_3 = a_{ij}, \quad i = 1, 2, 3, \tag{3.12}$$

$$= \begin{pmatrix} (\hat{e}_1' \cdot \hat{e}_1) & (\hat{e}_1' \cdot \hat{e}_2) & 0 \\ (\hat{e}_2' \cdot \hat{e}_1) & (\hat{e}_2' \cdot \hat{e}_2) & 0 \\ 0 & 0 & 1 \end{pmatrix} \tag{3.13}$$

$$= \begin{pmatrix} \cos\theta & \cos(\frac{\pi}{2} - \theta) & 0 \\ \cos(\frac{\pi}{2} + \theta) & \cos\theta & 0 \\ 0 & 0 & 1 \end{pmatrix} \tag{3.14}$$

$$= \begin{pmatrix} \cos\theta & \sin\theta & 0 \\ -\sin\theta & \cos\theta & 0 \\ 0 & 0 & 1 \end{pmatrix}. \tag{3.15}$$

3.1.2   Transformation Matrix and Orthogonality

General linear transformations between Cartesian coordinates can be written as

$$\boxed{x_i' = \sum_{j=1}^{3} a_{ij} x_j, \ i = 1, 2, 3,} \tag{3.16}$$

where the square array

$$S = a_{ij}, \quad i = 1, 2, 3, \tag{3.17}$$

$$= \begin{pmatrix} a_{11} & a_{12} & a_{13} \\ a_{21} & a_{22} & a_{23} \\ a_{31} & a_{32} & a_{33} \end{pmatrix} \tag{3.18}$$

is called the **transformation matrix**.

Let us now write the magnitude of $\vec{r}$ in the primed system as

$$\vec{r} \cdot \vec{r} = |r|^2 = \sum_{i=1}^{3} x'_i x'_i = x'^2_i + x'^2_2 + x'^2_3. \tag{3.19}$$

Using the transformation equations, we can write $|r|^2$ in the unprimed system as

$$\vec{r} \cdot \vec{r} = \sum_{i=1}^{3} \left[\sum_{j=1}^{3} a_{ij} x_j \right] \left[\sum_{j'=1}^{3} a_{ij'} x_{j'} \right] = \sum_{i=1}^{3} \sum_{j=1}^{3} \sum_{j'=1}^{3} a_{ij} a_{ij'} x_j x_{j'}. \tag{3.20}$$

Rearranging the triple sum, we write

$$\vec{r} \cdot \vec{r} = \sum_{j=1}^{3} \sum_{j'=1}^{3} \left[\sum_{i=1}^{3} a_{ij} a_{ij'} \right] x_j x_{j'}. \tag{3.21}$$

Since the orthogonal transformations preserve magnitudes of vectors, the transformation matrix has to satisfy

$$\boxed{\sum_{i=1}^{3} a_{ij} a_{ij'} = \delta_{jj'}, \quad j, j' = 1, 2, 3,} \tag{3.22}$$

which is called the **orthogonality condition**. Equation (3.21) now gives

$$\vec{r} \cdot \vec{r} = \sum_{j=1}^{3} x'_j x'_j = x'^2_i + x'^2_2 + x'^2_3 = \sum_{j=1}^{3} \sum_{j'=1}^{3} \delta_{jj'} x_j x_{j'} \tag{3.23}$$

$$= \sum_{j=1}^{3} x_j x_j = x^2_1 + x^2_2 + x^2_3. \tag{3.24}$$

3.1.3 Inverse Transformation Matrix

For the inverse transformation, we need to express x_i in terms of x'_i. This can only be done when the determinant of the transformation matrix does not vanish:

$$\det S = \det \begin{pmatrix} a_{11} & a_{12} & a_{13} \\ a_{21} & a_{22} & a_{23} \\ a_{31} & a_{32} & a_{33} \end{pmatrix} \neq 0. \tag{3.25}$$

Writing the orthogonality relation [Eq. (3.22)] explicitly as

$$\begin{pmatrix} a_{11} & a_{21} & a_{31} \\ a_{12} & a_{22} & a_{32} \\ a_{13} & a_{23} & a_{33} \end{pmatrix} \begin{pmatrix} a_{11} & a_{12} & a_{13} \\ a_{21} & a_{22} & a_{23} \\ a_{31} & a_{32} & a_{33} \end{pmatrix} = \begin{pmatrix} 1 & 0 & 0 \\ 0 & 1 & 0 \\ 0 & 0 & 1 \end{pmatrix}, \tag{3.26}$$

we can evaluate the determinant of S. The determinant of the left-hand side is the multiplication of the determinants of the individual matrices. Using the fact that interchanging rows and columns of a matrix does not change the value of its determinant, we can write

$$(\det S)^2 = 1, \tag{3.27}$$

which yields

$$\boxed{\det S = \pm 1.} \tag{3.28}$$

The negative sign corresponds to **improper transformations**, hence, the determinant of the transformation matrix provides a convenient tool to test whether a given transformation involves reflections or not. For a formal inversion, we multiply the transformation equation:

$$x_i' = \sum_{j=1}^{3} a_{ij} x_j, \quad i = 1, 2, 3, \tag{3.29}$$

with $a_{ij'}$ and sum over i to write

$$\sum_{i=1}^{3} a_{ij'} x_i' = \sum_{i=1}^{3} \sum_{j=1}^{3} a_{ij'} a_{ij} x_j = \sum_{j=1}^{3} \left[\sum_{i=1}^{3} a_{ij'} a_{ij} \right] x_j. \tag{3.30}$$

Substituting the orthogonality relation [Eq. (3.22)] for the sum inside the square brackets, we get

$$\sum_{i=1}^{3} a_{ij'} x_i' = \sum_{j=1}^{3} \delta_{j'j} x_j = x_{j'}, \quad j' = 1, 2, 3, \tag{3.31}$$

which, when written explicitly, gives

$$x_1 = a_{11} x_1' + a_{21} x_2' + a_{31} x_3', \tag{3.32}$$

$$x_2 = a_{12} x_1' + a_{22} x_2' + a_{32} x_3', \tag{3.33}$$

$$x_3 = a_{13} x_1' + a_{23} x_2' + a_{33} x_3'. \tag{3.34}$$

We now write the inverse transformation matrix as

$$S^{-1} = \begin{pmatrix} a_{11} & a_{21} & a_{31} \\ a_{12} & a_{22} & a_{32} \\ a_{13} & a_{23} & a_{33} \end{pmatrix}. \tag{3.35}$$

Comparing with S in Eq. (3.18), it is seen that

$$\boxed{S^{-1} = \tilde{S},} \tag{3.36}$$

where $\tilde{S}$ is called the **transpose** of S, which is obtained by interchanging the rows and the columns in S.

In summary, the inverse transformation matrix for orthogonal transformations is just the transpose of the transformation matrix. For rotations about the x_3-axis [Eq. (3.15)], the inverse transformation matrix is written as

$$S_3^{-1} = \begin{pmatrix} \cos\theta & -\sin\theta & 0 \\ \sin\theta & \cos\theta & 0 \\ 0 & 0 & 1 \end{pmatrix}. \tag{3.37}$$

Note that S_3^{-1} corresponds to a rotation in the opposite direction by the same amount:

$$\boxed{S_3^{-1}(\theta) = S_3(-\theta).} \tag{3.38}$$

3.2 CARTESIAN TENSORS

So far we have discussed the transformation properties of the position vector $\overrightarrow{r}$. We now extend this to an arbitrary vector $\overrightarrow{v}$ as

$$\overrightarrow{v}' = S\overrightarrow{v}, \tag{3.39}$$

where $S = a_{ij}$, $i, j = 1, 2, 3$, is the orthogonal transformation matrix [Eq. (3.18)]. In general, a given triplet of functions,

$$(v_1(\overrightarrow{r}), v_2(\overrightarrow{r}), v_3(\overrightarrow{r})), \tag{3.40}$$

cannot be used to define a vector:

$$\overrightarrow{v} = (v_1, v_2, v_3), \tag{3.41}$$

unless they transform as

$$v_i' = \sum_{j=1}^{3} a_{ij} v_j, \quad i = 1, 2, 3. \tag{3.42}$$

Under the orthogonal transformations a scalar function, $\Phi(x_1, x_2, x_3)$, transforms as

$$\Phi(x_1, x_2, x_3) = \Phi(x_1(\overrightarrow{r}'), x_2(\overrightarrow{r}'), x_3(\overrightarrow{r}')) = \Phi(x_1', x_2', x_3'). \tag{3.43}$$

Note that in the new coordinate system, Φ will naturally have a different functional dependence. However, the numerical value that Φ assumes at each point in space will be the same. It is for this reason that in Eq. (3.43), we have written Φ instead of Φ'. However, in order to indicate the coordinate system used, we may also write Φ'. For example, temperature is a scalar; hence, its value at a given point does not depend on the coordinate system used. A different choice of coordinate system assigns different coordinates (codes) to each point in space:

$$(x_1, x_2, x_3) \rightarrow (x'_1, x'_2, x'_3), \tag{3.44}$$

but the numerical value of the temperature at a given point will not change.

We now write the **scalar product** of two vectors, $\overrightarrow{x}$ and $\overrightarrow{y}$, in the primed coordinates by using the scalar product written in the unprimed coordinates:

$$\overrightarrow{x} \cdot \overrightarrow{y} = \sum_{i=1}^{3} x_i y_i = x_1 y_1 + x_2 y_2 + x_3 y_3, \tag{3.45}$$

Using orthogonal transformations:

$$x_i = \sum_{j=1}^{3} a_{ji} x'_j \quad \text{and} \quad y_i = \sum_{j'=1}^{3} a_{j'i} y'_{j'}, \tag{3.46}$$

we can write $\overrightarrow{x} \cdot \overrightarrow{y}$ as

$$\overrightarrow{x} \cdot \overrightarrow{y} = \sum_{i=1}^{3} \left[\sum_{j=1}^{3} a_{ji} x'_j \right] \left[\sum_{j'=1}^{3} a_{j'i} y'_{j'} \right] = \sum_{j=1}^{3} \sum_{j'=1}^{3} \left[\sum_{i=1}^{3} a_{j'i} a_{ji} \right] x'_j y'_{j'}. \tag{3.47}$$

Using the orthogonality relation [Eq. (3.22)]: $\widetilde{S}S = S\widetilde{S} = I$, this becomes

$$\overrightarrow{x} \cdot \overrightarrow{y} = \sum_{j=1}^{3} \sum_{j'=1}^{3} \delta_{j'j} x'_j y'_{j'} = \sum_{j=1}^{3} x'_j y'_j = x'_1 y'_1 + x'_2 y'_2 + x'_3 y'_3 = \overrightarrow{x}' \cdot \overrightarrow{y}'. \tag{3.48}$$

In other words, orthogonal transformations do not change the value of a scalar product. Properties of physical systems that preserve their value under coordinate transformations are called **invariants**. Identification of invariants in the study of nature is very important and plays a central role in both special and general theories of relativity.

In Chapter 2, we have defined vectors with respect to their geometric and algebraic properties. Definition of vectors with respect to their transformation properties under orthogonal transformations brings new levels into the subject and allows us to free the vector concept from being just a directed line segment drawn in space. Using the transformation properties, we can now define more complex objects called **tensors**. Tensors of **second rank**, T_{ij}, are among the most commonly encountered tensors in applications and have two indices. Vectors, v_i, have only one index, and they are tensors of **first rank**. Scalars, Φ, which have no indices, are tensors of **zeroth rank**. In general, tensors of higher ranks are written with the appropriate number of indices as

$$T = T_{ijkl...}, \quad i, j, k, \ldots = 1, 2, 3. \tag{3.49}$$

Each index of a tensor transforms like a vector:

$$\Phi(\overrightarrow{r}) = \Phi(\overrightarrow{r}'), \tag{3.50}$$

$$v'_i = \sum_{i'=1}^{3} a_{ii'} v_{i'}, \quad i = 1, 2, 3, \tag{3.51}$$

$$T'_{ij} = \sum_{i'=1}^{3} \sum_{j'=1}^{3} a_{ii'} a_{jj'} T_{i'j'}, \quad i, j = 1, 2, 3, \tag{3.52}$$

$$T'_{ijk} = \sum_{i'=1}^{3} \sum_{j'=1}^{3} \sum_{k'=1}^{3} a_{ii'} a_{jj'} a_{kk'} T_{i'j'k'}, \quad i, j, k = 1, 2, 3, \tag{3.53}$$

etc.

Tensors of second rank can be conveniently represented as 3×3 square matrices:

$$T_{ij} = \begin{pmatrix} t_{11} & t_{12} & t_{13} \\ t_{21} & t_{22} & t_{23} \\ t_{31} & t_{32} & t_{33} \end{pmatrix}. \tag{3.54}$$

Definition of tensors can be extended to n dimensions easily by taking the range of the indices from 1 to n. As we shall see shortly, tensors can also be defined in general coordinates. For the time being, we confine our discussion to **Cartesian tensors**, which are defined with respect to their transformation properties under orthogonal transformations.

3.2.1 Algebraic Properties of Tensors

Tensors of equal rank can be **added** or **subtracted** term by term and the result does not depend on the order of the tensors: For example, if A and B are two second-rank tensors, then their sum is

$$A + B = B + A = C, \tag{3.55}$$

$$C_{ij} = A_{ij} + B_{ij}, \quad i, j = 1, 2, 3. \tag{3.56}$$

Multiplication of a tensor with a **scalar**, α, is accomplished by multiplying all the component of that tensor with the same scalar. For a third-rank tensor A, we can write

$$\alpha A = \alpha A_{ijk}, \quad i, j, k = 1, 2, 3. \tag{3.57}$$

From the basic properties of matrices, second-rank tensors **do not commute** under multiplication:

$$\boxed{AB \neq BA,} \tag{3.58}$$

however, they **associate**:

$$\boxed{A(BC) = (AB)C,}$$
(3.59)

where A, B, C are second-rank tensors. **Antisymmetric** tensors satisfy

$$A_{ij} = -A_{ji}, \quad i, j = 1, 2, 3,$$
(3.60)

or

$$\boxed{\tilde{A} = -A,}$$
(3.61)

where $\tilde{A}$ is called the **transpose** of A, which is obtained by interchanging the rows and columns. Note that the diagonal terms, A_{11}, A_{22}, A_{33}, of an antisymmetric tensor are necessarily zero. If we set $i = j$ in Eq. (3.60), we obtain

$$A_{ii} = -A_{ii}, \quad i = 1, 2, 3,$$
(3.62)

$$2A_{ii} = 0, \quad i = 1, 2, 3,$$
(3.63)

$$A_{ii} = 0, \quad i = 1, 2, 3.$$
(3.64)

Symmetry and **antisymmetry** are invariant properties. If a second-rank tensor A is symmetric in one coordinate system,

$$A_{ij} = A_{ji}, \quad i, j = 1, 2, 3,$$
(3.65)

then A' is also symmetric. We first write

$$A'_{ij} = \sum_{i'=1}^{3} \sum_{j'=1}^{3} a_{ii'} a_{jj'} A_{i'j'}, \quad i, j = 1, 2, 3.$$
(3.66)

Since the components, a_{ij}, are constants or in general scalar functions, the order in which they are written in equations do not matter. Hence, we can write Eq. (3.66) as

$$A'_{ij} = \sum_{i'=1}^{3} \sum_{j'=1}^{3} a_{jj'} a_{ii'} A_{i'j'}, \quad i, j = 1, 2, 3.$$
(3.67)

Using the transformation property of second-rank tensors [Eq. (3.52)], for a symmetric second-rank tensor A this implies

$$A'_{ij} = A'_{ji}, \quad i, j = 1, 2, 3.$$
(3.68)

A similar proof can be given for the antisymmetric tensor. Any second-rank tensor can be written as the sum of a symmetric and an antisymmetric tensor:

$$A_{ij} = \frac{1}{2}(A_{ij} + A_{ji}) + \frac{1}{2}(A_{ij} - A_{ji}), \quad i,j = 1,2,3. \tag{3.69}$$

Using the components of two vectors, $\vec{a}$ and $\vec{b}$, we can construct a second-rank tensor A as

$$A = \begin{pmatrix} a_1b_1 & a_1b_2 & a_1b_3 \\ a_2b_1 & a_2b_2 & a_2b_3 \\ a_3b_1 & a_3b_2 & a_3b_3 \end{pmatrix}, \tag{3.70}$$

which is called the **outer product** or the **tensor product** of $\vec{a}$ and $\vec{b}$, and it is shown as

$$A = \vec{a}\,\vec{b} \tag{3.71}$$

or as

$$A = \vec{a} \otimes \vec{b}. \tag{3.72}$$

To justify that A is a second-rank tensor, we show that it obeys the correct transformation property; that is, it transforms like a second-rank tensor:

$$A'_{ij} = a'_i b'_j = \sum_{i'=1}^{3} a_{ii'} a_{i'} \sum_{j'=1}^{3} a_{jj'} b_{j'} = \sum_{i'=1}^{3} \sum_{j'=1}^{3} a_{ii'} a_{jj'} a_{i'} b_{j'} \tag{3.73}$$

$$= \sum_{i'=1}^{3} \sum_{j'=1}^{3} a_{ii'} a_{jj'} A_{i'j'}, \quad i,j = 1,2,3. \tag{3.74}$$

One can easily check that the outer product defined as $\vec{b} \otimes \vec{a}$ is the transpose of A. We remind the reader that even though we can construct a second-rank tensor from two vectors, the converse is not true. A second-rank tensor cannot always be written as the outer product of two vectors. Using the outer product, we can construct tensors of higher rank from tensors of lower rank:

$$A_{ijk} = B_{ij}a_k, \tag{3.75}$$

$$A_{ijk} = a_i b_j c_k, \tag{3.76}$$

$$A_{ijkl} = B_{ij}C_{kl}, \tag{3.77}$$

etc.,

where the indices take the values $1,2,3$.

For a given vector, there is only one invariant, namely, its magnitude. All the other invariants are functions of the magnitude. For a second-rank tensor, there are three invariants, one of which is the **spur** or the **trace**, which is defined as the sum of the diagonal elements:

$$\boxed{\operatorname{tr} A = A_{11} + A_{22} + A_{33}.}$$
(3.78)

We leave the proof as an exercise, but note that when A can be decomposed as the outer product of two vectors, the trace is the inner product of these vectors.

We can obtain a lower-rank tensor by summing over the pairs of indices. This operation is called **contraction**. Trace is obtained by contracting the two indices of a second-rank tensor as

$$\boxed{\operatorname{tr} A = \sum_{i=1}^{3} A_{ii}.}$$
(3.79)

Other examples of contraction are

$$T_{jk} = \sum_{i=1}^{3} T_{ijik}, \quad j, k = 1, 2, 3,$$
(3.80)

$$T = \sum_{i=1}^{3} \sum_{j=1}^{3} T_{iijj},$$
(3.81)

etc.

We can generalize the idea of the inner product by contracting the indices of a tensor with the indices of another tensor:

$$b_i = \sum_{j=1}^{3} T_{ij} a_j, \quad i = 1, 2, 3,$$
(3.82)

$$a_i = \sum_{j=1}^{3} \sum_{k=1}^{3} T_{ijk} A_{jk}, \quad i = 1, 2, 3,$$
(3.83)

$$d_{ik} = \sum_{j=1}^{3} \sum_{l=1}^{3} T_{ij} A_{jl} b_l c_k, \quad i, k = 1, 2, 3,$$
(3.84)

$$a = \sum_{i=1}^{3} \sum_{j=1}^{3} D_{ij} E_{ij},$$
(3.85)

etc.

The **rank** of the resulting tensor is equal to the number of the **free indices**, that is, the indices that are not summed over. Free indices take the values 1, 2, or 3. In this regard, we also write a tensor, say T_{ij}, $i, j = 1, 2, 3$, as simply T_{ij}. The indices that are summed over are called the **dummy indices**. Since dummy indices disappear in the final expression, we can always rename them.

3.2.2 Kronecker Delta and the Permutation Symbol

To check the tensor property of the **Kronecker delta**, δ_{ij}, we use the transformation equation for the second-rank tensors:

$$T'_{ij} = \sum_{i'=1}^{3} \sum_{j'=1}^{3} a_{ii'} a_{jj'} T_{i'j'}, \tag{3.86}$$

with $T_{i'j'} = \delta_{i'j'}$ and use the orthogonality relation [Eq. (3.22)] to write

$$\sum_{i'=1}^{3} \sum_{j'=1}^{3} a_{ii'} a_{jj'} \delta_{i'j'} = \sum_{i'=1}^{3} \sum_{j'=1}^{3} a_{ii'} a_{ji'} = \delta_{ij}. \tag{3.87}$$

In other words, the Kronecker delta is a symmetric second-rank tensor that transforms into itself under orthogonal transformations. It is also called the **identity tensor**, which is shown as **I**. Kronecker delta is the only tensor with this property.

Permutation symbol, ε_{ijk}, also called the **Levi-Civita symbol**, is defined as

$$\varepsilon_{ijk} = \begin{cases} 0 & \text{when any two indices are equal.} \\ 1 & \text{for even (cyclic) permutations: } 123, 231, 312. \\ -1 & \text{for odd (anticyclic) permutations: } 213, 321, 132. \end{cases} \tag{3.88}$$

Using the permutation symbol, we can write a **determinant** as

$$\det A_{i'j'} = \det \begin{pmatrix} A_{11} & A_{12} & A_{13} \\ A_{21} & A_{22} & A_{23} \\ A_{31} & A_{32} & A_{33} \end{pmatrix} = \sum_{ijk} A_{1i} A_{2j} A_{3k} \varepsilon_{ijk} \tag{3.89}$$

$$= A_{11}(A_{22}A_{33} - A_{32}A_{23}) - A_{12}(A_{21}A_{33} - A_{23}A_{31})$$
$$+ A_{13}(A_{21}A_{32} - A_{31}A_{22}). \tag{3.90}$$

Interchange any two of the indices of the permutation symbol [Eq. (3.89)], the determinant changes sign. This operation is equivalent to interchanging the corresponding rows and columns of a determinant. We now write the determinant of the transformation matrix $a_{i'j'}$ as

$$\det a_{i'j'} = -\sum_{ijk} a_{2j} a_{1i} a_{3k} \varepsilon_{jik}. \tag{3.91}$$

Renaming the dummy indices:

$$i \to j, \tag{3.92}$$

$$j \to i. \tag{3.93}$$

Equation (3.91) becomes

$$\det a_{i'j'} = -\sum_{ijk} a_{2i} a_{1j} a_{3k} \varepsilon_{ijk}. \tag{3.94}$$

From Eq. (3.28), we know that the determinant of the orthogonal transformation matrix is $\det a_{i'j'} = \mp 1$, hence, the component ε_{213} transforms as

$$(\mp 1)\varepsilon_{213} = \sum_{ijk} a_{2i} a_{1j} a_{3k} \varepsilon_{ijk}. \tag{3.95}$$

Similar arguments for the other components yields the transformation equation of ε_{lmn} as

$$(\mp 1)\varepsilon_{lmn} = \sum_{ijk} a_{li} a_{mj} a_{nk} \varepsilon_{ijk}. \tag{3.96}$$

The minus sign is for the improper transformations. In summary, ε_{ijk} transforms like a third-rank tensor for proper transformations, and a minus sign has to be inserted for improper transformations. Tensors that transform like this are called **tensor densities** or **pseudotensors**. Note that aside from the ∓ 1 factor, ε_{ijk} has the same constant components in all Cartesian coordinate systems. Permutation symbol is the only third-rank tensor with this property.

An important identity of ε_{ijk} is

$$\boxed{\sum_{i=1}^{3} \varepsilon_{ijk} \varepsilon_{ilm} = \delta_{jl}\delta_{km} - \delta_{jm}\delta_{kl}.} \tag{3.97}$$

Permutation symbol also satisfies

$$\boxed{\varepsilon_{ijk} = \varepsilon_{jki} = \varepsilon_{kij},} \tag{3.98}$$

for the cyclic permutations of the indices. For the anticyclic permutations, we write

$$\boxed{\varepsilon_{ijk} = -\varepsilon_{ikj} = -\varepsilon_{jik} = -\varepsilon_{kji}.} \tag{3.99}$$

Example 3.1. *Physical tensors:* Solid objects deform under stress to a certain extent. In general, forces acting on a solid can be described by a second-rank tensor called the **stress tensor**:

$$t_{ij} = \begin{pmatrix} t_{11} & t_{12} & t_{13} \\ t_{21} & t_{22} & t_{23} \\ t_{31} & t_{32} & t_{33} \end{pmatrix}. \tag{3.100}$$

Components of the stress tensor represent the forces acting on a unit test area when the normal is pointed in various directions. For example, t_{ij} is the ith component of the force when the normal is pointing in the jth direction. Since the stress tensor is a second-rank tensor, it transforms as

$$t'_{ij} = \sum_{k=1}^{3} \sum_{l=1}^{3} a_{ik} a_{jl} t_{kl}. \tag{3.101}$$

The amount of deformation is also described by a second rank tensor, σ_{ij}, called the **strain tensor**. The stress and the strain tensors are related by the equation

$$t_{ij} = \sum_{k=1}^{3} \sum_{l=1}^{3} C_{ijkl} \sigma_{kl}, \tag{3.102}$$

where the fourth-rank tensor C_{ijkl} represents the elastic constants of the solid. This is the most general expression that relates the deformation of a three-dimensional solid to the forces acting on it. For a long and thin solid sample, with cross section ΔA and with longitudinal loading F, Eq. (3.102) reduces to **Hook's law**:

$$\frac{F}{\Delta A} = Y \frac{\Delta l}{l}, \tag{3.103}$$

$$t = Y\sigma, \tag{3.104}$$

where t is the force per unit area, σ is the fractional change in length, $\Delta l/l$, and Y is **Young's modulus**.

Many of the scalar quantities in physics can be generalized as tensors of higher rank. In Newton's theory, mass of an object is defined as the proportionality constant, m, between the force acting on the object and the acceleration as

$$F_i = ma_i. \tag{3.105}$$

Mass is basically the ability of an object to resist acceleration, that is, its inertia. It is an experimental fact that mass does not depend on the direction in which we want to accelerate an object. Hence, it is defined as a scalar quantity. In some effective field theories, it may be advantageous to treat particles with a mass that depends on direction. In such cases, we can introduce effective mass as a second-rank tensor, m_{ij}, and write **Newton's second law** as

$$F_i = \sum_{j=1}^{3} m_{ij} a_j, \tag{3.106}$$

When the mass is isotropic, m_{ij} becomes

$$m_{ij} = m \delta_{ij}; \tag{3.107}$$

thus, Newton's second law reduces to its usual form.

3.3 GENERALIZED COORDINATES

3.3.1 Coordinate Curves and Surfaces

Before we introduce the **generalized coordinates**, which are also called the **curvilinear coordinates**, let us investigate some of the basic properties of the Cartesian coordinate system from a different perspective. In a Cartesian coordinate system at each point, there are three **planes** defined by the equations

$$x^1 = c_1, \quad x^2 = c_2, \quad x^3 = c_3. \tag{3.108}$$

These planes intersect at the point (c_1, c_2, c_3), which defines the **coordinates** of that point. In this section, we start by writing the coordinates with an upper index as x^i. There is no need for alarm: As far as the Cartesian coordinates are concerned there is no difference, that is, $x^i = x_i$. However, as we shall see shortly, this added richness in our notation is absolutely essential when we introduce the generalized coordinates. Treating (c_1, c_2, c_3) as parameters, the above equations define three mutually orthogonal families of surfaces, each of which is composed of infinitely many nonintersecting parallel planes. These surfaces are called the **coordinate surfaces** on which the corresponding coordinate has a fixed value (Figure 3.4). The coordinate surfaces intersect along the **coordinate curves**. For the Cartesian coordinate system, these curves are mutually orthogonal straight lines called the **coordinate axes** (Figure 3.4). Cartesian **basis vectors**, $\hat{e}_1, \hat{e}_2, \hat{e}_3$, are defined as the unit vectors along the coordinate axes. A unique property of the Cartesian coordinate system is that the basis vectors point in the same direction at every point in space (Figure 3.5).

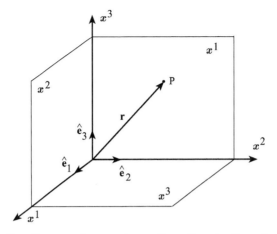

Figure 3.4   Coordinate surfaces and coordinate lines in Cartesian coordinates.

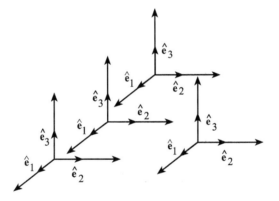

Figure 3.5   Basis vectors in Cartesian coordinates always point in the same direction.

We now introduce **generalized coordinates**, where the coordinate surfaces are defined in terms of the **Cartesian coordinates** (x^1, x^2, x^3) as three single-valued continuous functions with continuous partial derivatives:

$$\bar{c}_1 = \bar{x}^1(x^1, x^2, x^3), \tag{3.109}$$

$$\bar{c}_2 = \bar{x}^2(x^1, x^2, x^3), \tag{3.110}$$

$$\bar{c}_2 = \bar{x}^3(x^1, x^2, x^3). \tag{3.111}$$

Treating $\bar{c}_1, \bar{c}_2, \bar{c}_3$ as continuous variables, these give us three families of surfaces, where each family is composed of infinitely many nonintersecting surfaces (Figure 3.6). Using the fixed values that these functions, $\bar{x}^i(x^1, x^2, x^3)$,

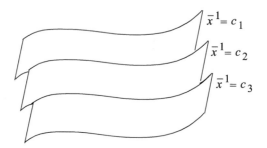

Figure 3.6 Coordinate surfaces for $\bar{x}^1 = $ const. in generalized coordinates.

$i = 1, 2, 3$, take on these surfaces, we define the generalized coordinates $(\bar{x}^1, \bar{x}^2, \bar{x}^3)$ as

$$\bar{x}^1 = \bar{x}^1(x^1, x^2, x^3), \tag{3.112}$$

$$\bar{x}^2 = \bar{x}^2(x^1, x^2, x^3), \tag{3.113}$$

$$\bar{x}^3 = \bar{x}^3(x^1, x^2, x^3). \tag{3.114}$$

Note that these equations are also the transformation equations between the Cartesian coordinates (x^1, x^2, x^3) and the generalized coordinates $(\bar{x}^1, \bar{x}^2, \bar{x}^3)$. For the new coordinates to be meaningful, the inverse transformations, $x^i = x^i(\bar{x}^j)$:

$$x^1 = x^1(\bar{x}^1, \bar{x}^2, \bar{x}^3), \tag{3.115}$$

$$x^2 = x^2(\bar{x}^1, \bar{x}^2, \bar{x}^3), \tag{3.116}$$

$$x^3 = x^3(\bar{x}^1, \bar{x}^2, \bar{x}^3), \tag{3.117}$$

should exist. In Chapter 1, we have seen that the necessary and the sufficient condition for the inverse transformation to exist, the **Jacobian**, J, of the transformation has to be different from zero. In other words, for a one-to-one correspondence between (x^1, x^2, x^3) and $(\bar{x}^1, \bar{x}^2, \bar{x}^3)$ we need to have

$$J = \frac{\partial(x^1, x^2, x^3)}{\partial(\bar{x}^1, \bar{x}^2, \bar{x}^3)} = \det \begin{pmatrix} \dfrac{\partial x^1}{\partial \bar{x}^1} & \dfrac{\partial x^2}{\partial \bar{x}^1} & \dfrac{\partial x^3}{\partial \bar{x}^1} \\[2ex] \dfrac{\partial x^1}{\partial \bar{x}^2} & \dfrac{\partial x^2}{\partial \bar{x}^2} & \dfrac{\partial x^3}{\partial \bar{x}^2} \\[2ex] \dfrac{\partial x^1}{\partial \bar{x}^3} & \dfrac{\partial x^2}{\partial \bar{x}^3} & \dfrac{\partial x^3}{\partial \bar{x}^3} \end{pmatrix} \neq 0 \tag{3.118}$$

or since $JK = 1$,

$$K = \frac{\partial(\bar{x}^1, \bar{x}^2, \bar{x}^3)}{\partial(x^1, x^2, x^3)} \neq 0. \tag{3.119}$$

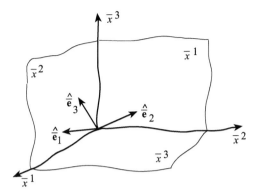

Figure 3.7 Generalized coordinates.

For the coordinate surfaces, given as

$$\overline{x}^1 = \overline{x}^1(x^1, x^2, x^3) = \overline{c}_1, \tag{3.120}$$

$$\overline{x}^2 = \overline{x}^2(x^1, x^2, x^3) = \overline{c}_2, \tag{3.121}$$

$$\overline{x}^3 = \overline{x}^3(x^1, x^2, x^3) = \overline{c}_3, \tag{3.122}$$

the intersection of the first two, $\overline{x}^1(x^1, x^2, x^3) = \overline{c}_1$ and $\overline{x}^2(x^1, x^2, x^3) = \overline{c}_2$, defines the coordinate curve along which $\overline{x}^3$ varies (Figure 3.7). We refer to this as the $\overline{x}^3$ curve, which can be parameterized in terms of $\overline{x}^3$ as $(x^1(\overline{x}^3), x^2(\overline{x}^3), x^3(\overline{x}^3))$. Similarly, two other curves exist for the $\overline{x}^1$ and the $\overline{x}^2$ coordinates. These curves are now the counterparts of the coordinate axes in Cartesian coordinates.

We now define the coordinate basis vectors $(\overrightarrow{e}_1, \overrightarrow{e}_2, \overrightarrow{e}_3)$ in terms of the Cartesian unit basis vectors $(\hat{e}_1, \hat{e}_2, \hat{e}_3)$ as the tangent vectors:

$$\overrightarrow{e}_1 = \frac{\partial x^1}{\partial \overline{x}^1}\hat{e}_1 + \frac{\partial x^2}{\partial \overline{x}^1}\hat{e}_2 + \frac{\partial x^3}{\partial \overline{x}^1}\hat{e}_3, \tag{3.123}$$

$$\overrightarrow{e}_2 = \frac{\partial x^1}{\partial \overline{x}^2}\hat{e}_1 + \frac{\partial x^2}{\partial \overline{x}^2}\hat{e}_2 + \frac{\partial x^3}{\partial \overline{x}^2}\hat{e}_3, \tag{3.124}$$

$$\overrightarrow{e}_3 = \frac{\partial x^1}{\partial \overline{x}^3}\hat{e}_1 + \frac{\partial x^2}{\partial \overline{x}^3}\hat{e}_2 + \frac{\partial x^3}{\partial \overline{x}^3}\hat{e}_3. \tag{3.125}$$

Note that $\overrightarrow{e}_i$ are in general neither orthogonal nor unit vectors. In fact, their magnitudes,

$$|\overrightarrow{e}_i|^2 = \left|\frac{\partial x^1}{\partial \overline{x}^i}\right|^2 + \left|\frac{\partial x^2}{\partial \overline{x}^i}\right|^2 + \left|\frac{\partial x^3}{\partial \overline{x}^i}\right|^2, \tag{3.126}$$

as well as their directions depend on their position. We define unit basis vectors in the direction of $\overrightarrow{\overline{e}}_i$ as

$$\widehat{\overline{e}}_i = \frac{\overrightarrow{\overline{e}}_i}{|\overrightarrow{\overline{e}}_i|}. \tag{3.127}$$

Coordinate basis vectors $\overrightarrow{\overline{e}}_i$ point in the direction of the change in the position vector, when we move an infinitesimal amount along the $\overline{x}^i$ curve. In other words, it is the tangent vector to the $\overline{x}^i$ curve at a given point. We can now interpret the condition $J \neq 0$ for a legitimate definition of generalized coordinates. We first write the Jacobian J as

$$J = \det \begin{pmatrix} \dfrac{\partial x^1}{\partial \overline{x}^1} & \dfrac{\partial x^2}{\partial \overline{x}^1} & \dfrac{\partial x^3}{\partial \overline{x}^1} \\[2mm] \dfrac{\partial x^1}{\partial \overline{x}^2} & \dfrac{\partial x^2}{\partial \overline{x}^2} & \dfrac{\partial x^3}{\partial \overline{x}^2} \\[2mm] \dfrac{\partial x^1}{\partial \overline{x}^3} & \dfrac{\partial x^2}{\partial \overline{x}^3} & \dfrac{\partial x^3}{\partial \overline{x}^3} \end{pmatrix} = \overrightarrow{\overline{e}}_1 \cdot (\overrightarrow{\overline{e}}_2 \times \overrightarrow{\overline{e}}_3). \tag{3.128}$$

Remembering that the triple product $\overrightarrow{\overline{e}}_1 \cdot (\overrightarrow{\overline{e}}_2 \times \overrightarrow{\overline{e}}_3)$ is the volume of the parallelepiped with the sides $\overrightarrow{\overline{e}}_1$, $\overrightarrow{\overline{e}}_2$, and $\overrightarrow{\overline{e}}_3$, the condition $J \neq 0$ for a legitimate definition of generalized coordinates means that the basis vectors have to be noncoplanar.

3.3.2   Why Upper and Lower Indices

Consider a particular generalized coordinate system with oblique axis on the plane (Figure 3.8). We now face a situation that we did not have with the Cartesian coordinates. We can define coordinates of a vector in two different ways, one of which is by drawing parallels to the coordinate axes and the other is by dropping perpendiculars to the axes (Figure 3.8). In general, these two

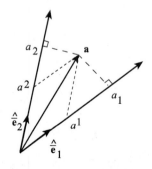

Figure 3.8   Covariant and contravariant components.

methods give different values for the coordinates. Coordinates found by drawing parallels are called the **contravariant components**, and we write them with an upper index as a^i. Now the vector $\vec{a}$ is expressed as

$$\vec{a} = a^1\hat{\vec{e}}_1 + a^2\hat{\vec{e}}_2,$$ (3.129)

where $\hat{\vec{e}}_1$ and $\hat{\vec{e}}_2$ are the unit basis vectors. Coordinates found by dropping perpendiculars to the coordinate axes are called the **covariant components**. They are written with a lower index as a_i, and their values are obtained as

$$a_1 = \vec{a} \cdot \hat{\vec{e}}_1, \ a_2 = \vec{a} \cdot \hat{\vec{e}}_2.$$ (3.130)

3.4   GENERAL TENSORS

Geometric interpretation of the covariant and the contravariant components demonstrates that the difference between the two types of coordinates is, in general, real. As in the case of Cartesian tensors, we can further enrich the concept of scalars and vectors by defining them with respect to their transformation properties under general coordinate transformations. We write the transformation equations between the **Cartesian coordinates** $x^i = (x^1, x^2, x^3)$ and the **generalized coordinates** $\overline{x}^i = (\overline{x}^1, \overline{x}^2, \overline{x}^3)$ as

$$\overline{x}^i = \overline{x}^i(x^j).$$ (3.131)

Similarly, we write the **inverse transformations** as

$$x^i = x^i(\overline{x}^j).$$ (3.132)

Note that each one of the above equations [Eqs. (3.131) and (3.132)] correspond to the three equations for $i = 1, 2, 3$. Even though we write our equations in three dimensions, they can be generalized to n dimensions by simply extending the range of the indices to n.

Using Eq. (3.131), we can write the transformation equation of the coordinate differentials as

$$d\overline{x}^i = \sum_{j=1}^{3} \frac{\partial \overline{x}^i}{\partial x^j} \, dx^j.$$ (3.133)

For a scalar function $\Phi(x^i)$, we can write the transformation equation of its gradient as

$$\frac{\partial \Phi}{\partial \overline{x}^i} = \sum_{j=1}^{3} \left(\frac{\partial \Phi}{\partial x^j} \right) \frac{\partial x^j}{\partial \overline{x}^i}.$$ (3.134)

We now generalize these to all vectors and define a **contravariant vector** as a vector that transforms like dx^j:

$$\bar{v}^i = \sum_{j=1}^{3} \frac{\partial \bar{x}^i}{\partial x^j} v^j \qquad (3.135)$$

and define a **covariant vector** as a vector that transforms like the gradient of a scalar function:

$$\bar{v}_i = \sum_{j=1}^{3} \frac{\partial x^j}{\partial \bar{x}^i} v_j . \qquad (3.136)$$

Analogous to Cartesian tensors, a second-rank **covariant tensor** T_{ij} is defined as

$$\bar{T}_{ij} = \sum_{i'=1}^{3} \sum_{j'=1}^{3} \frac{\partial x^{i'}}{\partial \bar{x}^i} \frac{\partial x^{j'}}{\partial \bar{x}^j} T_{i'j'} . \qquad (3.137)$$

Tensors with **contravariant** and **mixed indices** are also defined with respect to their transformation properties as

$$\bar{T}^{ij} = \sum_{i'=1}^{3} \sum_{j'=1}^{3} \frac{\partial \bar{x}^i}{\partial x^{i'}} \frac{\partial \bar{x}^j}{\partial x^{j'}} T^{i'j'} , \qquad (3.138)$$

$$\bar{T}^i_{\;j} = \sum_{i'=1}^{3} \sum_{j'=1}^{3} \frac{\partial \bar{x}^i}{\partial x^{i'}} \frac{\partial x^{j'}}{\partial \bar{x}^j} T^{i'}_{\;j'} . \qquad (3.139)$$

Note that the transformation equations between the coordinate differentials [Eq. (3.133)] are linear:

$$d\bar{x}^1 = \frac{\partial \bar{x}^1}{\partial x^1} dx^1 + \frac{\partial \bar{x}^1}{\partial x^2} dx^2 + \frac{\partial \bar{x}^1}{\partial x^3} dx^3 , \qquad (3.140)$$

$$d\bar{x}^2 = \frac{\partial \bar{x}^2}{\partial x^1} dx^1 + \frac{\partial \bar{x}^2}{\partial x^2} dx^2 + \frac{\partial \bar{x}^2}{\partial x^3} dx^3 , \qquad (3.141)$$

$$d\bar{x}^3 = \frac{\partial \bar{x}^3}{\partial x^1} dx^1 + \frac{\partial \bar{x}^3}{\partial x^2} dx^2 + \frac{\partial \bar{x}^3}{\partial x^3} dx^3 , \qquad (3.142)$$

hence, the elements of the **transformation matrix** A in $\bar{v} = Av$ [Eq. (3.136)] are given as

$$A = A^i_j = \frac{\partial \bar{x}^i}{\partial x^j} = \begin{pmatrix} \dfrac{\partial \bar{x}^1}{\partial x^1} & \dfrac{\partial \bar{x}^1}{\partial x^2} & \dfrac{\partial \bar{x}^1}{\partial x^3} \\[2mm] \dfrac{\partial \bar{x}^2}{\partial x^1} & \dfrac{\partial \bar{x}^2}{\partial x^2} & \dfrac{\partial \bar{x}^2}{\partial x^3} \\[2mm] \dfrac{\partial \bar{x}^3}{\partial x^1} & \dfrac{\partial \bar{x}^3}{\partial x^2} & \dfrac{\partial \bar{x}^3}{\partial x^3} \end{pmatrix} . \qquad (3.143)$$

If we apply this to orthogonal transformations between Cartesian coordinates defined in Eq. (3.10), we obtain the components of the transformation matrix as

$$A^i_j = A_{ij} = S_{ij} = \cos\theta_{ij}, \tag{3.144}$$

where θ_{ij} are the direction cosines, and we have used the fact that for Cartesian coordinates covariant and the contravariant components are equal.

Using the inverse transformation

$$dx^j = \sum_{k=1}^{3} \frac{\partial x^j}{\partial \overline{x}^k}\, d\overline{x}^k \tag{3.145}$$

in Eq. (3.133), we write

$$d\overline{x}^i = \sum_{k=1}^{3} \left[\sum_{j=1}^{3} \frac{\partial \overline{x}^i}{\partial x^j} \frac{\partial x^j}{\partial \overline{x}^k} \right] d\overline{x}^k = \sum_{k=1}^{3} \delta^i_k d\overline{x}^k = d\overline{x}^i, \tag{3.146}$$

to obtain the relation

$$\boxed{\frac{d\overline{x}^i}{d\overline{x}^k} = \sum_{j=1}^{3} \frac{\partial \overline{x}^i}{\partial x^j} \frac{\partial x^j}{\partial \overline{x}^k} = \delta^i_k.} \tag{3.147}$$

In general, we write the **transformation matrix** A and the **inverse** transformation matrix $\overline{A}$ as

$$A^i_j = \frac{\partial \overline{x}^i}{\partial x^j} \quad \text{and} \quad \overline{A}^i_j = \frac{\partial x^i}{\partial \overline{x}^j}, \tag{3.148}$$

respectively, which satisfy the relation

$$\boxed{\sum_{j=1}^{3} A^i_j \overline{A}^j_k = \delta^i_k.} \tag{3.149}$$

One should keep in mind that even though for ease in comparison we have identified the variables x^i as the Cartesian coordinates and we will continue to do so, the transformation equations represented by the transformation matrix in Eq. (3.143) could represent any transformation from one generalized coordinate system into another. We can also write the last equation [Eq. (3.149)] as

$$A\overline{A} = I, \tag{3.150}$$

$$A^{-1}A\overline{A} = A^{-1}I, \tag{3.151}$$

$$\overline{A} = A^{-1}, \tag{3.152}$$

thus showing that $\overline{A}$ is the inverse of $A = A^i_j$. If we apply Eq. (3.148) to the orthogonal transformations between Cartesian coordinates [Eq. (3.29)] and their inverse [Eq. (3.31)], we see that

$$\overline{A} = \tilde{A}. \tag{3.153}$$

We can now summarize the general transformation equations as

$$\overline{v}^i = \sum_{j=1}^{3} A^i_j v^j, \tag{3.154}$$

$$\overline{v}_i = \sum_{j=1}^{3} \overline{A}^j_i v_j, \tag{3.155}$$

$$\overline{T}^{ij} = \sum_{i'=1}^{3} \sum_{j'=1}^{3} A^i_{i'} A^j_{j'} T^{i'j'}, \tag{3.156}$$

$$\overline{T}_{ij} = \sum_{i'=1}^{3} \sum_{j'=1}^{3} \overline{A}^{i'}_i \overline{A}^{j'}_j T_{i'j'}, \tag{3.157}$$

$$\overline{T}^i_j = \sum_{i'=1}^{3} \sum_{j'=1}^{3} A^i_{i'} \overline{A}^{j'}_j T^{i'}_{j'}. \tag{3.158}$$

3.4.1   Einstein Summation Convention

From the above equations, we observe that whenever an index is repeated with one up and the other one down, it is summed over. We still have not shown how to raise or lower indices but from now on whenever there is a summation over two indices, we agree to write it with one up and the other down and omit the summation sign. It does not matter which index is written up or down. This is called the **Einstein summation convention**. Now the above transformation equations and their inverses can be written as

$$
\begin{array}{ll}
\overline{v}^i = A^i_j v^j & v^i = \overline{A}^i_j \overline{v}^j \\[4pt]
\overline{v}_i = \overline{A}^j_i v_j & v_i = A^j_i \overline{v}_j \\[4pt]
\overline{T}^{ij} = A^i_{i'} A^j_{j'} T^{i'j'} & T^{ij} = \overline{A}^i_{i'} \overline{A}^j_{j'} \overline{T}^{i'j'} \\[4pt]
\overline{T}_{ij} = \overline{A}^{i'}_i \overline{A}^{j'}_j T_{i'j'} & T_{ij} = A^{i'}_i A^{j'}_j \overline{T}_{i'j'} \\[4pt]
\overline{T}^i_j = A^i_{i'} \overline{A}^{j'}_j T^{i'}_{j'} & T^i_j = \overline{A}^i_{i'} A^{j'}_j \overline{T}^{i'}_{j'}.
\end{array}
\tag{3.159}
$$

A general tensor with **mixed indices** is defined with respect to the transformation rule

$$\boxed{T^{i_1 i_2 \ldots}_{j_1 j_2 \ldots} = \frac{\partial \overline{x}^{i_1}}{\partial x^{k_1}} \frac{\partial \overline{x}^{i_2}}{\partial x^{k_2}} \cdots \frac{\partial x^{l_1}}{\partial \overline{x}^{j_1}} \frac{\partial x^{l_2}}{\partial \overline{x}^{j_2}} \cdots T^{k_1 k_2 \ldots}_{l_1 l_2 \ldots}.} \tag{3.160}$$

To prove the tensor property of the **Kronecker delta** under general coordinate transformations, we use Eq. (3.149) to write

$$\overline{\delta}^i_j = A^i_{i'} \overline{A}^{j'}_j \delta^{i'}_{j'} = A^i_{i'} \overline{A}^{i'}_j = \delta^i_j. \tag{3.161}$$

Hence, δ^i_j is a second-rank tensor and has the same components in generalized coordinates. It is the only second-rank tensor with this property. Algebraic properties described for the Cartesian tensors are also valid for general tensors.

3.4.2   Line Element

We now write the line element in generalized coordinates, which gives the distance between two infinitesimally close points. We start with the line element in Cartesian coordinates, which is nothing but Pythagoras' theorem, which can be written in the following equivalent forms:

$$ds^2 = d\overrightarrow{r} \cdot d\overrightarrow{r} = (dx^1)^2 + (dx^2)^2 + (dx^3)^2, \tag{3.162}$$

$$ds^2 = \sum_{k=1}^{3} dx^k \, dx^k = dx^k \, dx_k = dx_k \, dx^k. \tag{3.163}$$

Using the inverse transformation:

$$dx^k = \frac{\partial x^k}{\partial \overline{x}^i} \, d\overline{x}^i \tag{3.164}$$

and the fact that ds is a scalar, we write the **line element** in generalized coordinates as

$$d\overline{s}^2 = ds^2 = \sum_{k=1}^{3} dx^k \, dx^k = \sum_{k=1}^{3} \frac{\partial x^k}{\partial \overline{x}^i} \, d\overline{x}^i \frac{\partial x^k}{\partial \overline{x}^j} \, d\overline{x}^j = \left[\sum_{k=1}^{3} \frac{\partial x^k}{\partial \overline{x}^i} \frac{\partial x^k}{\partial \overline{x}^j} \right] d\overline{x}^i d\overline{x}^j, \tag{3.165}$$

3.4.3   Metric Tensor

We now introduce a very important tensor, that is, the **metric tensor** g_{ij}, which is defined as

$$g_{ij} = \sum_{k=1}^{3} \frac{\partial x^k}{\partial \overline{x}^i} \frac{\partial x^k}{\partial \overline{x}^j}. \tag{3.166}$$

Note that the sum over k is written with both indices up. Hence, even though we still adhere to the Einstein summation convention, for these indices, we keep the summation sign. The metric tensor is the single most important second-rank tensor in tensor calculus, and the general theory of relativity. Now the **line element** in generalized coordinates becomes

$$ds^2 = g_{ij}d\bar{x}^i d\bar{x}^j. \qquad (3.167)$$

Needless to say, components of the metric tensor in Eq. (3.167) are all expressed in terms of the barred coordinates. Note that in **Cartesian coordinates**, the metric tensor is the identity tensor,

$$g_{ij} = \delta_{ij}; \qquad (3.168)$$

thus the line element in Cartesian coordinates becomes

$$ds^2 = \delta_{ij} \, dx^i \, dx^j = (dx^1)^2 + (dx^2)^2 + (dx^2)^2. \qquad (3.169)$$

3.4.4   How to Raise and Lower Indices

Given an arbitrary contravariant vector v^j, let us find how

$$[g_{ij}v^j] \qquad (3.170)$$

transforms. Using Eq. (3.159), we first write

$$g_{ij} = A_i^{i'} A_j^{j'} \bar{g}_{i'j'}, \qquad (3.171)$$

$$v^j = \overline{A}_k^j \bar{v}^k, \qquad (3.172)$$

and then substitute them into Eq. (3.170) to get

$$[g_{ij}v^j] = A_i^{i'} A_j^{j'} \overline{A}_k^j [\bar{g}_{i'j'}\bar{v}^k] \qquad (3.173)$$

$$= A_i^{i'} [A_j^{j'} \overline{A}_k^j][\bar{g}_{i'j'}\bar{v}^k] \qquad (3.174)$$

$$= A_i^{i'} [\delta_k^{j'}][\bar{g}_{i'j'}\bar{v}^k] \qquad (3.175)$$

$$= A_i^{i'} [\bar{g}_{i'k}\bar{v}^k]. \qquad (3.176)$$

Renaming the dummy variable k on the right-hand side as $k \to j$, we finally obtain

$$[g_{ij}v^j] = A_i^{i'} [\bar{g}_{i'j}\bar{v}^j]. \qquad (3.177)$$

Comparing with the corresponding equation in Eq. (3.159), it is seen that $g_{ij}v^j$ transforms like a covariant vector. We now define the **covariant component** of v^j as

$$v_i = g_{ij}v^j. \qquad (3.178)$$

We can also define the **metric tensor** with the **contravariant components** as

$$\boxed{g^{kl} = \sum_{i=1}^{3} \frac{\partial \bar{x}^k}{\partial x^i} \frac{\partial \bar{x}^l}{\partial x^i},}$$ (3.179)

where

$$g_{kl} g^{kl'} = \sum_{m=1}^{3} \frac{\partial x^m}{\partial \bar{x}^k} \frac{\partial x^m}{\partial \bar{x}^l} \sum_{n=1}^{3} \frac{\partial \bar{x}^k}{\partial x^n} \frac{\partial \bar{x}^{l'}}{\partial x^n}$$ (3.180)

$$= \sum_{m=1}^{3} \sum_{n=1}^{3} \left[\frac{\partial x^m}{\partial \bar{x}^k} \frac{\partial \bar{x}^k}{\partial x^n} \right] \left[\frac{\partial x^m}{\partial \bar{x}^l} \frac{\partial \bar{x}^{l'}}{\partial x^n} \right]$$ (3.181)

$$= \sum_{m=1}^{3} \sum_{n=1}^{3} \delta^m_n \left[\frac{\partial x^m}{\partial \bar{x}^l} \frac{\partial \bar{x}^{l'}}{\partial x^n} \right]$$ (3.182)

$$= \sum_{m=1}^{3} \left[\frac{\partial x^m}{\partial \bar{x}^l} \frac{\partial \bar{x}^{l'}}{\partial x^m} \right] = \delta^{l'}_l.$$ (3.183)

Note that in the above equations, in addition to the summation signs that come from the definition of the metric tensor, the Einstein summation convention is still in effect. Using the symmetry of the metric tensor, we can also write

$$\boxed{g_{lk} g^{kl'} = g^{l'}_l = \delta^{l'}_l.}$$ (3.184)

We now have a tool that can be used to raise and lower indices at will:

$$T_{ij} = g_{ii'} T^{i'}_j,$$ (3.185)

$$A^k_{ij} = g^{kk'} A_{ijk'},$$ (3.186)

$$C^j_{ik} = g^{jj'} g_{ii'} g_{kk'} C^{i'k'}_{j'},$$ (3.187)

etc.

Metric tensor in Cartesian coordinates is δ_{ij}, Using Eqs. (3.143) and (3.159), we can show that under the general coordinate transformations, it transforms into the metric tensor:

$$\bar{\delta}_{ij} = \bar{A}^{i'}_i \bar{A}^{j'}_j \delta_{i'j'} = \sum_{i'=1}^{3} \bar{A}^{i'}_i \bar{A}^{i'}_j = \sum_{i'=1}^{3} \frac{\partial x^{i'}}{\partial \bar{x}^i} \frac{\partial x^{i'}}{\partial \bar{x}^j} = g_{ij}.$$ (3.188)

3.4.5   Metric Tensor and the Basis Vectors

If we remember the definition of the basis vectors [Eqs. (3.123)–(3.125)]:

$$\vec{e}_i = \frac{\partial x^k}{\partial \overline{x}^i}, \quad i = 1, 2, 3 \tag{3.189}$$

$$= \frac{\partial x^1}{\partial \overline{x}^i} \widehat{e}_1 + \frac{\partial x^2}{\partial \overline{x}^i} \widehat{e}_2 + \frac{\partial x^3}{\partial \overline{x}^i} \widehat{e}_3, \tag{3.190}$$

which are tangents to the coordinate curves (Figure 3.7), we can write the metric tensor as the scalar product of the basis vectors as

$$g_{ij} = \sum_{k=1}^{3} \frac{\partial x^k}{\partial \overline{x}^i} \frac{\partial x^k}{\partial \overline{x}^j} = \vec{e}_i \cdot \vec{e}_j. \tag{3.191}$$

Note that the basis vectors $\vec{e}_i$ are given in terms of the unit basis vectors of the Cartesian coordinate system $\widehat{e}_i$. Similarly, using the definition of the metric tensor with the contravariant components:

$$g^{ij} = \sum_{k=1}^{3} \frac{\partial \overline{x}^i}{\partial x^k} \frac{\partial \overline{x}^j}{\partial x^k}, \tag{3.192}$$

we can define the new basis vectors $\vec{e}^i$ as

$$\vec{e}^i = \frac{\partial \overline{x}^i}{\partial x^k}, \quad i = 1, 2, 3, \tag{3.193}$$

which allows us to write the contravariant metric tensor as the scalar product

$$g^{ij} = \vec{e}^i \cdot \vec{e}^j. \tag{3.194}$$

The new basis vectors $\vec{e}^i$ are called the **inverse** basis vectors. Note that neither of the basis vectors, $\vec{e}^i$ or $\vec{e}_i$, are unit vectors and the indices do not refer to their components. Inverse basis vectors are actually the **gradients**:

$$\vec{e}^i = \vec{\nabla} \overline{x}^i(x^j). \tag{3.195}$$

Hence, they are perpendicular to the coordinate surfaces, while $\vec{e}_i$ are tangents to the coordinate curves. Usage of the upper or the lower indices for the basis vectors is justified by the fact that these indices can be lowered or raised by the metric tensor as

$$g_{ij}\,\overrightarrow{e}^{\,j} = \sum_{k=1}^{3} \frac{\partial x^k}{\partial \overline{x}^i} \frac{\partial x^k}{\partial \overline{x}^j} \frac{\partial \overline{x}^j}{\partial x^l} = \sum_{k=1}^{3} \frac{\partial x^k}{\partial \overline{x}^i} \left[\frac{\partial x^k}{\partial \overline{x}^j} \frac{\partial \overline{x}^j}{\partial x^l} \right] \tag{3.196}$$

$$= \sum_{k=1}^{3} \frac{\partial x^k}{\partial \overline{x}^i} \delta_l^k = \frac{\partial x^l}{\partial \overline{x}^i} = \overrightarrow{e}_i. \tag{3.197}$$

Similarly,

$$g^{ij}\,\overrightarrow{e}_j = \sum_{k=1}^{3} \frac{\partial \overline{x}^i}{\partial x^k} \frac{\partial \overline{x}^j}{\partial x^k} \frac{\partial x^l}{\partial \overline{x}^j} = \sum_{k=1}^{3} \frac{\partial \overline{x}^i}{\partial x^k} \left[\frac{\partial \overline{x}^j}{\partial x^k} \frac{\partial x^l}{\partial \overline{x}^j} \right] \tag{3.198}$$

$$= \sum_{k=1}^{3} \frac{\partial \overline{x}^i}{\partial x^k} \delta_k^l = \frac{\partial \overline{x}^i}{\partial x^l} = \overrightarrow{e}^{\,i}. \tag{3.199}$$

3.4.6 Displacement Vector

In generalized coordinates, the **displacement** vector between two infinitesimally close points is written as

$$\boxed{ d\overrightarrow{r} = \frac{\partial \overrightarrow{r}}{\partial \overline{x}^i} d\overline{x}^i = \overrightarrow{e}_i\, d\overline{x}^i = \overrightarrow{e}_1\, d\overline{x}^1 + \overrightarrow{e}_2\, d\overline{x}^2 + \overrightarrow{e}_3\, d\overline{x}^3. } \tag{3.200}$$

Using the displacement vector [Eq. (3.200)], we can write the **line element** as

$$\boxed{ ds^2 = d\overrightarrow{r} \cdot d\overrightarrow{r} = (\overrightarrow{e}_i \cdot \overrightarrow{e}_j) d\overline{x}^i d\overline{x}^j = g_{ij} d\overline{x}^i d\overline{x}^j. } \tag{3.201}$$

If we move along only on one of the coordinate curves, say $\overline{x}^1$, the distance covered is

$$ds_{\overline{x}^1} = \sqrt{g_{11}}\, d\overline{x}^1, \quad d\overline{x}^2 = d\overline{x}^3 = 0. \tag{3.202}$$

Similarly, for the displacements along the other axes, we obtain

$$ds_{\overline{x}^2} = \sqrt{g_{22}}\, d\overline{x}^2, \quad d\overline{x}^1 = d\overline{x}^3 = 0, \tag{3.203}$$

$$ds_{\overline{x}^3} = \sqrt{g_{33}}\, d\overline{x}^3, \quad d\overline{x}^1 = d\overline{x}^2 = 0. \tag{3.204}$$

For a general displacement, we have to use the line element [Eq. (3.201)]. For orthogonal generalized coordinates, where

$$(\overrightarrow{e}_i \cdot \overrightarrow{e}_j) = 0, \quad i \neq j, \tag{3.205}$$

the metric tensor has only the diagonal components and the line element reduces to

$$ds^2 = ds_{\overline{x}^1}^2 + ds_{\overline{x}^2}^2 + ds_{\overline{x}^3}^2 \tag{3.206}$$

$$= g_{11}(d\overline{x}^1)^2 + g_{22}(d\overline{x}^2)^2 + g_{33}(d\overline{x}^3)^2 \tag{3.207}$$

$$= (\overrightarrow{e}_1 \cdot \overrightarrow{e}_1)(d\overline{x}^1)^2 + (\overrightarrow{e}_2 \cdot \overrightarrow{e}_2)(d\overline{x}^2)^2 + (\overrightarrow{e}_3 \cdot \overrightarrow{e}_3)(d\overline{x}^3)^2. \tag{3.208}$$

3.4.7   Line Integrals

As in orthogonal transformations, value of a scalar function is invariant under generalized coordinate transformations, hence, we can write

$$\Phi(x^1, x^2, x^3) = \overline{\Phi}(\overline{x}^1, \overline{x}^2, \overline{x}^3) = \Phi(\overline{x}^1, \overline{x}^2, \overline{x}^3). \tag{3.209}$$

The scalar product of two vectors, $\overrightarrow{a}$ and $\overrightarrow{b}$, is also a scalar, thus preserving its value. In generalized coordinates, we write it as

$$\overrightarrow{a} \cdot \overrightarrow{b} = g_{ij}\overline{a}^i\overline{b}^j = \overline{a}^i\overline{b}_i. \tag{3.210}$$

Using the transformation equations:

$$\overline{a}^i = \frac{\partial \overline{x}^i}{\partial x^j}a^j, \tag{3.211}$$

$$\overline{b}_i = \frac{\partial x^k}{\partial \overline{x}^i}b_k, \tag{3.212}$$

it is clear that it has the same value that it has in Cartesian coordinates:

$$\overrightarrow{a} \cdot \overrightarrow{b} = \overline{a}^i\overline{b}_i = \frac{\partial \overline{x}^i}{\partial x^j}\frac{\partial x^k}{\partial \overline{x}^i}a^jb_k = \delta^k_j a^j b_k = a^k b_k = \overrightarrow{a} \cdot \overrightarrow{b}. \tag{3.213}$$

In the light of these, a given line integral in Cartesian coordinates:

$$I = \int \overrightarrow{V} \cdot d\overrightarrow{r} = \int V_1 \, dx^1 + V_2 \, dx^2 + V_3 \, dx^3, \tag{3.214}$$

can be written in generalized coordinates as

$$I = \int \overrightarrow{\overline{V}} \cdot d\overrightarrow{\overline{r}} = \int \overline{V}_i \, d\overline{x}^i = \int \overline{V}_1 \, d\overline{x}^1 + \overline{V}_2 \, d\overline{x}^2 + \overline{V}_3 \, d\overline{x}^3. \tag{3.215}$$

We can also write I as

$$\boxed{I = \int g_{ij}\overline{V}^i d\overline{x}^j.} \tag{3.216}$$

In orthogonal generalized coordinates, only the diagonal components of the metric tensor are nonzero, hence, I becomes

$$I = \int g_{11} \overline{V}^1 d\overline{x}^1 + g_{22} \overline{V}^2 d\overline{x}^2 + g_{33} \overline{V}^3 d\overline{x}^3. \tag{3.217}$$

It is important to keep in mind that a vector $\overrightarrow{V}$ exists independent of the coordinate system used to represent it. In other words, whether we write $\overrightarrow{V}$ in Cartesian coordinates as

$$\overrightarrow{V} = V^1 \widehat{e}_1 + V^2 \widehat{e}_2 + V^3 \widehat{e}_3 = V^i \widehat{e}_i, \tag{3.218}$$

or in generalized coordinates as

$$\overrightarrow{V} = \overline{V}^1 \overrightarrow{e}_1 + \overline{V}^2 \overrightarrow{e}_2 + \overline{V}^3 \overrightarrow{e}_3 = \overline{V}^i \overrightarrow{e}_i = \overline{V}_i \overrightarrow{e}^i, \tag{3.219}$$

it is the same vector. Hence, the bar on $\overrightarrow{V}$ is sometimes omitted. We remind the reader that $\overrightarrow{e}_i$ are not unit vectors in general. Covariant components of $\overrightarrow{V}$ are found as

$$\overrightarrow{V} \cdot \overrightarrow{e}_j = (\overline{V}^i \overrightarrow{e}_i) \cdot \overrightarrow{e}_j = \overline{V}^i (\overrightarrow{e}_i \cdot \overrightarrow{e}_j) = \overline{V}^i g_{ij} = \overline{V}_j. \tag{3.220}$$

Similarly, using the inverse basis vectors, $\overrightarrow{e}^i$, we can find the contravariant components as

$$\overrightarrow{V} \cdot \overrightarrow{e}^j = (\overline{V}_i \overrightarrow{e}^i) \cdot \overrightarrow{e}^j = \overline{V}_i (\overrightarrow{e}^i \cdot \overrightarrow{e}^j) = \overline{V}_i g^{ij} = \overline{V}^j. \tag{3.221}$$

The two types of components are related by

$$\overline{V}^j = g^{ij} \overline{V}_i. \tag{3.222}$$

We can now write the line integral [Eq. (3.214)] in the following equivalent ways:

$$I = \int \overrightarrow{V} \cdot (d\overline{x}^1 \overrightarrow{e}_1 + d\overline{x}^2 \overrightarrow{e}_2 + d\overline{x}^3 \overrightarrow{e}_3) \tag{3.223}$$

$$= \int (\overrightarrow{V} \cdot \overrightarrow{e}_1) d\overline{x}^1 + (\overrightarrow{V} \cdot \overrightarrow{e}_2) d\overline{x}^2 + (\overrightarrow{V} \cdot \overrightarrow{e}_3) d\overline{x}^3 \tag{3.224}$$

$$= \int \overline{V}_1 \, d\overline{x}^1 + \overline{V}_2 \, d\overline{x}^2 + \overline{V}_3 \, d\overline{x}^3 \tag{3.225}$$

$$= \int [\overline{V}^j g_{j1}] d\overline{x}^1 + [\overline{V}^j g_{j2}] d\overline{x}^2 + [\overline{V}^j g_{j3}] d\overline{x}^3. \tag{3.226}$$

3.4.8   Area Element in Generalized Coordinates

Using the expression for the area of a parallelogram defined by two vectors $\vec{a}$ and $\vec{b}$ as

$$\boxed{\text{area} = |\vec{a} \times \vec{b}|,}$$

(3.227)

we write the area element in generalized coordinates defined by the infinitesimal vectors $d\overline{x}^1\,\vec{e}_1$ and $d\overline{x}^2\,\vec{e}_2$ (Figure 3.9) as

$$d\overline{\sigma}_{\overline{x}^1\overline{x}^2} = \left|\vec{e}_1 \times \vec{e}_2\right| d\overline{x}^1 d\overline{x}^2,$$

(3.228)

Similarly, the other area elements are defined:

$$d\overline{\sigma}_{\overline{x}^1\overline{x}^3} = \left|\vec{e}_1 \times \vec{e}_3\right| d\overline{x}^1 d\overline{x}^3,$$

(3.229)

$$d\overline{\sigma}_{\overline{x}^2\overline{x}^3} = \left|\vec{e}_2 \times \vec{e}_3\right| d\overline{x}^2 d\overline{x}^3.$$

(3.230)

In orthogonal generalized coordinates, where the unit basis vectors, $\widehat{e}_i = \vec{e}_i / \left|\vec{e}_i\right|$, $i = 1, 2, 3$, satisfy

$$\widehat{e}_1 = \widehat{e}_2 \times \widehat{e}_3,$$

(3.231)

$$\widehat{e}_2 = \widehat{e}_3 \times \widehat{e}_1,$$

(3.232)

$$\widehat{e}_3 = \widehat{e}_1 \times \widehat{e}_2,$$

(3.233)

we can write

$$d\vec{\overline{\sigma}}_{\overline{x}^1\overline{x}^2} = \left|\vec{e}_1\right|\left|\vec{e}_2\right| d\overline{x}^1 d\overline{x}^2\,\widehat{e}_3 = \sqrt{g_{11}}\sqrt{g_{22}}\ d\overline{x}^1 d\overline{x}^2\,\widehat{e}_3,$$

(3.234)

where the area element is oriented in the $\widehat{e}_3$ direction. Similarly, we can write the other area elements as

$$d\vec{\overline{\sigma}}_{\overline{x}^3\overline{x}^1} = \left|\vec{e}_1\right|\left|\vec{e}_3\right| d\overline{x}^1 d\overline{x}^3\,\widehat{e}_2 = \sqrt{g_{11}}\sqrt{g_{33}}\ d\overline{x}^1 d\overline{x}^3\,\widehat{e}_2,$$

(3.235)

and

$$d\vec{\overline{\sigma}}_{\overline{x}^2\overline{x}^3} = \left|\vec{e}_2\right|\left|\vec{e}_3\right| d\overline{x}^2 d\overline{x}^3\,\widehat{e}_1 = \sqrt{g_{22}}\sqrt{g_{33}}\ d\overline{x}^2 d\overline{x}^3\,\widehat{e}_1.$$

(3.236)

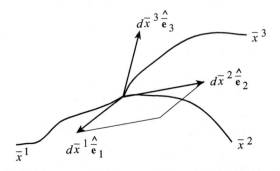

Figure 3.9   Area element in generalized coordinates.

3.4.9 Area of a Surface

A surface in three dimensional Cartesian space can be defined either as

$$x^3 = f(x^1, x^2)$$

or in terms of two coordinates (parameters), u and v, defined on the surface as (Figure 3.10)

$$x^1 = x^1(u, v), \tag{3.237}$$

$$x^2 = x^2(u, v), \tag{3.238}$$

$$x^3 = x^3(u, v). \tag{3.239}$$

The u and v coordinates are essentially the contravariant components of the coordinates that a two-dimensional observer living on this surface would use, that is,

$$\overline{x}^1 = u, \tag{3.240}$$

$$\overline{x}^2 = v. \tag{3.241}$$

We can write the infinitesimal Cartesian coordinate differentials (dx^1, dx^2, dx^3) corresponding to infinitesimal displacements on the surface, in terms of the surface coordinate differentials du and dv as

$$dx^1 = \frac{\partial x^1}{\partial u}\, du + \frac{\partial x^1}{\partial v}\, dv, \tag{3.242}$$

$$dx^2 = \frac{\partial x^2}{\partial u}\, du + \frac{\partial x^2}{\partial v}\, dv, \tag{3.243}$$

$$dx^3 = \frac{\partial x^3}{\partial u}\, du + \frac{\partial x^3}{\partial v}\, dv. \tag{3.244}$$

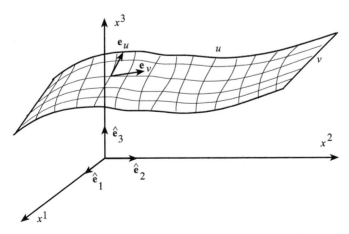

Figure 3.10 u and v coordinates defined on a surface.

We now write the distance ds between two infinitesimally close points on the surface entirely in terms of the surface coordinates u and v as

$$ds^2 = (dx^1)^2 + (dx^2)^2 + (dx^3)^2 \tag{3.245}$$

$$= \left[\left(\frac{\partial x^1}{\partial u}\right)^2 + \left(\frac{\partial x^2}{\partial u}\right)^2 + \left(\frac{\partial x^3}{\partial u}\right)^2\right] du^2$$

$$+ 2\left[\frac{\partial x^1}{\partial u}\frac{\partial x^1}{\partial v} + \frac{\partial x^2}{\partial u}\frac{\partial x^2}{\partial v} + \frac{\partial x^3}{\partial u}\frac{\partial x^3}{\partial v}\right] du\, dv$$

$$+ \left[\left(\frac{\partial x^1}{\partial v}\right)^2 + \left(\frac{\partial x^2}{\partial v}\right)^2 + \left(\frac{\partial x^3}{\partial v}\right)^2\right] dv^2 . \tag{3.246}$$

Comparing this with the line element for an observer living on the surface:

$$ds^2 = g_{ij}\, du\, dv, \quad i = 1, 2, \tag{3.247}$$

$$= g_{uu}\, du^2 + 2g_{uv}\, du\, dv + g_{vv}\, dv^2, \tag{3.248}$$

we obtain the components of the metric tensor as

$$g_{uu} = \left(\frac{\partial x^1}{\partial u}\right)^2 + \left(\frac{\partial x^2}{\partial u}\right)^2 + \left(\frac{\partial x^3}{\partial u}\right)^2, \tag{3.249}$$

$$g_{uv} = \frac{\partial x^1}{\partial u}\frac{\partial x^1}{\partial v} + \frac{\partial x^2}{\partial u}\frac{\partial x^2}{\partial v} + \frac{\partial x^3}{\partial u}\frac{\partial x^3}{\partial v}, \tag{3.250}$$

$$g_{vv} = \left(\frac{\partial x^1}{\partial v}\right)^2 + \left(\frac{\partial x^2}{\partial v}\right)^2 + \left(\frac{\partial x^3}{\partial v}\right)^2. \tag{3.251}$$

Since the metric tensor can also be written in terms of the surface basis vectors $\vec{e}_u$ and $\vec{e}_v$ as

$$ds^2 = (\vec{e}_u \cdot \vec{e}_u)du^2 + 2(\vec{e}_u \cdot \vec{e}_v)du\, dv + (\vec{e}_v \cdot \vec{e}_v)dv^2, \tag{3.252}$$

we can read $\vec{e}_u$ and $\vec{e}_v$ from Eqs. (3.249)–(3.251) as

$$\vec{e}_u = \frac{\partial x^1}{\partial u}\hat{e}_1 + \frac{\partial x^2}{\partial u}\hat{e}_2 + \frac{\partial x^3}{\partial u}\hat{e}_3, \tag{3.253}$$

$$\vec{e}_v = \frac{\partial x^1}{\partial v}\hat{e}_1 + \frac{\partial x^2}{\partial v}\hat{e}_2 + \frac{\partial x^3}{\partial v}\hat{e}_3. \tag{3.254}$$

Note that the surface basis vectors are given in terms of the Cartesian unit basis vectors $(\hat{e}_1, \hat{e}_2, \hat{e}_3)$. We can now write the **area element** of the surface in terms of the **surface coordinates** as

$$\boxed{d\vec{\sigma}_{uv} = \vec{e}_u du \times \vec{e}_v dv,} \tag{3.255}$$

which after substituting Eqs. (3.253) and (3.254) yields

$$d\vec{\sigma}_{uv} = \pm \left[\left(\frac{\partial x^2}{\partial u} \frac{\partial x^3}{\partial v} - \frac{\partial x^3}{\partial u} \frac{\partial x^2}{\partial v} \right) \hat{e}_1 \right.$$

$$+ \left(-\frac{\partial x^1}{\partial u} \frac{\partial x^3}{\partial v} + \frac{\partial x^3}{\partial u} \frac{\partial x^1}{\partial v} \right) \hat{e}_2$$

$$\left. + \left(\frac{\partial x^1}{\partial u} \frac{\partial x^2}{\partial v} - \frac{\partial x^2}{\partial u} \frac{\partial x^1}{\partial v} \right) \hat{e}_3 \right] du \; dv. \qquad (3.256)$$

This can also be written as

$$d\vec{\sigma}_{uv} = \pm \left[\frac{\partial(x^2, x^3)}{\partial(u, v)} \hat{e}_1 + \frac{\partial(x^3, x^1)}{\partial(u, v)} \hat{e}_2 + \frac{\partial(x^1, x^2)}{\partial(u, v)} \hat{e}_3 \right] du \; dv. \qquad (3.257)$$

The signs $\pm$ correspond to proper and improper transformations, respectively. Using Eq. (3.256), we can write the **magnitude** of the **area element** as

$$\boxed{|d\vec{\sigma}_{uv}| = \sqrt{EG - F^2} du \; dv,} \qquad (3.258)$$

where

$$E = \left(\frac{\partial x^1}{\partial u} \right)^2 + \left(\frac{\partial x^2}{\partial u} \right)^2 + \left(\frac{\partial x^3}{\partial u} \right)^2, \qquad (3.259)$$

$$F = \frac{\partial x^1}{\partial u} \frac{\partial x^1}{\partial v} + \frac{\partial x^2}{\partial u} \frac{\partial x^2}{\partial v} + \frac{\partial x^3}{\partial u} \frac{\partial x^3}{\partial v}, \qquad (3.260)$$

$$G = \left(\frac{\partial x^1}{\partial v} \right)^2 + \left(\frac{\partial x^2}{\partial v} \right)^2 + \left(\frac{\partial x^3}{\partial v} \right)^2. \qquad (3.261)$$

Integrating over the surface, we get the surface area

$$S = \int\int_{R_{uv}} |d\vec{\sigma}_{uv}| \qquad (3.262)$$

$$\boxed{S = \int\int_{R_{uv}} \sqrt{EG - F^2} du \; dv,} \qquad (3.263)$$

which is nothing but Eq. (2.178) we have written for the area of a surface in Chapter 2.

If the surface S is defined in Cartesian coordinates as $x^3 - f(x^1, x^2) = 0$, we can project the surface element $d\vec{\sigma}$ onto the $x^1 x^2$-plane as $dx^1 \; dx^2 = (\hat{n} \cdot \hat{e}_3) d\sigma = \cos \gamma d\sigma$, where $\hat{n} = \vec{n}/|\vec{n}|$ is the unit normal to the surface and integrate over the region $R_{x^1 x^2}$, which is the projection of S onto the $x^1 x^2$-plane

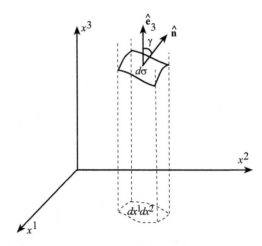

Figure 3.11   Projection of the surface area element.

(Figure 3.11). Since the **normal** is given as $\vec{n} = \left(-\frac{\partial f}{\partial x^1}, -\frac{\partial f}{\partial x^2}, 1\right)$, we write can write the **surface area** as

$$S = \int\int d\sigma = \int\int_{R_{x^1 x^2}} (1/\cos\gamma) dx^1\, dx^2, \tag{3.264}$$

or as

$$S = \int\int_{R_{x^1 x^2}} \sqrt{1 + \left(\frac{\partial f}{\partial x^1}\right)^2 + \left(\frac{\partial f}{\partial x^2}\right)^2}\, dx^1\, dx^2. \tag{3.265}$$

The two areas [Eqs. (3.263) and (3.265)] naturally agree.

Example 3.2.   *Curvilinear coordinates on the plane:* Transformations from the Cartesian to curvilinear coordinates on the plane, say the $x^1 x^2$-plane, is accomplished by the transformation equations

$$x^1 = x^1(u, v), \tag{3.266}$$

$$x^2 = x^2(u, v), \tag{3.267}$$

$$x^3 = 0. \tag{3.268}$$

Metric tensor can easily be constructed by using Eqs. (3.249)–(3.251). Area element is naturally in the x^3 direction and is given as [Eq. (3.257)]

$$d\vec{\sigma}_{uv} = \pm \left[\frac{\partial(x^1, x^2)}{\partial(u, v)}\, du\, dv\right] \hat{e}_3. \tag{3.269}$$

Taking the plus sign for proper transformations, we write the magnitude of the area element as

$$d\sigma_{uv} = \frac{\partial(x^1, x^2)}{\partial(u, v)} \, du \, dv. \tag{3.270}$$

In other words, under the above transformation [Eqs. (3.266)−(3.268)], the area element transforms as

$$dx^1 \, dx^2 \rightarrow \frac{\partial(x^1, x^2)}{\partial(u, v)} \, du \, dv. \tag{3.271}$$

Note that on the $x^1 x^2$-plane

$$\frac{\partial(x^2, x^3)}{\partial(u, v)} = \frac{\partial(x^1, x^3)}{\partial(u, v)} = 0. \tag{3.272}$$

Applying these to the plane polar coordinates defined by the transformation equations

$$x^1 = \rho \cos \phi, \tag{3.273}$$

$$x^2 = \rho \sin \phi, \tag{3.274}$$

where $u = \rho$ and $v = \phi$, we can write the line element as

$$ds^2 = d\rho^2 + \rho^2 \, d\phi^2. \tag{3.275}$$

Since

$$\frac{\partial(x^1, x^2)}{\partial(\rho, \phi)} = \begin{vmatrix} \cos \phi & -\rho \sin \phi \\ \sin \phi & \rho \cos \phi \end{vmatrix} = \rho, \tag{3.276}$$

the area element becomes

$$d\sigma = \rho \, d\rho \, d\phi. \tag{3.277}$$

3.4.10 Volume Element in Generalized Coordinates

In **Cartesian coordinates** the **volume element** is defined as the scalar

$$\boxed{d\tau = \hat{e}_1 \cdot (\hat{e}_2 \times \hat{e}_3) dx^1 \, dx^2 \, dx^3.} \tag{3.278}$$

Since the Cartesian basis vectors are mutually orthogonal and of unit magnitude, the infinitesimal volume element reduces to

$$d\tau = dx^1 \, dx^2 \, dx^3. \tag{3.279}$$

In generalized coordinates, we can write the volume element $d\tau'$, which is also equal to $d\tau$, as the volume of the infinitesimal parallelepiped with the sides defined by the vectors

$$\vec{e}_1 d\overline{x}^1, \quad \vec{e}_2 d\overline{x}^2, \quad \vec{e}_3 d\overline{x}^3 \tag{3.280}$$

as

$$d\tau' = (\vec{e}_1 d\overline{x}^1) \cdot (\vec{e}_2 d\overline{x}^2 \times \vec{e}_3 d\overline{x}^3) \tag{3.281}$$

$$= \vec{e}_1 \cdot (\vec{e}_2 \times \vec{e}_3) d\overline{x}^1\, d\overline{x}^2\, d\overline{x}^3. \tag{3.282}$$

Using Eq. (3.128), this can also be written as

$$d\tau' = \frac{\partial(x^1, x^2, x^3)}{\partial(\overline{x}^1, \overline{x}^2, \overline{x}^3)} d\overline{x}^1 d\overline{x}^2 d\overline{x}^3. \tag{3.283}$$

A tensor that transforms as

$$\overline{T}^{i_1 i_2 \cdots}_{j_1 j_2 \cdots} = \frac{\partial \overline{x}^{i_1}}{\partial x^{k_1}} \frac{\partial \overline{x}^{i_2}}{\partial x^{k_2}} \cdots \frac{\partial x^{l_1}}{\partial \overline{x}^{j_1}} \frac{\partial x^{l_2}}{\partial \overline{x}^{j_2}} \cdots T^{k_1 k_2 \cdots}_{l_1 l_2 \cdots} \left[\frac{\partial(x^1, x^2, x^3)}{\partial(\overline{x}^1, \overline{x}^2, \overline{x}^3)}\right]^w \tag{3.284}$$

is called a **tensor density** or a **pseudotensor** of **weight** w. Hence, the coordinate volume element, $d\overline{x}^1 d\overline{x}^2 d\overline{x}^3$, which transforms as

$$d\overline{x}^1\, d\overline{x}^2\, d\overline{x}^3 = dx^1\, dx^2\, dx^3 \left[\frac{\partial(x^1, x^2, x^3)}{\partial(\overline{x}^1, \overline{x}^2, \overline{x}^3)}\right]^{-1}, \tag{3.285}$$

is a scalar density of weight -1. The volume integral of a scalar function $\rho(\overline{x}^1, \overline{x}^2, \overline{x}^3)$ now transforms as

$$\iiint \rho(x^1, x^2, x^3)\, dx^1\, dx^2\, dx^2 = \iiint \rho(\overline{x}^1, \overline{x}^2, \overline{x}^3) \frac{\partial(x^1, x^2, x^3)}{\partial(\overline{x}^1, \overline{x}^2, \overline{x}^3)}\, d\overline{x}^1\, d\overline{x}^2\, d\overline{x}^3. \tag{3.286}$$

In **orthogonal** generalized coordinates, the volume element is given as

$$d\tau' = \left|\vec{e}_1\, d\overline{x}^1\right|\left|\vec{e}_2\, d\overline{x}^2\right|\left|\vec{e}_3\, d\overline{x}^3\right| \tag{3.287}$$

$$= \left|\vec{e}_1\right|\left|\vec{e}_2\right|\left|\vec{e}_3\right| d\overline{x}^1\, d\overline{x}^2\, d\overline{x}^3, \tag{3.288}$$

$$d\tau' = \sqrt{g_{11}}\sqrt{g_{22}}\sqrt{g_{33}} d\overline{x}^1 d\overline{x}^2 d\overline{x}^3. \tag{3.289}$$

3.4.11   Invariance and Covariance

We have seen that scalars preserve their value under general coordinate transformations that do not involve scale changes. Magnitudes of vectors and the traces of second-rank tensors are also other properties that do not change under such coordinate transformations. Properties that preserve their value under coordinate transformations are called **invariants**. Identification of invariants in nature is very important in understanding and developing new physical theories.

An important property of tensors is that tensor equations **preserve** their **form** under coordinate transformations. For example, a tensor equation given as

$$F_{ijk} = D^{lm} E_{lmijk} + A_l B^l_{ijk} + E_i G_{jk} + \cdots , \qquad (3.290)$$

transforms into

$$F'_{ijk} = D'^{lm} E'_{lmijk} + A'_l B'^l_{ijk} + E'_i G'_{jk} + \cdots . \qquad (3.291)$$

Even though the components of the individual tensors in a tensor equation change, the tensor equation itself preserves its form. This useful property is called **covariance**. Since the true laws of nature should not depend on the coordinate system we use, it should be possible to express them in coordinate independent formalism. In this regard, tensor calculus plays a very significant role in physics. In particular, it reaches its full potential with Einstein's special theory of relativity and the general theory of relativity [1, 2].

3.5   DIFFERENTIAL OPERATORS IN GENERALIZED COORDINATES

3.5.1   Gradient

We first write the differential of a scalar function $\Phi(\overline{x}^i)$ as

$$d\Phi = \frac{\partial \Phi}{\partial \overline{x}^1} \, d\overline{x}^1 + \frac{\partial \Phi}{\partial \overline{x}^2} \, d\overline{x}^2 + \frac{\partial \Phi}{\partial \overline{x}^3} \, d\overline{x}^3. \qquad (3.292)$$

Using the displacement vector written in terms of the generalized coordinates and the basis vectors $\overrightarrow{e}_i$ as

$$d\overrightarrow{r} = d\overline{x}^1 \overrightarrow{e}_1 + d\overline{x}^2 \overrightarrow{e}_2 + d\overline{x}^3 \overrightarrow{e}_3, \qquad (3.293)$$

we rewrite $d\Phi$ as

$$d\Phi = \overrightarrow{\nabla} \Phi \cdot d\overrightarrow{r} \qquad (3.294)$$

to get

$$d\Phi = \overrightarrow{\nabla}\Phi \cdot (d\overline{x}^1 \overrightarrow{e}_1 + d\overline{x}^2 \overrightarrow{e}_2 + d\overline{x}^3 \overrightarrow{e}_3) \tag{3.295}$$

$$= (\overrightarrow{\nabla}\Phi \cdot \overrightarrow{e}_1)d\overline{x}^1 + (\overrightarrow{\nabla}\Phi \cdot \overrightarrow{e}_2)d\overline{x}^2 + (\overrightarrow{\nabla}\Phi \cdot \overrightarrow{e}_3)d\overline{x}^3. \tag{3.296}$$

Comparing with Eq. (3.220), this gives the covariant components of the gradient in terms of the generalized coordinates as

$$\frac{\partial \Phi}{\partial \overline{x}^i} = \overrightarrow{\nabla}\Phi \cdot \overrightarrow{e}_i. \tag{3.297}$$

In orthogonal generalized coordinates, where the unit basis vectors are defined as

$$\widehat{e}_1 = \frac{\overrightarrow{e}_1}{|\overrightarrow{e}_1|} = \frac{\overrightarrow{e}_1}{\sqrt{g_{11}}}, \quad \widehat{e}_2 = \frac{\overrightarrow{e}_2}{\sqrt{g_{22}}}, \quad \widehat{e}_3 = \frac{\overrightarrow{e}_3}{\sqrt{g_{33}}}, \tag{3.298}$$

Equation (3.297) gives the gradient in terms of the generalized coordinates and their unit basis vectors:

$$\boxed{\overrightarrow{\nabla}\Phi = \frac{1}{\sqrt{g_{11}}}\frac{\partial \Phi}{\partial \overline{x}^1}\widehat{e}_1 + \frac{1}{\sqrt{g_{22}}}\frac{\partial \Phi}{\partial \overline{x}^2}\widehat{e}_2 + \frac{1}{\sqrt{g_{33}}}\frac{\partial \Phi}{\partial \overline{x}^3}\widehat{e}_3.} \tag{3.299}$$

3.5.2 Divergence

To obtain the divergence operator in generalized coordinates, we use the integral definition [Eq. (2.147)]:

$$\overrightarrow{\nabla} \cdot \overrightarrow{A} = \lim_{S,\Delta V \to 0} \frac{\oint_S d\overrightarrow{\sigma} \cdot \overrightarrow{A}}{\int_{\Delta V} d\tau}, \tag{3.300}$$

where S is a closed surface enclosing a small volume ΔV. We confine ourselves to orthogonal generalized coordinates so that the denominator can be taken as

$$\int_{\Delta V} d\tau' = \sqrt{g_{11}}\sqrt{g_{22}}\sqrt{g_{33}}\Delta\overline{x}^1\Delta\overline{x}^2\Delta\overline{x}^3. \tag{3.301}$$

For the numerator, we consider the flux through the closed surface enclosing the volume element shown in Figure 3.12. We first find the fluxes through the top and the bottom surfaces. We chose the location of the volume element such that the bottom surface is centered at $P_1 = (\overline{x}_0^1, \overline{x}_0^2, 0)$ and the top surface is centered at $P_2 = (\overline{x}_0^1, \overline{x}_0^2, \Delta\overline{x}^3)$. We write the flux through the bottom surface as

$$\int_{\text{bottom}} d\overrightarrow{\sigma} \cdot \overrightarrow{A} = -\sqrt{g_{11}}\Delta\overline{x}^1\sqrt{g_{22}}\overline{A}_3\Delta\overline{x}^2\Big|_{(\overline{x}_0^1, \overline{x}_0^2, 0)} \tag{3.302}$$

$$= -\sqrt{g_{11}}\sqrt{g_{22}}\overline{A}_3\Big|_{(\overline{x}_0^1, \overline{x}_0^2, 0)}\Delta\overline{x}^1\Delta\overline{x}^2, \tag{3.303}$$

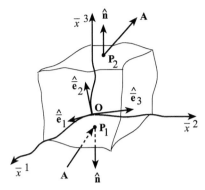

Figure 3.12   Volume element used in the derivation of the divergence operator.

where we used $\overline{\overline{A}}_3$ for the component of $\overrightarrow{A}$ along the unit basis vector $\widehat{\overline{e}}_3$, that is,

$$\overline{\overline{A}}_3 = \overrightarrow{A} \cdot \widehat{\overline{e}}_3. \tag{3.304}$$

The minus sign is due to the fact that $\overline{\overline{A}}_3$ and the normal to the surface are in opposite directions. Note that $\sqrt{g_{11}}$, $\sqrt{g_{22}}$, and $\overline{\overline{A}}_3$ are all functions of position. For the flux over the top surface, we write

$$\int_{\text{top}} d\overrightarrow{\sigma} \cdot \overrightarrow{A} = \sqrt{g_{11}}\Delta\overline{x}^1 \sqrt{g_{22}}\overline{\overline{A}}_3\Delta\overline{x}^2 \Big|_{(\overline{x}_0^1,\overline{x}_0^2,\Delta\overline{x}^3)} \tag{3.305}$$

$$= \sqrt{g_{11}}\sqrt{g_{22}}\overline{\overline{A}}_3 \Big|_{(\overline{x}_0^1,\overline{x}_0^2,\Delta\overline{x}^3)} \Delta\overline{x}^1\Delta\overline{x}^2. \tag{3.306}$$

We now have a plus sign, since $\overline{\overline{A}}_3$ and the normal are in the same direction. Since we use orthogonal generalized coordinates, the other components of $\overrightarrow{A}$ do not contribute to the flux through these surfaces. Since the right-hand side of Eq. (3.306) is to be evaluated at $(\overline{x}_0^1, \overline{x}_0^2, \Delta\overline{x}^3)$, we expand $(\sqrt{g_{11}}\sqrt{g_{22}}\overline{\overline{A}}_3)$ in Taylor series about $(\overline{x}_0^1, \overline{x}_0^2, 0)$ and keep only the first-order terms:

$$\sqrt{g_{11}}\sqrt{g_{22}}\overline{\overline{A}}_3 \Big|_{\overline{x}^1,\overline{x}^2,\Delta\overline{x}^3} = \sqrt{g_{11}}\sqrt{g_{22}}\overline{\overline{A}}_3 \Big|_{(\overline{x}_0^1,\overline{x}_0^2,0)}$$

$$+ \frac{\partial(\sqrt{g_{11}}\sqrt{g_{22}}\overline{\overline{A}}_3)}{\partial\overline{x}^3} \Big|_{(\overline{x}_0^1,\overline{x}_0^2,0)} \Delta\overline{x}^3. \tag{3.307}$$

Substituting this into Eq. (3.306), we obtain

$$\int_{\text{top}} d\overrightarrow{\sigma} \cdot \overrightarrow{A} = \sqrt{g_{11}}\sqrt{g_{22}}\overline{\overline{A}}_3 \Big|_{(\overline{x}_0^1,\overline{x}_0^2,0)} \Delta\overline{x}^1\Delta\overline{x}^2$$

$$+ \frac{\partial(\sqrt{g_{11}}\sqrt{g_{22}}\overline{\overline{A}}_3)}{\partial\overline{x}^3} \Big|_{(\overline{x}_0^1,\overline{x}_0^2,0)} \Delta\overline{x}^1\Delta\overline{x}^2\Delta\overline{x}^3. \tag{3.308}$$

Since the location of the volume element is arbitrary, we drop the subscripts and write the net flux through the top and the bottom surfaces as

$$\int_{\text{top+bottom}} d\vec{\sigma} \cdot \vec{A} = \frac{\partial(\sqrt{g_{11}}\sqrt{g_{22}}\overline{\overline{A}}_3)}{\partial \overline{x}^3} \Delta \overline{x}^1 \Delta \overline{x}^2 \Delta \overline{x}^3. \tag{3.309}$$

Similar terms are written for the other two pairs of surfaces, giving

$$\oint_S d\vec{\sigma} \cdot \vec{A} = \left[\frac{\partial(\sqrt{g_{22}}\sqrt{g_{33}}\overline{\overline{A}}_1)}{\partial \overline{x}^1} + \frac{\partial(\sqrt{g_{11}}\sqrt{g_{33}}\overline{\overline{A}}_2)}{\partial \overline{x}^2} \right.$$
$$\left. + \frac{\partial(\sqrt{g_{11}}\sqrt{g_{22}}\overline{\overline{A}}_3)}{\partial \overline{x}^3} \right] \Delta \overline{x}^1 \Delta \overline{x}^2 \Delta \overline{x}^3. \tag{3.310}$$

Substituting this into Eq. (3.300) with Eq. (3.301), we obtain the **divergence** in orthogonal generalized coordinates as

$$\vec{\nabla} \cdot \vec{A} = \frac{1}{\sqrt{g_{11}g_{22}g_{33}}} \left[\frac{\partial(\sqrt{g_{22}}\sqrt{g_{33}}\overline{\overline{A}}_1)}{\partial \overline{x}^1} + \frac{\partial(\sqrt{g_{11}}\sqrt{g_{33}}\overline{\overline{A}}_2)}{\partial \overline{x}^2} \right.$$
$$\left. + \frac{\partial(\sqrt{g_{11}}\sqrt{g_{22}}\overline{\overline{A}}_3)}{\partial \overline{x}^3} \right]. \tag{3.311}$$

3.5.3 Curl

We now find the curl operator in orthogonal generalized coordinates by using the integral definition [Eq. (2.277)]

$$(\vec{\nabla} \times \vec{A}) \cdot \lim_{\Delta S \to 0} \int_{\Delta S} d\vec{\sigma} = \lim_{C \to 0} \oint_C \vec{A} \cdot d\vec{r}, \tag{3.312}$$

where C is a small closed path bounding the oriented surface ΔS. The outward normal to $d\vec{\sigma}$ is found by the right-hand rule. We pick a single component of $\vec{\nabla} \times \vec{A}$ by pointing $d\vec{\sigma}$ in the desired direction, say $\hat{\vec{e}}_1$. In Figure 3.13, we show the outward unit normal $\hat{n}$ found by the right-hand rule, pointing in the direction of $\overline{x}^1$, that is, $\hat{n} = \hat{\vec{e}}_1$. We now write the complete line integral over C as

$$\oint_C \vec{A} \cdot d\vec{r} = \int_{C_a + C_b + C_c + C_d} \vec{A} \cdot d\vec{r} \tag{3.313}$$
$$= \int_{C_a + C_b + C_c + C_d} (\vec{A} \cdot \hat{\vec{e}}_1) d\overline{x}^1 + (\vec{A} \cdot \hat{\vec{e}}_2) d\overline{x}^2 + (\vec{A} \cdot \hat{\vec{e}}_3) d\overline{x}^3, \tag{3.314}$$

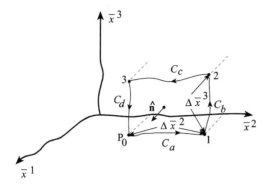

Figure 3.13 Closed path used in the definition of curl, where $\Delta \bar{x}^2$ and $\Delta \bar{x}^3$ represent the change in coordinates between the indicated points.

where we have used Eq. (3.224). We first consider the segments C_a and C_c. Along C_a, we write

$$\oint_{C_a} \vec{A} \cdot d\vec{r} = \int_{(\bar{x}_0^1, \bar{x}_0^2, \bar{x}_0^3)}^{(\bar{x}_0^1, \bar{x}_0^2 + \Delta \bar{x}^2, \bar{x}_0^3)} \sqrt{g_{22}} \overline{\overline{A}}_2 \, d\bar{x}^2, \tag{3.315}$$

where $\overline{\overline{A}}_2 = \vec{A} \cdot \hat{\bar{e}}_2$. We now write the Taylor series expansion of $\sqrt{g_{22}} \overline{\overline{A}}_2$ about $P_0 = (\bar{x}_0^1, \bar{x}_0^2, \bar{x}_0^3)$ with only the linear terms:

$$\sqrt{g_{22}} \overline{\overline{A}}_2 = \sqrt{g_{22}} \overline{\overline{A}}_2 \Big|_{(\bar{x}_0^1, \bar{x}_0^2, \bar{x}_0^3)} + \frac{\partial (\sqrt{g_{22}} \overline{\overline{A}}_2)}{\partial \bar{x}^1} \Big|_{(\bar{x}_0^1, \bar{x}_0^2, \bar{x}_0^3)} (\bar{x}^1 - \bar{x}_0^1)$$

$$+ \frac{\partial (\sqrt{g_{22}} \overline{\overline{A}}_2)}{\partial \bar{x}^2} \Big|_{(\bar{x}_0^1, \bar{x}_0^2, \bar{x}_0^3)} (\bar{x}^2 - \bar{x}_0^2)$$

$$+ \frac{\partial (\sqrt{g_{22}} \overline{\overline{A}}_2)}{\partial \bar{x}^3} \Big|_{(\bar{x}_0^1, \bar{x}_0^2, \bar{x}_0^3)} (\bar{x}^3 - \bar{x}_0^3). \tag{3.316}$$

Along C_a, we have $\bar{x}^3 = \bar{x}_0^3$ and $\bar{x}^1 = \bar{x}_0^1$; hence, Eq. (3.315) becomes

$$\oint_{C_a} \vec{A} \cdot d\vec{r} = \int_{(\bar{x}_0^1, \bar{x}_0^2, \bar{x}_0^3)}^{(\bar{x}_0^1, \bar{x}_0^2 + \Delta \bar{x}^2, \bar{x}_0^3)} \left[\sqrt{g_{22}} \overline{\overline{A}}_2 \Big|_{(\bar{x}_0^1, \bar{x}_0^2, \bar{x}_0^3)} \right.$$

$$\left. + \frac{\partial (\sqrt{g_{22}} \overline{\overline{A}}_2)}{\partial \bar{x}^2} \Big|_{(\bar{x}_0^1, \bar{x}_0^2, \bar{x}_0^3)} (\bar{x}^2 - \bar{x}_0^2) \right] d\bar{x}^2$$

$$= \sqrt{g_{22}} \overline{\overline{A}}_2 \Big|_{(\bar{x}_0^1, \bar{x}_0^2, \bar{x}_0^3)} \Delta \bar{x}^2 + \frac{\partial (\sqrt{g_{22}} \overline{\overline{A}}_2)}{\partial \bar{x}^2} \Big|_{(\bar{x}_0^1, \bar{x}_0^2, \bar{x}_0^3)} \frac{(\Delta \bar{x}^2)^2}{2}. \tag{3.317}$$

Next, to evaluate $\oint_{C_c} \overrightarrow{A} \cdot d\overrightarrow{r}$, we write

$$\oint_{C_c} \overrightarrow{A} \cdot d\overrightarrow{r} = \int_{(\overline{x}_0^1, \overline{x}_0^2 + \Delta\overline{x}^2, \overline{x}_0^3 + \Delta\overline{x}^3)}^{(\overline{x}_0^1, \overline{x}_0^2, \overline{x}_0^3 + \Delta\overline{x}^3)} \sqrt{g_{22}}\overline{\overline{A}}_2 \, d\overline{x}^2. \qquad (3.318)$$

We again use the Taylor series expansion [Eq.(3.316)] of $\sqrt{g_{22}}\overline{\overline{A}}_2$ about $P_0 = (\overline{x}_0^1, \overline{x}_0^2, \overline{x}_0^3)$ with only the linear terms. Along the path C_c, we have $\overline{x}^3 = \overline{x}_0^3 + \Delta\overline{x}_0^3$ and $\overline{x}^1 = \overline{x}_0^1$, which gives

$$\sqrt{g_{22}}\overline{\overline{A}}_2 = \sqrt{g_{22}}\overline{\overline{A}}_2\Big|_{(\overline{x}_0^1, \overline{x}_0^2, \overline{x}_0^3)} + \frac{\partial(\sqrt{g_{22}}\overline{\overline{A}}_2)}{\partial \overline{x}^2}\Big|_{(\overline{x}_0^1, \overline{x}_0^2, \overline{x}_0^3)} (\overline{x}^2 - \overline{x}_0^2)$$

$$+ \frac{\partial(\sqrt{g_{22}}\overline{\overline{A}}_2)}{\partial \overline{x}^3}\Big|_{(\overline{x}_0^1, \overline{x}_0^2, \overline{x}_0^3)} \Delta\overline{x}^3. \qquad (3.319)$$

Using this in Eq. (3.318), we obtain

$$\oint_{C_c} \overrightarrow{A} \cdot d\overrightarrow{r} = -\Bigg\{ \int_{(\overline{x}_0^1, \overline{x}_0^2, \overline{x}_0^3 + \Delta\overline{x}^3)}^{(\overline{x}_0^1, \overline{x}_0^2 + \Delta\overline{x}^2, \overline{x}_0^3 + \Delta\overline{x}^3)} \Big[\sqrt{g_{22}}\overline{\overline{A}}_2\Big|_{(\overline{x}_0^1, \overline{x}_0^2, \overline{x}_0^3)}$$

$$+ \frac{\partial(\sqrt{g_{22}}\overline{\overline{A}}_2)}{\partial \overline{x}^2}\Big|_{(\overline{x}_0^1, \overline{x}_0^2, \overline{x}_0^3)} (\overline{x}^2 - \overline{x}_0^2) + \frac{\partial(\sqrt{g_{22}}\overline{\overline{A}}_2)}{\partial \overline{x}^3}\Big|_{(\overline{x}_0^1, \overline{x}_0^2, \overline{x}_0^3)} \Delta\overline{x}^3 \Big] d\overline{x}^2 \Bigg\}$$

$$\qquad (3.320)$$

$$= -\Bigg\{ \sqrt{g_{22}}\overline{\overline{A}}_2\Big|_{(\overline{x}_0^1, \overline{x}_0^2, \overline{x}_0^3)} \Delta\overline{x}^2 + \frac{\partial(\sqrt{g_{22}}\overline{\overline{A}}_2)}{\partial \overline{x}^2}\Big|_{(\overline{x}_0^1, \overline{x}_0^2, \overline{x}_0^3)} \frac{(\Delta\overline{x}^2)^2}{2}$$

$$+ \frac{\partial(\sqrt{g_{22}}\overline{\overline{A}}_2)}{\partial \overline{x}^3}\Big|_{(\overline{x}_0^1, \overline{x}_0^2, \overline{x}_0^3)} \Delta\overline{x}^3 \Delta\overline{x}^2 \Bigg\}. \qquad (3.321)$$

This allows us to combine the integrals in Eqs. (3.317) and (3.321) to yield

$$\oint_{C_a + C_c} \overrightarrow{A} \cdot d\overrightarrow{r} = -\frac{\partial(\sqrt{g_{22}}\overline{\overline{A}}_2)}{\partial \overline{x}^3}\Big|_{(\overline{x}_0^1, \overline{x}_0^2, \overline{x}_0^3)} \Delta\overline{x}^2 \Delta\overline{x}^3. \qquad (3.322)$$

Since our choice of the point $(\overline{x}_0^1, \overline{x}_0^2, \overline{x}_0^3)$ is arbitrary, we write this for a general point as

$$\oint_{C_a + C_c} \overrightarrow{A} \cdot d\overrightarrow{r} = -\frac{\partial(\sqrt{g_{22}}\overline{\overline{A}}_2)}{\partial \overline{x}^3} \Delta\overline{x}^2 \Delta\overline{x}^3. \qquad (3.323)$$

A similar equation will be obtained from the other two segments as

$$\oint_{C_b+C_d} \vec{A} \cdot d\vec{r} = \frac{\partial(\sqrt{g_{33}}\overline{\overline{A}}_3)}{\partial \bar{x}^2} \Delta \bar{x}^2 \Delta \bar{x}^3. \tag{3.324}$$

Addition of Eqs. (3.323) and (3.324) yields

$$\oint_C \vec{A} \cdot d\vec{r} = \left[\frac{\partial(\sqrt{g_{33}}\overline{\overline{A}}_3)}{\partial \bar{x}^2} - \frac{\partial(\sqrt{g_{22}}\overline{\overline{A}}_2)}{\partial \bar{x}^3} \right] \Delta \bar{x}^2 \Delta \bar{x}^3. \tag{3.325}$$

Using this result in Eq. (3.312) gives the component of $\vec{\nabla} \times \vec{A}$ in the direction of $\widehat{e}_1$ as

$$(\vec{\nabla} \times \vec{A})_1 = \frac{\left[\frac{\partial(\sqrt{g_{33}}\overline{\overline{A}}_3)}{\partial \bar{x}^2} - \frac{\partial(\sqrt{g_{22}}\overline{\overline{A}}_2)}{\partial \bar{x}^3} \right] \Delta \bar{x}^2 \Delta \bar{x}^3}{\sqrt{g_{22}}\sqrt{g_{33}}\Delta \bar{x}^2 \Delta \bar{x}^3} \tag{3.326}$$

$$= \frac{1}{\sqrt{g_{22}}\sqrt{g_{33}}} \left[\frac{\partial(\sqrt{g_{33}}\overline{\overline{A}}_3)}{\partial \bar{x}^2} - \frac{\partial(\sqrt{g_{22}}\overline{\overline{A}}_2)}{\partial \bar{x}^3} \right]. \tag{3.327}$$

A similar procedure yields the other two components as

$$(\vec{\nabla} \times \vec{A})_2 = \frac{1}{\sqrt{g_{11}}\sqrt{g_{33}}} \left[\frac{\partial(\sqrt{g_{11}}\overline{\overline{A}}_1)}{\partial \bar{x}^3} - \frac{\partial(\sqrt{g_{33}}\overline{\overline{A}}_3)}{\partial \bar{x}^1} \right] \tag{3.328}$$

and

$$(\vec{\nabla} \times \vec{A})_3 = \frac{1}{\sqrt{g_{11}}\sqrt{g_{22}}} \left[\frac{\partial(\sqrt{g_{22}}\overline{\overline{A}}_2)}{\partial \bar{x}^1} - \frac{\partial(\sqrt{g_{11}}\overline{\overline{A}}_1)}{\partial \bar{x}^2} \right]. \tag{3.329}$$

The final expression for the **curl of a vector** field in orthogonal generalized coordinates can now be given as

$$\vec{\nabla} \times \vec{A} = \frac{1}{\sqrt{g_{22}}\sqrt{g_{33}}} \left[\frac{\partial(\sqrt{g_{33}}\overline{\overline{A}}_3)}{\partial \bar{x}^2} - \frac{\partial(\sqrt{g_{22}}\overline{\overline{A}}_2)}{\partial \bar{x}^3} \right] \widehat{e}_1$$

$$+ \frac{1}{\sqrt{g_{11}}\sqrt{g_{33}}} \left[\frac{\partial(\sqrt{g_{11}}\overline{\overline{A}}_1)}{\partial \bar{x}^3} - \frac{\partial(\sqrt{g_{33}}\overline{\overline{A}}_3)}{\partial \bar{x}^1} \right] \widehat{e}_2$$

$$+ \frac{1}{\sqrt{g_{11}}\sqrt{g_{22}}} \left[\frac{\partial(\sqrt{g_{22}}\overline{\overline{A}}_2)}{\partial \bar{x}^1} - \frac{\partial(\sqrt{g_{11}}\overline{\overline{A}}_1)}{\partial \bar{x}^2} \right] \widehat{e}_3, \tag{3.330}$$

which can also be expressed conveniently as

$$
\vec{\nabla} \times \vec{A} = \frac{1}{\sqrt{g_{11}} \sqrt{g_{22}} \sqrt{g_{33}}} \det \begin{pmatrix} \sqrt{g_{11}}\,\widehat{e}_1 & \sqrt{g_{22}}\,\widehat{e}_2 & \sqrt{g_{33}}\,\widehat{e}_3 \\ \dfrac{\partial}{\partial \overline{x}^1} & \dfrac{\partial}{\partial \overline{x}^2} & \dfrac{\partial}{\partial \overline{x}^3} \\ \sqrt{g_{11}}\,\overline{A}_1 & \sqrt{g_{22}}\,\overline{A}_2 & \sqrt{g_{33}}\,\overline{A}_3 \end{pmatrix}. \tag{3.331}
$$

3.5.4   Laplacian

Using the results for the gradient and the divergence operators [Eqs. (3.299) and (3.311)], we write the **Laplacian**, $\vec{\nabla}^2 = \vec{\nabla} \cdot \vec{\nabla}\Phi$, for orthogonal generalized coordinates as

$$
\vec{\nabla} \cdot \vec{\nabla}\Phi = \frac{1}{\sqrt{g_{11}} \sqrt{g_{22}} \sqrt{g_{33}}} \left[\frac{\partial}{\partial \overline{x}^1} \left(\frac{\sqrt{g_{22}} \sqrt{g_{33}}}{\sqrt{g_{11}}} \frac{\partial \Phi}{\partial \overline{x}^1} \right) \right.
$$
$$
\left. + \frac{\partial}{\partial \overline{x}^2} \left(\frac{\sqrt{g_{11}} \sqrt{g_{33}}}{\sqrt{g_{22}}} \frac{\partial \Phi}{\partial \overline{x}^2} \right) + \frac{\partial}{\partial \overline{x}^3} \left(\frac{\sqrt{g_{11}} \sqrt{g_{22}}}{\sqrt{g_{33}}} \frac{\partial \Phi}{\partial \overline{x}^3} \right) \right]. \tag{3.332}
$$

3.6   ORTHOGONAL GENERALIZED COORDINATES

The general formalism we have developed in the previous sections can be used to define new coordinate systems and to study their properties. Depending on the symmetries of a given system, certain coordinate systems may prove to be a lot easier to work through the mathematics. In this regard, many different coordinate systems have been designed. To name a few, Cartesian, cylindrical, spherical, paraboloidal, elliptic, toroidal, bipolar, and oblate spherical coordinate systems can be given. Among these, Cartesian, cylindrical, and spherical coordinate systems are the most frequently used ones, which we are going to discuss in detail.

Historically, the Cartesian coordinate system is the oldest and was introduced by Descartes in 1637. He labeled the coordinate axes as (x, y, z):

$$
\boxed{x^1 = x, \quad x^2 = y, \quad x^3 = z} \tag{3.333}
$$

and used $\widehat{i}, \widehat{j}, \widehat{k}$ for the unit basis vectors:

$$
\widehat{e}_1 = \widehat{i}, \quad \widehat{e}_2 = \widehat{j}, \quad \widehat{e}_3 = \widehat{k}. \tag{3.334}
$$

In Cartesian coordinates, motion of a particle is described by the radius or the position vector $\vec{r}(t)$ as

$$
\vec{r}(t) = x(t)\widehat{i} + y(t)\widehat{j} + z(t)\widehat{k}, \tag{3.335}
$$

where the parameter t is usually the time. Velocity $\overrightarrow{v}(t)$ and the acceleration $\overrightarrow{a}(t)$ are given as the first and the second derivatives of $\overrightarrow{r}(t)$:

$$\overrightarrow{v}(t) = \frac{d\overrightarrow{r}(t)}{dt} = \dot{x}\,\widehat{i} + \dot{y}\,\widehat{j} + \dot{z}\,\widehat{k}, \tag{3.336}$$

$$\overrightarrow{a}(t) = \frac{d\overrightarrow{v}}{dt} = \frac{d^2\overrightarrow{r}}{dt^2} = \ddot{x}\,\widehat{i} + \ddot{y}\,\widehat{j} + \ddot{z}\,\widehat{k}. \tag{3.337}$$

Example 3.3. *Circular motion:* Motion of a particle executing circular motion can be described by the following parametric equations:

$$x(t) = a_0 \cos \omega t, \quad y(t) = a_0 \sin \omega t, \quad z(t) = z_0. \tag{3.338}$$

Using the radius vector:

$$\overrightarrow{r}(t) = a_0 \cos \omega t\,\widehat{i} + a_0 \sin \omega t\,\widehat{j} + z_0\,\widehat{k}, \tag{3.339}$$

we write the velocity $\overrightarrow{v}(t)$:

$$\overrightarrow{v}(t) = -a_0\omega \sin \omega t\,\widehat{i} + a_0\omega \cos \omega t\,\widehat{j}, \tag{3.340}$$

and the acceleration $\overrightarrow{a}(t)$ as

$$\overrightarrow{a}(t) = -a_0\omega^2 \cos \omega t\,\widehat{i} - a_0\omega^2 \sin \omega t\,\widehat{j} = -\omega^2\overrightarrow{r}(t). \tag{3.341}$$

3.6.1 Cylindrical Coordinates

Cylindrical coordinates are defined by

$$\boxed{\overline{x}^1 = \rho, \quad \overline{x}^2 = \phi, \quad \overline{x}^3 = z.} \tag{3.342}$$

They are related to the Cartesian coordinates by the **transformation equations** (Figure 3.14)

$$\boxed{x = \rho \cos \phi, y = \rho \sin \phi, z = z,} \tag{3.343}$$

where $\rho \in [0, \infty]$, $\phi \in [0, 2\pi]$, $z \in [0, \infty]$. The **inverse** transformation equations are written as

$$\rho = \sqrt{x^2 + y^2}, \tag{3.344}$$

$$\phi = \tan^{-1}\left(\frac{y}{x}\right), \tag{3.345}$$

$$z = z. \tag{3.346}$$

We find the basis vectors [Eq. (3.189)] $\vec{e}_i = \frac{\partial x^k}{\partial \bar{x}^i}$ as

$$\vec{e}_1 = \vec{e}_\rho = \cos\phi\,\hat{i} + \sin\phi\,\hat{j}, \quad |\vec{e}_\rho| = 1, \tag{3.347}$$

$$\vec{e}_2 = \vec{e}_\phi = -\rho\sin\phi\,\hat{i} + \rho\cos\phi\,\hat{j}, \quad |\vec{e}_\phi| = \rho, \tag{3.348}$$

$$\vec{e}_3 = \vec{e}_z = \hat{k}, \quad |\vec{e}_z| = 1. \tag{3.349}$$

The unit basis vectors are now written as

$$\hat{e}_1 = \hat{e}_\rho = \cos\phi\,\hat{i} + \sin\phi\,\hat{j}, \tag{3.350}$$

$$\hat{e}_2 = \hat{e}_\phi = -\sin\phi\,\hat{i} + \cos\phi\,\hat{j}, \tag{3.351}$$

$$\hat{e}_3 = \hat{e}_z = \hat{k}. \tag{3.352}$$

It is easy to check that the basis vectors are mutually orthogonal Figure 3.14:

$$\begin{aligned}
\hat{e}_\rho &= \hat{e}_\phi \times \hat{e}_z, \\
\hat{e}_\phi &= \hat{e}_z \times \hat{e}_\rho, \\
\hat{e}_z &= \hat{e}_\rho \times \hat{e}_\phi.
\end{aligned} \tag{3.353}$$

It is important to note that the basis vectors $\vec{e}_i$ are mutually orthogonal; however, their direction and magnitude depends on position. We now write

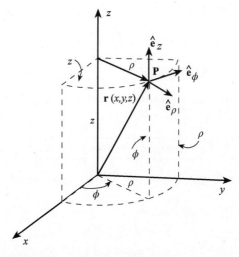

Figure 3.14 Cylindrical coordinates (ρ, ϕ, z): Coordinate surfaces for ρ, ϕ, z and the unit basis vectors.

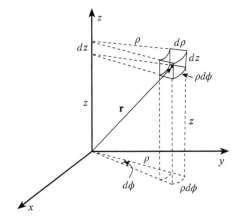

Figure 3.15   Infinitesimal displacements in cylindrical coordinates.

the position vector for the point $P(\rho, \phi, z)$ and the infinitesimal displacement
vector $d\vec{r}$ [Eq. (3.200)] in cylindrical coordinates as (Figures 3.14 and 3.15)

$$\vec{r} = \rho \cos \phi \, \hat{i} + \rho \sin \phi \, \hat{j} + z\hat{k}, \tag{3.354}$$

$$d\vec{r} = d\rho \, \vec{e}_\rho + d\phi \, \vec{e}_\phi + dz \, \vec{e}_z. \tag{3.355}$$

From the **line element**:

$$ds^2 = d\vec{r} \cdot d\vec{r} = d\rho^2(\vec{e}_\rho \cdot \vec{e}_\rho) + d\phi^2(\vec{e}_\phi \cdot \vec{e}_\phi) + dz^2(\vec{e}_z \cdot \vec{e}_z) \tag{3.356}$$

$$= g_{\rho\rho} \, d\rho^2 + g_{\phi\phi} \, d\phi^2 + g_{zz} \, dz^2, \tag{3.357}$$

$$\boxed{ds^2 = d\rho^2 + \rho^2 d\phi^2 + dz^2,} \tag{3.358}$$

we obtain the **metric tensor**:

$$g_{ij} = \vec{e}_i \cdot \vec{e}_j = \begin{pmatrix} 1 & 0 & 0 \\ 0 & \rho^2 & 0 \\ 0 & 0 & 1 \end{pmatrix}. \tag{3.359}$$

We can construct the **contravariant** metric tensor g^{ij} by using the inverse
basis vectors [Eq. (3.193)]:

$$\vec{e}^{\,i} = \frac{\partial \overline{x}^i}{\partial x^k} = \frac{\partial \overline{x}^i}{\partial x^1}\hat{e}_1 + \frac{\partial \overline{x}^i}{\partial x^2}\hat{e}_2 + \frac{\partial \overline{x}^i}{\partial x^3}\hat{e}_3 \tag{3.360}$$

$$= \frac{\partial \overline{x}^i}{\partial x}\hat{i} + \frac{\partial \overline{x}^i}{\partial y}\hat{j} + \frac{\partial \overline{x}^i}{\partial z}\hat{k}, \tag{3.361}$$

which are found by using the inverse transformation equations [Eqs. (3.344)–(3.346)] as

$$\vec{e}^{\,1} = \vec{e}^{\,\rho} = \frac{\partial \rho}{\partial x}\widehat{i} + \frac{\partial \rho}{\partial y}\widehat{j} + \frac{\partial \rho}{\partial z}\widehat{k} = \frac{x}{(x^2+y^2)^{1/2}}\widehat{i} + \frac{y}{(x^2+y^2)^{1/2}}\widehat{j} + 0\widehat{k} \quad (3.362)$$

$$= \frac{x}{\rho}\widehat{i} + \frac{y}{\rho}\widehat{j} = \cos\phi\,\widehat{i} + \sin\phi\,\widehat{j}, \quad (3.363)$$

$$\vec{e}^{\,2} = \vec{e}^{\,\phi} = \frac{\partial \phi}{\partial x}\widehat{i} + \frac{\partial \phi}{\partial y}\widehat{j} = \frac{-y}{(x^2+y^2)}\widehat{i} + \frac{x}{(x^2+y^2)}\widehat{j} \quad (3.364)$$

$$= \frac{1}{\rho}\left(-\frac{y}{\rho}\widehat{i} + \frac{x}{\rho}\widehat{j}\right) = \frac{1}{\rho}(-\sin\phi\,\widehat{i} + \cos\phi\,\widehat{j}), \quad (3.365)$$

$$\vec{e}^{\,3} = \vec{e}^{\,z} = \widehat{k}. \quad (3.366)$$

We can now write the **contravariant** components of the metric tensor as

$$g^{ij} = \vec{e}^{\,i} \cdot \vec{e}^{\,j} = \begin{pmatrix} 1 & 0 & 0 \\ 0 & \frac{1}{\rho^2} & 0 \\ 0 & 0 & 1 \end{pmatrix}. \quad (3.367)$$

Note that

$$\vec{e}^{\,\rho} = g^{\rho\rho}\vec{e}_{\,\rho}, \quad \vec{e}^{\,\phi} = g^{\phi\phi}\vec{e}_{\,\phi}, \quad \vec{e}^{\,z} = g^{zz}\vec{e}_{\,z}. \quad (3.368)$$

Line integrals in cylindrical coordinates are written as

$$I = \int \vec{A} \cdot d\vec{r} = \int \overline{A}_\rho\, d\rho + \overline{A}_\phi\, d\phi + \overline{A}_z\, dz, \quad (3.369)$$

where $\overline{A}_\rho = \vec{A} \cdot \vec{e}_\rho$, $\overline{A}_\phi = \vec{A} \cdot \vec{e}_\phi$, $\overline{A}_z = \vec{A} \cdot \vec{e}_z$. The area elements in cylindrical coordinates are given as (Figure 3.15)

$$d\sigma_{\rho\phi} = \rho d\rho\, d\phi, \quad d\sigma_{\rho z} = d\rho\, dz, d\sigma_{\phi z} = \rho d\phi dz, \quad (3.370)$$

while the **volume element** is

$$d\tau = d\rho(\rho d\phi)dz = \rho d\rho d\phi dz. \quad (3.371)$$

Applying our general results to cylindrical coordinates, we write the following differential operators:

Gradient [Eq. (3.299)]:

$$\vec{\nabla}\Phi(\rho, \phi, z) = \frac{\partial \Phi}{\partial \rho}\widehat{e}_\rho + \frac{1}{\rho}\frac{\partial \Phi}{\partial \phi}\widehat{e}_\phi + \frac{\partial \Phi}{\partial z}\widehat{k}. \quad (3.372)$$

Divergence [Eq. (3.311)]:

$$\vec{\nabla} \cdot \vec{A} = \frac{1}{\rho} \frac{\partial (\rho A_\rho)}{\partial \rho} + \frac{1}{\rho} \frac{\partial A_\phi}{\partial \phi} + \frac{\partial A_z}{\partial z}, \tag{3.373}$$

where $A_\rho = \vec{A} \cdot \hat{e}_\rho$, $A_\phi = \vec{A} \cdot \hat{e}_\phi$, $A_z = \vec{A} \cdot \hat{k}$.

Curl [Eq. (3.330)]:

$$\vec{\nabla} \times \vec{A} = \left[\frac{1}{\rho} \frac{\partial A_z}{\partial \phi} - \frac{\partial A_\phi}{\partial z} \right] \hat{e}_\rho + \left[\frac{\partial A_\rho}{\partial z} - \frac{\partial A_z}{\partial \rho} \right] \hat{e}_\phi$$

$$+ \frac{1}{\rho} \left[\frac{\partial (\rho A_\phi)}{\partial \rho} - \frac{\partial A_\rho}{\partial \phi} \right] \hat{k}, \tag{3.374}$$

where $A_\rho = \vec{A} \cdot \hat{e}_\rho$, $A_\phi = \vec{A} \cdot \hat{e}_\phi$, $A_z = \vec{A} \cdot \hat{k}$. Curl can also be conveniently expressed as

$$\vec{\nabla} \times \vec{A} = \frac{1}{\rho} \det \begin{pmatrix} \hat{e}_\rho & \rho\,\hat{e}_\phi & \hat{k} \\ \frac{\partial}{\partial \rho} & \frac{\partial}{\partial \phi} & \frac{\partial}{\partial z} \\ A_\rho & \rho A_\phi & A_z \end{pmatrix}. \tag{3.375}$$

Laplacian [Eq. (3.332)]:

$$\vec{\nabla}^2 \Phi(\rho, \phi, z) = \frac{1}{\rho} \frac{\partial}{\partial \rho} \left[\rho \frac{\partial \Phi}{\partial \rho} \right] + \frac{1}{\rho^2} \frac{\partial^2 \Phi}{\partial \phi^2} + \frac{\partial^2 \Phi}{\partial z^2}. \tag{3.376}$$

Example 3.4. *Acceleration in cylindrical coordinates:* In cylindrical coordinates the position vector is written as

$$\vec{r} = \rho \cos \phi\, \hat{i} + \rho \sin \phi\, \hat{j} + z\hat{k}. \tag{3.377}$$

Using the basis vectors $(\hat{e}_\rho, \hat{e}_\phi, \hat{k})$ [Eqs. (3.350)−(3.352)], we can also write this as

$$\vec{r} = (\vec{r} \cdot \hat{e}_\rho)\hat{e}_\rho + (\vec{r} \cdot \hat{e}_\phi)\hat{e}_\phi + (\vec{r} \cdot \hat{k})\hat{k} = \rho\hat{e}_\rho + z\hat{k}. \tag{3.378}$$

Since the basis vector $\hat{e}_\rho$ changes direction with position, velocity of a particle is written as

$$\vec{v} = \frac{d\vec{r}}{dt} = \dot{\rho}\hat{e}_\rho + \rho\dot{\hat{e}}_\rho + \dot{z}\hat{k}. \tag{3.379}$$

Using Eq. (3.350), we write the derivative of the basis vector, $\dot{\hat{e}}_\rho$, as

$$\dot{\hat{e}}_\rho = \dot{\phi}(-\sin \phi\, \hat{i} + \cos \phi\, \hat{j}) = \dot{\phi}\hat{e}_\phi, \tag{3.380}$$

thus obtaining the **velocity** vector:

$$\boxed{\vec{v} = \dot{\rho}\widehat{e}_\rho + \rho\dot{\phi}\widehat{e}_\phi + \dot{z}\widehat{k}.}$$ (3.381)

To write the acceleration, we also have to consider the change in the direction of $\widehat{e}_\phi$:

$$\vec{a} = \ddot{\rho}\widehat{e}_\rho + \dot{\rho}\widehat{\dot{e}}_\rho + \dot{\rho}\dot{\phi}\widehat{e}_\phi + \rho\ddot{\phi}\widehat{e}_\phi + \rho\dot{\phi}\widehat{\dot{e}}_\phi + \ddot{z}\widehat{k}.$$ (3.382)

Using Eq. (3.380) and

$$\widehat{\dot{e}}_\phi = \dot{\phi}(-\cos\phi\,\widehat{i} - \sin\phi\,\widehat{j}) = -\dot{\phi}\widehat{e}_\rho,$$ (3.383)

we finally write the **acceleration** vector as

$$\boxed{\vec{a} = (\ddot{\rho} - \rho\dot{\phi}^2)\widehat{e}_\rho + (\rho\ddot{\phi} + 2\dot{\rho}\dot{\phi})\widehat{e}_\phi + \ddot{z}\widehat{k}.}$$ (3.384)

3.6.2 Spherical Coordinates

Spherical coordinates (r, θ, ϕ) are related to the Cartesian coordinates by the transformation equations (Figure 3.16)

$$x = r\sin\theta\cos\phi,$$ (3.385)

$$y = r\sin\theta\sin\phi,$$ (3.386)

$$z = r\cos\theta,$$ (3.387)

where the ranges are given as $r \in [0, \infty]$, $\theta \in [0, \pi]$, $\phi \in [0, 2\pi]$. The inverse transformations are

$$r = \sqrt{x^2 + y^2 + z^2},$$ (3.388)

$$\theta = \tan^{-1}\left(\frac{\sqrt{x^2 + y^2}}{z}\right),$$ (3.389)

$$\phi = \tan^{-1}\left(\frac{y}{x}\right).$$ (3.390)

We now write the **radius vector** $\vec{r} = x\widehat{i} + y\widehat{j} + z\widehat{k}$ as

$$\boxed{\vec{r} = r\sin\theta\cos\phi\widehat{i} + r\sin\theta\sin\phi\widehat{j} + r\cos\theta\widehat{k}.}$$ (3.391)

Calling

$$\overline{x}^1 = r, \quad \overline{x}^2 = \theta, \quad \overline{x}^3 = \phi,$$ (3.392)

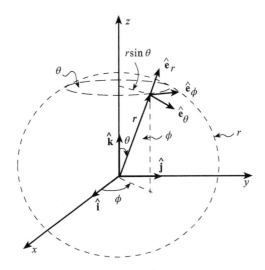

Figure 3.16 Spherical coordinates (r, θ, ϕ): Coordinate surfaces for r, θ, ϕ, and the unit basis vectors.

we obtain the basis vectors $\vec{e}_i = \frac{\partial x^j}{\partial \bar{x}^i}$:

$$\vec{e}_1 = \vec{e}_r = \sin\theta\cos\phi\,\hat{i} + \sin\theta\sin\phi\,\hat{j} + \cos\theta\,\hat{k}, \tag{3.393}$$

$$\vec{e}_2 = \vec{e}_\theta = r\cos\theta\cos\phi\,\hat{i} + r\cos\theta\sin\phi\,\hat{j} - r\sin\theta\,\hat{k}, \tag{3.394}$$

$$\vec{e}_3 = \vec{e}_\phi = -r\sin\theta\sin\phi\,\hat{i} + r\sin\theta\cos\phi\,\hat{j}. \tag{3.395}$$

Dividing with their respective magnitudes:

$$|\vec{e}_r| = 1, \quad |\vec{e}_\theta| = r, \quad |\vec{e}_\phi| = r\sin\theta, \tag{3.396}$$

gives us the unit basis vectors:

$$\hat{e}_r = \sin\theta\cos\phi\,\hat{i} + \sin\theta\sin\phi\,\hat{j} + \cos\theta\,\hat{k}, \tag{3.397}$$

$$\hat{e}_\theta = \cos\theta\cos\phi\,\hat{i} + \cos\theta\sin\phi\,\hat{j} - \sin\theta\,\hat{k}, \tag{3.398}$$

$$\hat{e}_\phi = -\sin\phi\,\hat{i} + \cos\phi\,\hat{j}, \tag{3.399}$$

which satisfy the relations

$$\begin{aligned}
\hat{e}_r &= \hat{e}_\theta \times \hat{e}_\phi, \\
\hat{e}_\theta &= \hat{e}_\phi \times \hat{e}_r, \\
\hat{e}_\phi &= \hat{e}_r \times \hat{e}_\theta.
\end{aligned} \tag{3.400}$$

Using the basis vectors, we construct the **metric tensor**:

$$g_{ij} = \overrightarrow{e}_i \cdot \overrightarrow{e}_j = \begin{pmatrix} 1 & 0 & 0 \\ 0 & r^2 & 0 \\ 0 & 0 & r^2\sin^2\theta \end{pmatrix}, \tag{3.401}$$

which gives the line element as

$$ds^2 = g_{ij}d\bar{x}^i d\bar{x}^j = dr^2 + r^2\,d\theta^2 + r^2\sin^2\theta\,d\phi^2. \tag{3.402}$$

The surface **area elements** (Figure 3.17) are now given as

$$d\sigma_{r\theta} = r\,dr\,d\theta, \quad d\sigma_{r\phi} = r\sin\theta\,dr\,d\phi, \quad d\sigma_{\theta\phi} = r^2\sin\theta\,d\theta\,d\phi, \tag{3.403}$$

while the **volume element** is

$$d\tau = r^2\sin\theta\,dr\,d\theta d\phi. \tag{3.404}$$

Following similar steps to cylindrical coordinates, we write the contravariant components of the metric tensor:

$$g^{ij} = \begin{pmatrix} 1 & 0 & 0 \\ 0 & \frac{1}{r^2} & 0 \\ 0 & 0 & \frac{1}{r^2\sin^2\theta} \end{pmatrix}. \tag{3.405}$$

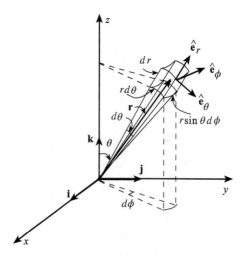

Figure 3.17 Infinitesimal displacements in spherical coordinates.

Using the metric tensor, we can now write the following differential operators for spherical polar coordinates:

Gradient [Eq. (3.299)]:

$$\vec{\nabla}\Phi(r,\theta,\phi) = \frac{\partial\Phi}{\partial r}\widehat{e}_r + \frac{1}{r}\frac{\partial\Phi}{\partial\theta}\widehat{e}_\theta + \frac{1}{r\sin\theta}\frac{\partial\Phi}{\partial\phi}\widehat{e}_\phi. \tag{3.406}$$

Divergence [Eq. (3.311)]:

$$\vec{\nabla}\cdot\vec{A} = \frac{1}{r^2\sin\theta}\left[\frac{\partial(r^2\sin\theta A_r)}{\partial r} + \frac{\partial(r\sin\theta A_\theta)}{\partial\theta} + \frac{\partial(rA_\phi)}{\partial\phi}\right] \tag{3.407}$$

$$= \frac{1}{r^2}\frac{\partial(r^2 A_r)}{\partial r} + \frac{1}{r\sin\theta}\frac{\partial(\sin\theta A_\theta)}{\partial\theta} + \frac{1}{r\sin\theta}\frac{\partial A_\phi}{\partial\phi}, \tag{3.408}$$

where

$$A_r = \vec{A}\cdot\widehat{e}_r, \quad A_\theta = \vec{A}\cdot\widehat{e}_\theta, \quad A_\phi = \vec{A}\cdot\widehat{e}_\phi. \tag{3.409}$$

Curl [Eq. (3.330)]:

$$\vec{\nabla}\times\vec{A} = \frac{1}{r\sin\theta}\left[\frac{\partial(\sin\theta A_\phi)}{\partial\theta} - \frac{\partial A_\theta}{\partial\phi}\right]\widehat{e}_r + \frac{1}{r}\left[\frac{1}{\sin\theta}\frac{\partial A_r}{\partial\phi} - \frac{\partial(rA_\phi)}{\partial r}\right]\widehat{e}_\theta$$

$$+ \frac{1}{r}\left[\frac{\partial(rA_\theta)}{\partial r} - \frac{\partial A_r}{\partial\theta}\right]\widehat{e}_\phi, \tag{3.410}$$

which can also be conveniently expressed as

$$\vec{\nabla}\times\vec{A} = \frac{1}{r^2\sin\theta}\det\begin{pmatrix} \widehat{e}_r & r\,\widehat{e}_\theta & r\sin\theta\,\widehat{e}_\phi \\ \dfrac{\partial}{\partial r} & \dfrac{\partial}{\partial\theta} & \dfrac{\partial}{\partial\phi} \\ A_r & rA_\theta & r\sin\theta A_\phi \end{pmatrix}, \tag{3.411}$$

where $A_r = \vec{A}\cdot\widehat{e}_r$, $A_\theta = \vec{A}\cdot\widehat{e}_\theta$, $A_\phi = \vec{A}\cdot\widehat{e}_\phi$.

Laplacian [Eq. (3.332)]:

$$\vec{\nabla}^2\Phi(r,\theta,\phi) = \frac{1}{r^2\sin\theta}\left[\frac{\partial}{\partial r}\left(r^2\sin\theta\frac{\partial\Phi}{\partial r}\right) + \frac{\partial}{\partial\theta}\left(\sin\theta\frac{\partial\Phi}{\partial\theta}\right) + \frac{\partial}{\partial\phi}\left(\frac{1}{\sin\theta}\frac{\partial\Phi}{\partial\phi}\right)\right] \tag{3.412}$$

$$= \frac{1}{r^2}\frac{\partial}{\partial r}\left(r^2\frac{\partial\Phi}{\partial r}\right) + \frac{1}{r^2\sin\theta}\frac{\partial}{\partial\theta}\left(\sin\theta\frac{\partial\Phi}{\partial\theta}\right) + \frac{1}{r^2\sin^2\theta}\frac{\partial^2\Phi}{\partial\phi^2}. \tag{3.413}$$

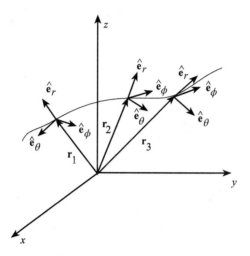

Figure 3.18   Basis vectors along the trajectory of a particle.

Example 3.5. *Motion in spherical coordinates:* In spherical coordinates, the position vector:

$$\vec{r} = x\widehat{i} + y\widehat{j} + z\widehat{k}, \tag{3.414}$$

is written as

$$\vec{r} = r\sin\theta\cos\phi\,\widehat{i} + r\sin\theta\sin\phi\,\widehat{j} + r\sin\theta\,\widehat{k} = r\widehat{e}_r. \tag{3.415}$$

For a particle in motion, the velocity vector is now given as

$$\vec{v} = \dot{r}\widehat{e}_r + r\dot{\widehat{e}}_r. \tag{3.416}$$

Since the unit basis vectors also change direction with position (Figure 3.18), we write the derivatives $\dot{\widehat{e}}_r$, $\dot{\widehat{e}}_\theta$, and $\dot{\widehat{e}}_\phi$ as

$$\dot{\widehat{e}}_r = (\cos\theta\cos\phi\,\dot\theta - \sin\theta\sin\phi\,\dot\phi)\widehat{i} + (\cos\theta\sin\phi\,\dot\theta + \sin\theta\cos\phi\,\dot\phi)\widehat{j}$$
$$- \sin\theta\,\dot\theta\widehat{k} \tag{3.417}$$

$$= \dot\theta\widehat{e}_\theta + \sin\theta\,\dot\phi\widehat{e}_\phi, \tag{3.418}$$

$$\dot{\widehat{e}}_\theta = (-\sin\theta\cos\phi\,\dot\theta - \cos\theta\sin\phi\,\dot\phi)\widehat{i} + (-\sin\theta\sin\phi\,\dot\theta + \cos\theta\cos\phi\,\dot\phi)\widehat{j}$$
$$- \cos\theta\,\dot\theta\widehat{k} \tag{3.419}$$

$$= -\dot\theta\widehat{e}_r + \cos\theta\,\dot\phi\widehat{e}_\phi, \tag{3.420}$$

$$\dot{\widehat{e}}_\phi = -\cos\phi\,\dot\phi\widehat{i} - \sin\phi\,\dot\phi\widehat{j} = -\sin\theta\,\dot\phi\widehat{e}_r - \cos\theta\,\dot\phi\widehat{e}_\theta. \tag{3.421}$$

Velocity [Eq. (3.416)] is now written as

$$\vec{v} = \dot{r}\widehat{e}_r + r\dot{\theta}\widehat{e}_\theta + r\sin\theta\,\dot{\phi}\widehat{e}_\phi, \tag{3.422}$$

which also leads to the acceleration

$$\vec{a} = (\ddot{r}\widehat{e}_r + \dot{r}\dot{\widehat{e}}_r) + (\dot{r}\dot{\theta}\widehat{e}_\theta + r\ddot{\theta}\widehat{e}_\theta + r\dot{\theta}\dot{\widehat{e}}_\theta)$$

$$+ (r\sin\theta\,\dot{\phi}\dot{\widehat{e}}_\phi + \dot{r}\sin\theta\,\dot{\phi}\widehat{e}_\phi + r\cos\theta\,\dot{\theta}\dot{\phi}\widehat{e}_\phi + r\sin\theta\,\ddot{\phi}\widehat{e}_\phi) \tag{3.423}$$

$$= (\ddot{r} - r\dot{\theta}^2 - r\sin\theta\,\dot{\phi}^2)\widehat{e}_r + (r\ddot{\theta} + 2\dot{r}\dot{\theta} - r\sin\theta\cos\theta\,\dot{\phi}^2)\widehat{e}_\theta$$

$$+ (r\sin\theta\,\ddot{\phi} + 2\dot{r}\dot{\phi}\sin\theta + 2r\cos\theta\,\dot{\theta}\dot{\phi})\widehat{e}_\phi. \tag{3.424}$$

REFERENCES

1. Hartle, J.B. (2003). *An Introduction to Einstein's General Relativity*. San Francisco, CA: Addison-Wesley.

2. Bayin, S.S. (2018). *Mathematical Methods in Science and Engineering*, 2e. Hoboken: NJ: Wiley.

PROBLEMS

1. Show that the transformation matrices for counterclockwise rotations through an angle of θ about the x^1-, x^2-, and x^3-axes, respectively, are given as

$$\begin{pmatrix} 1 & 0 & 0 \\ 0 & \cos\theta & \sin\theta \\ 0 & -\sin\theta & \cos\theta \end{pmatrix}, \quad \begin{pmatrix} \cos\theta & 0 & -\sin\theta \\ 0 & 1 & 0 \\ \sin\theta & 0 & \cos\theta \end{pmatrix}, \quad \begin{pmatrix} \cos\theta & \sin\theta & 0 \\ -\sin\theta & \cos\theta & 0 \\ 0 & 0 & 1 \end{pmatrix}.$$

2. Show that the tensor product defined as $\vec{b} \otimes \vec{a}$ is the transpose of $\vec{a} \otimes \vec{b}$.

3. Show that the trace of a second-rank tensor, $\text{tr}A = A_{11} + A_{22} + A_{33}$, is invariant under orthogonal transformations.

4. Using the permutation symbol, justify the formula

$$\det A_{ij} = \det \begin{pmatrix} A_{11} & A_{12} & A_{13} \\ A_{21} & A_{22} & A_{23} \\ A_{31} & A_{32} & A_{33} \end{pmatrix} = \sum_{ijk} A_{1i}A_{2j}A_{3k}\varepsilon_{ijk}.$$

5. Convert these tensor equations into index notation:

(i)

$$v_1 = A_{11}^1 + A_{12}^2 + A_{13}^3,$$
$$v_2 = A_{21}^1 + A_{22}^2 + A_{23}^3,$$
$$v_3 = A_{31}^1 + A_{32}^2 + A_{33}^3.$$

(ii)

$$A_{11} = B_1^1 C_{11} + B_1^2 C_{21},$$
$$A_{12} = B_1^1 C_{12} + B_1^2 C_{22},$$
$$A_{21} = B_2^1 C_{11} + B_2^2 C_{21},$$
$$A_{22} = B_2^1 C_{12} + B_2^2 C_{22}.$$

6. Write the components of the tensor equation $u_i = v_j v^j w_i$, $i, j = 1, 2, 3$, explicitly.

7. What are the ranks of the following tensors:

(i) $K^{ijkl} D_k B^m A_{ij}$,
(ii) $A^i B_i W^{jk} U_k$,
(iii) $A^{ij} A_{ij} B^k$.

8. Write the following tensors explicitly, take $i, j = 1, 2, 3$:

(i) $A^i B_i W^{jk} U_k$,
(ii) $A^{ij} A_{ij} B^k$.

9. Let A_{ij}, B_{ij}, and C_i be Cartesian tensors. Show that $\sum_k \sum_j A_{ij} B_{jk} C_k$ is a first-rank Cartesian tensor.

10. Show that the following matrices represent proper orthogonal transformations and interpret them geometrically:

(i) $\begin{pmatrix} \frac{2}{\sqrt{5}} & 0 & \frac{1}{\sqrt{5}} \\ 0 & 1 & 0 \\ -\frac{1}{\sqrt{5}} & 0 & \frac{2}{\sqrt{5}} \end{pmatrix}$ (ii) $\begin{pmatrix} \cos 30° & -\sin 30° & 0 \\ -\sin 30° & -\cos 30° & 0 \\ 0 & 0 & -1 \end{pmatrix}.$

11. Show that $g^{ij} g_{ik} = \delta_k^j$.

12. Show that in cylindrical coordinates the radius vector is written as

$$\vec{r} = \rho \hat{e}_\rho + z \hat{k}.$$

13. **Parabolic coordinates** $(\bar{x}^1, \bar{x}^2, \bar{x}^3)$ usually called (η, ξ, ϕ), are related to the Cartesian coordinates (x, y, z) by the transformation equations

$$x^1 = x = \eta \xi \cos \phi,$$
$$x^2 = y = \eta \xi \sin \phi,$$
$$x^2 = z = \frac{1}{2}(\xi^2 - \eta^2).$$

   (i) Show that this is an admissible coordinate system and write the inverse transformations.

   (ii) Find the basis vectors and construct the metric tensor g_{ij}.

   (iii) Write the line element.

   (iv) Write the inverse basis vectors and construct g^{ij}.

   (v) Verify the relation between the basis vectors and the inverse basis vectors [Eq. (3.197) and (3.199)].

   (vi) Write the differential operators, gradient, divergence, curl, and Laplacian, in parabolic coordinates.

   (vii) Write the velocity and acceleration in parabolic coordinates.

14. **Elliptic coordinates** (η, ξ, ϕ) are defined by

$$x^1 = x = \frac{a}{2} \cosh \eta \cos \xi,$$
$$x^2 = y = \frac{a}{2} \sinh \eta \sin \xi,$$
$$x^2 = z = z.$$

   (i) Show that this is an admissible coordinate system and write the inverse transformations.

   (ii) Find the basis vectors and construct the metric tensor g_{ij}.

   (iii) Write the line element.

   (iv) Write the inverse basis vectors and construct g^{ij}.

   (v) Verify the relations between the basis vectors and the inverse basis vectors [Eqs. (3.197) and (3.199)].

   (vi) Write the differential operators, gradient, divergence, curl, and Laplacian, in elliptic coordinates.

   (vii) Write the velocity and acceleration in elliptic coordinates.

15. Given the parametric equation of a surface:

$$x^1 = x^1(u, v),$$
$$x^2 = x^2(u, v),$$
$$x^3 = x^3(u, v),$$

and the surface area element

$$\vec{d\sigma}_{uv} = \pm \left[\frac{\partial(x^2, x^3)}{\partial(u, v)} \hat{e}_1 \right.$$
$$\left. + \frac{\partial(x^3, x^1)}{\partial(u, v)} \hat{e}_2 + \frac{\partial(x^1, x^2)}{\partial(u, v)} \hat{e}_3 \right] du\, dv,$$

show that the magnitude of the area element is given as

$$d\sigma_{uv} = \sqrt{EG - F^2}\, du\, dv,$$

where

$$E = \left(\frac{\partial x^1}{\partial u} \right)^2 + \left(\frac{\partial x^2}{\partial u} \right)^2 + \left(\frac{\partial x^3}{\partial u} \right)^2,$$
$$F = \frac{\partial x^1}{\partial u} \frac{\partial x^1}{\partial v} + \frac{\partial x^2}{\partial u} \frac{\partial x^2}{\partial v} + \frac{\partial x^3}{\partial u} \frac{\partial x^3}{\partial v},$$
$$G = \left(\frac{\partial x^1}{\partial v} \right)^2 + \left(\frac{\partial x^2}{\partial v} \right)^2 + \left(\frac{\partial x^3}{\partial v} \right)^2.$$

16. For the area of a surface in terms of the surface coordinates (u, v), we wrote

$$S = \int \int_{R_{uv}} d\sigma_{uv} = \int \int_{R_{uv}} \sqrt{EG - F^2}\, du\, dv$$

and said that in three dimensions and in Cartesian coordinates (x_1, x_2, x_3), S can also be written as

$$S = \int \int_{R_{x_1 x_2}} \sqrt{1 + \left(\frac{\partial x_3}{\partial x_1} \right)^2 + \left(\frac{\partial x_3}{\partial x_2} \right)^2}\, dx_1\, dx_2.$$

Is S a scalar? If your answer is yes, then scalar with respect to what? What happened to the third coordinate in the transformation?

17. Use the transformation

$$x + y = u,$$

$$y = uv$$

to evaluate

$$I = \int_{x=0}^{1} \int_{y=0}^{1-x} e^{y/(x+y)} \, dy dx.$$

18. Evaluate the integral

$$I = \int \int_{R} e^{(x-y)/(x+y)} \, dy dx$$

over the triangular region bounded by the lines $x = 0$, $y = 0$, and $x + y = 1$ by making a suitable coordinate transformation. Is this an orthogonal transformation?

19. Transform the following integral:

$$I = \int_{0}^{1} \int_{0}^{x} \log(1 + x^2 + y^2) \, dy dx$$

by using the substitution

$$x = u,$$

$$y = u - v.$$

Find the new basis vectors.

20. Evaluate the surface integral

$$I = \oint_{S} (x^2 + y^2) \, d\sigma$$

over the cone $z = \sqrt{3(x^2 + y^2)}$ between $z = 0$ and $z = 3$.

21. Express the differential operators

$$\frac{\partial}{\partial x}, \quad \frac{\partial}{\partial y}, \quad \frac{\partial}{\partial z}$$

in spherical coordinates and then show that

$$L_x = -i\left(y\frac{\partial}{\partial z} - z\frac{\partial}{\partial y}\right) = i\cot\theta\cos\phi\frac{\partial}{\partial\phi} + i\sin\phi\frac{\partial}{\partial\theta},$$

$$L_y = -i\left(z\frac{\partial}{\partial x} - x\frac{\partial}{\partial z}\right) = i\cot\theta\sin\phi\frac{\partial}{\partial\phi} - i\cos\phi\frac{\partial}{\partial\theta},$$

$$L_z = -i\left(x\frac{\partial}{\partial y} - y\frac{\partial}{\partial x}\right) = -i\frac{\partial}{\partial\phi},$$

where $\vec{L}$ is the angular momentum operator in quantum mechanics defined as $\vec{L} = -i(\vec{r}\times\vec{\nabla})$. We have set $\hbar = 1$.

Hint: Differential operators are meaningful only when they operate on a function. Hence, write

$$\frac{\partial\Psi(x,y,z)}{\partial x} = \frac{\partial\Psi}{\partial r}\frac{\partial r}{\partial x} + \frac{\partial\Psi}{\partial\theta}\frac{\partial\theta}{\partial x} + \frac{\partial\Psi}{\partial\phi}\frac{\partial\phi}{\partial x},$$

etc.

and proceed.

22. Quantum mechanical angular momentum is defined as the differential operator $\vec{L} = -i(\vec{r}\times\vec{\nabla})$. Show that

$$L_+ = L_x + iL_y = e^{i\phi}\left(\frac{\partial}{\partial\theta} + i\cot\theta\frac{\partial}{\partial\phi}\right),$$

$$L_- = L_x - iL_y = -e^{-i\phi}\left(\frac{\partial}{\partial\theta} - i\cot\theta\frac{\partial}{\partial\phi}\right),$$

where L_x and L_y are the Cartesian components of the angular momentum. Also show that

$$L^2 = L_x^2 + L_y^2 + L_z^2 = \frac{1}{2}(L_+L_- + L_-L_+) + L_z^2.$$

23. Show by using two different methods that in spherical coordinates

$$\vec{L} = -i(\vec{r}\times\vec{\nabla}) = i\left(\hat{e}_\theta\frac{1}{\sin\theta}\frac{\partial}{\partial\phi} - \hat{e}_\phi\frac{\partial}{\partial\theta}\right).$$

24. In spherical coordinates show that the L^2 operator in quantum mechanics, that is, $L^2 = L_x^2 + L_y^2 + L_z^2$, becomes

$$L^2 = -\left[\frac{1}{\sin\theta}\frac{\partial}{\partial\theta}\left(\sin\theta\frac{\partial}{\partial\theta}\right) + \frac{1}{\sin^2\theta}\frac{\partial^2}{\partial\phi^2}\right].$$

Hint: Use the result of Problem 21 and construct L^2.
Verify your result by finding L^2 as

$$L^2 = L^\theta L_\theta + L^\phi L_\phi.$$

25. Show that the cross product of two vectors, $\vec{a} \times \vec{b}$, is a pseudovector.

26. Show that the triple product of three vectors, $\vec{a} \cdot (\vec{b} \times \vec{c})$, is a pseudoscalar.

27. Prove that

$$\vec{\nabla} \left(\frac{1}{r^n} \right) = -\frac{n}{r^{n+2}} \vec{r}.$$

28. Show that the covariant components of velocity are given as

$$v_i = \frac{\partial}{\partial \dot{x}^i} \left(\frac{1}{2} v^2 \right), \quad i = 1, 2, 3,$$

where $v^2 = (\dot{x}^1)^2 + (\dot{x}^2)^2 + (\dot{x}^3)^2$. Justify this result by using spherical coordinates.

29. Given the following position vectors, find the rectangular components of velocity and acceleration:

(i) $\vec{r} = t^2 \hat{i} + 3t \hat{j} + (2t^3 + 4)\hat{k}$,
(ii) $\vec{r} = 2(t - \cos \omega t)\hat{i} + 2(1 - \sin \omega t)\hat{j} + t\hat{k}$.

30. Write the components of velocity and acceleration in polar coordinates for the two-dimensional position vector defined by

(i) $r = a/(b - \sin \phi)$, $\phi = wt$,
(ii) $r = a/t$, $\phi = bt$.

31. What are the components of velocity and acceleration in spherical coordinates with the position of the particle given as

$$r = a, \quad \theta = b \sin \omega t, \quad \phi = \omega t.$$

32. Find the expressions for the kinetic energy in
(i) cylindrical coordinates,
(ii) spherical coordinates.

33. Find the expression for the covariant components of velocity and acceleration in
(i) parabolic coordinates (Problem 13),
(ii) elliptic coordinates (Problem 14).

34. Given the expression

$$v^2 = Ax^2 + Bx\dot{x}\dot{\phi}\sin\phi + Cx^2\dot{\phi}^2$$

for the velocity squared in the following generalized coordinates:

$$\bar{x}_1 = x, \quad \bar{x}_2 = \phi,$$

find the covariant components of the velocity and acceleration.

35. Find the unit tangent vector to the intersection of the surfaces

$$\Psi_1(x, y, z) = x^2 + 3xy - y^2 + yz + z^2 = 5$$

and

$$\Psi_2(x, y, z) = 3x^2 - xy + y^2 = 3$$

at $(1,1,1)$.

Hint: First find the normals to the surfaces.

CHAPTER 4

DETERMINANTS AND MATRICES

In many areas of mathematical analysis, we encounter systems of ordered sets of elements, which could be sets of numbers, functions, or even equations. Determinants and matrices provide an elegant and an efficient computational tool for handling such systems. We already used some of the basic properties of matrices and determinants when we discussed coordinate systems and tensors. In this chapter, we give a formal treatment of matrices and determinants and their applications to systems of linear equations.

4.1 BASIC DEFINITIONS

We define a **rectangular matrix** of **dimension** $m \times n$ as an array [1]

$$
\mathbf{A} = \begin{pmatrix}
a_{11} & a_{12} & \cdots & a_{1n} \\
a_{21} & a_{22} & \cdots & a_{2n} \\
\vdots & \vdots & \ddots & \vdots \\
a_{m1} & a_{m2} & \cdots & a_{mn}
\end{pmatrix}, \tag{4.1}
$$

Essentials of Mathematical Methods in Science and Engineering, Second Edition. Selçuk Ş. Bayın.
© 2020 John Wiley & Sons, Inc. Published 2020 by John Wiley & Sons, Inc.

where the elements of $\mathbf{A}$ could be numbers, functions, or even other matrices. An alternate way to write a matrix is

$$\mathbf{A} = a_{ij}, \quad i = 1, \ldots, m, \quad j = 1, \ldots, n, \tag{4.2}$$

where a_{ij} is called the ijth **element** or the **component** of $\mathbf{A}$. The first subscript i denotes the **row number**, and the second subscript j denotes the **column number**. For the time being, we take a_{ij} as real numbers. When the number of rows is equal to the number of columns, $m = n$, we have a **square matrix** of order n. We define a **row matrix** with the dimension $1 \times n$ as

$$\mathbf{A} = \begin{pmatrix} a_{11} & a_{12} & \cdots & a_{1n} \end{pmatrix} \tag{4.3}$$

or simply as

$$\mathbf{A} = \begin{pmatrix} a_1 & a_2 & \cdots & a_n \end{pmatrix}. \tag{4.4}$$

Similarly, we define $m \times 1$ **column matrix** as

$$\mathbf{A} = \begin{pmatrix} a_{11} \\ a_{21} \\ \vdots \\ a_{m1} \end{pmatrix} \quad \text{or as} \quad \mathbf{A} = \begin{pmatrix} a_1 \\ a_2 \\ \vdots \\ a_m \end{pmatrix}. \tag{4.5}$$

4.2 OPERATIONS WITH MATRICES

The **transpose** of a matrix is obtained by interchanging its rows and columns, and it is denoted by $\mathbf{A}^T$ or $\widetilde{\mathbf{A}}$. The transpose of a matrix of dimension $m \times n$,

$$\mathbf{A} = \begin{pmatrix} a_{11} & a_{12} & \cdots & a_{1n} \\ a_{21} & a_{22} & \cdots & a_{2n} \\ \vdots & \vdots & \ddots & \vdots \\ a_{m1} & a_{m2} & \cdots & a_{mn} \end{pmatrix}, \tag{4.6}$$

is of dimension $n \times m$:

$$\mathbf{A}^T = \widetilde{\mathbf{A}} = \begin{pmatrix} a_{11} & a_{21} & \cdots & a_{m1} \\ a_{12} & a_{22} & \cdots & a_{m2} \\ \vdots & \vdots & \ddots & \vdots \\ a_{1n} & a_{2n} & \cdots & a_{mn} \end{pmatrix}. \tag{4.7}$$

The transpose of a column matrix is a row matrix:

$$\widetilde{\begin{pmatrix} a_1 \\ a_2 \\ \vdots \\ a_m \end{pmatrix}} = \begin{pmatrix} a_1 & a_2 & \cdots & a_n \end{pmatrix}, \tag{4.8}$$

and vice versa:

$$\widetilde{\begin{pmatrix} a_1 & a_2 & \cdots & a_n \end{pmatrix}} = \begin{pmatrix} a_1 \\ a_2 \\ \vdots \\ a_m \end{pmatrix}. \tag{4.9}$$

If the transpose of a square matrix is equal to itself, $\widetilde{\mathbf{A}} = \mathbf{A}$, it is called **symmetric**. When the transpose of a square matrix is equal to the negative of itself, $\widetilde{\mathbf{A}} = -\mathbf{A}$, it is called **antisymmetric** or **skew-symmetric**. If only the diagonal elements of a square matrix are nonzero:

$$\lambda = \begin{pmatrix} \lambda_1 & 0 & \cdots & 0 \\ 0 & \lambda_2 & \cdots & 0 \\ \vdots & \vdots & \ddots & \vdots \\ 0 & 0 & \cdots & \lambda_n \end{pmatrix}, \tag{4.10}$$

it is called **diagonal**. The **zero** or the **null matrix O** is defined as the square matrix with all of its elements zero:

$$\mathbf{O} = \begin{pmatrix} 0 & 0 & \cdots & 0 \\ 0 & 0 & \cdots & 0 \\ \vdots & \vdots & \ddots & \vdots \\ 0 & 0 & \cdots & 0 \end{pmatrix}. \tag{4.11}$$

Identity matrix I $= \delta_{ij}$, $i = 1, \ldots, n$, $j = 1, \ldots, n$, is defined as the square matrix

$$\mathbf{I} = \begin{pmatrix} 1 & 0 & \cdots & 0 \\ 0 & 1 & \cdots & 0 \\ \vdots & \vdots & \ddots & \vdots \\ 0 & 0 & \cdots & 1 \end{pmatrix}. \tag{4.12}$$

To indicate the dimension, we may also write $\mathbf{I}_n$. **Addition** of matrices is defined only when they have the same dimension. Two matrices, $\mathbf{A}$ and $\mathbf{B}$, of the same dimension $m \times n$ can be added term by term, with their sum being again a matrix of dimension $m \times n$:

$$\mathbf{C} = \mathbf{A} + \mathbf{B}, \tag{4.13}$$

$$c_{ij} = a_{ij} + b_{ij}, \quad i = 1, \ldots, m, \quad j = 1, \ldots, n. \tag{4.14}$$

Multiplication of a rectangular matrix with a constant α is accomplished by multiplying each element of the matrix with that constant:

$$\mathbf{C} = \alpha \mathbf{A}, \tag{4.15}$$

$$c_{ij} = \alpha a_{ij}, \quad i = 1, \ldots, m, \quad j = 1, \ldots, n. \tag{4.16}$$

Subtraction of two matrices can be accomplished by multiplying the matrix to be subtracted by -1 and then by adding to the other matrix as

$$\mathbf{A} - \mathbf{B} = \mathbf{A} + (-1)\mathbf{B}. \tag{4.17}$$

It is easy to verify that for three matrices, $\mathbf{A}, \mathbf{B}, \mathbf{C}$, of the same dimension, addition satisfies the following properties:

$$\mathbf{A} + \mathbf{B} = \mathbf{B} + \mathbf{A}, \tag{4.18}$$

$$(\mathbf{A} + \mathbf{B}) + \mathbf{C} = \mathbf{A} + (\mathbf{B} + \mathbf{C}). \tag{4.19}$$

Two matrices of the same dimension, $\mathbf{A}$ and $\mathbf{B}$, are equal if and only if all their corresponding components are equal, that is

$$a_{ij} = b_{ij} \text{ for all } i \text{ and } j. \tag{4.20}$$

Multiplication of rectangular matrices with constants satisfy

$$\alpha(\mathbf{A} + \mathbf{B}) = \alpha \mathbf{A} + \alpha \mathbf{B}, \tag{4.21}$$

$$(\alpha + \beta)\mathbf{A} = \alpha \mathbf{A} + \beta \mathbf{A}, \tag{4.22}$$

$$(\alpha\beta)\mathbf{A} = \alpha(\beta\mathbf{A}), \tag{4.23}$$

$$\alpha\mathbf{A} = \mathbf{A}\alpha, \tag{4.24}$$

where $\mathbf{A}$ and $\mathbf{B}$ have the same dimension and α and β are constants. **Multiplication** of rectangular matrices is defined only when the number of columns of the first matrix matches the number of rows of the second. Two matrices, $\mathbf{A}$ of dimension $m \times n$ and $\mathbf{B}$ of dimension $n \times p$, can be multiplied as

$$\mathbf{C} = \mathbf{AB}, \tag{4.25}$$

$$c_{ij} = \sum_{k=1}^{n} a_{ik} b_{kj}, \quad i = 1, \ldots, m, \quad j = 1, \ldots, p, \tag{4.26}$$

with the result being a matrix of dimension $m \times p$. In general, matrix multiplication is not **commutative**:

$$\boxed{\mathbf{AB} \neq \mathbf{BA},}$$ (4.27)

but it is **associative**:

$$\boxed{(\mathbf{AB})\mathbf{C} = \mathbf{A}(\mathbf{BC}).}$$ (4.28)

The **distributive property** of multiplication with respect to addition is true, that is

$$(\mathbf{A} + \mathbf{B})\mathbf{C} = \mathbf{AC} + \mathbf{BC},$$ (4.29)

$$\mathbf{A}(\mathbf{B} + \mathbf{C}) = \mathbf{AB} + \mathbf{AC}.$$ (4.30)

For a given square matrix $\mathbf{A}$ of order n, we say it is **invertible** or **nonsingular**, if there exists a matrix $\mathbf{B}$ such that

$$\mathbf{AB} = \mathbf{BA} = \mathbf{I}.$$ (4.31)

We call $\mathbf{B}$ the **inverse** of $\mathbf{A}$ and write

$$\mathbf{B} = \mathbf{A}^{-1}.$$ (4.32)

If two matrices can be multiplied, their transpose and inverse satisfy the relations

$$\widetilde{\mathbf{AB}} = \widetilde{\mathbf{B}}\widetilde{\mathbf{A}},$$ (4.33)

$$\widetilde{(\mathbf{A} + \mathbf{B})} = \widetilde{\mathbf{A}} + \widetilde{\mathbf{B}},$$ (4.34)

$$\widetilde{\alpha\mathbf{B}} = \alpha\widetilde{\mathbf{B}},$$ (4.35)

$$\widetilde{\mathbf{A}^{-1}} = (\widetilde{\mathbf{A}})^{-1},$$ (4.36)

$$(\mathbf{AB})^{-1} = \mathbf{B}^{-1}\mathbf{A}^{-1},$$ (4.37)

$$(\mathbf{A} + \mathbf{B})^{-1} = \mathbf{A}^{-1} + \mathbf{B}^{-1}.$$ (4.38)

The sum of the main diagonal elements of a square matrix is called the **spur** or the **trace** of the matrix:

$$\boxed{\operatorname{tr} \mathbf{A} = \sum_{i=1}^{n}\sum_{j=1}^{n} a_{ij}\delta_{ji} = a_{11} + a_{22} + \cdots + a_{nn}.}$$ (4.39)

Trace satisfies

$$\text{tr}(\mathbf{AB}) = \text{tr}(\mathbf{BA}), \tag{4.40}$$

$$\text{tr}(\mathbf{A}+\mathbf{B}) = \text{tr }\mathbf{A} + \text{tr }\mathbf{B}. \tag{4.41}$$

Example 4.1. *Operations with matrices:* Consider the following 2×3 matrix $\mathbf{A}$:

$$\mathbf{A} = \begin{pmatrix} 2 & 1 & 1 \\ 1 & 2 & -1 \end{pmatrix}, \tag{4.42}$$

which has two rows

$$\begin{pmatrix} 2 & 1 & 1 \end{pmatrix}, \begin{pmatrix} 1 & 2 & -1 \end{pmatrix} \tag{4.43}$$

and three columns

$$\begin{pmatrix} 2 \\ 1 \end{pmatrix}, \begin{pmatrix} 1 \\ 2 \end{pmatrix}, \begin{pmatrix} 1 \\ -1 \end{pmatrix}. \tag{4.44}$$

Its transpose is

$$\widetilde{\mathbf{A}} = \begin{pmatrix} 2 & 1 \\ 1 & 2 \\ 1 & -1 \end{pmatrix}. \tag{4.45}$$

Given two square matrices

$$\mathbf{A} = \begin{pmatrix} 2 & 1 \\ -1 & 3 \end{pmatrix} \text{ and } \mathbf{B} = \begin{pmatrix} 1 & 0 \\ -1 & 2 \end{pmatrix}, \tag{4.46}$$

we can write the following matrices

$$\mathbf{A}+\mathbf{B} = \begin{pmatrix} 2+1 & 1+0 \\ -1-1 & 3+2 \end{pmatrix} = \begin{pmatrix} 3 & 1 \\ -2 & 5 \end{pmatrix}, \tag{4.47}$$

$$2\mathbf{A} = \begin{pmatrix} 2.2 & 2.1 \\ 2.(-1) & 2.3 \end{pmatrix} = \begin{pmatrix} 4 & 2 \\ -2 & 6 \end{pmatrix}. \tag{4.48}$$

Now consider the following 3×3 matrices:

$$\mathbf{C} = \begin{pmatrix} 2 & 3 & -1 \\ 1 & 0 & 2 \\ 3 & 1 & 1 \end{pmatrix} \text{ and } \mathbf{D} = \begin{pmatrix} 1 & 2 & 1 \\ -1 & -1 & 0 \\ 0 & 0 & 1 \end{pmatrix}. \tag{4.49}$$

We can write the products

$$\mathbf{CD} = \begin{pmatrix} 2 & 3 & -1 \\ 1 & 0 & 2 \\ 3 & 1 & 1 \end{pmatrix} \begin{pmatrix} 1 & 2 & 1 \\ -1 & -1 & 0 \\ 0 & 0 & 1 \end{pmatrix} \tag{4.50}$$

$$= \begin{pmatrix} 2-3-0 & 4-3-0 & 2+0-1 \\ 1-0+0 & 2-0+0 & 1+0+2 \\ 3-1+0 & 6-1+0 & 3+0+1 \end{pmatrix} = \begin{pmatrix} -1 & 1 & 1 \\ 1 & 2 & 3 \\ 2 & 5 & 4 \end{pmatrix} \tag{4.51}$$

and

$$\mathbf{DC} = \begin{pmatrix} 1 & 2 & 1 \\ -1 & -1 & 0 \\ 0 & 0 & 1 \end{pmatrix} \begin{pmatrix} 2 & 3 & -1 \\ 1 & 0 & 2 \\ 3 & 1 & 1 \end{pmatrix} = \begin{pmatrix} 7 & 4 & 4 \\ -3 & -3 & -1 \\ 3 & 1 & 1 \end{pmatrix}. \tag{4.52}$$

Note that $\mathbf{DC} \neq \mathbf{CD}$. We can also write the following matrix multiplications:

$$\begin{pmatrix} 1 & 0 & 1 \\ -1 & 0 & 2 \\ 1 & 1 & 1 \end{pmatrix} \begin{pmatrix} 2 \\ 1 \\ 1 \end{pmatrix} = \begin{pmatrix} 2+0+1 \\ -2+0+2 \\ 2+1+1 \end{pmatrix} = \begin{pmatrix} 3 \\ 0 \\ 4 \end{pmatrix}, \tag{4.53}$$

$$\begin{pmatrix} 1 & 0 & 1 \\ -1 & 0 & 2 \\ 1 & 1 & 1 \end{pmatrix} \begin{pmatrix} 1 & 0 \\ 0 & 1 \\ 1 & -1 \end{pmatrix} = \begin{pmatrix} 1+0+1 & 0+0-1 \\ -1+0+2 & -0+0-2 \\ 1+0+1 & +0+1-1 \end{pmatrix} \tag{4.54}$$

$$= \begin{pmatrix} 2 & -1 \\ 1 & -2 \\ 2 & 0 \end{pmatrix}. \tag{4.55}$$

Note that the dimensions of the multiplied matrices and the product matrices [Eqs. (4.53) and (4.55)] satisfy the following relations, respectively:

$$(3 \times 3)(3 \times 1) = (3 \times 1), \tag{4.56}$$

$$(3 \times 3)(3 \times 2) = (3 \times 2). \tag{4.57}$$

The multiplication

$$\begin{pmatrix} 1 & 0 \\ 0 & 1 \\ 1 & -1 \end{pmatrix} \begin{pmatrix} 1 & 0 & 1 \\ -1 & 0 & 2 \\ 1 & 1 & 1 \end{pmatrix} \tag{4.58}$$

is not allowed, since the column number of the first matrix does not match the row number of the second one.

Example 4.2. *Multiplication with a diagonal matrix:* When a rectangular matrix is multiplied with a diagonal matrix from the right, we obtain

$$
\begin{pmatrix}
a_{11} & a_{12} & \cdots & a_{1n} \\
a_{21} & a_{22} & \cdots & a_{2n} \\
\vdots & \vdots & \ddots & \vdots \\
a_{m1} & a_{m2} & \cdots & a_{mn}
\end{pmatrix}
\begin{pmatrix}
\lambda_1 & 0 & \cdots & 0 \\
0 & \lambda_2 & \cdots & 0 \\
\vdots & \vdots & \ddots & \vdots \\
0 & 0 & \cdots & \lambda_n
\end{pmatrix}
\tag{4.59}
$$

$$
=
\begin{pmatrix}
a_{11}\lambda_1 & a_{12}\lambda_2 & \cdots & a_{1n}\lambda_n \\
a_{21}\lambda_1 & a_{22}\lambda_2 & \cdots & a_{2n}\lambda_n \\
\vdots & \vdots & \ddots & \vdots \\
a_{m1}\lambda_1 & a_{m2}\lambda_2 & \cdots & a_{mn}\lambda_n
\end{pmatrix}.
\tag{4.60}
$$

Similarly,

$$
\begin{pmatrix}
\lambda_1 & 0 & \cdots & 0 \\
0 & \lambda_2 & \cdots & 0 \\
\vdots & \vdots & \ddots & \vdots \\
0 & 0 & \cdots & \lambda_m
\end{pmatrix}
\begin{pmatrix}
a_{11} & a_{12} & \cdots & a_{1n} \\
a_{21} & a_{22} & \cdots & a_{2n} \\
\vdots & \vdots & \ddots & \vdots \\
a_{m1} & a_{m2} & \cdots & a_{mn}
\end{pmatrix}
\tag{4.61}
$$

$$
=
\begin{pmatrix}
\lambda_1 a_{11} & \lambda_1 a_{12} & \cdots & \lambda_1 a_{1n} \\
\lambda_2 a_{21} & \lambda_2 a_{22} & \cdots & \lambda_2 a_{2n} \\
\vdots & \vdots & \ddots & \vdots \\
\lambda_m a_{m1} & \lambda_m a_{m2} & \cdots & \lambda_m a_{mn}
\end{pmatrix}.
\tag{4.62}
$$

4.3 SUBMATRIX AND PARTITIONED MATRICES

Consider a given $(m \times n)$-dimensional matrix $\mathbf{A}$. If we delete r rows and s columns, we obtain a $[(m - r) \times (n - s)]$-dimensional matrix called the **submatrix** of $\mathbf{A}$. For example, given

$$
\mathbf{A} =
\begin{pmatrix}
a_{11} & a_{12} & a_{13} & a_{14} \\
a_{21} & a_{22} & a_{23} & a_{24} \\
a_{31} & a_{32} & a_{33} & a_{34} \\
a_{41} & a_{42} & a_{43} & a_{44}
\end{pmatrix},
\tag{4.63}
$$

we can write the 3×2 submatrix by deleting the fourth row and the third and the fourth columns as

$$
\mathbf{B} =
\begin{pmatrix}
a_{11} & a_{12} \\
a_{21} & a_{22} \\
a_{31} & a_{32}
\end{pmatrix}.
\tag{4.64}
$$

Even though there is no standard way of writing submatrices, we can write them either by indicating the **rows** and the **columns deleted** as

$$\boxed{\mathbf{B} = \mathbf{A}(4|3,4)}$$

(4.65)

or by indicating the **rows** and the **columns kept** as

$$\boxed{\mathbf{B} = \mathbf{A}[1,2,3|1,2].}$$

(4.66)

For a given $m \times n$ matrix, consider the submatrices

$$\mathbf{A}_{11} = \mathbf{A}[1,\ldots,m_1|1,\ldots,n_1],$$ (4.67)

$$\mathbf{A}_{12} = \mathbf{A}[1,\ldots,m_1|n_1+1,\ldots,n],$$ (4.68)

$$\mathbf{A}_{21} = \mathbf{A}[m_1+1,\ldots,m|1,\ldots,n_1],$$ (4.69)

$$\mathbf{A}_{22} = \mathbf{A}[m_1+1,\ldots,m|n_1+1,\ldots,n],$$ (4.70)

where m_1 and n_1 are integers satisfying $1 < m_1 < m$ and $1 < n_1 < n$. We now define the **partitioned matrix** in terms of the submatrices $\mathbf{A}_{ij}$ as

$$\boxed{\mathbf{A} = \begin{pmatrix} \mathbf{A}_{11} & \mathbf{A}_{12} \\ \mathbf{A}_{21} & \mathbf{A}_{22} \end{pmatrix}.}$$

(4.71)

Partitioned matrices are very important in applications. If $\mathbf{A} = \mathbf{A}_{ij}$ is a partitioned matrix, then multiplying $\mathbf{A}$ with a constant can be accomplished by multiplying each partition with the same constant. If $\mathbf{A}$ and $\mathbf{B}$ are two matrices of the same dimension and partitioned the same way, $\mathbf{A}_{ij}$ and $\mathbf{B}_{ij}$, $i = 1,\ldots,m$, $j = 1,\ldots,n$, they can be added blockwise with the result being another matrix with the same partition, that is

$$\mathbf{A} + \mathbf{B} = \mathbf{A}_{ij} + \mathbf{B}_{ij}, \quad i = 1,\ldots,m, \quad j = 1,\ldots,n.$$ (4.72)

Two partitioned matrices can be multiplied blockwise, if the multiplied partitions are **compatible**. For example, we can write

$$\boxed{\begin{pmatrix} \mathbf{A} & \mathbf{B} \\ \mathbf{C} & \mathbf{D} \end{pmatrix} \begin{pmatrix} \mathbf{E} & \mathbf{F} \\ \mathbf{G} & \mathbf{H} \end{pmatrix} = \begin{pmatrix} \mathbf{AE}+\mathbf{BG} & \mathbf{AF}+\mathbf{BH} \\ \mathbf{CE}+\mathbf{DG} & \mathbf{CF}+\mathbf{DH} \end{pmatrix},}$$

(4.73)

if the matrix multiplications $\mathbf{AE}, \mathbf{BG}, \mathbf{AF}, \ldots$ exist. Partitioned matrices that can be multiplied are called **conformable**.

Example 4.3. *Partitioned matrices:* Given the 4×4 matrices

$$\mathbf{A} = \begin{pmatrix} 1 & 0 & 1 & 1 \\ 0 & 1 & -1 & 0 \\ 1 & 1 & 0 & 1 \\ 0 & 0 & 1 & 0 \end{pmatrix} \text{ and } \mathbf{B} = \begin{pmatrix} -1 & 0 & 1 & 0 \\ 1 & 0 & 0 & 1 \\ -1 & -1 & 0 & 1 \\ 1 & 0 & 0 & 1 \end{pmatrix}, \tag{4.74}$$

the partitions

$$\mathbf{A} = \begin{pmatrix} \begin{pmatrix} 1 & 0 \\ 0 & 1 \end{pmatrix} & \begin{pmatrix} 1 & 1 \\ -1 & 0 \end{pmatrix} \\ \begin{pmatrix} 1 & 1 \\ 0 & 0 \end{pmatrix} & \begin{pmatrix} 0 & 1 \\ 1 & 0 \end{pmatrix} \end{pmatrix} \tag{4.75}$$

and

$$\mathbf{B} = \begin{pmatrix} \begin{pmatrix} -1 \\ 1 \\ -1 \\ 1 \end{pmatrix} & \begin{pmatrix} 0 \\ 0 \\ -1 \\ 0 \end{pmatrix} & \begin{pmatrix} 1 \\ 0 \\ 0 \\ 0 \end{pmatrix} & \begin{pmatrix} 0 \\ 1 \\ 1 \\ 1 \end{pmatrix} \end{pmatrix} \tag{4.76}$$

are not conformable. However, the partitioned matrices:

$$\mathbf{A} = \begin{pmatrix} \mathbf{A}_{11} & \mathbf{A}_{12} \\ \mathbf{A}_{21} & \mathbf{A}_{22} \end{pmatrix} = \begin{pmatrix} \begin{pmatrix} 1 & 0 \\ 0 & 1 \end{pmatrix} & \begin{pmatrix} 1 & 1 \\ -1 & 0 \end{pmatrix} \\ \begin{pmatrix} 1 & 1 \\ 0 & 0 \end{pmatrix} & \begin{pmatrix} 0 & 1 \\ 1 & 0 \end{pmatrix} \end{pmatrix}, \tag{4.77}$$

$$\mathbf{B} = \begin{pmatrix} \mathbf{B}_{11} & \mathbf{B}_{12} \\ \mathbf{B}_{21} & \mathbf{B}_{22} \end{pmatrix} \begin{pmatrix} \begin{pmatrix} -1 & 0 \\ 1 & 0 \end{pmatrix} & \begin{pmatrix} 1 & 0 \\ 0 & 1 \end{pmatrix} \\ \begin{pmatrix} -1 & -1 \\ 1 & 0 \end{pmatrix} & \begin{pmatrix} 0 & 1 \\ 0 & 1 \end{pmatrix} \end{pmatrix} \tag{4.78}$$

are conformable and can be multiplied blockwise to yield the partitioned matrix

$$\mathbf{AB} = \begin{pmatrix} \mathbf{A}_{11} & \mathbf{A}_{12} \\ \mathbf{A}_{21} & \mathbf{A}_{22} \end{pmatrix} \begin{pmatrix} \mathbf{B}_{11} & \mathbf{B}_{12} \\ \mathbf{B}_{21} & \mathbf{B}_{22} \end{pmatrix} \tag{4.79}$$

$$= \begin{pmatrix} \mathbf{A}_{11}\mathbf{B}_{11} + \mathbf{A}_{12}\mathbf{B}_{21} & \mathbf{A}_{11}\mathbf{B}_{12} + \mathbf{A}_{12}\mathbf{B}_{22} \\ \mathbf{A}_{21}\mathbf{B}_{11} + \mathbf{A}_{22}\mathbf{B}_{21} & \mathbf{A}_{21}\mathbf{B}_{12} + \mathbf{A}_{22}\mathbf{B}_{22} \end{pmatrix} \tag{4.80}$$

$$= \begin{pmatrix} \begin{pmatrix} -1 & -1 \\ 2 & 1 \end{pmatrix} & \begin{pmatrix} 1 & 2 \\ 0 & 0 \end{pmatrix} \\ \begin{pmatrix} 1 & 0 \\ -1 & -1 \end{pmatrix} & \begin{pmatrix} 1 & 2 \\ 0 & 1 \end{pmatrix} \end{pmatrix}, \tag{4.81}$$

a result that can be verified by direct multiplication of the matrices.

4.4 SYSTEMS OF LINEAR EQUATIONS

In many scientific and engineering problems, we encounter systems of linear equations to be solved simultaneously. In general, such systems can be written in terms of n equations and n unknowns, $x_1, x_2, \ldots, x_n$, as

$$
\begin{aligned}
a_{11}x_1 + a_{12}x_2 + \cdots + a_{1n}x_n &= y_1, \\
a_{21}x_1 + a_{22}x_2 + \cdots + a_{2n}x_n &= y_2, \\
&\vdots \\
a_{n1}x_1 + a_{n2}x_2 + \cdots + a_{nn}x_n &= y_n,
\end{aligned}
\tag{4.82}
$$

where a_{ij} and y_i are given constants. Introducing the matrices

$$
\mathbf{A} = \begin{pmatrix} a_{11} & a_{12} & \cdots & a_{1n} \\ a_{21} & a_{22} & \cdots & a_{2n} \\ \vdots & \vdots & \ddots & \vdots \\ a_{n1} & a_{n2} & \cdots & a_{nn} \end{pmatrix}, \quad \mathbf{x} = \begin{pmatrix} x_1 \\ x_2 \\ \vdots \\ x_n \end{pmatrix}, \quad \mathbf{y} = \begin{pmatrix} y_1 \\ y_2 \\ \vdots \\ y_n \end{pmatrix},
\tag{4.83}
$$

the aforementioned system can be written as

$$
\boxed{\mathbf{A}\mathbf{x} = \mathbf{y}.}
\tag{4.84}
$$

Solution of this system is the $x_1, x_2, \ldots, x_n$ values that satisfy all the n equations simultaneously. To demonstrate the basic method of solution, let us start with a simple system with two equations and two unknowns:

$$
a_{11}x_1 + a_{12}x_2 = y_1,
\tag{4.85}
$$

$$
a_{21}x_1 + a_{22}x_2 = y_2.
\tag{4.86}
$$

Since these are linear equations, using their linear combinations, we can write infinitely many equivalent sets that will yield the same solution with the original set. Now, a method of solution appears. We simply try to reduce the aforementioned set into an equivalent set with the hope that the new set will yield the solution immediately. We first multiply Eq. (4.85) by a_{21}/a_{11} and then subtract from Eq. (4.86) to get the equivalent set

$$
a_{11}x_1 + a_{12}x_2 = y_1,
\tag{4.87}
$$

$$
\left(a_{21} - \frac{a_{21}}{a_{11}}a_{11}\right)x_1 + \left(a_{22} - \frac{a_{21}}{a_{11}}a_{12}\right)x_2 = y_2 - \frac{a_{21}}{a_{11}}y_1,
\tag{4.88}
$$

or

$$
a_{11}x_1 + a_{12}x_2 = y_1,
\tag{4.89}
$$

$$
\left(a_{22} - \frac{a_{21}}{a_{11}}a_{12}\right)x_2 = y_2 - \frac{a_{21}}{a_{11}}y_1.
\tag{4.90}
$$

Since a_{ij} and y_i are known numbers, Eq. (4.90) gives the value of one of the unknowns, x_2, directly, while Eq. (4.89) gives the remaining unknowns, x_1, in terms of x_2. Substituting the value of x_2 obtained from the second equation into the first, we obtain the value of x_1, thereby completing the solution.

4.5   GAUSS'S METHOD OF ELIMINATION

Consider three equations with three unknowns:

$$x_1 + 2x_2 + x_3 = 1, \tag{4.91}$$

$$2x_1 + 3x_2 - 2x_3 = 1, \tag{4.92}$$

$$3x_1 + 4x_2 - 3x_3 = -3. \tag{4.93}$$

To obtain an equivalent set, we start with Eq. (4.91) and use it to eliminate x_1 from Eqs. (4.92) and (4.93). To obtain the second equation of the new set, we multiply the first equation by 2 and then subtract from the second. For the third equation, we multiply the first equation by 3 and then subtract from the third, thus obtaining the following equivalent set:

$$x_1 + 2x_2 + x_3 = 1, \tag{4.94}$$

$$[2 - 2(1)]x_1 + [3 - 2(2)]x_2 + [-2 - 2(1)]x_3 = 1 - 2(1), \tag{4.95}$$

$$[3 - 3(1)]x_1 + [4 - 3(2)]x_2 + [-3 - 3(1)]x_3 = -3 - 3(1), \tag{4.96}$$

that is,

$$x_1 + 2x_2 + x_3 = 1, \tag{4.97}$$

$$-x_2 - 4x_3 = -1, \tag{4.98}$$

$$-2x_2 - 6x_3 = -6. \tag{4.99}$$

We now use Eq. (4.98) to eliminate x_2 from the Eqs. (4.97) and (4.99). We first multiply the second equation by 2 and then add to the first equation and then multiply the second equation by 2 and subtract from the third equation to obtain another equivalent set:

$$x_1 + [2 - 2(1)]x_2 + [1 - 2(4)]x_3 = 1 - 2(1), \tag{4.100}$$

$$-x_2 - 4x_3 = -1, \tag{4.101}$$

$$[-2 - 2(-1)]x_2 + [-6 - 2(-4)]x_3 = -6 - 2(-1), \tag{4.102}$$

which is

$$x_1 - 7x_3 = -1, \tag{4.103}$$

$$-x_2 - 4x_3 = -1, \tag{4.104}$$

$$2x_3 = -4. \tag{4.105}$$

From the last equation, the value of x_3 can be read immediately. For the other two unknowns, x_1 and x_2, the method allows us to express them in terms of the already determined x_3 as

$$x_1 = -1 + 7x_3, \tag{4.106}$$

$$x_2 = 1 - 4x_3, \tag{4.107}$$

thus yielding the solution as

$$x_1 = -15, \ x_2 = 9, \ x_3 = -2. \tag{4.108}$$

It can be checked easily that this solution satisfies all three of the equivalent sets. This method is called the **Gauss's method of elimination**, also called the **Gauss-Jordan reduction**. When the original set of equations are compatible, Gauss's method of elimination always yields the solution. Furthermore, the algorithm of the Gauss's method is very convenient to be carried out by computers.

Of course, an important issue about a given system of n equations with n unknowns is the **existence** of solution, that is the **compatibility** of the system. Consider the system

$$2x_1 + 3x_2 - 6x_3 = 1, \tag{4.109}$$

$$5x_1 - x_2 + 2x_3 = 2, \tag{4.110}$$

$$3x_1 + 2x_2 - 4x_3 = 2. \tag{4.111}$$

Proceeding with Gauss's method of elimination, we use the first equation to eliminate x_1 from the other two to get the equivalent set

$$2x_1 + 3x_2 - 6x_3 = 1, \tag{4.112}$$

$$-17x_2 + 34x_3 = -1, \tag{4.113}$$

$$-5x_2 + 10x_3 = 1. \tag{4.114}$$

We now use the second equation of the new set to eliminate x_2 from the other two equations of the new set to write the final reduced set as

$$2x_1 = \frac{14}{17}, \tag{4.115}$$

$$-17x_2 + 34x_3 = -1, \tag{4.116}$$

$$0 = \frac{22}{17}. \tag{4.117}$$

Obviously, the last equation is a contradiction. This indicates that the original set is incompatible, hence does not have a solution. Could we have seen this from the beginning? Before we give the answer, let us go back to the first case where we have two equations and two unknowns [Eqs. (4.85) and (4.86)] and write it as

$$a_1 x + b_1 y = c_1, \tag{4.118}$$

$$a_2 x + b_2 y = c_2. \tag{4.119}$$

We first solve the first equation for x in terms of y as

$$x = \frac{c_1}{a_1} - \frac{b_1}{a_1} y, \tag{4.120}$$

which, when substituted back into Eq. (4.119) and solved for y gives

$$y = \frac{c_2}{b_2} - \frac{a_2}{b_2} x \tag{4.121}$$

$$= \frac{c_2}{b_2} - \frac{a_2}{b_2} \left(\frac{c_1}{a_1} - \frac{b_1}{a_1} y \right) \tag{4.122}$$

$$= \frac{\frac{1}{b_2} \left(c_2 - \frac{a_2}{a_1} c_1 \right)}{\frac{1}{b_2} (b_2 - \frac{a_2}{a_1} b_1)}. \tag{4.123}$$

Using this in Eq. (4.120), the final solution can be written as

$$x = \frac{c_1 b_2 - b_1 c_2}{b_2 a_1 - a_2 b_1}, \tag{4.124}$$

$$y = \frac{c_2 a_1 - a_2 c_1}{b_2 a_1 - a_2 b_1}. \tag{4.125}$$

In general, the solution exists if the denominator is different from zero, that is when

$$b_2 a_1 - a_2 b_1 \neq 0, \tag{4.126}$$

$$\frac{b_1}{a_1} \neq \frac{b_2}{a_2}. \tag{4.127}$$

Geometrically, this result can be understood by the fact that the two equations, $a_1 x + b_1 y = c_1$ and $a_2 x + b_2 y = c_2$, correspond to two straight lines in the xy-plane, and the existence of solution implies their intersection at a finite point. When the two ratios are equal:

$$\frac{b_1}{a_1} = \frac{b_2}{a_2}, \tag{4.128}$$

the two lines have the same slope. Hence, they are parallel, and they never intersect.

We now introduce **determinants** that play a very important role in the theory of **matrices**. We first write Eqs. (4.118) and (4.119) as a matrix equation:

$$\begin{pmatrix} a_1 & b_1 \\ a_2 & b_2 \end{pmatrix} \begin{pmatrix} x \\ y \end{pmatrix} = \begin{pmatrix} c_1 \\ c_2 \end{pmatrix}, \tag{4.129}$$

$$\mathbf{A}\mathbf{x} = \mathbf{y}, \tag{4.130}$$

where the matrices $\mathbf{A}, \mathbf{x}$, and $\mathbf{y}$ are defined as

$$\mathbf{A} = \begin{pmatrix} a_1 & b_1 \\ a_2 & b_2 \end{pmatrix}, \; \mathbf{x} = \begin{pmatrix} x \\ y \end{pmatrix}, \; \mathbf{y} \begin{pmatrix} c_1 \\ c_2 \end{pmatrix}. \tag{4.131}$$

We now write the solution [Eqs. (4.124) and (4.125)] in terms of **determinants** as

$$x = \frac{\det \begin{pmatrix} c_1 & b_1 \\ c_2 & b_2 \end{pmatrix}}{\det \begin{pmatrix} a_1 & b_1 \\ a_2 & b_2 \end{pmatrix}}, y = \frac{\det \begin{pmatrix} a_1 & c_1 \\ a_2 & c_2 \end{pmatrix}}{\det \begin{pmatrix} a_1 & b_1 \\ a_2 & b_2 \end{pmatrix}}, \tag{4.132}$$

where the scalar in the denominator is called the **determinant of the coefficients**,

$$\det \mathbf{A} = \begin{vmatrix} a_1 & b_1 \\ a_2 & b_2 \end{vmatrix} = a_1 b_2 - a_2 b_1, \tag{4.133}$$

In this regard, solution of the linear system [Eqs. (4.118) and (4.119)] exists if the determinant of the coefficients does not vanish. Before we generalize this result to a system of n equations with n unknowns, we give a formal introduction to determinants.

4.6 DETERMINANTS

Determinants are **scalars** associated with **square matrices**. We have already introduced the determinant of a 2×2 square matrix [2]:

$$\mathbf{A} = \begin{pmatrix} a_{11} & a_{12} \\ a_{21} & a_{22} \end{pmatrix} \tag{4.134}$$

as

$$\det \mathbf{A} = \begin{vmatrix} a_{11} & a_{12} \\ a_{21} & a_{22} \end{vmatrix} = a_{11}a_{22} - a_{21}a_{12}, \tag{4.135}$$

which is also written as $D(\mathbf{A})$. The order of a square matrix is also the order of its determinant. There are several different but equivalent ways to define a determinant. For $n = 1$, the determinant of a scalar is the scalar itself. For $n = 2$, we have defined the determinant as in Eq. (4.135). In general, we define determinants of order $n > 1$ as

$$\det \mathbf{A} = \sum_{\nu=1}^{n} (-1)^{1+\nu} a_{1\nu} \det \mathbf{A}(1|\nu), \tag{4.136}$$

where $\det \mathbf{A}(1|\nu)$ is called the **minor** of $a_{1\nu}$, which is the determinant of the submatrix obtained by deleting the first row and the νth column of $\mathbf{A}$. When $n = 2$, Eq. (4.136) gives

$$\begin{vmatrix} a_{11} & a_{12} \\ a_{21} & a_{22} \end{vmatrix} = (-1)^{1+1} a_{11} \det \mathbf{A}(1|1) + (-1)^{1+2} a_{12} \det \mathbf{A}(1|2) \tag{4.137}$$

$$= a_{11} a_{22} - a_{12} a_{21} \tag{4.138}$$

and for $n = 3$, the determinant becomes

$$\begin{vmatrix} a_{11} & a_{12} & a_{13} \\ a_{21} & a_{22} & a_{23} \\ a_{31} & a_{32} & a_{33} \end{vmatrix} = a_{11} \begin{vmatrix} a_{22} & a_{23} \\ a_{32} & a_{33} \end{vmatrix} - a_{12} \begin{vmatrix} a_{21} & a_{23} \\ a_{31} & a_{33} \end{vmatrix} + a_{13} \begin{vmatrix} a_{21} & a_{22} \\ a_{31} & a_{32} \end{vmatrix}. \tag{4.139}$$

In Eq. (4.136), we have used the first row to expand the determinant in terms of its minors. However, any row can be used, with the result being the same for all. In other words, for a **square matrix A** of **order** n, the **determinant** can be written as

$$\det \mathbf{A} = \sum_{\nu=1}^{n} (-1)^{i+\nu} a_{i\nu} \det \mathbf{A}(i|\nu), \tag{4.140}$$

where i is any one of the numbers $i = 1, \ldots, n$. From this definition, it is seen that the determinant of a square matrix of order n can be expanded in terms of the determinants of its submatrices of order $n - 1$. This is called the **Laplace development**. Determinants can also be found by expanding with respect to a column, hence we can also write

$$\det \mathbf{A} = \sum_{\nu=1}^{n} (-1)^{\nu+i} a_{\nu i} \det \mathbf{A}(\nu|i), \tag{4.141}$$

where i is anyone of the integers $i = 1, \ldots, n$. Formal proofs are obtained by induction. The **rank** of a square matrix is the largest number among the orders of the minors with nonvanishing determinants.

Example 4.4. *Evaluation of determinants:* Given the matrix

$$\mathbf{A} = \begin{pmatrix} 0 & 1 & 2 \\ 1 & -1 & 1 \\ 1 & 0 & 1 \end{pmatrix}, \tag{4.142}$$

let us evaluate the determinant by using the Laplace development with respect to the first row:

$$\det \mathbf{A} = (-1)^{1+1}(0) \begin{vmatrix} -1 & 1 \\ 0 & 1 \end{vmatrix} + (-1)^{1+2}(1) \begin{vmatrix} 1 & 1 \\ 1 & 1 \end{vmatrix} + (-1)^{1+3}(2) \begin{vmatrix} 1 & -1 \\ 1 & 0 \end{vmatrix} \tag{4.143}$$

$$= 0(-1) - 1(1-1) + 2(1) = 2. \tag{4.144}$$

To check we also expand with respect to the third row:

$$\det \mathbf{A} = (-1)^{3+1}(1) \begin{vmatrix} 1 & 2 \\ -1 & 1 \end{vmatrix} + (-1)^{3+2}(0) \begin{vmatrix} 0 & 2 \\ 1 & 1 \end{vmatrix} + (-1)^{3+3}(1) \begin{vmatrix} 0 & 1 \\ 1 & -1 \end{vmatrix} \tag{4.145}$$

$$= 1(1+2) + 1(-1) = 2 \tag{4.146}$$

and the second column:

$$\det \mathbf{A} = (-1)^{1+2}(1) \begin{vmatrix} 1 & 1 \\ 1 & 1 \end{vmatrix} + (-1)^{2+2}(-1) \begin{vmatrix} 0 & 2 \\ 1 & 1 \end{vmatrix} + (-1)^{3+2}(0) \begin{vmatrix} 0 & 2 \\ 1 & 1 \end{vmatrix} \tag{4.147}$$

$$= -1(1-1) + (-1)(-2) - 0(-2) = 2. \tag{4.148}$$

A convenient way to determine the signs in front of the minors is to use the following rule:

$$\begin{vmatrix} + & - & + & - & + & \cdots \\ - & + & - & + & \cdots \\ + & - & + & \cdots \\ - & + & \cdots \\ + & \cdots \\ \vdots \end{vmatrix}. \tag{4.149}$$

Advantage of the Laplace development is that sometimes most of the elements in a row or column may be zero, thus expanding with respect to that row or column simplifies the calculations significantly. For example, for the matrix

$$\mathbf{A} = \begin{pmatrix} 1 & 2 & 1 & -1 \\ 1 & 2 & 0 & 3 \\ 1 & 0 & 0 & 1 \\ 0 & 0 & 0 & 2 \end{pmatrix}, \tag{4.150}$$

evaluation of the determinant is considerably simplified if we use the third column and then the second column of the 3×3 minor as

$$\det \mathbf{A} = +(1) \begin{vmatrix} 1 & 2 & 3 \\ 1 & 0 & 1 \\ 0 & 0 & 2 \end{vmatrix} = -(2) \begin{vmatrix} 1 & 1 \\ 0 & 2 \end{vmatrix} = -2(2) = -4. \qquad (4.151)$$

4.7 PROPERTIES OF DETERMINANTS

Evaluation of determinants by using the basic definition is practical only for small orders, usually up to 3 or 4. In this regard, the following properties are extremely useful in matrix operations:

Theorem 4.1. The determinant of the transpose of a square matrix is equal to the determinant of the matrix itself:

$$\boxed{\det \mathbf{A} = \det \widetilde{\mathbf{A}}} \qquad (4.152)$$

For example, the determinant of the transpose of the matrix $\mathbf{A}$ in Eq. (4.150) is again -4:

$$\det \widetilde{\mathbf{A}} = \begin{vmatrix} 1 & 1 & 1 & 0 \\ 2 & 2 & 0 & 0 \\ 1 & 0 & 0 & 0 \\ -1 & 3 & 1 & 2 \end{vmatrix} = (+1) \begin{vmatrix} 1 & 1 & 0 \\ 2 & 0 & 0 \\ 3 & 1 & 2 \end{vmatrix} = (-1)(2) \begin{vmatrix} 1 & 0 \\ 1 & 2 \end{vmatrix} = -4.$$

$$(4.153)$$

Theorem 4.2. If each element of one row or column is multiplied by a constant, then the value of the determinant is also multiplied by the same constant.

For example, if

$$\begin{vmatrix} 1 & 0 & 1 \\ 0 & 1 & 1 \\ 1 & 0 & 2 \end{vmatrix} = 2, \qquad (4.154)$$

then

$$\begin{vmatrix} (c_1)1 & (c_1)0 & (c_1)(c_0)1 \\ 0 & 1 & (c_0)1 \\ 1 & 0 & (c_0)2 \end{vmatrix} = 2(c_1 c_0), \qquad (4.155)$$

where c_0 and c_1 are constants.

Theorem 4.3. If two rows or columns of a square matrix are proportional, then its determinant is zero. A special case of this theorem says, if two rows or columns of a square matrix are identical then its determinant is zero.

Consider the square matrix

$$\mathbf{A} = \begin{pmatrix} 2 & 1 & 0 \\ 4 & 2 & 1 \\ 2 & 1 & 0 \end{pmatrix}, \tag{4.156}$$

where the first two columns are proportional, the determinant is

$$\det \mathbf{A} = 2 \begin{vmatrix} 2 & 1 \\ 1 & 0 \end{vmatrix} - 1 \begin{vmatrix} 4 & 1 \\ 2 & 0 \end{vmatrix} = -2 + 2 = 0. \tag{4.157}$$

Theorem 4.4. If two rows or columns of a determinant are interchanged, then the value of the determinant reverses sign.

For example, if we interchange the first two rows in the determinant

$$\begin{vmatrix} 1 & 0 & 1 \\ 0 & 1 & 1 \\ 1 & 0 & 2 \end{vmatrix} = 1, \tag{4.158}$$

we get

$$\begin{vmatrix} 0 & 1 & 1 \\ 1 & 0 & 1 \\ 1 & 0 & 2 \end{vmatrix} = -1. \tag{4.159}$$

Theorem 4.5. The value of a determinant is unchanged, if we add to each element of one row or column a constant times the corresponding element of another row or column.

For example, consider

$$\mathbf{A} = \begin{pmatrix} 1 & 0 & 1 \\ 1 & 1 & 0 \\ 0 & 1 & 1 \end{pmatrix} \tag{4.160}$$

with $\det \mathbf{A} = 2$. We now add three times the first column to the second column:

$$\mathbf{A}' = \begin{pmatrix} 1 & 0+3 & 1 \\ 1 & 1+3 & 0 \\ 0 & 1+0 & 1 \end{pmatrix} = \begin{pmatrix} 1 & 3 & 1 \\ 1 & 4 & 0 \\ 0 & 1 & 1 \end{pmatrix} \tag{4.161}$$

and add twice the first row of $\mathbf{A}'$ to its third row to write

$$\mathbf{A}'' = \begin{pmatrix} 1 & 3 & 1 \\ 1 & 4 & 0 \\ 0+2 & 1+6 & 1+2 \end{pmatrix} = \begin{pmatrix} 1 & 3 & 1 \\ 1 & 4 & 0 \\ 2 & 7 & 3 \end{pmatrix}, \tag{4.162}$$

where the determinants of $\mathbf{A}$, $\mathbf{A}'$, and $\mathbf{A}''$ are all equal:

$$\det \mathbf{A} = \det \mathbf{A}' = \det \mathbf{A}'' = 2. \tag{4.163}$$

Theorem 4.6. Determinant of the product of two square matrices, $\mathbf{A}$ and $\mathbf{B}$, is equal to the product of their determinants:

$$\boxed{\det(\mathbf{AB}) = (\det \mathbf{A})(\det \mathbf{B}).} \tag{4.164}$$

Consider two square matrices

$$\mathbf{A} = \begin{pmatrix} 1 & 0 & 1 \\ 1 & 1 & 0 \\ 0 & 1 & 1 \end{pmatrix} \text{ and } \mathbf{B} = \begin{pmatrix} 1 & 0 & 0 \\ 0 & 1 & 1 \\ 1 & -1 & 1 \end{pmatrix} \tag{4.165}$$

with the determinants $\det \mathbf{A} = 2$ and $\det \mathbf{B} = 2$. The determinant of their product is

$$\det \mathbf{AB} = \begin{vmatrix} 2 & -1 & 1 \\ 1 & 1 & 1 \\ 1 & 0 & 2 \end{vmatrix} = (\det \mathbf{A})(\det \mathbf{B}) = 4. \tag{4.166}$$

Theorem 4.7. The rank of the product of two rectangular matrices does not exceed the ranks of the matrices multiplied.

4.8 CRAMER'S RULE

We now consider the case of three **linear equations** with three **unknowns** to be solved simultaneously:

$$a_{11}x_1 + a_{12}x_2 + a_{13}x_3 = b_1, \tag{4.167}$$

$$a_{21}x_1 + a_{22}x_2 + a_{23}x_3 = b_2, \tag{4.168}$$

$$a_{31}x_1 + a_{32}x_2 + a_{33}x_3 = b_3. \tag{4.169}$$

Following Gauss's method of elimination, we use the first equation to eliminate x_1 from the other two equations. We multiply the first equation by a_{21}/a_{11}, $a_{11} \neq 0$ and subtract from the second. Similarly, we multiply the first equation by a_{31}/a_{11}, $a_{11} \neq 0$ and subtract from the third to get the equivalent set

$$a_{11}x_1 + a_{12}x_2 + a_{13}x_3 = b_1, \tag{4.170}$$

$$\left(a_{22} - \frac{a_{21}}{a_{11}} a_{12} \right) x_2 + \left(a_{23} - \frac{a_{21}}{a_{11}} a_{13} \right) x_3 = \left(b_2 - \frac{a_{21}}{a_{11}} b_1 \right), \tag{4.171}$$

$$\left(a_{32} - \frac{a_{31}}{a_{11}} a_{12} \right) x_2 + \left(a_{33} - \frac{a_{31}}{a_{11}} a_{13} \right) x_3 = \left(b_3 - \frac{a_{31}}{a_{11}} b_1 \right). \tag{4.172}$$

We now use the second equation of the new set to eliminate x_2 from the other two equations. We first multiply the second equation with

$$\frac{a_{12}}{\left(a_{22} - \dfrac{a_{21}}{a_{11}}a_{12}\right)}, \quad a_{22} \neq \frac{a_{21}}{a_{11}}a_{12}, \qquad (4.173)$$

and subtract from the first, and then we multiply the second equation by

$$\frac{\left(a_{32} - \dfrac{a_{31}}{a_{11}}a_{12}\right)}{\left(a_{22} - \dfrac{a_{21}}{a_{11}}a_{12}\right)}, \quad a_{22} \neq \frac{a_{21}}{a_{11}}a_{12}, \qquad (4.174)$$

and subtract from the third to get the final equivalent set:

$$a_{11}x_1 - \left[a_{13} - \frac{a_{12}\left(a_{23} - \dfrac{a_{21}}{a_{11}}a_{13}\right)}{\left(a_{22} - \dfrac{a_{21}}{a_{11}}a_{12}\right)}\right]x_3 = \left[b_1 - \frac{a_{12}\left(b_2 - \dfrac{a_{21}}{a_{11}}b_1\right)}{\left(a_{22} - \dfrac{a_{21}}{a_{11}}a_{12}\right)}\right], \quad (4.175)$$

$$\left(a_{22} - \frac{a_{21}}{a_{11}}a_{12}\right)x_2 + \left(a_{23} - \frac{a_{21}}{a_{11}}a_{13}\right)x_3 = \left(b_2 - \frac{a_{21}}{a_{11}}b_1\right), \qquad (4.176)$$

$$\left[\left(a_{33} - \frac{a_{31}}{a_{11}}a_{13}\right) - \frac{\left(a_{32} - \dfrac{a_{31}}{a_{11}}a_{12}\right)\left(a_{23} - \dfrac{a_{21}}{a_{11}}a_{13}\right)}{\left(a_{22} - \dfrac{a_{21}}{a_{11}}a_{12}\right)}\right]x_3$$

$$= \left[\left(b_3 - \frac{a_{31}}{a_{11}}b_1\right) - \frac{\left(a_{32} - \dfrac{a_{31}}{a_{11}}a_{12}\right)\left(b_2 - \dfrac{a_{21}}{a_{11}}b_1\right)}{\left(a_{22} - \dfrac{a_{21}}{a_{11}}a_{12}\right)}\right]. \qquad (4.177)$$

Equation (4.177) gives the value of x_3 directly, while Eqs. (4.175) and (4.176) give the values of x_1 and x_2 in terms of x_3, respectively. Thus completing the solution. We have assumed that the coefficients a_{11} and $\left(a_{22} - \frac{a_{21}}{a_{11}}a_{12}\right)$ to be different from zero. In cases where we encounter zeros in these coefficients, we simply rename the variables and equations so that they are not zeros.

We now concentrate on Eq. (4.177), which can be rearranged as

$$x_3 = \frac{a_{11}(a_{22}b_3 - a_{32}b_2) - a_{12}(a_{21}b_3 - a_{31}b_2) + b_1(a_{21}a_{32} - a_{31}a_{22})}{a_{11}(a_{22}a_{33} - a_{32}a_{23}) - a_{12}(a_{21}a_{33} - a_{31}a_{23}) + a_{13}(a_{21}a_{32} - a_{31}a_{22})}.$$

$$(4.178)$$

Using determinants, this can be written as

$$x_3 = \frac{\begin{vmatrix} a_{11} & a_{12} & b_1 \\ a_{21} & a_{22} & b_2 \\ a_{31} & a_{32} & b_3 \end{vmatrix}}{\begin{vmatrix} a_{11} & a_{12} & a_{13} \\ a_{21} & a_{22} & a_{23} \\ a_{31} & a_{32} & a_{33} \end{vmatrix}}. \tag{4.179}$$

Values of the remaining two variables, x_1 and x_2, can be obtained from Eq. (4.178) by **cyclic permutations** of the indices, that is

$$1 \to 2 \to 3 \to 1 \to 2 \to \cdots, \tag{4.180}$$

as

$$x_1 = \frac{a_{22}(a_{33}b_1 - a_{13}b_3) - a_{23}(a_{32}b_1 - a_{12}b_3) + b_2(a_{32}a_{13} - a_{12}a_{33})}{a_{11}(a_{22}a_{33} - a_{32}a_{23}) - a_{12}(a_{21}a_{33} - a_{31}a_{23}) + a_{13}(a_{21}a_{32} - a_{31}a_{22})}, \tag{4.181}$$

and

$$x_2 = \frac{a_{33}(a_{11}b_2 - a_{21}b_1) - a_{31}(a_{13}b_2 - a_{23}b_1) + b_3(a_{13}a_{21} - a_{23}a_{11})}{a_{11}(a_{22}a_{33} - a_{32}a_{23}) - a_{12}(a_{21}a_{33} - a_{31}a_{23}) + a_{13}(a_{21}a_{32} - a_{31}a_{22})}. \tag{4.182}$$

In terms of determinants, these can be expressed as

$$x_1 = \frac{1}{\det \mathbf{A}} \begin{vmatrix} b_1 & a_{12} & a_{13} \\ b_2 & a_{22} & a_{23} \\ b_3 & a_{32} & a_{33} \end{vmatrix}, \quad x_2 = \frac{1}{\det \mathbf{A}} \begin{vmatrix} a_{11} & b_1 & a_{13} \\ a_{21} & b_2 & a_{23} \\ a_{31} & b_3 & a_{33} \end{vmatrix}, \tag{4.183}$$

where $\det \mathbf{A}$ is the determinant of the coefficients:

$$\det \mathbf{A} = \begin{vmatrix} a_{11} & a_{12} & a_{13} \\ a_{21} & a_{22} & a_{23} \\ a_{31} & a_{32} & a_{33} \end{vmatrix}. \tag{4.184}$$

Note that for the system to yield a finite solution, the determinant of the coefficients, $\det \mathbf{A}$, has to be different from zero. Geometrically, Eqs. (4.167)–(4.169) correspond to three planes in the $x_1x_2x_3$-space with their normals given as the vectors

$$\mathbf{n}_1 = (a_{11}, a_{12}, a_{13}), \tag{4.185}$$

$$\mathbf{n}_2 = (a_{21}, a_{22}, a_{23}), \tag{4.186}$$

$$\mathbf{n}_3 = (a_{31}, a_{32}, a_{33}). \tag{4.187}$$

The condition that $\det \mathbf{A} \neq 0$ is equivalent to requiring that $\mathbf{n}_1$, $\mathbf{n}_2$, $\mathbf{n}_3$ be **noncoplanar**:

$$\boxed{\mathbf{n}_1 \cdot (\mathbf{n}_2 \times \mathbf{n}_3) \neq 0.} \tag{4.188}$$

The aforementioned procedure can be generalized to n **linear equations** with n **unknowns**:

$$a_{11}x_1 + a_{12}x_2 + \cdots + a_{1n}x_n = b_1,$$
$$a_{21}x_1 + a_{22}x_2 + \cdots + a_{2n}x_n = b_2,$$
$$\vdots$$
$$a_{n1}x_1 + a_{n2}x_2 + \cdots + a_{nn}x_n = b_n, \tag{4.189}$$

which can be conveniently written as

$$\boxed{\mathbf{A}\mathbf{x} = \mathbf{b},} \tag{4.190}$$

where

$$\mathbf{A} = \begin{pmatrix} a_{11} & a_{12} & \cdots & a_{1n} \\ a_{21} & a_{22} & \cdots & a_{2n} \\ \vdots & \vdots & \ddots & \vdots \\ a_{n1} & a_{n2} & \cdots & a_{nn} \end{pmatrix}, \quad \mathbf{x} = \begin{pmatrix} x_1 \\ x_2 \\ \vdots \\ x_n \end{pmatrix}, \quad \mathbf{b} = \begin{pmatrix} b_1 \\ b_2 \\ \vdots \\ b_n \end{pmatrix}. \tag{4.191}$$

We can also write $\mathbf{A}$ in terms of the column matrices

$$\mathbf{A}_j = \begin{pmatrix} a_{1j} \\ a_{2j} \\ \vdots \\ a_{nj} \end{pmatrix} \tag{4.192}$$

as

$$\mathbf{A} = \mathbf{A}(\mathbf{A}_1, \mathbf{A}_2, \ldots, \mathbf{A}_n). \tag{4.193}$$

Cramer's rule says that the solution of the linear system $\mathbf{A}\mathbf{x} = \mathbf{b}$ is unique and is given as

$$x_j = \frac{\det \mathbf{A}(\mathbf{A}_1, \mathbf{A}_2, \ldots, \mathbf{A}_{j-1}, \mathbf{b}, \mathbf{A}_{j+1}, \ldots, \mathbf{A}_n)}{\det \mathbf{A}}, \tag{4.194}$$

where

$$\det \mathbf{A} = \begin{vmatrix} a_{11} & a_{12} & \cdots & a_{1n} \\ a_{21} & a_{22} & \cdots & a_{2n} \\ \vdots & \vdots & \ddots & \vdots \\ a_{n1} & a_{n2} & \cdots & a_{nn} \end{vmatrix} \tag{4.195}$$

is the determinant of the coefficients and

$$
\det \mathbf{A}(\mathbf{A}_1, \mathbf{A}_2, \ldots, \mathbf{A}_{j-1}, \mathbf{b}, \mathbf{A}_{j+1}, \ldots, \mathbf{A}_n)
$$

$$
= \begin{vmatrix}
a_{11} & \cdots & a_{1(j-1)} & b_1 & a_{1(j+1)} & \cdots & a_{1n} \\
a_{21} & \cdots & a_{2(j-1)} & b_2 & a_{2(j+1)} & \cdots & a_{2n} \\
\vdots & \vdots & \vdots & \vdots & \vdots & \cdots & \vdots \\
a_{n1} & \cdots & a_{n(j-1)} & b_n & a_{n(j+1)} & \cdots & a_{nn}
\end{vmatrix} \tag{4.196}
$$

is the determinant obtained from $\mathbf{A}$ by replacing the jth column with $\mathbf{b}$. The solution exists and is unique when the determinant of the coefficients is different from zero, that is

$$
\det \mathbf{A} \neq 0. \tag{4.197}
$$

Using determinants, the proof is rather straightforward. We substitute Eq. $\mathbf{Ax} = \mathbf{b}$ into Eq. (4.194) to write

$$
(\det \mathbf{A})x_j = \det \mathbf{A}(\mathbf{A}_1, \ldots, \mathbf{A}_{j-1}, (\mathbf{A}_1 x_1 + \cdots + \mathbf{A}_n x_n), \mathbf{A}_{j+1}, \ldots, \mathbf{A}_n). \tag{4.198}
$$

Laplace development with respect to the jth column gives

$$
(\det \mathbf{A})x_j = \det \mathbf{A}(\mathbf{A}_1, \ldots, \mathbf{A}_{j-1}, \mathbf{A}_1 x_1, \mathbf{A}_{j+1} \ldots, \mathbf{A}_n)
$$
$$
+ \cdots + \det \mathbf{A}(\mathbf{A}_1, \ldots, \mathbf{A}_{j-1}, \mathbf{A}_n x_n, \mathbf{A}_{j+1}, \ldots, \mathbf{A}_n), \tag{4.199}
$$

which after using Theorem 4.2 becomes

$$
(\det \mathbf{A})x_j = x_1 \det \mathbf{A}(\mathbf{A}_1, \ldots, \mathbf{A}_{j-1}, \mathbf{A}_1, \mathbf{A}_{j+1}, \ldots, \mathbf{A}_n)
$$
$$
+ \cdots + x_j \det \mathbf{A}(\mathbf{A}_1, \ldots, \mathbf{A}_{j-1}, \mathbf{A}_j, \mathbf{A}_{j+1}, \ldots, \mathbf{A}_n)
$$
$$
+ \cdots + x_n \det \mathbf{A}(\mathbf{A}_1, \ldots, \mathbf{A}_{j-1}, \mathbf{A}_n, \mathbf{A}_{j+1}, \ldots, \mathbf{A}_n). \tag{4.200}
$$

Except the jth term, all the determinants on the right-hand side have two identical columns, hence by Theorem 4.3 they vanish, leaving only the jth term giving

$$
(\det \mathbf{A})x_j = x_j \det \mathbf{A}(\mathbf{A}_1, \ldots, \mathbf{A}_{j-1}, \mathbf{A}_j, \mathbf{A}_{j+1}, \ldots, \mathbf{A}_n), \tag{4.201}
$$
$$
(\det \mathbf{A}) = (\det \mathbf{A}), \tag{4.202}
$$

thereby proving the theorem.

4.9 INVERSE OF A MATRIX

Another approach to solving systems of linear equations, $\mathbf{Ax} = \mathbf{b}$, is by finding the inverse matrix $\mathbf{A}^{-1}$, which satisfies

$$\mathbf{A}^{-1}\mathbf{A} = \mathbf{A}\mathbf{A}^{-1} = \mathbf{I}. \tag{4.203}$$

The solution can now be found as

$$\mathbf{A}^{-1}\mathbf{Ax} = \mathbf{A}^{-1}\mathbf{b}, \tag{4.204}$$

$$\mathbf{x} = \mathbf{A}^{-1}\mathbf{b}. \tag{4.205}$$

We return to Eq. (4.178) and rewrite it as

$$x_3 = \frac{b_1(a_{21}a_{32} - a_{31}a_{22}) - b_2(a_{11}a_{32} - a_{12}a_{31}) + b_3(a_{11}a_{22} - a_{12}a_{21})}{\det \mathbf{A}}. \tag{4.206}$$

We immediately identify the terms in the parentheses as determinants of 2×2 matrices and write

$$x_3 = \frac{1}{\det \mathbf{A}} \left[b_1 \begin{vmatrix} a_{21} & a_{22} \\ a_{31} & a_{32} \end{vmatrix} - b_2 \begin{vmatrix} a_{11} & a_{12} \\ a_{31} & a_{32} \end{vmatrix} + b_3 \begin{vmatrix} a_{11} & a_{12} \\ a_{21} & a_{22} \end{vmatrix} \right]. \tag{4.207}$$

Similarly, by cyclic permutations of the indices, we write

$$x_1 = \frac{1}{\det \mathbf{A}} \left[b_2 \begin{vmatrix} a_{32} & a_{33} \\ a_{12} & a_{13} \end{vmatrix} - b_3 \begin{vmatrix} a_{22} & a_{23} \\ a_{12} & a_{13} \end{vmatrix} + b_1 \begin{vmatrix} a_{22} & a_{23} \\ a_{32} & a_{33} \end{vmatrix} \right] \tag{4.208}$$

and

$$x_2 = \frac{1}{\det \mathbf{A}} \left[b_3 \begin{vmatrix} a_{13} & a_{11} \\ a_{23} & a_{21} \end{vmatrix} - b_1 \begin{vmatrix} a_{33} & a_{31} \\ a_{23} & a_{21} \end{vmatrix} + b_2 \begin{vmatrix} a_{33} & a_{31} \\ a_{13} & a_{11} \end{vmatrix} \right]. \tag{4.209}$$

Comparing with $\mathbf{x} = \mathbf{A}^{-1}\mathbf{b}$,

$$\begin{pmatrix} x_1 \\ x_2 \\ x_3 \end{pmatrix} = \begin{pmatrix} a_{11}^{-1} & a_{12}^{-1} & a_{13}^{-1} \\ a_{21}^{-1} & a_{22}^{-1} & a_{23}^{-1} \\ a_{31}^{-1} & a_{32}^{-1} & a_{33}^{-1} \end{pmatrix} \begin{pmatrix} b_1 \\ b_2 \\ b_3 \end{pmatrix}, \tag{4.210}$$

we obtain the components of the inverse matrix as

$$
a_{31}^{-1} = \frac{\begin{vmatrix} a_{21} & a_{22} \\ a_{31} & a_{32} \end{vmatrix}}{\det \mathbf{A}}, \quad a_{32}^{-1} = -\frac{\begin{vmatrix} a_{11} & a_{12} \\ a_{31} & a_{32} \end{vmatrix}}{\det \mathbf{A}}, \quad a_{33}^{-1} = \frac{\begin{vmatrix} a_{11} & a_{12} \\ a_{21} & a_{22} \end{vmatrix}}{\det \mathbf{A}}, \ldots.
$$

(4.211)

In short, we can write this result as

$$
a_{ij}^{-1} = (-1)^{i+j} \frac{\det \mathbf{A}(j|i)}{\det \mathbf{A}},
$$

(4.212)

where $\det \mathbf{A}(j|i)$ is the minor obtained by deleting the jth row and the ith column. Notice that the indices in $\det \mathbf{A}(j|i)$ are reversed. Furthermore, this result is also valid for $n \times n$ matrices. In general, for a given square matrix $\mathbf{A}$, if $\det \mathbf{A} \neq 0$, we say $\mathbf{A}$ is **nonsingular** and a **unique inverse exists** as

$$
\boxed{a_{ij}^{-1} = (-1)^{i+j} \frac{\det \mathbf{A}(j|i)}{\det \mathbf{A}}, \quad i = 1, \ldots, n, \ j = 1, \ldots, n.}
$$

(4.213)

Usually the signed minors are called **cofactors**:

$$
(\text{cofactor } \mathbf{A})_{ij} = (-1)^{i+j} \det \mathbf{A}(i|j).
$$

(4.214)

Hence

$$
\boxed{a_{ij}^{-1} = \frac{(\text{cofactor } \mathbf{A})_{ji}}{\det \mathbf{A}}.}
$$

(4.215)

Another way to write the inverse is

$$
\boxed{\mathbf{A}^{-1} = \frac{\text{adj } \mathbf{A}}{\det \mathbf{A}},}
$$

(4.216)

where the **adjoint matrix**, adj$\mathbf{A}$, is defined as the matrix, whose components are the **cofactors** of the **transposed** matrix, that is,

$$
\boxed{\text{adj } \mathbf{A} = \widetilde{\text{cofactor } \mathbf{A}} = \text{cofactor } \widetilde{\mathbf{A}}.}
$$

(4.217)

Inverse operation satisfies

$$
(\mathbf{A} + \mathbf{B})^{-1} = (\mathbf{B} + \mathbf{A})^{-1},
$$

(4.218)

$$
(\mathbf{AB})^{-1} = \mathbf{B}^{-1}\mathbf{A}^{-1},
$$

(4.219)

$$
(\alpha \mathbf{A})^{-1} = \frac{1}{\alpha} \mathbf{A}^{-1}.
$$

(4.220)

Example 4.5. *Inverse matrix:* Consider the square matrix

$$\mathbf{A} = \begin{pmatrix} 1 & 0 & 2 \\ 0 & 1 & 1 \\ 1 & 1 & 0 \end{pmatrix}. \tag{4.221}$$

Since the determinant of $\mathbf{A}$ is different from zero, $\det \mathbf{A} = -3$, its inverse exists. To find $\mathbf{A}^{-1}$, we first write the transpose:

$$\widetilde{\mathbf{A}} = \begin{pmatrix} 1 & 0 & 1 \\ 0 & 1 & 1 \\ 2 & 1 & 0 \end{pmatrix} \tag{4.222}$$

and then construct the adjoint matrix term by term by finding the cofactors:

$$\text{adj } \mathbf{A} = \begin{pmatrix} (-1)^{1+1} \begin{vmatrix} 1 & 1 \\ 1 & 0 \end{vmatrix} & (-1)^{1+2} \begin{vmatrix} 0 & 1 \\ 2 & 0 \end{vmatrix} & (-1)^{1+3} \begin{vmatrix} 0 & 1 \\ 2 & 1 \end{vmatrix} \\ (-1)^{2+1} \begin{vmatrix} 0 & 1 \\ 1 & 0 \end{vmatrix} & (-1)^{2+2} \begin{vmatrix} 1 & 1 \\ 2 & 0 \end{vmatrix} & (-1)^{2+3} \begin{vmatrix} 1 & 0 \\ 2 & 1 \end{vmatrix} \\ (-1)^{3+1} \begin{vmatrix} 0 & 1 \\ 1 & 1 \end{vmatrix} & (-1)^{3+2} \begin{vmatrix} 1 & 1 \\ 0 & 1 \end{vmatrix} & (-1)^{3+3} \begin{vmatrix} 1 & 0 \\ 0 & 1 \end{vmatrix} \end{pmatrix} \tag{4.223}$$

$$= \begin{pmatrix} -1 & 2 & -2 \\ 1 & -2 & -1 \\ -1 & -1 & 1 \end{pmatrix}. \tag{4.224}$$

Finally, we write the inverse as

$$\mathbf{A}^{-1} = -\frac{1}{3} \begin{pmatrix} -1 & 2 & -2 \\ 1 & -2 & -1 \\ -1 & -1 & 1 \end{pmatrix}. \tag{4.225}$$

It is always a good idea to check:

$$\mathbf{A}^{-1}\mathbf{A} = -\frac{1}{3} \begin{pmatrix} -1 & 2 & -2 \\ 1 & -2 & -1 \\ -1 & -1 & 1 \end{pmatrix} \begin{pmatrix} 1 & 0 & 2 \\ 0 & 1 & 1 \\ 1 & 1 & 0 \end{pmatrix} = -\frac{1}{3} \begin{pmatrix} -3 & 0 & 0 \\ 0 & -3 & 0 \\ 0 & 0 & -3 \end{pmatrix} \tag{4.226}$$

$$= \mathbf{I} = \mathbf{A}\mathbf{A}^{-1}. \tag{4.227}$$

4.10   HOMOGENEOUS LINEAR EQUATIONS

When the right-hand sides of all the equations in a linear system of equations are zero:

$$a_{11}x_1 + \cdots + a_{1n}x_n = 0,$$

$$\vdots \tag{4.228}$$

$$a_{n1}x_1 + \cdots + a_{nn}x_n = 0,$$

we say the system is **homogeneous**. When the determinant of the coefficients, $\det \mathbf{A}$, is different from zero, there is only the **trivial solution:**

$$x_1 = x_2 = \cdots = x_n = 0. \tag{4.229}$$

When the determinant of the coefficients vanishes,

$$\det \mathbf{A} = \begin{vmatrix} a_{11} & a_{12} & \cdots & a_{1n} \\ a_{21} & a_{22} & \cdots & a_{2n} \\ \vdots & \vdots & \ddots & \vdots \\ a_{n1} & a_{n2} & \cdots & a_{nn} \end{vmatrix} = 0, \tag{4.230}$$

then there are infinitely many solutions.

Example 4.6. *Homogeneous equations:* Let us consider the following homogeneous system:

$$2x + 3y - 6z = 0, \tag{4.231}$$

$$5x - y + 2z = 0, \tag{4.232}$$

$$3x + 2y - 4z = 0. \tag{4.233}$$

For this system, the determinant of the coefficients is zero:

$$\begin{vmatrix} 2 & 3 & -6 \\ 5 & -1 & 2 \\ 3 & 2 & -4 \end{vmatrix} = 0. \tag{4.234}$$

We now seek a solution using Gauss's method of elimination. Using the first equation, we eliminate x from the other two equations to obtain the equivalent set

$$2x + 3y - 6z = 0, \tag{4.235}$$

$$-\frac{17}{2}y + 17z = 0, \tag{4.236}$$

$$-\frac{5}{2}y + 5z = 0. \tag{4.237}$$

Since the last two equations are identical, this yields the solution

$$x = 0, \tag{4.238}$$

$$y = 2z. \tag{4.239}$$

For the infinitely many values that z could take this represents an infinite number of solutions to the homogeneous system of linear equations [Eqs. (4.231)–(4.233)].

Whether the equation is homogeneous or not Gauss's method of elimination can always be used with m equations and n unknowns. In general, the method will yield all the solutions. In cases where there are no solutions, the elimination process will lead to a contradiction. In case we obtain $0 = 0$ for an equation, we disregard it. In cases where there are fewer equations than unknowns, $n > m$, the system cannot have a unique solution. In this case, m of the variables can be solved in terms of the remaining $n - m$ variables. For matrices larger than 4×4, the method becomes too cumbersome to follow. However, using the variations of Gauss's method, there exists package programs written in various computer languages. Also, programs like Matlab, Mathematica, Maple, etc., include matrix diagonalization and inverse matrix calculation routines.

REFERENCES

1. Gantmacher, F.R. (1960). *The Theory of Matrices*. New York: Chelsea Publishing Company.

2. Spiegel, M.R. (1971). *Advanced Mathematics for Engineers and Scientists: Schaum's Outline Series in Mathematics*. New York: McGraw-Hill.

PROBLEMS

1. Given $\mathbf{A}$, $\mathbf{B}$, and $\mathbf{C}$ as

$$\mathbf{A} = \begin{pmatrix} -1 & 1 & 0 \\ 1 & 2 & 3 \end{pmatrix}, \quad \mathbf{B} = \begin{pmatrix} 4 & 1 & -1 \\ 2 & 1 & 1 \end{pmatrix}, \quad \mathbf{C} = \begin{pmatrix} 1 & 0 \\ 2 & 1 \\ 1 & -1 \end{pmatrix},$$

write the following matrices:

   (i) $\mathbf{A} + \mathbf{B}$, $\mathbf{B} + \mathbf{A}$, $\mathbf{A} + 2\mathbf{B}$, $\mathbf{B} - 2\mathbf{A}$, $\mathbf{B} - \mathbf{C}$.

   (ii) $\tilde{\mathbf{A}}$, $\widetilde{\mathbf{AC}}$, $\tilde{\mathbf{C}}$, $\widetilde{\mathbf{A} + \mathbf{B}}$, $\tilde{\mathbf{A}}\mathbf{C}$, $(\tilde{\mathbf{A}}\mathbf{A} + \mathbf{B})$.

   (iii) $\mathbf{AB}$, $\mathbf{BA}$, $\tilde{\mathbf{A}}\mathbf{A}$, $\tilde{\mathbf{A}}\mathbf{B}$, $\mathbf{CA}$, $\mathbf{AC}$.

   (iv) Write the row and the column matrices for $\mathbf{A}$, $\mathbf{B}$, and $\mathbf{C}$.

2. If $\mathbf{A}$ and $\mathbf{B}$ are two $m \times n$ matrices and α is a constant, show that

$$(\widetilde{\mathbf{A} + \alpha \mathbf{B}}) = \tilde{\mathbf{A}} + \alpha \tilde{\mathbf{B}}.$$

3. For a symmetric matrix, show that $\tilde{\mathbf{A}} = \mathbf{A}$. Also show that $\mathbf{A} + \tilde{\mathbf{A}}$ is symmetric.

4. Using the index notation, prove that $\mathbf{A}(\mathbf{BC}) = (\mathbf{AB})\mathbf{C}$.

5. Calculate the product $\mathbf{AB}$ for
   (i)
   $$\mathbf{A} = \begin{pmatrix} -2 & 1 \\ 4 & 1 \end{pmatrix}, \ \mathbf{B} = \begin{pmatrix} 1 & 0 & 1 & 1 & 1 \\ 0 & 1 & 2 & 1 & 3 \end{pmatrix}.$$

   (ii)
   $$\mathbf{A} = \begin{pmatrix} 1 & 2 & 1 \\ 1 & 1 & 0 \\ 2 & 0 & 1 \end{pmatrix}, \ \mathbf{B} = \begin{pmatrix} 1 & 2 & 1 & -1 & 1 \\ 2 & 1 & 1 & 1 & 2 \\ 1 & 0 & 1 & -1 & 1 \end{pmatrix}.$$

6. Use Gauss's method of elimination to solve the following linear systems:
   (i)
   $$3x_1 + 2x_2 = 1,$$
   $$5x_1 + 6x_2 = 2.$$

   (ii)
   $$5x_1 - 2x_2 = 3,$$
   $$x_1 + x_2 = 2.$$

7. For a square matrix $\mathbf{A}$, prove the following relations:
   (a) $\mathbf{A}^2 - \mathbf{I} = (\mathbf{A} + \mathbf{I})(\mathbf{A} - \mathbf{I})$,
   (b) $\mathbf{A}^3 - \mathbf{I} = (\mathbf{A} - \mathbf{I})(\mathbf{A}^2 + \mathbf{A} + \mathbf{I})$,
   (c) $\mathbf{A}^2 - 4\mathbf{A} + 3\mathbf{I} = (\mathbf{A} - 3\mathbf{I})(\mathbf{A} - \mathbf{I})$.

8. Solve the following system of linear equations by using Gauss's method of elimination and check your answer via the Cramer's rule:
   $$3x + 2y + 4z = 2,$$
   $$2x - y + z = 1,$$
   $$x + 2y + 3z = 0.$$

9. Given
   $$\mathbf{A} = \begin{pmatrix} 2 & 1 & 0 & -1 & 1 & 0 \\ 4 & 2 & 1 & 2 & -3 & 3 \\ 3 & 1 & 2 & 1 & 1 & 1 \\ 0 & -1 & 1 & 2 & 0 & 1 \end{pmatrix},$$

write

$$\mathbf{A}(1|1,3,5), \quad \mathbf{A}(1,2,3|1,3,5),$$
$$\mathbf{A}[1,2|3,5], \quad \mathbf{A}[2,3|3,5,6].$$

10. Using the following partitioned matrices, find $\mathbf{AB}$ and $\mathbf{A}+\mathbf{B}[1,2|1,2]$:

$$\mathbf{A} = \begin{pmatrix} \mathbf{I}_2 & 0 \\ 0 & \mathbf{A}_1 \end{pmatrix}, \quad \mathbf{B} = \begin{pmatrix} \mathbf{B}_1 & 0 & \mathbf{B}_2 \\ 0 & \mathbf{I}_1 & \mathbf{B}_3 \end{pmatrix},$$

where

$$\mathbf{I}_2 = \begin{pmatrix} 1 & 0 \\ 0 & 1 \end{pmatrix}, \quad \mathbf{A}_1 = 2, \quad \mathbf{B}_1 = \begin{pmatrix} 1 & 1 \\ 2 & 1 \end{pmatrix},$$

$$\mathbf{B}_2 = \begin{pmatrix} 2 \\ 1 \end{pmatrix}, \quad \mathbf{B}_3 = 2, \quad \mathbf{I}_1 = 1.$$

11. Evaluate the following determinants:

(i) $\begin{vmatrix} 1 & 13 \\ 5 & 5 \end{vmatrix}$,

(ii) $\begin{vmatrix} 1 & 2 & 0 & 0 \\ 0 & 3 & 4 & 0 \\ 0 & 0 & 5 & 6 \\ 1 & 2 & 3 & 4 \end{vmatrix}$,

(iii) $\begin{vmatrix} 0 & 1 & 0 & 1 & 2 \\ 2 & 1 & 0 & -1 & 2 \\ 0 & 0 & 1 & 0 & 2 \\ -1 & 1 & 1 & 2 & 1 \\ 2 & 1 & 0 & 1 & 0 \end{vmatrix}$.

12. Find the ranks of the following matrices:

(i) $\begin{pmatrix} 3 & -2 & 0 & 0 \\ 1 & 4 & 2 & 1 \\ 2 & 1 & 0 & 3 \\ 1 & 1 & 2 & 1 \end{pmatrix}$,

(ii) $\begin{pmatrix} 0 & 1 & 0 & 1 & 2 \\ 1 & 1 & 0 & -1 & 2 \\ 0 & 0 & 1 & 0 & 1 \\ -1 & 1 & 1 & -1 & 1 \\ 2 & 1 & 0 & 1 & 0 \end{pmatrix}$.

13. Find the values for which the following determinant is singular:

$$\begin{vmatrix} x - 1 & 4 \\ 2 & x + 1 \end{vmatrix}.$$

14. By using Gauss's method of elimination, solve the following linear systems:

    (i)

$$2x_1 + 5x_2 = 1,$$
$$x_1 + 3x_2 = 2.$$

    (ii)

$$2x_1 + 2x_2 + 2x_3 = 2,$$
$$x_1 + x_2 + 3x_3 = 2,$$
$$x_1 - x_2 + x_1 = 1.$$

    For each case, verify your solution by finding $\mathbf{A}^{-1}$.

15. First check that the inverse exists and then find the inverses of the following matrices. Verify your answers.

    (i) $\mathbf{A} = \begin{pmatrix} 1 & 2 \\ 3 & 4 \end{pmatrix},$

    (ii) $\mathbf{A} = \begin{pmatrix} 1 & 2 & 0 \\ 0 & 2 & 1 \\ 1 & 1 & -1 \end{pmatrix},$

    (iii) $\mathbf{A} = \begin{pmatrix} 1 & 0 & 2 \\ 2 & 1 & 3 \\ 3 & 0 & 4 \end{pmatrix},$

    (iv) $\mathbf{A} = \begin{pmatrix} 1 & 0 & 2 & 2 \\ 1 & 2 & 3 & 1 \\ 3 & 0 & 4 & 1 \\ 2 & -1 & 0 & 2 \end{pmatrix}.$

16. Simplify

    (i) $[(\mathbf{AB})^{-1}\mathbf{A}^{-1}]^{-1},$

    (ii) $(\mathbf{ABC})^{-1}(\mathbf{C}^{-1}\mathbf{B}^{-1}\mathbf{A}^{-1})^{-1},$

    (iii) $[(\mathbf{AB})^T A^T]^T.$

17. If $\mathbf{A}$ and $\mathbf{B}$ are two nonsingular matrices with the property $\mathbf{AB} = \mathbf{BA}$, then show that
$$\mathbf{A}^{-1}\mathbf{B}^{-1} = \mathbf{B}^{-1}\mathbf{A}^{-1}.$$

18. Show that
$$(\mathbf{AB})^{-1} = \mathbf{B}^{-1}\mathbf{A}^{-1}.$$

19. If $\mathbf{ABC} = \mathbf{I}$, then show that
$$\mathbf{BCA} = \mathbf{CAB} = \mathbf{I}.$$

20. Find the inverses of the matrices
$$\mathbf{A} = \begin{pmatrix} 1 & 2 & 4 \\ 2 & 0 & 2 \\ 3 & -1 & 2 \end{pmatrix}$$
and
$$\mathbf{B} = \begin{pmatrix} 2 & 1 & 2 \\ 4 & 0 & 2 \\ 6 & -1 & 3 \end{pmatrix}.$$

Check the relation $(\mathbf{AB})^{-1} = \mathbf{B}^{-1}\mathbf{A}^{-1}$ for these matrices.

21. Given the linear system of equations:
$$2x_1 - 2x_2 + x_3 = 2,$$
$$x_1 + 2x_2 - x_3 = 1,$$
$$5x_1 - 5x_2 + 4x_3 = 4.$$

   (i) Solve by using Gauss's method of elimination.
   (ii) Solve by using Cramer's rule.
   (iii) Solve by finding the inverse matrix of the matrix of the coefficients.

22. For the homogeneous system
$$x_1 - x_2 + x_3 + x_4 = 0,$$
$$x_1 + 2x_2 - x_3 - x_4 = 0,$$
$$3x_1 - x_2 - x_3 + 2x_4 = 0,$$
$$x_1 + 3x_2 + x_3 - 2x_4 = 0,$$

show that the determinant of the coefficients is zero and find the solution.

23. Show the determinant

$$\begin{vmatrix} 1 & x_1 & x_1^2 \\ 1 & x_2 & x_2^2 \\ 1 & x_3 & x_3^2 \end{vmatrix} = (x_2 - x_1)(x_3 - x_1)(x_3 - x_2).$$

24. By using Laplace construction with respect to a row and a column of your choice evaluate the determinant

$$\begin{vmatrix} -1 & 1 & 2 & 0 \\ 0 & 3 & 2 & 0 \\ 0 & 2 & 1 & 1 \\ 1 & 2 & 0 & 1 \end{vmatrix}.$$

25. Solve the following system of linear equations by a method of your choice:
    (i)

$$3x_1 + x_2 - x_3 = 1,$$
$$x_1 + x_2 + x_3 = 1,$$
$$x_2 - x_3 = 0.$$

(ii)

$$4x_1 - 2x_2 + 2x_3 = 1,$$
$$x_1 + 3x_2 - 2x_3 = 1,$$
$$4x_1 - 3x_2 + x_3 = 1.$$

26. Write the inverse of a diagonal matrix.

27. Solve the following system of linear equations by using Cramer's rule and interpret your results geometrically:
    (i)

$$3x_1 - 2x_2 + 2x_3 = 10,$$
$$x_1 + 2x_2 - 3x_3 = -1,$$
$$4x_1 + x_2 + 2x_3 = 3.$$

(ii)

$$2x_1 + 5x_2 - 3x_3 = 0,$$
$$x_1 - 2x_2 + x_3 = 0,$$
$$7x_1 + 4x_2 - 3x_3 = 0.$$

(iii)

$$2x_1 + 5x_2 - 3x_3 = 3,$$
$$x_1 - 2x_2 + x_3 = 2,$$
$$7x_1 + 4x_2 - 3x_3 = 12.$$

(iv)

$$3x_1 - 2x_2 + 2x_3 = 0,$$
$$x_1 + 2x_2 - 3x_3 = 0,$$
$$4x_1 + x_2 + 2x_3 = 0.$$

CHAPTER 5

LINEAR ALGEBRA

Vectors are usually introduced with their geometric definition as directed line segments. Introduction of a coordinate system allows the concept of vector to be extended to a much broader class of objects called tensors, which are defined in terms of their transformation properties. As a tensor, vectors are now classified as first-rank tensors. In n dimensions, a given vector can be written as the linear combination of n linearly independent basis vectors. Linear algebra is essentially the branch of mathematics that uses the concept of linear combination to extend the vector concept to a much broader class of objects. In this chapter, we discuss abstract vector spaces, which paves the way to many scientific and engineering applications of linear algebra.

5.1 FIELDS AND VECTOR SPACES

We start with the basic definitions used throughout this chapter [1–3]. As usual, a collection of objects is called a **set**. The set of all real numbers is denoted by $\mathbb{R}$ and the set of all complex numbers by $\mathbb{C}$. The set of all n-tuples of real numbers,

$$X = (x_1, x_2, \ldots, x_n), \tag{5.1}$$

is shown by $\mathbb{R}^n$. Similarly, the set of n-tuples of complex numbers is shown by $\mathbb{C}^n$. An essential part of linear algebra consists of the definitions of **field** and **vector space**, which is also called the **linear space**.

Definition 5.1. A field K is a set of objects that satisfy the following conditions:

I. If α, β are elements of K, then their sum, $\alpha + \beta$, and their multiplication, $\alpha\beta$, are also elements of K.

II. If α is an element of K, then $-\alpha$ such that $\alpha + (-\alpha) = 0$ is also an element of K. Furthermore, if $\alpha \neq 0$, then α^{-1}, where $\alpha(\alpha^{-1}) = 1$ is also an element of K.

III. The elements 0 and 1 are also elements of K.

The set of all real numbers, $\mathbb{R}$, and the set of all complex numbers, $\mathbb{C}$, are fields. Since the condition II is not satisfied, the set of all integers is not a field. Notice that a field is essentially a set of objects, elements of which can be added and multiplied according to the ordinary rules of arithmetic and that can be divided by nonzero elements. We use K to denote any field. Elements of K will be called **numbers** or **scalars**.

We now introduce the concept of **linear space** or as usually called the **vector space**. A vector space is defined in conjunction with a **field** and its elements are called **vectors**.

Definition 5.2. A vector space V over a field K is a set of objects that can be added with the result being an element of V and the product of an element of V with an element of K is again an element of V. A vector space also satisfies the following properties:

I. Given any three elements, u, v, w, of V, addition is associative:

$$u + (v + w) = (u + v) + w. \tag{5.2}$$

II. There is a unique element 0 of V, such that

$$0 + u = u + 0 = u, \quad u \in V. \tag{5.3}$$

III. Let u be an element of V, then there exists a unique vector $-u$ in V such that

$$u + (-u) = 0. \tag{5.4}$$

IV. For any two elements, u, v, of V, addition is commutative:

$$u + v = v + u. \tag{5.5}$$

V. If α is a number belonging to the field K, then

$$\alpha(u + v) = \alpha u + \alpha v, \quad u, v \in V. \tag{5.6}$$

VI. If α, β are two numbers in K, then we have

$$(\alpha + \beta)v = \alpha v + \beta v, \quad v \in V. \tag{5.7}$$

VII. For any two numbers α and β in K, we have

$$(\alpha\beta)v = \alpha(\beta v), \quad v \in V. \tag{5.8}$$

VIII. For all u of V, we have

$$1.u = u, \tag{5.9}$$

where 1 is the number one.

As we shall demonstrate in the following examples, use of the word *vector* for the elements of a vector space is largely a matter of convenience. These are essentially linear spaces, where their elements can be many things such as ordinary vectors, matrices, tensors, functions.

Example 5.1. *The n-tuple space* X^n: For a given field K, Let X^n be the set of all n-tuples

$$x = (x_1, x_2, \ldots, x_n), \tag{5.10}$$

where x_i are numbers in K. If

$$y = (y_1, y_2, \ldots, y_n) \tag{5.11}$$

is another element of X^n with the sum $x + y$:

$$x + y = (x_1 + y_1, x_2 + y_2, \ldots, x_n + y_n) \tag{5.12}$$

and the product αx is defined as

$$\alpha x = (\alpha x_1, \alpha x_2, \ldots, \alpha x_n), \quad \alpha \in K, \tag{5.13}$$

then it can easily be checked that X^n is a vector space.

Example 5.2. *The space of* $m \times n$ *matrices:* The sum of two $m \times n$ matrices, $\mathbf{A}$ and $\mathbf{B}$, is defined as

$$\mathbf{A} + \mathbf{B} = a_{ij} + b_{ij}, \quad i = 1, \ldots, m, \qquad j = 1, \ldots, n, \tag{5.14}$$

while their multiplication with a scalar α is defined as

$$\alpha\mathbf{A} = \alpha a_{ij}, \quad i = 1, \ldots, m, \qquad j = 1, \ldots, n. \tag{5.15}$$

It is again obvious that the eight conditions in the definition of a vector field are satisfied. In this sense, the set of $m \times n$ matrices defined over some field K is a vector space and its elements are vectors.

Example 5.3. *Functions as vectors:* Consider the set V of all continuous functions of $\mathbb{R}$ into $\mathbb{R}$. It is obvious that V is a vector space. A subset of V is the set V' of all differentiable functions of $\mathbb{R}$ into $\mathbb{R}$. If f and g are two differentiable functions, then $f + g$ is also differentiable. If α is a scalar in K, then αf is also a differentiable function. Since the zero function is differentiable, V' also forms a vector space. We say V' is a **subspace** of V. Similarly, if a subset K' of K also satisfies the conditions to be a field, it is a called a **subfield**.

5.2 LINEAR COMBINATIONS, GENERATORS, AND BASES

Let $v_1, v_2, \ldots, v_n$ be the elements of a vector space V defined over the field K and let $\alpha_1, \alpha_2, \ldots, \alpha_n$ be numbers. An expression like

$$\alpha_1 v_1 + \alpha_2 v_2 + \cdots + \alpha_n v_n \tag{5.16}$$

is called a **linear combination** of $v_1, v_2, \ldots, v_n$. As can be shown easily, the set V' of all linear combinations is a subspace of V and the vectors

$$v_1, v_2, \ldots, v_n \tag{5.17}$$

are called the **generators** of V'. If $V = V'$, we say $v_1, v_2, \ldots, v_n$ generate V over K. Consider a vector space V defined over the field K. If $v_1, v_2, \ldots, v_n$ are elements of V, then $v_1, v_2, \ldots, v_n$ are said to be **linearly dependent** over K; if there exist numbers, $\alpha_1, \alpha_2, \ldots, \alpha_n$, in K, not all of them are zero, such that

$$\alpha_1 v_1 + \alpha_2 v_2 + \cdots + \alpha_n v_n = 0. \tag{5.18}$$

If such a set of numbers, $\alpha_1, \alpha_2, \ldots, \alpha_n$, cannot be found, then we say the vectors $v_1, v_2, \ldots, v_n$ are **linearly independent**.

Example 5.4. *Linear independence:* Consider the vector space $V = \mathbb{R}^n$ and the vectors

$$\begin{aligned} e_1 &= (1, 0, \ldots, 0), \\ e_2 &= (0, 1, \ldots, 0), \\ &\ \ \vdots \\ e_n &= (0, 0, \ldots, 1), \end{aligned} \tag{5.19}$$

Since we cannot have

$$\alpha_1 e_1 + \alpha_2 e_2 + \cdots + \alpha_n e_n = 0 \tag{5.20}$$

unless all α_i are zero, $e_1, e_2, \ldots, e_n$ are linearly independent.

Example 5.5. *Linear independence of functions:* Consider the functions $\cos t$ and $\sin t$, where t is real. For these functions to be linearly independent, it should be impossible to find two real numbers, a and b, both nonzero, such that the equation

$$a \cos t + b \sin t = 0 \tag{5.21}$$

is satisfied for all t. Differentiating the above equation gives another equation,

$$-a \sin t + b \cos t = 0, \tag{5.22}$$

which, when combined with the original equation, gives

$$a^2 = -b^2. \tag{5.23}$$

Naturally, this cannot be satisfied in the real domain unless

$$a = b = 0. \tag{5.24}$$

Hence, $\cos t$ and $\sin t$ are linearly independent.

In general, let V be the vector space of all functions of the real variable t; then the linear independence of any given number of functions,

$$f_1(t), f_2(t), \ldots, f_n(t), \tag{5.25}$$

can be checked by showing that the equation

$$\alpha_1 f_1(t) + \alpha_2 f_2(t) + \cdots + \alpha_n f_n(t) = 0 \tag{5.26}$$

cannot be satisfied unless all α_i are zero.

Consider an arbitrary vector space V defined over the field K. Let $v_1, v_2, \ldots, v_n$ be linearly independent vectors in V. Suppose that we have two linear combinations that are equal:

$$\alpha_1 v_1 + \alpha_2 v_2 + \cdots + \alpha_n v_n = \beta_1 v_1 + \beta_2 v_2 + \cdots + \beta_n v_n, \tag{5.27}$$

then

$$\alpha_i = \beta_i, \quad i = 1, \ldots, n. \tag{5.28}$$

The proof is simple: We write Eq. (5.27) as

$$(\alpha_1 - \beta_1) v_1 + (\alpha_2 - \beta_2) v_2 + \cdots + (\alpha_n - \beta_n) v_n = 0. \tag{5.29}$$

Since $v_1, v_2, \ldots, v_n$ are linearly independent, the only way to satisfy the above equation is by having

$$\alpha_i = \beta_i, i = 1, \ldots, n. \tag{5.30}$$

We now define **basis vectors** of V over K as the set of elements $\{v_1, v_2, \ldots, v_n\}$ of V, which generate V and that are linearly independent. In this regard, the set of vectors $\{e_1, e_2, \ldots, e_n\}$ defined in Eq. (5.19), forms a basis for $\mathbb{R}^n$ over $\mathbb{R}$. In general, a basis for V is a set of linearly independent vectors $\{v_1, v_2, \ldots, v_n\}$ that **spans** V. In other words, every **element** v of V can be written as a **linear combination** of the **basis vectors** $\{v_1, v_2, \ldots, v_n\}$ as

$$\boxed{v = \alpha_1 v_1 + \alpha_2 v_2 + \cdots + \alpha_n v_n, \alpha_i \in K.} \tag{5.31}$$

A vector space is **finite-dimensional** if it has finite basis, where the **dimension** is defined as the number of the elements in its basis $\{v_1, v_2, \ldots, v_n\}$. In a vector space V, if one set of basis has n elements and another basis has m elements, then $n = m$. For a given vector space V over the field K, if U and W are **subspaces** such that

$$U + W = V \tag{5.32}$$

and

$$U \cap W = \{0\}, \tag{5.33}$$

that is, the intersection of U and W is the null set, then V is the **direct sum** of U with W, which is written as

$$\boxed{V = U \oplus W.} \tag{5.34}$$

Dimension of V is equal to the sum of the dimensions of U and W.

5.3   COMPONENTS

One of the advantages of introducing basis vectors $\{v_1, v_2, \ldots, v_n\}$ in n-dimensional space V is that we can define coordinates or components analogous to the coordinates of vectors in Cartesian space. **Components** of a **vector** v in V are the **scalars** $\alpha_1, \alpha_2, \ldots, \alpha_n$, which are used to express v as a **linear** combination of the **basis vectors** as

$$\boxed{v = \alpha_1 v_1 + \alpha_2 v_2 + \cdots + \alpha_n v_n = \sum_{i=1}^{n} \alpha_i v_i.} \tag{5.35}$$

For a given vector v and the basis $B = \{v_1, v_2, \ldots, v_n\}$, the scalars α_i, $i = 1, \ldots, n$, are unique. A given vector v can be conveniently represented as the

column matrix

$$\mathbf{v} = \begin{pmatrix} \alpha_1 \\ \alpha_2 \\ \vdots \\ \alpha_n \end{pmatrix}. \tag{5.36}$$

To indicate the **matrix representation,** we use **boldface,** that is, v and $\mathbf{v}$ represent the same element in V. In the matrix representation order of the numbers, α_i follow the order in which the basis vectors are written.

In a given n-dimensional space V, let

$$B = \{v_1, v_2, \ldots, v_n\} \tag{5.37}$$

and

$$B' = \{v_1', v_2', \ldots, v_n'\} \tag{5.38}$$

be two sets of bases. Then there exists unique scalars S_{ij}, $i = 1, 2, \ldots, n$, $j = 1, 2, \ldots, n$, such that

$$v_i' = \sum_{j=1}^{n} S_{ij} v_j. \tag{5.39}$$

Let α_i', $i = 1, \ldots, n$, be the components in terms of B' of a given v in V:

$$v = \alpha_1' v_1' + \alpha_2' v_2' + \cdots + \alpha_n' v_n' = \sum_{i=1}^{n} \alpha_i' v_i'. \tag{5.40}$$

Using Eq. (5.39), we can write Eq. (5.40) as

$$v = \sum_{i=1}^{n} \alpha_i' \left(\sum_{j=1}^{n} S_{ij} v_j \right) = \sum_{i=1}^{n} \sum_{j=1}^{n} S_{ij} v_j \alpha_i' = \sum_{j=1}^{n} \left(\sum_{i=1}^{n} S_{ij} \alpha_i' \right) v_j. \tag{5.41}$$

Since in the unprimed basis v is written as

$$v = \sum_{j=1}^{n} \alpha_j v_j, \tag{5.42}$$

we obtain the relation between the components found with respect to the primed and the unprimed bases as

$$\boxed{\alpha_j = \sum_{i=1}^{n} S_{ij} \alpha_i'.} \tag{5.43}$$

Representing S_{ij}, $i = 1, 2, \ldots, n$, $j = 1, 2, \ldots, n$, in Eq. (5.39) as an $n \times n$ matrix,

$$
\mathbf{S} = \begin{pmatrix} S_{11} & S_{12} & \cdots & S_{1n} \\ S_{21} & S_{22} & \cdots & S_{2n} \\ \vdots & \vdots & \ddots & \vdots \\ S_{n1} & S_{n2} & \cdots & S_{nn} \end{pmatrix},
\tag{5.44}
$$

we can write Eq. (5.43) as

$$
\boxed{\mathbf{v} = \widetilde{\mathbf{S}}\mathbf{v}',}
\tag{5.45}
$$

where $\mathbf{v}$ and $\mathbf{v}'$ are the column matrices

$$
\mathbf{v} = \begin{pmatrix} \alpha_1 \\ \alpha_2 \\ \vdots \\ \alpha_n \end{pmatrix}, \quad \mathbf{v}' = \begin{pmatrix} \alpha_1' \\ \alpha_2' \\ \vdots \\ \alpha_n' \end{pmatrix}
\tag{5.46}
$$

and $\widetilde{\mathbf{S}}$ is the transpose of $\mathbf{S}$. We now write Eq. (5.43) explicitly as

$$
\begin{aligned}
S_{11}\alpha_1' + S_{21}\alpha_2' + \cdots + S_{n1}\alpha_n' &= \alpha_1, \\
S_{12}\alpha_1' + S_{22}\alpha_2' + \cdots + S_{n2}\alpha_n' &= \alpha_2,
\end{aligned}
$$

$$
\vdots
$$

$$
S_{1n}\alpha_1' + S_{2n}\alpha_2' + \cdots + S_{nn}\alpha_n' = \alpha_n.
\tag{5.47}
$$

When $\mathbf{v} = 0$, that is,

$$
\alpha_1 = \alpha_2 = \cdots = \alpha_n = 0,
\tag{5.48}
$$

the above system of linear equations [Eq. (5.47)] is homogeneous, hence the only solution that it has is the trivial solution:

$$
\mathbf{v}' = 0
\tag{5.49}
$$

or

$$
\alpha_1' = \alpha_2' = \cdots = \alpha_n' = 0.
\tag{5.50}
$$

This means that the determinant of $\widetilde{\mathbf{S}}$ is different from zero. Remembering that for a square matrix the determinant of $\mathbf{S}$ is equal to the determinant of $\widetilde{\mathbf{S}}$, we conclude that both $\mathbf{S}$ and $\widetilde{\mathbf{S}}$ are **invertible**, hence, we can write

$$
\boxed{\mathbf{v}' = \widetilde{\mathbf{S}}^{-1}\mathbf{v}.}
\tag{5.51}
$$

We remind the reader that $\mathbf{v}'$ and $\mathbf{v}$ are essentially the same vector in V. They are just different representations of the same vector in terms of the primed, B', and the unprimed, B, bases, respectively [Eqs. (5.40) and (5.42)]. Note that if we prefer to write Eq. (5.39) as $v_i' = \sum_{j=1}^{n} S_{ji} v_j$, Eq. (5.45) becomes $\mathbf{v} = \mathbf{S}\mathbf{v}'$.

5.4   LINEAR TRANSFORMATIONS

Linear transformations, which are also called **linear operators**, help us to relate the elements of two vector spaces defined over the same field. A **linear transformation** T is defined as a function that transforms the elements of one vector space, V, into the elements of another, W, which are both defined over the field K and such that

$$\boxed{T(\alpha u + \beta v) = \alpha T(u) + \beta T(v)} \tag{5.52}$$

for all u, v in V and for all scalars α, β in K.

Example 5.6. *Space of polynomials:* Consider the differential operator $T = d/dx$ acting on the space V of polynomials. Given a polynomial $f(x)$ of order n:

$$f(x) = a_0 + a_1 x + a_2 x^2 + \cdots + a_n x^n, \tag{5.53}$$

the action of T on $f(x)$ is to produce another polynomial of order $n - 1$ as

$$\frac{d}{dx} f(x) = a_1 + 2a_2 x + \cdots + na_n x^{n-1}. \tag{5.54}$$

It is easy to check that T is a linear transformation from V into V.

Example 5.7. *Linear transformations:* Consider the matrices

$$\mathbf{A} = \begin{pmatrix} a_{11} & a_{12} & \cdots & a_{1n} \\ a_{21} & a_{22} & \cdots & a_{2n} \\ \vdots & \vdots & \ddots & \vdots \\ a_{m1} & a_{m2} & \cdots & a_{mn} \end{pmatrix}, \quad \mathbf{x} = \begin{pmatrix} x_1 \\ x_2 \\ \vdots \\ x_n \end{pmatrix} \tag{5.55}$$

with elements in the field K. The function T defined by the equation

$$T(x) = \mathbf{A}\mathbf{x} \tag{5.56}$$

is a linear transformation from the $(n \times 1)$-tuple space K^n to the $(m \times 1)$-tuple space K^m.

Example 5.8. *Linear Transformations:* Let V be the space of continuous functions defined over the field of real numbers $\mathbb{R}$. The transformation defined as

$$(Tf)(x) = \int_0^x f(x)\, dt, \tag{5.57}$$

is a linear transformation from V into V, where $f(x)$ is an element of V. Linearity of the transformation follows from the properties of the Riemann integral.

An important property of linear transformations is that they preserve linear combinations, that is, if $u_1, u_2, \ldots, u_n$ are vectors in V, then

$$T(c_1 u_1 + c_2 u_2 + \cdots + c_n u_n) = c_1 T(u_1) + c_2 T(u_2) + \cdots + c_n T(u_n). \quad (5.58)$$

We now state an important theorem:

Theorem 5.1. In a given finite-dimensional vector space V defined over the field K, let the basis vectors be $\{v_1, v_2, \ldots, v_n\}$. In another vector space W defined over the same field K, let $\{w_1, w_2, \ldots, w_n\}$ be any set of vectors. Then, there is precisely one linear transformation T from V into W such that

$$T(v_i) = w_i, \quad i = 1, \ldots, n. \quad (5.59)$$

The proof of this theorem can be found in Hoffman and Kunze [1]. Its importance is in formally stating the central role that linear transformations play among many possible transformations from V into W.

5.5 MATRIX REPRESENTATION OF TRANSFORMATIONS

Consider an n-dimensional vector space V over the field K with the basis $B_V = \{v_1, v_2, \ldots, v_n\}$. Similarly, let W be an m-dimensional vector space over the same field K with the basis $B_W = \{w_1, w_2, \ldots, w_m\}$. If T is a given linear transformation from V into W over the field K, then the effect of T on a vector of V will be to convert it into a vector in W. In other words, the transformed vector, Tv, can be expressed uniquely as a linear combination of the basis vectors of W. Similarly, each one of the transformed basis vectors, Tv_j, can be uniquely expressed as a linear combination:

$$\boxed{Tv_j = \sum_{i=1}^{m} A_{ij} w_i,} \quad (5.60)$$

in terms of the basis vectors $\{w_1, w_2, \ldots, w_m\}$, where the scalars A_{ij} correspond to the components of Tv_j in the W space in terms of the basis B_W. In other words, the $m \times n$ matrix, A_{ij}, $i = 1, \ldots, m$, $j = 1, \ldots, n$, uniquely defines the effect of T on the vectors of V. Hence it is called the **transformation matrix** of T with respect to the bases B_V and B_W.

Consider an arbitrary vector v in V, which we write in terms of the basis B_V as

$$v = \alpha_1 v_1 + \alpha_2 v_2 + \cdots + \alpha_n v_n = \sum_{j=1}^{n} \alpha_j v_j. \quad (5.61)$$

If we act on v by T, we obtain

$$Tv = T \sum_{j=1}^{n} \alpha_j v_j = \sum_{j=1}^{n} \alpha_j (Tv_j). \tag{5.62}$$

Using Eq. (5.60), we write this as

$$Tv = \sum_{j=1}^{n} \alpha_j \sum_{i=1}^{m} A_{ij} w_i = \sum_{i=1}^{m} \left[\sum_{j=1}^{n} A_{ij} \alpha_j \right] w_i. \tag{5.63}$$

In other words,

$$\boxed{\beta_i = \sum_{j=1}^{n} A_{ij} \alpha_j} \tag{5.64}$$

are the components, β_i, $i = 1, 2, \ldots m$, of Tv in terms of the basis B_W of the space W. Using the matrix representations of **v**, **w**, and **A**:

$$\mathbf{v} = \begin{pmatrix} \alpha_1 \\ \alpha_2 \\ \vdots \\ \alpha_n \end{pmatrix}, \quad \mathbf{w} = \begin{pmatrix} \beta_1 \\ \beta_2 \\ \vdots \\ \beta_m \end{pmatrix}, \tag{5.65}$$

$$\mathbf{A} = \begin{pmatrix} A_{11} & A_{12} & \cdots & A_{1n} \\ A_{21} & A_{22} & \cdots & A_{2n} \\ \vdots & \vdots & \ddots & \vdots \\ A_{m1} & A_{m2} & \cdots & A_{mn} \end{pmatrix}, \tag{5.66}$$

we can write Eq. (5.64) as

$$\boxed{\mathbf{w} = \mathbf{A}\mathbf{v}.} \tag{5.67}$$

We now summarize this result formally as a theorem:

Theorem 5.2. Let V be an n-dimensional vector space over the field K with the basis vectors $B_V = \{v_1, v_2, \ldots, v_n\}$ and let W be an m-dimensional vector space over the same field with the basis vectors $B_W = \{w_1, w_2, \ldots, w_m\}$. For each linear transformation T from V into W, we can write an $m \times n$ matrix **A** with the entries in the field K such that

$$[Tv]_{B_W} = \mathbf{A}\mathbf{v} = \mathbf{A}[v]_{B_V}, \tag{5.68}$$

where v is any element in V. Furthermore, the transformation from $T \rightarrow A$ is a one-to-one correspondence between the set of all linear transformations from V into W and the set of all $m \times n$ matrices over the field K. We call the matrix

A the transformation matrix of T with respect to the bases B_V and B_W. To make this dependence explicit, we write the matrix that represents T as

$$\mathbf{A} = [T]_{B_W B_V}. \tag{5.69}$$

We have written the subscripts as $B_W B_V$, since the row dimension is determined by W and the column dimension is determined by V.

5.6   ALGEBRA OF TRANSFORMATIONS

Let T and U be two **linear** transformations from V into W, where V is a vector space of dimension n, and W is a vector space of dimension m, both of which are defined over the field K. Let $\mathbf{A}$ and $\mathbf{B}$ be the transformation matrices of T and U with respect to the bases B_V and B_W, respectively. If P is the transformation written as the linear combination

$$P = c_1 T + c_2 U, \tag{5.70}$$

where c_1 and c_2 are scalars in field K, the transformation matrix of P with respect to the bases B_V and B_W is the same linear combination of the transformation matrices of T and U with respect to the same bases:

$$\boxed{[P]_{B_W B_V} = c_1 [T]_{B_W B_V} + c_2 [U]_{B_W B_V}.} \tag{5.71}$$

For the **proof** we first write

$$Tv_j = \sum_{i=1}^{m} A_{ij} w_i, \tag{5.72}$$

$$Uv_j = \sum_{i=1}^{m} B_{ij} w_i, \tag{5.73}$$

where v_j belongs to the basis $B_V = \{v_1, v_2, \ldots, v_n\}$ and w_j belongs to $B_W = \{w_1, w_2, \ldots, w_m\}$. We then write

$$Pv_j = (c_1 T + c_2 U)v_j = c_1(Tv_j) + c_2(Uv_j) \tag{5.74}$$

$$= c_1 \left[\sum_{i=1}^{m} A_{ij} w_i \right] + c_2 \left[\sum_{i=1}^{m} B_{ij} w_i \right] \tag{5.75}$$

$$= \sum_{i=1}^{m} [c_1 A_{ij} + c_2 B_{ij}] w_i, \tag{5.76}$$

thus obtaining

$$[P]_{B_W B_V} = c_1 A_{ij} + c_2 B_{ij}, \quad i = 1, \ldots, m, \quad j = 1, \ldots, n. \tag{5.77}$$

Let us now consider the **product** of transformations. Given three finite-dimensional vector spaces, V, W, Z, defined over the field K, let T be a linear transformation from V into W and let U be a linear transformation from W into Z. Also let $B_V = \{v_1, v_2, \ldots, v_n\}$, $B_W = \{w_1, w_2, \ldots, w_m\}$, and $B_Z = \{z_1, z_2, \ldots, z_p\}$ be the bases for the spaces V, W, and Z, respectively. We can write

$$(UT)(v_j) = U(Tv_j) \tag{5.78}$$

$$= U\left(\sum_{k=1}^{m} A_{kj} w_k\right) = \sum_{k=1}^{m} A_{kj}(U w_k) \tag{5.79}$$

$$= \sum_{k=1}^{m} A_{kj}\left(\sum_{i=1}^{p} B_{ik} z_i\right) = \sum_{i=1}^{p}\left(\sum_{k=1}^{m} B_{ik} A_{kj}\right) z_i, \tag{5.80}$$

where A_{kj}, $k = 1, \ldots, m$, $j = 1, \ldots, n$, is the transformation matrix of T with respect to the bases B_V and B_W, and B_{ik}, $i = 1, \ldots, p$, $k = 1, \ldots, m$, is the transformation matrix of U with respect to the bases B_W and B_Z. In other words, the **matrix** of the **product transformation**:

$$\boxed{[UT]_{B_Z B_V} = [U]_{B_Z B_W} [T]_{B_W B_V},} \tag{5.81}$$

is the **product** of the transformation matrices of U and T.

When T and U are two linear transformations in the same space V with respect to the basis B_V, we write

$$[UT]_{B_V} = [U]_{B_V} [T]_{B_V}. \tag{5.82}$$

The **inverse** transformation is defined as

$$\boxed{[T^{-1}T]_{B_V} = [TT^{-1}]_{B_V} = \mathbf{I}.} \tag{5.83}$$

Using Eq. (5.81) we can conclude that

$$[T^{-1}]_{B_V} [T]_{B_V} = [T]_{B_V} [T^{-1}]_{B_V} = \mathbf{I}, \tag{5.84}$$

$$[T^{-1}]_{B_V} = [T]_{B_V}^{-1}. \tag{5.85}$$

That is, the matrix of the inverse transformation is the inverse matrix of the matrix of the transformation. The one-to-one correspondence that B_V establishes between transformations and matrices guarantees that a linear transformation is invertible, if and only if the transformation matrix is **invertible**, that is, **nonsingular**:

$$\boxed{\det [T]_{B_V} \neq 0.} \tag{5.86}$$

5.7 CHANGE OF BASIS

We have seen that the matrix representation of a given transformation depends on the bases used. Since there are infinitely many possible bases for a given n-dimensional vector space, we would like to find how the matrix elements change under a change of bases.

Let T be a given linear transformation or an operator in n-dimensional vector space V and let $B_V = \{v_1, v_2, \ldots, v_n\}$ and $B'_V = \{v'_1, v'_2, \ldots, v'_n\}$ be two possible sets of bases for V. We need a relation between the matrices

$$[T]_{B_V} \text{ and } [T]_{B'_V}. \tag{5.87}$$

Let $\mathbf{S}$ be the unique $n \times n$ matrix, S_{ij}, $i = 1, 2, \ldots, n$, $j = 1, 2, \ldots, n$, relating the components of a vector v in V in terms of the two bases as

$$\boxed{[v]_{B_V} = \mathbf{S}[v]_{B'_V},} \tag{5.88}$$

where

$$[v]_{B_V} = \begin{pmatrix} \alpha_1 \\ \alpha_2 \\ \vdots \\ \alpha_n \end{pmatrix}, \quad [v]_{B'_V} = \begin{pmatrix} \alpha'_1 \\ \alpha'_2 \\ \vdots \\ \alpha'_n \end{pmatrix}. \tag{5.89}$$

By definition, for the linear transformation T we can write

$$[Tv]_{B_V} = [T]_{B_V}[v]_{B_V}. \tag{5.90}$$

Applying Eq. (5.88) to the vector Tv, we obtain

$$[Tv]_{B_V} = \mathbf{S}[Tv]_{B'_V}, \tag{5.91}$$

which, when substituted in Eq. (5.90), gives

$$[T]_{B_V}[v]_{B_V} = \mathbf{S}[Tv]_{B'_V}. \tag{5.92}$$

We now substitute Eq. (5.88) into the left-hand side of the above equation to write

$$[T]_{B_V}\mathbf{S}[v]_{B'_V} = \mathbf{S}[Tv]_{B'_V}. \tag{5.93}$$

Multiplying both sides by the inverse $\mathbf{S}^{-1}$, we obtain

$$\mathbf{S}^{-1}[T]_{B_V}\mathbf{S}[v]_{B'_V} = \mathbf{S}^{-1}\mathbf{S}[Tv]_{B'_V}, \tag{5.94}$$

which we write as

$$[Tv]_{B'_V} = (\mathbf{S}^{-1}[T]_{B_V}\mathbf{S})[v]_{B'_V}. \tag{5.95}$$

Comparing with the corresponding equation [Eq. (5.90)] written in the B'_V basis:

$$[Tv]_{B'_V} = [T]_{B'_V}[v]_{B'_V}, \tag{5.96}$$

we obtain the expression for the transformation matrix of T in the B'_V basis, $[T]_{B'_V}$, in terms of $[T]_{B_V}$ as

$$\boxed{[T]_{B'_V} = \mathbf{S}^{-1}[T]_{B_V}\mathbf{S}.} \tag{5.97}$$

Equation (5.88) means that the components of v in the bases B_V and B'_V are related as

$$\boxed{\alpha_i = \sum_{j=1}^{n} S_{ij}\alpha'_j.} \tag{5.98}$$

In terms of the basis B_V, v is written as $v = \sum_i \alpha_i v_i$. Using Eq. (5.98) we can write it in the basis B'_V as

$$v = \sum_{i=1}^{n}\sum_{j=1}^{n} S_{ij}\alpha'_j v_i = \sum_{j=1}^{n}\left[\sum_{i=1}^{n} S_{ij}v_i\right]\alpha'_j = \sum_j \alpha'_j v'_j. \tag{5.99}$$

In other words, the transpose of the $n \times n$ matrix $\mathbf{S}$ relates the two sets of basis vectors as

$$\boxed{\sum_{i=1}^{n} S_{ij}v_i = v'_j.} \tag{5.100}$$

If two $n \times n$ matrices, $\mathbf{A}$ and $\mathbf{B}$, defined over a field K, can be related by an invertible $n \times n$ matrix $\mathbf{S}$ over K as

$$\boxed{\mathbf{B} = \mathbf{S}^{-1}\mathbf{A}\mathbf{S},} \tag{5.101}$$

we say that $\mathbf{B}$ is **similar** to $\mathbf{A}$ over K. In conjunction with our result in Eq. (5.97), when we say $[T]_{B'_V}$ is similar to $[T]_{B_V}$ means that on each n-dimensional vector space V over K, the matrices $[T]_{B_V}$ and $[T]_{B'_V}$ represent the same transformation, or operator, in terms of their respective bases B_V and B'_V, defined in V. Note that if $\mathbf{B}$ is similar to $\mathbf{A}$, then $\mathbf{A}$ is similar to $\mathbf{B}$ through

$$\boxed{\mathbf{A} = \mathbf{S}\mathbf{B}\mathbf{S}^{-1}.} \tag{5.102}$$

5.8   INVARIANTS UNDER SIMILARITY TRANSFORMATIONS

Under similarity transformations, the **determinant** and the **trace** of a matrix remains unchanged. If $\mathbf{A}$ and $\mathbf{B}$ are two similar matrices, $\mathbf{A} = \mathbf{S}\mathbf{B}\mathbf{S}^{-1}$, then their determinants are equal. The proof follows from the properties of determinants:

$$\det \mathbf{A} = \det \left(\mathbf{SBS}^{-1} \right) = \det \mathbf{S} \det \mathbf{B} \det \mathbf{S}^{-1}. \tag{5.103}$$

Since $\det \mathbf{S}^{-1} = 1/\det \mathbf{S}$, we get $\det \mathbf{A} = \det \mathbf{B}$. If $\mathbf{A}$ and $\mathbf{B}$ are two similar matrices, then their traces are equal:

$$\mathrm{tr}\mathbf{B} = \sum_{i=1}^{n} [B]_{ii} = \sum_{i=1}^{n} [\mathbf{S}^{-1}\mathbf{AS}]_{ii} = \sum_{i=1}^{n}\sum_{j=1}^{n}\sum_{k=1}^{n} [\mathbf{S}^{-1}]_{ij}[\mathbf{A}]_{jk}[\mathbf{S}]_{ki} \tag{5.104}$$

$$= \sum_{i=1}^{n}\sum_{j=1}^{n}\sum_{k=1}^{n} [\mathbf{S}]_{ki}[\mathbf{S}^{-1}]_{ij}[\mathbf{A}]_{jk} = \sum_{j=1}^{n}\sum_{k=1}^{n} [\mathbf{SS}^{-1}]_{kj}[\mathbf{A}]_{jk} \tag{5.105}$$

$$= \sum_{j=1}^{n}\sum_{k=1}^{n} \delta_{kj}[\mathbf{A}]_{jk} = \sum_{k=1}^{n} [\mathbf{A}]_{kk}. \tag{5.106}$$

5.9 EIGENVALUES AND EIGENVECTORS

We have seen that matrix representations of linear operators depend on the bases used. Since we have also established the fact that two different representations of a given operator are related by a similarity transformation, we are ready to search for a basis that presents the greatest advantage. Since diagonal matrices are the simplest matrices, whose rank and determinant can be read at a glance, our aim is to find a basis in which a given linear operator T is represented by a diagonal matrix. The question that needs to be answered is, Can we always represent a linear operator by a diagonal matrix? If the answer is yes, then how do we find such a basis? Since in a given basis, $B = \{v_1, v_2, \ldots, v_n\}$, the matrix representation of T is **diagonal**:

$$[T]_B = \lambda_i \delta_{ij}, i, j = 1, 2, \ldots n = \begin{pmatrix} \lambda_1 & 0 & \cdots & 0 \\ 0 & \lambda_2 & \cdots & 0 \\ \vdots & \vdots & \ddots & \vdots \\ 0 & 0 & \cdots & \lambda_n \end{pmatrix}, \tag{5.107}$$

if and only if

$$Tv_i = \sum_{j=1}^{n}(\lambda_i \delta_{ij})v_j = \lambda_i v_i, \tag{5.108}$$

we start by searching for vectors that are sent to multiples of themselves by T.

Definition 5.3. Let T be a linear operator on a vector space V defined over the field K. An **eigenvalue**, also called the **characteristic value**, is a **scalar**,

λ, such that there is a nonzero **vector** v in V with

$$Tv = \lambda v, \tag{5.109}$$

where v is called the **eigenvector** of T corresponding to the **eigenvalue** λ.

In an n-dimensional vector space a linear operator is represented by an $n \times n$ matrix, hence we concentrate on determining the eigenvalues and the eigenvectors of a given square matrix $\mathbf{A}$. We can write the eigenvalue equation (5.109) as the matrix equation

$$\boxed{(\mathbf{A} - \lambda \mathbf{I})\mathbf{v} = 0,} \tag{5.110}$$

where $\mathbf{A} = A_{ij}$, $i, j = 1, \ldots, n$ is an $n \times n$ matrix. Eigenvector $\mathbf{v}$, corresponding to the eigenvalue λ, is represented by the $n \times 1$ column matrix:

$$\mathbf{v} = \begin{pmatrix} a_1 \\ a_2 \\ \vdots \\ a_n \end{pmatrix}, \tag{5.111}$$

where a_i are the components. The $n \times n$ identity matrix is written as $\mathbf{I}$. When Eq. (5.110) is written explicitly, we obtain a linear system of homogeneous equations:

$$(A_{11} - \lambda)a_1 + A_{12}a_2 + \cdots + A_{1n}a_n = 0,$$
$$A_{21}a_1 + (A_{22} - \lambda)a_2 + \cdots + A_{2n}a_n = 0,$$
$$\vdots$$
$$A_{n1}a_1 + A_{n2}a_2 + \cdots + (A_{nn} - \lambda)a_n = 0, \tag{5.112}$$

which has a nontrivial solution if and only if the determinant of the coefficients vanishes:

$$\boxed{|\mathbf{A} - \lambda \mathbf{I}| = \begin{vmatrix} (A_{11} - \lambda) & A_{12} & \cdots & A_{1n} \\ A_{21} & (A_{22} - \lambda) & \cdots & A_{2n} \\ \vdots & \vdots & \ddots & \vdots \\ A_{n1} & A_{n2} & \cdots & (A_{nn} - \lambda) \end{vmatrix} = 0.} \tag{5.113}$$

This gives an nth-order polynomial in λ called the **characteristic equation**. A polynomial of order n has n roots, not necessarily all distinct, in the complex field $\mathbb{C}$. The set of eigenvalues $\{\lambda_i\}$ is called the **spectrum** of $\mathbf{A}$. When some of the roots are multiple, the spectrum is called **degenerate**. In general, a given polynomial may not have real roots, hence from now on, unless otherwise stated, we work in the complex field $\mathbb{C}$.

To identify the eigenvalues, we write the eigenvalue equation as

$$\boxed{\mathbf{A}\mathbf{v}_k = \lambda_k \mathbf{v}_k,} \tag{5.114}$$

where $\mathbf{v}_k$ is the eigenvector belonging to the kth eigenvalue λ_k.

Theorem 5.3. If $\mathbf{A}$ and $\mathbf{B}$ are two **similar matrices**, then they have the same characteristic equation and the same eigenvalues, that is, the **same spectrum**.

Proof is given as follows: Since $\mathbf{A}$ and $\mathbf{B}$ are similar matrices:

$$\mathbf{S}^{-1}\mathbf{A}\mathbf{S} = \mathbf{B}, \tag{5.115}$$

we can write

$$\det(\mathbf{B} - \lambda\mathbf{I}) = \det(\mathbf{S}^{-1}\mathbf{A}\mathbf{S} - \lambda\mathbf{I}) \tag{5.116}$$

$$= \det(\mathbf{S}^{-1}\mathbf{A}\mathbf{S} - \mathbf{S}^{-1}\lambda\mathbf{I}\mathbf{S}) \tag{5.117}$$

$$= \det(\mathbf{S}^{-1}(\mathbf{A} - \lambda\mathbf{I})\mathbf{S}) \tag{5.118}$$

$$= \det\mathbf{S}^{-1}\det(\mathbf{A} - \lambda\mathbf{I})\det\mathbf{S} \tag{5.119}$$

$$= \det(\mathbf{A} - \lambda\mathbf{I}), \tag{5.120}$$

thereby proving that $\mathbf{A}$ and $\mathbf{B}$ have the same characteristic equation, thus the same eigenvalues. We now write the matrix equation [Eq. (5.114)] as

$$\sum_{j=1}^{n} A_{ij} a_{jk} = \lambda_k a_{ik}, \tag{5.121}$$

where a_{jk} is the jth component of the kth eigenvector. Consider the $n \times n$ matrix $\mathbf{S}$ whose columns are the eigenvectors $\mathbf{v}_k$, $k = 1, 2, \ldots, n$:

$$\mathbf{S} = \left(\begin{pmatrix} \\ \mathbf{v}_1 \\ \end{pmatrix} \begin{pmatrix} \\ \mathbf{v}_2 \\ \end{pmatrix} \cdots \begin{pmatrix} \\ \mathbf{v}_n \\ \end{pmatrix} \right), \tag{5.122}$$

where

$$\mathbf{v}_k = \begin{pmatrix} a_{1k} \\ a_{2k} \\ \vdots \\ a_{nk} \end{pmatrix}. \tag{5.123}$$

If the eigenvectors are linearly independent, then

$$\det\mathbf{S} \neq 0. \tag{5.124}$$

Hence, the inverse of S exists. We can now write Eq. (5.121) as

$$\sum_{j=1}^{n} A_{ij} S_{jk} = \lambda_k S_{ik} \tag{5.125}$$

$$= \sum_{l=1}^{n} S_{il}(\delta_{lk}\lambda_k) \tag{5.126}$$

$$= \sum_{l=1}^{n} S_{il} D_{lk}, \quad i, k = 1, \ldots, n, \tag{5.127}$$

where we have introduced the diagonal matrix D_{lk} as

$$\mathbf{D} = \begin{pmatrix} \lambda_1 & 0 & \cdots & 0 \\ 0 & \lambda_2 & \cdots & 0 \\ \vdots & \vdots & \ddots & \vdots \\ 0 & 0 & \cdots & \lambda_n \end{pmatrix}. \tag{5.128}$$

We can also write Eq. (5.127) as

$$\mathbf{AS} = \mathbf{SD}, \tag{5.129}$$

which, after multiplying both sides by the inverse $\mathbf{S}^{-1}$, becomes

$$\mathbf{S}^{-1}\mathbf{AS} = \mathbf{S}^{-1}\mathbf{SD} = \mathbf{D}. \tag{5.130}$$

In other words, the matrix $\mathbf{S}$, columns of which are the eigenvectors of $\mathbf{A}$, diagonalizes $\mathbf{A}$ by the similarity transformation

$$\mathbf{S}^{-1}\mathbf{AS} = \mathbf{D}. \tag{5.131}$$

We now express these results formally in terms of the following theorems:

Theorem 5.4. Let V be an n-dimensional vector space defined over the field K and let A be a linear operator or transformation from V into V. Let $v_1, v_2, \ldots, v_n$ be the eigenvectors of A corresponding to the eigenvalues $\lambda_1, \lambda_1, \ldots, \lambda_n$, respectively. If the eigenvalues are **distinct**:

$$\lambda_i \neq \lambda_j, \quad \text{when } i \neq j, \tag{5.132}$$

then the corresponding eigenvectors, $v_1, v_2, \ldots, v_n$, are **linearly independent** and span V.

Theorem 5.5. Let V be an n-dimensional vector space over the field K and let A be a linear operator from V into V. Assume that there exists a basis $B = \{v_1, v_2, \ldots, v_n\}$ that spans V consisting of the eigenvectors of A, with the eigenvalues $\lambda_1, \lambda_1, \ldots, \lambda_n$, respectively, then the matrix of A with respect to this basis is the diagonal matrix

$$[A]_B = \mathbf{D} = \begin{pmatrix} \lambda_1 & 0 & \cdots & 0 \\ 0 & \lambda_2 & \cdots & 0 \\ \vdots & \vdots & \ddots & \vdots \\ 0 & 0 & \cdots & \lambda_n \end{pmatrix}. \tag{5.133}$$

Example 5.9. *Characteristic equation and the eigenvalues:* Consider the matrix

$$\mathbf{A} = \begin{pmatrix} 1 & 0 & 1 \\ 0 & 1 & 0 \\ 1 & 0 & 1 \end{pmatrix}, \tag{5.134}$$

where the characteristic equation is

$$\det(\mathbf{A} - \lambda\mathbf{I}) = \begin{vmatrix} (1-\lambda) & 0 & 1 \\ 0 & (1-\lambda) & 0 \\ 1 & 0 & (1-\lambda) \end{vmatrix} \tag{5.135}$$

$$= (1-\lambda)(1-\lambda)^2 + (1)(-1)(1-\lambda) \tag{5.136}$$

$$= (1-\lambda)[(1-\lambda)^2 - 1] \tag{5.137}$$

$$= (1-\lambda)(1 - 2\lambda + \lambda^2 - 1) \tag{5.138}$$

$$= \lambda(1-\lambda)(\lambda - 2) = 0. \tag{5.139}$$

Roots of the characteristic equation gives the three real and distinct eigenvalues:

$$\lambda_1 = 0, \quad \lambda_2 = 1, \quad \lambda_3 = 2. \tag{5.140}$$

Example 5.10. *Characteristic equation and the eigenvalues:* Consider

$$\mathbf{A} = \begin{pmatrix} 0 & 2 \\ -2 & 0 \end{pmatrix}, \tag{5.141}$$

where the characteristic equation:

$$\det(\mathbf{A} - \lambda\mathbf{I}) = \begin{vmatrix} -\lambda & 2 \\ -2 & -\lambda \end{vmatrix} = \lambda^2 + 4 = 0, \tag{5.142}$$

gives two complex eigenvalues as $\lambda_{1,2} = \pm 2i$.

Example 5.11. *Eigenvalues and eigenvectors:* Consider the following 3×3 matrix:

$$\mathbf{A} = \begin{pmatrix} 1 & \sqrt{2} & 0 \\ \sqrt{2} & 0 & 0 \\ 0 & 0 & 0 \end{pmatrix}. \tag{5.143}$$

From the roots of the characteristic equation:

$$\det(\mathbf{A} - \lambda\mathbf{I}) = \begin{vmatrix} (1-\lambda) & \sqrt{2} & 0 \\ \sqrt{2} & (0-\lambda) & 0 \\ 0 & 0 & (0-\lambda) \end{vmatrix} \quad (5.144)$$

$$= (1-\lambda)(-\lambda)(-\lambda) - \sqrt{2}(-\lambda)\sqrt{2} \quad (5.145)$$

$$= (1-\lambda)\lambda^2 + 2\lambda \quad (5.146)$$

$$= \lambda\{(1-\lambda)\lambda + 2\} \quad (5.147)$$

$$= -\lambda(\lambda^2 - \lambda - 2) \quad (5.148)$$

$$= -\lambda(\lambda - 2)(\lambda + 1) = 0, \quad (5.149)$$

we obtain the eigenvalues as

$$\lambda_1 = 0, \quad \lambda_2 = 2, \quad \lambda_3 = -1. \quad (5.150)$$

We now find the eigenvectors one by one, by using the eigenvalue equation

$$\mathbf{A}\mathbf{v}_k = \lambda_k\mathbf{v}_k. \quad (5.151)$$

For $\lambda_1 = 0$ this gives

$$\begin{pmatrix} 1 & \sqrt{2} & 0 \\ \sqrt{2} & 0 & 0 \\ 0 & 0 & 0 \end{pmatrix} \begin{pmatrix} a_{11} \\ a_{21} \\ a_{31} \end{pmatrix} = 0 \begin{pmatrix} a_{11} \\ a_{21} \\ a_{31} \end{pmatrix}, \quad (5.152)$$

which leads to the equations to be solved for the components as

$$a_{11} + \sqrt{2}a_{21} = 0, \quad (5.153)$$

$$\sqrt{2}a_{11} + 0a_{31} = 0, \quad (5.154)$$

$$0a_{31} = 0. \quad (5.155)$$

The last equation leaves the third component arbitrary. Hence, we take it as

$$a_{31} = c_1, \quad c_1 \neq 0, \quad (5.156)$$

where c_1 is any constant. The first two equations determine the remaining components as

$$a_{11} = a_{21} = 0, \quad (5.157)$$

thus giving the first eigenvector as

$$\mathbf{v}_1 = c_1 \begin{pmatrix} 0 \\ 0 \\ 1 \end{pmatrix}. \quad (5.158)$$

Similarly, the other eigenvectors are found. For $\lambda_2 = 2$, we write

$$\begin{pmatrix} 1 & \sqrt{2} & 0 \\ \sqrt{2} & 0 & 0 \\ 0 & 0 & 0 \end{pmatrix} \begin{pmatrix} a_{12} \\ a_{22} \\ a_{32} \end{pmatrix} = 2 \begin{pmatrix} a_{12} \\ a_{22} \\ a_{32} \end{pmatrix}, \tag{5.159}$$

$$a_{12} + \sqrt{2}a_{22} = 2a_{12}, \tag{5.160}$$

$$\sqrt{2}a_{12} = 2a_{22}, \tag{5.161}$$

$$0 = 2a_{32}, \tag{5.162}$$

and obtain the second eigenvector as

$$a_{12} = \sqrt{2}c_2, \quad a_{22} = c_2, \quad a_{32} = 0, \tag{5.163}$$

$$\mathbf{v}_2 = c_2 \begin{pmatrix} \sqrt{2} \\ 1 \\ 0 \end{pmatrix}, \quad c_2 \neq 0, \tag{5.164}$$

where c_2 is an arbitrary constant. For $\lambda_3 = -1$, we write

$$\begin{pmatrix} 1 & \sqrt{2} & 0 \\ \sqrt{2} & 0 & 0 \\ 0 & 0 & 0 \end{pmatrix} \begin{pmatrix} a_{13} \\ a_{23} \\ a_{33} \end{pmatrix} = -1 \begin{pmatrix} a_{13} \\ a_{23} \\ a_{33} \end{pmatrix}, \tag{5.165}$$

$$a_{13} + \sqrt{2}a_{23} = -a_{13}, \tag{5.166}$$

$$\sqrt{2}a_{13} + 0a_{23} = -a_{23}, \tag{5.167}$$

$$0 = -a_{33} \tag{5.168}$$

and obtain the third eigenvector as

$$\mathbf{v}_3 = c_3 \begin{pmatrix} 1 \\ -\sqrt{2} \\ 0 \end{pmatrix}, \quad c_3 \neq 0, \tag{5.169}$$

where c_3 is an arbitrary constant. For the time being, we leave these constants in the eigenvectors arbitrary. Using these eigenvectors, we construct the transformation matrix $\mathbf{S}$ as

$$\mathbf{S} = \left(\begin{pmatrix} \\ \mathbf{v}_1 \\ \end{pmatrix} \begin{pmatrix} \\ \mathbf{v}_2 \\ \end{pmatrix} \begin{pmatrix} \\ \mathbf{v}_3 \\ \end{pmatrix} \right) \tag{5.170}$$

$$= \begin{pmatrix} 0 & \sqrt{2}c_2 & c_3 \\ 0 & c_2 & -\sqrt{2}c_3 \\ c_1 & 0 & 0 \end{pmatrix}. \tag{5.171}$$

Since the determinant of $\mathbf{S}$ is nonzero:

$$\det \mathbf{S} = -3c_1 c_2 c_3 \neq 0, \tag{5.172}$$

the inverse transformation, $\mathbf{S}^{-1}$, exists and its components can be found by using the formula [Eq. (4.213)]

$$\boxed{S_{ij}^{-1} = (-1)^{i+j} \frac{\det \mathbf{S}(j|i)}{\det \mathbf{S}},} \tag{5.173}$$

where $\det \mathbf{S}(j|i)$ is the minor obtained by deleting the jth row and the ith column. Note that the order of the indices in $\det \mathbf{S}(j|i)$ are reversed; hence, we first write the transpose of $\mathbf{S}$ as

$$\tilde{\mathbf{S}} = \begin{pmatrix} 0 & 0 & c_1 \\ \sqrt{2}c_2 & c_2 & 0 \\ c_3 & -\sqrt{2}c_3 & 0 \end{pmatrix} \tag{5.174}$$

and then construct the inverse by finding the minors one by one as

$$
\begin{aligned}
\mathbf{S}^{-1} &= \frac{1}{\det \mathbf{S}}
\begin{pmatrix}
(+)\begin{vmatrix} c_2 & 0 \\ -\sqrt{2}c_3 & 0 \end{vmatrix} & (-)\begin{vmatrix} \sqrt{2}c_2 & 0 \\ c_3 & 0 \end{vmatrix} & (+)\begin{vmatrix} \sqrt{2}c_2 & c_2 \\ c_3 & -\sqrt{2}c_3 \end{vmatrix} \\[2mm]
(-)\begin{vmatrix} 0 & c_1 \\ -\sqrt{2}c_3 & 0 \end{vmatrix} & (+)\begin{vmatrix} 0 & c_1 \\ c_3 & 0 \end{vmatrix} & (-)\begin{vmatrix} 0 & 0 \\ c_3 & -\sqrt{2}c_3 \end{vmatrix} \\[2mm]
(+)\begin{vmatrix} 0 & c_1 \\ c_2 & 0 \end{vmatrix} & (-)\begin{vmatrix} 0 & c_1 \\ \sqrt{2}c_2 & 0 \end{vmatrix} & (+)\begin{vmatrix} 0 & 0 \\ \sqrt{2}c_2 & c_2 \end{vmatrix}
\end{pmatrix} \\[4mm]
&= \frac{1}{-3c_1 c_2 c_3}
\begin{pmatrix}
0 & 0 & -3c_2 c_3 \\
-\sqrt{2}c_1 c_3 & -c_1 c_3 & 0 \\
-c_1 c_2 & \sqrt{2}c_1 c_2 & 0
\end{pmatrix} \\[4mm]
&= \begin{pmatrix}
0 & 0 & \dfrac{1}{c_1} \\[2mm]
\dfrac{\sqrt{2}}{3c_2} & \dfrac{1}{3c_2} & 0 \\[2mm]
\dfrac{1}{3c_3} & -\dfrac{\sqrt{2}}{3c_3} & 0
\end{pmatrix}.
\end{aligned} \tag{5.175}
$$

We can easily check that $\mathbf{S}^{-1}\mathbf{S} = \mathbf{I}$:

$$\mathbf{S}^{-1}\mathbf{S} = \begin{pmatrix} 0 & 0 & \dfrac{1}{c_1} \\ \dfrac{\sqrt{2}}{3c_2} & \dfrac{1}{3c_2} & 0 \\ \dfrac{1}{3c_3} & -\dfrac{\sqrt{2}}{3c_3} & 0 \end{pmatrix} \begin{pmatrix} 0 & \sqrt{2}c_2 & c_3 \\ 0 & c_2 & -\sqrt{2}c_3 \\ c_1 & 0 & 0 \end{pmatrix} \qquad (5.176)$$

$$= \begin{pmatrix} 1 & 0 & 0 \\ 0 & 1 & 0 \\ 0 & 0 & 1 \end{pmatrix} = \mathbf{S}^{-1}\mathbf{S}. \qquad (5.177)$$

We finally construct the transformed matrix

$$\mathbf{A}' = \mathbf{S}^{-1}\mathbf{A}\mathbf{S} \qquad (5.178)$$

as

$$\mathbf{A}' = \begin{pmatrix} 0 & 0 & \dfrac{1}{c_1} \\ \dfrac{\sqrt{2}}{3c_2} & \dfrac{1}{3c_2} & 0 \\ \dfrac{1}{3c_3} & -\dfrac{\sqrt{2}}{3c_3} & 0 \end{pmatrix} \begin{pmatrix} 1 & \sqrt{2} & 0 \\ \sqrt{2} & 0 & 0 \\ 0 & 0 & 0 \end{pmatrix} \begin{pmatrix} 0 & \sqrt{2}c_2 & c_3 \\ 0 & c_2 & -\sqrt{2}c_3 \\ c_1 & 0 & 0 \end{pmatrix}$$

$$= \begin{pmatrix} 0 & 0 & \dfrac{1}{c_1} \\ \dfrac{\sqrt{2}}{3c_2} & \dfrac{1}{3c_2} & 0 \\ \dfrac{1}{3c_3} & -\dfrac{\sqrt{2}}{3c_3} & 0 \end{pmatrix} \begin{pmatrix} 0 & 2\sqrt{2}c_2 & -c_3 \\ 0 & 2c_2 & \sqrt{2}c_3 \\ 0 & 0 & 0 \end{pmatrix} = \begin{pmatrix} 0 & 0 & 0 \\ 0 & 2 & 0 \\ 0 & 0 & -1 \end{pmatrix},$$

$$(5.179)$$

which is in diagonal form with the diagonal elements being equal to the eigenvalues:

$$A'_{11} = \lambda_1 = 0, \qquad (5.180)$$

$$A'_{22} = \lambda_2 = 2, \qquad (5.181)$$

$$A'_{33} = \lambda_3 = -1. \qquad (5.182)$$

5.10 MOMENT OF INERTIA TENSOR

In rotation problems, we can view the **moment of inertia tensor I** as an operator, or a transformation, in the equation

$$\boxed{\vec{L} = \mathbf{I}\vec{\omega},}$$ (5.183)

which relates the **angular momentum** $\vec{L}$ and the **angular velocity** $\vec{\omega}$ vectors. The moment of inertia tensor is represented by the $n \times n$ matrix [4]:

$$\boxed{I_{ij} = \int_V \rho(\mathbf{r})(\mathbf{r}^2\delta_{ij} - r_i r_j)d\tau,}$$ (5.184)

where $\rho(\mathbf{r})$ is the mass density and r_i, $i = 1, 2, 3$, stand for the Cartesian components of the position vector $\mathbf{r}$. The angular momentum and the angular velocity vectors are represented by the following column matrices:

$$\vec{L} = \begin{pmatrix} L_x \\ L_y \\ L_z \end{pmatrix} \text{ and } \vec{\omega} = \begin{pmatrix} \omega_x \\ \omega_y \\ \omega_z \end{pmatrix}.$$ (5.185)

We can now write the transformation equation [Eq. (5.183)] as

$$\boxed{L_i = \sum_{j=1}^{3} I_{ij}\omega_j,}$$ (5.186)

which, when written explicitly, becomes

$$L_x = I_{xx}\omega_x + I_{xy}\omega_y + I_{xz}\omega_z,$$ (5.187)

$$L_y = I_{yx}\omega_x + I_{yy}\omega_y + I_{yz}\omega_z,$$ (5.188)

$$L_z = I_{zx}\omega_x + I_{zy}\omega_y + I_{zz}\omega_z.$$ (5.189)

It is desirable to orient the Cartesian axes so that in terms of the new coordinates, the moment of inertia tensor $\mathbf{I}'$ is represented by the diagonal matrix

$$\mathbf{I}' = \begin{pmatrix} I_1 & 0 & 0 \\ 0 & I_2 & 0 \\ 0 & 0 & I_3 \end{pmatrix},$$ (5.190)

where I_1, I_2, and I_3 are called the **principal moments of inertia**. Directions of the new axes denoted by the subscripts 1, 2, 3 are called the **principal directions** of the object and the corresponding Cartesian coordinate system is called the **principal coordinates**. In the principal coordinates, Eqs. (5.187)–(5.189) simplify to

$$L_1 = I_1\omega_1, \tag{5.191}$$

$$L_2 = I_2\omega_2, \tag{5.192}$$

$$L_3 = I_3\omega_3, \tag{5.193}$$

where (L_1, L_2, L_3) and $(\omega_1, \omega_2, \omega_3)$ are the components of the angular momentum and the angular velocity along the principal axes, respectively. Finding the principal directions, which always exist, and the corresponding transformation matrix requires the solution of an eigenvalue problem for the moment of inertia tensor. This is demonstrated in the following example:

Example 5.12. *Moment of inertia tensor:* Let us now consider the moment of inertia tensor of a uniform flat rigid body with the density σ and in the shape of a 45° right triangle as shown in Figure 5.1. Components of the moment of inertia tensor [Eq. (5.184)] are evaluated as follows: Since $r^2 = x^2 + y^2$, we first find I_{xx} as

$$I_{xx} = \int_0^a \int_0^{a-x} \sigma(r^2 - x^2) \, dy \, dx \tag{5.194}$$

$$= \sigma \int_0^a \int_0^{a-x} y^2 \, dy \, dx = \frac{\sigma}{3} \int_0^a (a - x)^3 \, dx \tag{5.195}$$

$$= \frac{\sigma}{3} \int_0^a (a^3 - 3a^2 x + 3ax^2 - x^3) \, dx \tag{5.196}$$

$$= \frac{\sigma}{3} \left(a^4 - 3a^2 \frac{a^2}{2} + 3a \frac{a^3}{3} - \frac{a^4}{4} \right) = \sigma \frac{a^4}{12}. \tag{5.197}$$

Since the mass of the object is $M = \sigma a^2/2$, we write $I_{xx} = Ma^2/6$. Similarly, we obtain the other diagonal terms as

$$I_{yy} = \frac{Ma^2}{6}, \quad I_{zz} = \frac{Ma^2}{3}. \tag{5.198}$$

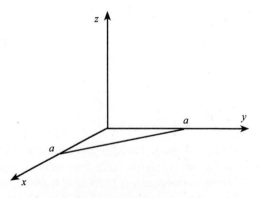

Figure 5.1   Moment of inertia tensor of a flat uniform triangular mass distribution.

For the off-diagonal terms, we find

$$I_{xy} = I_{yx} = -\int_0^a \int_0^{a-x} \sigma xy \, dy \, dx \tag{5.199}$$

$$= -\frac{2M}{a^2} \int_0^a \int_0^{a-x} xy \, dy \, dx = -\frac{2M}{a^2} \int_0^a \frac{x(a-x)^2}{2} \, dx \tag{5.200}$$

$$= -\frac{2M}{2a^2} \int_0^a (a^2 x - 2ax^2 + x^3) \, dx = -\frac{Ma^2}{12}, \tag{5.201}$$

$$I_{xz} = I_{zx} = 0, \quad I_{yz} = I_{zy} = 0, \tag{5.202}$$

thus obtaining the complete moment of inertia tensor as

$$\mathbf{I} = \begin{pmatrix} \dfrac{Ma^2}{6} & -\dfrac{Ma^2}{12} & 0 \\[2mm] -\dfrac{Ma^2}{12} & \dfrac{Ma^2}{6} & 0 \\[2mm] 0 & 0 & \dfrac{Ma^2}{3} \end{pmatrix}. \tag{5.203}$$

The principal moments of inertia are found by solving the characteristic equation:

$$\begin{vmatrix} \left(\dfrac{Ma^2}{6} - I\right) & -\dfrac{Ma^2}{12} & 0 \\[2mm] -\dfrac{Ma^2}{12} & \left(\dfrac{Ma^2}{6} - I\right) & 0 \\[2mm] 0 & 0 & \left(\dfrac{Ma^2}{3} - I\right) \end{vmatrix} = 0, \tag{5.204}$$

$$\left(\frac{Ma^2}{6} - I\right)^2 \left(\frac{Ma^2}{3} - I\right) + \frac{Ma^2}{12}\left(-\frac{Ma^2}{12}\right)\left(\frac{Ma^2}{3} - I\right) = 0, \tag{5.205}$$

$$\left(\frac{Ma^2}{3} - I\right)\left[\left(\frac{Ma^2}{6} - I\right)^2 - \left(\frac{Ma^2}{12}\right)^2\right] = 0, \tag{5.206}$$

which yields the principal moments of inertia as

$$I_1 = \frac{Ma^2}{3}, \quad I_2 = \frac{Ma^2}{4}, \quad I_3 = \frac{Ma^2}{12}. \tag{5.207}$$

Orientation of the principal axes, which are along the eigenvectors, are found as follows: For the principle value $I_1 = \frac{Ma^2}{3}$, we write the eigenvalue equation as

$$\frac{Ma^2}{3} \begin{pmatrix} 1/2 & -1/4 & 0 \\ -1/4 & 1/2 & 0 \\ 0 & 0 & 1 \end{pmatrix} \begin{pmatrix} a_{11} \\ a_{21} \\ a_{31} \end{pmatrix} = \frac{Ma^2}{3} \begin{pmatrix} a_{11} \\ a_{21} \\ a_{31} \end{pmatrix}, \tag{5.208}$$

which gives

$$\frac{Ma^2}{6}a_{11} - \frac{Ma^2}{12}a_{21} = \frac{Ma^2}{3}a_{11}, \tag{5.209}$$

$$-\frac{Ma^2}{12}a_{11} + \frac{Ma^2}{6}a_{21} = \frac{Ma^2}{3}a_{21}, \tag{5.210}$$

$$\frac{Ma^2}{3}a_{31} = \frac{Ma^2}{3}a_{31}. \tag{5.211}$$

The first two equations reduce to

$$a_{21} = -2a_{11}, \tag{5.212}$$

$$a_{21} = -\frac{1}{2}a_{11}, \tag{5.213}$$

which cannot be satisfied unless

$$a_{11} = a_{21} = 0. \tag{5.214}$$

The third equation [Eq. (5.211)] gives the only nonzero component of the first eigenvector, $a_{31} = c_1$, thus yielding

$$\mathbf{v}_1 = c_1 \begin{pmatrix} 0 \\ 0 \\ 1 \end{pmatrix}, \tag{5.215}$$

where c_1 is any constant different from zero. Similarly, the other eigenvectors are found as

$$\mathbf{v}_2 = c_2 \begin{pmatrix} 1 \\ -1 \\ 0 \end{pmatrix}, \quad \mathbf{v}_3 = c_3 \begin{pmatrix} 1 \\ 1 \\ 0 \end{pmatrix}. \tag{5.216}$$

Again the constants c_2 and c_3 are arbitrary and different from zero. We now construct the transformation matrix $\mathbf{S}$ as

$$\mathbf{S} = \left(\begin{pmatrix} \\ \mathbf{v}_1 \\ \end{pmatrix} \begin{pmatrix} \\ \mathbf{v}_2 \\ \end{pmatrix} \begin{pmatrix} \\ \mathbf{v}_3 \\ \end{pmatrix} \right) \tag{5.217}$$

$$= \begin{pmatrix} 0 & c_2 & c_3 \\ 0 & -c_2 & c_3 \\ c_1 & 0 & 0 \end{pmatrix}. \tag{5.218}$$

Since $\det \mathbf{S} = 2c_1 c_2 c_3 \neq 0$, the inverse exists and is found by using the formula [Eq. (4.213)]

$$S_{ij}^{-1} = (-1)^{i+j} \frac{\det \mathbf{S}(j|i)}{\det \mathbf{S}}, \tag{5.219}$$

where $\det \mathbf{S}(j|i)$ is the minor obtained by deleting the jth row and the ith column. This yields the inverse as

$$
\mathbf{S}^{-1} = \frac{1}{\det \mathbf{S}} \begin{pmatrix} 0 & 0 & 2c_2c_3 \\ c_1c_3 & -c_1c_3 & 0 \\ c_1c_2 & c_1c_2 & 0 \end{pmatrix} = \frac{1}{2c_1c_2c_3} \begin{pmatrix} 0 & 0 & 2c_2c_3 \\ c_1c_3 & -c_1c_3 & 0 \\ c_1c_2 & c_1c_2 & 0 \end{pmatrix}
$$

$$
= \begin{pmatrix} 0 & 0 & \dfrac{1}{c_1} \\ \dfrac{1}{2c_2} & -\dfrac{1}{2c_2} & 0 \\ \dfrac{1}{2c_3} & \dfrac{1}{2c_3} & 0 \end{pmatrix}. \tag{5.220}
$$

The inverse can be checked easily as

$$
\mathbf{S}^{-1}\mathbf{S} = \begin{pmatrix} 0 & 0 & \dfrac{1}{c_1} \\ \dfrac{1}{2c_2} & -\dfrac{1}{2c_2} & 0 \\ \dfrac{1}{2c_3} & \dfrac{1}{2c_3} & 0 \end{pmatrix} \begin{pmatrix} 0 & c_2 & c_3 \\ 0 & -c_2 & c_3 \\ c_1 & 0 & 0 \end{pmatrix} = \begin{pmatrix} 1 & 0 & 0 \\ 0 & 1 & 0 \\ 0 & 0 & 1 \end{pmatrix} = \mathbf{S}\mathbf{S}^{-1}.
$$

In terms of the new basis vectors $(\mathbf{v}_1, \mathbf{v}_2, \mathbf{v}_3)$, we can now write the moment of inertia tensor as

$$
\mathbf{S}^{-1}\mathbf{I}\mathbf{S} = \begin{pmatrix} 0 & 0 & \dfrac{1}{c_1} \\ \dfrac{1}{2c_2} & -\dfrac{1}{2c_2} & 0 \\ \dfrac{1}{2c_3} & \dfrac{1}{2c_3} & 0 \end{pmatrix} \times \frac{Ma^2}{3} \begin{pmatrix} \dfrac{1}{2} & -\dfrac{1}{4} & 0 \\ -\dfrac{1}{4} & \dfrac{1}{2} & 0 \\ 0 & 0 & 1 \end{pmatrix} \begin{pmatrix} 0 & c_2 & c_3 \\ 0 & -c_2 & c_3 \\ c_1 & 0 & 0 \end{pmatrix}
$$
$$
\tag{5.221}
$$

$$
= \frac{Ma^2}{3} \begin{pmatrix} 0 & 0 & \dfrac{1}{c_1} \\ \dfrac{1}{2c_2} & -\dfrac{1}{2c_2} & 0 \\ \dfrac{1}{2c_3} & \dfrac{1}{2c_3} & 0 \end{pmatrix} \begin{pmatrix} 0 & \dfrac{3c_2}{4} & \dfrac{c_3}{4} \\ 0 & -\dfrac{3c_2}{4} & \dfrac{c_3}{4} \\ c_1 & 0 & 0 \end{pmatrix}, \tag{5.222}
$$

$$S^{-1}IS = \begin{pmatrix} \dfrac{Ma^2}{3} & 0 & 0 \\ 0 & \dfrac{Ma^2}{4} & 0 \\ 0 & 0 & \dfrac{Ma^2}{12} \end{pmatrix} = \begin{pmatrix} I_1 & 0 & 0 \\ 0 & I_2 & 0 \\ 0 & 0 & I_3 \end{pmatrix}. \tag{5.223}$$

5.11   INNER PRODUCT SPACES

So far our discussion of vector spaces was rather abstract with limited potential for applications. We now introduce the missing element, that is, the **inner product**, which defines multiplication of vectors with vectors. This added richness allows us to define **magnitude** of vectors, **distances**, and **angles** and also allows us to determine the matrix representation of linear transformations. We will also be able to evaluate the proportionality constants, which were left undetermined in eigenvector calculations by normalizing their magnitudes to unity [Eqs. (5.158), (5.164), and (5.169) and Eqs. (5.215) and (5.216)].

5.12   THE INNER PRODUCT

For ordinary vectors in three-dimensional real space, the scalar product is defined as

$$\vec{a} \cdot \vec{b} = \sum_{i=1}^{3} a_i b_i, \tag{5.224}$$

where $\vec{a}$, $\vec{b}$ are two vectors given in terms of their Cartesian components:

$$\vec{a} = (a_1, a_2, a_3), \quad \vec{b} = (b_1, b_2, b_3). \tag{5.225}$$

We find the magnitude a of $\vec{a}$ as

$$a = |\vec{a}| = (\vec{a} \cdot \vec{a})^{1/2} = \left(\sum_{i=1}^{3} a_i^2 \right)^{1/2}, \tag{5.226}$$

while the angle between $\vec{a}$ and $\vec{b}$ is defined as

$$\theta_{ab} = \cos^{-1} \left[\frac{\vec{a} \cdot \vec{b}}{ab} \right]. \tag{5.227}$$

We have seen that in terms of matrices, vectors can be written as column matrices, which we denote by boldface characters:

$$\vec{a} = \mathbf{a} = \begin{pmatrix} a_1 \\ a_2 \\ a_3 \end{pmatrix}. \tag{5.228}$$

Using the transpose $\tilde{\mathbf{a}}$, we can write the scalar product of $\mathbf{a}$ with itself as

$$\vec{a} \cdot \vec{a} = \tilde{\mathbf{a}}\mathbf{a} = \begin{pmatrix} a_1 & a_2 & a_3 \end{pmatrix} \begin{pmatrix} a_1 \\ a_2 \\ a_3 \end{pmatrix} = \sum_{i=1}^{3} a_i a_i. \qquad (5.229)$$

Similarly, the scalar product of $\vec{a}$ and $\vec{b}$, is defined as

$$\vec{a} \cdot \vec{b} = \tilde{\mathbf{a}}\mathbf{b} = \begin{pmatrix} a_1 & a_2 & a_3 \end{pmatrix} \begin{pmatrix} b_1 \\ b_2 \\ b_3 \end{pmatrix} = \sum_{i=1}^{3} a_i b_i. \qquad (5.230)$$

Note that the scalar product is symmetric:

$$\vec{b} \cdot \vec{a} = \tilde{\mathbf{b}}\mathbf{a} = \begin{pmatrix} b_1 & b_2 & b_3 \end{pmatrix} \begin{pmatrix} a_1 \\ a_2 \\ a_3 \end{pmatrix} = \sum_{i=1}^{3} b_i a_i = \vec{a} \cdot \vec{b}. \qquad (5.231)$$

Generalization to n dimensions is simply accomplished by letting the sums run from 0 to n.

For a given abstract n-dimensional vector space V defined over the field K, we can also represent a vector v as the column matrix:

$$\mathbf{v} = \begin{pmatrix} \alpha_1 \\ \alpha_2 \\ \vdots \\ \alpha_n \end{pmatrix}, \qquad (5.232)$$

where α_i stands for the coefficients in the linear combination

$$v = \alpha_1 v_1 + \alpha_2 v_2 + \cdots + \alpha_n v_n, \qquad (5.233)$$

in terms of the basis

$$B = \{v_1, v_2, \ldots, v_n\}. \qquad (5.234)$$

We now introduce the **inner product** for abstract vector spaces as

$$\boxed{(v, w) = \sum_{i=1}^{n} \alpha_i \beta_i,} \qquad (5.235)$$

where v and w belong to V with the components

$$\mathbf{v} = \begin{pmatrix} \alpha_1 \\ \alpha_2 \\ \vdots \\ \alpha_n \end{pmatrix}, \quad \mathbf{w} = \begin{pmatrix} \beta_1 \\ \beta_2 \\ \vdots \\ \beta_n \end{pmatrix}. \qquad (5.236)$$

We can also write the **inner product** [Eq. (5.235)] in the following equivalent ways:

$$(v, w) = \widetilde{\mathbf{v}}\mathbf{w} = \widetilde{\mathbf{w}}\mathbf{v}. \tag{5.237}$$

The **angle** between v and w is now defined as

$$\theta_{vw} = \cos^{-1}\left[\frac{(v, w)}{(v, v)^{1/2}(w, w)^{1/2}}\right], \tag{5.238}$$

where (v, v) and (w, w) stand for the magnitude squares $|u|^2$ and $|w|^2$, respectively.

This definition of the inner product works fine for vector fields defined over the **real field** $\mathbb{R}$. For vector fields defined over the complex field $\mathbb{C}$, it presents problems. For example, consider the one-dimensional vector $v = i\alpha$, where its magnitude:

$$|v| = (i\alpha, i\alpha)^{1/2} = (-\alpha^2)^{1/2}, \tag{5.239}$$

is imaginary for real α. Since the inner product represents **lengths** or **distances**, it has to be real, hence this is not acceptable. However, there is a simple cure. We modify the definition of the inner product over the **complex field** $\mathbb{C}$ as

$$(v, w) = \sum_{i=1}^{n} \alpha_i^* \beta_i, \tag{5.240}$$

where the "$*$" indicates that the **complex conjugate** has to be taken. With this definition, the **magnitude** of a vector defined over the complex field becomes

$$|v| = (v, v)^{1/2} = \left[\sum_{i=1}^{n} \alpha_i^* \alpha_i\right]^{1/2} = \left[\sum_{i=1}^{n} |\alpha_i|^2\right]^{1/2}. \tag{5.241}$$

Since $|\alpha_i|^2 \geq 0$, magnitude is always real. It is important to keep in mind that for vector fields defined over the complex field, the inner product is not symmetric:

$$(v, w) \neq (w, v). \tag{5.242}$$

However, the following is true:

$$(v, w) = (w, v)^* \tag{5.243}$$

Definition 5.4. For a given vector field V, defined over the real field $\mathbb{R}$, or the complex field $\mathbb{C}$, an **inner product** is a rule that associates a scalar-valued function to any ordered pairs of vectors, v and w, in V such that

I.

$$(v, w) = (w, v)^*, \tag{5.244}$$

II.

$$(v + w, u) = (v, w) + (w, u),$$ (5.245)

where u, v, and w are vectors in V.

III. If α is a scalar in the underlying field K, then

$$(\alpha v, w) = \alpha^*(v, w),$$ (5.246)

$$(v, \alpha w) = \alpha(v, w),$$ (5.247)

where v and w are vectors in V.

IV. For any v in V, we have $(v, v) \geq 0$ and $(v, v) = 0$ if and only if $v = 0$. The scalar $|v| = (v, v)^{1/2}$ is called the **norm** or the **magnitude** of v.

A vector space with an inner product definition is called an **inner product space**. In the aforementioned definition, if the underlying field is real, the complex conjugation becomes superfluous. A Euclidean space is a real vector space with an inner product, which is also called the **dot** or the **scalar product**. An inner product space defined over the complex field $\mathbb{C}$ is called a **unitary space**.

Example 5.13. *Definition of inner product:* One can easily show that the set of all continuous functions defined over the complex field $\mathbb{C}$ and in the closed interval $[0, 1]$ forms a vector space. In this space, we can define an inner product as

$$(f(x), g(x)) = \int_0^1 f^*(x)g(x) \, dx,$$ (5.248)

where $f(x)$ and $g(x)$ are complex-valued functions of the real argument x.

5.13   ORTHOGONALITY AND COMPLETENESS

After the introduction of the concepts such as distance, magnitude/norm, and angle, applications to physics and engineering becomes easier. Among the remaining concepts that are needed for applications are the **orthogonality** and the **completeness** of sets of vectors that form a vector space.

We say that two vectors, v and w, are **orthogonal**, if and only if their inner product vanishes:

$$(v, w) = 0.$$ (5.249)

Note that the condition of orthogonality is symmetric, that is, if $(v, w) = 0$, then $(w, v) = 0$. For ordinary vectors, orthogonality implies the two vectors being perpendicular to each other. Eq. (5.249) extends the concept of orthogonality to abstract vector spaces, whether they are finite- or infinite-dimensional. A set of vectors $\{v_1, v_2, \ldots, v_n\}$ is said to be **orthonormal** if for all i and j the following relation holds:

$$\boxed{(v_i, v_j) = \delta_{ij}.}$$
(5.250)

In a finite-dimensional vector space V, an orthonormal set is **complete**, if every vector v in V can be expressed as a linear combination of the vectors in the set. A set of vectors $\{v_1, v_2, \ldots, v_n\}$ is said to be **linearly independent** if the equation

$$c_1 v_1 + c_2 v_2 + \cdots + c_n v_n = 0,$$
(5.251)

where $c_1, c_2, \ldots, c_n$ are scalars, cannot be satisfied unless all the coefficients are zero: $c_1 = c_2 = \cdots = c_n = 0$. To obtain a formal expression for the criteria of linear independence, we form the inner products of Eq. (5.251) by $v_1, v_2, \ldots, v_n$ and use the linearity of the inner product to write

$$c_1(v_1, v_1) + c_2(v_1, v_2) + \cdots + c_n(v_1, v_n) = 0,$$
$$c_1(v_2, v_1) + c_2(v_2, v_2) + \cdots + c_n(v_2, v_n) = 0,$$
$$\vdots$$
$$c_1(v_n, v_1) + c_2(v_n, v_2) + \cdots + c_n(v_n, v_n) = 0.$$
(5.252)

This gives a set of n linear equations to be solved for the coefficients $c_1, c_2, \ldots, c_n$ simultaneously. Unless the determinant of the coefficients vanishes, these equations do not have a solution besides the trivial solution, that is, $c_1 = c_2 = \cdots = c_n = 0$. Hence, we can write the condition of linear independence for the set $\{v_1, v_2, \ldots, v_n\}$ as

$$G = \begin{vmatrix} (v_1, v_1) & (v_1, v_2) & \cdots & (v_1, v_n) \\ (v_2, v_1) & (v_2, v_2) & \cdots & (v_2, v_n) \\ \vdots & \vdots & \ddots & \vdots \\ (v_n, v_1) & (v_n, v_2) & \cdots & (v_n, v_n) \end{vmatrix} \neq 0.$$
(5.253)

The determinant G is called the **Gramian**. For an orthonormal set in n dimensions the Gramian reduces to

$$G = \begin{vmatrix} 1 & 0 & \cdots & 0 \\ 0 & 1 & \cdots & 0 \\ \vdots & \vdots & \ddots & \vdots \\ 0 & 0 & \cdots & 1 \end{vmatrix} = 1.$$
(5.254)

Hence, an orthonormal set is linearly independent. For a given orthonormal set $\{v_1, v_2, \ldots, v_n\}$ in an n-dimensional vector space, the following statements are equivalent:

1. The set $\{v_1, v_2, \ldots, v_n\}$ is complete.
2. If $(v_i, v) = 0$ for all i, then $v = 0$.
3. The set $\{v_1, v_2, \ldots, v_n\}$ spans V.
4. If v belongs to V, then

$$v = \sum_{i=1}^{n} c_i v_i, \tag{5.255}$$

where $c_i = (v_i, v)$ are called the **components** of v with respect to the basis $B = \{v_1, v_2, \ldots, v_n\}$.

5.14 GRAM–SCHMIDT ORTHOGONALIZATION

It is often advantageous to work with an orthonormal set of n linearly independent vectors. Given a set of linearly independent vectors $\{v_1, v_2, \ldots, v_n\}$, we can always construct an orthonormal set. We start with any one of the vectors in the set, say v_1, call it $w_1 = v_1$, and normalize it as

$$\widehat{e}_1 = \frac{w_1}{l(w_1)}, \tag{5.256}$$

where $l(w_1) = (w_1, w_1)^{1/2}$ is the norm of w_1. Next we choose another element from the original set, say v_2, and subtract a constant, c, times $\widehat{e}_1$ from it to write

$$w_2 = v_2 - c\widehat{e}_1. \tag{5.257}$$

We now determine c such that w_2 is orthogonal to $\widehat{e}_1$:

$$(\widehat{e}_1, (v_2 - c\widehat{e}_1)) = (\widehat{e}_1, v_2) - c(\widehat{e}_1, \widehat{e}_1) = 0. \tag{5.258}$$

Since $(\widehat{e}_1, \widehat{e}_1) = 1$, this gives c as

$$c = (\widehat{e}_1, v_2). \tag{5.259}$$

Hence, w_2 becomes

$$w_2 = v_2 - (\widehat{e}_1, v_2)\widehat{e}_1. \tag{5.260}$$

We now normalize w_2 to obtain the second member of the orthonormal set as

$$\widehat{e}_2 = \frac{w_2}{l(w_2)}, \tag{5.261}$$

where

$$l(w_2) = (w_2, w_2)^{1/2}. \tag{5.262}$$

We continue the process with a third vector from the original set, say v_3, and write

$$w_3 = v_3 - c_1\widehat{e}_1 - c_2\widehat{e}_2. \tag{5.263}$$

Requiring w_3 to be orthogonal to both $\widehat{e}_1$ and $\widehat{e}_2$ gives us two equations to be solved simultaneously for c_1 and c_2, which yields the coefficients as $c_1 = (\widehat{e}_1, v_3)$ and $c_2 = (\widehat{e}_2, v_3)$. The third orthonormal vector is now written as

$$\widehat{e}_3 = \frac{w_3}{l(w_3)}, \tag{5.264}$$

where $l(w_3) = (w_3, w_3)^{1/2}$. Continuing this process, we finally obtain the **last member** of the orthonormal set $\{\widehat{e}_1, \widehat{e}_2, \ldots, \widehat{e}_n\}$ as

$$\widehat{e}_n = \frac{w_n}{l(w_n)}, \tag{5.265}$$

where

$$w_n = v_n - \sum_{i=1}^{n-1}(\widehat{e}_i, v_n)\widehat{e}_i \tag{5.266}$$

and

$$l(w_n) = (w_n, w_n)^{1/2}. \tag{5.267}$$

This method is called the **Gram-Schmidt** orthogonalization procedure.

5.15 EIGENVALUE PROBLEM FOR REAL SYMMETRIC MATRICES

In many physically interesting cases linear operators or transformations are represented by real symmetric matrices. Consider the following **eigenvalue problem**:

$$\mathbf{A}\mathbf{v}_\lambda = \lambda\mathbf{v}_\lambda, \tag{5.268}$$

where $\mathbf{A}$ is a symmetric $n \times n$ matrix with real components and $\mathbf{v}_\lambda$ is the $n \times 1$ column matrix representing the eigenvector corresponding to the eigenvalue λ. Eigenvalues are the roots of the **characteristic equation**:

$$|\mathbf{A} - \lambda\mathbf{I}| = \begin{vmatrix} (a_{11} - \lambda) & a_{12} & \cdots & a_{1n} \\ a_{21} & (a_{22} - \lambda) & \cdots & a_{2n} \\ \vdots & \vdots & \ddots & \vdots \\ a_{n1} & a_{n2} & \cdots & (a_{nn} - \lambda) \end{vmatrix} = 0, \tag{5.269}$$

which gives an nth-order polynomial in λ. Even though the matrix $\mathbf{A}$ is real, roots of the characteristic equation could be complex, and in turn the corresponding eigenvectors could be complex. Consider two eigenvalues, λ_i and λ_j, and write the corresponding eigenvalue equations as

$$\mathbf{A}\mathbf{v}_i = \lambda_i \mathbf{v}_i, \tag{5.270}$$

$$\mathbf{A}\mathbf{v}_j = \lambda_j \mathbf{v}_j. \tag{5.271}$$

We multiply the first equation from the left by $\widetilde{\mathbf{v}}_j^*$ and write

$$\widetilde{\mathbf{v}}_j^* \mathbf{A}\mathbf{v}_i = \lambda_i \widetilde{\mathbf{v}}_j^* \mathbf{v}_i. \tag{5.272}$$

We now consider the second equation [Eq. (5.271)] and take its transpose, and then take its complex conjugate to write

$$\widetilde{\mathbf{v}}_j^* \widetilde{\mathbf{A}}^* = \lambda_j^* \widetilde{\mathbf{v}}_j^*. \tag{5.273}$$

Since $\mathbf{A}$ is real and symmetric, $\widetilde{\mathbf{A}}^* = \mathbf{A}$, Eq. (5.273) is also equal to

$$\widetilde{\mathbf{v}}_j^* \mathbf{A} = \lambda_j^* \widetilde{\mathbf{v}}_j^*. \tag{5.274}$$

Multiplying Eq. (5.274) with $\mathbf{v}_i$ from the right, we obtain another equation:

$$\widetilde{\mathbf{v}}_j^* \mathbf{A}\mathbf{v}_i = \lambda_j^* \widetilde{\mathbf{v}}_j^* \mathbf{v}_i. \tag{5.275}$$

Subtracting Eq. (5.272) from Eq. (5.275), we obtain

$$0 = (\lambda_j^* - \lambda_i)\widetilde{\mathbf{v}}_j^* \mathbf{v}_i, \tag{5.276}$$

which leads us to the following conclusions:

I. When $j = i$, the expression $\widetilde{\mathbf{v}}_i^* \mathbf{v}_i = (v_i, v_i) = |\mathbf{v}_i|^2$ is the square of the norm of the ith eigenvector, which is always positive and different from zero, hence we obtain

$$\boxed{\lambda_i^* = \lambda_i.} \tag{5.277}$$

That is, the **eigenvalues** are **real**. In fact, this important property of real and symmetric matrices was openly displayed in Examples 5.9–5.12.

II. When $j \neq i$ and when the **eigenvalues** are **distinct**, $\lambda_j \neq \lambda_i$, Eq. (5.276) implies

$$\boxed{\widetilde{\mathbf{v}}_j^* \mathbf{v}_i = (v_j, v_i) = 0.} \tag{5.278}$$

In other words, the **eigenvectors** corresponding to the **distinct eigenvalues** are **orthogonal**.

III. When $i \neq j$ but $\lambda_i = \lambda_j$, that is, when the characteristic equation has multiple roots, for the root with the multiplicity s, $s < n$, there always exist s linearly independent eigenvectors [5]. We should make a note that this statement about multiple roots does not in general hold for nonsymmetric matrices. Eigenvalues corresponding to multiple roots are called **degenerate**.

5.16   PRESENCE OF DEGENERATE EIGENVALUES

To demonstrate how one handles cases with degenerate eigenvalues, we use an example. Consider an operator represented by the following real and symmetric matrix in a three-dimensional real inner product space:

$$\mathbf{A} = \begin{pmatrix} 1 & 1 & 1 \\ 1 & 1 & 1 \\ 1 & 1 & 1 \end{pmatrix}. \tag{5.279}$$

We write the characteristic equation as

$$|\mathbf{A} - \lambda\mathbf{I}| = \begin{vmatrix} (1-\lambda) & 1 & 1 \\ 1 & (1-\lambda) & 1 \\ 1 & 1 & (1-\lambda) \end{vmatrix} = \lambda^2(3-\lambda) = 0, \tag{5.280}$$

which has the roots

$$\lambda_1 = \lambda_2 = 0 \quad \text{and} \quad \lambda_3 = 3. \tag{5.281}$$

We first find the eigenvector corresponding to the multiple root, $\lambda_1 = \lambda_2 = 0$, by writing the corresponding eigenvalue equation as

$$\begin{pmatrix} 1 & 1 & 1 \\ 1 & 1 & 1 \\ 1 & 1 & 1 \end{pmatrix} \begin{pmatrix} a_{11} \\ a_{21} \\ a_{31} \end{pmatrix} = 0 \begin{pmatrix} a_{11} \\ a_{21} \\ a_{31} \end{pmatrix}, \tag{5.282}$$

which gives a single equation:

$$a_{11} + a_{21} + a_{31} = 0. \tag{5.283}$$

In this case, we can only solve for one of the components, say a_{11}, in terms of the remaining two, a_{21} and a_{31}, as

$$a_{11} = -a_{21} - a_{31}. \tag{5.284}$$

Since a_{21} and a_{31} are not fixed, we are free to choose them at will. For one of the eigenvectors, we choose

$$a_{31} = c_1, \quad a_{21} = 0, \quad c_1 \neq 0, \tag{5.285}$$

which gives $a_{11} = -c_1$, thus obtaining the first eigenvector as

$$\mathbf{v}_1 = \begin{pmatrix} -c_1 \\ 0 \\ c_1 \end{pmatrix} = c_1 \begin{pmatrix} -1 \\ 0 \\ 1 \end{pmatrix}. \tag{5.286}$$

For the second eigenvector corresponding to the degenerate eigenvalue $\lambda_{1,2} = 0$, we choose

$$a_{21} = c_2, \quad a_{31} = 0, \quad c_2 \neq 0, \tag{5.287}$$

which gives $a_{11} = -c_2$, thus yielding the second and linearly independent eigenvector as

$$\mathbf{v}_2 = \begin{pmatrix} -c_2 \\ c_2 \\ 0 \end{pmatrix} = c_2 \begin{pmatrix} -1 \\ 1 \\ 0 \end{pmatrix}. \tag{5.288}$$

At this point, aside from the fact that the eigenvectors look relatively simple, there is no reason for making these choices. Any other choice would be equally good as long as the two eigenvectors are not constant multiples of each other. Finally, for the third eigenvalue, $\lambda_3 = 3$, we write

$$\begin{pmatrix} 1 & 1 & 1 \\ 1 & 1 & 1 \\ 1 & 1 & 1 \end{pmatrix} \begin{pmatrix} a_{13} \\ a_{23} \\ a_{33} \end{pmatrix} = 3 \begin{pmatrix} a_{13} \\ a_{23} \\ a_{33} \end{pmatrix}, \tag{5.289}$$

$$a_{13} + a_{23} + a_{33} = 3a_{13}, \tag{5.290}$$

$$a_{13} + a_{23} + a_{33} = 3a_{23}, \tag{5.291}$$

$$a_{13} + a_{23} + a_{33} = 3a_{33}, \tag{5.292}$$

which implies

$$a_{13} = a_{23} = a_{33} = c_3, \quad c_3 \neq 0. \tag{5.293}$$

This yields $\mathbf{v}_3$ up to an arbitrary constant c_3 as

$$\mathbf{v}_3 = c_3 \begin{pmatrix} 1 \\ 1 \\ 1 \end{pmatrix}. \tag{5.294}$$

Linear independence of the eigenvectors $(\mathbf{v}_1, \mathbf{v}_2, \mathbf{v}_3)$ can be checked by showing that the **Gramian** is different from zero:

$$G = c_1 c_2 c_3 \begin{vmatrix} -1 & -1 & 1 \\ 0 & 1 & 1 \\ 1 & 0 & 1 \end{vmatrix} = -3c_1 c_2 c_3 \neq 0. \tag{5.295}$$

We now construct the transformation matrix $\mathbf{S}$ as

$$\mathbf{S} = \left(\left(\mathbf{v}_1 \right) \left(\mathbf{v}_2 \right) \left(\mathbf{v}_3 \right) \right) = \begin{pmatrix} -c_1 & -c_2 & c_3 \\ 0 & c_2 & c_3 \\ c_1 & 0 & c_3 \end{pmatrix}. \tag{5.296}$$

Since the determinant of $\mathbf{S}$, $\det \mathbf{S} = -3c_1 c_2 c_3$, is different from zero, the inverse transformation matrix exists and can be found by using the formula

$$S_{ij}^{-1} = (-1)^{i+j} \frac{\det \mathbf{S}(j|i)}{\det \mathbf{S}}, \tag{5.297}$$

where $\det \mathbf{S}(j|i)$ is the minor obtained by deleting the jth row and the ith column. Notice that the indices in $\det \mathbf{S}(j|i)$ are reversed [Eq. (4.213)]. We find $\mathbf{S}^{-1}$ as

$$\mathbf{S}^{-1} = \begin{pmatrix} -1/3c_1 & -1/3c_1 & 2/3c_1 \\ -1/3c_2 & 2/3c_2 & -1/3c_2 \\ 1/3c_3 & 1/3c_3 & 1/3c_3 \end{pmatrix}. \tag{5.298}$$

It can easily be checked that $\mathbf{S}\mathbf{S}^{-1} = \mathbf{S}^{-1}\mathbf{S} = \mathbf{I}$ and

$$\mathbf{S}^{-1}\mathbf{A}\mathbf{S} = \begin{pmatrix} \lambda_1 & 0 & 0 \\ 0 & \lambda_2 & 0 \\ 0 & 0 & \lambda_3 \end{pmatrix} = \begin{pmatrix} 0 & 0 & 0 \\ 0 & 0 & 0 \\ 0 & 0 & 3 \end{pmatrix}. \tag{5.299}$$

We now analyze the eigenvectors for this case more carefully. Using the definition of the inner product [Eq. (5.235)], it is seen that the two eigenvectors corresponding to the degenerate eigenvalues are not orthogonal, that is, $(v_1, v_2) \neq 0$. However, the third eigenvector is orthogonal to both of them:

$$(v_1, v_3) = (v_2, v_3) = 0. \tag{5.300}$$

Since we picked two of the components of the eigenvectors corresponding to the degenerate eigenvalue randomly, were we just extremely lucky to have them orthogonal to v_3? If we look at the eigenvalue equation for the degenerate eigenvalues [Eq. (5.283)], $\lambda_{1,2} = 0$, we see that we have only one equation for the three components, hence, we can write the corresponding eigenvectors as

$$v_{\lambda_{1,2}} = \begin{pmatrix} -a_{21} - a_{31} \\ a_{21} \\ a_{31} \end{pmatrix}. \tag{5.301}$$

For reasons to be clear shortly, we introduce the constants c_1, c_2, b, and c as

$$\left\{ \begin{array}{l} a_{21} = c_2 + c_1 b, \\ a_{31} = c_1 + c_2 c \end{array} \right\} \tag{5.302}$$

and write

$$\mathbf{v}_{\lambda_{1,2}} = \begin{pmatrix} -c_1(1+b) - c_2(1+c) \\ c_2 + c_1 b \\ c_1 + c_2 c \end{pmatrix} = c_1 \begin{pmatrix} -(1+b) \\ b \\ 1 \end{pmatrix} + c_2 \begin{pmatrix} -(1+c) \\ 1 \\ c \end{pmatrix}.$$

(5.303)

In other words, eigenvectors $\mathbf{v}_{\lambda_{1,2}}$ for the doubly degenerate eigenvalues $\lambda_{1,2} = 0$ can be written as the linear combination of two vectors as

$$\mathbf{v}_{\lambda_{1,2}} = c_1 \mathbf{w}_1 + c_2 \mathbf{w}_2,$$

(5.304)

where

$$\mathbf{w}_1 = \begin{pmatrix} -(1+b) \\ b \\ 1 \end{pmatrix} \quad \text{and} \quad \mathbf{w}_2 = \begin{pmatrix} -(1+c) \\ 1 \\ c \end{pmatrix}.$$

(5.305)

It can easily be checked that $\mathbf{w}_1$ and $\mathbf{w}_2$ satisfy the eigenvalue equations for the matrix $\mathbf{A}$ with the eigenvalue $\lambda = 0$ [Eq. (5.282)], that is,

$$\mathbf{A}\mathbf{w}_1 = 0\mathbf{w}_1,$$

(5.306)

$$\mathbf{A}\mathbf{w}_2 = 0\mathbf{w}_2.$$

(5.307)

In general, any linear combination of the eigenvectors belonging to the same eigenvalue is also an eigenvector with the same eigenvalue. The proof is simple. Let $\mathbf{w}_1$ and $\mathbf{w}_2$ be the eigenvectors of $\mathbf{B}$ with the eigenvalue λ:

$$\mathbf{B}\mathbf{w}_1 = \lambda\mathbf{w}_1,$$

(5.308)

$$\mathbf{B}\mathbf{w}_2 = \lambda\mathbf{w}_2.$$

(5.309)

Then,

$$\mathbf{B}(c_0\mathbf{w}_1 + c_1\mathbf{w}_2) = \mathbf{B}(c_0\mathbf{w}_1) + \mathbf{B}(c_1\mathbf{w}_2) = c_0\mathbf{B}\mathbf{w}_1 + c_1\mathbf{B}\mathbf{w}_2$$

(5.310)

$$= c_0\lambda\mathbf{w}_1 + c_1\lambda\mathbf{w}_2 = \lambda(c_0\mathbf{w}_1 + c_1\mathbf{w}_2).$$

(5.311)

The converse of this result is also true. If we can write an eigenvector $\mathbf{w}$ corresponding to the eigenvalue λ:

$$\mathbf{B}\mathbf{w} = \lambda\mathbf{w},$$

(5.312)

as the linear combination of two linearly independent vectors $\mathbf{w}_1$ and $\mathbf{w}_2$:

$$\mathbf{w} = c_0\mathbf{w}_1 + c_1\mathbf{w}_2,$$

(5.313)

where both c_0 and c_1 are different from zero, then both $\mathbf{w}_1$ and $\mathbf{w}_2$ are eigenvectors of $\mathbf{B}$ corresponding to the eigenvalue λ. The proof is as follows:

$$\mathbf{B}(c_0\mathbf{w}_1 + c_1\mathbf{w}_2) = \lambda(c_0\mathbf{w}_1 + c_1\mathbf{w}_2),$$

(5.314)

$$c_0 \mathbf{B} \mathbf{w}_1 + c_1 \mathbf{B} \mathbf{w}_2 = c_0 \lambda \mathbf{w}_1 + c_1 \lambda \mathbf{w}_2, \tag{5.315}$$

$$c_0 (\mathbf{B} \mathbf{w}_1 - \lambda \mathbf{w}_1) + c_1 (\mathbf{B} \mathbf{w}_2 - \lambda \mathbf{w}_2) = 0. \tag{5.316}$$

Since in general c_0 and c_1 are different from zero, the only way to satisfy this equation for all c_0 and c_1 is to have

$$\mathbf{B} \mathbf{w}_1 = \lambda \mathbf{w}_1, \tag{5.317}$$

$$\mathbf{B} \mathbf{w}_2 = \lambda \mathbf{w}_2, \tag{5.318}$$

which completes the proof.

As we stated earlier, for symmetric matrices of order n and for multiple roots of order s $(s < n)$, there always exists s linearly independent eigenvectors. For the matrix in Eq. (5.279) n is 3 and s is 2, where the degenerate eigenvalues $\lambda_{1,2}$ are both 0. For the the degenerate eigenvalues, we can now take the two linearly independent eigenvectors as [Eq. (5.303)]:

$$\mathbf{v}_1 = c_1 \begin{pmatrix} -(1+b) \\ b \\ 1 \end{pmatrix}, \quad \mathbf{v}_2 = c_2 \begin{pmatrix} -(1+c) \\ 1 \\ c \end{pmatrix}, \tag{5.319}$$

where the constants b and c are still arbitrary. In other words, there are infinitely many possibilities for the eigenvectors of the degenerate eigenvalue $\lambda_{1,2} = 0$. However, notice that all these vectors satisfy

$$(v_1, v_3) = 0, \quad (v_2, v_3) = 0. \tag{5.320}$$

That is, they are all perpendicular to the remaining eigenvector v_3, corresponding to $\lambda_3 = 3$. Using the freedom in the choice of b and c, we can also choose v_1 and v_2 as perpendicular:

$$(v_1, v_2) = 0, \tag{5.321}$$

which gives a relation between b and c as

$$(1+b)(1+c) + b + c = 0, \tag{5.322}$$

$$b = -\frac{1+2c}{2+c}. \tag{5.323}$$

Choosing c as 1, Eq. (5.323) gives $b = -1$, thus obtaining the orthogonal set of eigenvectors of $\mathbf{A}$ [Eq. (5.279)] as

$$\mathbf{v}_1 = c_1 \begin{pmatrix} 0 \\ -1 \\ 1 \end{pmatrix}, \quad \mathbf{v}_2 = c_2 \begin{pmatrix} -2 \\ 1 \\ 1 \end{pmatrix}, \quad \mathbf{v}_3 = c_3 \begin{pmatrix} 1 \\ 1 \\ 1 \end{pmatrix}. \tag{5.324}$$

In all the eigenvalue problems, the eigenvectors are determined up to a multiplicative constant. This follows from the fact that any constant multiple of an eigenvector is also an eigenvector; that is, if

$$\mathbf{A}\mathbf{v} = \lambda \mathbf{v}, \tag{5.325}$$

then $\alpha \mathbf{v}$, where α is a scalar, is also an eigenvector:

$$\mathbf{A}(\alpha \mathbf{v}) = \lambda(\alpha \mathbf{v}). \tag{5.326}$$

Having defined the inner product of two vectors, we can now fix the constants left arbitrary in the definition of the eigenvectors [Eq. (5.324)] by normalizing their norms to unity. Thus, obtaining an orthonormal set of eigenvectors as

$$\hat{\mathbf{e}}_1 = \frac{1}{\sqrt{2}}\begin{pmatrix} 0 \\ -1 \\ 1 \end{pmatrix}, \quad \hat{\mathbf{e}}_2 = \frac{1}{\sqrt{6}}\begin{pmatrix} -2 \\ 1 \\ 1 \end{pmatrix}, \quad \hat{\mathbf{e}}_3 = \frac{1}{\sqrt{3}}\begin{pmatrix} 1 \\ 1 \\ 1 \end{pmatrix}. \tag{5.327}$$

Geometrically, the degeneracy can be understood from Figure 5.2, where all the vectors lying on the plane perpendicular to the third eigenvector v_3 are possible eigenvectors to $\lambda_{1,2} = 0$. Now the transformation matrix $\mathbf{S}$ can be written as

$$\mathbf{S} = \begin{pmatrix} 0 & -\frac{2}{\sqrt{6}} & \frac{1}{\sqrt{3}} \\ -\frac{1}{\sqrt{2}} & \frac{1}{\sqrt{6}} & \frac{1}{\sqrt{3}} \\ \frac{1}{\sqrt{2}} & \frac{1}{\sqrt{6}} & \frac{1}{\sqrt{3}} \end{pmatrix}, \tag{5.328}$$

we can easily check that the inverse is given as

$$\mathbf{S}^{-1} = \begin{pmatrix} 0 & -\frac{1}{\sqrt{2}} & \frac{1}{\sqrt{2}} \\ -\frac{2}{\sqrt{6}} & \frac{1}{\sqrt{6}} & \frac{1}{\sqrt{6}} \\ \frac{1}{\sqrt{3}} & \frac{1}{\sqrt{3}} & \frac{1}{\sqrt{3}} \end{pmatrix}. \tag{5.329}$$

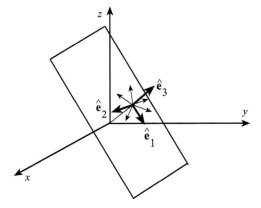

Figure 5.2   Eigenvectors corresponding to the degenerate eigenvalues.

Note that the inverse transformation matrix is the transpose of $\mathbf{S}$. In other words, $\mathbf{S}$ represents an orthogonal transformation.

5.17   QUADRATIC FORMS

Let $\mathbf{A}$ be an operator defined in an n-dimensional inner product space V over the field of real numbers $\mathbb{R}$. Also let $\mathbf{A}$ be represented by a symmetric $n \times n$ matrix

$$\mathbf{A} = A_{ij}, \quad i, j = 1, \ldots, n, \tag{5.330}$$

in terms of the basis vectors $B = \{v_1, v_2, \ldots, v_n\}$. In this space, we can construct scalar quantities like

$$\boxed{Q = \tilde{\mathbf{v}}\mathbf{A}\mathbf{v},} \tag{5.331}$$

where $\mathbf{v}$ is a vector in V represented by the column matrix

$$\mathbf{v} = \begin{pmatrix} \alpha_1 \\ \alpha_2 \\ \vdots \\ \alpha_n \end{pmatrix}. \tag{5.332}$$

If we write Q explicitly, we obtain the quadratic expression:

$$Q = A_{11}\alpha_1\alpha_1 + A_{22}\alpha_2\alpha_2 + A_{33}\alpha_3\alpha_3 + \cdots + A_{nn}\alpha_n\alpha_n \tag{5.333}$$
$$+ 2A_{12}\alpha_1\alpha_2 + 2A_{13}\alpha_1\alpha_3 + \cdots + 2A_{n-1,n}\alpha_{n-1}\alpha_n. \tag{5.334}$$

Such expressions are called **quadratic forms**, which are frequently encountered in applications. For example, the quadratic form I constructed from the moment of inertia tensor $\mathbf{I}$ [Eq. (5.184)]:

$$\boxed{I = \tilde{\hat{\mathbf{n}}}\mathbf{I}\hat{\mathbf{n}},} \tag{5.335}$$

where $\hat{\mathbf{n}}$ is the unit vector in the direction of the angular velocity $\vec{\omega} = \omega\hat{\mathbf{n}}$, is a scalar called the **moment of inertia**. In terms of I, the **kinetic energy** [4]:

$$T = \frac{1}{2}I\omega^2, \tag{5.336}$$

is written as the quadratic form

$$T = I_{xx}\omega_x^2 + I_{yy}\omega_y^2 + I_{zz}\omega_z^2 + 2I_{xy}\omega_x\omega_y + 2I_{xz}\omega_x\omega_z + 2I_{yz}\omega_y\omega_z, \tag{5.337}$$

where $(\omega_x, \omega_y, \omega_z)$ are the components of the angular velocity in Cartesian coordinates. It is naturally desirable to pick the orientation of our Cartesian axes so that the moment of inertia tensor is represented by a diagonal matrix: $\mathbf{I} = I_i\delta_{ij}$, $i, j = 1, 2, 3$, hence, the kinetic energy simplifies to

$$T = \frac{1}{2}I_1\omega_1^2 + \frac{1}{2}I_2\omega_2^2 + \frac{1}{2}I_3\omega_3^2. \tag{5.338}$$

In this equation, I_1, I_2, I_3 are the principal moments of inertia and $(\omega_1, \omega_2, \omega_3)$ are the components of the angular momentum along the principal axes.

This is naturally an eigenvalue problem, where we look for a new set of basis $B' = \{v'_1, v'_2, \ldots, v'_n\}$, where $\mathbf{A}$ [Eq. (5.330)] is represented by a diagonal matrix. In Eq. (5.88), we have written the transformation equation between two vectors, v and V', as $v = SV'$. Since the inverse transformation exists, we can also write this as $V' = S^{-1}v$.

For the new basis vectors, we first solve the eigenvalue problem $A\omega = \lambda\omega$ and find the orthogonal set of eigenvectors $(w_1, w_2, \ldots, w_n)$. Then by using these eigenvectors, we construct the new orthonormal basis, which also spans V as

$$\widehat{B}' = \{\widehat{e}_1, \widehat{e}_2, \ldots, \widehat{e}_n\}, \tag{5.339}$$

where $(\widehat{e}_i, \widehat{e}_j) = \delta_{ij}$ and

$$\widehat{e}_i = \frac{w_i}{|w_i|}, \quad Aw_i = \lambda_i w_i, \quad i = 1, 2, \ldots, n. \tag{5.340}$$

We have demonstrated that for real and symmetric matrices, the eigenvectors corresponding to distinct eigenvalues are always orthogonal and for the degenerate eigenvalues, eigenvectors can always be arranged as orthogonal. Hence, the set $\widehat{B}'$ can always be constructed. We can now write the $n \times n$ transformation matrix $\mathbf{S}$ as

$$\mathbf{S} = \left(\begin{pmatrix} \\ \widehat{e}_1 \end{pmatrix} \begin{pmatrix} \\ \widehat{e}_2 \end{pmatrix} \cdots \begin{pmatrix} \\ \widehat{e}_n \end{pmatrix} \right), \tag{5.341}$$

where $\begin{pmatrix} \\ \widehat{e}_i \end{pmatrix}$ is the $n \times 1$ column matrix representation of the eigenvector $\widehat{e}_i$. Taking the transpose of $\mathbf{S}$:

$$\widetilde{\mathbf{S}} = \begin{pmatrix} \begin{pmatrix} \widetilde{\widehat{e}}_1 \end{pmatrix} \\ \begin{pmatrix} \widetilde{\widehat{e}}_2 \end{pmatrix} \\ \vdots \\ \begin{pmatrix} \widetilde{\widehat{e}}_n \end{pmatrix} \end{pmatrix}, \tag{5.342}$$

we construct the product

$$\widetilde{\mathbf{S}}\mathbf{S} = \begin{pmatrix} (\widehat{e}_1, \widehat{e}_1) & (\widehat{e}_1, \widehat{e}_2) & \cdots & (\widehat{e}_1, \widehat{e}_n) \\ (\widehat{e}_2, \widehat{e}_1) & (\widehat{e}_2, \widehat{e}_2) & \cdots & (\widehat{e}_2, \widehat{e}_n) \\ \vdots & \vdots & \ddots & \vdots \\ (\widehat{e}_n, \widehat{e}_1) & (\widehat{e}_n, \widehat{e}_2) & \cdots & (\widehat{e}_n, \widehat{e}_n) \end{pmatrix}. \tag{5.343}$$

Using the orthonormality relation of the new set $\widehat{B}'$:

$$(\widehat{e}_i, \widehat{e}_j) = \delta_{ij}, \tag{5.344}$$

Equation (5.343) becomes $\widetilde{\mathbf{S}}\mathbf{S} = \mathbf{I}$. In other words, the inverse transformation is the transpose of the transformation matrix: $\mathbf{S}^{-1} = \widetilde{\mathbf{S}}$. Matrices satisfying this property are called **orthogonal matrices**. Transformations represented by orthogonal matrices preserve norms of vectors. This can be easily seen by using $\mathbf{v} = \mathbf{S}\mathbf{v}'$ to write $|\mathbf{v}|^2$ in terms of the new basis $\widehat{B}'$ as

$$|\mathbf{v}|^2 = \widetilde{\mathbf{v}}\mathbf{v} = \widetilde{(\mathbf{S}\mathbf{v}')}\mathbf{S}\mathbf{v}' = \widetilde{\mathbf{v}}'\widetilde{\mathbf{S}}\mathbf{S}\mathbf{v}' = \widetilde{\mathbf{v}}'\mathbf{I}\mathbf{v}' = \widetilde{\mathbf{v}}'\mathbf{v}' = |\mathbf{v}'|^2. \tag{5.345}$$

The quadratic form $Q = \widetilde{\mathbf{v}}\mathbf{A}\mathbf{v}$ can now be written in terms of the new bases. Using the transformation equation $\mathbf{v} = \mathbf{S}\mathbf{v}'$, we can write

$$Q = \widetilde{(\mathbf{S}\mathbf{v}')}\mathbf{A}\mathbf{S}\mathbf{v}' = \widetilde{\mathbf{v}}'\widetilde{\mathbf{S}}\mathbf{A}\mathbf{S}\mathbf{v}' = \widetilde{\mathbf{v}}'(\mathbf{S}^{-1}\mathbf{A}\mathbf{S})\mathbf{v}', \tag{5.346}$$

where $\mathbf{S}^{-1}\mathbf{A}\mathbf{S}$ is the matrix representation of the operator $\mathbf{A}$ in terms of the new bases:

$$\mathbf{A}' = \mathbf{S}^{-1}\mathbf{A}\mathbf{S}. \tag{5.347}$$

It is important to note that orthogonal transformations preserve the values of quadratic forms, in other words,

$$Q = \widetilde{\mathbf{v}}\mathbf{A}\mathbf{v} = \widetilde{\mathbf{v}}'\mathbf{A}'\mathbf{v}' = Q'. \tag{5.348}$$

Writing the transformation matrix [Eq. (5.341)] $\mathbf{S} = \widehat{e}_{ij}$, $i, j = 1, \ldots, n$, explicitly:

$$\mathbf{S} = \begin{pmatrix} \widehat{e}_{11} & \widehat{e}_{12} & \cdots & \widehat{e}_{1n} \\ \widehat{e}_{21} & \widehat{e}_{22} & \cdots & \widehat{e}_{2n} \\ \vdots & \vdots & \ddots & \vdots \\ \widehat{e}_{n1} & \widehat{e}_{n2} & \cdots & \widehat{e}_{nn} \end{pmatrix}, \tag{5.349}$$

where $\widehat{e}_{ij}$ is the ith component of the jth normalized eigenvector, we form the product

$$\mathbf{A}\mathbf{S} = \begin{pmatrix} A_{11} & A_{12} & \cdots & A_{1n} \\ A_{21} & A_{22} & \cdots & A_{2n} \\ \vdots & \vdots & \ddots & \vdots \\ A_{n1} & A_{n2} & \cdots & A_{nn} \end{pmatrix} \begin{pmatrix} \widehat{e}_{11} & \widehat{e}_{12} & \cdots & \widehat{e}_{1n} \\ \widehat{e}_{21} & \widehat{e}_{22} & \cdots & \widehat{e}_{2n} \\ \vdots & \vdots & \ddots & \vdots \\ \widehat{e}_{n1} & \widehat{e}_{n2} & \cdots & \widehat{e}_{nn} \end{pmatrix}$$

$$= \begin{pmatrix} \lambda_1\widehat{e}_{11} & \lambda_2\widehat{e}_{12} & \cdots & \lambda_n\widehat{e}_{1n} \\ \lambda_1\widehat{e}_{21} & \lambda_2\widehat{e}_{22} & \cdots & \lambda_n\widehat{e}_{2n} \\ \vdots & \vdots & \ddots & \vdots \\ \lambda_1\widehat{e}_{n1} & \lambda_2\widehat{e}_{n2} & \cdots & \lambda_n\widehat{e}_{nn} \end{pmatrix}. \tag{5.350}$$

In the above equation, we have used the eigenvalue equation [Eq. (5.340)]:

$$\sum_{j=1}^{n} A_{ij}\widehat{e}_{jk} = \sum_{l=1}^{n} \widehat{e}_{il}\delta_{lk}\lambda_k.$$

(5.351)

Equation (5.350) can also be written as

$$\mathbf{AS} = \mathbf{S} \begin{pmatrix} \lambda_1 & 0 & \cdots & 0 \\ 0 & \lambda_2 & \cdots & 0 \\ \vdots & \vdots & \ddots & \vdots \\ 0 & 0 & \cdots & \lambda_n \end{pmatrix},$$

(5.352)

which leads to

$$\mathbf{S}^{-1}\mathbf{AS} = \begin{pmatrix} \lambda_1 & 0 & \cdots & 0 \\ 0 & \lambda_2 & \cdots & 0 \\ \vdots & \vdots & \ddots & \vdots \\ 0 & 0 & \cdots & \lambda_n \end{pmatrix};$$

(5.353)

that is, in the new bases $\mathbf{A}$ is represented by a diagonal matrix with the diagonal elements being the eigenvalues.

5.18 HERMITIAN MATRICES

We now consider matrices in the **complex field** $\mathbb{C}$. The **adjoint** of a matrix, $\mathbf{A}^\dagger$, is defined as $\widetilde{\mathbf{A}}^*$, where the transpose of $\mathbf{A}$ along with the complex conjugate of each element is taken. If the adjoint of a matrix is equal to itself, $\mathbf{A}^\dagger = \mathbf{A}$, it is called **self-adjoint**. In real inner product spaces, self-adjoint matrices are simply symmetric matrices. In complex inner product spaces, self-adjoint matrices are called **Hermitian**. Adjoint operation has the following properties:

$$(\mathbf{A} + \mathbf{B})^\dagger = \mathbf{A}^\dagger + \mathbf{B}^\dagger,$$

(5.354)

$$(\mathbf{AB})^\dagger = \mathbf{B}^\dagger\mathbf{A}^\dagger,$$

(5.355)

$$(\alpha\mathbf{A})^\dagger = \alpha^*\mathbf{A}^\dagger,$$

(5.356)

$$(\mathbf{A}^\dagger)^\dagger = \mathbf{A}.$$

(5.357)

Hermitian matrices have very useful properties. Consider the eigenvalue problem for an $n \times n$ Hermitian matrix $\mathbf{H}$:

$$\mathbf{Hu}_\lambda = \lambda\mathbf{u}_\lambda,$$

(5.358)

where $\mathbf{u}_\lambda$ is the $n \times 1$ column matrix representing the eigenvector corresponding to the eigenvalue λ. For two eigenvalues, λ_i and λ_j, we write the corresponding eigenvalue equations, respectively, as

$$\mathbf{Hu}_i = \lambda_i\mathbf{u}_i$$

(5.359)

and

$$\mathbf{Hu}_j = \lambda_j \mathbf{u}_j. \tag{5.360}$$

We now multiply Eq. (5.359) with $\widetilde{\mathbf{u}}_j^*$ from the left and write

$$\widetilde{\mathbf{u}}_j^* \mathbf{Hu}_i = \lambda_i \widetilde{\mathbf{u}}_j^* \mathbf{u}_i. \tag{5.361}$$

Next we take the adjoint of Eq. (5.360):

$$(\mathbf{Hu}_j)^\dagger = \lambda_j^* \mathbf{u}_j^\dagger, \tag{5.362}$$

$$\widetilde{\mathbf{u}}_j^* \mathbf{H}^\dagger = \lambda_j^* \widetilde{\mathbf{u}}_j^* \tag{5.363}$$

and multiply by $\mathbf{u}_i$ from the right:

$$\widetilde{\mathbf{u}}_j^* \mathbf{H}^\dagger \mathbf{u}_i = \lambda_j^* \widetilde{\mathbf{u}}_j^* \mathbf{u}_i. \tag{5.364}$$

Using the fact that for Hermitian matrices $\mathbf{H}^\dagger = \mathbf{H}$, Eq. (5.364) becomes

$$\widetilde{\mathbf{u}}_j^* \mathbf{Hu}_i = \lambda_j^* \widetilde{\mathbf{u}}_j^* \mathbf{u}_i. \tag{5.365}$$

We now subtract Eq. (5.365) from Eq. (5.361) to write

$$0 = (\lambda_i - \lambda_j^*)\widetilde{\mathbf{u}}_j^* \mathbf{u}_i. \tag{5.366}$$

As in the case of real symmetric matrices, there are basically three cases:
I. When $i = j$, Eq. (5.366) becomes

$$0 = (\lambda_i - \lambda_i^*)\widetilde{\mathbf{u}}_i^* \mathbf{u}_i = (\lambda_i - \lambda_i^*)|\mathbf{u}_i|^2. \tag{5.367}$$

Since $|\mathbf{u}_i|^2$ is always positive, this implies that the **eigenvalues** of a **Hermitian matrix** are always **real**, that is,

$$\boxed{\lambda_i = \lambda_i^*.} \tag{5.368}$$

In **quantum mechanics** eigenvalues correspond to directly measurable quantities in laboratory, hence, **observables** are represented by Hermitian matrices.
II. When $i \neq j$ and the eigenvalues are distinct, $\lambda_i \neq \lambda_j$, the corresponding eigenvectors are orthogonal in the Hermitian sense, that is,

$$\boxed{(u_j, u_i) = \widetilde{\mathbf{u}}_j^* \mathbf{u}_i = 0.} \tag{5.369}$$

III. As in the case of real and symmetric matrices, for repeated roots of order s there exists a set of s linearly independent eigenvectors. Hence, for a Hermitian $n \times n$ matrix $\mathbf{H}$, we can construct a set of linearly independent orthonormal basis vectors $\{\widehat{u}_1, \widehat{u}_2, \ldots, \widehat{u}_n\}$, which spans the n-dimensional inner product space V defined over the complex field $\mathbb{C}$. It then follows that any vector v in V can be written as the linear combination

$$v = \sum_{i=1}^{n} \alpha_i \widehat{u}_i, \qquad (5.370)$$

where the components α_i are numbers in the complex field $\mathbb{C}$. Using the definition of the inner product and the orthogonality of the normalized basis vectors, we can evaluate α_j as

$$(\widehat{u}_j, v) = \sum_{i=1}^{n} (\widehat{u}_j, \alpha_i \widehat{u}_i) = \sum_{i=1}^{n} \alpha_i (\widehat{u}_j, \widehat{u}_i) = \sum_{i=1}^{n} \alpha_i \delta_{ij} = \alpha_j. \qquad (5.371)$$

In this complex vector space, Hermitian operators, H, transform vectors in V into other vectors in V. Again, it is desirable to find a new set of basis vectors, where H is represented by a diagonal Hermitian matrix. Steps to follow are similar to the procedure described for the real and symmetric matrices. We define the transformation matrix $\mathbf{U}$ as

$$\mathbf{U} = \left(\left(\begin{matrix} \\ \widehat{\mathbf{u}}_1 \\ \end{matrix} \right) \left(\begin{matrix} \\ \widehat{\mathbf{u}}_2 \\ \end{matrix} \right) \cdots \left(\begin{matrix} \\ \widehat{\mathbf{u}}_n \\ \end{matrix} \right) \right), \qquad (5.372)$$

where $\widehat{\mathbf{u}}_i$, $i = 1, 2, \ldots, n$, are the unit eigenvectors represented by the column matrices

$$\widehat{\mathbf{u}}_i = \begin{pmatrix} \widehat{u}_{11} \\ \vdots \\ \widehat{u}_{n1} \end{pmatrix}. \qquad (5.373)$$

Here, $\widehat{u}_{ji}$ corresponds to the jth component of the ith eigenvector $\widehat{\mathbf{u}}_i$, satisfying the eigenvalue equation

$$\mathbf{H}\widehat{\mathbf{u}}_i = \lambda_i \widehat{\mathbf{u}}_i. \qquad (5.374)$$

Due to the orthonormality of the eigenvectors,

$$(\widehat{u}_j, \widehat{u}_i) = \widetilde{\widehat{\mathbf{u}}}_j^{*} \widehat{\mathbf{u}}_i = \delta_{ij}, \qquad (5.375)$$

we can write

$$\widetilde{\mathbf{U}}^{*}\mathbf{U} = \begin{pmatrix} \left(\quad \widetilde{\widehat{\mathbf{u}}}_1^{*} \quad \right) \\ \left(\quad \widetilde{\widehat{\mathbf{u}}}_2^{*} \quad \right) \\ \vdots \\ \left(\quad \widetilde{\widehat{\mathbf{u}}}_n^{*} \quad \right) \end{pmatrix} \left(\left(\begin{matrix} \\ \widehat{\mathbf{u}}_1 \\ \end{matrix} \right) \left(\begin{matrix} \\ \widehat{\mathbf{u}}_2 \\ \end{matrix} \right) \cdots \left(\begin{matrix} \\ \widehat{\mathbf{u}}_n \\ \end{matrix} \right) \right), \qquad (5.376)$$

$$\mathbf{U}^\dagger\mathbf{U} = \begin{pmatrix} 1 & 0 & \cdots & 0 \\ 0 & 1 & \cdots & 0 \\ \vdots & \vdots & \ddots & \vdots \\ 0 & 0 & \cdots & 1 \end{pmatrix} = \mathbf{I}, \tag{5.377}$$

hence

$$\mathbf{U}^{-1} = \mathbf{U}^\dagger. \tag{5.378}$$

A matrix whose **inverse** is equal to its adjoint is called a **unitary matrix**. Similarly, a transformation represented by a unitary matrix is called a **unitary transformation**. Unitary transformations are the counterpart of orthogonal transformations in complex inner product spaces, and they play a very important role in quantum mechanics. A unitary transformation, $\mathbf{U}^\dagger\mathbf{U} = \mathbf{I}$, preserves the norms of vectors. This can be proven by using

$$\mathbf{v} = \mathbf{U}\mathbf{v}' \tag{5.379}$$

to write

$$\widetilde{\mathbf{v}}^*\mathbf{v} = \widetilde{(\mathbf{U}\mathbf{v}')}^*\mathbf{U}\mathbf{v}' = \widetilde{\mathbf{v}'}^*\widetilde{\mathbf{U}}^*\mathbf{U}\mathbf{v}' = \widetilde{\mathbf{v}'}^*\mathbf{U}^\dagger\mathbf{U}\mathbf{v}' = \widetilde{\mathbf{v}'}^*\mathbf{v}', \tag{5.380}$$

which is the desired result:
$$|\mathbf{v}|^2 = |\mathbf{v}'|^2. \tag{5.381}$$

Consider a Hermitian matrix $\mathbf{H}$, representing a Hermitian operator H, which transforms a vector $\mathbf{v}$ in V into another vector $\mathbf{w}$ in V defined over the complex field $\mathbb{C}$ as
$$\mathbf{w} = \mathbf{H}\mathbf{v}. \tag{5.382}$$

Vectors $\mathbf{v}$ and $\mathbf{w}$ transform under unitary transformations as

$$\mathbf{w} = \mathbf{U}\mathbf{w}', \tag{5.383}$$
$$\mathbf{v} = \mathbf{U}\mathbf{v}'. \tag{5.384}$$

Using these in Eq. (5.382), we obtain

$$\mathbf{U}\mathbf{w}' = \mathbf{H}\mathbf{U}\mathbf{v}', \tag{5.385}$$
$$\mathbf{U}^{-1}\mathbf{U}\mathbf{w}' = (\mathbf{U}^{-1}\mathbf{H}\mathbf{U})\mathbf{v}', \tag{5.386}$$
$$\mathbf{w}' = (\mathbf{U}^{-1}\mathbf{H}\mathbf{U})\mathbf{v}'. \tag{5.387}$$

In conclusion, in terms of the new bases defined by the unitary transformation $\mathbf{U}$, $\mathbf{H}$ is expressed as
$$\mathbf{H}' = \mathbf{U}^{-1}\mathbf{H}\mathbf{U} = \mathbf{U}^\dagger\mathbf{H}\mathbf{U}. \tag{5.388}$$

A unitary transformation constructed by using the eigenvectors of **H**, as described in the argument leading to Eq. (5.353), diagonalizes **H** as

$$
\mathbf{H}' = \begin{pmatrix} \lambda_1 & 0 & \cdots & 0 \\ 0 & \lambda_2 & \cdots & 0 \\ \vdots & \vdots & \ddots & \vdots \\ 0 & 0 & \cdots & \lambda_n \end{pmatrix}. \tag{5.389}
$$

5.19 MATRIX REPRESENTATION OF HERMITIAN OPERATORS

Having defined the inner product spaces, we can now show how the matrix representation of a given transformation or operator can be found. Consider an n-dimensional vector space V, spun by the orthonormal set $B = \{\widehat{e}_1, \widehat{e}_2, \ldots, \widehat{e}_n\}$ and defined over the field K. Let T be a linear transformation from V into V, that is,

$$
w = Tv, \tag{5.390}
$$

where v and w are elements in V. Using the orthonormal basis B, we can write the vectors v and w as

$$
v = \sum_{i=1}^{n} \alpha_i \widehat{e}_i, \tag{5.391}
$$

$$
w = \sum_{j=1}^{n} \beta_j \widehat{e}_j, \tag{5.392}
$$

where the components α_i and β_j are evaluated by using the inner products

$$
\alpha_i = (\widehat{e}_i, v) \text{ and } \beta_j = (\widehat{e}_j, w). \tag{5.393}
$$

We now write Eq. (5.390) as

$$
\sum_{j=1}^{n} \beta_j \widehat{e}_j = T \sum_{i=1}^{n} \alpha_i \widehat{e}_i = \sum_{i=1}^{n} \alpha_i T \widehat{e}_i. \tag{5.394}
$$

If we take the inner product of both sides with $\widehat{e}_k$ and use the orthogonality condition, $(\widehat{e}_k, \widehat{e}_j) = \delta_{kj}$, we obtain

$$
\sum_{j=1}^{n} \beta_j (\widehat{e}_k, \widehat{e}_j) = \sum_{i=1}^{n} \alpha_i (\widehat{e}_k, T \widehat{e}_i), \tag{5.395}
$$

$$
\sum_{j=1}^{n} \beta_j \delta_{kj} = \sum_{i=1}^{n} \alpha_i (\widehat{e}_k, T \widehat{e}_i), \tag{5.396}
$$

$$
\beta_k = \sum_{i=1}^{n} (\widehat{e}_k, T \widehat{e}_i) \alpha_i. \tag{5.397}
$$

Comparing with $\beta_i = \sum_{j=1}^{n} A_{ij}\alpha_j$ [Eq. (5.64)], we see that the elements of the transformation matrix are obtained as

$$\boxed{A_{ki} = (\widehat{e}_k, T\widehat{e}_i).}$$
(5.398)

For any linear operator T acting on V, we can define another linear operator $T^\dagger$ called the **adjoint** of T, which for any two vectors v and w in V has the property

$$\boxed{(Tv, w) = (v, T^\dagger w).}$$
(5.399)

If we write v and w in terms of the basis $B = \{\widehat{e}_1, \widehat{e}_2, \ldots, \widehat{e}_n\}$, we obtain

$$\boxed{(T\widehat{e}_i, \widehat{e}_j) = (\widehat{e}_i, T^\dagger\widehat{e}_j) = (\widehat{e}_j, T\widehat{e}_i)^*.}$$
(5.400)

In other words, the matrix $[T^\dagger]_{ij} = A^\dagger_{ij}$, representing the adjoint operator $T^\dagger$, is obtained from the matrix $[T]_{ij} = A_{ij}$, which represents T by complex conjugation and transposition. A linear operator satisfying

$$\boxed{T^\dagger = T}$$
(5.401)

is called a **Hermitian operator** or a **self-adjoint operator**. Hermitian operators are represented by **Hermitian matrices**.

5.20   FUNCTIONS OF MATRICES

We are now ready to state an important result from Gantmacher [6]:

Theorem 5.6. If a scalar function $f(x)$ can be expanded in power series:

$$f(x) = \sum_{n=0}^{\infty} c_n (x - x_0)^n, \quad |x - x_0| < r,$$
(5.402)

where r is the radius of convergence, then this expansion remains valid when x is replaced by a matrix $\mathbf{A}$ whose eigenvalues are within the radius of convergence.

Using this theorem, we can write the expansions:

$$e^{\mathbf{A}} = \sum_{n=0}^{\infty} \frac{\mathbf{A}^n}{n!},$$
(5.403)

$$\cos \mathbf{A} = \sum_{n=0}^{\infty} \frac{(-1)^n}{(2n)!} \mathbf{A}^{2n},$$
(5.404)

$$\sin \mathbf{A} = \sum_{n=0}^{\infty} \frac{(-1)^n}{(2n+1)!} \mathbf{A}^{2n+1},$$
(5.405)

An important consequence of this theorem is the **Baker –Hausdorf formula**:

$$e^{i\mathbf{A}}\mathbf{H}e^{-i\mathbf{A}} = \mathbf{H} + i[\mathbf{A},\mathbf{H}] - \frac{1}{2}[\mathbf{A},[\mathbf{A},\mathbf{H}]] + \cdots , \qquad (5.406)$$

where $\mathbf{A}$ and $\mathbf{H}$ are $n \times n$ matrices and

$$[\mathbf{A},\mathbf{H}] = \mathbf{AH} - \mathbf{HA} \qquad (5.407)$$

is called the **commutator** of $\mathbf{A}$ with $\mathbf{H}$. For the proof, we first expand $e^{i\mathbf{A}}$ and $e^{-i\mathbf{A}}$ in power series and then substitute the results into $e^{i\mathbf{A}}\mathbf{H}e^{-i\mathbf{A}}$ and collect similar terms.

Another very useful result is the **trace formula**:

$$\det e^{\mathbf{A}} = e^{\mathrm{tr}\mathbf{A}}, \qquad (5.408)$$

where $\mathbf{A}$ is a Hermitian matrix. Since unitary transformations, $\mathbf{U}^{\dagger}\mathbf{U} = \mathbf{I}$, preserve values of determinants (Section 5.8), we can write

$$\det e^{\mathbf{A}} = \det(\mathbf{U}^{\dagger}e^{\mathbf{A}}\mathbf{U}) = \det e^{\mathbf{U}^{\dagger}\mathbf{A}\mathbf{U}}.$$

Using a unitary transformation that diagonalizes $\mathbf{A}$ as

$$\mathbf{U}^{\dagger}\mathbf{A}\mathbf{U} = \mathbf{D}, \qquad (5.409)$$

$$\mathbf{D} = \begin{pmatrix} \lambda_1 & 0 & \cdots & 0 \\ 0 & \lambda_2 & \cdots & 0 \\ \vdots & \vdots & \ddots & \vdots \\ 0 & 0 & \cdots & \lambda_n \end{pmatrix}, \qquad (5.410)$$

where λ_i are the eigenvalues of $\mathbf{A}$, we can write

$$\det e^{\mathbf{A}} = \det e^{\mathbf{D}}. \qquad (5.411)$$

Since the exponential of a diagonal matrix, $\mathbf{D} = \lambda_i \delta_{ik}$, $i,k = 1, \ldots, n$, is

$$e^{\mathbf{D}} = e^{\lambda_i}\delta_{ik}, \quad i,k = 1, \ldots, n, \qquad (5.412)$$

we also write

$$\det e^{\mathbf{A}} = \det(e^{\lambda_i}\delta_{ik}), \quad i,k = 1, \ldots, n. \qquad (5.413)$$

Since the determinant of a diagonal matrix is equal to the product of its diagonal terms:

$$\det e^{\mathbf{A}} = \prod_{i=1}^{n} e^{\lambda_i}, \qquad (5.414)$$

and since unitary transformations, $\mathbf{U}^\dagger\mathbf{U} = \mathbf{I}$, preserve trace (Section 5.8), we finally obtain the trace formula as

$$\det e^{\mathbf{A}} = e^{(\lambda_1+\lambda_2+\cdots+\lambda_n)} = e^{\text{tr}\mathbf{D}} = e^{\text{tr}\mathbf{A}}. \tag{5.415}$$

5.21 FUNCTION SPACE AND HILBERT SPACE

We now define a new vector space L_2, whose elements are complex valued square integrable functions of the real variable x, defined in the closed interval $[a, b]$. A function is **square integrable** if the integral

$$\int_a^b f^*(x)f(x)dx = \int_a^b |f(x)|^2 dx \tag{5.416}$$

exists and is finite. With the **inner product** of two square integrable functions, f_1 and f_2, defined as

$$(f_1, f_2) = \int_a^b f_1^* f_2 dx, \tag{5.417}$$

the resulting infinite-dimensional inner product space is called the **Hilbert space**. Concepts of orthogonality and normalization in Hilbert space are defined as before. In a finite-dimensional subspace V of the Hilbert space, a set of square integrable functions $\{g_1, g_2, \ldots, g_n\}$ is called an **orthonormal** set if its elements satisfy

$$\int_a^b g_i^* g_j dx = \delta_{ij}, \quad i, j = 1, \ldots, n. \tag{5.418}$$

Furthermore, we say that the set $\{g_1, g_2, \ldots, g_n\}$ is **complete** if it spans V; that is, any **square integrable** function $g(x)$ in V can be written as

$$g(x) = \sum_{i=0}^n c_i g_i(x). \tag{5.419}$$

Using Eq. (5.418), we can evaluate the **expansion coefficients** c_i as

$$c_i = (g_i, g) = \int_a^b g_i^* g dx. \tag{5.420}$$

Using a linear operator A acting in the Hilbert space, we can define an eigenvalue problem as

$$Af_i(x) = \lambda_i f_i(x), \quad i = 1, 2, \ldots, \tag{5.421}$$

where $f_i(x)$ is the eigenfunction corresponding to the eigenvalue λ_i. We have seen that in a finite-dimensional vector space V, Hermitian operators are represented by Hermitian matrices, which have real eigenvalues and their eigenvectors form a complete orthonormal set that spans V. For the infinite-dimensional Hilbert space, the proof of completeness is beyond the scope of this book, a discussion of this point can be found in Byron and Fuller [7]. The state of a system in quantum mechanics is described by a square integrable function in Hilbert space, which is called the **state function** or the **wave function**. Observables in quantum mechanics are Hermitian operators with real eigenvalues acting on square integrable functions in Hilbert space.

5.22 DIRAC'S BRA AND KET VECTORS

A different notation introduced by Dirac has advantages when we consider eigenvalue problems and Hermitian and unitary operators in quantum mechanics. For two vectors in Hilbert space, the inner product is **not symmetric**:

$$(\Psi_1, \Psi_2) = \int \Psi_1^* \Psi_2 \, dx \neq (\Psi_2, \Psi_1) = \int \Psi_2^* \Psi_1 \, dx. \tag{5.422}$$

However, from the relations

$$(\Psi_1, \Psi_2) = (\Psi_2, \Psi_1)^*, \tag{5.423}$$

$$(\Psi_1 + \Psi_2, \Psi_3) = (\Psi_1, \Psi_3) + (\Psi_2, \Psi_3), \tag{5.424}$$

$$(\Psi_1, \Psi_2 + \Psi_3) = (\Psi_1, \Psi_2) + (\Psi_1, \Psi_3), \tag{5.425}$$

$$(\Psi_1, \alpha\Psi_2) = \alpha(\Psi_1, \Psi_2), \tag{5.426}$$

$$(\alpha\Psi_1, \Psi_2) = \alpha^*(\Psi_1, \Psi_2), \tag{5.427}$$

where α is a number in the complex field $\mathbb{C}$, we see that the inner product is nonlinear with respect to the prefactor. This apparent asymmetry can be eliminated if we think of both vectors as belonging to different spaces – that is, the space of *prefactor* vectors and the space of *postfactor* vectors, where each space is linear within itself but related to each other in a nonlinear manner through the definition of the inner product. Hence, they are called **dual spaces.** Dirac called the prefactor vectors **bra** and showed them as $\langle\Psi|$, and called the postfactor vectors **ket** and showed them as $|\Psi\rangle$. For each bra, there exists a ket in its **dual space**, that is,

$$|\Psi\rangle \leftrightarrow \langle\Psi|, \tag{5.428}$$

$$|\Psi_1 + \Psi_2\rangle \leftrightarrow \langle\Psi_1 + \Psi_2|, \tag{5.429}$$

$$\alpha|\Psi\rangle \leftrightarrow \alpha^*\langle\Psi|. \tag{5.430}$$

Each space on its own is a vector space. The connection between them is established through the definition of the inner product as

$$\langle \Psi_1 | \Psi_2 \rangle = \int \Psi_1^* \Psi_2 \, dx. \tag{5.431}$$

Obviously,

$$\langle \Psi_2 | \Psi_1 \rangle = \int \Psi_2^* \Psi_1 \, dx = \langle \Psi_1 | \Psi_2 \rangle^*. \tag{5.432}$$

Note that generally, we write $\langle \Psi_1 | \Psi_2 \rangle$ rather than $\langle \Psi_1 || \Psi_2 \rangle$. A linear operator A associates a ket with another ket:

$$A | \Psi \rangle = | \Phi \rangle, \tag{5.433}$$

$$A | \Psi_1 + \Psi_2 \rangle = A | \Psi_1 \rangle + A | \Psi_2 \rangle, \tag{5.434}$$

$$A(\alpha | \Psi \rangle) = \alpha(A | \Psi \rangle). \tag{5.435}$$

By writing $\langle \Psi_1 | A | \Psi_2 \rangle$ as

$$\langle \Psi_1 | [A | \Psi_2 \rangle] = [\langle \Psi_1 | A] | \Psi_2 \rangle \tag{5.436}$$

we may define a bra, $\langle \Psi_1 | A$, that allows us to use A to act on either the ket or the bra as

$$(\langle \Psi_1 | + \langle \Psi_2 |)A = \langle \Psi_1 | A + \langle \Psi_2 | A, \tag{5.437}$$

$$(\langle \Psi | \alpha)A = \alpha(\langle \Psi | A). \tag{5.438}$$

In other words, A is a linear operator in both the bra and the ket space. In terms of the bra-ket notation, the definition of **Hermitian operators** become

$$\boxed{\langle \Psi_1 | A^\dagger | \Psi_2 \rangle = \langle \Psi_2 | A | \Psi_1 \rangle^*.} \tag{5.439}$$

In general, we can establish the correspondence

$$A | \Psi \rangle = | \Phi \rangle \leftrightarrow \langle \Phi | = \langle \Psi | A^\dagger. \tag{5.440}$$

REFERENCES

1. Hoffman, K. and Kunze, R. (1971). *Linear Algebra*, 2e. Upper Saddle River, NJ: Prentice Hall.

2. Kaye, R. and Wilson, R. (1998). *Linear Algebra*. New York: Oxford University Press.

3. Lang, S. (1966). *Linear Algebra*. Reading, MA: Addison-Wesley.

4. Goldstein, H., Poole, C., and Safko, J. (2002). *Classical Mechanics*, 3e. San Francisco, CA: Addison-Wesley.

5. Hildebrand, F.B. (1992). *Methods of Applied Mathematics*, 2nd reprint edition. New York: Dover Publications.

6. Gantmacher, F.R. (1960). *The Theory of Matrices*. New York: Chelsea Publishing Company.

7. Byron, F.W. Jr. and Fuller, R.W. (1992). *Mathematics of Classical and Quantum Physics*. New York: Dover Publications.

PROBLEMS

1. Is the set of integers, $\ldots, -1, 0, 1, \ldots$, a subfield of $\mathbb{C}$?
2. Is the set of rational numbers a subfield of $\mathbb{C}$? Explain.
3. Write three vectors that are linearly dependent in $\mathbb{R}^3$ but such that any two of them are linearly independent.
4. Let V be the vector space of all 2×2 matrices over the field K. Show that V has the dimension 4 by writing a basis for V which has four elements.
5. Let V be the vector space of all polynomials from $\mathbb{R}$ into $\mathbb{R}$ of degree 2 or less. In other words, the space of all functions of the form

$$f(x) = a_0 + a_1 x + a_2 x^2.$$

Show that $B = \{g_1, g_2, g_3\}$, where

$$g_1(x) = 1, \quad g_2(x) = x + 1, \quad g_3(x) = (x + 2)^2,$$

forms a basis for V. If $f(x) = a_0 + a_1 x + a_2 x^2$, find the components with respect to B.

6. Show that the vectors

$$v_1 = (1, 1, 0, 0),$$
$$v_2 = (1, 0, 0, 2),$$
$$v_3 = (0, 0, 1, 1),$$
$$v_4 = (0, 0, 0, 1)$$

form a basis for $\mathbb{R}^4$.

7. Which of the following transformations, $T(x_1, x_2)$, from $\mathbb{R}^2$ into $\mathbb{R}^2$ are linear transformations?

$$
\begin{array}{ll}
\text{(i)} & T(x_1, x_2) = (x_1, 1 + x_2), \\
\text{(ii)} & T(x_1, x_2) = (x_2, x_1), \\
\text{(iii)} & T(x_1, x_2) = (x_1, x_2^2), \\
\text{(iv)} & T(x_1, x_2) = (x_1 x_2, 0).
\end{array}
$$

8. If T and U are linear operators on $\mathbb{R}^2$ defined as
$$T(x_1, x_2) = (x_2, x_1),$$
$$U(x_1, x_2) = (x_1, 0),$$

give a geometric interpretation of T and U. Also write the following transformations explicitly:
$$(U + T), \quad UT, \quad TU, \quad T^2, \quad U^2.$$

9. Let V be the vector space of all $n \times n$ matrices over the field K. Show that
$$T(A) = AB - BA,$$

where A belongs to V and B is any fixed $n \times n$ matrix, is a linear transformation.

10. Let V be the space of polynomials of degree three or less:
$$f(x) = a_0 + a_1 x + a_2 x^2 + a_3 x^3.$$

The differential operator $D = d/dx$ maps V into V, since its effect on $f(x)$ is to lower its degree by one.

(i) Let
$$B = \{f_1, f_2, f_3, f_4\} = \{1, x, x^2, x^3\}$$

be the basis for V. Find the matrix representation of D, that is, $[D]_B$.

(ii) Show that the set $B' = \{g_1, g_2, g_3, g_4\}$, where
$$g_1 = f_1,$$
$$g_2 = f_1 + f_2,$$
$$g_3 = f_1 + 2f_2 + f_3,$$
$$g_4 = f_1 + 3f_2 + 3f_3 + f_4,$$

also forms a basis for V.

(iii) Find the matrix representation of D in terms of the basis B'.

11. For a distribution of m point masses, the moment of inertia tensor is written as
$$I_{ij} = \sum_{k=1}^{m} m_k [r_k^2 \delta_{ij} - (\mathbf{r}_k)_i (\mathbf{r}_k)_j],$$

where k stands for the kth particle and i and j refer to the Cartesian coordinates of the kth particle. Consider 11 equal points located at the following points:

$(a, 0, 0), \quad (0, a, 0), \quad (0, 0, a), \quad (-a, 0, 0), \quad (0, -a, 0), \quad (0, 0, -a),$

$(a, a, 0), \quad (a, -a, 0), \quad (-a, a, 0), \quad (-a, -a, 0), \quad (2a, 2a, 0),$

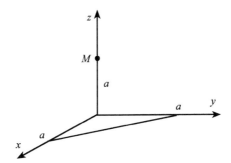

Figure 5.3   Mass distribution in Problem 12.

(i) Find the principal moments of inertia.

(ii) Find the principal axes and plot them.

(iii) Write the transformation matrix S and show that it corresponds to an orthogonal transformation.

(iv) Show explicitly that S diagonalizes the moment of inertia tensor.

(v) Write the quadratic form corresponding to the moment of inertia, I, in both coordinates.

12. Consider a rigid body with uniform density of mass $2M$ in the shape of a $45°$ right triangle lying on the xy-plane and with a point mass M on the z-axis as shown in Figure 5.3.
Find the principal moments of inertia and the eigenvectors.

13. Find the eigenvalues and the eigenvectors of the following matrices:

(i)

$$\mathbf{A} = \begin{pmatrix} 1 & \sqrt{2} & 0 \\ \sqrt{2} & 0 & 0 \\ 0 & 0 & 0 \end{pmatrix},$$

(ii)

$$\mathbf{A} = \begin{pmatrix} 2 & 0 & 0 \\ 0 & 1 & 1 \\ 0 & 1 & 1 \end{pmatrix}.$$

14. Write the quadratic form

$$Q = 5x_1^2 - 2x_2^2 - 3x_3^2 + 12x_1x_2 - 8x_1x_3 + 20x_2x_3$$

in matrix form $Q = \tilde{\mathbf{x}}\mathbf{A}\mathbf{x}$ and then

(i) Find the eigenvalues and the eigenvectors of A (*Hint:* one of the eigenvalues is 6).

(ii) Construct the transformation matrix that diagonalizes the above quadratic form and show that this transformation is orthogonal and diagonalizes Q.

15. Find the transformation that removes the xy term in

$$x^2 + xy + y^2 = 16$$

and interpret your answer geometrically.

16. Given the quadratic form

$$Q = 2x_1^2 + 5x_2^2 + 2x_3^2 + 4x_1x_3.$$

(i) Express Q in matrix form as

$$Q = \tilde{\mathbf{x}}\mathbf{A}\mathbf{x}.$$

(ii) Find the eigenvalues and the eigenvectors of A.

(iii) Find the transformation matrix that diagonalizes the above quadratic form and show that this transformation is orthogonal.

(iv) Write Q in diagonal form.

17. **Schwarz inequality**: Since the cosine of any angle lies between -1 and $+1$, it is clear that geometric vectors satisfy the Schwarz inequality:

$$|\vec{A} \cdot \vec{B}| \leq |\vec{A}||\vec{B}|.$$

Show the analog of this inequality for the general inner product spaces:

$$|(u, v)|^2 \leq (u, u)(v, v).$$

18. **Triangle inequality**: For the inner product spaces prove the triangle inequality

$$|u + v| \leq |u| + |v|.$$

19. Show that the space of square integrable functions forms a vector space.

20. Prove the **Baker–Hausdorf** formula:

$$e^{i\mathbf{A}}\mathbf{H}e^{-i\mathbf{A}} = \mathbf{H} + i[\mathbf{A}, \mathbf{H}] - \frac{1}{2}[\mathbf{A}, [\mathbf{A}, \mathbf{H}]] + \cdots,$$

where $\mathbf{A}$ and $\mathbf{H}$ are $n \times n$ matrices and $[\mathbf{A}, \mathbf{H}] = \mathbf{AH} - \mathbf{HA}$.

21. In bra-ket notation Hermitian operators are defined as

$$\langle \Psi_1 | A^\dagger | \Psi_2 \rangle = \langle \Psi_2 | A | \Psi_1 \rangle^*.$$

Establish the correspondence

$$A|\Psi\rangle = |\Phi\rangle \leftrightarrow \langle \Phi| = \langle \Psi | A^\dagger.$$

CHAPTER 6

PRACTICAL LINEAR ALGEBRA

In the previous chapter, we have introduced the basic and somewhat abstract concepts of linear algebra based on the definition of linear vector spaces and their properties. Here, we introduce the practical side of linear algebra used in many of its real-life applications. We first discuss linear systems with m equations and n unknowns in detail. We then introduce row echelon forms, row, column and null spaces, and rank and nullity of matrices. Linear independence, basis vectors, linear transformations and their representations are among the other topics we discuss. Finally, we introduce some of the frequently used numerical methods of linear algebra, where partial pivoting, LU-factorization, interpolation, and numerical methods of finding solutions are among the topics covered.

Essentials of Mathematical Methods in Science and Engineering, Second Edition. Selçuk Ş. Bayın. **293**
© 2020 John Wiley & Sons, Inc. Published 2020 by John Wiley & Sons, Inc.

6.1 SYSTEMS OF LINEAR EQUATIONS

In general, a linear system with m equations and n unknowns $(x_1, x_2, \ldots, x_n)$ is written as

$$
\begin{aligned}
a_{11}x_1 + a_{12}x_2 + \cdots + a_{1n}x_n &= b_1, \\
a_{21}x_1 + a_{22}x_2 + \cdots + a_{2n}x_n &= b_2, \\
&\vdots \\
a_{m1}x_1 + a_{m2}x_2 + \cdots + a_{mn}x_n &= b_m.
\end{aligned}
\tag{6.1}
$$

We can interpret this system as m planes in n dimensional real space $\mathbb{R}^n$, which may have a **unique, infinitely many**, or **no solution** at all. If such a system does not have a solution, we say it is **inconsistent**, or **incompatible**. A solution, $\vec{x} = (x_1, x_2, \ldots, x_n)$, is a point in $\mathbb{R}^n$ that is common to all the planes. **Gauss's method of elimination**, which is also called **row-reduction**, can always be used to solve linear systems whether they are consistent or not (Chapter 4).

Gauss's method involves certain basic operations that leads to an equivalent system, which has the same solution as the original system, but where the solution can be obtained easily by a method called **back-substitution**. These operations consist of

  I. Interchanging two equations.

  II. Multiplying an equation with a nonzero constant.

  III. Adding a constant multiple of an equation to another.

It is obvious that the first two operations do not change the solution. To understand the effect of the third operation, we write the jth equation of the set as

$$
a_{j1}x_1 + a_{j2}x_2 + \cdots + a_{jn}x_n - b_j = 0
\tag{6.2}
$$

and multiply it with a nonzero constant c_0:

$$
c_0 a_{j1}x_1 + c_0 a_{j2}x_2 + \cdots + c_0 a_{jn}x_n - c_0 b_j = 0.
\tag{6.3}
$$

Adding this to the kth equation of the set, which is equivalent to adding zero, we write

$$
\begin{aligned}
c_0 a_{j1}x_1 + c_0 a_{j2}x_2 + \cdots + c_0 a_{jn}x_n - c_0 b_j \\
+ a_{k1}x_1 + a_{k2}x_2 + \cdots + a_{kn}x_n - b_k = 0,
\end{aligned}
\tag{6.4}
$$

to obtain

$$
(c_0 a_{j1} + a_{k1})x_1 + (c_0 a_{j2} + a_{k2})x_2 + \cdots + (c_0 a_{jn} + a_{kn})x_n = (c_0 b_j + b_k).
\tag{6.5}
$$

Obviously, the new set of equations obtained by replacing the kth equation of the original set [Eq. (6.1)] with Eq. (6.5) will have the same solution. Furthermore, these operations can be performed as many times as desired, each time producing an equivalent set of equations that has the same solution as the original one.

6.1.1 Matrices and Elementary Row Operations

Using matrices we can write a general linear system [Eq. (6.1)] as

$$\boxed{\mathbf{Ax} = \mathbf{b},} \tag{6.6}$$

where

$$\mathbf{A} = \begin{pmatrix} a_{11} & a_{12} & \cdots & a_{1n} \\ a_{21} & a_{22} & \cdots & a_{2n} \\ \vdots & \vdots & \ddots & \vdots \\ a_{m1} & a_{m2} & \cdots & a_{mn} \end{pmatrix}, \ \mathbf{x} = \begin{pmatrix} x_1 \\ x_2 \\ \vdots \\ x_n \end{pmatrix}, \ \mathbf{b} = \begin{pmatrix} b_1 \\ b_2 \\ \vdots \\ b_m \end{pmatrix}. \tag{6.7}$$

The matrix $\mathbf{A}$ is called the matrix of coefficients. We now introduce the **augmented matrix** $\widetilde{\mathbf{A}}$, which is defined as

$$\boxed{\widetilde{\mathbf{A}} = (\mathbf{A}|\mathbf{b}) = \begin{pmatrix} a_{11} & a_{12} & \cdots & a_{1n} & b_1 \\ a_{21} & a_{22} & \cdots & a_{2n} & b_2 \\ \vdots & \vdots & \ddots & \vdots & \vdots \\ a_{m1} & a_{m2} & \cdots & a_{mn} & b_m \end{pmatrix}.} \tag{6.8}$$

Note that $\widetilde{\mathbf{A}}$ does not mean the transpose of $\mathbf{A}$. The augmented matrix carries all the relevant information about the linear system in Eq. (6.1). The basic operations of the **Gauss's method of elimination** now corresponds to the following elementary **row operations**:

    I. Swapping rows.

    II. Multiplying a row with a nonzero constant.

    III. Adding a constant multiple of one row to another.

In general two matrices that can be obtained from each other by a finite number of elementary row operations are called **row-equivalent**.

6.1.2 Gauss-Jordan Method

Let us consider the following linear system:

$$\begin{array}{l} x_1 + 2x_2 + 3x_3 = 1 \\ 3x_1 + 2x_2 + x_3 = 2 \\ x_1 + x_2 - x_3 = -2 \end{array} \ \rightarrow \widetilde{\mathbf{A}} = \left(\begin{array}{ccc|c} 1 & 2 & 3 & 1 \\ 3 & 2 & 1 & 2 \\ 1 & 1 & -1 & -2 \end{array} \right), \tag{6.9}$$

where on the right-hand side we have written the corresponding augmented matrix. The systematics of the Gauss's method of elimination consists of starting with the first equation and then using it to eliminate x_1 from the remaining equations. This equation is called the **pivot equation** and a_{11}, the coefficient of x_1, is called the **pivot**. If $a_{11} = 0$, we can always swap the first equation with the one that has nonzero a_{11}. For the above system, we multiply the first equation by 3 and then subtract from the second equation, thus eliminating x_1 from the second equation. Next, we multiply the first equation by 1 and then subtract from the third equation to obtain

$$\begin{array}{rcl} x_1 + 2x_2 + 3x_3 & = & 1 \\ -4x_2 - 8x_3 & = & -1 \\ -x_2 - 4x_3 & = & -3 \end{array} \quad \rightarrow \tilde{\mathbf{A}} = \left(\begin{array}{ccc|c} 1 & 2 & 3 & 1 \\ 0 & -4 & -8 & -1 \\ 0 & -1 & -4 & -3 \end{array} \right). \tag{6.10}$$

In general, this step requires division of the first equation by the pivot, a_{11}, followed by a multiplication with a_{21} and then subtraction from the second equation. This will eliminate x_1 from the second equation. Similarly, we eliminate x_1 from the subsequent equations by using the pivot equation. In the next step, we use the second equation as our pivot equation and a_{22} as our pivot, and eliminate x_2 from the equations lying below the pivot equation to obtain the following equivalent system:

$$\begin{array}{rcl} x_1 + 2x_2 + 3x_3 & = & 1 \\ -4x_2 - 8x_3 & = & -1 \\ -2x_3 & = & -11/4 \end{array} \quad \rightarrow \tilde{\mathbf{A}} = \left(\begin{array}{ccc|c} 1 & 2 & 3 & 1 \\ 0 & -4 & -8 & -1 \\ 0 & 0 & -2 & -11/4 \end{array} \right). \tag{6.11}$$

From the last equation, we read the value of x_3 as

$$x_3 = 11/8. \tag{6.12}$$

Substituting this into the previous equation gives the value of x_2:

$$x_2 = -5/2, \tag{6.13}$$

which along with x_3, when substituted into the first equation yields the value of the final unknown as

$$x_1 = 15/8. \tag{6.14}$$

This method can be used with any number of equations and unknowns. For consistent systems, it yields the last unknown immediately and the rest are obtained by the method called **back-substitution**. Note that the **matrix of coefficients** of the system in Eq. (6.9) is **row-reduced** as

$$\mathbf{A} = \left(\begin{array}{ccc} 1 & 2 & 3 \\ 3 & 2 & 1 \\ 1 & 1 & -1 \end{array} \right) \rightarrow \left(\begin{array}{ccc} 1 & 2 & 3 \\ 0 & -4 & -8 \\ 0 & 0 & -2 \end{array} \right), \tag{6.15}$$

which is in **upper-triangular** form. A square matrix in upper triangular form
is given as

$$
\begin{pmatrix}
* & * & * & * & * \\
0 & * & * & * & * \\
\vdots & \vdots & \ddots & \vdots & \vdots \\
0 & 0 & 0 & * & * \\
0 & 0 & 0 & 0 & *
\end{pmatrix},
\tag{6.16}
$$

where all the terms below the main diagonal are zero. Due to the **wedge
shape** of the nonzero entrees in a row-reduced $m \times n$ matrix, it is called the
row-echelon form. In the row-reduced augmented matrix [Eq. (6.11)], the first
nonzero terms that have all zeros below them, 1 and -4, are called the **leading
terms**.

The augmented matrix [Eq. (6.11)] can be row-reduced further by **normalizing** the leading terms to 1 by multiplying the corresponding rows with the
appropriate constants as

$$
\widetilde{\mathbf{A}} = \left(
\begin{array}{ccc|c}
1 & 2 & 3 & 1 \\
0 & 1 & 2 & 1/4 \\
0 & 0 & 1 & 11/8
\end{array}
\right).
\tag{6.17}
$$

The **normalized** row-reduced form of a matrix is called the **reduced
row-echelon** form. In general, the procedure that leads to the reduced
row-echelon form of a matrix is called the **Gauss-Jordan method** or the
Gauss-Jordan row-reduction.

In the row-echelon form of a matrix,
  (i) in a row that does not consist entirely of zeros, the first nonzero term
      is called the **leading term**,
 (ii) the rows that consist entirely of zeros are grouped at the bottom,
(iii) in any two successive rows, the leading term of the lower row is to the
      right of the leading term of the previous row.
Examples of matrices in row-echelon form can be given as

$$
\begin{pmatrix}
* & * & * & * \\
0 & * & * & * \\
0 & 0 & 0 & * \\
0 & 0 & 0 & 0
\end{pmatrix},
\begin{pmatrix}
* & * & 0 & * \\
0 & * & * & 0 \\
0 & 0 & * & * \\
0 & 0 & 0 & *
\end{pmatrix},
\begin{pmatrix}
* & 0 & 0 & * & * \\
0 & * & * & * & * \\
0 & 0 & 0 & 0 & 0 \\
0 & 0 & 0 & 0 & 0
\end{pmatrix},
\tag{6.18}
$$

where the stars indicate the nonzero entrees. In addition to the aforementioned
three conditions, if the leading terms are also **normalized** to 1, we say the
matrix is in **reduced row-echelon** form:

$$
\begin{pmatrix}
1 & * & * & * \\
0 & 1 & * & * \\
0 & 0 & 0 & 1 \\
0 & 0 & 0 & 0
\end{pmatrix},
\begin{pmatrix}
1 & 0 & 0 & * & * & * \\
0 & 0 & 1 & * & * & * \\
0 & 0 & 0 & 1 & 0 & * \\
0 & 0 & 0 & 0 & 0 & 1
\end{pmatrix},
\begin{pmatrix}
1 & * & 0 & * \\
0 & 1 & * & * \\
0 & 0 & 0 & 0 \\
0 & 0 & 0 & 0
\end{pmatrix}
\tag{6.19}
$$

If we now start from the last row of the reduced row-echelon form [Eq. (6.17)] and continue the reduction process **upwards** by eliminating the terms above the leading terms, we obtain

$$\tilde{A} = \left(\begin{array}{ccc|c} 1 & 2 & 3 & 1 \\ 0 & 1 & 2 & 1/4 \\ 0 & 0 & 1 & 11/8 \end{array} \right) \rightarrow \left(\begin{array}{ccc|c} 1 & 2 & 3 & 1 \\ 0 & 1 & 0 & -5/2 \\ 0 & 0 & 1 & 11/8 \end{array} \right)$$

$$\rightarrow \left(\begin{array}{ccc|c} 1 & 2 & 0 & -25/8 \\ 0 & 1 & 0 & -5/2 \\ 0 & 0 & 1 & 11/8 \end{array} \right) \rightarrow \left(\begin{array}{ccc|c} 1 & 0 & 0 & 15/8 \\ 0 & 1 & 0 & -5/2 \\ 0 & 0 & 1 & 11/8 \end{array} \right). \qquad (6.20)$$

Now the final form of the augmented matrix represents the following equivalent linear system:

$$x_1 = 15/8,$$
$$x_2 = -5/2, \qquad (6.21)$$
$$x_3 = 11/8,$$

which is nothing but the desired solution.

Example 6.1. *Gauss Jordan method:* Interpret the following reduced row-echelon forms of the augmented matrices corresponding to linear systems of equations:

$$\left(\begin{array}{ccc|c} 1 & 0 & 0 & 1 \\ 0 & 1 & 0 & 2 \\ 0 & 0 & 1 & 3 \end{array} \right), \left(\begin{array}{cccc|c} 1 & 0 & 0 & 3 & -1 \\ 0 & 1 & 2 & 0 & 2 \\ 0 & 0 & 1 & 1 & 3 \end{array} \right), \left(\begin{array}{ccc|c} 1 & 0 & 0 & 1 \\ 0 & 1 & 2 & 2 \\ 0 & 0 & 0 & 3 \end{array} \right). \quad (6.22)$$

Substituting the variables, the first augmented matrix corresponds to the system

$$x_1 = 1,$$
$$x_2 = 2, \qquad (6.23)$$
$$x_3 = 3,$$

which has a unique solution. The second augmented matrix corresponds to the system

$$x_1 + 3x_4 = -1, \qquad (6.24)$$
$$x_2 + 2x_3 = 2, \qquad (6.25)$$
$$x_3 + x_4 = 3, \qquad (6.26)$$

which has infinitely many solutions. Treating $x_4 = s$ as a parameter, we obtain the solution as

$$x_1 = -1 - 3s, \qquad (6.27)$$
$$x_2 = -4 + 2s, \qquad (6.28)$$

$$x_3 = 3 - s, \tag{6.29}$$

$$x_4 = s. \tag{6.30}$$

Finally, the last augmented matrix corresponds to the system

$$x_1 = 1, \tag{6.31}$$

$$x_2 + 2x_3 = 2, \tag{6.32}$$

$$0 = 3. \tag{6.33}$$

The last equation is a contradiction; hence, this system is inconsistent and has no solution. This is also confirmed by the fact that the determinant of coefficients is zero (Chapter 4):

$$\det \mathbf{A} = \begin{vmatrix} 1 & 0 & 0 \\ 0 & 1 & 2 \\ 0 & 0 & 0 \end{vmatrix} = 0. \tag{6.34}$$

On the other hand, the augmented matrix

$$\left(\begin{array}{ccc|c} 1 & 0 & 0 & 1 \\ 0 & 1 & 2 & 2 \\ 0 & 0 & 1 & 0 \end{array} \right), \tag{6.35}$$

which corresponds to the system

$$x_1 = 1, \tag{6.36}$$

$$x_2 + 2x_3 = 2, \tag{6.37}$$

$$x_3 = 0, \tag{6.38}$$

has a unique solution:

$$x_1 = 1, \tag{6.39}$$

$$x_2 = 2, \tag{6.40}$$

$$x_3 = 0 \tag{6.41}$$

Example 6.2. *Gauss-Jordan method:* When the matrix of coefficients of a linear system of equations is row reduced by Gauss-Jordan method, the following reduced row-echelon form is obtained:

$$\mathbf{A} = \begin{pmatrix} 1 & 2 & 0 & 0 & 1 \\ 0 & 0 & 1 & 0 & 2 \\ 0 & 0 & 0 & 1 & 0 \\ 0 & 0 & 0 & 0 & 0 \end{pmatrix}. \tag{6.42}$$

Let us find the solution of the homogeneous equation:

$$\mathbf{Ax} = \mathbf{0}. \tag{6.43}$$

We first write the augmented matrix:

$$\tilde{\mathbf{A}} = \left(\begin{array}{ccccc|c} 1 & 2 & 0 & 0 & 1 & 0 \\ 0 & 0 & 1 & 0 & 2 & 0 \\ 0 & 0 & 0 & 1 & 0 & 0 \\ 0 & 0 & 0 & 0 & 0 & 0 \end{array} \right), \tag{6.44}$$

and the corresponding homogeneous linear system of equations:

$$x_1 + 2x_2 + x_5 = 0, \tag{6.45}$$

$$x_3 + 2x_5 = 0, \tag{6.46}$$

$$x_4 = 0. \tag{6.47}$$

The nontrivial solutions in terms of x_2 and x_5 are obtained as

$$x_1 = -2x_2 - x_5, \tag{6.48}$$

$$x_3 = -2x_5, \tag{6.49}$$

$$x_4 = 0, \tag{6.50}$$

where x_2 and x_5 are called the **free variables** or the **parameters** of the solution.

6.1.3   Information From the Row-Echelon Form

After using the Gauss-Jordan method, the augmented matrix of a linear system with m equations and n unknowns takes the following **row-echelon** form:

$$\tilde{\mathbf{A}} = \left(\begin{array}{cccccccc|c} a_{11} & a_{12} & \cdots & \cdots & \cdots & \cdots & a_{1n-1} & a_{1n} & \tilde{b}_1 \\ 0 & c_{22} & \cdots & \cdots & \cdots & \cdots & c_{2n-1} & c_{2n} & \tilde{b}_2 \\ 0 & 0 & d_{33} & \cdots & \cdots & \cdots & d_{3n-1} & d_{3n} & \tilde{b}_3 \\ \vdots & \vdots & 0 & \ddots & \vdots & \vdots & \vdots & \vdots & \vdots \\ 0 & 0 & 0 & 0 & k_{rr} & \cdots & \cdots & k_{rn} & \tilde{b}_r \\ 0 & 0 & 0 & 0 & 0 & 0 & 0 & 0 & \tilde{b}_{r+1} \\ \vdots & \vdots & \vdots & \vdots & \vdots & \vdots & \vdots & \vdots & \vdots \\ 0 & 0 & 0 & 0 & 0 & 0 & 0 & 0 & \tilde{b}_m \end{array} \right), \tag{6.51}$$

where $r \leq m$, $a_{11} \neq 0$, $c_{22} \neq 0, \ldots, k_{rr} \neq 0$ and in a given column all the entrees between the zeros are zero. If we also **normalize** all the left most entrees to

1, we have the **reduced row-echelon** form. We can now write the following cases:

I. When $r = n$ and $\tilde{b}_{r+1}, \ldots, \tilde{b}_m$ are all zero, then we solve the nth equation for x_n as

$$\boxed{x_n = \tilde{b}_n / k_{nn}} \tag{6.52}$$

and obtain the remaining unknowns by **back-substitution**. In this case, we have a **unique solution**.

II. If $r < n$ and $\tilde{b}_{r+1}, \ldots, \tilde{b}_m$ are all zero, then we can obtain the values of the r unknowns, $x_1, \ldots, x_r$, in terms of the remaining variables

$$\boxed{x_{r+1}, \ldots, x_n.} \tag{6.53}$$

In this case, there are **infinitely many** solutions, where $x_{r+1}, \ldots, x_n$ are the **free variables**.

III. When $r < m$ and any one of the entrees $\tilde{b}_{r+1}, \ldots, \tilde{b}_m$ is different from zero, the system is **inconsistent** and we have **no solutions**.

6.1.4 Elementary Matrices

An $n \times n$ matrix $\mathbf{E}$ that can be obtained from the identity matrix $\mathbf{I}_n$ by a single elementary row operation is called an **elementary matrix**. For example,

$$\mathbf{E}_1 = \begin{pmatrix} 0 & 0 & 1 \\ 0 & 1 & 0 \\ 1 & 0 & 0 \end{pmatrix}, \ \mathbf{E}_2 = \begin{pmatrix} 3 & 0 & 0 \\ 0 & 1 & 0 \\ 0 & 0 & 1 \end{pmatrix} \tag{6.54}$$

are elementary matrices, since $\mathbf{E}_1$ can be obtained from $\mathbf{I}_3$ by swapping the first and the third rows, while $\mathbf{E}_2$ can be obtained by multiplying the first row of $\mathbf{I}_3$ with 3.

Theorem 6.1. Let $\mathbf{E}$ be an elementary matrix obtained from $\mathbf{I}_m$ by a certain elementary row operation and let $\mathbf{A}$ be an $m \times n$ matrix, then the product $\mathbf{EA}$ is the matrix that results when the same row operation is performed on $\mathbf{A}$.

For example, for

$$\mathbf{E}_1 = \begin{pmatrix} 0 & 0 & 1 \\ 0 & 1 & 0 \\ 1 & 0 & 0 \end{pmatrix} \text{ and } \mathbf{A} = \begin{pmatrix} 2 & 0 & 1 & 1 \\ 1 & 1 & 0 & 2 \\ 1 & 0 & 3 & 4 \end{pmatrix}, \tag{6.55}$$

the product of $\mathbf{E}_1$ with $\mathbf{A}$ swaps the first and the third rows of $\mathbf{A}$:

$$\mathbf{E}_1 \mathbf{A} = \begin{pmatrix} 1 & 0 & 3 & 4 \\ 1 & 1 & 0 & 2 \\ 2 & 0 & 1 & 1 \end{pmatrix}. \tag{6.56}$$

It can be easily shown that every elementary matrix is invertible and its inverse is also an elementary matrix. For example, the inverse of

$$\mathbf{E} = \begin{pmatrix} 3 & 0 & 0 \\ 0 & 1 & 0 \\ 0 & 0 & 1 \end{pmatrix} \tag{6.57}$$

is

$$\mathbf{E}^{-1} = \begin{pmatrix} 1/3 & 0 & 0 \\ 0 & 1 & 0 \\ 0 & 0 & 1 \end{pmatrix}, \tag{6.58}$$

thus $\mathbf{E}\mathbf{E}^{-1} = \mathbf{E}^{-1}\mathbf{E} = \mathbf{I}_3$.

For an $n \times n$ matrix $\mathbf{A}$, we can write the following equivalent statements:
(i) $\mathbf{A}$ is invertible.
(ii) $\mathbf{A}\mathbf{x} = \mathbf{0}$ has only the trivial solution.
(iii) The reduced row-echelon form of $\mathbf{A}$ is $\mathbf{I}_n$.
(iv) $\mathbf{A}$ is expressible as the product of elementary matrices.
We leave the proof of the equivalence of these statements as an exercise. From the last statement, we can write

$$\boxed{\mathbf{E}_k \cdots \mathbf{E}_2 \mathbf{E}_1 \mathbf{A} = \mathbf{I}_n,} \tag{6.59}$$

thus the inverse of $\mathbf{A}$ is

$$\boxed{\mathbf{A}^{-1} = \mathbf{E}_1^{-1} \mathbf{E}_2^{-1} \cdots \mathbf{E}_k^{-1}.} \tag{6.60}$$

6.1.5   Inverse by Gauss-Jordan Row-Reduction

Using the above statement, we see that an $n \times n$ matrix is invertible if and only if it is row equivalent to the identity matrix. For example, for the inverse of the matrix

$$\mathbf{A} = \begin{pmatrix} 1 & 2 & 0 \\ 0 & 1 & 0 \\ 1 & 1 & 1 \end{pmatrix}, \tag{6.61}$$

we write

$$(\mathbf{A}|\mathbf{I}_3) \tag{6.62}$$

and then apply elementary row operations until we obtain the identity matrix on the left-hand side as

$$(\mathbf{I}_3|\mathbf{A}^{-1}) : \tag{6.63}$$

$$\left(\begin{array}{ccc|ccc} 1 & 2 & 0 & 1 & 0 & 0 \\ 0 & 1 & 0 & 0 & 1 & 0 \\ 1 & 1 & 1 & 0 & 0 & 1 \end{array} \right) \xrightarrow{(\text{Row3} - \text{Row1})} \left(\begin{array}{ccc|ccc} 1 & 2 & 0 & 1 & 0 & 0 \\ 0 & 1 & 0 & 0 & 1 & 0 \\ 0 & -1 & 1 & -1 & 0 & 1 \end{array} \right)$$

$$\xrightarrow{\text{(Row2 + Row3)}} \begin{pmatrix} 1 & 2 & 0 & | & 1 & 0 & 0 \\ 0 & 1 & 0 & | & 0 & 1 & 0 \\ 0 & 0 & 1 & | & -1 & 1 & 1 \end{pmatrix}$$

$$\xrightarrow{\text{(Row1 − 2Row2)}} \begin{pmatrix} 1 & 0 & 0 & | & 1 & -2 & 0 \\ 0 & 1 & 0 & | & 0 & 1 & 0 \\ 0 & 0 & 1 & | & -1 & 1 & 1 \end{pmatrix}, \tag{6.64}$$

thus the inverse is obtained as

$$\mathbf{A}^{-1} = \begin{pmatrix} 1 & -2 & 0 \\ 0 & 1 & 0 \\ -1 & 1 & 1 \end{pmatrix}. \tag{6.65}$$

Generalization to an $n \times n$ matrix $\mathbf{A}$ is obvious:

$$\boxed{(\mathbf{A}|\mathbf{I}_n) \to (\mathbf{I}_n|\mathbf{A}^{-1}).} \tag{6.66}$$

6.1.6 Row Space, Column Space, and Null Space

Given an $m \times n$ matrix,

$$\mathbf{A} = \begin{pmatrix} a_{11} & a_{12} & \cdots & a_{1n} \\ a_{21} & a_{22} & \cdots & a_{2n} \\ \vdots & \vdots & \ddots & \vdots \\ a_{m1} & a_{m2} & \cdots & a_{mn} \end{pmatrix}, \tag{6.67}$$

we can define the following row vectors:

$$\mathbf{r}_1 = \begin{pmatrix} a_{11} & a_{12} & \cdots & a_{1n} \end{pmatrix}, \tag{6.68}$$

$$\mathbf{r}_2 = \begin{pmatrix} a_{21} & a_{22} & \cdots & a_{2n} \end{pmatrix}, \tag{6.69}$$

$$\vdots \tag{6.70}$$

$$\mathbf{r}_m = \begin{pmatrix} a_{m1} & a_{m2} & \cdots & a_{mn} \end{pmatrix} \tag{6.71}$$

and the column vectors:

$$\mathbf{c}_1 = \begin{pmatrix} a_{11} \\ a_{21} \\ \vdots \\ a_{m1} \end{pmatrix}, \quad \mathbf{c}_2 = \begin{pmatrix} a_{12} \\ a_{22} \\ \vdots \\ a_{m2} \end{pmatrix}, \dots, \quad \mathbf{c}_n = \begin{pmatrix} a_{1n} \\ an \\ \vdots \\ a_{mn} \end{pmatrix}. \tag{6.72}$$

Definition 6.1. For a given $m \times n$ matrix $\mathbf{A}$, the subspace of $\mathbb{R}^n$ spanned by the row vectors is called the **row space** of $\mathbf{A}$ and the subspace of $\mathbb{R}^m$ spanned by the column vectors is called the **column space** of $\mathbf{A}$.

Definition 6.2. The solution space of the **homogeneous equation**

$$\boxed{Ax = 0}$$

(6.73)

is a subspace of $\mathbb{R}^n$ and is called the **null space** of A.

Using A [Eq. (6.67)] and the column vector

$$x = \begin{pmatrix} x_1 \\ x_2 \\ \vdots \\ x_n \end{pmatrix},$$

(6.74)

we can write $Ax = 0$ as

$$\boxed{Ax = x_1 c_1 + x_2 c_2 + \cdots + x_n c_n.}$$

(6.75)

Thus, the linear system $Ax = b$, is consistent, if and only if b is expressible as a linear combination of the column vectors of A, that is, if and only if b is in the column space of A.

Theorem 6.2. Let x_p be a particular (single) solution of

$$\boxed{Ax = b}$$

(6.76)

and let

$$v_1, v_2, \ldots, v_n$$

(6.77)

be a basis for the null space of A, then the **general solution** of $Ax = b$ can be written as

$$\boxed{x = x_p + x_h,}$$

(6.78)

where

$$\boxed{x_h = c_1 v_1 + c_2 v_2 + \cdots + c_n v_n}$$

(6.79)

is the solution of the homogeneous equation $Ax = 0$.

Note that given two solutions, x_1 and x_2, of the homogeneous equation $Ax = 0$:

$$Ax_1 = 0 \text{ and } Ax_2 = 0,$$

(6.80)

then their linear combination

$$c_0 x_1 + c_1 x_2,$$

(6.81)

is also a solution:

$$A(c_0 x_1 + c_1 x_2) = 0.$$

(6.82)

Example 6.3. *Null space of a matrix:* Consider the following matrix:

$$
\mathbf{A} = \begin{pmatrix} 1 & 2 & -2 & 0 & 1 & 0 \\ 1 & 3 & -2 & 1 & 2 & 1 \\ 0 & 0 & 2 & 3 & 0 & 1 \\ 2 & 2 & 0 & 4 & 4 & 3 \end{pmatrix}. \tag{6.83}
$$

To find the null space of $\mathbf{A}$, we reduce the augmented matrix:

$$
\tilde{\mathbf{A}} = \left(\begin{array}{cccccc|c} 1 & 2 & -2 & 0 & 1 & 0 & 0 \\ 1 & 3 & -2 & 1 & 2 & 1 & 0 \\ 0 & 0 & 2 & 3 & 0 & 1 & 0 \\ 2 & 2 & 0 & 4 & 4 & 3 & 0 \end{array} \right), \tag{6.84}
$$

of the homogeneous equation, $\mathbf{Ax} = \mathbf{0}$, via the Gauss-Jordan reduction to obtain the reduced row-echelon form as

$$
\left(\begin{array}{cccccc|c} 1 & 0 & 0 & 1 & 0 & -1/4 & 0 \\ 0 & 1 & 0 & 1 & 0 & 1/4 & 0 \\ 0 & 0 & 1 & 3/2 & 0 & 1/2 & 0 \\ 0 & 0 & 0 & 0 & 1 & 3/4 & 0 \end{array} \right). \tag{6.85}
$$

Substituting the variables, the corresponding system of equations become

$$
x_1 + x_4 - \frac{1}{4}x_6 = 0, \tag{6.86}
$$

$$
x_2 + x_4 + \frac{1}{4}x_6 = 0, \tag{6.87}
$$

$$
x_3 + \frac{3}{2}x_4 + \frac{1}{2}x_6 = 0, \tag{6.88}
$$

$$
x_5 + \frac{3}{4}x_6 = 0. \tag{6.89}
$$

Using $x_4 = s$ and $x_6 = t$ as parameters, we can write the solution as

$$
x_1 = -s + \frac{1}{4}t, \tag{6.90}
$$

$$
x_2 = -s - \frac{1}{4}t \tag{6.91}
$$

$$
x_3 = -\frac{3}{2}s - \frac{1}{2}t, \tag{6.92}
$$

$$
x_5 = -\frac{3}{4}t. \tag{6.93}
$$

There are infinitely many solutions in terms of the parameters s and t as

$$\begin{pmatrix} x_1 \\ x_2 \\ x_3 \\ x_4 \\ x_5 \\ x_6 \end{pmatrix} = s \begin{pmatrix} -1 \\ -1 \\ -3/2 \\ 1 \\ 0 \\ 0 \end{pmatrix} + t \begin{pmatrix} 1/4 \\ -1/4 \\ -1/2 \\ 0 \\ -3/4 \\ 1 \end{pmatrix}. \tag{6.94}$$

In other words, the **null space** of $\mathbf{A}$ is two-dimensional and it is spanned by the vectors

$$\mathbf{v}_1 = \begin{pmatrix} -1 \\ -1 \\ -3/2 \\ 1 \\ 0 \\ 0 \end{pmatrix} \text{ and } \mathbf{v}_2 = \begin{pmatrix} 1/4 \\ -1/4 \\ -1/2 \\ 0 \\ -3/4 \\ 1 \end{pmatrix}. \tag{6.95}$$

To find the general solution of $\mathbf{A}\mathbf{x} = \mathbf{b}$, where

$$\mathbf{b} = \begin{pmatrix} 0 \\ -1 \\ 1 \\ 6 \end{pmatrix}, \tag{6.96}$$

we use the Gauss-Jordan method to find the reduced row-echelon form of the corresponding augmented matrix:

$$\tilde{\mathbf{A}} = \left(\begin{array}{cccccc|c} 1 & 2 & -2 & 0 & 1 & 0 & 0 \\ 1 & 3 & -2 & 1 & 2 & 1 & -1 \\ 0 & 0 & 2 & 3 & 0 & 1 & 1 \\ 2 & 2 & 0 & 4 & 4 & 3 & 6 \end{array} \right), \tag{6.97}$$

which is

$$\tilde{\mathbf{A}} = \left(\begin{array}{cccccc|c} 1 & 0 & 0 & 1 & 0 & -1/4 & 7/2 \\ 0 & 1 & 0 & 1 & 0 & 1/4 & -3/2 \\ 0 & 0 & 1 & 3/2 & 0 & 1/2 & 1/2 \\ 0 & 0 & 0 & 0 & 1 & 3/4 & 1/2 \end{array} \right). \tag{6.98}$$

Substituting the variables, this gives the system

$$x_1 = \frac{7}{2} - x_4 + \frac{1}{4}x_6, \tag{6.99}$$

$$x_2 = -\frac{3}{2} - x_4 - \frac{1}{4}x_6, \tag{6.100}$$

$$x_3 = \frac{1}{2} - \frac{3}{2}x_4 - \frac{1}{2}x_6, \tag{6.101}$$

$$x_5 = \frac{1}{2} - \frac{3}{4}x_6. \tag{6.102}$$

Treating x_4 and x_6 as parameters and writing $x_4 = s$ and $x_6 = t$, we write the solution of $\mathbf{Ax} = \mathbf{b}$ as

$$\begin{pmatrix} x_1 \\ x_2 \\ x_3 \\ x_4 \\ x_5 \\ x_6 \end{pmatrix} = \begin{pmatrix} 7/2 \\ -3/2 \\ 1/2 \\ 0 \\ 1/2 \\ 0 \end{pmatrix} + s \begin{pmatrix} -1 \\ -1 \\ -3/2 \\ 1 \\ 0 \\ 0 \end{pmatrix} + t \begin{pmatrix} 1/4 \\ -1/4 \\ -1/2 \\ 0 \\ -3/4 \\ 1 \end{pmatrix}. \tag{6.103}$$

Comparing with Eqs. (6.94) and (6.78), we see that it is of the form $\mathbf{x} = \mathbf{x}_p + \mathbf{x}_h$, thus the **particular solution** is found as

$$\mathbf{x}_p = \begin{pmatrix} 7/2 \\ -3/2 \\ 1/2 \\ 0 \\ 1/2 \\ 0 \end{pmatrix}. \tag{6.104}$$

6.1.7 Bases for Row, Column, and Null Spaces

It can be seen easily that the elementary row operations do not change the row space and the null space of a matrix. However, the column space changes. Consider the following 2×2 matrix:

$$\begin{pmatrix} 1 & 2 \\ 2 & 4 \end{pmatrix}. \tag{6.105}$$

The second row is twice the first row; thus, the row space is the space of all vectors that lie along the line in the direction of

$$\mathbf{r}_1 = \begin{pmatrix} 1 & 2 \end{pmatrix}. \tag{6.106}$$

Multiplying the first row by 2 and then subtracting from the second row gives

$$\begin{pmatrix} 1 & 2 \\ 0 & 0 \end{pmatrix}. \tag{6.107}$$

The new row space is still all vectors along the same line pointing along the direction of $\mathbf{r}_1$.

On the other hand, the second column vector in the first matrix [Eq. (6.105)] is twice the first column vector, which is still true for the second matrix [Eq. (6.107)]. However, the column space of the first matrix, which consists of all vectors lying on the line pointing in the direction of

$$\mathbf{c}_1 = \begin{pmatrix} 1 \\ 2 \end{pmatrix},$$

(6.108)

in the row-reduced matrix, is rotated and now consists of all vectors lying on the line pointing in the $\begin{pmatrix} 1 \\ 0 \end{pmatrix}$ direction. In general, even though the elementary row operations can change the column space of a matrix, it preserves relations of linear dependence or independence.

Theorem 6.3. If two matrices, $\mathbf{A}$ and $\mathbf{B}$, are row equivalent, then the set of column vectors of $\mathbf{A}$ is linearly independent if and only if the corresponding column vectors of $\mathbf{B}$ are linearly independent. Also, a given set of column vectors of $\mathbf{A}$ forms a basis for the column space of $\mathbf{A}$, if and only if the corresponding set of column vectors of $\mathbf{B}$ forms a basis for the column space of $\mathbf{B}$.

The proof of this theorem simply follows from the fact that the homogeneous equations $\mathbf{A}\mathbf{x} = \mathbf{0}$ and $\mathbf{B}\mathbf{x} = \mathbf{0}$ have the same solution set. We now give a very useful theorem in finding the row/column spaces of matrices.

Theorem 6.4. Given a matrix $\mathbf{A}$ in reduced row-echelon form $\mathbf{R}$, the row vectors corresponding to the leading 1s in $\mathbf{R}$ form a basis for the row space of $\mathbf{R}$ and similarly, the column vectors corresponding to the leading 1s in $\mathbf{R}$ form a basis for the column space of $\mathbf{R}$.

Example 6.4. *Basis for the column space:* In the following matrix, which is already in reduced row-echelon form:

$$\mathbf{R} = \begin{pmatrix} 1 & 0 & 0 & 1 & 2 & 0 \\ 0 & 0 & 1 & 1 & 0 & 1 \\ 0 & 0 & 0 & 0 & 1 & 0 \\ 0 & 0 & 0 & 0 & 0 & 0 \end{pmatrix},$$

(6.109)

the row vectors:

$$\mathbf{r}_1 = \begin{pmatrix} 1 & 0 & 0 & 1 & 2 & 0 \end{pmatrix},$$ (6.110)

$$\mathbf{r}_2 = \begin{pmatrix} 0 & 0 & 1 & 1 & 0 & 1 \end{pmatrix},$$ (6.111)

$$\mathbf{r}_3 = \begin{pmatrix} 0 & 0 & 0 & 0 & 1 & 0 \end{pmatrix},$$ (6.112)

form a basis for the row space of $\mathbf{R}$. Similarly, the corresponding column vectors:

$$\mathbf{c}_1 = \begin{pmatrix} 1 \\ 0 \\ 0 \\ 0 \end{pmatrix}, \mathbf{c}_2 = \begin{pmatrix} 0 \\ 1 \\ 0 \\ 0 \end{pmatrix}, \mathbf{c}_3 = \begin{pmatrix} 2 \\ 0 \\ 1 \\ 0 \end{pmatrix},$$

(6.113)

form a basis for the column space of $\mathbf{R}$.

Example 6.5. ***Bases for the row and column spaces:*** To find bases for the row and the column spaces of the following matrix:

$$
\mathbf{A} = \begin{pmatrix} 1 & -3 & 4 & -2 & 5 & 4 \\ 0 & 0 & 1 & 3 & -2 & -6 \\ 2 & -6 & 9 & -1 & 9 & 7 \\ 2 & -6 & 8 & -4 & 10 & 8 \end{pmatrix}, \tag{6.114}
$$

we first put $\mathbf{A}$ in reduced row-echelon form as

$$
\mathbf{R} = \begin{pmatrix} 1 & -3 & 4 & -2 & 5 & 4 \\ 0 & 0 & 1 & 3 & -2 & -6 \\ 0 & 0 & 0 & 0 & 1 & 5 \\ 0 & 0 & 0 & 0 & 0 & 0 \end{pmatrix}. \tag{6.115}
$$

The basis for the row space of $\mathbf{R}$ is now obtained as the rows containing the leading 1s as

$$
\mathbf{r}_1 = \begin{pmatrix} 1 & -3 & 4 & -2 & 5 & 4 \end{pmatrix}, \tag{6.116}
$$

$$
\mathbf{r}_2 = \begin{pmatrix} 0 & 0 & 1 & 3 & -2 & -6 \end{pmatrix}, \tag{6.117}
$$

$$
\mathbf{r}_3 = \begin{pmatrix} 0 & 0 & 0 & 0 & 1 & 5 \end{pmatrix}, \tag{6.118}
$$

while the basis for the column space of $\mathbf{R}$ is obtained as the columns containing the leading 1s:

$$
\mathbf{c}_1' = \begin{pmatrix} 1 \\ 0 \\ 0 \\ 0 \end{pmatrix}, \ \mathbf{c}_3' = \begin{pmatrix} 4 \\ 1 \\ 0 \\ 0 \end{pmatrix}, \ \mathbf{c}_5' = \begin{pmatrix} 5 \\ -2 \\ 1 \\ 0 \end{pmatrix}. \tag{6.119}
$$

We have seen that $\mathbf{A}$ and $\mathbf{R}$ may have different column spaces, in other words, we can not find the column space of $\mathbf{A}$ directly from $\mathbf{R}$. However, Theorem 6.3 says that the corresponding column vectors of $\mathbf{A}$ will form a basis for the column space of $\mathbf{A}$, that is, the first, third, and fifth columns of $\mathbf{A}$:

$$
\mathbf{c}_1 = \begin{pmatrix} 1 \\ 0 \\ 2 \\ 2 \end{pmatrix}, \ \mathbf{c}_3 = \begin{pmatrix} 4 \\ 1 \\ 9 \\ 8 \end{pmatrix}, \ \mathbf{c}_5 = \begin{pmatrix} 5 \\ -2 \\ 9 \\ 10 \end{pmatrix}. \tag{6.120}
$$

Note that even though the set $\{\mathbf{r}_1, \mathbf{r}_2, \mathbf{r}_3\}$ is not composed of the rows of $\mathbf{A}$, it still spans the row space of $\mathbf{A}$. If we want to find a basis for the row space of $\mathbf{A}$ entirely in terms of the rows of $\mathbf{A}$, we first take the transpose of $\mathbf{A}$, thus converting the row space of $\mathbf{A}$ into the column space of $\mathbf{A}^T$:

$$\mathbf{A}^T = \begin{pmatrix} 1 & 0 & 2 & 2 \\ -3 & 0 & -6 & -6 \\ 4 & 1 & 9 & 8 \\ -2 & 3 & -1 & -4 \\ 5 & -2 & 9 & 10 \\ 4 & -6 & 7 & 8 \end{pmatrix}, \tag{6.121}$$

and then find its reduced row-echelon form:

$$\mathbf{R}' = \begin{pmatrix} 1 & 0 & 2 & 2 \\ 0 & 1 & 1 & 0 \\ 0 & 0 & 1 & 0 \\ 0 & 0 & 0 & 0 \\ 0 & 0 & 0 & 0 \\ 0 & 0 & 0 & 0 \end{pmatrix}. \tag{6.122}$$

Since the first, second, and third column vectors span the column space of $\mathbf{R}'$, a basis for the column space of $\mathbf{A}^T$ is found by writing the corresponding columns of $\mathbf{A}^T$:

$$\mathbf{c}_1 = \begin{pmatrix} 1 \\ -3 \\ 4 \\ -2 \\ 5 \\ 4 \end{pmatrix}, \ \mathbf{c}_2 = \begin{pmatrix} 0 \\ 0 \\ 1 \\ 3 \\ -2 \\ -6 \end{pmatrix}, \ \mathbf{c}_3 = \begin{pmatrix} 2 \\ -6 \\ 9 \\ -1 \\ 9 \\ 7 \end{pmatrix}. \tag{6.123}$$

We now take their transpose and obtain a basis for the row space of $\mathbf{A}$ entirely in terms of the rows of $\mathbf{A}$ as

$$\mathbf{a}_1 = \begin{pmatrix} 1 & -3 & 4 & -2 & 5 & 4 \end{pmatrix}, \tag{6.124}$$

$$\mathbf{a}_2 = \begin{pmatrix} 0 & 0 & 1 & 3 & -2 & -6 \end{pmatrix}, \tag{6.125}$$

$$\mathbf{a}_3 = \begin{pmatrix} 2 & -6 & 9 & -1 & 9 & 7 \end{pmatrix}. \tag{6.126}$$

6.1.8   Vector Spaces Spanned by a Set of Vectors

We now use these ideas to find a basis for a vector space spanned by a given set of vectors. We demonstrate the general procedure with an example. Consider a vector space spanned by the following set of vectors:

$$\mathbf{v}_1 = \begin{pmatrix} 1 & 2 & 3 \end{pmatrix}, \tag{6.127}$$

$$\mathbf{v}_2 = \begin{pmatrix} 1 & 1 & 1 \end{pmatrix}, \tag{6.128}$$

$$\mathbf{v}_3 = (\begin{array}{ccc} 2 & 1 & 0 \end{array}), \tag{6.129}$$

$$\mathbf{v}_4 = (\begin{array}{ccc} 1 & 0 & 1 \end{array}), \tag{6.130}$$

$$\mathbf{v}_5 = (\begin{array}{ccc} 1 & -1 & 1 \end{array}). \tag{6.131}$$

We first form the matrix, the columns of which are $\mathbf{v}_i$, $i = 1, \ldots, 4$:

$$\mathbf{A} = \begin{pmatrix} 1 & 1 & 2 & 1 & 1 \\ 2 & 1 & 1 & 0 & -1 \\ 3 & 1 & 0 & 1 & 1 \end{pmatrix}, \tag{6.132}$$

and then find its reduced row-echelon form:

$$\mathbf{R} = \begin{pmatrix} 1 & 1 & 2 & 1 & 1 \\ 0 & 1 & 3 & 2 & 3 \\ 0 & 0 & 0 & 1 & 2 \end{pmatrix}. \tag{6.133}$$

Even though this is sufficient to identify the bases for the row and the column spaces of $\mathbf{R}$, we continue the reduction process with the Gauss-Jordan method and eliminate the entrees above the leading terms also, to get

$$\mathbf{R}' = \begin{pmatrix} 1 & 0 & -1 & 0 & 0 \\ 0 & 1 & 3 & 0 & -1 \\ 0 & 0 & 0 & 1 & 2 \end{pmatrix}. \tag{6.134}$$

Now the columns of the leading 1s of $\mathbf{R}'$:

$$\mathbf{w}_1 = \begin{pmatrix} 1 \\ 0 \\ 0 \end{pmatrix}, \ \mathbf{w}_2 = \begin{pmatrix} 0 \\ 1 \\ 0 \end{pmatrix}, \ \mathbf{w}_4 = \begin{pmatrix} 0 \\ 0 \\ 1 \end{pmatrix}, \tag{6.135}$$

form a basis for the column space of the reduced matrix $\mathbf{R}'$. The corresponding columns of $\mathbf{A}$:

$$\mathbf{v}_1 = \begin{pmatrix} 1 \\ 2 \\ 3 \end{pmatrix}, \ \mathbf{v}_2 = \begin{pmatrix} 1 \\ 1 \\ 1 \end{pmatrix}, \ \mathbf{v}_4 = \begin{pmatrix} 1 \\ 0 \\ 1 \end{pmatrix}, \tag{6.136}$$

form a basis for the column space of $\mathbf{A}$. This means that the remaining vectors, $\mathbf{v}_3$ and $\mathbf{v}_5$, of the set [Eqs. (6.127)–(6.131)] can be expressed in terms of these basis vectors.

To express $\mathbf{v}_3$ and $\mathbf{v}_5$ in terms of the basis vectors $(\mathbf{v}_1, \mathbf{v}_2, \mathbf{v}_5)$, we use the basis vectors of the column space of $\mathbf{R}'$, where by inspection we can easily write the following linear combinations:

$$\mathbf{w}_3 = -\mathbf{w}_1 + 3\mathbf{w}_2, \tag{6.137}$$

$$\mathbf{w}_5 = -\mathbf{w}_2 + 2\mathbf{w}_4. \tag{6.138}$$

Since the elementary row operations preserve linear combinations, we deduce that

$$\mathbf{v}_3 = -\mathbf{v}_1 + 3\mathbf{v}_2, \tag{6.139}$$

$$\mathbf{v}_5 = -\mathbf{v}_2 + 2\mathbf{v}_4, \tag{6.140}$$

which can be checked easily. This procedure can be used to find a subset of a set of vectors, S, containing k vectors in $\mathbb{R}^n$:

$$S = \{\mathbf{v}_1, \mathbf{v}_2, \dots, \mathbf{v}_k\}, \tag{6.141}$$

that spans S.

6.1.9   Rank and Nullity

For a given $m \times n$ matrix $\mathbf{A}$, there are six basic vector spaces:

Row space of $\mathbf{A}$	Row space of $\mathbf{A}^T$
Column space of $\mathbf{A}$	Column space of $\mathbf{A}^T$
Null space of $\mathbf{A}$	Null space of $\mathbf{A}^T$

$$\tag{6.142}$$

However, since the row space of $\mathbf{A}$ and the column space of $\mathbf{A}^T$ are the same, and also the column space of $\mathbf{A}$ and the row space of $\mathbf{A}^T$ are the same, there are only four fundamental vector spaces:

Row space of $\mathbf{A}$	Column space of $\mathbf{A}$
Null space of $\mathbf{A}$	Null space of $\mathbf{A}^T$

$$\tag{6.143}$$

The row space and the null space of $\mathbf{A}$ are subspaces of $\mathbb{R}^n$ and the column space and the row space of $\mathbf{A}^T$ are subspaces of $\mathbb{R}^m$. If $\mathbf{A}$ is any $m \times n$ matrix, we have seen that the key factor in determining the bases for the row and the column spaces is the number of leading 1s in the reduced row-echelon form, hence, both the row space and the column space have the same dimension.

Definition 6.3. The common dimension of the row and the column spaces is called the **rank** of $\mathbf{A}$. The dimension of the null space of $\mathbf{A}$ is called the **nullity** of $\mathbf{A}$.

The reduced row-echelon form for

$$\mathbf{A} = \begin{pmatrix} 1 & -3 & 4 & -2 & 5 & 4 \\ 0 & 0 & 1 & 3 & -2 & -6 \\ 2 & -6 & 9 & -1 & 9 & 7 \\ 2 & -6 & 8 & -4 & 10 & 8 \end{pmatrix} \tag{6.144}$$

is given as

$$\mathbf{R} = \begin{pmatrix} 1 & -3 & 4 & -2 & 5 & 4 \\ 0 & 0 & 1 & 3 & -2 & -6 \\ 0 & 0 & 0 & 0 & 1 & 5 \\ 0 & 0 & 0 & 0 & 0 & 0 \end{pmatrix}. \tag{6.145}$$

Since there are three leading 1s, both the row space and the column spaces have the dimension 3, hence,

$$\text{rank}(\mathbf{A}) = 3. \tag{6.146}$$

Note that this definition of rank is in agreement with our previous definition (Chapter 4), which is the order of the largest nonzero determinant of the submatrices of $\mathbf{A}$.

For the nullity of $\mathbf{A}$, one has to solve the homogeneous equation $\mathbf{Ax} = \mathbf{0}$. For example, consider the following set of equations:

$$x_1 - 2x_3 + 4x_4 - 3x_5 + x_6 = 0, \tag{6.147}$$

$$x_2 - 3x_4 + 4x_5 + 2x_6 = 0. \tag{6.148}$$

Nontrivial solutions exist and the solution can be written as

$$x_1 = 2x_3 - 4x_4 + 3x_5 - x_6, \tag{6.149}$$

$$x_2 = 3x_4 - 4x_5 - 2x_6. \tag{6.150}$$

Treating x_3, x_4, x_5, x_6 as parameters:

$$x_3 = r, \ x_4 = s, \ x_5 = t, \ x_6 = u \tag{6.151}$$

we can write the solution as

$$\begin{pmatrix} x_1 \\ x_2 \\ x_3 \\ x_4 \\ x_5 \\ x_6 \end{pmatrix} = r \begin{pmatrix} 2 \\ 0 \\ 1 \\ 0 \\ 0 \\ 0 \end{pmatrix} + s \begin{pmatrix} -4 \\ 3 \\ 0 \\ 1 \\ 0 \\ 0 \end{pmatrix} + t \begin{pmatrix} 3 \\ -4 \\ 0 \\ 0 \\ 1 \\ 0 \end{pmatrix} + u \begin{pmatrix} -1 \\ -2 \\ 0 \\ 0 \\ 0 \\ 1 \end{pmatrix}, \tag{6.152}$$

Hence, the nullity of the matrix of coefficients:

$$\mathbf{A} = \begin{pmatrix} 1 & 0 & -2 & 4 & -3 & 1 \\ 0 & 1 & 0 & -3 & 4 & 2 \\ 0 & 0 & 0 & 0 & 0 & 0 \\ 0 & 0 & 0 & 0 & 0 & 0 \\ 0 & 0 & 0 & 0 & 0 & 0 \\ 0 & 0 & 0 & 0 & 0 & 0 \end{pmatrix}, \tag{6.153}$$

is 4, that is,

$$\text{nullity}(\mathbf{A}) = 4. \tag{6.154}$$

Rank and nullity have the following properties:

(i) The transpose of a matrix has the same rank as the matrix itself:

$$\text{rank}(\mathbf{A}) = \text{rank}(\mathbf{A}^T). \tag{6.155}$$

This follows from the fact that the determinant of a matrix is the same as the determinant of its transpose.

(ii) If a matrix $\mathbf{A}$ has n columns, then

$$\text{rank}(\mathbf{A}) + \text{nullity}(\mathbf{A}) = n. \tag{6.156}$$

This follows from the fact that the number of columns in a linear system is equal to the number of unknowns and the number of rows is equal to the number of equations. Also, the number of leading 1s in the reduced row-echelon form of $\mathbf{A}$ gives the number of linearly independent equations, that is, the number of variables that can be determined in terms of the free variables. Thus, the **nullity** is the **number of free parameters** in the system. In other words,

$$\text{no. of leading variables} + \text{no. of free variables} = n. \tag{6.157}$$

(iii) The maximum value for rank is

$$\text{rank}(\mathbf{A}) \leq \min(m, n). \tag{6.158}$$

We now summarize these results in terms of a theorem:

Theorem 6.5. If $\mathbf{Ax} = \mathbf{b}$ is a linear system with m equations and n unknowns, where

$$\mathbf{A} = \begin{pmatrix} a_{11} & a_{12} & \cdots & a_{1n} \\ a_{21} & a_{22} & \cdots & a_{2n} \\ \vdots & \vdots & \ddots & \vdots \\ a_{m1} & a_{m2} & \cdots & a_{mn} \end{pmatrix}, \ \mathbf{x} = \begin{pmatrix} x_1 \\ x_2 \\ \vdots \\ x_n \end{pmatrix}, \ \mathbf{b} = \begin{pmatrix} b_1 \\ b_2 \\ \vdots \\ b_m \end{pmatrix}. \tag{6.159}$$

I. The system is **consistent** if and only if the matrix of coefficients $\mathbf{A}$ and the augmented matrix $\tilde{\mathbf{A}} = (\mathbf{A}|\mathbf{b})$ have the same rank.

II. The system has a **unique** solution if and only if the common rank r of $\mathbf{A}$ and $\tilde{\mathbf{A}}$ is equal to n.

III. The system has **infinitely many** solutions if the common rank r satisfies $r < n$. Since basically the rank gives the number of linearly independent equations, r of the variables can be determined in terms of the remaining $n - r$ variables treated as **free parameters**.

IV. If $r > n$, the system is **overdetermined**, thus **inconsistent**.

For all these cases, the **Gauss-Jordan method** can be used and yields the solution when it exists and also reveals when the system is inconsistent.

6.1.10   Linear Transformations

We now look at linear transformations from a slightly different perspective. A linear transformation T from $\mathbb{R}^n$ to $\mathbb{R}^m$ is shown as

$$T\colon \mathbb{R}^n \to \mathbb{R}^m. \tag{6.160}$$

If $\mathbf{x}$ and $\mathbf{y}$ are two points in $\mathbb{R}^n$ and $\mathbb{R}^m$, respectively, we write

$$T(\mathbf{x}) = \mathbf{y}. \tag{6.161}$$

Picking a basis, $\{\mathbf{e}_1, \mathbf{e}_2, \ldots, \mathbf{e}_n\}$, in $\mathbb{R}^n$, we can represent $\mathbf{x}$ in terms of its coordinates, x_i, $i = 1, \ldots, n$, as

$$\mathbf{x} = x_1 \mathbf{e}_1 + x_2 \mathbf{e}_2 + \cdots + x_n \mathbf{e}_n, \tag{6.162}$$

which can also be written as a column matrix:

$$\mathbf{x} = \begin{pmatrix} x_1 \\ x_2 \\ \vdots \\ x_n \end{pmatrix}. \tag{6.163}$$

Since T is a linear operator, we can write

$$\mathbf{y} = T(\mathbf{x}) = x_1 T(\mathbf{e}_1) + x_2 T(\mathbf{e}_2) + \cdots + x_n T(\mathbf{e}_n), \tag{6.164}$$

thus T is determined uniquely by its action on the basis vectors. For $\mathbb{R}^n$, let us choose the **standard basis**:

$$\widehat{\mathbf{e}}_1 = \begin{pmatrix} 1 \\ 0 \\ \vdots \\ 0 \end{pmatrix}, \; \widehat{\mathbf{e}}_2 = \begin{pmatrix} 0 \\ 1 \\ \vdots \\ 0 \end{pmatrix}, \; \cdots, \; \widehat{\mathbf{e}}_n = \begin{pmatrix} 0 \\ 0 \\ \vdots \\ 1 \end{pmatrix}. \tag{6.165}$$

We now determine an $m \times n$ matrix $\mathbf{A}$:

$$\mathbf{A} = \begin{pmatrix} a_{11} & a_{12} & \cdots & a_{1n} \\ a_{21} & a_{22} & \cdots & a_{2n} \\ \vdots & \vdots & \ddots & \vdots \\ a_{m1} & a_{m2} & \cdots & a_{mn} \end{pmatrix}, \tag{6.166}$$

such that for every $\mathbf{x} \in \mathbb{R}^n$ the image $\mathbf{y} \in \mathbb{R}^m$ is written as

$$\mathbf{y} = T(\mathbf{x}) = \mathbf{A}\mathbf{x}. \tag{6.167}$$

To denote that we are representing T with the matrix $\mathbf{A}$, we could also write $\mathbf{A} = [T_A]$.

Let us now find the images of the standard basis as

$$\widehat{\mathbf{w}}_1 = \mathbf{A}\widehat{\mathbf{e}}_1, \ \widehat{\mathbf{w}}_2 = \mathbf{A}\widehat{\mathbf{e}}_2, \ldots, \ \widehat{\mathbf{w}}_n = \mathbf{A}\widehat{\mathbf{e}}_n. \tag{6.168}$$

If we write $\widehat{\mathbf{w}}_1 = \mathbf{A}\widehat{\mathbf{e}}_1$ explicitly as

$$\begin{pmatrix} \widehat{\mathbf{w}}_{11} \\ \widehat{\mathbf{w}}_{12} \\ \vdots \\ \widehat{\mathbf{w}}_{1m} \end{pmatrix} = \begin{pmatrix} a_{11} & a_{12} & \cdots & a_{1n} \\ a_{21} & a_{22} & \cdots & a_{2n} \\ \vdots & \vdots & \ddots & \vdots \\ a_{m1} & a_{m2} & \cdots & a_{mn} \end{pmatrix} \begin{pmatrix} 1 \\ 0 \\ \vdots \\ 0 \end{pmatrix} \tag{6.169}$$

$$= \begin{pmatrix} a_{11} \\ a_{21} \\ \vdots \\ a_{m1} \end{pmatrix}, \tag{6.170}$$

we see that the image of $\widehat{\mathbf{e}}_1$ gives the first column of $\mathbf{A}$, that is, $\mathbf{A}_1$. Hence, we can construct $\mathbf{A}$ column wise by finding the images of the standard basis vectors,

$$\mathbf{A}_1 = T(\widehat{\mathbf{e}}_1), \ \mathbf{A}_2 = T(\widehat{\mathbf{e}}_2), \ldots, \mathbf{A}_n = T(\widehat{\mathbf{e}}_n), \tag{6.171}$$

as

$$\mathbf{A} = (T(\widehat{\mathbf{e}}_1)|T(\widehat{\mathbf{e}}_2)| \cdots |T(\widehat{\mathbf{e}}_1)|T(\widehat{\mathbf{e}}_n)). \tag{6.172}$$

The matrix representation, $\mathbf{A}$, of the transformation T constructed via the standard basis is called the **standard matrix** of T. For one-to-one transformations, the inverse exists and is represented by the inverse matrix $\mathbf{A}^{-1}$. There is a one-to-one correspondence with matrices and transformations. Every $m \times n$ matrix can be interpreted as a linear transformation and vice-versa.

Example 6.6. *Linear transformations:* Consider orthogonal **projections** onto the xy-plane, which is $T: \mathbb{R}^3 \to \mathbb{R}^3$. The images of the standard basis vectors are now given as

$$T(\widehat{\mathbf{e}}_1) = \widehat{\mathbf{e}}_1 = \begin{pmatrix} 1 \\ 0 \\ 0 \end{pmatrix}, \tag{6.173}$$

$$T(\widehat{\mathbf{e}}_2) = \widehat{\mathbf{e}}_2 = \begin{pmatrix} 0 \\ 1 \\ 0 \end{pmatrix}, \tag{6.174}$$

$$T(\widehat{\mathbf{e}}_3) = \mathbf{0} = \begin{pmatrix} 0 \\ 0 \\ 0 \end{pmatrix}, \tag{6.175}$$

thus **A** is constructed as

$$\mathbf{A} = \begin{pmatrix} 1 & 0 & 0 \\ 0 & 1 & 0 \\ 0 & 0 & 0 \end{pmatrix}. \tag{6.176}$$

Note that the projection operator is **many-to-one**, hence, it is not invertible, which is also apparent from its matrix representation, which has a zero determinant.

Example 6.7. *Linear transformations:* For an operator performing counterclockwise rotation about the z-axis by θ, the images of the standard basis vectors are

$$T(\widehat{\mathbf{e}}_1) = \begin{pmatrix} \cos\theta \\ \sin\theta \\ 0 \end{pmatrix}, \tag{6.177}$$

$$T(\widehat{\mathbf{e}}_2) = \begin{pmatrix} -\sin\theta \\ \cos\theta \\ 0 \end{pmatrix}, \tag{6.178}$$

$$T(\widehat{\mathbf{e}}_3) = \begin{pmatrix} 0 \\ 0 \\ 1 \end{pmatrix}, \tag{6.179}$$

thus

$$\mathbf{A} = \begin{pmatrix} \cos\theta & -\sin\theta & 0 \\ \sin\theta & \cos\theta & 0 \\ 0 & 0 & 1 \end{pmatrix}. \tag{6.180}$$

Since $\det \mathbf{A} = 1$, this transformation is reversible and the inverse transformation is represented by

$$\mathbf{A}^{-1} = \begin{pmatrix} \cos\theta & \sin\theta & 0 \\ -\sin\theta & \cos\theta & 0 \\ 0 & 0 & 1 \end{pmatrix}. \tag{6.181}$$

6.2   NUMERICAL METHODS OF LINEAR ALGEBRA

6.2.1   Gauss-Jordan Row-Reduction and Partial Pivoting

A general linear system with m equations and n unknowns $(x_1, x_2, \ldots, x_n)$ is written as

$$\begin{aligned} E_1 : \quad & a_{11}x_1 + a_{12}x_2 + \cdots + a_{1n}x_n = b_1, \\ E_2 : \quad & a_{21}x_1 + a_{22}x_2 + \cdots + a_{2n}x_n = b_2, \\ & \vdots \qquad\qquad\qquad \vdots \\ E_m : \quad & a_{m1}x_1 + a_{m2}x_2 + \cdots + a_{mn}x_n = b_m, \end{aligned} \tag{6.182}$$

which can be conveniently represented in terms of matrices as

$$\mathbf{Ax} = \mathbf{b}, \tag{6.183}$$

where

$$\mathbf{A} = \begin{pmatrix} a_{11} & a_{12} & \cdots & a_{1n} \\ a_{21} & a_{22} & \cdots & a_{2n} \\ \vdots & \vdots & \ddots & \vdots \\ a_{m1} & a_{m2} & \cdots & a_{mn} \end{pmatrix}, \; \mathbf{x} = \begin{pmatrix} x_1 \\ x_2 \\ \vdots \\ x_n \end{pmatrix}, \; \mathbf{b} = \begin{pmatrix} b_1 \\ b_2 \\ \vdots \\ b_m \end{pmatrix}. \tag{6.184}$$

The matrix $\mathbf{A}$ is called the **matrix of coefficients** and $\widetilde{\mathbf{A}}$ is the **augmented matrix** defined as

$$\widetilde{\mathbf{A}} = (\mathbf{A}|\mathbf{b}) = \left(\begin{array}{cccc|c} a_{11} & a_{12} & \cdots & a_{1n} & b_1 \\ a_{21} & a_{22} & \cdots & a_{2n} & b_2 \\ \vdots & \vdots & \ddots & \vdots & \vdots \\ a_{m1} & a_{m2} & \cdots & a_{mn} & b_m \end{array} \right). \tag{6.185}$$

The augmented matrix carries all the relevant information about a linear system and since computers can handle data entered in the form of matrices, it plays a central role in numerical applications of linear systems.

Gauss-Jordan Row-Reduction

The systematics of Gauss-Jordan row-reduction involves the following basic steps:

- I. Use the first equation, E_1, and eliminate x_1 from the remaining equations, $E_2, E_3, \ldots, E_m$, to obtain a new set $E_1, \overline{E}_2, \overline{E}_3, \ldots, \overline{E}_m$. The first equation E_1 is called the **pivot equation** and a_{11} is called the **pivot**.
- II. Next use the second equation, $\overline{E}_2$, as the pivot equation and eliminate x_2 in $\overline{E}_3, \ldots, \overline{E}_m$.
- III. Continue this way until the entire matrix is reduced and a triangular form for the augmented matrix is obtained. The system can now be solved by **back-substitution**.

All these steps being well defined, in practice there are critical points that one needs to be careful. Consider the following linear system:

$$\begin{array}{rcl} -\varepsilon x_1 + x_2 + \varepsilon x_3 &=& 1 \\ x_1 + 2x_2 + x_3 &=& 4 \\ x_1 + \varepsilon x_2 - \varepsilon x_3 &=& 1 \end{array} \;\; \rightarrow \widetilde{\mathbf{A}} = \left(\begin{array}{ccc|c} -\varepsilon & 1 & \varepsilon & 1 \\ 1 & 2 & 1 & 4 \\ 1 & \varepsilon & -\varepsilon & 1 \end{array} \right), \tag{6.186}$$

where on the right-hand side we have written the corresponding augmented matrix. The number ε is very small so that the approximations

$$(1+\varepsilon) \simeq 1 \text{ and } \left(1 + \frac{1}{\varepsilon}\right) \simeq \frac{1}{\varepsilon} \tag{6.187}$$

can be used. We first eliminate x_1 using the first equation to write

$$-\varepsilon x_1 + x_2 + \varepsilon x_3 = 1$$
$$(2 + 1/\varepsilon)x_2 + 2x_3 = (4 + 1/\varepsilon) \quad \rightarrow$$
$$(\varepsilon + 1/\varepsilon)x_2 + (-\varepsilon + 1)x_3 = (1 + 1/\varepsilon)$$

$$\tilde{\mathbf{A}} = \left(\begin{array}{ccc|c} -\varepsilon & 1 & \varepsilon & 1 \\ 0 & (2 + 1/\varepsilon) & 2 & (4 + 1/\varepsilon) \\ 0 & (1 + 1/\varepsilon) & (-\varepsilon + 1) & (1 + 1/\varepsilon) \end{array} \right). \tag{6.188}$$

Applying the approximations given in Eq. (6.187), the above system becomes

$$-\varepsilon x_1 + x_2 + \varepsilon x_3 \simeq 1$$
$$(1/\varepsilon)x_2 + 2x_3 \simeq (1/\varepsilon) \quad \rightarrow \tilde{\mathbf{A}} \simeq \left(\begin{array}{ccc|c} -\varepsilon & 1 & \varepsilon & 1 \\ 0 & (1/\varepsilon) & 2 & (1/\varepsilon) \\ 0 & (1/\varepsilon) & 1 & (1/\varepsilon) \end{array} \right). \tag{6.189}$$
$$(1/\varepsilon)x_2 + x_3 \simeq (1/\varepsilon)$$

In the next step, we reduce the second column and obtain the following row-echelon form of the augmented matrix :

$$-\varepsilon x_1 + x_2 + \varepsilon x_3 \simeq 1$$
$$(1/\varepsilon)x_2 + 2x_3 \simeq (1/\varepsilon) \quad \rightarrow \tilde{\mathbf{A}} \simeq \left(\begin{array}{ccc|c} -\varepsilon & 1 & \varepsilon & 1 \\ 0 & (1/\varepsilon) & 2 & (1/\varepsilon) \\ 0 & 0 & -1 & 0 \end{array} \right). \tag{6.190}$$
$$-x_3 \simeq 0$$

Finally, back-substitution yields the approximate solution:

$$x_1 = 0, \tag{6.191}$$

$$x_2 = 1, \tag{6.192}$$

$$x_3 = 0. \tag{6.193}$$

It can easily be checked that the exact solution of this system is

$$x_1 = 1, \tag{6.194}$$

$$x_2 = 1, \tag{6.195}$$

$$x_3 = 1. \tag{6.196}$$

It is important to note that the determinant of coefficients of this system is

$$\det(\mathbf{A}) = \left| \begin{array}{ccc} -\varepsilon & 1 & \varepsilon \\ 1 & 2 & 1 \\ 1 & \varepsilon & -\varepsilon \end{array} \right| = 1 - \varepsilon + 4\varepsilon^2 \simeq 1, \tag{6.197}$$

Hence, the problem is well defined to the order we are considering.

The point that one needs to be careful with the Gauss-Jordan method of row-reduction is that since the first step involves division of the pivot equation by the pivot, the pivot has to be different from zero. Actually, as in the above example, even when the pivot is not zero but just small, the **roundoff errors** could cause serious problems. The cure is to reorder the equations so that the pivot equation is the one with a large pivot term, preferably the largest in terms of absolute value. This is called **partial pivoting**. The **total pivoting** is reserved for cases where both rows and columns are swapped, however, it is hardly used in practice. Another method that is not as common as partial pivoting is **row scaling**.

Let us now apply partial pivoting to the above example and swap the first and the second rows:

$$
\begin{aligned}
x_1 + 2x_2 + x_3 &= 4 \\
-\varepsilon x_1 + x_2 + \varepsilon x_3 &= 1 \\
x_1 + \varepsilon x_2 - \varepsilon x_3 &= 1
\end{aligned}
\quad \rightarrow \tilde{\mathbf{A}} = \left(\begin{array}{ccc|c}
1 & 2 & 1 & 4 \\
-\varepsilon & 1 & \varepsilon & 1 \\
1 & \varepsilon & -\varepsilon & 1
\end{array} \right).
\tag{6.198}
$$

Reducing the first column, we get

$$
\begin{aligned}
x_1 + 2x_2 + x_3 &= 4 \\
(1 + 2\varepsilon)x_2 + 2\varepsilon x_3 &= (1 + 4\varepsilon) \quad \rightarrow \\
(\varepsilon - 2)x_2 + (-\varepsilon - 1)x_3 &= -3
\end{aligned}
$$

$$
\tilde{\mathbf{A}} = \left(\begin{array}{ccc|c}
1 & 2 & 1 & 4 \\
0 & (1 + 2\varepsilon) & 2\varepsilon & (1 + 4\varepsilon) \\
0 & (\varepsilon - 2) & (-\varepsilon - 1) & -3
\end{array} \right).
\tag{6.199}
$$

Using the same approximations [Eq. (6.187)], we can write

$$
\begin{aligned}
x_1 + 2x_2 + x_3 &\simeq 4 \\
x_2 + 2\varepsilon x_3 &\simeq 1 \\
-2x_2 + -x_3 &\simeq -3
\end{aligned}
\quad \rightarrow \tilde{\mathbf{A}} \simeq \left(\begin{array}{ccc|c}
1 & 2 & 1 & 4 \\
0 & 1 & 2\varepsilon & 1 \\
0 & -2 & -1 & -3
\end{array} \right),
\tag{6.200}
$$

which after reduction of the second column yields

$$
\begin{aligned}
x_1 + 2x_2 + x_3 &\simeq 4 \\
x_2 + 2\varepsilon x_3 &\simeq 1 \\
(-1 + 4\varepsilon)x_3 &\simeq -1
\end{aligned}
\quad \rightarrow \tilde{\mathbf{A}} \simeq \left(\begin{array}{ccc|c}
1 & 2 & 1 & 4 \\
0 & 1 & 2\varepsilon & 1 \\
0 & 0 & (-1 + 4\varepsilon) & -1
\end{array} \right).
\tag{6.201}
$$

To the same order of approximation, this becomes

$$
\begin{aligned}
x_1 + 2x_2 + x_3 &\simeq 4 \\
x_2 + 2\varepsilon x_3 &\simeq 1 \\
-x_3 &\simeq -1
\end{aligned}
\quad \rightarrow \tilde{\mathbf{A}} \simeq \left(\begin{array}{ccc|c}
1 & 2 & 1 & 4 \\
0 & 1 & 2\varepsilon & 1 \\
0 & 0 & -1 & -1
\end{array} \right).
\tag{6.202}
$$

After back-substitution, we obtain the approximate solution as

$$
x_1 \simeq 1,
\tag{6.203}
$$

$$x_2 \simeq 1, \tag{6.204}$$

$$x_3 \simeq 1. \tag{6.205}$$

Note that in Eq. (6.202), instead of using back-substitution, we could continue the elimination process in the upward direction with the Gauss-Jordan reduction and put the augmented matrix into the following form

$$\tilde{\mathbf{A}} \simeq \left(\begin{array}{ccc|c} 1 & 0 & 0 & 1 \\ 0 & 1 & 0 & 1 \\ 0 & 0 & 1 & 1 \end{array} \right), \tag{6.206}$$

where the answer can be read by inspection from the last column. However, since each elementary operation (addition, multiplication and division) in a computer takes a certain amount of time and generates heat, in numerical analysis, the number of elementary operations that a computer has to perform becomes a major source of concern as the size of the matrices grow. It can be shown that for an $n \times n$ system, for large n, the number of elementary operations for the Gauss-Jordan elimination process goes as n^3, while for the back-substitution it goes only as n^2. In the accompanying file we present some codes for Matlab applications of Gauss-Jordan row reduction and partial pivoting.

6.2.2   LU-Factorization

For a given linear system of equations, $\mathbf{Ax} = \mathbf{b}$, where $\mathbf{A}$ is the $n \times n$ matrix of coefficients, we now introduce three methods that requires fewer operations. They are all modifications of the Gauss-Jordan method of row-reduction and use a method called the **LU-factorization**, where a square matrix $\mathbf{A}$ is written as the product of a **lower triangular** matrix $\mathbf{L}$ and an **upper triangular** matrix $\mathbf{U}$ as

$$\mathbf{A} = \mathbf{LU}. \tag{6.207}$$

For $n = 3$, the $\mathbf{L}$ and the $\mathbf{U}$ matrices as given as

$$\mathbf{L} = \left(\begin{array}{ccc} L_{11} & 0 & 0 \\ L_{21} & L_{22} & 0 \\ L_{31} & L_{32} & L_{33} \end{array} \right), \quad \mathbf{U} = \left(\begin{array}{ccc} U_{11} & U_{12} & U_{13} \\ 0 & U_{22} & U_{23} \\ 0 & 0 & U_{33} \end{array} \right). \tag{6.208}$$

Writing $\mathbf{A} = \mathbf{LU}$ explicitly:

$$\left(\begin{array}{ccc} a_{11} & a_{12} & a_{13} \\ a_{21} & a_{22} & a_{23} \\ a_{31} & a_{32} & a_{33} \end{array} \right) = \left(\begin{array}{ccc} L_{11} & 0 & 0 \\ L_{21} & L_{22} & 0 \\ L_{31} & L_{32} & L_{33} \end{array} \right) \left(\begin{array}{ccc} U_{11} & U_{12} & U_{13} \\ 0 & U_{22} & U_{23} \\ 0 & 0 & U_{33} \end{array} \right), \tag{6.209}$$

we see that in general a 3×3 matrix has 9 components, thus we can only determine 9 of the $6 + 6 = 12$ components of the $\mathbf{L}$ and the $\mathbf{U}$ matrices. This gives us the freedom to choose any 3 of the components of **L** and **U**. In this regard, the **Doolittle method** chooses to set the diagonal components of **L** to 1:

$$L_{11} = L_{22} = L_{33} = 1, \tag{6.210}$$

while the **Crout method** sets the diagonal components of **U** to 1:

$$U_{11} = U_{22} = U_{33} = 1. \tag{6.211}$$

Since the two methods are very similar, we only discuss the Doolittle method.

Doolittle Method

Using the LU-factorization, we can write the linear system **Ax** = **b** as

$$\mathbf{LUx} = \mathbf{b}, \tag{6.212}$$

$$\mathbf{L(Ux)} = \mathbf{b}. \tag{6.213}$$

Calling

$$\mathbf{Ux} = \mathbf{y}, \tag{6.214}$$

we write

$$\mathbf{Ly} = \mathbf{b} : \tag{6.215}$$

$$\begin{pmatrix} 1 & 0 & 0 \\ L_{21} & 1 & 0 \\ L_{31} & L_{32} & 1 \end{pmatrix} \begin{pmatrix} y_1 \\ y_2 \\ y_3 \end{pmatrix} = \begin{pmatrix} b_1 \\ b_2 \\ b_3 \end{pmatrix}, \tag{6.216}$$

to obtain the following set of equations:

$$y_1 = b_1, \tag{6.217}$$

$$L_{21}y_1 + y_2 = b_2, \tag{6.218}$$

$$L_{31}y_1 + L_{32}y_2 + y_3 = b_3. \tag{6.219}$$

This yields **y** as

$$y_1 = b_1, \tag{6.220}$$

$$y_2 = b_2 - L_{21}b_1, \tag{6.221}$$

$$y_3 = b_3 - L_{31}b_1 - L_{32}(b_2 - L_{21}b_1). \tag{6.222}$$

Once **y** is determined, we turn to Eq. (6.214):

$$\begin{pmatrix} U_{11} & U_{12} & U_{13} \\ 0 & U_{22} & U_{23} \\ 0 & 0 & U_{33} \end{pmatrix} \begin{pmatrix} x_1 \\ x_2 \\ x_3 \end{pmatrix} = \begin{pmatrix} y_1 \\ y_2 \\ y_3 \end{pmatrix} \tag{6.223}$$

and read the values of **x** from

$$U_{11}x_1 + U_{12}x_2 + U_{13}x_3 = y_1, \tag{6.224}$$

$$U_{22}x_2 + U_{23}x_3 = y_2, \tag{6.225}$$

$$U_{33}x_3 = y_3, \tag{6.226}$$

as

$$x_1 = \frac{y_1}{U_{11}} - \frac{U_{12}}{U_{11}U_{22}}\left(y_2 - U_{23}\frac{y_3}{U_{33}}\right) - \frac{U_{13}y_3}{U_{11}U_{33}}, \tag{6.227}$$

$$x_2 = \left(y_2 - U_{23}\frac{y_3}{U_{33}}\right)\frac{1}{U_{22}}, \tag{6.228}$$

$$x_3 = \frac{y_3}{U_{33}}. \tag{6.229}$$

To find the components of **L** and **U** in terms of the components of **A**, we write the product **LU**:

$$\mathbf{LU} = \begin{pmatrix} 1 & 0 & 0 \\ L_{21} & 1 & 0 \\ L_{31} & L_{32} & 1 \end{pmatrix} \begin{pmatrix} U_{11} & U_{12} & U_{13} \\ 0 & U_{22} & U_{23} \\ 0 & 0 & U_{33} \end{pmatrix}, \tag{6.230}$$

explicitly as

$$\mathbf{LU} = \begin{pmatrix} U_{11} & U_{12} & U_{13} \\ L_{21}U_{11} & L_{21}U_{12} + U_{22} & L_{21}U_{13} + U_{23} \\ L_{31}U_{11} & L_{31}U_{12} + L_{32}U_{22} & L_{31}U_{13} + L_{32}U_{23} + U_{33} \end{pmatrix} \tag{6.231}$$

and compare with the components of **A**:

$$\mathbf{A} = \begin{pmatrix} a_{11} & a_{12} & a_{13} \\ a_{21} & a_{22} & a_{23} \\ a_{31} & a_{32} & a_{33} \end{pmatrix}. \tag{6.232}$$

This yields the components of **L** as

$$L_{21} = \frac{a_{21}}{a_{11}}, \tag{6.233}$$

$$L_{31} = \frac{a_{31}}{a_{11}}, \tag{6.234}$$

$$L_{32} = \left(a_{32} - \frac{a_{31}}{a_{11}}a_{12}\right) \Big/ \left(a_{22} - \frac{a_{21}}{a_{11}}a_{12}\right), \tag{6.235}$$

while the components of $\mathbf{U}$ are found as

$$U_{11} = a_{11}, \tag{6.236}$$

$$U_{12} = a_{12}, \tag{6.237}$$

$$U_{13} = a_{13}, \tag{6.238}$$

$$U_{22} = a_{22} - \frac{a_{21}}{a_{11}} a_{12}, \tag{6.239}$$

$$U_{23} = a_{23} - \frac{a_{21}}{a_{11}} a_{13}, \tag{6.240}$$

$$U_{33} = a_{33} - \frac{a_{31}}{a_{11}} a_{13} - \frac{\left(a_{32} - \dfrac{a_{31}a_{12}}{a_{11}} \right) \left(a_{23} - \dfrac{a_{21}a_{13}}{a_{11}} \right)}{\left(a_{22} - \dfrac{a_{21}a_{12}}{a_{11}} \right)}. \tag{6.241}$$

For a **general** $n \times n$ **matrix** $\mathbf{A}$, the components of $\mathbf{L}$ and $\mathbf{U}$ are given as

$$U_{1k} = a_{1k}, \quad k = 1, \ldots, n, \tag{6.242}$$

$$L_{j1} = \frac{a_{j1}}{U_{11}}, \quad j = 2, \ldots, n, \tag{6.243}$$

$$U_{jk} = a_{jk} - \sum_{s=1}^{j-1} L_{js}U_{sk}, \quad k = j, \ldots, n; \quad j \geq 2, \tag{6.244}$$

$$L_{jk} = \frac{1}{U_{kk}} \left(a_{jk} - \sum_{s=1}^{k-1} L_{js}U_{sk} \right), \quad j = k+1, \ldots, n; \quad k \geq 2. \tag{6.245}$$

Cholesky's Method

For **symmetric** and **positive definite** matrices, that is, when

$$\mathbf{A} = \mathbf{A}^T \text{ and } \mathbf{x}^T \mathbf{A} \mathbf{x} > 0 \text{ for all } \mathbf{x} \neq 0, \tag{6.246}$$

we can choose the $\mathbf{L}$ and the $\mathbf{U}$ matrices as

$$\mathbf{U} = \mathbf{L}^T, \tag{6.247}$$

$$\mathbf{L} = \mathbf{L}. \tag{6.248}$$

Thus, $\mathbf{A} = \mathbf{L}\mathbf{U}$ becomes

$$\mathbf{A} = \mathbf{L}\mathbf{L}^T. \tag{6.249}$$

For a 3×3 matrix of coefficients, we can take $\mathbf{L}$ as

$$\mathbf{L} = \begin{pmatrix} L_{11} & 0 & 0 \\ L_{21} & L_{22} & 0 \\ L_{31} & L_{32} & L_{33} \end{pmatrix}. \tag{6.250}$$

Note that since $\mathbf{A}$ is symmetric, for a 3×3 matrix, we have only six components: $L_{11}, L_{21}, L_{22}, L_{31}, L_{32}, L_{33}$, to determine. In general, for an $n \times n$ matrix, the Cholesky's method yields the components of $\mathbf{L}$ as

$$L_{11} = (a_{11})^{1/2}, \tag{6.251}$$

$$L_{j1} = \frac{a_{j1}}{L_{11}}, \quad j = 2, \ldots, n, \tag{6.252}$$

$$L_{jj} = \left(a_{jj} - \sum_{s=1}^{j-1} L_{js}^2 \right)^{1/2}, \quad j = 2, \ldots, n, \tag{6.253}$$

$$L_{pj} = \frac{1}{L_{jj}} \left(a_{pj} - \sum_{s=1}^{j-1} L_{js} L_{ps} \right), \quad p = j+1, \ldots, n; \quad j \geq 2. \tag{6.254}$$

Once $\mathbf{L}$ is determined in terms of the components of $\mathbf{A}$, we write

$$\mathbf{A}\mathbf{x} = \mathbf{b}, \tag{6.255}$$

$$\mathbf{L}(\mathbf{L}^T\mathbf{x}) = \mathbf{b}. \tag{6.256}$$

Calling

$$\mathbf{L}^T\mathbf{x} = \mathbf{y}, \tag{6.257}$$

we write Eq. (6.256) as

$$\mathbf{L}\mathbf{y} = \mathbf{b}, \tag{6.258}$$

which allows us to determine $\mathbf{y}$. For $n = 3$, $\mathbf{y}$ is given as

$$y_1 = b_1/L_{11}, \tag{6.259}$$

$$y_2 = (b_2 - L_{21}b_1/L_{11})/L_{22}, \tag{6.260}$$

$$y_3 = [b_3 - L_{31}b_1/L_{11} - L_{32}(b_2 - L_{21}b_1/L_{11})/L_{22}]/L_{33}. \tag{6.261}$$

Substituting $\mathbf{y}$ into

$$\mathbf{L}^T\mathbf{x} = \mathbf{y}, \tag{6.262}$$

yields $\mathbf{x}$ as in the Doolittle method.

6.2.3   Solutions of Linear Systems by Iteration

As straightforward as the Gauss-Jordan method of elimination seems, sometimes it is more practical to try an **iterative approach**, where we start with a **guess** for the solution and hope to approach the exact solution to any desired level of accuracy. Iterative methods are naturally very helpful when the **convergence** is rapid. Iterative methods are usually preferred for matrices with large diagonal elements. They are also useful for **sparse systems**, which involve large matrices with many zeros for their entrees.

Gauss–Seidal Iteration

We demonstrate the Gauss–Seidal method on a linear system $\mathbf{Ax} = \mathbf{b}$ with three equations and three unknowns:

$$\begin{aligned}
a_{11}x_1 + a_{12}x_2 + a_{13}x_3 &= b_1, \\
a_{21}x_1 + a_{22}x_2 + a_{23}x_3 &= b_2, \\
a_{31}x_1 + a_{32}x_2 + a_{33}x_3 &= b_3.
\end{aligned} \tag{6.263}$$

Assuming that the diagonal terms are all nonzero:

$$a_{11} \neq 0, \ a_{22} \neq 0, \ a_{33} \neq 0, \tag{6.264}$$

we can write the above system as

$$\begin{aligned}
x_1 + \left(\frac{a_{12}}{a_{11}}\right)x_2 + \left(\frac{a_{13}}{a_{11}}\right)x_3 &= \left(\frac{b_1}{a_{11}}\right), \\
\left(\frac{a_{21}}{a_{22}}\right)x_2 \qquad + x_2 + \left(\frac{a_{23}}{a_{22}}\right)x_3 &= \left(\frac{b_2}{a_{22}}\right), \\
\left(\frac{a_{31}}{a_{33}}\right)x_3 + \left(\frac{a_{32}}{a_{33}}\right)x_2 \qquad + x_3 &= \left(\frac{b_3}{a_{33}}\right).
\end{aligned} \tag{6.265}$$

If some of the diagonal terms are zero, we can rearrange the equations or the variables so that none of the diagonal terms are zero. Introducing the following matrices:

$$\mathbf{L} = \begin{pmatrix} 0 & 0 & 0 \\ L_{21} & 0 & 0 \\ L_{31} & L_{32} & 0 \end{pmatrix}, \ \mathbf{U} = \begin{pmatrix} 0 & U_{12} & U_{13} \\ 0 & 0 & U_{23} \\ 0 & 0 & 0 \end{pmatrix}, \ \tilde{\mathbf{b}} = \begin{pmatrix} \tilde{b}_1 \\ \tilde{b}_2 \\ \tilde{b}_3 \end{pmatrix}, \tag{6.266}$$

where the components are given as

$$L_{21} = \left(\frac{a_{21}}{a_{22}}\right), \ L_{31} = \left(\frac{a_{31}}{a_{33}}\right), \ L_{32} = \left(\frac{a_{32}}{a_{33}}\right), \tag{6.267}$$

$$U_{12} = \left(\frac{a_{12}}{a_{11}}\right), \ U_{13} = \left(\frac{a_{13}}{a_{11}}\right), \ U_{23} = \left(\frac{a_{23}}{a_{22}}\right), \tag{6.268}$$

$$\tilde{b}_1 = \left(\frac{b_1}{a_{11}}\right), \ \tilde{b}_2 = \left(\frac{b_2}{a_{22}}\right), \ \tilde{b}_3 = \left(\frac{b_3}{a_{33}}\right), \tag{6.269}$$

we can write the system of equations in Eq. (6.265) as

$$(\mathbf{L} + \mathbf{I} + \mathbf{U})\mathbf{x} = \tilde{\mathbf{b}}. \tag{6.270}$$

We start with a trial solution:

$$x_1 = x_1^{(0)}, \ x_1 = x_2^{(0)}, \ x_3 = x_3^{(0)}, \tag{6.271}$$

and substitute into Eq. (6.265) to write

$$
x_1^{(1)} = -\left(\frac{a_{12}}{a_{11}}\right) x_2^{(0)} - \left(\frac{a_{13}}{a_{11}}\right) x_3^{(0)} + \left(\frac{b_1}{a_{11}}\right),
$$

$$
x_2^{(1)} = -\left(\frac{a_{21}}{a_{22}}\right) x_1^{(1)} - \left(\frac{a_{23}}{a_{22}}\right) x_3^{(0)} + \left(\frac{b_2}{a_{22}}\right), \qquad (6.272)
$$

$$
x_3^{(1)} = -\left(\frac{a_{31}}{a_{33}}\right) x_1^{(1)} - \left(\frac{a_{32}}{a_{33}}\right) x_2^{(1)} + \left(\frac{b_3}{a_{33}}\right).
$$

Note that starting with the first equation, we substitute the trial solution. But in the subsequent equations, we use the most recent value as soon as it is available. For example, after the first equation, $x_1^{(1)}$ becomes available, hence, we use $x_1^{(1)}$ in the second equation rather than $x_1^{(0)}$. Similarly, in the third equation, we use $x_1^{(1)}$ and $x_2^{(1)}$.

The above system of equations [Eq. (6.272)] can be generalized as

$$
\mathbf{x}^{(m+1)} = \widetilde{\mathbf{b}} - \mathbf{L}\mathbf{x}^{(m+1)} - \mathbf{U}\mathbf{x}^{(m)}, \qquad (6.273)
$$

which when rearranged:

$$
(\mathbf{I} + \mathbf{L})\mathbf{x}^{(m+1)} = \widetilde{\mathbf{b}} - \mathbf{U}\mathbf{x}^{(m)}, \qquad (6.274)
$$

yields the **iteration formula**:

$$
\boxed{\mathbf{x}^{(m+1)} = \mathbf{C}^{-1}(\widetilde{\mathbf{b}} - \mathbf{U}\mathbf{x}^{(m)}),} \qquad (6.275)
$$

where

$$
\mathbf{C} = (\mathbf{I} + \mathbf{L}). \qquad (6.276)
$$

A sufficient condition for convergence is $||\mathbf{C}|| < 1$, where $||\mathbf{C}||$ is the **norm** of $\mathbf{C}$, which is defined as

$$
\boxed{||\mathbf{C}|| = \sqrt{\sum_{j=1}^{n}\sum_{k=1}^{n} C_{jk}^2.}} \qquad (6.277)
$$

$||\mathbf{C}||$ is also called the **Frobenius norm**.

Residual and the Error Margin

For the exact solution, we can write

$$
\mathbf{A}\mathbf{x}_{\text{exact}} = \mathbf{b}, \qquad (6.278)
$$

Hence, we define the **residual** $\mathbf{r}^{(m)}$ as the difference

$$\boxed{\mathbf{r}^{(m)} = \mathbf{b} - \mathbf{A}\mathbf{x}^{(m)}} \tag{6.279}$$

and the **error margin** ε as

$$\boxed{|\mathbf{r}^{(m)}| < \varepsilon,} \tag{6.280}$$

which is

$$\sqrt{(r_1^{(m)})^2 + (r_2^{(m)})^2 + \cdots + (r_n^{(m)})^2} < \varepsilon. \tag{6.281}$$

It can be proven that for any positive definite matrix $\mathbf{A}$, $\mathbf{x}^T\mathbf{A}\mathbf{x} > 0$ for $\mathbf{x} > 0$, Gauss–Seidel iteration **converges** independent of our choice of $\mathbf{x}^{(0)}$ [1].

Jacobi Iteration

In the Gauss–Seidal iteration, we use the most recent value of a variable as soon as it is available. In parallel processors, it is now possible to solve all n equations simultaneously at each step of the iteration, hence, it is desirable to use iteration formulas where $\mathbf{x}^{(m+1)}$ is not used until a step is completed. In the **Jacobi iteration**, the general iteration formula is given as

$$\boxed{\mathbf{x}^{(m+1)} = \mathbf{b} + (\mathbf{I} - \mathbf{A})\mathbf{x}^{(m)},} \tag{6.282}$$

where until an iteration is completed, the new value is not used. This iteration formula converges for all choices of $\mathbf{x}^{(0)}$, if and only if the spectral radius of $(\mathbf{I} - \mathbf{A})$ lies within the unit circle. The proof follows from the expansion of $\mathbf{x}$ in terms of the eigenvectors of $(\mathbf{I} - \mathbf{A})$. Note that **spectral radius** means the eigenvalue with the largest absolute value.

6.2.4   Interpolation

Vandermonde Matrix

We now introduce another important application of linear algebra. Consider a set of $n + 1$ data points:

$$\boxed{\{(x_0, y_0), (x_1, y_1), \ldots, (x_n, y_n)\}.} \tag{6.283}$$

We would like to **interpolate** this data with a **polynomial of minimum order**:

$$\boxed{y(x) = a_m x^m + a_{m-1} x^{m-1} + \cdots + a_0,} \tag{6.284}$$

such that each data point satisfies the equations

$$y_0 = a_m x_0^m + a_{m-1} x_0^{m-1} + \cdots + a_0,$$
$$y_1 = a_m x_1^m + a_{m-1} x_1^{m-1} + \cdots + a_0,$$
$$\vdots$$
$$y_n = a_m x_n^m + a_{m-1} x_n^{m-1} + \cdots + a_0, \tag{6.285}$$

respectively. It is not essential that polynomials are used for interpolation. In principle, any finite set of functions can be used [1]. The linear system in Eq. (6.285) is set of $(n+1)$ equations for the $(m+1)$ unknowns a_i, $i = 0, 1, \ldots, m$. In terms of matrices we can write this set as

$$\boxed{\mathbf{X a} = \mathbf{y},} \tag{6.286}$$

where

$$\mathbf{X} = \begin{pmatrix} 1 & x_0 & \cdots & x_0^m \\ 1 & x_1 & \cdots & x_1^m \\ \vdots & \vdots & \vdots & \vdots \\ 1 & x_n & \cdots & x_n^m \end{pmatrix}, \quad \mathbf{a} = \begin{pmatrix} a_0 \\ a_1 \\ \vdots \\ a_m \end{pmatrix}, \quad \mathbf{y} = \begin{pmatrix} y_0 \\ y_1 \\ \vdots \\ y_m \end{pmatrix}. \tag{6.287}$$

If $m = n$, then the matrix of coefficients, $\mathbf{X}$, becomes a square matrix, called the **Vandermonde matrix**:

$$\mathbf{X} = \begin{pmatrix} 1 & x_0 & \cdots & x_0^n \\ 1 & x_1 & \cdots & x_1^n \\ \vdots & \vdots & \vdots & \vdots \\ 1 & x_n & \cdots & x_n^n \end{pmatrix} \tag{6.288}$$

and the corresponding linear system is called the **Vandermonde system**. Note that the jth column of the Vandermonde matrix is the second column raised to the $(j-1)$th power.

The Vandermonde system can be solved by Gauss-Jordan's row reduction. However, let us see what happens if we replace the interpolating polynomial, which is called the **standard form**:

$$\boxed{y(x) = a_n x^n + a_{n-1} x^{n-1} + \cdots + a_0,} \tag{6.289}$$

by the **Newton form**:

$$\boxed{\begin{aligned} y(x) = {}& b_n (x - x_{n-1})(x - x_{n-2}) \cdots (x - x_0) \\ & + b_{n-1}(x - x_{n-2})(x - x_{n-3}) \cdots (x - x_0) \\ & + \cdots + b_1(x - x_0) + b_0. \end{aligned}} \tag{6.290}$$

Now, the first condition [Eq. (6.283)]: $y(x_0) = y_0$, immediately yields the value of b_0 as $b_0 = y_0$. Similarly, we write the other conditions to obtain the linear system

$$y(x_0) = b_0,$$
$$y(x_1) = b_1 h_1 + b_0,$$
$$y(x_2) = b_2(h_1 + h_2)h_2 + b_0(h_1 + h_2) + b_0,$$
$$y(x_3) = b_3(h_1 + h_2 + h_3)(h_2 + h_3)h_3 + b_2(h_1 + h_2 + h_3)(h_2 + h_3)$$
$$+ b_1(h_1 + h_2 + h_3) + b_0,$$
$$\vdots$$
$$y(x_n) = \cdots , \tag{6.291}$$

where $h_i = x_i - x_{i-1}$, $i = 1, 2, \ldots, n$.

In general, this gives us a **lower triangular** matrix for the matrix of coefficients. For example, for $n = 3$ we obtain

$$\begin{pmatrix} 1 & 0 & 0 & 0 \\ 1 & h_1 & 0 & 0 \\ 1 & h_1 + h_2 & (h_1 + h_2)h_2 & 0 \\ 1 & (h_1 + h_2 + h_3) & (h_1 + h_2 + h_3)(h_2 + h_3) & (h_1 + h_2 + h_3)(h_2 + h_3)h_3 \end{pmatrix}$$

$$\times \begin{pmatrix} b_0 \\ b_1 \\ b_2 \\ b_3 \end{pmatrix} = \begin{pmatrix} y_0 \\ y_1 \\ y_2 \\ y_3 \end{pmatrix}, \tag{6.292}$$

which is much simpler than the Vandermonde system, and can be solved by **forward-substitution**. If we have equally spaced points, where

$$h = h_i = (x_i - x_{i-1}) > 0, \quad i = 1, 2, \ldots, n, \tag{6.293}$$

we obtain

$$\begin{pmatrix} 1 & 0 & 0 & 0 \\ 1 & h & 0 & 0 \\ 1 & 2h & 2h^2 & 0 \\ 1 & 3h & 6h^2 & 6h^3 \end{pmatrix} \begin{pmatrix} b_0 \\ b_1 \\ b_2 \\ b_3 \end{pmatrix} = \begin{pmatrix} y_0 \\ y_1 \\ y_2 \\ y_3 \end{pmatrix}. \tag{6.294}$$

Conversion Between Two Forms

We have seen that the Newton form offers certain advantages in finding the interpolating polynomial; however, the standard form is still much simpler. Conversion between the two forms is possible and can be done by a simple comparison of the equal powers of x between the two forms. For a **third-order** polynomial, the **standard form** is given as

$$y(x) = a_3 x^3 + a_2 x^2 + a_1 x + a_0, \tag{6.295}$$

where the **Newton form** is

$$y(x) = b_3(x - x_2)(x - x_1)(x - x_0) + b_2(x - x_1)(x - x_0) + b_1(x - x_0) + b_0. \tag{6.296}$$

Equating the two forms and comparing their equal powers of x, yields the transformation equation

$$\begin{pmatrix} a_0 \\ a_1 \\ a_2 \\ a_3 \end{pmatrix} = \begin{pmatrix} 1 & -x_0 & x_0 x_1 & -x_0 x_1 x_2 \\ 0 & 1 & -(x_0 + x_1) & x_0 x_1 + x_0 x_2 + x_1 x_2 \\ 0 & 0 & 1 & -(x_0 + x_1 + x_2) \\ 0 & 0 & 0 & 1 \end{pmatrix} \begin{pmatrix} b_0 \\ b_1 \\ b_2 \\ b_3 \end{pmatrix}. \tag{6.297}$$

The significance of this result is that we can obtain the standard form from the Newton form by solving an upper-triangular system via the back-substitution. For large matrices, the number of basic operations that a computer has to do for back-substitution and also for forward-substitution goes as n^2. On the other hand, using the Gauss-Jordan's reduction process, the number of operations for a general Vandermonde system goes as n^3. In other words, instead of solving the Vandermonde system to obtain the standard form of the interpolating formula, it is much faster to find the Newton form and then to convert it into the standard form.

6.2.5   Power Method for Eigenvalues

We now introduce a simple **iterative method** to find the dominant eigenvalue, that is, the **largest eigenvalue** in terms of absolute value, of any real symmetric $n \times n$ matrix $\mathbf{A}$. The method starts with any $n \times 1$ vector $\mathbf{x}_0$ and computes successively

$$\mathbf{x}_1 = \mathbf{A}\mathbf{x}_0, \ \mathbf{x}_2 = \mathbf{A}\mathbf{x}_1, \ \ldots, \ \mathbf{x}_j = \mathbf{A}\mathbf{x}_{j-1}, \ldots. \tag{6.298}$$

In the jth iteration:

$$\boxed{\mathbf{x}_j = \mathbf{A}\mathbf{x}_{j-1},} \tag{6.299}$$

if we define three scalars, m_{j-1}, m_j, p_j, as

$$m_{j-1} = \mathbf{x}_{j-1}^T \mathbf{x}_{j-1}, \ m_j = \mathbf{x}_j^T \mathbf{x}_j, \ p_j = \mathbf{x}_{j-1}^T \mathbf{x}_j, \tag{6.300}$$

the **quotient**, also called the **Rayleigh quotient**:

$$q_j = \frac{p_j}{m_{j-1}}, \tag{6.301}$$

gives the jth **approximation** q_j to the **dominant eigenvalue** λ.

Furthermore, if we define the **error** ε_j as

$$\lambda - q_j = \varepsilon_j, \tag{6.302}$$

the Rayleigh quotient gives the error of the jth iteration as

$$|\varepsilon_j| \leq \delta_j = \sqrt{\frac{m_j}{m_{j-1}} - q_j^2}. \tag{6.303}$$

Let us now prove these statements. Since $\mathbf{A}$ is real and symmetric, it has a complete set of orthogonal eigenvectors (Sections 5.13 and 5.15, and [2]):

$$\mathbf{v}_1, \mathbf{v}_2, \ldots, \mathbf{v}_n, \tag{6.304}$$

corresponding to the eigenvalues

$$\lambda_1, \lambda_2, \ldots, \lambda_n, \tag{6.305}$$

respectively. Thus, we can write $\mathbf{x}_{j-1}$ as the linear combination

$$\mathbf{x}_{j-1} = a_1 \mathbf{v}_1 + a_2 \mathbf{v}_2 + \cdots + a_n \mathbf{v}_n. \tag{6.306}$$

Since the eigenvectors satisfy the eigenvalue equation:

$$\mathbf{A}\mathbf{v}_i = \lambda_i \mathbf{v}_i, \quad i = 1, 2, \ldots, n, \tag{6.307}$$

we can also write Eq. (6.299) as

$$\mathbf{x}_j = \mathbf{A}\mathbf{x}_{j-1} = a_1 \lambda_1 \mathbf{v}_1 + a_2 \lambda_2 \mathbf{v}_2 + \cdots + a_n \lambda_n \mathbf{v}_n. \tag{6.308}$$

We now write the difference

$$\mathbf{x}_j - q_j \mathbf{x}_{j-1} = a_1(\lambda_1 - q_j)\mathbf{v}_1 + a_2(\lambda_2 - q_j)\mathbf{v}_2 + \cdots + a_n(\lambda_n - q_j)\mathbf{v}_n \tag{6.309}$$

and use the orthogonality of the eigenvectors, $\mathbf{v}_i^T \mathbf{v}_j = \delta_{ij}$, $i, j = 1, 2, \ldots, n$, to obtain the scalar product

$$(\mathbf{x}_j - q_j \mathbf{x}_{j-1})^T (\mathbf{x}_j - q_j \mathbf{x}_{j-1}) = a_1^2(\lambda_1 - q_j)^2 + \cdots + a_n^2(\lambda_n - q_j). \tag{6.310}$$

For the same scalar product, we can also write

$$(\mathbf{x}_j - q_j \mathbf{x}_{j-1})^T (\mathbf{x}_j - q_j \mathbf{x}_{j-1}) = m_j - 2q_j p_j + q_j^2 m_{j-1} \tag{6.311}$$

$$= m_j - q_j^2 m_{j-1} \tag{6.312}$$

$$= \delta_j^2 m_{j-1}, \tag{6.313}$$

where in the last step, we have defined δ_j as

$$\delta_j = \sqrt{\frac{m_j}{m_{j-1}} - q_j^2} \,. \tag{6.314}$$

Equating Eqs. (6.310) and (6.313), we now get

$$\delta_j^2 m_{j-1} = a_1^2(\lambda_1 - q_j)^2 + a_2^2(\lambda_2 - q_j)^2 + \cdots + a_n^2(\lambda_n - q_j)^2. \tag{6.315}$$

Let us assume that $\tilde{\lambda}$ is one of the eigenvalues of $\mathbf{A}$ that q_j comes closest, hence, we can write

$$(\tilde{\lambda} - q_j)^2 \le (\lambda_j - q_j)^2, \ j = 1, 2, \ldots, n. \tag{6.316}$$

Using this in Eq. (6.315) gives

$$\delta_j^2 m_{j-1} \ge (\tilde{\lambda} - q_j)^2(a_1^2 + a_2^2 + \cdots + a_n^2) \tag{6.317}$$

$$\ge (\tilde{\lambda} - q_j)^2 m_{j-1}, \tag{6.318}$$

which yields the following inequality for δ_j:

$$\delta_j \ge (\tilde{\lambda} - q_j). \tag{6.319}$$

Using Eq. (6.314) and the definition of error [Eq. (6.302)], we finally obtain the desired result as

$$|\varepsilon_j| \le \delta_j = \sqrt{\frac{m_j}{m_{j-1}} - q_j^2} \,, \tag{6.320}$$

where

$$|\varepsilon_j| = (\tilde{\lambda} - q_j). \tag{6.321}$$

One can improve the convergence significantly by shifting the eigenvalues, which can be achieved by using the matrix $(\mathbf{A} - p\mathbf{I})$ instead of $\mathbf{A}$, with a suitably chosen p [1]. The reader can show that $(\mathbf{A} - p\mathbf{I})$ and $\mathbf{A}$ have the same eigenvectors, while the eigenvalues are shifted from λ_i to $\lambda_i - p$.

Note: It is important to note that the iteration process applies to any $n \times n$ matrix that has a dominant eigenvalue. However, the error bound in Eq. (6.320) is valid only for symmetric matrices.

6.2.6 Solution of Equations

Solution of an equation given as

$$f(x) = 0, \tag{6.322}$$

is probably one of the most commonly encountered problems in numerical analysis. In this section, we introduce some of the available methods for finding the solution of a single equation, where $f(x)$ is an **algebraic** or a **transcendental** function of the real variable x. A solution is called the **zero** or the **root** of $f(x)$, which is the value of the argument:

$$x = a, \qquad (6.323)$$

that satisfies the equation:

$$f(a) = 0. \qquad (6.324)$$

Even though we confine ourselves to the real domain, $f(x)$ could also be defined over the field of complex numbers. As in the case of quadratic polynomials, sometimes a general formula for the solutions can be written. However, in cases where no exact expressions can be found, we resort to the available approximation methods. These methods can be divided into two groups as **iterative** and **always-convergent**. Among the always-convergent methods, we have the **bisection**, **reguli falsi** and the **Brent's methods**, which under certain conditions guarantee the convergence to the required root, while the iterative methods, like the **fixed-point iteration**, **Newton–Raphson method**, and the **secant iteration**, may or may not converge [1].

Fixed-Point Iteration

In the fixed-point iteration, if the equation $f(x) = 0$ can be written as

$$x = g(x) \qquad (6.325)$$

and if an approximate value for the root is assumed, say x_n, we can hope to improve it by iterating the above formula as

$$\boxed{x_{n+1} = g(x_n).} \qquad (6.326)$$

This simple iterative approach, converges under certain conditions.

To investigate the **convergence properties** of the fixed-point iteration, we first let the exact value of the root be a and write Eq. (6.326) as

$$x_{n+1} - a = g(x_n) - g(a). \qquad (6.327)$$

If $g(x)$ is a continuous function, the **mean-value** theorem can be used to write

$$x_{n+1} - a = (x_n - a)g'(\xi_n), \qquad (6.328)$$

where ξ_n is a point in the interval $[a, x_n]$. We start with an initial guess x_0, which is in some interval J that also contains a, and iterate the formula [Eq. (6.328)]

n times. Assuming that the inequality

$$\boxed{|g'(x)| \leq K < 1}$$ (6.329)

holds for any interval J that contains a, which is also a **sufficient condition** for convergence, we obtain the inequality

$$|x_n - a| = |x_{n-1} - a||g'(\xi_{n-1})|$$ (6.330)

$$\leq |x_{n-2} - a||g'(\xi_{n-1})||g'(\xi_{n-2})| \leq \cdots \leq |x_0 - a|K^n.$$ (6.331)

For $K < 1$, as $n \to \infty$, the following limit holds:

$$\lim_{n\to\infty} |x_n - a| \to 0,$$ (6.332)

hence, the iteration converges for any x_0 in J.

Let ε_n denote the error of the nth iteration:

$$\varepsilon_n = x_n - a.$$ (6.333)

Assuming that the iteration is convergent and converges to a:

$$\lim_{n\to\infty} x_n \to a,$$ (6.334)

we can also write

$$g'(\xi_n) \to g'(a).$$ (6.335)

Hence, using Eqs. (6.328) and (6.331), we can write

$$\lim_{n\to\infty} \frac{\varepsilon_{n+1}}{\varepsilon_n} = g'(a),$$ (6.336)

and

$$\lim_{n\to\infty} \varepsilon_n \simeq A|g'(a)|^n,$$ (6.337)

where A is some constant. For the convergence of the sequence, it is **necessary** that $|g'(a)| < 1$ be satisfied [1]. In summary, for the convergence of the fixed-point iteration, in the neighborhood of the intersection point, $x \simeq a$, of the two curves, $y_1 = x$ and $y_2 = g(x)$, the slope of y_2 has to be less than the slope of y_1.

Example 6.8. *Roots by fixed point iteration:* Given the equation

$$f(x) = x^2 - 3.2x + 1.5 = 0,$$ (6.338)

we write it as

$$x = g(x) = \frac{1}{3.2}(x^2 + 1.5), \qquad (6.339)$$

thus the corresponding iteration formula becomes

$$x_{n+1} = \frac{1}{3.2}(x_n^2 + 1.5). \qquad (6.340)$$

Starting with $x_0 = 1$, after the first 12 iterations, we obtain

n	0	1	2	3	4	5	$\cdots$	12
x_n	1.0	0.7813	0.6595	0.6047	0.5830	0.5750	$\cdots$	0.5704

hence the iteration converges to the smaller root 0.5704 where the exact roots are 0.5704 and 2.6296. For another choice, $x_0 = 1.5$, we again obtain another convergent sequence that converges to the smaller root but less rapidly (Figure 6.1):

n	0	1	2	3	4	5	$\cdots$	14
x_n	1.5	1.1719	0.8979	0.7207	0.6311	0.5932	$\cdots$	0.5704

However, the iteration for $x_0 = 3$ diverges:

n	0	1	2	3	4	5	$\cdots$
x_n	3.0	3.2813	3.8333	5.0607	8.4721	22.8991	$\cdots$

Notice that for an initial point closer to the root, we get a sequence that converges much faster:

n	0	1	2	3	4	5	$\cdots$
x_n	0.570	0.5703	0.5704	0.5704	0.5704	0.5704	$\cdots$

Let us now consider the same equation [Eq. (6.338)] with the following choice for $g(x)$:

$$g(x) = 3.2 - \frac{1.5}{x}. \qquad (6.341)$$

Now for $x_0 = 10$ the first seven iterations yield

n	0	1	2	3	4	5	6	7
x_n	10.00	3.0500	2.7082	2.6461	2.6331	2.6303	2.6297	2.6296

which converges to the larger root 2.6296. The graph of x with $g(x)$ is given in Figure 6.2. In the light of the convergence properties of the

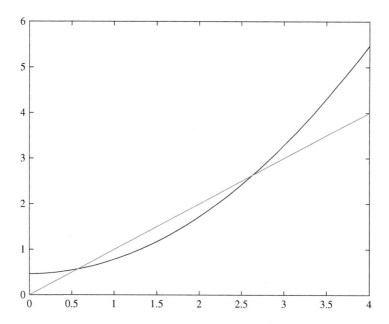

Figure 6.1   Plot of $y = x$ and $g(x) = (\frac{1}{3.2})(x^2 + 1.5)$.

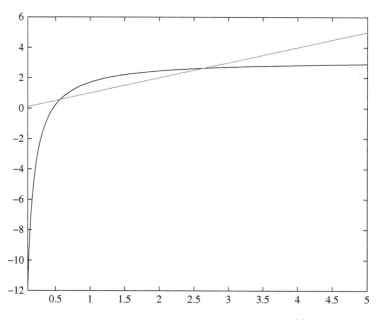

Figure 6.2   Plot of $y = x$ vs. $g(x) = 3.2 - \frac{1.5}{x}$.

fixed-point iteration method, for the first choice of $g(x)$ [Eq. (6.339)] the derivative $g'(x)$ is

$$g'(x) = \frac{2x}{(3.2)} = 0.625x, \tag{6.342}$$

where $|g'(x)| < 1$ for all $x < 1.6$. Hence, for the initial points, $x_0 = 1$ and $x_0 = 1.5$, we have obtained convergent sequences that yielded the smaller root but not for $x_0 = 3$. For the second choice of $g(x)$ [Eq. (6.341)], the inequality $|g'(x)| < 1$ is satisfied for $x > 1.22$. In fact, for $x_0 = 10$, we have obtained the larger root. Also note that in the light of Eq. (6.329), for the first choice of $g(x)$, the two exact roots, 0.5704 and 2.6296, give

$$|g'(0.5704)| = 0.3565 < 1, \tag{6.343}$$

$$|g'(2.6296)| = 1.6435 > 1, \tag{6.344}$$

while for the second choice of $g(x)$ we have

$$|g'(0.5704)| = 4.6103 > 1, \tag{6.345}$$

$$|g'(2.6296)| = 0.2169 < 1. \tag{6.346}$$

Example 6.9. *Roots by fixed point iteration:* Consider the following equation:

$$f(x) = x^3 + 1.5x - 1.2 = 0 \tag{6.347}$$

and write it as

$$x = g(x) = \frac{1.2}{x^2 + 1.5}, \tag{6.348}$$

which has the iteration formula

$$x_{n+1} = \frac{1.2}{x_n^2 + 1.5}. \tag{6.349}$$

Using the starting value $x_0 = 1$, we obtain the sequence

n	0	1	2	3	4	5	$\cdots$	12
x_n	1.00	0.4800	0.6935	0.6058	0.6428	0.6272	$\cdots$	0.6318

where the exact value is 0.6318. Note that for this choice of $g(x)$ [Eq. (6.349)] the inequality $|g(x)| < 1$ is always satisfied (Figure 6.3). In general, for an appropriate choice of $g(x)$, a certain amount of **experimentation** is needed.

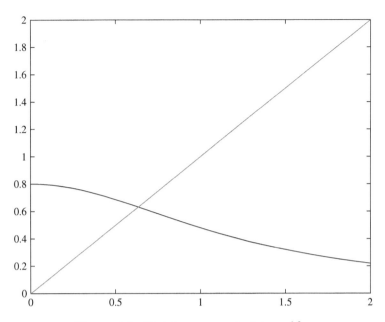

Figure 6.3 Plot for $y = x$ and $g(x) = \frac{1.2}{x^2 + 1.5}$.

Newton–Raphson Method

When a function $f(x)$ has a continuous derivative, the Newton–Raphson method can be used. It is based on approximating a given function by suitably chosen tangents. The Newton–Raphson method is usually preferred for its **simplicity** and **speed**. Starting with an initial value x_0 for the zero of the function $f(x)$, we draw the tangent line at $x = x_0$, the intercept of which with the x-axis gives another point x_1. Using the definition of the slope, we write (Figure 6.4)

$$\tan \theta = f'(x_0) = \frac{f(x_0)}{x_0 - x_1}, \tag{6.350}$$

which gives the value of x_1 as

$$x_1 = x_0 - \frac{f(x_0)}{f'(x_0)}. \tag{6.351}$$

We now continue by drawing the tangent at x_1 and by finding its intercept at x_2 as (Figure 6.4)

$$x_2 = x_1 - \frac{f(x_1)}{f'(x_1)}. \tag{6.352}$$

Repeating this process, we obtain the **iteration formula** of the Newton–Raphson method as

$$\boxed{x_{n+1} = x_n - \frac{f(x_n)}{f'(x_n)}.} \tag{6.353}$$

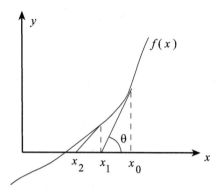

Figure 6.4 Newton–Raphson method.

This formula can also be obtained by approximating a function with the first two terms of its **Taylor series** expansion:

$$f(x) = f(x_0) + f'(x_0)(x - x_0) + \frac{1}{2}f''(x_0)(x - x_0)^2 + \cdots, \tag{6.354}$$

as

$$f(x) \approx f(x_0) + f'(x_0)(x - x_0), \tag{6.355}$$

which gives us the **approximate** value of the **root** as

$$x \approx x_0 + \frac{f(x_0)}{f'(x_0)}. \tag{6.356}$$

Iterating this formula, we hope to approach the exact root.

Error in the Newton–Raphson Method

For a general iteration formula:

$$x_{n+1} = g(x_n), \tag{6.357}$$

we can write the Taylor expansion of $g(x_n)$ about the root $x = a$ as

$$g(x_{n+1}) = g(a) + g'(a)(x_n - a) + \frac{1}{2}g''(a)(x_n - a)^2 + \cdots. \tag{6.358}$$

Using the iteration formula [Eq. (6.357)] and the fact that for the root $x = a$, $a = g(a)$, we write

$$x_{n+1} = a + g'(a)(x_n - a) + \frac{1}{2}g''(a)(x_n - a)^2 + \cdots. \tag{6.359}$$

Since the error of the nth iteration is defined as $\varepsilon_n = a - x_n$, Eq. (6.359) can also be written as

$$x_{n+1} - a = g'(a)(x_n - a) + \frac{1}{2}g''(a)(x_n - a)^2 + \cdots , \qquad (6.360)$$

$$\varepsilon_{n+1} = g'(a)\varepsilon_n - \frac{1}{2}g''(a)\varepsilon_n^2 + \cdots . \qquad (6.361)$$

For the Newton–Raphson method, $g_n(x_n)$ is given as [Eq. (6.353)]

$$g_n(x_n) = x_n - \frac{f(x_n)}{f'(x_n)}, \qquad (6.362)$$

which when differentiated and substituted into Eq. (6.361), yields the **error** of the Newton–Raphson method as

$$\boxed{\varepsilon_{n+1} \simeq \frac{1}{2}\frac{f''(x_n)}{f(x_n)}\varepsilon_n^2.} \qquad (6.363)$$

This shows that in the Newton–Raphson method, the error in x_{n+1} is to **second-order**. Of course, it is needless to say that in the Newton–Raphson method, it is assumed that all the required derivatives exist and to avoid round-off errors, one has to watch out for cases where $|f'(x)|$ is small near a solution.

Example 6.10. *Newton–Raphson method:* We now reconsider the function in Example 6.9:

$$f(x) = x^3 + 1.5x - 1.2 = 0, \qquad (6.364)$$

The iteration formula now becomes

$$x_{n+1} = x_n - \frac{x^3 + 1.5x - 1.2}{3x^2 + 1.5}. \qquad (6.365)$$

For the same initial point $x_0 = 1$, and for the 10th iteration, we obtain

n	0	1	2	3	4	$\cdots$	10
x_n	1.00	0.7111	0.6361	0.6319	0.6318	$\cdots$	0.6318

Notice the rapid convergence.

Example 6.11. *Newton–Raphson method:* Consider the transcendental equation

$$f(x) = x - 2.7\sin x = 0,$$

which has the iteration formula

$$x_{n+1} = x_n - \frac{x_n - 2.7\sin x}{1 - 2.7\cos x} \tag{6.366}$$

$$= \frac{2.7(\sin x - x\cos x)}{1 - 2.7\cos x}. \tag{6.367}$$

For $x_0 = 2$, we obtain

n	0	1	2	3	$\cdots$	10
x_n	2.00	2.2143	2.1936	2.1934	$\cdots$	2.1934

The Secant Method

Newton–Raphson method is very practical and effective when it works, but sometimes to write the derivative of a function could not only be cumbersome but also be computationally inefficient. In this regard, the secant method replaces the derivative in the iteration formula with (Figure 6.5)

$$f'(x_k) \approx \frac{f(x_k) - f(x_{k-1})}{x_k - x_{k-1}}, \tag{6.368}$$

which yields the **iteration formula**

$$\boxed{x_{k+1} = x_k - f(x_k)\frac{x_k - x_{k-1}}{f(x_k) - f(x_{k-1})}.} \tag{6.369}$$

We now need two initial values x_0 and x_1.

Example 6.12. *The secant method:* For comparison, we use the function in the previous example with the secant method and write the corre-

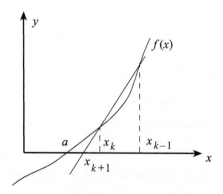

Figure 6.5   Secant method.

sponding iteration formula as

$$x_{k+1} = x_k - \frac{(x - 2.7\sin x)(x_k - x_{k-1})}{x_k - x_{k-1} + 2.7(\sin x_{k-1} - \sin x_k)}. \tag{6.370}$$

Using the initial values $x_0 = 1.7$ and $x_1 = 2.1$, we obtain

n	0	1	2	3	4	5	6
x_k	1.7000	2.1000	2.2235	2.1921	2.1934	2.1934	2.1934

With the accompanying Mathlab files, the reader can experiment with different functions and different choices for the $g(x)$ functions.

6.2.7 Numerical Integration

Numerical integration is one of the oldest topics of numerical analysis, which dates back to Archimedes who tried to calculate the area of a circle. Numerical integration is both simple and difficult. In most cases, it is possible to get a reasonable answer even with the simplest techniques and yet in some other cases, it may require complex techniques with huge amount of computer capacity. Nowadays, computer programs such as Maple, Mathematica are capable of performing both numerical and analytical integration. We usually resort to the numerical techniques when the analytic methods fail. However, since the numerical techniques give only the value of an integral with the specific end points substituted, to get a global feeling of the integral it is always advisable to push the problem analytically as far as possible. This may not only save computer time but also may allow us to get a better feeling of the integral on the various parameters of the problem. In this section, we discuss the basic techniques like the **rectangular rule** and the **trapezoidal rule**. Other techniques with detailed error analysis can be found in Refs. [1], [3], and [4].

Rectangular Rule

In one dimension, the integral

$$I = \int_a^b f(x)dx$$

is geometrically the area under the function $f(x)$ between the limits a and b. The simplest method we can use is the rectangular rule, where we divide the interval $[a, b]$ into n subintervals of equal length:

$$h = \frac{b - a}{n}, \tag{6.371}$$

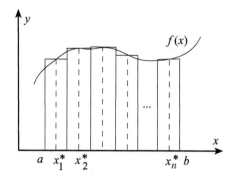

Figure 6.6   Rectangular method.

and approximate the value of the integral with the sum of the areas of the rectangles as

$$I \approx h[f(x_1^*) + f(x_2^*) + \cdots + f(x_n^*)]. \tag{6.372}$$

The height of the rectangles, $f(x_i^*)$, are in general taken as one of the three possibilities: the values of the function at the left end points, the right end points and the mid-points of the intervals (Figure 6.6).

Trapezoidal Rule

The trapezoidal rule is in general more accurate. Instead of rectangles, we use trapezoidal segments, where the upper side of the of the trapezoids are the chords (Figure 6.7). Areas of the segments are now given, respectively, as

$$\frac{1}{2}[f(a) + f(x_1)]h, \ \frac{1}{2}[f(x_1) + f(x_2)]h, \ldots, \frac{1}{2}[f(x_{n-1}) + f(b)]h. \tag{6.373}$$

Their sum approximates the integral as

$$I \approx h\left[\frac{f(a)}{2} + f(x_1) + f(x_2) + \cdots + f(x_{n-1}) + \frac{1}{2}f(b)\right]. \tag{6.374}$$

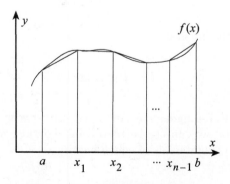

Figure 6.7   The trapezoidal rule.

A large class of functions can be integrated by using interpolating functions between the end points of the intervals. Usually, these interpolating functions are taken as polynomials. The rectangular or the trapezoidal rules are two examples of cases where the interpolating functions are linear, hence the function is represented by straight lines between the end points of the intervals. In general, interpolation with polynomials between equally spaced points is called the **Newton–Cotes formulas**. The **Simpson's rule**, which is based on interpolation by polynomials of second order, is also a Newton–Cotes formula.

Euler–Maclaurin Sum Formula

We now introduce the Euler–Maclaurin sum formula, which is a very useful and a versatile tool in numerical calculations. To derive the Euler–Maclaurin sum formula, we use the properties of the Bernoulli polynomials, where the first five polynomials are given below [2]:

$$
\begin{array}{c}
\textbf{Bernoulli Polynomials} \\[4pt]
B_0(x) = 1, \\[6pt]
B_1(x) = x - \dfrac{1}{2}, \\[6pt]
B_2(x) = x^2 - x + \dfrac{1}{6}, \\[6pt]
B_3(x) = x\left(x - \dfrac{1}{2}\right)(x - 1), \\[6pt]
B_4(x) = x^4 - 2x^3 + x^2 - \dfrac{1}{30}, \\[6pt]
B_5(x) = x\left(x - \dfrac{1}{2}\right)(x - 1)\left(x^2 - x - \dfrac{1}{3}\right).
\end{array}
\tag{6.375}
$$

Bernoulli numbers are defined as $B_i = B_i(0)$:

$$
\begin{array}{ccccc}
B_0 = 1, & B_1 = -\dfrac{1}{2}, & B_2 = \dfrac{1}{6}, & B_3 = 0, & B_4 = -\dfrac{1}{30}, \\[8pt]
B_5 = 0, & B_6 = \dfrac{1}{42}, & B_7 = 0, & B_8 = -\dfrac{1}{30}, & \cdots
\end{array}
\tag{6.376}
$$

Bernoulli polynomials have the following useful properties:

$$
B_s'(x) = sB_{s-1}(x), \quad \int_0^1 B_s(x)dx = 0, \quad s \geq 1, \tag{6.377}
$$

$$
B_s(1 - x) = (-1)^s B_s(x). \tag{6.378}
$$

We now consider the integral

$$\int_0^1 f(x)dx. \tag{6.379}$$

Since the first Bernoulli polynomial is given as

$$B_0(x) = 1, \tag{6.380}$$

we can write this as

$$\int_0^1 f(x)dx = \int_0^1 f(x)B_0(x)dx. \tag{6.381}$$

Next, the following property of the Bernoulli polynomials:

$$B_1'(x) = B_0(x) = 1, \tag{6.382}$$

allows us to write this as

$$\int_0^1 f(x)dx = \int_0^1 f(x)B_0(x)dx = \int_0^1 f(x)B_1'(x)dx. \tag{6.383}$$

Integrating by parts yields

$$\int_0^1 f(x)dx = f(x)B_1(x)|_0^1 - \int_0^1 f'(x)B_1(x)dx \tag{6.384}$$

$$= \frac{1}{2}[f(1) + f(0)] - \int_0^1 f'(x)B_1(x)dx, \tag{6.385}$$

where we have used the special values

$$B_1(1) = \frac{1}{2} \text{and } B_1(0) = -\frac{1}{2}. \tag{6.386}$$

We now make use of the property

$$B_1(x) = \frac{1}{2}B_2'(x) \tag{6.387}$$

and integrate by parts again to obtain

$$\int_0^1 f(x)dx = \frac{1}{2}[f(1) + f(0)] - \frac{1}{2!}[f'(1)B_2(1) - f'(0)B_2(0)]$$

$$+ \frac{1}{2!}\int_0^1 f''(x)B_2(x)dx. \tag{6.388}$$

Using the special values of the Bernoulli polynomials:

$$B_{2n}(1) = B_{2n}(0) = B_{2n}, \qquad n = 0, 1, 2, \ldots,$$

$$B_{2n+1}(1) = B_{2n+1}(0) = 0, \qquad n = 1, 2, 3, \ldots, \tag{6.389}$$

and continuing like this we obtain

$$\int_0^1 f(x)dx = \frac{1}{2}[f(1) + f(0)] - \sum_{p=1}^q \frac{B_{2p}}{(2p)!}[f^{(2p-1)}(1) - f^{(2p-1)}(0)]$$

$$+ \frac{1}{(2q)!} \int_0^1 f^{(2q)}(x)B_{2q}(x)dx. \tag{6.390}$$

This equation is called the **Euler–Maclaurin sum formula**. We have assumed that all the necessary derivatives of $f(x)$ exist and q is an integer greater than one.

To change the limits of the integral from 0–1 to 1–2, we write

$$\int_1^2 f(x)dx = \int_0^1 f(y + 1)dy \tag{6.391}$$

and repeat the same steps in the derivation of the Euler–Maclaurin sum formula for $f(y + 1)$ to write

$$\int_0^1 f(y + 1)dy = \frac{1}{2}[f(2) + f(1)]$$

$$- \sum_{p=1}^q \frac{B_{2p}}{(2p)!}[f^{(2p-1)}(2) - f^{(2p-1)}(1)]$$

$$+ \frac{1}{(2q)!} \int_0^1 f^{(2q)}(y + 1)P_{2q}(y)dy. \tag{6.392}$$

In the above result, we have employed the Bernoulli periodic function, where the **Bernoulli periodic function:**

$$P_s(x) = B_s(x - [x]), \tag{6.393}$$

is continuous and has the period 1 and $[x]$ means the greatest integer in the interval $(x - 1, x]$. The Bernoulli periodic function also satisfies the relations

$$P_s'(x) = sP_{s-1}(x), \quad s = 1, 2, 3, \ldots \tag{6.394}$$

and

$$P_s(1) = (-1)^s P_s(0), \quad s = 0, 1, 2, 3, \ldots. \tag{6.395}$$

Making the transformation $y + 1 = x$, we can now write

$$\int_1^2 f(x)dx = \int_0^1 f(y+1)dy \tag{6.396}$$

$$= \frac{1}{2}[f(2) + f(1)]$$

$$- \sum_{p=1}^q \frac{B_{2p}}{(2p)!}[f^{(2p-1)}(2) - f^{(2p-1)}(1)]$$

$$+ \frac{1}{(2q)!}\int_0^1 f^{(2q)}(x)P_{2q}(x-1)dx. \tag{6.397}$$

Repeating this for the interval $[2,3]$, we write

$$\int_2^3 f(x)dx = \frac{1}{2}[f(3) + f(2)]$$

$$- \sum_{p=1}^q \frac{B_{2p}}{(2p)!}[f^{(2p-1)}(3) - f^{(2p-1)}(2)]$$

$$+ \frac{1}{(2q)!}\int_2^3 f^{(2q)}(x)P_{2q}(x-2)dx. \tag{6.398}$$

Integrals for the other intervals can be written similarly. Since the integral for the interval $[0,n]$ can be written as

$$\int_0^n f(x)dx = \int_0^1 f(x)dx + \int_1^2 f(x)dx + \cdots + \int_{n-1}^n f(x)dx, \tag{6.399}$$

we substitute the formulas found above into the right-hand side to obtain

$$\int_0^n f(x)dx = \frac{1}{2}f(0) + f(1) + f(2) + \cdots + \frac{1}{2}f(n)$$

$$- \sum_{p=1}^q \frac{B_{2p}}{(2p)!}[f^{(2p-1)}(n) - f^{(2p-1)}(0)]$$

$$+ \frac{1}{(2q)!}\int_0^n f^{(2q)}(x)P_{2q}(x)dx. \tag{6.400}$$

In this derivation, we have used the fact that the function $P_{2q}(x)$ is periodic with the period one. Rearranging this, we can write

$$\sum_{j=0}^n f(j) = \int_0^n f(x)dx + \frac{1}{2}[f(0) + f(n)]$$

$$+ \sum_{p=1}^{q-1} \frac{B_{2p}}{(2p)!} [f^{(2p-1)}(n) - f^{(2p-1)}(0)]$$

$$+ \frac{B_{2q}}{(2q)!} [f^{(2q-1)}(n) - f^{(2q-1)}(0)] - \frac{1}{(2q)!} \int_0^n f^{(2q)}(x) P_{2q}(x) dx.$$

$$(6.401)$$

The last two terms on the right-hand side can be written under the same integral sign, which gives us the final form of the Euler–Maclaurin sum formula as

$$\sum_{j=0}^{n} f(j) = \int_0^n f(x) dx + \frac{1}{2} [f(0) + f(n)]$$

$$+ \sum_{p=1}^{q-1} \frac{B_{2p}}{(2p)!} [f^{(2p-1)}(n) - f^{(2p-1)}(0)]$$

$$+ \int_0^n \frac{[B_{2q} - B_{2q}(x - [x])]}{(2q)!} f^{(2q)}(x) dx. \qquad (6.402)$$

In this derivation, we have assumed that $f(x)$ is continuous and has all the required derivatives. This is a very versatile formula that can be used in several ways. When q is chosen as a finite number, it allows us to evaluate a given series as an integral plus some correction terms. When q is chosen as infinity, it could allow us to replace a slowly converging series with a rapidly converging one. When we take the integral, $\int_0^n f(x) dx$, to the left-hand side and if the integral in the last term is manageable, it can also be used for numerical evaluation of integrals:

$$\int_0^n f(x) dx = \sum_{j=0}^{n} f(j) - \frac{1}{2} [f(0) + f(n)]$$

$$- \sum_{p=1}^{q-1} \frac{B_{2p}}{(2p)!} [f^{(2p-1)}(n) - f^{(2p-1)}(0)]$$

$$- \int_0^n \frac{[B_{2q} - B_{2q}(x - [x])]}{(2q)!} f^{(2q)}(x) dx. \qquad (6.403)$$

REFERENCES

1. Antia, H.M. (2002). *Numerical Methods for Scientists and Engineers*. Birkhäuser.

2. Bayin, S.S. (2018). *Mathematical Methods in Science and Engineering*, 2e. Wiley.

3. Davis, P.J. and Rabinowitz, P. (1984). *Methods of Numerical Integration*, 2e. Academic Press.

4. Koenig, H.A. (1998). *Modern Computational Methods*. Taylor and Francis.

PROBLEMS

1. Use the Gauss-Jordan reduction method to obtain the reduced row-echelon form of

$$
\mathbf{A} = \left(\begin{array}{cccccc|c}
1 & 2 & -2 & 0 & 1 & 0 & 0 \\
1 & 3 & -2 & 1 & 2 & 1 & 0 \\
0 & 0 & 2 & 3 & 0 & 1 & 0 \\
2 & 2 & 0 & 4 & 4 & 3 & 0
\end{array} \right)
$$

as

$$
\mathbf{R} = \left(\begin{array}{cccccc|c}
1 & 0 & 0 & 1 & 0 & -1/4 & 0 \\
0 & 1 & 0 & 1 & 0 & 1/4 & 0 \\
0 & 0 & 1 & 3/2 & 0 & 1/2 & 0 \\
0 & 0 & 0 & 0 & 1 & 3/4 & 0
\end{array} \right).
$$

2. Show that the reduced row-echelon form of

$$
\mathbf{A} = \left(\begin{array}{cccc}
1 & 0 & 2 & 2 \\
-3 & 0 & -6 & -6 \\
4 & 1 & 9 & 8 \\
-2 & 3 & -1 & -4 \\
5 & -2 & 9 & 10 \\
4 & -6 & 7 & 8
\end{array} \right)
$$

is

$$
\mathbf{R} = \left(\begin{array}{cccc}
1 & 0 & 2 & 2 \\
0 & 1 & 1 & 0 \\
0 & 0 & 1 & 0 \\
0 & 0 & 0 & 0 \\
0 & 0 & 0 & 0 \\
0 & 0 & 0 & 0
\end{array} \right).
$$

3. Find the reduced row-echelon form of $\mathbf{A}$:

$$
\mathbf{A} = \left(\begin{array}{cccccc}
1 & -3 & 4 & -2 & 5 & 4 \\
0 & 0 & 1 & 3 & -2 & -6 \\
2 & -6 & 9 & -1 & 9 & 7 \\
2 & -6 & 8 & -4 & 10 & 8
\end{array} \right).
$$

4. For an operator performing counterclockwise rotation about the z-axis by θ, show that the images of the standard basis vectors are

$$
T(\hat{\mathbf{e}}_1) = \left(\begin{array}{c}
\cos \theta \\
\sin \theta \\
0
\end{array} \right),
$$

$$T(\widehat{e}_2) = \begin{pmatrix} -\sin\theta \\ \cos\theta \\ 0 \end{pmatrix},$$

$$T(\widehat{e}_3) = \begin{pmatrix} 0 \\ 0 \\ 1 \end{pmatrix}.$$

5. Show that the standard matrix of the transformation that projects a point in $\mathbb{R}^2$ onto the line that goes through the origin and makes an angle θ with the x-axis is

$$\mathbf{A} = \begin{pmatrix} \cos^2\theta & \sin\theta\cos\theta \\ \sin\theta\cos\theta & \cos^2\theta \end{pmatrix}.$$

Is this transformation one-to-one? Is it reversible?

6. For the following 3×5 matrix:

$$\mathbf{A} = \begin{pmatrix} 2 & 1 & -2 & 3 & 1 \\ 1 & 2 & -1 & 2 & 1 \\ 4 & 5 & -4 & 7 & 3 \end{pmatrix},$$

(i) Find a basis for the column space.

(ii) Find a subset of the rows of $\mathbf{A}$ forming a basis for the row space of $\mathbf{A}$.

(iii) Find the null space of $\mathbf{A}$.

(iv) If a solution of the system

$$\mathbf{Av} = \mathbf{b}, \ \mathbf{b} = \begin{pmatrix} 5 \\ 5 \\ 15 \end{pmatrix}$$

is

$$\mathbf{x} = \begin{pmatrix} 1 \\ 1 \\ 1 \\ 1 \\ 1 \end{pmatrix},$$

write the general solution.

7. Given

$$\mathbf{A} = \begin{pmatrix} 1 & 2 & 3 \\ 3 & 2 & 1 \\ 0 & 1 & 1 \end{pmatrix},$$

(i) Find the elementary matrices $\mathbf{E}_1$ and $\mathbf{E}_2$ such that

$$\mathbf{E}_2\mathbf{E}_1\mathbf{A}$$

is upper triangular, that is, in row-echelon form.

(ii) Write $\mathbf{A}$ as the product of elementary matrices.

8. Let T be a linear transformation from $\mathbb{R}^3 \to \mathbb{R}^3$ that rotates a point counterclockwise about the z-axis by $30°$, followed by a reflection about the y-axis by $90°$. Find the standard matrix for T.

9. Consider the following linear system of equations:

$$y + az = 1,$$
$$x + y + z = b,$$
$$x - z = a.$$

Find the values of a and b such that
   (i) There is a unique solution.
   (ii) The system is inconsistent.
   (iii) There are infinitely many solutions.

10. Find the inverse of

$$\mathbf{A} = \begin{pmatrix} 1 & 0 & 2 \\ 1 & 1 & 3 \\ 1 & 1 & 0 \end{pmatrix}$$

by using the **Gauss-Jordan method**. Check your answer by using the adjoint matrix method.

11. Suppose the reduced row-echelon form of the matrix $\mathbf{A}$ is obtained as

$$\mathbf{R} = \begin{pmatrix} 0 & 1 & 7 & 0 & 0 \\ 0 & 0 & 0 & 1 & 0 \\ 0 & 0 & 0 & 0 & 1 \\ 0 & 0 & 0 & 0 & 0 \end{pmatrix}.$$

Find the null space of $\mathbf{A}$. What is the nullity of $\mathbf{A}$?

12. Given the matrix

$$\mathbf{A} = \begin{pmatrix} 2 & 1 & 3 & 2 \\ 4 & 1 & 8 & 5 \\ -2 & -3 & 1 & -2 \\ 8 & 2 & 16 & 16 \end{pmatrix},$$

(i) Find its rank.

(ii) What is the dimension of the null space, that is, the nullity of **A** ?

(iii) Is there a **b** such that the system **Ax** = **b** does not have a solution?

(iv) Find the nontrivial solutions for **Ax** = **0**, that is, the null space of **A**.

(v) Are the row vectors of **A** linearly independent?

13. Use the **triangulation method** to evaluate the determinant of

$$
\mathbf{A} = \begin{pmatrix} 1 & 1 & 1 & 1 & 1 \\ 1 & 2 & 2 & 2 & 2 \\ 1 & 2 & 3 & 3 & 3 \\ 1 & 2 & 3 & 4 & 4 \\ 1 & 2 & 3 & 4 & 5 \end{pmatrix}.
$$

Hint: Write **A** in row-echelon form, be careful which row operations you use and their effect on the value of the determinant.

CHAPTER 7

APPLICATIONS OF LINEAR ALGEBRA

Linear algebra has many important applications to real-life problems. In this chapter, we concentrate on some of its applications to science and engineering. We consider applications to chemistry and chemical engineering, linear programming, Leontief input-output model, geometry, elimination theory, coding theory and cryptography, and finally graph theory. Besides these, linear algebra also has interesting applications to image processing and computer graphics, networks, genetics, coupled linear oscillations, Markov chains, etc.

7.1 CHEMISTRY AND CHEMICAL ENGINEERING

In chemistry, we are often interested in finding a set of **independent reactions** among a given set of reactions.

Essentials of Mathematical Methods in Science and Engineering, Second Edition. Selçuk Ş. Bayın.
© 2020 John Wiley & Sons, Inc. Published 2020 by John Wiley & Sons, Inc.

7.1.1 Independent Reactions and Stoichiometric Matrix

In determining the composition of an equilibrium system that involves several reactions, it is important to find the **minimum** number of reactions that produces the same result. Consider the following set of reactions:

$$C + O_2 \rightarrow CO_2, \tag{7.1}$$

$$2C + O_2 \rightarrow 2CO, \tag{7.2}$$

$$3C + 2O_2 \rightarrow CO_2 + 2CO, \tag{7.3}$$

$$2CO + O_2 \rightarrow 2CO_2. \tag{7.4}$$

It is clear that the third reaction can be obtained by adding the first two reactions, while the sum of the second and the fourth reactions gives twice the first reaction. We now write the above set as

$$C + O_2 - CO_2 = 0, \tag{7.5}$$

$$2C + O_2 - 2CO = 0, \tag{7.6}$$

$$3C + 2O_2 - CO_2 - 2CO = 0, \tag{7.7}$$

$$2CO + O_2 - 2CO_2 = 0. \tag{7.8}$$

and call

$$A_1 = C, \ A_2 = O_2, \ A_3 = CO_2, \ A_4 = CO, \tag{7.9}$$

to write

$$A_1 + A_2 - A_3 = 0, \tag{7.10}$$

$$2A_1 + A_2 - 2A_4 = 0, \tag{7.11}$$

$$3A_1 + 2A_2 - A_3 - 2A_4 = 0, \tag{7.12}$$

$$A_2 - 2A_3 + 2A_4 = 0. \tag{7.13}$$

These equations constitute a linear system of homogeneous equations for the unknowns A_i, $i = 1, \ldots 4$, which can be written as the matrix equation

$$\mathbf{SA} = 0, \tag{7.14}$$

where

$$\mathbf{S} = \begin{pmatrix} 1 & 1 & -1 & 0 \\ 2 & 1 & 0 & -2 \\ 3 & 2 & -1 & -2 \\ 0 & 1 & -2 & 2 \end{pmatrix}, \ \mathbf{A} \begin{pmatrix} A_1 \\ A_2 \\ A_3 \\ A_4 \end{pmatrix}. \tag{7.15}$$

The matrix S is called the **stoichiometric matrix**. Employing the Gauss's row reduction to S, we obtain the following **reduced row-echelon** form (Chapter 6):

$$rref(S) = \begin{pmatrix} 1 & 1 & -1 & 0 \\ 0 & 1 & -2 & 2 \\ 0 & 0 & 0 & 0 \\ 0 & 0 & 0 & 0 \end{pmatrix}. \tag{7.16}$$

The number of linearly independent equations/reactions is equal to the **rank** or the number of the leading 1s in the reduced row-echelon form of the stoichiometric matrix, which in this case is 2. Other interesting examples of applications of linear algebra to chemistry and chemical engineering can be found in Tosun [1].

Example 7.1. *Independent reaction set:* Consider the following set of reactions suggested for the steam reacting with coal at high temperatures to form hydrogen, carbon monoxide, carbon dioxide, and methane:

$$C + 2H_2O \rightleftharpoons CO_2 + 2H_2, \tag{7.17}$$

$$C + H_2O \rightleftharpoons CO + H_2, \tag{7.18}$$

$$C + CO_2 \rightleftharpoons 2CO, \tag{7.19}$$

$$C + 2H_2 \rightleftharpoons CH_4, \tag{7.20}$$

$$CO + H_2O \rightleftharpoons CO_2 + H_2. \tag{7.21}$$

To find the set of independent reactions, we first let

$$A_1 = C, \ A_2 = H_2O, \ A_3 = CO_2, \ A_4 = H_2, \ A_5 = CO, \ A_6 = CH_4 \tag{7.22}$$

and express the given reactions as the following system of linear homogeneous equations:

$$-A_1 - 2A_2 + A_3 + 2A_4 = 0, \tag{7.23}$$

$$-A_1 - A_2 + A_4 + A_5 = 0, \tag{7.24}$$

$$-A_1 - A_3 + 2A_5 = 0, \tag{7.25}$$

$$-A_1 - 2A_4 + A_6 = 0, \tag{7.26}$$

$$-A_2 + A_3 + A_4 - A_5 = 0. \tag{7.27}$$

The corresponding stoichiometric matrix S is

$$S = \begin{pmatrix} -1 & -2 & 1 & 2 & 0 & 0 \\ -1 & -1 & 0 & 1 & 1 & 0 \\ -1 & 0 & -1 & 0 & 2 & 0 \\ -1 & 0 & 0 & -2 & 0 & 1 \\ 0 & -1 & 1 & 1 & -1 & 0 \end{pmatrix}, \tag{7.28}$$

the **reduced row-echelon** form of which is

$$rref(\mathbf{S}) = \begin{pmatrix} 1 & 0 & 0 & 2 & 0 & -1 \\ 0 & 1 & 0 & -3 & -1 & 1 \\ 0 & 0 & 1 & -2 & -2 & 1 \\ 0 & 0 & 0 & 0 & 0 & 0 \\ 0 & 0 & 0 & 0 & 0 & 0 \end{pmatrix}. \tag{7.29}$$

The number of nonzero rows indicate that there are three independent reactions. To find an independent set of reactions is equivalent to finding a basis for the row space of $\mathbf{S}$ in terms of the row vectors of $\mathbf{S}$. Thus, we first take the transpose of $\mathbf{S}$:

$$\widetilde{\mathbf{S}} = \begin{pmatrix} -1 & -1 & -1 & -1 & 0 \\ -2 & -1 & 0 & 0 & -1 \\ 1 & 0 & -1 & 0 & 1 \\ 2 & 1 & 0 & -2 & 1 \\ 0 & 1 & 2 & 0 & -1 \\ 0 & 0 & 0 & 1 & 0 \end{pmatrix} \tag{7.30}$$

and then find its reduced row-echelon form:

$$\begin{pmatrix} 1 & 1 & 1 & 1 & 0 \\ 0 & 1 & 2 & 2 & -1 \\ 0 & 0 & 0 & 1 & 0 \\ 0 & 0 & 0 & 0 & 0 \\ 0 & 0 & 0 & 0 & 0 \\ 0 & 0 & 0 & 0 & 0 \end{pmatrix}. \tag{7.31}$$

The leading ones appear in columns 1, 2, and 4. Hence, we write the corresponding columns of $\widetilde{\mathbf{S}}$:

$$\begin{pmatrix} -1 \\ -2 \\ 1 \\ 2 \\ 0 \\ 0 \end{pmatrix}, \begin{pmatrix} -1 \\ -1 \\ 0 \\ 1 \\ 1 \\ 0 \end{pmatrix}, \begin{pmatrix} -1 \\ 0 \\ 0 \\ -2 \\ 0 \\ 1 \end{pmatrix} \tag{7.32}$$

and then take their transpose to obtain the set of independent row vectors as

$$\begin{pmatrix} -1 & -2 & 1 & 2 & 0 & 0 \end{pmatrix},$$
$$\begin{pmatrix} -1 & -1 & 0 & 1 & 1 & 0 \end{pmatrix}, \tag{7.33}$$
$$\begin{pmatrix} -1 & 0 & 0 & -2 & 0 & 1 \end{pmatrix}.$$

These correspond to the reactions

$$C + 2H_2O \rightleftharpoons CO_2 + 2H_2, \qquad (7.34)$$

$$C + H_2O \rightleftharpoons H_2 + CO, \qquad (7.35)$$

$$C + 2H_2 \rightleftharpoons CH_4, \qquad (7.36)$$

which are sufficient to describe the process represented by the five reactions in Eqs. (7.17)–(7.21). Note that since there are only three leading ones, the rank of the stoichiometric metric is three.

7.1.2 Independent Reactions from a Set of Species

In the previous section, we have seen how to find a set of independent reactions among a given set of possible reactions. We now address a different problem, where we aim to find a set of independent reactions among a set of given **chemical species**. In this process, the key concepts are the **conservation of mass** and the **conservation of the elements** involved in a reaction. In a chemical reaction, the number of molecules change, hence, the number of molecules is not conserved.

Consider the complete combustion of methane to carbon dioxide and water vapor. Actors of this reaction are

$$CH_4, \ O_2, \ CO_2, \ H_2O. \qquad (7.37)$$

The basic reaction is of the form

$$CH_4 + O_2 \rightarrow CO_2 + H_2O, \qquad (7.38)$$

where appropriate coefficients (amounts) of the chemical species have to be found. Conservation of mass allows us to write

$$\alpha_1 A_1 + \alpha_2 A_2 + \alpha_3 A_3 + \alpha_4 A_4 = 0, \qquad (7.39)$$

where we have substituted

$$A_1 = CH_4, \ A_2 = O_2, \ A_3 = CO_2, \ A_4 = H_2O. \qquad (7.40)$$

To find the coefficients, we first use the conservation of elements to write

C balance $1\alpha_1 + 0\alpha_2 + 1\alpha_3 + 0\alpha_4 = 0,$

H balance $4\alpha_1 + 0\alpha_2 + 0\alpha_3 + 2\alpha_4 = 0,$ (7.41)

O balance $0\alpha_1 + 2\alpha_2 + 2\alpha_3 + 1\alpha_4 = 0.$

This can be written as a matrix equation:

$$
\begin{array}{cccc}
\text{CH}_4 & \text{O}_2 & \text{CO}_2 & \text{H}_2\text{O}
\end{array}
$$

$$
\begin{array}{c} \text{C} \\ \text{H} \\ \text{O} \end{array}
\begin{pmatrix}
1 & 0 & 1 & 0 \\
4 & 0 & 0 & 2 \\
0 & 2 & 2 & 1
\end{pmatrix}
\begin{pmatrix}
\alpha_1 \\ \alpha_2 \\ \alpha_3 \\ \alpha_4
\end{pmatrix} = 0. \tag{7.42}
$$

The matrix of coefficients:

$$
\begin{array}{cccc}
\text{CH}_4 & \text{O}_2 & \text{CO}_2 & \text{H}_2\text{O}
\end{array}
$$

$$
\begin{array}{c} \text{C} \\ \text{H} \\ \text{O} \end{array}
\begin{pmatrix}
1 & 0 & 1 & 0 \\
4 & 0 & 0 & 2 \\
0 & 2 & 2 & 1
\end{pmatrix}, \tag{7.43}
$$

is also called the **element-species** matrix [1]. Reducing the coefficient matrix yields

$$
\begin{array}{cccc}
\text{CH}_4 & \text{O}_2 & \text{CO}_2 & \text{H}_2\text{O}
\end{array}
$$

$$
\begin{array}{c} \text{C} \\ \text{H} \\ \text{O} \end{array}
\begin{pmatrix}
1 & 0 & 1 & 0 \\
0 & 1 & 1 & 1/2 \\
0 & 0 & 1 & -1/2
\end{pmatrix}
\begin{pmatrix}
\alpha_1 \\ \alpha_2 \\ \alpha_3 \\ \alpha_4
\end{pmatrix} = 0, \tag{7.44}
$$

which corresponds to the following linear system of equations:

$$
\alpha_1 + \alpha_3 = 0, \tag{7.45}
$$

$$
\alpha_2 + \alpha_3 + \frac{1}{2}\alpha_4 = 0, \tag{7.46}
$$

$$
\alpha_3 - \frac{1}{2}\alpha_4 = 0, \tag{7.47}
$$

where we have three homogeneous equations in four unknowns. Choosing α_4 as a parameter, we can write the remaining variables as

$$
\alpha_1 = -\alpha_4/2, \tag{7.48}
$$

$$
\alpha_2 = -\alpha_4, \tag{7.49}
$$

$$
\alpha_3 = \alpha_4/2, \tag{7.50}
$$

thus obtaining the solution

$$
\begin{pmatrix}
\alpha_1 \\ \alpha_2 \\ \alpha_3 \\ \alpha_4
\end{pmatrix} = \alpha_4
\begin{pmatrix}
-1/2 \\ -1 \\ 1/2 \\ 1
\end{pmatrix}. \tag{7.51}
$$

Using Eq. (7.39):

$$(A_1 \quad A_2 \quad A_3 \quad A_4) \begin{pmatrix} \alpha_1 \\ \alpha_2 \\ \alpha_3 \\ \alpha_4 \end{pmatrix} = 0, \tag{7.52}$$

that is,

$$(CH_4 \quad O_2 \quad CO_2 \quad H_2O) \begin{pmatrix} -1/2 \\ -1 \\ 1/2 \\ 1 \end{pmatrix} = 0, \tag{7.53}$$

we finally obtain the reaction as

$$-\tfrac{1}{2}CH_4 \quad -O_2 \quad +\tfrac{1}{2}CO_2 \quad H_2O = 0, \tag{7.54}$$

or as

$$CH_4 + 2O_2 \rightarrow CO_2 + 2H_2O. \tag{7.55}$$

Example 7.2. _Independent reactions:_ Let us now consider another example, where carbon monoxide is hydrogenized to produce methanol (CH_3OH). In this process, carbon monoxide reacts with hydrogen to produce methanol. However, there is also a side reaction producing water vapor and methane. Actors in this process can be taken as [1]

$$\{(CO, \; H_2, \; H_2O, \; CH_4, \; CH_3OH), \; (C, H, O)\}, \tag{7.56}$$

where in the first parenthesis, we list the **chemical species**, while in the second parenthesis, we display the **chemical elements**. We now write the **element-species** matrix:

$$\begin{array}{c} \\ C \\ H \\ O \end{array} \begin{array}{ccccc} CO & H_2 & H_2O & CH_4 & CH_3OH \end{array} \\ \begin{pmatrix} 1 & 0 & 0 & 1 & 1 \\ 0 & 2 & 2 & 4 & 4 \\ 1 & 0 & 1 & 0 & 1 \end{pmatrix}, \tag{7.57}$$

which in reduced row-echelon form becomes

$$\begin{array}{c} \\ C \\ H \\ O \end{array} \begin{array}{ccccc} CO & H_2 & H_2O & CH_4 & CH_3OH \end{array} \\ \begin{pmatrix} 1 & 0 & 0 & 1 & 1 \\ 0 & 1 & 0 & 3 & 2 \\ 0 & 0 & 1 & -1 & 0 \end{pmatrix}. \tag{7.58}$$

This gives the following linear system of equations for α_i, $i = 1, 2, 3, 4$:

$$\alpha_1 + \alpha_4 + \alpha_5 = 0, \tag{7.59}$$

$$\alpha_2 + 3\alpha_4 + 2\alpha_5 = 0, \tag{7.60}$$

$$\alpha_3 - \alpha_4 = 0. \tag{7.61}$$

Since we have three equations and five unknowns, we choose α_4 and α_5 as the parameters of the solution and write the solution as

$$\begin{pmatrix} \alpha_1 \\ \alpha_2 \\ \alpha_3 \\ \alpha_4 \\ \alpha_5 \end{pmatrix} = \alpha_4 \begin{pmatrix} -1 \\ -3 \\ 1 \\ 1 \\ 0 \end{pmatrix} + \alpha_5 \begin{pmatrix} -1 \\ -2 \\ 0 \\ 0 \\ 1 \end{pmatrix}. \tag{7.62}$$

This yields the two reactions as

$$\begin{pmatrix} CO & H_2 & H_2O & CH_4 & CH_3OH \end{pmatrix} \begin{pmatrix} -1 \\ -3 \\ 1 \\ 1 \\ 0 \end{pmatrix} = 0 \tag{7.63}$$

and

$$\begin{pmatrix} CO & H_2 & H_2O & CH_4 & CH_3OH \end{pmatrix} \begin{pmatrix} -1 \\ -2 \\ 0 \\ 0 \\ 1 \end{pmatrix} = 0, \tag{7.64}$$

or as

$$CO + 3H_2 \rightarrow H_2O + CH_4, \tag{7.65}$$

$$CO + 2H_2 \rightarrow CH_3OH. \tag{7.66}$$

7.2 LINEAR PROGRAMMING

Many real-life situations involving planning and decision-making require optimization, maximization, or minimization of a **target function** or the **objective function** that represents the gain or the loss in the process. Linear programming is a very useful tool in solving such problems, which are frequently encountered in industry and science. In linear programming, the objective function is a linear expression of a number of variables that needs to be optimized under a certain number of constraints expressed in general as inequalities.

7.2.1   The Geometric Method

The general problem involves a linear **objective function** r, which in two variables can be expressed as

$$r = ax_1 + bx_2 \tag{7.67}$$

that has to be optimized subject to the following **linear constraints**:

$$
\begin{aligned}
&a_1 x_1 + b_1 x_2 (\geqslant, \leqslant, =) c_1, \\
&a_2 x_1 + b_2 x_2 (\geqslant, \leqslant, =) c_2, \\
&\quad \vdots \\
&a_n x_1 + b_n x_2 (\geqslant, \leqslant, =) c_n
\end{aligned}
\tag{7.68}
$$

and

$$
\begin{aligned}
&x_1 \geqslant 0, \\
&x_2 \geqslant 0.
\end{aligned}
\tag{7.69}
$$

In the constraints [Eq. (7.68)], any one of the symbols $(\geqslant, \leqslant, =)$ can be used. If the objective function satisfies all the constraints, then it is called a **feasible solution**. The set of all feasible solutions define the **feasible region** in the $x_1 x_2$-plane. Each constraint in the form $ax_1 + bx_2 = c$ is a straight-line in the $x_1 x_2$-plane, thus each constraint of the form $ax_1 + bx_2 (\geqslant, \leqslant) c$ defines a half plane including the boundary. In this regard, the feasible region, which could be **bounded** or **unbounded**, is the intersection of these half planes (Figure 7.1). If the feasible region contains no points, then the constraints are **inconsistent**, and the problem has **no solution**. The crucial point of the solution is to identify the points in the feasible region that extremizes the objective function. For this, the following theorem plays a key role [2].

Theorem 7.1. In linear programming, if the feasible region is bounded, then the objective function attains both a **maximum** and a **minimum** value, and these occur at the **extreme points** of the feasible region. If the feasible

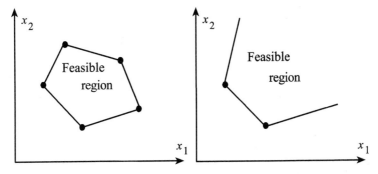

Figure 7.1   Bounded and unbounded regions and the extreme points.

region is unbounded, then the objective function may or may not possess a maximum or a minimum value. But if it does, then these values occur at the extreme points of the feasible region.

We now demonstrate the method through a number of specific examples.

Example 7.3. *Geometric method:* A merchant has 150 units of product 1 and 70 units of product 2. He decides to market them as two mixtures, say A and B. A unit of A contains 1/3 product 1 and 2/3 product 2 and will sell for \$5 per unit, while the mixture B contains 2/3 product 1 and 1/3 product 2 and will sell for \$8 per unit. The problem is how many units of each type should this merchant prepare to maximize the revenue.

Let x be the units of A to be prepared and y be the units of B to be prepared so that the **revenue function** r:

$$r = 5x + 8y, \tag{7.70}$$

is a **maximum**. Since the total units of product 1 cannot exceed 150, can write

$$\frac{1}{3}x + \frac{2}{3}y \leqslant 150. \tag{7.71}$$

Similarly, the total number of units of product 2 cannot exceed 70; hence, we write

$$\frac{2}{3}x + \frac{1}{3}y \leqslant 70. \tag{7.72}$$

Since x and y cannot be negative, we write the complete set of constraints as

$$\frac{1}{3}x + \frac{2}{3}y \leqslant 150, \tag{7.73}$$

$$\frac{2}{3}x + \frac{1}{3}y \leqslant 70, \tag{7.74}$$

$$x \geqslant 0, \tag{7.75}$$

$$y \geqslant 0. \tag{7.76}$$

Now the feasible region defined by these constraints is bounded and is the triangular region defined by the points 1–3 in Figure 7.2. Using Theorem (7.1), we check the revenue at the extreme points (vertices) to find where the **maximum** is attained:

Pt. 1 $r_1 = 5(0) + 8(210) = \$1680$,

Pt. 2 $r_2 = 5(105) + 8(0) = \$525$,   (7.77)

Pt. 3 $r_3 = 5(0) + 8(0) = \$0$.

Since the maximum revenue is obtained at point 1, this merchant should sell all of its goods as mixture B composed of 2/3 product 1 and 1/3 product 2 for the maximum profit of \$1680.

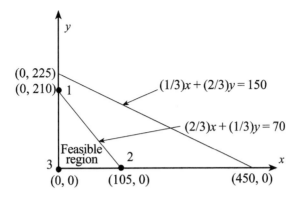

Figure 7.2 Extreme points in Case 1.

Example 7.4. **_Geometric method:_** Let us now change the numbers in Example 7.3 so that the merchant has 100 units of product 1 and 180 units of product 2. Let a unit of A be 1/3 product 1 and 2/3 product 2 which will sell for $6.5 per unit, while a unit of B consists of 2/5 product 1 and 3/5 product 2, which will sell for $7 per unit.

Now the revenue function is written as

$$r = 6.5x + 7y, \tag{7.78}$$

where the constraints become

$$\frac{1}{3}x + \frac{2}{5}y \leqslant 100, \tag{7.79}$$

$$\frac{2}{3}x + \frac{3}{5}y \leqslant 180, \tag{7.80}$$

$$x \geqslant 0, \tag{7.81}$$

$$y \geqslant 0. \tag{7.82}$$

In this case, the feasible region is still bounded and defined by the trapezoid with the vertices at points 1–4 (Figure 7.3). Checking the values of the revenue function at the extreme points:

$$
\begin{aligned}
Pt.\ 1 \quad & r_1 = 6.5(0) + 7(250) = \$1750, \\
Pt.\ 2 \quad & r_2 = 6.5(180) + 7(100) = \$1870, \\
Pt.\ 3 \quad & r_3 = 6.5(270) + 6(0) = \$1755, \\
Pt.\ 4 \quad & r_4 = 5(0) + 6(0) = \$0,
\end{aligned}
\tag{7.83}
$$

we see that the **maximum revenue** is obtained when the merchant markets its products as 180 units of mixture A and 100 units of mixture B.

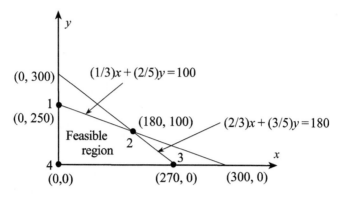

Figure 7.3 Extreme points of the feasible region in Case 2.

Example 7.5. *Geometric method:* An auto parts producer has two factories. The operation cost of factory 1 is $8000 for a week and on the average can produce 5 parts of type 1 and 6 parts of type 2 per week. The second factory costs $10 000 to operate for a week and produces 4 parts of type 1 and 10 parts of type 2 per week. The producer commits itself to deliver 150 type 1 parts and 250 type 2 parts for a year.

The problem is how many weeks should this producer run each of its factories to minimize costs and still meet its commitment. To minimize the operating cost of two factories, the producer has to look for the **minimum** of the **cost function** c:

$$c = 8000x + 10\ 000y, \tag{7.84}$$

where x and y stand for the number of weeks that each factory will be working, respectively, to meet the demand. The constraints can now be written as

$$5x + 4y \geqslant 150, \tag{7.85}$$

$$6x + 10y \geqslant 250, \tag{7.86}$$

$$x \geqslant 0, \tag{7.87}$$

$$y \geqslant 0. \tag{7.88}$$

The feasible region is the open region shown in Figure 7.4. Checking the cost function at the extreme points:

Pt. 1 $r_1 = 8000(0) + 10\ 000(37.5) = \$375\ 000,$

Pt. 2 $r_2 = 8000(19.2) + 10\ 000(13.5) = \$288\ 600,$ (7.89)

Pt. 3 $r_3 = 8000(41.7) + 10\ 000(0) = \$333\ 600,$

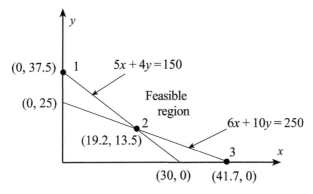

Figure 7.4   Extreme points in Case 3.

we see that to **minimize** the cost of operation, this producer should commit factory 1 for 19.2 weeks and factory 2 for 13.5 weeks to produce the promised parts of types 1 and 2. Note that the region is unbounded, hence, we take the other extreme points at infinity. Since we are looking for a minimum, these points are not going to be optimal.

7.2.2   The Simplex Method

The **geometric method** works in two dimensions, but as the number of variables increases, it becomes rather cumbersome. In 1947, Dantzig introduced an elegant algorithm known as the **simplex method**. It is a simple and an efficient method of solving linear programs encountered in planning and decision-making processes. The goal of the simplex method is to find the optimum, maximum, or minimum, value of a linear objective function under a system of linear constraints expressed in general as inequalities. In n dimensions, the system of linear inequalities define a region, a convex polytope, as a feasible region for the optimum value. In two dimensions, starting from a vertex, the algorithm moves along the edges until the vertex with the optimum value is reached.

The general form of a linear programming problem is to **maximize** or **minimize** the **objective function** r:

$$r = c_1 x_1 + c_2 x_2 + \cdots + c_n x_n, \tag{7.90}$$

subject to the **constraints**

$$a_{11} x_1 + a_{12} x_2 + \cdots + a_{1n} x_n (\leqslant, =, \geqslant) b_1, \tag{7.91}$$

$$a_{21} x_1 + a_{22} x_2 + \cdots + a_{2n} x_n (\leqslant, =, \geqslant) b_2, \tag{7.92}$$

$$\vdots \tag{7.93}$$

$$a_{m1} x_1 + a_{m2} x_2 + \cdots + a_{mn} x_n (\leqslant, =, \geqslant) b_m, \tag{7.94}$$

where the constants a_{ij}, b_i, and c_j are real. The constants c_j are basically the **cost** or the **price** coefficients of the unknowns x_i. When all the inequalities in the constraints are $\leqslant$ and the unknowns satisfy

$$x_i \geqslant 0, \quad i = 1, \ldots, n, \tag{7.95}$$

we call the form of the problem **canonical**. Furthermore, if

$$b_j \geqslant 0, \quad j = 1, \ldots, m, \tag{7.96}$$

we call the form **feasible canonical**.

For the **simplex method** to be applicable, we have to convert the problem into the **standard form**, where we seek the **maximum** of the objective function:

$$r = c_1 x_1 + c_2 x_2 + \cdots + c_n x_n, \tag{7.97}$$

subject to the constraints

$$a_{11} x_1 + a_{12} x_2 + \cdots + a_{1n} x_n = b_1, \tag{7.98}$$

$$a_{21} x_1 + a_{22} x_2 + \cdots + a_{2n} x_n = b_2, \tag{7.99}$$

$$\vdots \tag{7.100}$$

$$a_{m1} x_1 + a_{m2} x_2 + \cdots + a_{mn} x_n = b_m, \tag{7.101}$$

$$x_i \geqslant 0, \quad i = 1, \ldots, n, \tag{7.102}$$

where b_j are also assumed to be nonnegative:

$$b_j \geqslant 0, \ j = 1, \ldots, m. \tag{7.103}$$

One can always convert a linear programming problem into **canonical** or the **standard** form by using the following procedures:

(i) If the original problem demands the minimization of

$$r = c_1 x_1 + c_2 x_2 + \cdots + c_n x_n, \tag{7.104}$$

we can always maximize the equivalent objective function

$$r' = (-c_1) x_1 + (-c_2) x_2 + \cdots + (-c_n) x_n = -r. \tag{7.105}$$

(ii) If a variable, x_j, is not restricted to nonnegative values, we can replace it with

$$x_j = x_j^+ - x_j^-, \tag{7.106}$$

where the first column numbers the equations. This is called the initial **simplex tableau**, where $\langle .. \rangle$ stands for the **slack variables**, which are also called the **basic variables**. The corresponding columns are called the **basic columns**. Note that in the basic columns, all the entries are zeros except the one that is equal to one. This is just like the augmented matrix in Gauss-Jordan elimination method with the exception that only the **feasible solutions** are allowed. An obvious solution is when all the nonbasic variables are zero:

$$x_1 = 0, \ x_2 = 0, \ x_3 = 0, \tag{7.131}$$

and the basic variables take the values

$$x_4 = 5, \ x_5 = 7, \ x_6 = 25, \tag{7.132}$$

with $r = 0$. In other words, $(x_1, x_2, x_3, x_4, x_5, x_6) = (0,0,0,5,7,25)$ is a feasible solution yielding 0 for the objective function. This is called the **basic feasible solution**, which is zero in this case. Our aim is to find an **optimal feasible solution**, that is, the one that is **maximum**. The key to this process is the following theorem.

Theorem 7.2. If a standard linear programming problem has a solution, then there is a basic feasible solution that yields the maximum value of the objective function. Such feasible solutions are called **optimal** [2].

Since in Eq. (7.116), the coefficients of x_1 and x_2 are positive, by increasing x_1 or x_2 we can improve r. However, it is better to increase x_2 since its coefficient is greater. Thus, we modify the tableau so that x_2 is the new basic variable. Now, x_2 is also called the **entering variable** and its column is called the **pivot column**:

$$\begin{bmatrix} & x_1 & x_2 & x_3 & \langle x_4 \rangle & \langle x_5 \rangle & \langle x_6 \rangle & r & - \\ 1 & 2 & [1] & 1 & 1 & 0 & 0 & 0 & 5 \\ 2 & 1 & -2 & 2 & 0 & 1 & 0 & 0 & 7 \\ 3 & 3 & 4 & 6 & 0 & 0 & 1 & 0 & 25 \\ 4 & -1 & -2 & 3 & 0 & 0 & 0 & 1 & 0 \end{bmatrix} \quad \begin{matrix} - \\ 5/1 \\ - \\ 25/4 \\ - \end{matrix}, \tag{7.133}$$

The next crucial step in the algorithm is to convert the pivot column into a **basic column**. That is, all zeros except one, which will be set to 1. But which one? For this, we need to choose a **pivot**. We do not choose the last row, since we do not want to disturb r. However, any of the other rows with **positive terms** can be chosen. Since the pivot has to be positive, we choose the row that produces the smallest number when divided into the entry in the last column [Eq. (7.133)]. In this case, the first entry in the pivot column produces the smallest number, $5/1 = 5$, compared to the entry in the third row, $25/4 = 6.3$, thus, the first term is our pivot.

We show the pivot as [..]. Since the pivot is already one, we do not need to normalize it to one.

Using the first row, we perform Gauss-Jordan row reduction to obtain

$$
\begin{bmatrix}
 & x_1 & \langle x_2 \rangle & x_3 & x_4 & \langle x_5 \rangle & \langle x_6 \rangle & r & - \\
1 & 2 & [1] & 1 & 1 & 0 & 0 & 0 & 5 \\
2 & 5 & 0 & 4 & 2 & 1 & 0 & 0 & 17 \\
3 & -5 & 0 & 2 & -4 & 0 & 1 & 0 & 5 \\
4 & 3 & 0 & 5 & 2 & 0 & 0 & 1 & 10
\end{bmatrix}. \qquad (7.134)
$$

Note that x_4 is no longer a basic variable; hence, it is called the **departing variable**. Setting the non **basic variables**, x_1, x_3, x_4, to zero, the new feasible solution is obtained as

$$ x_2 = 5, \ x_5 = 17, \ x_6 = 5; \ r = 10. \qquad (7.135) $$

Obviously, $r = 10$ is better than zero. The new last row, that is, the new objective function, becomes

$$ r = 10 - 3x_1 - 5x_3 - 2x_4. \qquad (7.136) $$

Since all the coefficients of the nonbasic variables, x_1, x_3, x_4, are negative, we can no longer improve r; hence, we claim that $r = 10$ is the optimal maximum solution with $(0, 5, 0, 0, 17, 5)$.

Example 7.7. *Simplex method:* Let us now maximize the objective function

$$ r = 3x_1 + 2x_2 - 3x_3, \qquad (7.137) $$

subject to the constraints

$$ 2x_1 + x_2 + x_3 \leqslant 6, \qquad (7.138) $$

$$ x_1 - 2x_2 + 2x_3 \leqslant 8, \qquad (7.139) $$

$$ 3x_1 + 4x_2 + 6x_3 \leqslant 27, \qquad (7.140) $$

$$ x_1, x_2, x_3 \geqslant 0. \qquad (7.141) $$

Introducing slack variables, x_4, x_5, x_6, we write r:

$$ r = 3x_1 + 2x_2 - 3x_3 + 0x_4 + 0x_5 + 0x_6, \qquad (7.142) $$

and the constraints as

$$ 2x_1 + x_2 + x_3 + x_4 + 0x_5 + 0x_6 = 6, \qquad (7.143) $$

$$x_1 - 2x_2 + 2x_3 + 0x_4 + x_5 + 0x_6 = 8, \qquad (7.144)$$

$$3x_1 + 4x_2 + 6x_3 + 0x_4 + 0x_5 + x_6 = 27, \qquad (7.145)$$

$$x_1, x_2, x_3, x_4, x_5, x_6 \geqslant 0. \qquad (7.146)$$

Using the slack variables, we write r as the fourth equation:

$$2x_1 + x_2 + x_3 + x_4 + 0x_5 + 0x_6 = 6, \qquad (7.147)$$

$$x_1 - 2x_2 + 2x_3 + 0x_4 + x_5 + 0x_6 = 8, \qquad (7.148)$$

$$3x_1 + 4x_2 + 6x_3 + 0x_4 + 0x_5 + x_6 = 27, \qquad (7.149)$$

$$-3x_1 - 2x_2 + 3x_3 + 0x_4 + 0x_5 + 0x_6 + r = 0. \qquad (7.150)$$

The augmented matrix of this system is now written as

$$\begin{bmatrix} & x_1 & x_2 & x_3 & \langle x_4 \rangle & \langle x_5 \rangle & \langle x_6 \rangle & r & - \\ 1 & 2 & 1 & 1 & 1 & 0 & 0 & 0 & 6 \\ 2 & 1 & -2 & 2 & 0 & 1 & 0 & 0 & 8 \\ 3 & 3 & 4 & 6 & 0 & 0 & 1 & 0 & 27 \\ 4 & -3 & -2 & 3 & 0 & 0 & 0 & 1 & 0 \end{bmatrix}. \qquad (7.151)$$

From Eq. (7.137), since the coefficient of x_1 is larger than x_2, we can do better by increasing x_1, thus, we choose the pivot column as the first column and since the pivot must be positive with the smallest ratio, we write

$$\begin{bmatrix} & x_1 & x_2 & x_3 & \langle x_4 \rangle & \langle x_5 \rangle & \langle x_6 \rangle & r & - \\ 1 & [2] & 1 & 1 & 1 & 0 & 0 & 0 & 6 \\ 2 & 1 & -2 & 2 & 0 & 1 & 0 & 0 & 8 \\ 3 & 3 & 4 & 6 & 0 & 0 & 1 & 0 & 27 \\ 4 & -3 & -2 & 3 & 0 & 0 & 0 & 1 & 0 \end{bmatrix} \begin{array}{l} - \\ 6/2 = 3 \\ 8/1 = 8 \\ 27/3 = 9 \\ - \end{array}, \qquad (7.152)$$

and use the first entry shown as [..] as our pivot. Dividing the first row with 2, we normalize the pivot to 1 to write

$$\begin{bmatrix} & \langle x_1 \rangle & x_2 & x_3 & x_4 & \langle x_5 \rangle & \langle x_6 \rangle & r & - \\ 1 & 1 & 1/2 & 1/2 & 1/2 & 0 & 0 & 0 & 3 \\ 2 & 1 & -2 & 2 & 0 & 1 & 0 & 0 & 8 \\ 3 & 3 & 4 & 6 & 0 & 0 & 1 & 0 & 27 \\ 4 & -3 & -2 & 3 & 0 & 0 & 0 & 1 & 0 \end{bmatrix}. \qquad (7.153)$$

Using the first row, we perform row reduction to obtain

$$
\begin{bmatrix}
 & \langle x_1 \rangle & x_2 & x_3 & x_4 & \langle x_5 \rangle & \langle x_6 \rangle & r & - \\
1 & 1 & 1 & 1/2 & 1/2 & 1/2 & 0 & 0 & 0 & 3 \\
2 & 0 & -5/2 & 3/2 & -1/2 & 1 & 0 & 0 & 5 \\
3 & 0 & 5/2 & 9/2 & -3/2 & 0 & 1 & 0 & 18 \\
4 & 0 & -1/2 & 9/2 & 3/2 & 0 & 0 & 1 & 9
\end{bmatrix}. \tag{7.154}
$$

The new objective function becomes

$$
r = 9 + \frac{1}{2}x_2 - \frac{9}{2}x_3 - \frac{3}{2}x_4, \tag{7.155}
$$

where 9 is an improvement over 0, but we can still improve on this by increasing x_2. Seeking for the smallest ratio among the positive terms of the new basic column corresponding to x_2

$$
\begin{bmatrix}
 & \langle x_1 \rangle & x_2 & x_3 & x_4 & \langle x_5 \rangle & \langle x_6 \rangle & r & - \\
1 & 1 & 1/2 & 1/2 & 1/2 & 0 & 0 & 0 & 3 \\
2 & 0 & -5/2 & 3/2 & -1/2 & 1 & 0 & 0 & 5 \\
3 & 0 & 5/2 & 9/2 & -3/2 & 0 & 1 & 0 & 18 \\
4 & 0 & -1/2 & 4 & 1 & 0 & 0 & 1 & 6
\end{bmatrix}
\quad
\begin{matrix}
- \\
3/(1/2) = 6 \\
- \\
18/(5/2) = 7.2 \\
-
\end{matrix}, \tag{7.156}
$$

we choose the pivot term as $1/2$ and normalize the pivot equation by multiplying with 2:

$$
\begin{bmatrix}
 & x_1 & \langle x_2 \rangle & x_3 & x_4 & \langle x_5 \rangle & \langle x_6 \rangle & r & - \\
1 & 2 & [1] & 1 & 1 & 0 & 0 & 0 & 6 \\
2 & 0 & -5/2 & 3/2 & -1/2 & 1 & 0 & 0 & 5 \\
3 & 0 & 5/2 & 9/2 & -3/2 & 0 & 1 & 0 & 18 \\
4 & 0 & -1/2 & 9/2 & 3/2 & 0 & 0 & 1 & 9
\end{bmatrix}. \tag{7.157}
$$

Performing row reduction, we obtain

$$
\begin{bmatrix}
 & x_1 & \langle x_2 \rangle & x_3 & x_4 & \langle x_5 \rangle & \langle x_6 \rangle & r & - \\
1 & 2 & 1 & 1 & 1 & 0 & 0 & 0 & 6 \\
2 & 5 & 0 & 4 & 2 & 1 & 0 & 0 & 20 \\
3 & -5 & 0 & 2 & -4 & 0 & 1 & 0 & 3 \\
4 & 1 & 0 & 5 & 2 & 0 & 0 & 1 & 12
\end{bmatrix}. \tag{7.158}
$$

Now the last equation gives the new objective function as

$$
r = 12 - x_1 - 5x_3 - 2x_4. \tag{7.159}
$$

Since 12 is better than 9 and since this cannot be improved any further, $(0, 6, 0, 0, 20, 3)$ represents the optimum feasible solution with $r = 12$ as the maximum value.

Since the simplex method runs along the polytope edges, as the number of edges increase convergence becomes an issue. In this regard, a more efficient method called the **interior point method** was introduced by Karmarker in 1987, which goes through the interior points of the feasible region [5].

7.3 LEONTIEF INPUT–OUTPUT MODEL OF ECONOMY

In 1930s, Leontief divided the US economy into 500 sectors and introduced an economic model known as the **input–output** model. Leontief aimed to predict the proper output levels of each sector so that the total demand is met within the economy. In this model, to produce something, each sector consumes some of its own output as well as the outputs of the other sectors. The main objective is to balance the total amount of goods produced by the economy with the total demand. With Leontief's model, it was possible to determine the proper production levels of each sector so that the total demand from the other sectors within the economy as well as the outside demand are met. It also became possible to assess whether an economy is productive or nonproductive. Leontief's model was a great step in modelling economies using real data. Over the years, this model has been used by various institutions and governments for economic production planning. For his contributions, Leontief was awarded the 1973 Nobel Prize in economy [6].

7.3.1 Leontief Closed Model

Assume that an economy consists of n independent sectors:

$$S_1, S_2, \ldots, S_n, \tag{7.160}$$

where to produce, each sector consumes some of the outputs of the other sectors as well as some of its own. If no other products leave or enter into this system, we say it is a **closed economy**. To model such an economy, let m_{ij} be the number of units produced by the sector S_i and necessary to produce one unit of sector S_j. If p_j is the **production level** of the sector S_j, then $m_{ij}p_j$ represents the number of units produced by the sector S_i and consumed by the sector S_j. In a closed system, since all the goods or services produced by the sector S_i are consumed by the other sectors, the total number of units produced by the sector S_i is the sum

$$m_{i1}p_1 + m_{i2}p_2 + \cdots + m_{in}p_n, \tag{7.161}$$

which is naturally equal the production level p_i of the sector S_i:

$$p_i = m_{i1}p_1 + m_{i2}p_2 + \cdots + m_{in}p_n, \tag{7.162}$$

which leads to the following linear system of equations:

$$p_1 = m_{11}p_1 + m_{12}p_2 + \cdots + m_{1n}p_n, \tag{7.163}$$

$$p_2 = m_{21}p_1 + m_{22}p_2 + \cdots + m_{2n}p_n, \tag{7.164}$$

$$\vdots \tag{7.165}$$

$$p_n = m_{n1}p_1 + m_{n2}p_2 + \cdots + m_{nn}p_n, \tag{7.166}$$

In matrix form, we can write this system as

$$\boxed{\mathbf{p} = \mathbf{Ap},} \tag{7.167}$$

where $\mathbf{A}$:

$$\mathbf{A} = \begin{pmatrix} m_{11} & m_{12} & \cdots & m_{1n} \\ m_{21} & m_{22} & \cdots & m_{2n} \\ \cdots & \cdots & \ddots & \cdots \\ m_{n1} & m_{n2} & \cdots & m_{nn} \end{pmatrix}, \tag{7.168}$$

is called the **input–output matrix** and $\mathbf{p}$:

$$\mathbf{p} = \begin{pmatrix} p_1 \\ p_2 \\ \vdots \\ p_n \end{pmatrix}, \tag{7.169}$$

is the **production vector**. The purpose of the model is to find the vector $\mathbf{p}$ that satisfies Eq. (7.167) with **nonnegative** components, where at least one component is positive. In Leontief closed models, there is no external demand, hence, whatever is produced by the economy is consumed within the economy. In this regard, each column in the input–output matrix has to add up to 1. In Eq. (7.167), we are basically looking for a vector $\mathbf{p}$ that remains unchanged when multiplied with $\mathbf{A}$. In order for an economy to balance, that is, production match consumption, we need to guarantee that a steady state exists. In fact, when $\mathbf{A}$ is a Markov matrix, we can guarantee that a solution exists. An $n \times n$ matrix that has all nonnegative entries and the sum of each column vector is equal to 1 is called a **Markov matrix**. In an eigenvalue problem:

$$\boxed{\mathbf{Ax} = \lambda \mathbf{x},} \tag{7.170}$$

when $\mathbf{A}$ is a Markov matrix, there is always an eigenvalue equal to 1 and all the other eigenvalues are smaller or equal to 1.

To clarify these points, let us now assume that the economy of a certain region depends on three sectors, I, II, and III. Over a period of a year, it is observed that

(1) To produce 1 unit of its product, sector I consumes 0.4 units of its own production, 0.3 units of sector II's product and 0.4 units of sector III's product.

(2) To produce 1 unit of its product, sector II uses 0.4 units of sector I's product, 0.2 units of its own production and 0.4 units of sector III' s product.

(3) During this period, sector III uses 0.2 units of sector I's product, 0.5 units of sector II's product and 0.2 units of its own production.

We can write the input–output matrix of this economy as

$$\mathbf{A} = \begin{pmatrix} 0.4 & 0.3 & 0.4 \\ 0.4 & 0.2 & 0.4 \\ 0.2 & 0.5 & 0.2 \end{pmatrix}. \tag{7.171}$$

Note that in a closed Leontief models, since no goods enter or leave the economy, each column of $\mathbf{A}$ adds up to 1. Since $\mathbf{p}$ satisfies

$$(\mathbf{A} - \mathbf{I})\mathbf{p} = \mathbf{0}, \tag{7.172}$$

$$\begin{pmatrix} -0.6 & 0.3 & 0.4 \\ 0.4 & -0.8 & 0.4 \\ 0.2 & 0.5 & -0.8 \end{pmatrix} \begin{pmatrix} p_1 \\ p_2 \\ p_n \end{pmatrix} = 0 \tag{7.173}$$

and $\det(\mathbf{A} - \mathbf{I}) = 0$, this homogeneous system of linear equations has infinitely many solutions. Writing the corresponding augmented matrix (Chapter 6):

$$\tilde{\mathbf{A}} = \left(\begin{array}{ccc|c} -0.6 & 0.3 & 0.4 & 0 \\ 0.4 & -0.8 & 0.4 & 0 \\ 0.2 & 0.5 & -0.8 & 0 \end{array} \right) = 0, \tag{7.174}$$

which when reduced becomes

$$\left(\begin{array}{ccc|c} 1 & 0 & -0.11 & 0 \\ 0 & 1 & -1.2 & 0 \\ 0 & 0 & 0 & 0 \end{array} \right). \tag{7.175}$$

To write the final solution, we let

$$p_3 = t, \tag{7.176}$$

where t is a parameter, thus yielding the solution as

$$p_1 = 0.11t, \tag{7.177}$$

$$p_2 = 1.2t, \tag{7.178}$$

$$p_3 = t. \tag{7.179}$$

To make sense, p_1, p_2, p_3 must be nonnegative, hence, $t \geqslant 0$. Taking $t = 10$ units, we obtain the production levels for each sector so that the demands of the economy is met:

$$p_1 = 1.1 \text{ units}, \tag{7.180}$$

$$p_2 = 12 \text{ units}, \tag{7.181}$$

$$p_3 = 10 \text{ units}. \tag{7.182}$$

7.3.2 Leontief Open Model

In Leontief closed models, no goods enter or leave the system. However, we usually have outside demand. Let d_i represent the outside demand for the ith sector. We now modify Eq. (7.162) as

$$p_i = m_{i1}p_1 + m_{i2}p_2 + \cdots + m_{in}p_n + d_i, \ i = 1, \ldots, n. \tag{7.183}$$

This gives the following linear system of equations:

$$\boxed{\mathbf{p} = \mathbf{A}\mathbf{p} + \mathbf{d},} \tag{7.184}$$

where $\mathbf{A}$ and $\mathbf{p}$ are defined as in Eqs. (7.168) and (7.169), and

$$\boxed{\mathbf{d} = \begin{pmatrix} d_1 \\ d_2 \\ \vdots \\ d_n \end{pmatrix}} \tag{7.185}$$

is the **demand vector**. The solution of this linear system can be found by writing

$$\mathbf{p} = \mathbf{A}\mathbf{p} + \mathbf{d}, \tag{7.186}$$

$$(\mathbf{I} - \mathbf{A})\mathbf{p} = \mathbf{d}, \tag{7.187}$$

$$(\mathbf{I} - \mathbf{A})^{-1}(\mathbf{I} - \mathbf{A})\mathbf{p} = (\mathbf{I} - \mathbf{A})^{-1}\mathbf{d} \tag{7.188}$$

as

$$\boxed{\mathbf{p} = (\mathbf{I} - \mathbf{A})^{-1}\mathbf{d}.} \tag{7.189}$$

Of course, this assumes that the matrix $(\mathbf{I} - \mathbf{A})$ is invertible, which some-times may not be the case, and in such cases, it is impossible to solve for the production vector. For an acceptable solution, we need a positive production vector. Since the demand vector is always positive, we require that the product $(\mathbf{I} - \mathbf{A})^{-1}\,\mathbf{d}$ also be a positive vector. This means that $(\mathbf{I} - \mathbf{A})^{-1}$ is a **positive definite** matrix. Note that from basic economics, when the demand increases, $\Delta \mathbf{d} > 0$, a positive definite $(\mathbf{I} - \mathbf{A})^{-1}$ in Eq. (7.189) implies an increase in the total output, that is, $\Delta \mathbf{p} > 0$. In such cases, we call the economy **productive**. To check whether an economy is productive or not we can use the Hawkins–Simon condition.

Hawkins–Simon Condition

If all the leading principal minors of a symmetric matrix are positive, then the inverse exists and is nonnegative.

The kth leading principal minor of an $n \times n$ matrix is defined as the determinant defined by deleting the last $n - k$ rows and columns of the matrix. For a symmetric matrix to be positive definite, it is necessary and sufficient that all the leading principal minors, Δ_k, be positive. For a given $n \times n$ symmetric matrix:

$$\begin{pmatrix} a_{11} & a_{12} & a_{13} & \cdots & a_{1n} \\ a_{21} & a_{22} & a_{23} & \cdots & a_{2n} \\ \vdots & & \vdots & & \vdots \\ a_{n1} & a_{n2} & a_{n3} & \cdots & a_{nn} \end{pmatrix}, \tag{7.190}$$

the first, the second, and the third principal minors are defined, respectively, as

$$|a_{11}|, \quad \begin{vmatrix} a_{11} & a_{12} \\ a_{21} & a_{22} \end{vmatrix}, \quad \begin{vmatrix} a_{11} & a_{12} & a_{13} \\ a_{21} & a_{22} & a_{23} \\ a_{31} & a_{32} & a_{33} \end{vmatrix}. \tag{7.191}$$

The higher-order minors are defined similarly. In general, a positive definite matrix $\mathbf{A}$ satisfies $\mathbf{x}^T \mathbf{A} \mathbf{x} > 0$ for all $\mathbf{x} \neq 0$. Being a positive definite matrix, $(\mathbf{I} - \mathbf{A})^{-1}$ or $(\mathbf{I} - \mathbf{A})$, assures that economy can meet any given demand.

To demonstrate Leontief open models, consider that the economy of a certain region depends on three sectors: I, II, and III. Over a period of a year it is observed that

(1) To produce 1 unit of product, sector I must purchase 0.2 units of its own production, 0.35 units of sector II's product, and 0.15 units of sector III's product.

(2) To produce 1 unit of product, sector II must purchase 0.5 units of sector I's product, 0.2 units of its product, and 0.1 units of sector III's product.

(3) Finally, to produce 1 unit of product, sector III must purchase 0.4 units of sector I's product, 0.3 units of sector II's product and 0.1 units of its product.

We write the input–output matrix of this economy as

$$\mathbf{A} = \begin{pmatrix} 0.20 & 0.50 & 0.40 \\ 0.35 & 0.20 & 0.30 \\ 0.15 & 0.10 & 0.10 \end{pmatrix} \tag{7.192}$$

and take the demand vector as

$$\mathbf{d} = \begin{pmatrix} 10\ 000 \\ 15\ 000 \\ 40\ 000 \end{pmatrix}. \tag{7.193}$$

In an open economy, elements of the input–output matrix are positive and its columns adds up to a number less than 1, which means that some of the outputs are used to meet the outside demand. We now write

$$(\mathbf{I} - \mathbf{A}) = \begin{pmatrix} 0.80 & -0.50 & -0.40 \\ -0.35 & 0.80 & -0.30 \\ -0.15 & -0.10 & 0.90 \end{pmatrix} \tag{7.194}$$

and evaluate the inverse as

$$(\mathbf{I} - \mathbf{A})^{-1} = \begin{pmatrix} 2.2258 & 1.5806 & 1.5161 \\ 1.1613 & 2.1290 & 1.2258 \\ 0.5000 & 0.5000 & 1.5000 \end{pmatrix}, \tag{7.195}$$

which yields the production vector

$$\mathbf{p} = (\mathbf{I} - \mathbf{A})^{-1}\mathbf{d} = \begin{pmatrix} 1.0661 \times 10^5 \\ 0.9258 \times 10^5 \\ 0.7250 \times 10^5 \end{pmatrix}. \tag{7.196}$$

To meet the total demand, the output of sector I must be 1.0661×10^5 units, the output of sector II must be 0.9258×10^5 units, and the output of sector III must be 0.7250×10^5 units. To check whether the economy is productive or not, we evaluate the principal minors of $(\mathbf{I} - \mathbf{A})$. Since the first principal minor:

$$|0.80| = 0.8 > 0, \tag{7.197}$$

the second principal minor:

$$\begin{vmatrix} 0.80 & -0.50 \\ -0.35 & 0.80 \end{vmatrix} = 0.465 > 0, \tag{7.198}$$

and the third principle minor:

$$\begin{vmatrix} 0.80 & -0.50 & -0.40 \\ -0.35 & 0.80 & -0.30 \\ -0.15 & -0.10 & 0.90 \end{vmatrix} = 0.31 > 0, \tag{7.199}$$

are all positive, by the Hawkins–Simon criteria, we conclude that the economy is productive, thus any given demand can be met.

Originally, Leontief divided the US economy into 500 sectors, thus, leading to a 500×500 input–output matrix, which gave a linear system of 500 equations with 500 variables. Since computers at that that time could not handle such large matrices, he had to narrow his final analysis to a 42×42 system. Leontief models approaches economic problems with the methods of linear algebra and allows one to use real data for production planning and predicting whether an economy can meet certain demands. The models we have discussed here are static, where in real life interdependence of different sectors could change with time. Leontief models can also be modified to apply to dynamic situations.

7.4 APPLICATIONS TO GEOMETRY

Practical problems often involve finding a geometric figure passing through a given number of points. For example, let

$$P_1 = (x_1, y_1), \tag{7.200}$$

$$P_2 = (x_2, y_2), \tag{7.201}$$

be two points on the xy-plane. Equation of a line in this plane can be written as

$$ax + by + c = 0. \tag{7.202}$$

For the equation of a line passing through the points P_1 and P_2, the constants (a, b, c) in Eq. (7.202) have to satisfy the equations

$$ax_1 + by_1 + c = 0, \tag{7.203}$$

$$ax_2 + by_2 + c = 0. \tag{7.204}$$

Together with Eq. (7.202), these define a system of three homogeneous linear equations for the three unknowns (a, b, c) as

$$xa + yb + c = 0, \tag{7.205}$$

$$x_1 a + y_1 b + c = 0, \tag{7.206}$$

$$x_2 a + y_2 b + c = 0. \tag{7.207}$$

Since we are sure that there is a line that passes through these points, this system will have at least one solution. However, since for a given solution (a, b, c), $k(a, b, c)$, where k is a constant, will also be a solution, there are infinitely many solutions, hence, the determinant of the coefficients must be zero:

$$\begin{vmatrix} x & y & 1 \\ x_1 & y_1 & 1 \\ x_2 & y_2 & 1 \end{vmatrix} = 0. \tag{7.208}$$

This gives the equation of the line as

$$x(y_1 - y_2) + y(x_2 - x_1) + (x_1 y_2 - x_2 y_1) = 0. \tag{7.209}$$

7.4.1   Orbit Calculations

Due to the nature of the **central force** problem orbits of particles about the force center are always **conic sections**. The general equation of a conic section can be written as

$$\boxed{Ax^2 + Bxy + Cy^2 + Dx + Ey + F = 0,} \tag{7.210}$$

where (A, B, C, D, E, F) are constants, and the type of the conic section is given as

$B^2 - 4AC$	$-$
< 0	ellipse,
$= 0$	parabola,
> 0	hyperbola.

$$(7.211)$$

Another important property of the central force problem is that the orbit lies in a plane. Let us assume that an astronomer after setting up a Cartesian coordinate system in the plane of the orbit, makes five different observations, which give five different points of the orbit as

$$(x_i, y_i), \quad i = 1, \ldots, 5. \tag{7.212}$$

We can now write the following six linear homogeneous equations to be solved for the constants (A, B, C, D, E, F) as

$$x^2 A + xy B + y^2 C + x D + y E + F = 0, \tag{7.213}$$

$$x_1^2 A + x_1 y_1 B + y_1^2 C + x_1 D + y_1 E + F = 0, \tag{7.214}$$

$$x_2^2 A + x_2 y_2 B + y_2^2 C + x_2 D + y_2 E + F = 0, \tag{7.215}$$

$$x_3^2 A + x_3 y_3 B + y_3^2 C + x_3 D + y_3 E + F = 0, \tag{7.216}$$

$$x_4^2 A + x_4 y_4 B + y_4^2 C + x_4 D + y_4 E + F = 0, \tag{7.217}$$

$$x_5^2 A + x_5 y_5 B + y_5^2 C + x_5 D + y_5 E + F = 0. \tag{7.218}$$

Since we are certain that the orbit exists, this system has a solution. However, if (A, B, C, D, E, F) is a solution, then $k(A, B, C, D, E, F)$, where k is a constant, is also a solution. In other words, there are infinitely many solutions, which implies that the determinant of the coefficients is zero:

$$
\begin{vmatrix}
x^2 & xy & y^2 & x & y & 1 \\
x_1^2 & x_1 y_1 & y_1^2 & x_1 & y_1 & 1 \\
x_2^2 & x_2 y_2 & y_2^2 & x_2 & y_2 & 1 \\
x_3^2 & x_3 y_3 & y_3^2 & x_3 & y_3 & 1 \\
x_4^2 & x_4 y_4 & y_4^2 & x_4 & y_4 & 1 \\
x_5^2 & x_5 y_5 & y_5^2 & x_5 & y_5 & 1
\end{vmatrix} = 0.
\tag{7.219}
$$

Evaluation of this determinant:

$$
x^2
\begin{vmatrix}
x_1 y_1 & y_1^2 & x_1 & y_1 & 1 \\
x_2 y_2 & y_2^2 & x_2 & y_2 & 1 \\
x_3 y_3 & y_3^2 & x_3 & y_3 & 1 \\
x_4 y_4 & y_4^2 & x_4 & y_4 & 1 \\
x_5 y_5 & y_5^2 & x_5 & y_5 & 1
\end{vmatrix}
- xy
\begin{vmatrix}
x_1^2 & y_1^2 & x_1 & y_1 & 1 \\
x_2^2 & y_2^2 & x_2 & y_2 & 1 \\
x_3^2 & y_3^2 & x_3 & y_3 & 1 \\
x_4^2 & y_4^2 & x_4 & y_4 & 1 \\
x_5^2 & y_5^2 & x_5 & y_5 & 1
\end{vmatrix}
$$

$$
+ y^2
\begin{vmatrix}
x_1^2 & x_1 y_1 & x_1 & y_1 & 1 \\
x_2^2 & x_2 y_2 & x_2 & y_2 & 1 \\
x_3^2 & x_3 y_3 & x_3 & y_3 & 1 \\
x_4^2 & x_4 y_4 & x_4 & y_4 & 1 \\
x_5^2 & x_5 y_5 & x_5 & y_5 & 1
\end{vmatrix}
- x
\begin{vmatrix}
x_1^2 & x_1 y_1 & y_1^2 & y_1 & 1 \\
x_2^2 & x_2 y_2 & y_2^2 & y_2 & 1 \\
x_3^2 & x_3 y_3 & y_3^2 & y_3 & 1 \\
x_4^2 & x_4 y_4 & y_4^2 & y_4 & 1 \\
x_5^2 & x_5 y_5 & y_5^2 & y_5 & 1
\end{vmatrix}
$$

$$
+ y
\begin{vmatrix}
x_1^2 & x_1 y_1 & y_1^2 & x_1 & 1 \\
x_2^2 & x_2 y_2 & y_2^2 & x_2 & 1 \\
x_3^2 & x_3 y_3 & y_3^2 & x_3 & 1 \\
x_4^2 & x_4 y_4 & y_4^2 & x_4 & 1 \\
x_5^2 & x_5 y_5 & y_5^2 & x_5 & 1
\end{vmatrix}
-
\begin{vmatrix}
x_1^2 & x_1 y_1 & y_1^2 & x_1 & y_1 \\
x_2^2 & x_2 y_2 & y_2^2 & x_2 & y_2 \\
x_3^2 & x_3 y_3 & y_3^2 & x_3 & y_3 \\
x_4^2 & x_4 y_4 & y_4^2 & x_4 & y_4 \\
x_5^2 & x_5 y_5 & y_5^2 & x_5 & y_5
\end{vmatrix} = 0,
\tag{7.220}
$$

yields the equation of the orbit in the xy-plane in terms of the coordinates of the five observed points.

7.5 ELIMINATION THEORY

In practice, we often encounter situations where we have to decide whether two polynomials share a **common root** or not. Also, we may need the conditions under which they share a **common factor**. Naturally, as the order of the polynomials increase, this becomes a daunting task even for computers. Elimination theory is a valuable tool in checking the existence of a common solution for two polynomials without actually solving equations. It is also useful in reducing the number of variables in a system of polynomial equations to get an equivalent but a simpler system.

7.5.1 Quadratic Equations and the Resultant

Given two quadratic functions:

$$f(x) = a_2 x^2 + a_1 x + a_0, \qquad (7.221)$$

$$g(x) = b_2 x^2 + b_1 x + b_0, \qquad (7.222)$$

let us find under what conditions the two equations $f(x) = 0$ and $g(x) = 0$, have a common solution. Naturally, if they have a common solution, they must have a common linear factor $h(x)$, so that we can write

$$f(x) = q_1(x)h(x), \qquad (7.223)$$

$$g(x) = q_2(x)h(x), \qquad (7.224)$$

where

$$q_1(x) = A_1 x + A_0, \qquad (7.225)$$

$$q_2(x) = -B_1 x - B_0. \qquad (7.226)$$

The minus signs in $q_2(x)$ are introduced for future convenience and A_1, A_0, B_1, B_0 are constants. Using Eqs. (7.223) and (7.224), we can write

$$\frac{f(x)}{q_1(x)} = \frac{g(x)}{q_2(x)} = h(x), \qquad (7.227)$$

thus

$$f(x)q_2(x) = g(x)q_1(x), \qquad (7.228)$$

$$(a_2 x^2 + a_1 x + a_0)(-B_1 x - B_0) = (b_2 x^2 + b_1 x + b_0)(A_1 x - A_0), \qquad (7.229)$$

which when expanded and like terms collected yields

$$(a_2 B_1 + b_2 A_1)x^3 + (b_1 A_1 + a_1 B_1 + a_2 B_0 + b_2 A_0)x^2$$
$$+ (a_0 B_1 + b_0 A_1 + b_1 A_0 + a_1 B_0)x + (a_0 B_0 + b_0 A_0) = 0. \qquad (7.230)$$

For this equation to be satisfied for all x, we set the coefficients of all the powers of x to zero:

$$a_2 B_1 + b_2 A_1 = 0, \qquad (7.231)$$

$$b_1 A_1 + a_1 B_1 + a_2 B_0 + b_2 A_0 = 0, \qquad (7.232)$$

$$a_0 B_1 + b_0 A_1 + b_1 A_0 + a_1 B_0 = 0, \qquad (7.233)$$

$$a_0 B_0 + b_0 A_0 = 0. \qquad (7.234)$$

We rewrite this as

$$a_2 B_1 + 0 B_0 + b_2 A_1 + 0 A_0 = 0, \tag{7.235}$$

$$a_1 B_1 + a_2 B_0 + b_1 A_1 + b_2 A_0 = 0, \tag{7.236}$$

$$a_0 B_1 + a_1 B_0 + b_0 A_1 + b_1 A_0 = 0, \tag{7.237}$$

$$0 B_1 + a_0 B_0 + 0 A_1 + b_0 A_0 = 0. \tag{7.238}$$

This is a system of four linear homogeneous equations with four unknowns:

$$B_1, \ B_0, \ A_1, \ A_0. \tag{7.239}$$

For this system to have a nontrivial solution, determinant of the coefficients must vanish:

$$\begin{vmatrix} a_2 & 0 & b_2 & 0 \\ a_1 & a_2 & b_1 & b_2 \\ a_0 & a_1 & b_0 & b_1 \\ 0 & a_0 & 0 & b_0 \end{vmatrix} = 0. \tag{7.240}$$

Since taking the transpose of a matrix does not change the value of its determinant, we take the transpose and rewrite it as

$$\begin{vmatrix} a_2 & a_1 & a_0 & 0 \\ 0 & a_2 & a_1 & a_0 \\ b_2 & b_1 & b_0 & 0 \\ 0 & b_2 & b_1 & b_0 \end{vmatrix} = 0. \tag{7.241}$$

This determinant is called the (**Sylvester**) **resultant** of $f(x)$ and $g(x)$ with respect to x and gives the relation between the coefficients of $f(x)$ and $g(x)$ so that the two equations, $f(x) = 0$ and $g(x) = 0$, have a common solution. The matrix of coefficients:

$$Syl(f, g, x) = \begin{pmatrix} a_2 & a_1 & a_0 & 0 \\ 0 & a_2 & a_1 & a_0 \\ b_2 & b_1 & b_0 & 0 \\ 0 & b_2 & b_1 & b_0 \end{pmatrix}, \tag{7.242}$$

is called the **Sylvester matrix** of $f(x)$ and $g(x)$ with respect to x.

Keeping track of the special locations of the zeros in the Sylvester matrix, we can generalize this result to the case of two polynomials of degrees m and n, respectively:

$$f(x) = a_m x^m + \cdots + a_0, \ a_m \neq 0, \tag{7.243}$$

$$g(x) = b_n x^n + \cdots + b_0, \ b_n \neq 0, \tag{7.244}$$

as

$$
Syl(f,g,x) =
\begin{pmatrix}
a_m & a_{m-1} & \cdots & a_1 & a_0 & 0 & \cdots & 0 \\
0 & a_m & a_{m-1} & \cdots & a_1 & a_0 & \cdots & 0 \\
\vdots & \ddots & \ddots & \ddots & \vdots & \ddots & \ddots & \vdots \\
0 & \cdots & 0 & a_m & a_{m-1} & \cdots & a_1 & a_0 \\
b_n & b_{n-1} & \cdots & b_1 & b_0 & 0 & \cdots & 0 \\
0 & b_n & b_{n-1} & \cdots & b_1 & b_0 & \cdots & 0 \\
\vdots & \ddots & \ddots & \ddots & \vdots & \ddots & \ddots & \vdots \\
0 & \cdots & 0 & b_n & b_{n-1} & \cdots & b_1 & b_0
\end{pmatrix}, \quad (7.245)
$$

where the **Sylvester matrix** is defined as an $m+n$ dimensional **square matrix**. We now present the following theorem.

Theorem 7.3. Given two polynomials:

$$f(x) = a_m x^m + \cdots + a_0, \ a_m \neq 0, \tag{7.246}$$

$$g(x) = b_n x^n + \cdots + b_0, \ b_n \neq 0, \tag{7.247}$$

the equations:

$$f(x) = 0, \tag{7.248}$$

$$g(x) = 0, \tag{7.249}$$

have a common solution if and only if the **resultant** $\mathrm{Res}(f,g,x)$, which is the **determinant** of the Sylvester matrix of $f(x)$ and $g(x)$ with respect to x is zero:

$$\boxed{\det(Syl(f,g,x)) = \mathrm{Res}(f,g,x) = 0.} \tag{7.250}$$

Example 7.8. *Resultant:* Given the following two polynomials:

$$f(x) = 3x^3 - 14x^2 + 20x - 8, \tag{7.251}$$

$$g(x) = 3x^2 - 8x + 4. \tag{7.252}$$

Without solving the equations $f(x) = 0$ and $g(x) = 0$, we can check if they have a common root by calculating the resultant. For this case, the Sylvester matrix is the following five dimensional square matrix:

$$
Syl(f,g,x) =
\begin{pmatrix}
3 & -14 & 20 & -8 & 0 \\
0 & 3 & -14 & 20 & -8 \\
3 & -8 & 4 & 0 & 0 \\
0 & 3 & -8 & 4 & 0 \\
0 & 0 & 3 & -8 & 4
\end{pmatrix}. \tag{7.253}
$$

Evaluating its determinant gives the resultant as

$$\text{Res}(f, g, x) = \det(Syl(f, g, x)) = 0. \qquad (7.254)$$

Using **Theorem (7.3)**, we can conclude that they have a common root. Indeed, we can write

$$f(x) = (3x^2 - 8x + 4)(x - 2), \qquad (7.255)$$

$$g(x) = (x - 2/3)(x - 2), \qquad (7.256)$$

to see that the common root is 2.

Example 7.9. *Resultant:* Let us now consider the following polynomials:

$$f(x) = 3x^4 + x^3 - 14x^2 + 20x - 8, \qquad (7.257)$$

$$g(x) = 3x^3 + 8x^2 + x + 4. \qquad (7.258)$$

The square Sylvester matrix is seven-dimensional, and the resultant is calculated as

$$\text{Res}(f, g, x) = \begin{vmatrix} 3 & 1 & -14 & 20 & -8 & 0 & 0 \\ 0 & 3 & 1 & -14 & 20 & -8 & 0 \\ 0 & 0 & 3 & 1 & -14 & 20 & -8 \\ 3 & 8 & 1 & 4 & 0 & 0 & 0 \\ 0 & 3 & 8 & 1 & 4 & 0 & 0 \\ 0 & 0 & 3 & 8 & 1 & 4 & 0 \\ 0 & 0 & 0 & 3 & 8 & 1 & 4 \end{vmatrix} = -2.9 \times 10^5. \qquad (7.259)$$

In this example, since $\text{Res}(f, g, x)$ is different from 0, the polynomials $f(x)$ and $g(x)$ do not have a common root.

Example 7.10. *Resultant:* Let us now consider the following two equations:

$$y^2 + 8y + 36x^2 - 36x - 11 = 0, \qquad (7.260)$$

$$y^2 + 4(4x^2 - 9) = 0, \qquad (7.261)$$

which we can consider as two polynomials in y with coefficients as functions of x. For a common solution, we evaluate the resultant to write

$$\text{Res}(f, g, x) = \begin{vmatrix} 1 & 8 & (36x^2 - 36x - 11) & 0 \\ 0 & 1 & 8 & (36x^2 - 36x - 11) \\ 1 & 0 & (16x^2 - 36) & 0 \\ 0 & 1 & 0 & (16x^2 - 36) \end{vmatrix}, \qquad (7.262)$$

$$= 400x^4 - 1440x^3 + 3320x^2 - 1800x - 1679. \qquad (7.263)$$

Setting the resultant to zero:

$$400x^4 - 1440x^3 + 3320x^2 - 1800x - 1679 = 0, \qquad (7.264)$$

gives the equation for the x coordinates of the intersection points of the two curves [Eqs. (7.260) and (7.261)] in the xy-plane. Numerical evaluation of the roots yield two real roots:

$$x_1 = -0.46, \quad x_2 = 1.43. \qquad (7.265)$$

7.6   CODING THEORY

Transmitted messages or data are always subject to some noise. One way to guard against errors is to repeat the message several times. However, resending and copying the message several times on some electronic media is not only inefficient but also requires extra memory. Coding theory deals with the development of **error detecting** and **correcting** methods for a reliable transmission of information through noisy channels. **Linear codes** is a class of codes, where the concepts of linear algebra are proven to be extremely useful. Since digital messages are sent in binary coding as sequences of 0s and 1s, we start by reviewing the basic concepts of linear algebra and binary arithmetic [7].

7.6.1   Fields and Vector Spaces

A collection of objects is called a **set**. The set of all real numbers is denoted by $\mathbb{R}$ and the set of all complex numbers by $\mathbb{C}$. The set of all n-tuples of real numbers:

$$X = (x_1, x_2, \ldots, x_n), \qquad (7.266)$$

is shown by $\mathbb{R}^n$. Similarly, the set of n-tuples of complex numbers is shown by $\mathbb{C}^n$. In coding theory, we use the special set $\mathbb{Z}_2$, which consists of two numbers $\{0, 1\}$ that obeys the following rules of **modulo-2** arithmetic:

$$0 + 0 = 0, \ 1 + 0 = 1, \ 0 + 1 = 1, \ 1 + 1 = 0, \qquad (7.267)$$

$$0 \cdot 0 = 0, \ 1 \cdot 0 = 0, \ 0 \cdot 1 = 0, \ 1 \cdot 1 = 1. \qquad (7.268)$$

Note that addition and subtraction are the same in $\mathbb{Z}_2$. An essential part of linear algebra consists of the definition of **vector space**, which is also called the **linear space**.

Definition 7.1. A **field** K is a set of objects that satisfy the following conditions:

I. If α, β are the elements of K, then their sum, $\alpha + \beta$, and their product, $\alpha\beta$, are also the elements of K.

II. If α is an element of K, then $-\alpha$ such that $\alpha + (-\alpha) = 0$ is also an element of K. Furthermore, if $\alpha \neq 0$, then α^{-1}, where $\alpha(\alpha^{-1}) = 1$ is also an element of K.

III. The elements 0 and 1 are also the elements of K.

In this regard, according to the rules given in Eqs. (7.267) and (7.268), $\mathbb{Z}_2$ is a **field**. In general, the elements of a field are called **numbers** or **scalars**. We now introduce the concept of **linear space**, or as usually called the **vector space**. A vector space is defined in conjunction with a field and its elements are called **vectors**.

Definition 7.2. A **vector space** V over a field K is a set of objects that can be added with the result being again an element of V. Also, the product of an element of V with an element of K is again an element of V. A vector space also satisfies the following properties:

I. For all u, v, w of V, **addition** is **associative**:

$$u + (v + w) = (u + v) + w. \tag{7.269}$$

II. There is a unique element 0 of V, such that

$$0 + u = u + 0 = u, \ u \in V. \tag{7.270}$$

III. Let u be an element of V, then there exists a unique vector $-u$ in V such that
$$u + (-u) = 0. \tag{7.271}$$

IV. For any two elements u and v of V, **addition** is **commutative**:

$$u + v = v + u. \tag{7.272}$$

V. If α is a number belonging to the field K, then

$$\alpha(u + v) = \alpha u + \alpha v, \ u, v \in V. \tag{7.273}$$

VI. If α, β are two numbers in K,, then we have

$$(\alpha + \beta)v = \alpha v + \beta v, \ v \in V. \tag{7.274}$$

VII. For any two numbers α and β in K,, we have

$$(\alpha\beta)v = \alpha(\beta v), \ v \in V. \tag{7.275}$$

VIII. For all u of V, we have
$$1.u = u, \tag{7.276}$$

where 1 is the number one.

Just like the n-dimensional vectors over the field of real numbers, $\mathbf{v} \in \mathbb{R}^n$:

$$\mathbf{v} = (x_1, \ldots, x_n), \tag{7.277}$$

we can use the above structure to define a vector $\mathbf{v} \in \mathbb{Z}_2^n$ in $\mathbb{Z}_2$. For example, a vector in $\mathbb{Z}_2^5$ could be given as

$$\mathbf{v} = (0, 1, 1, 0, 1), \tag{7.278}$$

or in $\mathbb{Z}_2^6$ as

$$\mathbf{v} = (1, 1, 1, 0, 1, 1). \tag{7.279}$$

A major difference between vectors in $\mathbb{R}^n$ and $\mathbb{Z}_2^n$ is that $\mathbb{Z}_2^n$ contains a **finite**, 2^n, number of vectors.

7.6.2 Hamming (7,4) Code

In 1950s a linear **single error correcting** code introduced by Hamming is known as the **Hamming (7,4) code**. We now discuss this code through its connections with linear algebra. Given two integers k and n, $k \leqslant n$, a k dimensional subspace of $\mathbb{Z}_2^n$ is called an (n, k) **linear code**, where the elements of this subspace are called the **encoded words**. Elements of an (n, k) code corrects any single error if all the elements of its **parity-check matrix** are distinct and nonzero.

We can construct such a $(3, 7)$-dimensional **parity-check matrix H** for the $(7, 4)$ code by taking all its seven columns as the binary representations of the numbers

$$c_1 = 1,\ c_2 = 2,\ c_3 = 3,\ c_4 = 4,\ c_5 = 5,\ c_6 = 6,\ c_7 = 7 \tag{7.280}$$

as

$$\mathbf{H} = \begin{pmatrix} 0 & 0 & 0 & 1 & 1 & 1 & 1 \\ 0 & 1 & 1 & 0 & 0 & 1 & 1 \\ 1 & 0 & 1 & 0 & 1 & 0 & 1 \end{pmatrix}. \tag{7.281}$$

The null space, Null($\mathbf{H}$), of $\mathbf{H}$ is called the **Hamming (7,4) code**. Since Null($\mathbf{H}$) is the solution set of the homogeneous equation (Chapter 6):

$$\mathbf{Hx} = 0, \tag{7.282}$$

we can either use the Gauss-Jordan method with the binary arithmetic, or simply in this case swap rows 1 and 3, to write the following row-echelon form of $\mathbf{H}$:

$$\text{ref}(\mathbf{H}) = \begin{pmatrix} 1 & 0 & 1 & 0 & 1 & 0 & 1 \\ 0 & 1 & 1 & 0 & 0 & 1 & 1 \\ 0 & 0 & 0 & 1 & 1 & 1 & 1 \end{pmatrix}. \tag{7.283}$$

Since the **rank** of $\mathbf{H}$ is 3, the **dimension** of $\text{Null}(\mathbf{H})$ is

$$7 - 3 = 4. \tag{7.284}$$

We now write $\mathbf{Hx} = 0$ explicitly as

$$x_1 + x_3 + x_5 + x_7 = 0, \tag{7.285}$$
$$x_2 + x_3 + x_6 + x_7 = 0, \tag{7.286}$$
$$x_4 + x_5 + x_6 + x_7 = 0. \tag{7.287}$$

Using the parametrization

$$x_1 = s, \tag{7.288}$$
$$x_2 = t, \tag{7.289}$$
$$x_3 = v, \tag{7.290}$$
$$x_4 = w, \tag{7.291}$$

we obtain

$$x_5 = -t - v + w, \tag{7.292}$$
$$x_6 = s + v - w, \tag{7.293}$$
$$x_7 = -t - s + w. \tag{7.294}$$

This gives the solution of the homogeneous equation $\mathbf{Hx} = 0$ as

$$\mathbf{x} = s \begin{pmatrix} 1 \\ 0 \\ 0 \\ 0 \\ 0 \\ 1 \\ 1 \end{pmatrix} + t \begin{pmatrix} 0 \\ 1 \\ 0 \\ 0 \\ 1 \\ 0 \\ 1 \end{pmatrix} + v \begin{pmatrix} 0 \\ 0 \\ 1 \\ 0 \\ 1 \\ 1 \\ 0 \end{pmatrix} + w \begin{pmatrix} 0 \\ 0 \\ 0 \\ 1 \\ 1 \\ 1 \\ 1 \end{pmatrix}, \tag{7.295}$$

which yields the basis vectors of the null space, $\text{Null}(\mathbf{H})$, over $\mathbb{Z}_2$ as

$$(1, 0, 0, 0, 0, 1, 1), \tag{7.296}$$
$$(0, 1, 0, 0, 1, 0, 1), \tag{7.297}$$
$$(0, 0, 1, 0, 1, 1, 0), \tag{7.298}$$
$$(0, 0, 0, 1, 1, 1, 1). \tag{7.299}$$

We now use these basis vectors as the rows of a (4×7)-dimensional matrix, called the **generator matrix G** of the Hamming $(7, 4)$ code:

$$\mathbf{G} = \begin{pmatrix} 1 & 0 & 0 & 0 & 0 & 1 & 1 \\ 0 & 1 & 0 & 0 & 1 & 0 & 1 \\ 0 & 0 & 1 & 0 & 1 & 1 & 0 \\ 0 & 0 & 0 & 1 & 1 & 1 & 1 \end{pmatrix}. \tag{7.300}$$

Note that **G** has the following form called the **systematic form**:

$$\mathbf{G} = (\mathbf{I_4}|\mathbf{P}), \tag{7.301}$$

where $\mathbf{I_4}$ is the 4×4 identity matrix and **P** is a 4×3 matrix called the **check bits matrix**:

$$\mathbf{I_4} = \begin{pmatrix} 1 & 0 & 0 & 0 \\ 0 & 1 & 0 & 0 \\ 0 & 0 & 1 & 0 \\ 0 & 0 & 0 & 1 \end{pmatrix}, \ \mathbf{P} = \begin{pmatrix} 0 & 1 & 1 \\ 1 & 0 & 1 \\ 1 & 1 & 0 \\ 1 & 1 & 1 \end{pmatrix}. \tag{7.302}$$

Each **word**:

$$\mathbf{u} = (u_1, u_2, u_3, u_4), \tag{7.303}$$

can be **encoded** as

$$\mathbf{v} = \mathbf{uG} = (u_1, u_2, u_3, u_4, u_5, u_6, u_7), \tag{7.304}$$

where the first 4 digits, called the **information digits**, are the same as **u**. The remaining 3 digits, u_5, u_6, u_7, are called the **check digits**.

Before we go any further, let $\{\mathbf{e}_i, 1, \dots, 7\}$:

$$\mathbf{e_1} = \begin{pmatrix} 1 \\ 0 \\ 0 \\ 0 \\ 0 \\ 0 \\ 0 \end{pmatrix}, \ \mathbf{e_2} = \begin{pmatrix} 0 \\ 1 \\ 0 \\ 0 \\ 0 \\ 0 \\ 0 \end{pmatrix}, \ \mathbf{e_3} = \begin{pmatrix} 0 \\ 0 \\ 1 \\ 0 \\ 0 \\ 0 \\ 0 \end{pmatrix}, \ \mathbf{e_4} = \begin{pmatrix} 0 \\ 0 \\ 0 \\ 1 \\ 0 \\ 0 \\ 0 \end{pmatrix},$$

$$\mathbf{e_5} = \begin{pmatrix} 0 \\ 0 \\ 0 \\ 0 \\ 1 \\ 0 \\ 0 \end{pmatrix}, \ \mathbf{e_6} = \begin{pmatrix} 0 \\ 0 \\ 0 \\ 0 \\ 0 \\ 1 \\ 0 \end{pmatrix}, \ \mathbf{e_7} = \begin{pmatrix} 0 \\ 0 \\ 0 \\ 0 \\ 0 \\ 0 \\ 1 \end{pmatrix}, \tag{7.305}$$

be the **standard basis** of $\mathbb{Z}_2^7$. It can be checked by inspection that

$$H\mathbf{e}_i = c_i, \tag{7.306}$$

where c_i is the binary representation of the numbers given in Eq. (7.280). Thus, $\mathbf{e}_i$, $i = 1, \ldots, 7$, is not in the null space of $\mathbf{H}$.

We now have the following **remarks:**

   I. If $\mathbf{v} \in \mathbb{Z}_2^7$ is a vector in the null space of $\mathbf{H}$, that is, $Null(\mathbf{H})$, then $\mathbf{v} + \mathbf{e}_i$, $i = 1, \ldots, 7$, is not in $Null(\mathbf{H})$.

  II. If $\mathbf{w} \in \mathbb{Z}_2^7$ is a vector that satisfies $\mathbf{Hw} = c_j$ for some j, where j could take any value from 1 to 7, then $H(\mathbf{w} + \mathbf{e}_j) = 0$, thus $\mathbf{w} + \mathbf{e}_j \in Null(\mathbf{H})$. For all $i \neq j$, $\mathbf{w} + \mathbf{e}_i$ does not belong to $Null(\mathbf{H})$.

7.6.3 Hamming Algorithm for Error Correction

Suppose we send a message with four binary digits:

$$\mathbf{u} = (u_1, u_2, u_3, u_4), \tag{7.307}$$

that might get scrambled by noise. Hamming algorithm can be described as follows:

  (1) Using the basis vectors in Eq. (7.295), or the matrix multiplication

$$\mathbf{v} = \mathbf{uG} = (u_1, u_2, u_3, u_4) \begin{pmatrix} 1 & 0 & 0 & 0 & 0 & 1 & 1 \\ 0 & 1 & 0 & 0 & 1 & 0 & 1 \\ 0 & 0 & 1 & 0 & 1 & 1 & 0 \\ 0 & 0 & 0 & 1 & 1 & 1 & 1 \end{pmatrix}, \tag{7.308}$$

we can write the encoded message $\mathbf{v}$ as a seven-digit vector:

$$\mathbf{v} = (v_1, v_2, v_3, v_4, v_5, v_6, v_7), \tag{7.309}$$

where the first four digits is the message:

$$\mathbf{v} = (u_1, u_2, u_3, u_4, v_5, v_6, v_7), \tag{7.310}$$

and the remaining 3 digits are the check digits.

  (2) However, instead of $\mathbf{v}$, let us assume that the scrambled message $\mathbf{w}$ is received. Our purpose is to find any errors it may contain and to correct them. We now compute $\mathbf{Hw}$, where $\mathbf{H}$ is the parity-check matrix [Eq. (7.281)].

  (3) If $\mathbf{Hw} = 0$, then by the first remark, $\mathbf{w}$ is in $Null(\mathbf{H})$, hence, the received message does not have any errors and the message is the first four digits of $\mathbf{w}$.

(4) If $\mathbf{Hw} = c_i$ for some i, then the second remark, $(\mathbf{w} + \mathbf{e}_i) \in \text{Null}(\mathbf{H})$, tells us that by changing the ith component of $\mathbf{w}$, we can correct the error. In other words, if the ith component is 0, we change it to 1, and vice versa, to get the corrected vector $\mathbf{w}'$ whose first four digits represent the transmitted message $\mathbf{u}$.

Example 7.11. ***Hamming code:*** Suppose that a message encoded by the Hamming code is received as

$$
\mathbf{w} = \begin{pmatrix} 1 \\ 1 \\ 0 \\ 1 \\ 0 \\ 0 \\ 0 \end{pmatrix}. \tag{7.311}
$$

We suspect that it contains a single **error**. To find the original Message, we calculate $\mathbf{Hw}$ as

$$
\mathbf{Hw} = \begin{pmatrix} 0 & 0 & 0 & 1 & 1 & 1 & 1 \\ 0 & 1 & 1 & 0 & 0 & 1 & 1 \\ 1 & 0 & 1 & 0 & 1 & 0 & 1 \end{pmatrix} \begin{pmatrix} 1 \\ 1 \\ 0 \\ 1 \\ 0 \\ 0 \\ 0 \end{pmatrix} = \begin{pmatrix} 1 \\ 1 \\ 1 \end{pmatrix}. \tag{7.312}
$$

Since this is equal to the seventh column of $\mathbf{H}$, changing the last digit of the received message, we obtain the encoded word as

$$
\mathbf{v} = \begin{pmatrix} 1 \\ 1 \\ 0 \\ 1 \\ 0 \\ 0 \\ 1 \end{pmatrix}, \tag{7.313}
$$

where the first four digits gives the original message as

$$
\begin{pmatrix} 1 \\ 1 \\ 0 \\ 1 \end{pmatrix}. \tag{7.314}
$$

Example 7.12. *Hamming code:* Suppose the received message is

$$\mathbf{w} = \begin{pmatrix} 0 \\ 1 \\ 0 \\ 1 \\ 0 \\ 1 \\ 0 \end{pmatrix}. \tag{7.315}$$

Again, calculating $\mathbf{Hw}$:

$$\mathbf{Hw} = \begin{pmatrix} 0 & 0 & 0 & 1 & 1 & 1 & 1 \\ 0 & 1 & 1 & 0 & 0 & 1 & 1 \\ 1 & 0 & 1 & 0 & 1 & 0 & 1 \end{pmatrix} \begin{pmatrix} 0 \\ 1 \\ 0 \\ 1 \\ 0 \\ 1 \\ 0 \end{pmatrix}, \tag{7.316}$$

we find

$$\mathbf{Hw} = \begin{pmatrix} 0 \\ 0 \\ 0 \end{pmatrix}, \tag{7.317}$$

which indicates that the received message contains no errors, that is, $\mathbf{w} = \mathbf{v}$.

Example 7.13. *Hamming code:* Let us now consider a case where the message $\mathbf{v}$ in Eq. (7.315) is received with two errors:

$$\mathbf{w} = \begin{pmatrix} 0 \\ 1 \\ 1 \\ 0 \\ 0 \\ 1 \\ 0 \end{pmatrix}, \tag{7.318}$$

where the third and the fourth positions are in error. Calculating $\mathbf{Hw}$:

$$\mathbf{Hw} = \begin{pmatrix} 0 & 0 & 0 & 1 & 1 & 1 & 1 \\ 0 & 1 & 1 & 0 & 0 & 1 & 1 \\ 1 & 0 & 1 & 0 & 1 & 0 & 1 \end{pmatrix} \begin{pmatrix} 0 \\ 1 \\ 1 \\ 0 \\ 0 \\ 1 \\ 0 \end{pmatrix} = \begin{pmatrix} 1 \\ 1 \\ 1 \end{pmatrix}, \tag{7.319}$$

we see that it points to an error in the seventh position, thus changing
the seventh position from 0 to 1 gives

$$\mathbf{v}' = \begin{pmatrix} 0 \\ 1 \\ 1 \\ 0 \\ 0 \\ 1 \\ 1 \end{pmatrix}. \tag{7.320}$$

Comparing with the message sent, $\mathbf{v}$, we see that instead of correcting the
existing two errors, we have made things worse by adding a third error.
In fact, the Hamming (7,4) code works only for cases with a single error.
After Shannon's original paper people have been looking for codes with
low probability of error and that could be coded and decoded with ease.
Over the years, many different codes with elegant and intricate schemes
have been developed [7]. Here, we discussed the simplest one of such
codes to demonstrate the links of linear coding theory with the methods
of linear algebra.

7.7   CRYPTOGRAPHY

Cryptography is an essential tool in protecting sensitive information communicated in the presence of adversaries. In particular, it aims to develop and
investigate protocols that blocks third parties. Modern cryptography is an interdisciplinary science at the crossroads of mathematics, computer science, and
electrical engineering with applications to ATM machines, electronic passwords,
electronic banking, etc. One of the key concepts of cryptography is **encryption**,
which is the transformation of information into a form that is unintelligible by
the adversaries. **Decryption** is the inverse of encryption, which converts the
encrypted information into intelligible form. Obviously, encryption and decryption involves some **secret key** shared by the parties who want to communicate
privately. Today, governments and companies use sophisticated protocols that
protect their communication from eavesdropping. Modern cryptography can
be discussed under two main categories as **symmetric-key** cryptography and
asymmetric or **public-key** cryptography. Until mid seventies symmetric-key
cryptography was the dominant tool publicly used, where the sender and the
receiver used the same key.

7.7.1   Single-Key Cryptography

One of the methods of **single-key** cryptography is to use a large matrix called
the **encoding matrix** as the key to encode a message, where the receiver

decodes the encrypted message by using the inverse of the encoding matrix. As an example, let us use the following 3×3 encoding matrix:

$$\mathbf{E} = \begin{pmatrix} 5 & 1 & 3 \\ -1 & 3 & 4 \\ 2 & -4 & 1 \end{pmatrix} \qquad (7.321)$$

to send the message

$$\text{ACCEPT ALL TERMS AND SIGN} \qquad (7.322)$$

Assigning a number to each letter of the alphabet:

$$
\begin{array}{cccccccccccccc}
\text{A} & \text{B} & \text{C} & \text{D} & \text{E} & \text{F} & \text{G} & \text{H} & \text{I} & \text{J} & \text{K} & \text{L} & \text{M} & \text{N} \\
1 & 2 & 3 & 4 & 5 & 6 & 7 & 8 & 9 & 10 & 11 & 12 & 13 & 14 \\
\text{O} & \text{P} & \text{Q} & \text{R} & \text{S} & \text{T} & \text{U} & \text{V} & \text{W} & \text{X} & \text{Y} & \text{Z} \\
15 & 16 & 17 & 18 & 19 & 20 & 21 & 22 & 23 & 24 & 25 & 26 & 27
\end{array} \qquad (7.323)
$$

where we have assigned the number 27 to space, we can code this message as

$$
\begin{array}{cccccccccccc}
\text{A} & \text{C} & \text{C} & \text{E} & \text{P} & \text{T} & * & \text{A} & \text{L} & \text{L} & * \\
1 & 3 & 3 & 5 & 16 & 20 & 27 & 1 & 12 & 12 & 27 \\
\text{T} & \text{E} & \text{R} & \text{M} & \text{S} & * & \text{A} & \text{N} & \text{D} & * & \text{S} & \text{I} & \text{G} & \text{N} \\
20 & 5 & 18 & 13 & 19 & 27 & 1 & 14 & 4 & 27 & 19 & 9 & 7 & 14
\end{array} \qquad (7.324)
$$

Since we are using a 3×3 encoding matrix, we break the coded message, which is going to be transmitted as a string of numbers:

$$
\begin{array}{ccccccccccc}
1 & 3 & 3 & 5 & 16 & 20 & 27 & 1 & 12 & 12 & 27 & 20 & 5 \\
18 & 13 & 19 & 27 & 1 & 14 & 4 & 27 & 19 & 9 & 7 & 14
\end{array} \qquad (7.325)
$$

into a sequence of 3×1 column matrices:

$$
\begin{pmatrix} 1 \\ 3 \\ 3 \end{pmatrix}, \begin{pmatrix} 5 \\ 16 \\ 20 \end{pmatrix}, \begin{pmatrix} 27 \\ 1 \\ 12 \end{pmatrix}, \begin{pmatrix} 12 \\ 27 \\ 20 \end{pmatrix},
$$

$$
\begin{pmatrix} 5 \\ 18 \\ 13 \end{pmatrix}, \begin{pmatrix} 19 \\ 27 \\ 1 \end{pmatrix}, \begin{pmatrix} 14 \\ 4 \\ 27 \end{pmatrix}, \begin{pmatrix} 19 \\ 9 \\ 7 \end{pmatrix}, \begin{pmatrix} 14 \\ 27 \\ 27 \end{pmatrix}, \qquad (7.326)
$$

where in the last matrix, we have added two spaces to complete the matrix. Using these column matrices as its columns, we construct the following 3×9 matrix:

$$
\mathbf{A} = \begin{pmatrix} 1 & 5 & 27 & 12 & 5 & 19 & 14 & 19 & 14 \\ 3 & 16 & 1 & 27 & 18 & 27 & 4 & 9 & 27 \\ 3 & 20 & 12 & 20 & 13 & 1 & 27 & 7 & 27 \end{pmatrix}. \qquad (7.327)
$$

To encode this message, we multiply $\mathbf{A}$ with the encoding matrix $\mathbf{E}$:

$$\widetilde{\mathbf{A}} = \mathbf{EA} \tag{7.328}$$

$$= \begin{pmatrix} 5 & 1 & 3 \\ -1 & 3 & 4 \\ 2 & -4 & 1 \end{pmatrix} \begin{pmatrix} 1 & 5 & 27 & 12 & 5 & 19 & 14 & 19 & 14 \\ 3 & 16 & 1 & 27 & 18 & 27 & 4 & 9 & 27 \\ 3 & 20 & 12 & 20 & 13 & 1 & 27 & 7 & 27 \end{pmatrix}, \tag{7.329}$$

to obtain the encoded message:

$$\widetilde{\mathbf{A}} = \begin{pmatrix} 17 & 101 & 172 & 147 & 82 & 125 & 155 & 125 & 178 \\ 20 & 123 & 24 & 149 & 101 & 66 & 106 & 36 & 175 \\ -7 & -34 & 62 & -64 & -49 & -69 & 39 & 9 & -53 \end{pmatrix}. \tag{7.330}$$

The columns of this matrix forms the encoded message sent and received in the following linear form:

17 20 −7 101 123 −34 172 24 62 147 149 −64 82 101

−49 125 66 −69 155 106 39 125 36 9 178 175 −53

$$\tag{7.331}$$

Now the receiver has to decode this message, which is done via the **inverse** of the **encoding matrix**:

$$\widetilde{\mathbf{E}} = \begin{pmatrix} 0.1939 & -0.1327 & -0.0510 \\ 0.0918 & -0.0102 & -0.2347 \\ 0.0204 & 0.2245 & 0.1633 \end{pmatrix}. \tag{7.332}$$

Thus, the decoded message is

$$\mathbf{A} = \widetilde{\mathbf{E}}\widetilde{\mathbf{A}} = \begin{pmatrix} 0.1939 & -0.1327 & -0.0510 \\ 0.0918 & -0.0102 & -0.2347 \\ 0.0204 & 0.2245 & 0.1633 \end{pmatrix} \tag{7.333}$$

$$\times \begin{pmatrix} 17 & 101 & 172 & 147 & 82 & 125 & 155 & 125 & 178 \\ 20 & 123 & 24 & 149 & 101 & 66 & 106 & 36 & 175 \\ -7 & -34 & 62 & -64 & -49 & -69 & 39 & 9 & -53 \end{pmatrix}$$

$$= \begin{pmatrix} 1 & 5 & 27 & 12 & 5 & 19 & 14 & 19 & 14 \\ 3 & 16 & 1 & 27 & 18 & 27 & 4 & 9 & 27 \\ 3 & 20 & 12 & 20 & 13 & 1 & 27 & 7 & 27 \end{pmatrix}. \tag{7.334}$$

Now the receiver can write this in linear form:

1 3 3 5 16 20 27 1 12 12 27 20 5 $\qquad$ (7.335)

18 13 19 27 1 14 4 27 19 9 7 14 $\qquad$ (7.336)

and using the coding in Eq. (7.323) read the message as

$$\text{ACCEPT ALL TERMS AND SIGN.} \qquad (7.337)$$

Depending on the size of the encoding matrix, the single-key cryptography could be rather difficult to break, however, the obvious weakness of this method is the management of keys, which require extreme care. In this regard, after mid-1970s **public-key** cryptography is developed and found wide spread use. In public-key cryptography, there is both a public and a private key. Even though both keys are mathematically related, they are designed such that computationally it is not feasible to construct the private-key from the public-key. Both keys are essentially generated secretly as an interrelated pair.

7.8   GRAPH THEORY

Graphs can be used to visualize many different types of relations and processes in nature. In this regard, it is not surprising to see that graph theory has found a wide range of applications in engineering and as well as in physical, biological, and social sciences. Computer science and information technologies are other areas, where the graph theory and the tools of linear algebra are widely used. In mathematics, graphs are used in geometry and in certain branches of topology like the knot theory and the curve theory with deep and insightful results. Graph theory also played a crucial role in the solution of the four color problem. Group theory and algebraic graph theory are also closely related areas of mathematics. What follows is a brief introduction to this intriguing and far-reaching branch of mathematics. We concentrate on the basic definitions of graphs and their matrix representations, which are the starting points for the implementation of the highly versatile tools of linear algebra [8–11].

7.8.1   Basic Definition

Even though there is no universal agreement with some of the definitions in graph theory, here we introduce some of the more basic and agreed upon aspects of defining graphs and the corresponding mathematical objects.

A **graph** is a collection of points called **vertices** and lines called **edges** that connect these vertices (Figure 7.5).

Formally, a graph $G = G(V, E)$ can be given as a pair of sets, where V is the set of **vertices**:

$$V = \{p_1, p_2, \ldots\}, \qquad (7.338)$$

and E is the set of **edges**,

$$E = \{e_1, e_2, \ldots\}. \qquad (7.339)$$

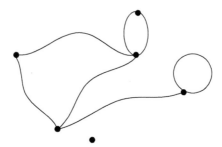

Figure 7.5   A graph is a collection of points connected with lines.

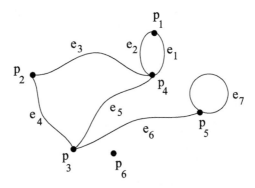

Figure 7.6   Vertices and edges of a graph.

We can also show E as the pairs of vertices that connect the edges. For example, for the graph in Figure 7.6, we can write

$$E = \{(p_1, p_4), (p_4, p_1), (p_4, p_2), (p_2, p_3), (p_3, p_4), (p_3, p_5), (p_5, p_5)\}. \qquad (7.340)$$

This also called the **list description** of a graph.

7.8.2   Terminology

1. A graph with finite number of vertices and edges is called a **finite graph**. In theory and applications, almost all graphs are finite.
2. It is not important whether the **edges** are drawn as straight-lines or curves.
3. In Eq. (7.340), (p_i, p_j) stands for the edge beginning at the vertex p_i and ending at p_j.
4. Edges with the same vertices are called **parallel**, e.g. e_1 and e_2 in Figure 7.6.
5. An edge (p_i, p_i) that starts and ends at the same vertex is called a **loop**, e.g. e_7 in Figure 7.6.
6. A graph that does not have any parallel edges or loops, is called a **simple graph**.

7. A graph is **empty** if it has no edges.
8. A graph is a **null graph** if it has no vertices and no edges.
9. Edges that share a common end vertex are called **adjacent edges**, e.g. e_4 and e_5 in Figure 7.6.
10. Two vertices connected by an edge are called **adjacent vertices**, e.g. p_3 and p_4 in Figure 7.6.
11. When p_i is the end vertex of an edge e_j, we call p_i and e_j as **incident** to each other.
12. A vertex that has no incident edge is called an **isolated vertex**, e.g. p_6 in Figure 7.6.
13. The **degree** of a vertex, $d(p)$, is the number of edges that ends at p. In this regard, a loop counts twice and parallel edges contribute separately. An **isolated vertex** has degree zero. A vertex whose degree is 1 is called a **pendant vertex**.
14. Since in a finite graph every edge has two end vertices, for a graph with n vertices and m edges we have

$$\boxed{\sum_{i=1}^{n} d(p_i) = 2m.}$$ (7.341)

For the graph in Figure 7.6, there are seven edges and six vertices, hence,

$$\sum_{i=1}^{6} d(p_i) = 2 + 2 + 3 + 4 + 3 + 0 = 2(7) = 14.$$ (7.342)

15. If there is a path connecting any two vertices, the graph is called **connected**, otherwise, it is **disconnected** (Figure 7.7). The graph in Figure 7.7 is disconnected with two parts. In general, a disconnected graph will have two or more parts called **components**.
16. A simple graph that contains every possible edge is called **complete** (Figure 7.8).

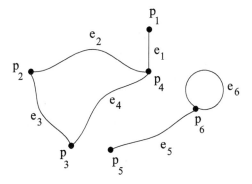

Figure 7.7 A disconnected graph.

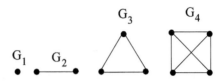

Figure 7.8   Examples of complete graphs.

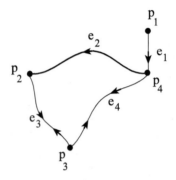

Figure 7.9   When the edges are directed, we have a digraph.

17. A graph $G_2(V_2, E_2)$ is a **subgraph** of $G_1(V_1, E_1)$ if $V_2 \subseteq V_1$ and every edge of G_2 is also an edge of G_1.

18. When the edges of a graph are directed, it is called a **directed graph** or a **digraph** (Figure 7.9).

19. Two graphs G and G' are called **isomorphic**, if there is a one-to-one correspondence between their vertices and edges such that the incidence relation is preserved. Except for the labelling of their vertices and edges, isomorphic graphs are equivalent. However, showing that two graphs are isomorphic is by all means not an easy task that could be settled by inspection. There are various algorithms developed for checking isomorphism via computers.

20. Two or more subgraphs of a graph are called **edge-disjoint** if they do not have any edges in common. Note that edge-disjoint graphs may still have common vertices.

21. Two or more subgraphs that do not have common vertices are called **vertex-disjoint**. If two graphs do not have any common vertices, then they cannot possibly have common edges.

7.8.3   Walks, Trails, Paths and Circuits

A **walk** is a sequence of vertices and edges, where the end points of each edge are the preceding and the following vertices in the sequence. A walk in $G(V, E)$ is the finite sequence

$$p_{i_0}, e_{j_1}, p_{i_1}, e_{j_2}, \ldots, e_{j_k}, p_{i_k}, \tag{7.343}$$

consisting of alternating vertices and edges. The walk starts at a vertex p_{i_0} called the **initial vertex**, and ends at the **terminal vertex** p_{i_k}, where k is the **length** of the walk which gives the number of edges traversed. A walk is **closed**, if the starting and the terminal vertices are the same, and **open** if they are different. A zero-length walk is just a single vertex. In a walk, it is allowed to visit a vertex or to traverse an edge more than once.

A **trail** is a walk where all edges are distinct, that is, when any edge is traversed only once. A closed trail is called a **tour** or a **circuit**. Even though it is not universal, an open walk is also called a **path**, however, it is common to use path for cases where no vertices or edges are repeated.

The graph in Figure 7.10 can be given as

$$G = G(V, E), \tag{7.344}$$

$$V = \{1, 2, 3, 4, 5, 6\}, \tag{7.345}$$

$$E = \{(1,2), (2,3), (2,4), (3,5), (4,5), (4,6), (5,6)\}. \tag{7.346}$$

In this graph,

$$1 - 2 - 4 - 5 - 4 - 6 \tag{7.347}$$

is an open walk with length 5, however,

$$3 - 5 - 4 - 5 \tag{7.348}$$

is an open walk with length 3, while

$$2 - 4 - 5 - 6 - 4 - 2 \tag{7.349}$$

is a closed walk of length 5 and

$$2 - 4 - 5 - 3 - 2 \tag{7.350}$$

is a circuit.

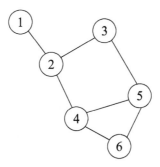

Figure 7.10   A walk is a sequence of vertices and edges.

7.8.4   Trees and Fundamental Circuits

Trees are among the most important concepts in graph theory. A **tree** is a connected graph without a circuit (Figure 7.11). In applications, communication networks, power distribution networks, mail sorting according to zip codes, decision processes, etc., can all be represented by trees. Some of the main properties of trees can be given as follows:

1. There is one and only one path between every pair of vertices in a tree.
2. A tree with n vertices has $n - 1$ edges.
3. A tree T is called a **spanning tree** of a connected graph G, if it is a subgraph of G and includes all the vertices of G (Figure 7.12).
4. A connected graph has at least one spanning tree.
5. A connected graph is a tree, if and only if adding an edge between any two vertices creates exactly one circuit.
6. A circuit created by adding an edge to a spanning tree is called a **fundamental circuit**. A graph may have more than one fundamental circuit (Figure 7.13).
7. An edge in a spanning tree T is called a **branch** of T.
8. An edge of G that is not in a given spanning tree T is called a **chord**.

7.8.5   Graph Operations

It is possible to consider a large graph as composed of smaller ones by using the basic graph operations. Given two graphs $G_1(V_1, E_1)$ and $G_2(V_2, E_2)$, their

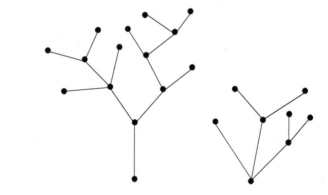

Figure 7.11   Trees.

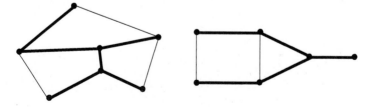

Figure 7.12   Spanning trees.

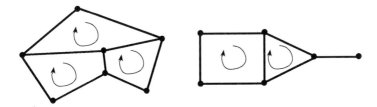

Figure 7.13 Fundamental circuits of the spanning trees in Figure 7.12.

union $G_1 \cup G_2$ is defined as

$$G_1 \cup G_2 = G(V_1 \cup V_2, E_1 \cup E_2). \tag{7.351}$$

Similarly, their **intersection** $G_1 \cap G_2$ is defined as

$$G_1 \cap G_2 = G(V_1 \cap V_2, E_1 \cap E_2). \tag{7.352}$$

The intersection of two graphs consists of the vertices and the edges that are common to both graphs. The **sum** $G_1 \oplus G_2$ of two graphs also called the **ring sum**, consists of the vertex set $V_1 \cup V_2$, and the edges that are either in G_1 or G_2 but not in both (Figure 7.14). Note that these three operations are **commutative**:

$$G_1 \cup G_2 = G_2 \cup G_1, \tag{7.353}$$

$$G_1 \cap G_2 = G_2 \cap G_1, \tag{7.354}$$

$$G_1 \oplus G_2 = G_2 \oplus G_1. \tag{7.355}$$

7.8.6 Cut Sets and Fundamental Cut Sets

In a connected graph G a **cut set** is defined as the set of edges, removal of which leaves G **disconnected**, provided no proper subset of it does not do so. Since a set of edges along with their end vertices is a subgraph of G, a cut set is also a subgraph of G. For example, In Figure 7.15a, the set $\{a, e, f, c\}$ is a cut set that leaves G disconnected (Figure 7.15b). The set $\{i\}$ with one edge and $\{a, b, h\}$ are also cut sets, but $\{b, c, g, h\}$ is not a cut set, since one of its proper subsets $\{b, c, g\}$ is a cut set.

Cut sets have very important applications in communication and transportation networks. For example, let the vertices in Figure 7.15a represent six centers in a transportation network. We need to find the weak spots so that new roads can be opened to strengthen the system. For this, we look at all the cut sets and in particular look for the one with the smallest number of edges, which will be the most vulnerable. From Figure 7.15a, we see that the transportation to or from the city represented by the vertex 5 to the other cities in the network can be halted by shutting down a single edge.

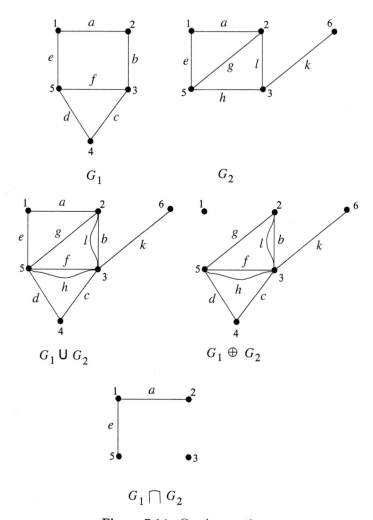

Figure 7.14   Graph operations.

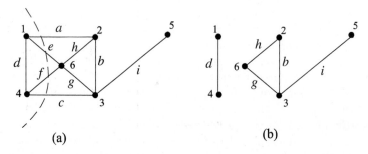

Figure 7.15   The cut set $\{a, e, f, c\}$ separates the graph into two.

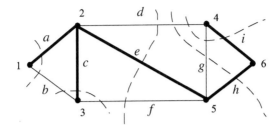

Figure 7.16 Fundamental cut sets of a graph with respect to a spanning tree.

Some of the properties of **cut sets** are
1. A cut set always separates a graph into two.
2. Every edge of a tree is a cut set.
3. A cut set is a subgraph of G.

Let T be a spanning tree of a connected graph G (Figure 7.16). A **cut set** that contains exactly **one branch** of T is called a **fundamental cut set** of G with respect to T. In Figure 7.16, the set $\{a, b, f\}$ is a fundamental cut set with respect to T, since it contains one branch, a, of T. However, the set $\{b, c, e, f\}$ is not a fundamental cut set, since it contains two edges, c and e, of T. A vertex of a connected graph is called a **cut vertex**, if its deletion with the edges incident on it disconnects the graph.

7.8.7 Vector Space Associated with a Graph

In Section 7.6.1, we have defined **vector spaces** over the **field** $\mathbb{Z}_2$, which consists of two numbers $\{0, 1\}$ and that obeys the rules of **binary arithmetic**:

$$0 + 0 = 0, \ 1 + 0 = 1, \ 0 + 1 = 1, \ 1 + 1 = 0, \tag{7.356}$$

$$0 \cdot 0 = 0, \ 1 \cdot 0 = 0, \ 0 \cdot 1 = 0, \ 1 \cdot 1 = 1. \tag{7.357}$$

We now consider the graph G in Figure 7.17 with four vertices and five edges. Any subgraph of G can be represented by a five-tuple

$$\mathbf{X} = (x_1, x_2, x_3, x_4, x_5), \tag{7.358}$$

where x_i are defined as

$$\begin{aligned} x_i &= 1 \quad \text{when } e_i \text{ is in the subgraph,} \\ x_i &= 0 \quad \text{when } e_i \text{ is not in the subgraph.} \end{aligned} \tag{7.359}$$

For example, the subgraph G_1 is represented by $\mathbf{X}_1 = (1, 1, 0, 0, 1)$ and G_2 is represented by $\mathbf{X}_2 = (0, 0, 1, 0, 1)$. Including the **zero vector** $(0, 0, 0, 0, 0)$ and the graph $(1, 1, 1, 1, 1)$, all together, we have $2^5 = 32$ such five-tuples possible.

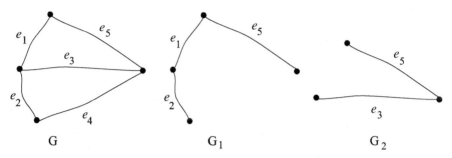

Figure 7.17   Graph G and its two subgraphs G_1 and G_2.

The sum of these five-tuples are done according to the rules of binary arithmetic given in Eqs. (7.356) and (7.357). For example,

$$G_1 \oplus G_2 = X_1 + X_2, \tag{7.360}$$

$$(1,1,0,0,1) \oplus (0,0,1,0,1) = (1,1,1,0,0). \tag{7.361}$$

We are now ready to associate a **vector space** W_G for every graph G over the field $\mathbb{Z}_2$, consisting of 2^n n-tuples, where n is the number of edges in G. Let $\mathbf{X} = (x_1, \ldots, x_n)$, $\mathbf{Y} = (y_1, \ldots, y_n)$ and $\mathbf{Z} = (z_1, \ldots, z_n)$ be any three vectors in the vector space W_G, where in W_G

1. **Summation** is defined as

$$\mathbf{X} \oplus \mathbf{Y} = (x_1 + y_1, \ldots, x_n + y_n), \tag{7.362}$$

where $\mathbf{X} \oplus \mathbf{Y}$ is also an element of W_G:

$$\mathbf{X} \oplus \mathbf{Y} \in W_G, \tag{7.363}$$

and satisfies the **commutative** and the **associative** rules:

$$\mathbf{X} \oplus \mathbf{Y} = \mathbf{Y} \oplus \mathbf{X}, \tag{7.364}$$

$$\mathbf{X} \oplus (\mathbf{Y} \oplus \mathbf{Z}) = (\mathbf{X} \oplus \mathbf{Y}) \oplus \mathbf{Z}. \tag{7.365}$$

2. **Multiplication** with a scalar $c \in \mathbb{Z}_2$ is defined as

$$c \cdot \mathbf{X} = (cx_1, \ldots, cx_n), \tag{7.366}$$

where

$$c \cdot \mathbf{X} \in W_G. \tag{7.367}$$

We also have the **distributive rule**:

$$c\mathbf{Z} \cdot (\mathbf{X} \oplus \mathbf{Y}) = c\mathbf{Z} \cdot \mathbf{X} \oplus c\mathbf{Z} \cdot \mathbf{Y}. \tag{7.368}$$

3. There exists a unique **zero element 0** in W_G, such that for any vector $\mathbf{X} \in W_G$,

$$0 \cdot \mathbf{X} = \mathbf{0}, \tag{7.369}$$

$$\mathbf{0} \oplus \mathbf{X} = \mathbf{X}. \tag{7.370}$$

4. For each element $\mathbf{X} \in W_G$, there exists a unique **inverse** $-\mathbf{X} \in W_G$, such that

$$\mathbf{X} \oplus (-\mathbf{X}) = \mathbf{0}. \tag{7.371}$$

We have used $\oplus$ and $\cdot$ to indicate that addition and multiplication in $\mathbb{Z}_2$ are to be performed according to the rules of binary arithmetic given in Eqs. (7.356) and (7.357).

Linear independence and **basis vectors** in W_G are defined in the same way as done in ordinary vector spaces. For example, the **natural basis** for the graph G in Figure 7.17 is the following five vectors:

$$\begin{pmatrix} 1 \\ 0 \\ 0 \\ 0 \\ 0 \end{pmatrix}, \begin{pmatrix} 0 \\ 1 \\ 0 \\ 0 \\ 0 \end{pmatrix}, \begin{pmatrix} 0 \\ 0 \\ 1 \\ 0 \\ 0 \end{pmatrix}, \begin{pmatrix} 0 \\ 0 \\ 0 \\ 1 \\ 0 \end{pmatrix}, \begin{pmatrix} 0 \\ 0 \\ 0 \\ 0 \\ 1 \end{pmatrix}. \tag{7.372}$$

Any one of the $2^5 = 32$ vectors, $\mathbf{w} \in W_G$, can be represented in terms of linear combinations of these five **basis vectors** as

$$\mathbf{w} = c_1 \begin{pmatrix} 1 \\ 0 \\ 0 \\ 0 \\ 0 \end{pmatrix} + c_2 \begin{pmatrix} 0 \\ 1 \\ 0 \\ 0 \\ 0 \end{pmatrix} + c_3 \begin{pmatrix} 0 \\ 0 \\ 1 \\ 0 \\ 0 \end{pmatrix} + c_4 \begin{pmatrix} 0 \\ 0 \\ 0 \\ 1 \\ 0 \end{pmatrix} + c_5 \begin{pmatrix} 0 \\ 0 \\ 0 \\ 0 \\ 1 \end{pmatrix} \tag{7.373}$$

by a suitable choice of the constants $c_i \in \mathbb{Z}_2$, $i = 1, \ldots, 5$.

7.8.8 Rank and Nullity

We have seen that in specifying a graph, there are three critical numbers, n, e, and k, where n is the number of **vertices**, e is the number of **edges**, and k is the number of **components**. For a connected graph $k = 1$. We also have relations among these numbers. Since every component of a graph must have at least one vertex, we have $n \geqslant k$. Also, since the number of edges in a component cannot be less than $n - 1$, we can write $e \geqslant n - k$. Aside from these constraints:

$$\boxed{n - k \geqslant 0,} \tag{7.374}$$

$$\boxed{e - n + k \geqslant 0,} \tag{7.375}$$

the three numbers, n, e, and k, are independent. From these three numbers, we can define two additional very important parameters as **rank** and **nullity**:

$$\text{rank;} \quad \boxed{r = n - k,}$$

$$\text{nullity;} \quad \boxed{\mu = e - n + k.} \qquad (7.376)$$

We could also determine rank and nullity by the following relations:

$$r = \text{no of branches in any spanning tree of } G, \qquad (7.377)$$

$$\mu = \text{no of chords in } G, \qquad (7.378)$$

$$r + \mu = \text{no of edges in } G. \qquad (7.379)$$

7.8.9   Subspaces in W_G

We have seen that for every graph G, we can associate a vector space W_G. In W_G a **circuit vector** represents either a circuit or the union, or the sum, of edge-disjoint circuits. Similarly, a **cut set vector** represents a cut set or a union, or the sum, of edge disjoint cut sets.

We now state the following results without proof:

1. The set of all circuit vectors, W_r, including **0**, is a subspace of W_G, where the set of fundamental circuits with respect to a spanning tree forms a basis. The dimension of W_r is equal to μ, that is, the nullity of G. The number of circuit vectors in W_r is 2^μ.
2. The set of all cut-set vectors, W_S, including **0**, is a also a subspace of W_G, where the set of fundamental cut sets with respect to a spanning tree of G is a basis for W_S. The dimension of W_S is equal to the rank r of G. The number of vectors in W_S is 2^r.

For the graph G in Figure 7.17, using the spanning tree in Figure 7.18, we determine the critical parameters as

$$e = 5, \ n = 4, \ k = 1; \ r = 3, \ \mu = 2. \qquad (7.380)$$

The number of circuit vectors in W_r is $2^2 = 4$ (Figure 7.19):

$$\begin{pmatrix} 0 \\ 1 \\ 1 \\ 1 \\ 0 \end{pmatrix}, \begin{pmatrix} 1 \\ 1 \\ 0 \\ 1 \\ 1 \end{pmatrix}, \begin{pmatrix} 1 \\ 0 \\ 1 \\ 0 \\ 1 \end{pmatrix}, \begin{pmatrix} 0 \\ 0 \\ 0 \\ 0 \\ 0 \end{pmatrix}. \qquad (7.381)$$

Note that the first two vectors above corresponds to the fundamental circuits of the spanning tree in Figure 7.18, thus they form a basis for W_r. In fact, the remaining two vectors can be written as their linear combination.

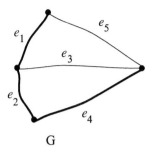

Figure 7.18 Spanning tree for G.

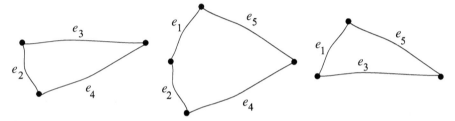

Figure 7.19 Circuit vectors of G in Figure 7.18.

The dimension of the cut set space W_S is 3, hence, there are $2^3 = 8$ vectors in W_S:

$$
\begin{pmatrix} 1 \\ 0 \\ 0 \\ 0 \\ 1 \end{pmatrix},
\begin{pmatrix} 0 \\ 1 \\ 0 \\ 1 \\ 0 \end{pmatrix},
\begin{pmatrix} 0 \\ 0 \\ 1 \\ 1 \\ 1 \end{pmatrix},
\begin{pmatrix} 1 \\ 1 \\ 1 \\ 0 \\ 0 \end{pmatrix},
\begin{pmatrix} 1 \\ 1 \\ 0 \\ 1 \\ 1 \end{pmatrix},
\begin{pmatrix} 1 \\ 0 \\ 1 \\ 1 \\ 1 \end{pmatrix},
\begin{pmatrix} 0 \\ 1 \\ 1 \\ 0 \\ 1 \end{pmatrix},
\begin{pmatrix} 0 \\ 0 \\ 0 \\ 0 \\ 0 \end{pmatrix},
$$

$$(7.382)$$

which are shown in Figure 7.20. Note that the three cut sets in Figure 7.20, (a), (c), (g), are the **fundamental cut sets** with respect to the spanning tree in Figure 7.18, while (b), (d), (f), (e) are their vector sums (g) $\oplus$ (c), (a) $\oplus$ (g), (a) $\oplus$ (c), (a) $\oplus$ (g) $\oplus$ (c), respectively. Therefore, (a), (c), (g) serve as the basis vectors for W_S, where the rest can be written as their linear combinations.

7.8.10   Dot Product and Orthogonal vectors

Dot product in the vector space W_G of a graph G is defined the same way as in ordinary vector spaces. Let $\mathbf{X} = (x_1, \ldots, x_n)$ and $\mathbf{Y} = (y_1, \ldots, y_n)$ be any two vectors in W_G, then the **dot product** of these vectors is defined as

$$\mathbf{X} \cdot \mathbf{Y} = x_1 y_1 + \cdots + x_n y_n, \tag{7.383}$$

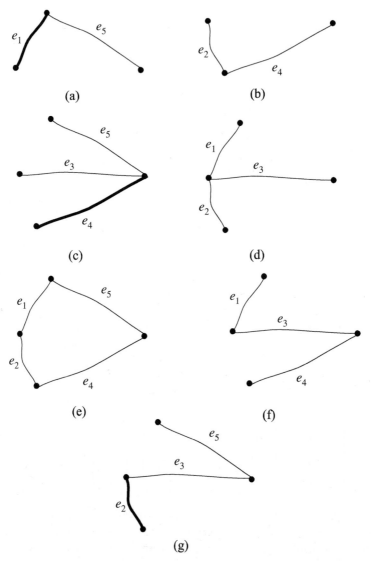

Figure 7.20   Cut set vectors in W_S of G in Figure 7.18.

where the multiplication and addition is done according to the rules of **binary arithmetic**. For example,

$$(1,0,0,0,1) \cdot (0,1,0,1,0) = 1 \cdot 0 + 0 \cdot 1 + 0 \cdot 0 + 0 \cdot 1 + 1 \cdot 0 \qquad (7.384)$$

$$= 0 + 0 + 0 + 0 + 0 = 0. \qquad (7.385)$$

When the dot product of two vectors is zero; they are called **orthogonal** to each other. If every vector in subspace is orthogonal to every other vector in another subspace, we call them **orthogonal subspaces**. We leave it to the reader to verify that the circuit and the cut set spaces W_r and W_S of G are orthogonal. We now state the following useful theorems:

Theorem 7.4. Let G_1 and G_2 be two subgraphs of G, the dot product of their corresponding vectors is 0 if the number of their common edges is even, and 1 if the number of their common edges is odd.

Theorem 7.5. In the vector space W_G of a graph, the circuit subspace W_r and the cut set subspace W_S are orthogonal.

7.8.11 Matrix Representation of Graphs

Even though the pictorial description of graphs are useful, for large and complicated graphs, we need something more practical and efficient. In particular, we need a representation that can be fed into a computer that will allow graph operations in an electronic environment. In this regard, matrix representations of graphs are very useful. Since graphs are completely defined by giving either their adjacency or their incidence structure, the **adjacency** or the **incidence** matrices are the starting points in implementing the versatile tools of linear algebra and also the way to define operations on graphs with computers.

The Adjacency Matrix

Given a graph $G = G(V, E)$ with vertices $V = \{p_1, p_2, \ldots, p_n\}$, the **adjacency matrix** $\mathbf{A} = a_{ij}$, $i, j = 1, \ldots, n$, is defined as the following $n \times n$ matrix:

$$a_{ij} = \begin{cases} 1 & \text{if there is an edge connecting } p_i \text{ to } p_j \\ 0 & \text{otherwise} \end{cases} \tag{7.386}$$

The adjacency matrix of the graph in Figure 7.21 with the vertices $p_i, i = 1, \ldots, 6$ is written as

$$\mathbf{A} = \begin{array}{c} p_i \backslash p_j \\ 1 \\ 2 \\ 3 \\ 4 \\ 5 \\ 6 \end{array} \begin{array}{cccccc} 1 & 2 & 3 & 4 & 5 & 6 \\ \begin{pmatrix} 0 & 0 & 0 & 1 & 0 & 0 \\ 0 & 0 & 1 & 1 & 0 & 0 \\ 0 & 1 & 0 & 1 & 1 & 0 \\ 1 & 1 & 1 & 0 & 0 & 0 \\ 0 & 0 & 1 & 0 & 0 & 1 \\ 0 & 0 & 0 & 0 & 1 & 0 \end{pmatrix} \end{array} \tag{7.387}$$

We can make the following observations about the adjacency matrix:

1. The adjacency matrix for undirected simple graphs are symmetric. It is not necessary for the adjacency matrix of a digraph to be symmetric.

2. For a simple graph, the adjacency matrix is composed of 0s and 1s and with 0 on its diagonal.

We now state a useful theorem:

Theorem 7.6. Given the adjacency matrix $\mathbf{A}$ of a graph, the ijth element of $\mathbf{A}^k$ gives the number of distinct k-walks from the vertex p_i to p_j.

Calculating $\mathbf{A}^2$ for the graph in Figure 7.21, we get

$$
\mathbf{A}^2 = \begin{array}{c} p_i\backslash p_j \\ 1 \\ 2 \\ 3 \\ 4 \\ 5 \\ 6 \end{array} \begin{array}{c} \begin{array}{cccccc} 1 & 2 & 3 & 4 & 5 & 6 \end{array} \\ \left(\begin{array}{cccccc} 1 & 1 & 1 & 0 & 0 & 0 \\ 1 & 2 & 1 & 1 & 1 & 0 \\ 1 & 1 & 3 & 1 & 0 & 1 \\ 0 & 1 & 1 & 3 & 1 & 0 \\ 0 & 1 & 0 & 1 & 2 & 0 \\ 0 & 0 & 1 & 0 & 0 & 1 \end{array}\right) \end{array}, \tag{7.388}
$$

This matrix gives the number of distinct walks from a given vertex to another by travelling only over two edges. For example, there are no two-walks between the vertices p_i and p_j, where $i \neq j$. Similarly, we calculate $\mathbf{A}^3$:

$$
\mathbf{A}^3 = \begin{array}{c} p_i\backslash p_j \\ 1 \\ 2 \\ 3 \\ 4 \\ 5 \\ 6 \end{array} \begin{array}{c} \begin{array}{cccccc} 1 & 2 & 3 & 4 & 5 & 6 \end{array} \\ \left(\begin{array}{cccccc} 0 & 1 & 1 & 3 & 1 & 0 \\ 1 & 2 & 4 & 4 & 1 & 1 \\ 1 & 4 & 2 & 5 & 4 & 0 \\ 3 & 4 & 5 & 2 & 1 & 1 \\ 1 & 1 & 4 & 1 & 0 & 2 \\ 0 & 1 & 0 & 1 & 2 & 0 \end{array}\right) \end{array}, \tag{7.389}
$$

which gives the number of distinct three-walks. As can be verified from the graph, there are four distinct paths between p_3 and p_5.

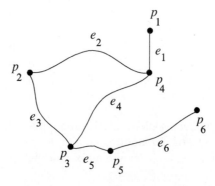

Figure 7.21 Adjacency matrix.

Example 7.14. *Adjacency matrix of a digraph:* We write the adjacency matrix for the digraph in Figure 7.9 as

$$
\mathbf{A} = \begin{array}{c} p_i \backslash p_j \\ 1 \\ 2 \\ 3 \\ 4 \end{array} \begin{array}{cccc} 1 & 2 & 3 & 4 \\ \left(\begin{array}{cccc} 0 & 0 & 0 & 1 \\ 0 & 0 & 1 & 0 \\ 0 & 1 & 0 & 1 \\ 0 & 1 & 1 & 0 \end{array} \right) \end{array}. \tag{7.390}
$$

Consider the digraph in Figure 7.22, representing the delivery route of a company servicing five cities, $p_1, \ldots, p_5$.

We write the adjacency matrix as

$$
\mathbf{A} = \begin{array}{c} p_i \backslash p_j \\ 1 \\ 2 \\ 3 \\ 4 \\ 5 \end{array} \begin{array}{ccccc} 1 & 2 & 3 & 4 & 5 \\ \left(\begin{array}{ccccc} 0 & 1 & 0 & 0 & 0 \\ 1 & 0 & 1 & 1 & 1 \\ 0 & 1 & 0 & 1 & 1 \\ 0 & 0 & 1 & 0 & 0 \\ 0 & 1 & 0 & 1 & 0 \end{array} \right) \end{array} \tag{7.391}
$$

and its powers $\mathbf{A}^2$ and $\mathbf{A}^3$ as

$$
\mathbf{A}^2 = \begin{array}{c} p_i \backslash p_j \\ 1 \\ 2 \\ 3 \\ 4 \\ 5 \end{array} \begin{array}{ccccc} 1 & 2 & 3 & 4 & 5 \\ \left(\begin{array}{ccccc} 1 & 0 & 1 & 1 & 1 \\ 0 & 3 & 1 & 2 & 1 \\ 1 & 1 & 2 & 2 & 1 \\ 0 & 1 & 0 & 1 & 1 \\ 1 & 0 & 2 & 1 & 1 \end{array} \right) \end{array}, \tag{7.392}
$$

$$
\mathbf{A}^3 = \begin{array}{c} p_i \backslash p_j \\ 1 \\ 2 \\ 3 \\ 4 \\ 5 \end{array} \begin{array}{ccccc} 1 & 2 & 3 & 4 & 5 \\ \left(\begin{array}{ccccc} 0 & 3 & 1 & 2 & 1 \\ 3 & 2 & 5 & 5 & 4 \\ 1 & 4 & 2 & 4 & 3 \\ 1 & 1 & 2 & 2 & 1 \\ 0 & 4 & 1 & 3 & 2 \end{array} \right) \end{array}. \tag{7.393}
$$

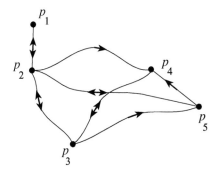

Figure 7.22 Delivery route of a company.

Note that for transportation from p_1 to p_2, there is only 1-step connection, no 2-step connections, but there are three 3-step connections. Also note that even though the edges (p_2, p_5) and (p_3, p_4) appear to intersect, they are not connected by a vertex. Such edges are not coplanar.

The Incidence Matrix

The incidence matrix, also called the **all-vertex incidence** matrix, of a nonempty and loopless graph $G(V, E)$ is an $n \times m$ matrix $\mathbf{A} = a_{ij}$, $i = 1, \ldots, n$, $j = 1, \ldots, m$, where n is the number of vertices and m is the number of edges:

$$a_{ij} = \begin{cases} 1 & \text{if } p_i \text{ is an end vertex of } e_j \\ 0 & \text{otherwise} \end{cases} . \tag{7.394}$$

We can write the incidence matrix of the graph in Figure 7.23 as

$$\mathbf{A} = \begin{array}{c} p_i \backslash e_j \\ 1 \\ 2 \\ 3 \\ 4 \\ 5 \end{array} \begin{array}{cccccc} 1 & 2 & 3 & 4 & 5 & 6 \\ \left(\begin{array}{cccccc} 1 & 0 & 0 & 0 & 0 & 0 \\ 1 & 1 & 1 & 0 & 0 & 1 \\ 0 & 1 & 1 & 1 & 0 & 0 \\ 0 & 0 & 0 & 1 & 1 & 1 \\ 0 & 0 & 0 & 0 & 1 & 0 \end{array} \right) \end{array} \tag{7.395}$$

The incidence matrix of a nonempty and directed loopless graph is defined as

$$a_{ij} = \begin{cases} 1 & \text{if } p_i \text{ is the initial vertex of } e_j \\ -1 & \text{if } p_i \text{ is a terminal vertex of } e_j \\ 0 & \text{otherwise} \end{cases} . \tag{7.396}$$

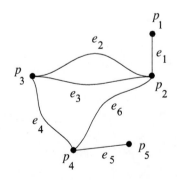

Figure 7.23   Incidence matrix.

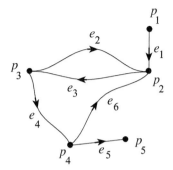

Figure 7.24 Incidence matrix of a directed graph.

For the graph in Figure 7.24, we write the incidence matrix as

$$
\mathbf{A} = \begin{array}{c} p_i\backslash e_j \\ 1 \\ 2 \\ 3 \\ 4 \\ 5 \end{array}
\begin{pmatrix}
\begin{array}{cccccc}
1 & 2 & 3 & 4 & 5 & 6 \\
1 & 0 & 0 & 0 & 0 & 0 \\
-1 & -1 & 1 & 0 & 0 & -1 \\
0 & 1 & -1 & 1 & 0 & 0 \\
0 & 0 & 0 & -1 & 1 & 1 \\
0 & 0 & 0 & 0 & -1 & 0
\end{array}
\end{pmatrix}. \tag{7.397}
$$

We can make the following observations about the incidence matrix:
1. The incidence matrix is a **binary matrix** that contains only 0s and 1s.
2. Since every edge connects two vertices, each column of the incidence matrix has two 1s.
3. The number of 1s in each row gives the **degree** of the corresponding vertex.
4. A row with all zeros corresponds to an **isolated vertex**.
5. Edges that are **parallel** yield identical columns in the incidence matrix.
6. The **incidence matrix** of a disconnected graph, $A(G)$, consisting of two disconnected parts, $A(G_1)$ and $A(G_2)$, can be written in block diagonal form as

$$
A(G) = \begin{pmatrix} A(G_1) & 0 \\ 0 & A(G_2) \end{pmatrix}. \tag{7.398}
$$

7. Permutations of any two rows or columns of the of the incidence matrix corresponds to renaming the vertices and the edges.

7.8.12   Dominance Directed Graphs

In social sciences we are often interested in studying relations in a group. Using graph theory, we can assign a vertex to each member of a group and draw an edge that points toward the member of the group that is thought to be **dominated**, or **influenced**, by the member represented by the preceding vertex (Figure 7.25).

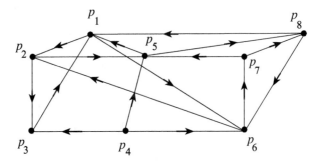

Figure 7.25 Application to sociology via dominance directed graphs.

The adjacency matrix of the group in Figure 7.25 can be written as

$$
\mathbf{A} =
\begin{array}{c}
p_i \backslash p_j \\
1 \\
2 \\
3 \\
4 \\
5 \\
6 \\
7 \\
8
\end{array}
\begin{array}{cccccccc}
1 & 2 & 3 & 4 & 5 & 6 & 7 & 8 \\
\left(\begin{array}{cccccccc}
0 & 1 & 0 & 0 & 0 & 1 & 0 & 0 \\
0 & 0 & 1 & 0 & 1 & 0 & 0 & 0 \\
1 & 0 & 0 & 0 & 0 & 0 & 0 & 0 \\
0 & 0 & 1 & 0 & 1 & 1 & 0 & 0 \\
1 & 0 & 0 & 0 & 0 & 0 & 0 & 1 \\
0 & 1 & 0 & 0 & 0 & 0 & 1 & 0 \\
0 & 0 & 0 & 0 & 1 & 0 & 0 & 1 \\
1 & 0 & 0 & 0 & 0 & 1 & 0 & 0
\end{array}\right)
\end{array}.
\qquad (7.399)
$$

The row that has the most number of 1s corresponds to the most influential member of the group. In this case, it is the member number 4, which has three direct connections. In other words, number 4 has direct access to more members of the group than the others. Calculating $\mathbf{A}^2$ and $\mathbf{A}^3$:

$$
\mathbf{A}^2 =
\begin{array}{c}
p_i \backslash p_j \\
1 \\
2 \\
3 \\
4 \\
5 \\
6 \\
7 \\
8
\end{array}
\begin{array}{cccccccc}
1 & 2 & 3 & 4 & 5 & 6 & 7 & 8 \\
\left(\begin{array}{cccccccc}
0 & 1 & 1 & 0 & 1 & 0 & 1 & 0 \\
2 & 0 & 0 & 0 & 0 & 0 & 0 & 1 \\
0 & 1 & 0 & 0 & 0 & 1 & 0 & 0 \\
2 & 1 & 0 & 0 & 0 & 0 & 1 & 1 \\
1 & 1 & 0 & 0 & 0 & 2 & 0 & 0 \\
0 & 0 & 1 & 0 & 2 & 0 & 0 & 1 \\
2 & 0 & 0 & 0 & 0 & 1 & 0 & 1 \\
0 & 2 & 0 & 0 & 0 & 1 & 1 & 0
\end{array}\right)
\end{array},
\qquad (7.400)
$$

$$
\mathbf{A}^3 =
\begin{array}{c}
p_i \backslash p_j \\
1 \\
2 \\
3 \\
4 \\
5 \\
6 \\
7 \\
8
\end{array}
\begin{array}{cccccccc}
1 & 2 & 3 & 4 & 5 & 6 & 7 & 8 \\
\left(\begin{array}{cccccccc}
2 & 0 & 1 & 0 & 2 & 0 & 0 & 2 \\
1 & 2 & 0 & 0 & 0 & 3 & 0 & 0 \\
0 & 1 & 1 & 0 & 1 & 0 & 1 & 0 \\
1 & 2 & 1 & 0 & 2 & 3 & 0 & 1 \\
0 & 3 & 1 & 0 & 1 & 1 & 2 & 0 \\
4 & 0 & 0 & 0 & 0 & 1 & 0 & 2 \\
1 & 3 & 0 & 0 & 0 & 3 & 1 & 0 \\
0 & 1 & 2 & 0 & 3 & 0 & 1 & 1
\end{array}\right)
\end{array},
\qquad (7.401)
$$

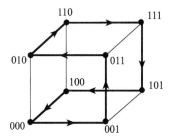

Figure 7.26   A 3-bit Gray code corresponds to a circuit on a 3-cube.

we see that the member number 4 is still the most influential member with five 2-stage and ten 3-stage influences on the group.

7.8.13   Gray Codes in Coding Theory

When information is converted into binary form, it usually consists of a list of m-tuples such that each entry differs from the preceding one in just one or more digits or coordinates. Let us consider a case where the angular position of a shaft is determined by means of three brushes on a commutator. As the shaft rotates, the angle changes from $0°$ to $360°$. Using the activation sequence of the switches in the commutator, we record the information about the angular position of the shaft as the following list of three-tuples:

$$
\begin{array}{rcl}
0\text{–}45° & \rightarrow & 000 \\
45\text{–}90° & \rightarrow & 001 \\
90\text{–}135° & \rightarrow & 011 \\
135\text{–}180° & \rightarrow & 010 \\
180\text{–}225° & \rightarrow & 110 \\
225\text{–}270° & \rightarrow & 111 \\
270\text{–}315° & \rightarrow & 101 \\
315\text{–}360° & \rightarrow & 100
\end{array}
\tag{7.402}
$$

Note that the last state rolls over to

$$
0\text{–}45° \quad \rightarrow \quad 000
\tag{7.403}
$$

with the change of one digit, that is, with one switch flip. Such a code where two successive values differ in only one bit is called a **gray code, reflected binary code, circuit code**, or **cyclic code**. Gray codes are widely used in linear or rotary position encoders and also in digital communication for error correction and in some cable TV systems. An m-bit Gray code corresponds to a circuit in a m-cube, where for $m = 3$ the corresponding circuit is shown with thick lines in Figure 7.26.

REFERENCES

1. Tosun, I. (2012). *The Thermodynamics of Phase and Reaction Equilibria*. Springer-Verlag.

2. Gass, S.I. (2010). *Linear Programming: Methods and Applications*, 5e. Dover Publications.

3. Dongerra, J. and Sullivan, F. (2000). Guest Editors' Introdroduction: the top 10 algorithms. *Comput. Sci. Eng.* 2: 22–23.

4. Nash, J.C. (2000). The (Dantzig) simplex method for linear prgramming. *Comput. Sci. Eng.* 2: 29–31.

5. Karmarkar, N.K. (1991). Interior methods in optimization. *Proceedings of the 2nd International Conference on Industrial and Applied Mathematics*, SIAM, p. 160181.

6. Leontief, V. (1986). *Input-Output Economics*, 2e. Oxford University Press.

7. Jones, G.A. and Jones, J.M. (2006). *Information and Coding Theory*. Springer-Verlag.

8. Buckley, F. and Lewinter, M. (2013). *Introduction to Graph Theory With Applications*. Waveland Press.

9. Chartrand, G. and Zang, P. (2012). *A First Course in Graph Theory*. Dover Publications.

10. Deo, N. (1974). *Graph Theory With Applications to Engineering and Computer Science*. Prentice Hall of India.

11. Yadav, S.K. (2010). *Elements of Graph Theory*. Ane Books Pvt. Ltd.

PROBLEMS

1. The following eight reactions are suggested for the steam reforming of glycerol ($C_3H_8O_3$):

    1. $C_3H_8O_3 + 3H_2O \rightleftharpoons 7H_2 + 3CO_2$
    2. $C_3H_8O_3 \rightleftharpoons 4H_2 + 3\,CO$
    3. $C + H_2O \rightleftharpoons H_2 + CO$
    4. $CO + H_2O \rightleftharpoons H_2 + CO_2$
    5. $C + 2H_2 \rightleftharpoons CH_4$
    6. $CO + 3H_2 \rightleftharpoons CH_4 + H_2O$
    7. $CO_2 + 4H_2 \rightleftharpoons CH_4 + 2H_2O$
    8. $C + CO_2 \rightleftharpoons 2CO.$

    Determine the number of independent reactions representing this set. Also, find a set of independent reactions.

2. Given a mixture consisting of $KmnO_4$, HCl, Cl_2, $MnCl_2$, H_2O, KCl, and NaCl, show that the following reaction:

    $$2KMnO_4 + 16\,HCl \rightleftharpoons 5Cl_2 + 2MnCl_2 + 8H_2O + 2\,KCl,$$

    represents this system. Keep in mind that NaCl does not take part in the reaction.

3. Balance the following reaction using the techniques of linear algebra:

$$a_1 \text{ Pb } (\text{N}_3)_2 + a_2 \text{ Cr } (\text{MnO}_4)_2 + a_3\text{Cr}_2\text{O}_3 + a_4\text{MnO}_2 + a_5\text{NO}$$
$$+ a_6\text{Pb}_3\text{O}_4 = 0.$$

Hint: First construct the 5×6 element-species matrix $\mathbf{A}$ and the 6×1 column matrix $\mathbf{a}$:

$$\mathbf{A} = \begin{pmatrix} 1 & 0 & 0 & 0 & 0 & 3 \\ 6 & 0 & 0 & 0 & 1 & 0 \\ 0 & 1 & 2 & 0 & 0 & 0 \\ 0 & 2 & 0 & 1 & 0 & 0 \\ 0 & 8 & 3 & 2 & 1 & 4 \end{pmatrix}, \quad \mathbf{a} = \begin{pmatrix} a_1 \\ a_2 \\ a_3 \\ a_4 \\ a_5 \\ a_6 \end{pmatrix},$$

and then solve the homogeneous equation $\mathbf{Aa} = 0$. Discuss what all this means.

4. Maximize the objective function

$$r = 2x_1 + 2x_2 - 2x_3,$$

subject to the constraints

$$x_1 + 2x_2 + 2x_3 \leqslant 8,$$
$$3x_1 - x_2 + 2x_3 \leqslant 12,$$
$$2x_1 + 3x_2 + 5x_3 \leqslant 15,$$
$$x_1, x_2, x_3 \geqslant 0.$$

5. Maximize the following revenue function:

$$r = 10x_1 + 5x_2$$

subject to

$$3x_1 + x_2 \leqslant 8,$$
$$x_1 - x_2 \leqslant 1,$$
$$x_1, x_2 \geqslant 0,$$

both with the geometric method and the simplex method. Discuss your answers.

6. Prove that a Markov matrix always has an eigenvalue equal to 1 and all other eigenvalues are in absolute value less than or equal to 1.

7. Given the following 2×2 matrix:

$$\mathbf{A} = \begin{pmatrix} a & b \\ 1-a & 1-b \end{pmatrix}.$$

If $\mathbf{A}$ is a Markov matrix, then show that one of the eigenvalues is 1 and the other is less than 1.

8. For Leontief closed models show that $\det(\mathbf{I} - \mathbf{A}) = \mathbf{0}$.

9. In a Leontief open model, the input–output matrix is given

$$\mathbf{A} = \begin{pmatrix} 0.1 & 0.2 & 0.3 \\ 0.1 & 0.1 & 0.4 \\ 0.3 & 0.2 & 0.1 \end{pmatrix},$$

Is this economy productive? Explain.

10. Find the radius and the center of the circle passing through the points $(-1, 2), (1, 0)$, and $(3, 1)$. Discuss your result.

11. Find the equation of a plane that passes through three points: (x_1, y_1, z_1), (x_2, y_2, z_2), (x_3, y_3, z_3), not on the same line.

12. For the following polynomials:

$$f = y^2 x - 3yx^2 + y^2 - 3yx,$$
$$g = y^3 x + y^3 - 4x^2 - 3x + 1,$$

(i) Compute $\text{Res}(f, g, y)$ and $\text{Res}(f, g, x)$.

(ii) From the first part what can you say about f and g?

13. Evaluate the following operations in $\mathbb{Z}_2^6$:

$$(1, 0, 1, 1, 1) + (1, 1, 0, 0, 1),$$
$$0 \cdot (1, 1, 0, 1, 0),$$
$$1 \cdot (1, 0, 0, 1, 1),$$
$$(1, 1, 0, 0, 1) \cdot (0, 1, 1, 1, 0).$$

14. Given that the received message in Hamming $(7,4)$ code is $\mathbf{w} = (1, 1, 0, 1, 0, 1)$. Assuming that only one error exists, find the original message. Check your answer.

15. The following message is received in Hamming $(7,4)$ code: $(1, 0, 0, 0, 0, 1, 1)$. Does it contain any errors?

16. Verify that Eq. (7.306): $\mathbf{H}\mathbf{e}_i = c_i$, is true.

17. Express the numbers 11, 12, 15, 33, 105 in binary notation.

18. Evaluate the sum $16 + 30$ and the product 12×55 in binary notation and convert your answer back to decimal notation.

19. Use the following 3×3 encoding matrix:

$$
\mathbf{E} = \begin{pmatrix} 3 & 1 & 5 \\ 2 & -1 & 1 \\ 1 & 0 & 2 \end{pmatrix},
$$

    to send the message *DO NOT NEGOTIATE*. Find the encoded message and also show how to decode it.

20. Find the incidence matrix of the Graph G_1 in Figure 7.27.

21. Write the adjacency matrix for Graph G_2 in Figure 7.28. How many distinct two-walks exist between P_5 and P_2, and P_5 and P_3.. Show them on the graph.

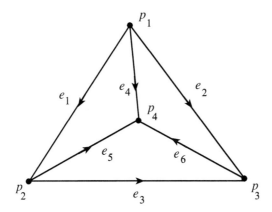

Figure 7.27   Graph G_1 for problem 20.

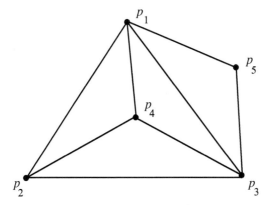

Figure 7.28   Graph G_2 for Problem 21.

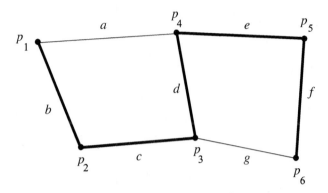

Figure 7.29   Problems 23 and 24.

22. For the spanning tree $T = \{a, c, d, e, f\}$ in Figure 7.29, which one of the following cut sets is fundamental cut set: $\{a, b, c\}$, $\{a, b, g\}$, $\{a, e, g\}$.

23. For a different spanning tree of the graph in Figure 7.29, write all the vectors in the cut space. Which ones are the fundamental cut sets. Verify your answer.

24. Find the circuit vectors in the circuit vector space of the graph in Figure 7.29. Which ones are the fundamental circuits.

CHAPTER 8

SEQUENCES AND SERIES

Sequences and series have found a wide range of applications in both pure and applied mathematics. A sequence is a succession of numbers or functions that may or may not converge to a limit value. A natural way to generate sequences is to use partial sums of series. In this regard, the convergence of a series is synonymous with the convergence of the corresponding sequence of partial sums. In this chapter, we introduce the basic concepts of sequences and series and discuss their most commonly used properties. We can name the evaluation of definite integrals, calculation of limits, and series approximation of functions to any desired level of accuracy as being among the most frequently used techniques with series. Series solutions of differential and integral equations and perturbative techniques are other important applications of series.

Essentials of Mathematical Methods in Science and Engineering, Second Edition. Selçuk Ş. Bayın. **425**
© 2020 John Wiley & Sons, Inc. Published 2020 by John Wiley & Sons, Inc.

8.1 SEQUENCES

An **infinite sequence** is defined by assigning a number S_n to each positive integer n and then by writing them in a definite order:

$$\{S_1, S_2, \ldots, S_n, S_{n+1}, \ldots\},$$
(8.1)

or as

$$\{S_n\}, n = 1, 2, \ldots.$$
(8.2)

A **finite sequence** has a finite number of elements $\{S_1, S_2, \ldots, S_n\}$. In general, a sequence is defined by a rule that allows us to generate the sequence, which is usually given as the nth term of the sequence. For example, the sequences:

$$\left\{\frac{1}{2}, \frac{1}{4}, \frac{1}{8}, \ldots\right\},$$
(8.3)

$$\left\{2, \left(\frac{3}{2}\right)^2, \left(\frac{4}{3}\right)^3, \ldots\right\},$$
(8.4)

$$\{-1, 1, -1, \ldots\},$$
(8.5)

$$\left\{1, \left(1 + \frac{1}{2}\right), \left(1 + \frac{1}{2} + \frac{1}{3}\right), \ldots\right\},$$
(8.6)

are generated by the following nth terms, respectively:

$$S_n = \frac{1}{2^n}, \quad S_n = \left(\frac{n+1}{n}\right)^n, \quad S_n = (-1)^n, \quad S_n = \sum_{k=1}^{n} \frac{1}{k}.$$
(8.7)

These sequences can also be written, respectively, as

$$\left\{\frac{1}{2^n}\right\}, \quad \left\{\left(\frac{n+1}{n}\right)^n\right\}, \quad \{(-1)^n\}, \quad \left\{\sum_{k=1}^{n} \frac{1}{k}\right\}, \quad n = 1, 2, \ldots.$$
(8.8)

A sequence $\{S_n\}$ is said to **converge** to the number S, that is,

$$\lim_{n \to \infty} S_n \to S,$$
(8.9)

if for every positive number ε we can find an N such that for all $n > N$ the inequality

$$|S_n - S| < \varepsilon, \quad n > N,$$
(8.10)

is satisfied. The limit, when it exists, is unique.

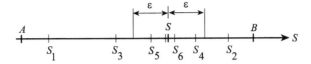

Figure 8.1   Convergent sequence.

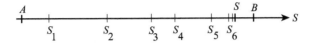

Figure 8.2   Convergent monotonic increasing sequence.

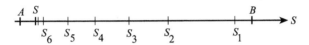

Figure 8.3   Convergent monotonic decreasing sequence.

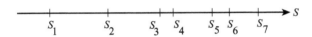

Figure 8.4   Divergent monotonic increasing sequence.

Figure 8.5   Divergent bounded sequence.

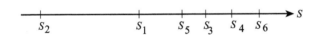

Figure 8.6   Divergent unbounded sequence.

In Figures 8.1–8.6, we show various possibilities for sequences. **Convergence** of a sequence can also be decided without the knowledge of its limit by using the **Cauchy criteria**:

Theorem 8.1. *Cauchy criteria*: For a given infinite sequence $\{S_n\}$, the limit

$$\boxed{\lim_{n\to\infty} S_n = S} \tag{8.11}$$

exists, if and only if the following condition is satisfied: For every $\varepsilon > 0$, there exists an integer N, such that for all n and m greater than N

$$|S_n - S_m| < \varepsilon, \; n \neq m, \tag{8.12}$$

is true.

A sequence is called **divergent** if it is not convergent. Sequences in Eqs. (8.3) and (8.4) converge since they have the limits

$$\lim_{n\to\infty} \frac{1}{2^n} \to 0, \tag{8.13}$$

$$\lim_{n\to\infty} \left(\frac{n+1}{n}\right)^n \to e, \tag{8.14}$$

while the sequences in Eqs. (8.5) and (8.6) diverge. A **monotonic increasing** sequence satisfies (Figure 8.2)

$$\boxed{S_1 \le S_2 \le S_3 \le \cdots \le S_n \le \cdots .} \tag{8.15}$$

Similarly, a **monotonic decreasing** sequence is defined as (Figure 8.3)

$$\boxed{S_1 \ge S_2 \ge S_3 \ge \cdots \ge S_n \ge \cdots .} \tag{8.16}$$

A sequence is **bounded** if there are two numbers, A and B, such that $A \le S_n \le B$ for all n. Sequences in Figures 8.4 and 8.6 are unbounded. It can be proven that every bounded monotonic decreasing sequence is convergent and every unbounded sequence is divergent [1]. Sequences that are bounded but not monotonic can diverge or converge. For example, the sequence

$$\{-1, 1, -1, 1, \ldots, (-1)^n, \ldots\} \tag{8.17}$$

diverges by oscillating between -1 and 1. Another sequence

$$\left\{\cos\frac{n\pi}{2}\right\}, \quad n = 1, 2, \ldots, \tag{8.18}$$

oscillates between the values $0, 1$, and -1:

$$\{0, -1, 0, 1, 0, -1, \ldots\}. \tag{8.19}$$

In other words, a bounded series may diverge not because it does not have a limit value but because it has too many.

For bounded sequences, we define an **upper limit** as

$$\boxed{\overline{\lim}_{n\to\infty} S_n \to U} \tag{8.20}$$

if for every $\varepsilon > 0$ there exists a number N such that

$$|S_n - U| < \varepsilon \tag{8.21}$$

is true for infinitely many $n > N$ and if no number larger than U has this property. Similarly, we define the **lower limit** as

$$\underline{\lim}_{n \to \infty} S_n \to L,$$ (8.22)

if for every $\varepsilon > 0$ there exists a number M such that

$$|S_n - L| < \varepsilon$$ (8.23)

is true for infinitely many $n > M$ and if no number less than L has this property. A bounded sequence has an upper and a lower bound such that $U > L$. For unbounded sequences, we can take $+\infty$ and $-\infty$ as possible limit "values."

For a sequence of real numbers $\{S_n\}$, we can say the following:

(i)

$$\overline{\lim}_{n \to \infty} S_n \geq \underline{\lim}_{n \to \infty} S_n.$$ (8.24)

(ii) A sequence converges if and only if both limits are finite and equal:

$$\overline{\lim}_{n \to \infty} S_n = \underline{\lim}_{n \to \infty} S_n = S.$$ (8.25)

(iii) A sequence diverges to $+\infty$ if and only if

$$\overline{\lim}_{n \to \infty} S_n = \underline{\lim}_{n \to \infty} S_n = +\infty.$$ (8.26)

(iv) A sequence diverges to $-\infty$ if and only if

$$\overline{\lim}_{n \to \infty} S_n = \underline{\lim}_{n \to \infty} S_n = -\infty.$$ (8.27)

(v) A sequence where we have

$$\overline{\lim}_{n \to \infty} S_n \neq \underline{\lim}_{n \to \infty} S_n$$ (8.28)

is said to oscillate. By simple graphing one can verify the following limits:

$$
\begin{array}{lll}
S_n = (-1)^n \left(1 - \tfrac{1}{n}\right); & \underline{\lim}_{n \to \infty} S_n = -1, & \overline{\lim}_{n \to \infty} S_n = +1, \\
S_n = 1 + (-1)^n/3^n; & \underline{\lim}_{n \to \infty} S_n = 1, & \overline{\lim}_{n \to \infty} S_n = 1, \\
S_n = (-1)^n n^2; & \underline{\lim}_{n \to \infty} S_n = -\infty, & \overline{\lim}_{n \to \infty} S_n = +\infty, \quad (8.29) \\
S_n = n \sin^2(\tfrac{1}{2} n\pi); & \underline{\lim}_{n \to \infty} S_n = 0, & \overline{\lim}_{n \to \infty} S_n = +\infty, \\
S_n = \sin(\tfrac{1}{2} n\pi); & \underline{\lim}_{n \to \infty} S_n = -1, & \overline{\lim}_{n \to \infty} S_n = +1.
\end{array}
$$

8.2   INFINITE SERIES

An infinite series of real numbers is defined as the **infinite sum**

$$a_1 + a_2 + a_3 + \cdots = \sum_{n=1}^{\infty} a_n. \tag{8.30}$$

When there is no room for confusion, we also write an infinite series as $\sum_n a_n$ or simply as $\sum a_n$. With each infinite sum, we can associate a sequence of partial sums,

$$\{S_1, S_2, S_3, \ldots, S_n, \ldots\}, \tag{8.31}$$

where

$$S_1 = a_1,$$
$$S_2 = a_1 + a_2,$$
$$S_3 = a_1 + a_2 + a_3,$$
$$\vdots$$
$$S_n = \sum_{i=1}^{n} a_i. \tag{8.32}$$

The series $\sum_{i=1}^{\infty} a_i$ is said to converge if the sequence of partial sums $\{S_n\}$ converges and the series diverges if the sequence of partial sums diverges. When the sequence of partial sums converges to S, we say the **series converges** and call S the sum of the series and write

$$\sum_{i=1}^{\infty} a_i = S. \tag{8.33}$$

In other words,

$$\lim_{n \to \infty} \sum_{i=1}^{n} a_i = \sum_{i=1}^{\infty} a_i = S. \tag{8.34}$$

When the limit exists and if $\lim_{n \to \infty} S_n = \pm\infty$, we say the series is (properly) **divergent**. Let $S_a = \sum_{n=1}^{\infty} a_n$ and $S_b = \sum_{n=1}^{\infty} b_n$ be two **convergent series** and let α and β be two constants, then we can write

$$\sum_{n=1}^{\infty} [\alpha a_n + \beta b_n] = \alpha \sum_{n=1}^{\infty} a_n + \beta \sum_{n=1}^{\infty} b_n, = \alpha S_a + \beta S_b. \tag{8.35}$$

8.3   ABSOLUTE AND CONDITIONAL CONVERGENCE

A series $\sum a_n$ is called **absolutely convergent** if the series constructed by taking the absolute value of each term, $\sum |a_n|$, converges. A series that does not converge absolutely but converges otherwise is called **conditionally convergent**. For a convergent series, the nth term a_n necessarily goes to zero as n goes to infinity. This is obvious from the difference

$$S_n - S_{n-1} = a_n, \tag{8.36}$$

which by the **Cauchy criteria** must approach to zero as n goes to infinity. However, the n **th term test**,

$$\boxed{\lim_{n \to \infty} a_n \to 0,} \tag{8.37}$$

is a **necessary** but **not a sufficient** condition for the convergence of a series. In what follows, we present some of the most commonly used tests, which are also sufficient for absolutely convergent series or series with positive terms. For series with positive terms, the absolute value signs are superfluous.

8.3.1   Comparison Test

Given two infinite series $\sum a_n$ and $\sum b_n$ with $b_n > 0$ and $|a_n| \le b_n$ for all n, then the convergence of $\sum b_n$ implies the absolute convergence of $\sum a_n$.

8.3.2   Limit Comparison Test

Assume that $b_n > 0$ for $n = 1, 2, \ldots$ and suppose that

$$\lim_{n \to \infty} \left| \frac{a_n}{b_n} \right| = c, \ c \neq 0, \tag{8.38}$$

then the series $\sum a_n$ converges absolutely, if and only if $\sum b_n$ converges. When the limit $\lim_{n \to \infty} |a_n/b_n| \to 0$ is true, we can only conclude that the convergence of $\sum b_n$ implies the convergence of $\sum |a_n|$.

8.3.3   Integral Test

Let $f(x)$ be a positive decreasing function in the interval $[1, \infty]$ satisfying the following conditions:
   (i) $f(x)$ is continuous in $[1, \infty]$,
   (ii) $f(x)$ is monotonic decreasing and $\lim_{n \to \infty} f(x) \to 0$,
   (iii) $f(n) = a_n$.
   Then the series $\sum a_n$ converges if the integral $\int_1^\infty f(x)dx$ converges and diverges if the integral diverges.

8.3.4   Ratio Test

Given an infinite series $\sum a_n$, if $\lim_{n\to\infty} \left| \frac{a_{n+1}}{a_n} \right| = L$, then for

$$
\begin{aligned}
&L < 1; &&\sum a_n \text{ is absolutely convergent,}\\
&L > 1; &&\sum a_n \text{ is divergent,}\\
&L = 1; &&\text{the test is inconclusive.}
\end{aligned}
\tag{8.39}
$$

8.3.5   Root Test

For a given infinite series $\sum a_n$, let $R = \lim_{n\to\infty} \sqrt[n]{|a_n|}$, then for

$$
\begin{aligned}
&R < 1; &&\sum a_n \text{ is absolutely convergent,}\\
&R > 1; &&\sum a_n \text{ is divergent,}\\
&R = 1; &&\text{the test is inconclusive.}
\end{aligned}
\tag{8.40}
$$

The strategy in establishing the convergence of an unknown series is to start with the simplest test, that is, the comparison test. In order to use the comparison test effectively, we need to have at our disposal some series of known behavior. For this purpose, we often use the **harmonic series** or the **geometric series**:

Theorem 8.2. The **harmonic series** of order p:

$$
\sum_{n=1}^{\infty} \frac{1}{n^p} = 1 + \frac{1}{2^p} + \frac{1}{3^p} + \cdots ,
\tag{8.41}
$$

converges for $p > 1$ and diverges for $0 < p \leq 1$.

We can establish this result most easily by using the integral test. For $p > 0$, we write the integral

$$
\int_1^{\infty} \frac{1}{x^p}\, dx = \lim_{b\to\infty} \int_1^b \frac{1}{x^p}\, dx = \lim_{b\to\infty} \left[\frac{1}{(p-1)} \left(1 - \frac{1}{b^{p-1}} \right) \right].
\tag{8.42}
$$

When $p > 1$, the above limit exists and takes the value $1/(p-1)$, hence, the series converges. For $0 < p < 1$, the integral diverges, hence, the series also diverges. For $p = 1$, we have

$$
\int_1^{\infty} \frac{1}{x}\, dx = \lim_{b\to\infty} (\ln b) = \infty
\tag{8.43}
$$

and the harmonic series diverges again. Finally, for $p < 0$, the nth term readily establishes the divergence, thus completing the proof.

Theorem 8.3. The **geometric series:**

$$\sum_{n=0}^{\infty} x^n = 1 + x + x^2 + \cdots ,$$ (8.44)

converges for $-1 < x < 1$ as

$$\sum_{n=0}^{\infty} x^n = \frac{1}{1-x}, \ |x| < 1.$$ (8.45)

The proof can be established by the ratio test. From the limit

$$\lim_{n \to \infty} \left| \frac{x^{n+1}}{x^n} \right| = |x|,$$ (8.46)

we need $|x| < 1$ for convergence. The series obviously diverges for $x = 1$ by the nth term test.

Example 8.1. *Convergence tests:* Consider the series

$$\sum_{n=1}^{\infty} \frac{n+2}{4n^2 + n + 1}.$$ (8.47)

Since the nth term converges to zero, it is of no help. For large n, the nth term behaves as $1/4n$, which suggests comparison with $1/4n$. Since the inequality

$$\frac{n+2}{4n^2 + n + 1} > \frac{1}{4n}$$ (8.48)

is satisfied for all $n \geq 1$, by comparison with the harmonic series, $\sum_n 1/n$, we conclude that the above series [Eq. (8.47)] also diverges. Note that for this case, the ratio test fails:

$$\lim_{n \to \infty} \left[\frac{n+1+2}{4(n+1)^2 + n + 2} \cdot \frac{4n^2 + n + 1}{n+2} \right] \to 1.$$ (8.49)

Example 8.2. *Convergence tests:* The series

$$\sum_{n=2}^{\infty} \frac{1}{n \ln n}$$ (8.50)

diverges by the integral test:

$$\int_2^{\infty} \frac{dx}{x \ln x} = \lim_{b \to \infty} [\ln(\ln x)]_2^b \to \infty.$$ (8.51)

Note that comparison with the harmonic series is of no help since

$$\frac{1}{n \ln n} < \frac{1}{n}, \quad n = 2, 3, \ldots. \tag{8.52}$$

Example 8.3. *Convergence tests:* For the series

$$\sum_{n=1}^{\infty} \frac{\ln n}{n}, \tag{8.53}$$

we can use the comparison test with the harmonic series:

$$\frac{\ln n}{n} > \frac{1}{n}, \quad n = 3, 4, \ldots, \tag{8.54}$$

to conclude that the series is divergent.

Example 8.4. *Convergence tests:* Given the series

$$\sum_{n=1}^{\infty} \frac{c^n}{n!}, \tag{8.55}$$

where c is a positive constant, we can use the ratio test:

$$\lim_{n \to \infty} \left[\frac{c^{n+1}}{(n+1)!} \frac{n!}{c^n} \right] = \lim_{n \to \infty} \left[\frac{c}{n+1} \right] \to 0, \tag{8.56}$$

to conclude that the series converges. Similarly, for the series $\sum_{n=1}^{\infty} \frac{n^n}{n!}$, we apply the ratio test,

$$\lim_{n \to \infty} \left[\frac{(n+1)^{n+1}}{(n+1)!} \frac{n!}{n^n} \right] = \lim_{n \to \infty} \left[1 + \frac{1}{n} \right]^n \to e > 1, \tag{8.57}$$

to conclude that the series diverges. In fact, in this case, we can decide quicker via the nth term test:

$$\lim_{n \to \infty} \frac{n^n}{n!} \to \lim_{n \to \infty} \frac{e^n}{\sqrt{2\pi n}} \to \infty, \tag{8.58}$$

where we have used the Stirling's approximation $n! \simeq n^n e^{-n} \sqrt{2\pi n}$ for large n.

Example 8.5. *Convergence tests:* We can write the series

$$\sum_{n=1}^{\infty} \frac{n^2 + 3^n}{3^n n^2} \tag{8.59}$$

as

$$\sum_{n=1}^{\infty} \left(\frac{1}{3^n} + \frac{1}{n^2} \right), \tag{8.60}$$

where the first series converges by the ratio test:

$$\lim_{n \to \infty} \frac{1/3^{n+1}}{1/3^n} \to \frac{1}{3} < 1, \tag{8.61}$$

while the second series converges by the integral test:

$$\int_1^{\infty} \frac{dx}{x^2} = 1, \tag{8.62}$$

thus the original series [Eq. (8.59)] converges. We should make a note that the sum of two convergent series is convergent and the sum of a divergent series with a convergent series is divergent. However, the sum of two divergent series may be convergent.

Example 8.6. *Convergence tests:* Given the series

$$\sum_{n=2}^{\infty} \left(\frac{n}{n^2 + 1} \right)^n, \tag{8.63}$$

the ratio test fails:

$$\lim_{n \to \infty} \left[\frac{n+1}{(n+1)^2 + 1} \right]^{n+1} \left[\frac{n^2 + 1}{n} \right]^n \to \lim_{n \to \infty} \left[\frac{n+1}{n} \right]^n \left[\frac{n^2 + 1}{(n+1)^2 + 1} \right]^n \to 1. \tag{8.64}$$

However, the root test:

$$\lim_{n \to \infty} \sqrt[n]{a_n} \to \lim_{n \to \infty} \frac{n}{n^2 + 1} \to 0, \tag{8.65}$$

yields that the series is convergent. In general, the root test is more powerful than the ratio test. When the ratio test is inconclusive, usually a result can be obtained via the root test.

The above tests work for the absolutely convergent series or series with positive terms. For conditionally convergent series, we can use the following theorem:

Theorem 8.4. An alternating series:

$$\sum_{n=0}^{\infty} (-1)^n a_n, \ a_n > 0, \tag{8.66}$$

converges when the following two conditions are met:
(i) a_n are monotonic decreasing, that is,

$$a_{n+1} \le a_n, \ n = 0, 1, 2, \ldots. \tag{8.67}$$

(ii)

$$\lim_{n \to \infty} a_n \to 0. \tag{8.68}$$

Example 8.7. *Alternating series:* The alternating series

$$\sum_{n=1}^{\infty} (-1)^n \frac{\ln n}{n} \tag{8.69}$$

converges since $\frac{\ln n}{n}$ is monotonic, decreasing for $n = 2, 3, \ldots$ and the limit

$$\lim_{n \to \infty} \frac{\ln n}{n} \to 0 \tag{8.70}$$

is true.

8.4   OPERATIONS WITH SERIES

We have already pointed out that convergent series can be added, subtracted, and multiplied by a constant:

$$\sum_{n=1}^{\infty} (\alpha a_n) + \sum_{n=1}^{\infty} (\beta b_n) = \alpha \sum_{n=1}^{\infty} a_n + \beta \sum_{n=1}^{\infty} b_n, \tag{8.71}$$

where $\sum a_n$ and $\sum b_n$ are two convergent series and α and β are constants. In addition to these there are three other operations like **grouping**, **rearrangement**, and **multiplication** that can be performed on series. For **absolutely convergent** series or series with positive terms, grouping – that is, inserting parentheses – yields a new convergent series having the same sum as the original series. For the sum

$$\sum_{n=1}^{\infty} a_n = a_1 + a_2 + a_3 + a_4 + \cdots, \tag{8.72}$$

inserting parentheses:

$$(a_1 + a_2 + a_3) + (a_4 + a_5 + a_6) + (a_7 + a_8 + a_9) + \cdots , \quad (8.73)$$

while preserving the order of terms, will only cause one to skip some of the terms in the sequence of partial sums of the original series; that is,

$$\{S_1, S_2, S_3, S_4, \ldots\} \quad (8.74)$$

will be replaced by

$$\{S_3, S_6, S_9, \ldots\}. \quad (8.75)$$

If the first sequence [Eq. (8.74)] converges to S, then the second sequence [Eq. (8.75)] obtained by skipping some of the terms will also converge to the same value. If the original series is properly divergent, then insertion of parentheses yields again a properly divergent series. Since a series with positive terms is either convergent or properly divergent, we can insert parentheses freely without affecting the sum. Divergent series with variable signs can occasionally produce convergent series. For example, the divergent series

$$\sum_{n=0}^{\infty} (-1)^n = +1 - 1 + 1 - 1 + 1 - \cdots \quad (8.76)$$

can be written as

$$\sum_{n=0}^{\infty} (-1)^n = +1 + (-1 + 1) + (-1 + 1) + \cdots , \quad (8.77)$$

where all the pairs in parentheses give zeros, thus yielding 1 as the sum.

Another operation with the series is **rearrangement** of its terms. For absolutely convergent series converging to S, every rearrangement of its terms will also converge absolutely and will yield the same sum. For conditionally convergent series, no such guarantees can be given. In fact, by a suitable arrangement of its terms, a conditionally convergent series can be made to converge to any value or even to diverge. To this effect, we present the following theorem without proof [2]:

Theorem 8.5. Let $\sum a_i$ be a conditionally convergent series and also let a, b, $a < b$, be any two numbers in the closed interval $[-\infty, +\infty]$; then there exists a rearrangement $\sum b_i$ of $\sum a_i$ such that

$$\underline{\lim}_{n \to \infty} \sum b_n = a \text{ and } \overline{\lim}_{n \to \infty} \sum b_n = b. \quad (8.78)$$

To demonstrate Theorem 8.5, consider the alternating series

$$\sum_{k=1}^{\infty} \frac{(-1)^{k+1}}{k} = 1 - \frac{1}{2} + \frac{1}{3} - \frac{1}{4} + \frac{1}{5} - \cdots , \quad (8.79)$$

which converges to $\ln 2$. We rearrange its terms as

$$\sum_{k=1}^{\infty} \frac{(-1)^{k+1}}{k} = 1 + \frac{1}{3} + \frac{1}{5} + \cdots - \frac{1}{2} - \frac{1}{4} - \frac{1}{6} - \cdots \qquad (8.80)$$

$$= 1 + \frac{1}{3} + \frac{1}{5} + \cdots - \frac{1}{2}\left[1 + \frac{1}{2} + \frac{1}{3} + \cdots\right]. \qquad (8.81)$$

We first add and then subtract $\frac{1}{2}\left[1 + \frac{1}{2} + \frac{1}{3} + \cdots\right]$ to the above series to obtain

$$\sum_{k=1}^{\infty} \frac{(-1)^{k+1}}{k} = 1 + \frac{1}{3} + \frac{1}{5} + \cdots - \frac{1}{2}\left[1 + \frac{1}{2} + \frac{1}{3} + \cdots\right]$$

$$+ \frac{1}{2}\left[1 + \frac{1}{2} + \frac{1}{3} + \cdots\right] - \frac{1}{2}\left[1 + \frac{1}{2} + \frac{1}{3} + \cdots\right] \qquad (8.82)$$

$$= \left[1 + \frac{1}{2} + \frac{1}{3} + \cdots\right] - \left[1 + \frac{1}{2} + \frac{1}{3} + \cdots\right] = 0. \qquad (8.83)$$

Note that when we rearrange the terms of a series, there is a one-to-one correspondence between the terms of the rearranged series and the original series.

So far we have emphasized the central role that absolute convergence plays in operations with series. In fact, **multiplication** is another operation defined only for the absolutely convergent series. If $\sum_{n=1}^{\infty} a_n$ and $\sum_{n=1}^{\infty} b_n$ are two absolutely convergent series, then their product, $(\sum a_n)(\sum b_n) = \sum c_n$, is also absolutely convergent. Furthermore, if $\sum a_n$ converges to A and $\sum b_n$ converges to B, then their product converges to AB. Since the product series, $\sum c_n$, is absolutely convergent, a particular rearrangement of its terms is known as the **Cauchy product**:

$$\left(\sum_{n=0}^{\infty} a_n\right)\left(\sum_{n=0}^{\infty} b_n\right) = \sum_{n=0}^{\infty} c_n \qquad (8.84)$$

$$= a_0 b_0 + (a_0 b_1 + a_1 b_0) + (a_0 b_2 + a_1 b_1 + a_2 b_0) + \cdots , \qquad (8.85)$$

where in general, we can write c_n as

$$c_n = \sum_{k=0}^{n} a_k b_{n-k}, \quad n = 0, 1, 2, \ldots . \qquad (8.86)$$

8.5 SEQUENCES AND SERIES OF FUNCTIONS

So far we have confined our discussion to sequences and series of numbers. One can also define **sequences of functions** as

$$\boxed{\{f_n(x)\}, \quad n = 1, 2, \ldots ,} \qquad (8.87)$$

where $f_n(x)$ are functions defined over the same interval, which can be infinite along the x-axis. Similarly, we can define **series of functions** as

$$\sum_{n=1}^{\infty} u_n(x), x \in [L_1, L_2],$$ (8.88)

where the nth **partial sum** is defined as

$$S_n(x) = \sum_{i=1}^{n} u_i(x).$$ (8.89)

At each point of their interval of definition, sequences or series of functions reduce to sequences and series of numbers, hence, their convergence or divergence is defined the same way. However, there is one subtle question that needs to be answered, that is, a given series or sequence may converge at each point of its interval, but does it converge at the same rate? For this, we introduce the concept of **uniform convergence**:

Definition 8.1. A series, $\sum_{n=1}^{\infty} u_n(x)$, is said to converge uniformly to $S(x)$ in the interval $[L_1, L_2]$ if, for a given $\varepsilon > 0$, an integer N can be found such that

$$|S_n(x) - S(x)| < \varepsilon$$ (8.90)

for all $n \geq N$ and for all x in $[L_1, L_2]$.

This definition of uniform convergence works equally well for sequences, that is, a sequence $\{f_n(x)\}$ is uniformly convergent to $f(x)$ in the interval $[L_1, L_2]$ if, for a given $\varepsilon > 0$, an integer N can be found such that

$$|f_n(x) - f(x)| < \varepsilon$$ (8.91)

for all $n \geq N$. In other words, the uniform convergence of a series is equivalent to the uniform convergence of the sequence of its partial sums. Note that in this definition, ε basically stands for the error that one makes by approximating a uniformly convergent series with its nth partial sum. Uniform convergence assures that the error will remain in the predetermined margin:

$$f(x) - \varepsilon < f_n(x) < f(x) + \varepsilon,$$ (8.92)

for all x in $[L_1, L_2]$.

Example 8.8. *The geometric series:* Consider the geometric series

$$\sum_{n=0}^{\infty} x^n, |x| < 1,$$ (8.93)

where the nth partial sum is given as

$$S_n(x) = 1 + x + x^2 + \cdots + x^{n-1} = \frac{1 - x^n}{1 - x} \qquad (8.94)$$

and the sum is

$$S(x) = \frac{1}{1 - x}. \qquad (8.95)$$

Absolute value of the error, ε_n, committed by approximating the geometric series with its nth partial sum is

$$\varepsilon_n = \left| \frac{1}{1 - x} - \frac{1 - x^n}{1 - x} \right| = \frac{|x^n|}{|1 - x|}. \qquad (8.96)$$

Consider the interval $[-\frac{1}{3}, \frac{1}{3}]$. Since the largest error always occurs at the end points, for an error tolerance of $\varepsilon_n = 10^{-3}$, we need to sum at least seven terms:

$$10^{-3} = \frac{1/3^n}{1 - 1/3}, \qquad (8.97)$$

$$3^n = \frac{3}{2} 10^3, \qquad (8.98)$$

$$n \log 3 = \log \left[\frac{3 \times 10^3}{2} \right], \qquad (8.99)$$

$$n \simeq 7. \qquad (8.100)$$

As we approach the end points of the interval $(-1, 1)$, the number of terms to be added increases dramatically. For example, for the same error margin, $\varepsilon_n = 10^{-3}$, but this time in the interval $(-0.99, 0.99)$, the number of terms to be added is 1146. It is seen from Eq. (8.96) that as n goes to infinity, the error ε_n, which is also equal to the remainder of the series, goes to zero:

$$\lim_{n \to \infty} \frac{|x^n|}{|1 - x|} \to 0 \text{ for } |x| < 1. \qquad (8.101)$$

Adding 1146 terms assures us that the fractional error, ε_n/S, committed in the interval $[-0.99, 0.99]$ is always less than

$$\frac{10^{-3}}{\left[\frac{1}{1 - 0.99} \right]} = \frac{0.001}{100} = 10^{-5}. \qquad (8.102)$$

We write the geometric series approximated by its first 1146 terms as

$$S(x) \simeq S_{1146}(x) + 0(10^{-5}), \quad [-0.99, 0.99]. \qquad (8.103)$$

In this notation, $0(10^{-5})$ means that the partial sum $S_{1146}(x)$ approximates the exact sum $S(x)$ accurate up to the fifth digit.

8.6 M-TEST FOR UNIFORM CONVERGENCE

A commonly used test for uniform convergence is the **Weierstrass M-test** or
in short, the M-**test**:

Theorem 8.6. Let $\sum_{n=1}^{\infty} u_n(x)$ be a series of functions defined in the interval
$[L_1, L_2]$. If we can find a **convergent** series of positive constants, $\sum_{n=1}^{\infty} M_n$,
such that the inequality

$$|u_n(x)| \leq M_n \tag{8.104}$$

holds for all x in the interval $[L_1, L_2]$, then the series $\sum_{n=1}^{\infty} u_n(x)$ converges
uniformly and absolutely in the interval $[L_1, L_2]$.

Example 8.9. *Weierstrass M-test:* Consider the series $\sum_{n=1}^{\infty} \frac{x^n}{n^p}$, which
reduces to the geometric series for $p = 0$. To find the values of p for which
this series converges, we first apply the ratio test:

$$\lim_{n \to \infty} \left| \frac{a_{n+1}}{a_n} \right| = \lim_{n \to \infty} \left| \frac{x^{n+1}}{(n+1)^p} \frac{n^p}{x^n} \right| = \lim_{n \to \infty} \left(\frac{n}{n+1} \right)^p |x| = |x|, \tag{8.105}$$

which says that independent of the value of p the series converges abso-
lutely for $|x| < 1$. At one of the end points, $x = 1$, the series converges
for $p > 1$ and diverges for $0 < p \leq 1$ [Eq. (8.41)]. At the other end point,
$x = -1$, for $p > 1$ the series becomes alternating series, which converges
by Theorem 8.4. To check for uniform convergence, we use the M-test with
the series $\sum_{n=1}^{\infty} \frac{1}{n^p}$, which converges for $p > 1$. Comparing with $\sum_{n=1}^{\infty} \frac{x^n}{n^p}$,
we write

$$\left| \frac{x^n}{n^p} \right| \leq \frac{1}{n^p} = M_n, \tag{8.106}$$

which holds for $|x| \leq 1$. Hence, we conclude that the series $\sum_{n=1}^{\infty} \frac{x^n}{n^p}$ con-
verges uniformly and absolutely in the interval $|x| \leq 1$ for $p > 1$. Keep in
mind that uniform convergence and absolute convergence are two independent concepts, neither of them implies the other. Weierstrass M-test
checks for both uniform and absolute convergence.

8.7   PROPERTIES OF UNIFORMLY CONVERGENT SERIES

Uniform convergence is very important in dealing with series of functions. Three
of the most important and frequently used properties of **uniformly convergent** series are given below in terms of three theorems:

Theorem 8.7. The sum of a uniformly convergent series of continuous
functions is continuous. In other words, if $u_n(x)$ are continuous functions in
the interval $[L_1, L_2]$, then their sum, $f(x) = \sum_{n=1}^{\infty} u_n(x)$, is also continuous
provided that the series $\sum_{n=1}^{\infty} u_n(x)$ is uniformly convergent in the interval
$[L_1, L_2]$.

From this theorem, we see that any series of functions which has discontinuities in the interval $[L_1, L_2]$ cannot be uniformly convergent.

Theorem 8.8. A uniformly convergent series of continuous functions can be integrated term by term in the interval of uniform convergence. In other words, in the interval of uniform convergence, $[L_1, L_2]$, we can interchange the integral and the summation signs:

$$\int_{L_1}^{L_2} dx f(x) = \int_{L_1}^{L_2} dx \left[\sum_{n=1}^{\infty} u_n(x) \right] = \sum_{n=1}^{\infty} \left[\int_{L_1}^{L_2} dx u_n(x) \right]. \qquad (8.107)$$

Theorem 8.9. If $u'_n(x) = \frac{du_n}{dx}$ are continuous in $[L_1, L_2]$ and the series $\sum_{n=1}^{\infty} u_n(x)$ converges to $f(x)$ in $[L_1, L_2]$, and the series $\sum_{n=1}^{\infty} u'_n(x)$ converges uniformly in $[L_1, L_2]$, then we can interchange the order of the differentiation and the summation signs:

$$\frac{df}{dx} = \frac{d}{dx} \left[\sum_{n=1}^{\infty} u_n(x) \right] = \sum_{n=1}^{\infty} u'_n(x). \qquad (8.108)$$

Other operations like addition and subtraction as well as multiplication with a continuous function $h(x)$ of the uniformly convergent series $f(x) = \sum_{n=1}^{\infty} u_n(x)$ and $g(x) = \sum_{n=1}^{\infty} v_n(x)$ can be performed as follows:

$$f(x) \pm g(x) = \sum_{n=1}^{\infty} [u_n(x) \pm v_n(x)], \quad [L_1, L_2], \qquad (8.109)$$

$$h(x)g(x) = \sum_{n=1}^{\infty} h(x)v_n(x), \quad [L_1, L_2], \qquad (8.110)$$

where the results are again uniformly convergent in $[L_1, L_2]$.

Example 8.10. *Uniformly convergent series:* Differentiability and integrability of uniformly convergent series gives us a powerful tool in obtaining new series sums from the existing ones. For example, the geometric series:

$$\frac{1}{1-x} = \sum_{n=0}^{\infty} x^n = 1 + x + x^2 + x^3 + \cdots, \quad |x| < 1, \qquad (8.111)$$

after differentiation gives

$$\frac{d}{dx} \left[\frac{1}{1-x} \right] = \frac{d}{dx} \sum_{n=0}^{\infty} x^n = \sum_{n=0}^{\infty} \frac{d}{dx} x^n, \qquad (8.112)$$

$$\frac{1}{(1-x)^2} = \sum_{n=1}^{\infty} n x^{n-1} = 1 + 2x + 3x^2 + \cdots, \quad |x| < 1. \qquad (8.113)$$

Differentiating once more, we obtain

$$\frac{d}{dx}\left[\frac{1}{(1-x)^2}\right] = \frac{d}{dx}\sum_{n=1}^{\infty} nx^{n-1} = \sum_{n=1}^{\infty} \frac{d}{dx}(nx^{n-1}), \quad |x| < 1, \qquad (8.114)$$

$$\frac{2}{(1-x)^3} = \sum_{n=2}^{\infty} n(n-1)x^{n-2} = 2 + 6x + \cdots, \quad |x| < 1, \qquad (8.115)$$

Similarly, by integration we obtain the series

$$\int_0^x dx \left[\frac{1}{1-x}\right] = \int_0^x dx \left[\sum_{n=0}^{\infty} x^n\right] = \sum_{n=0}^{\infty} \int_0^x dx\, x^n, \qquad (8.116)$$

$$\ln\left[\frac{1}{1-x}\right] = \sum_{n=0}^{\infty} \frac{x^{n+1}}{n+1} = x + \frac{x^2}{2} + \frac{x^3}{3} + \cdots, \quad |x| < 1. \qquad (8.117)$$

8.8   POWER SERIES

An important and frequently encountered class of uniformly convergent series is the **power series** of the form

$$\sum_{n=0}^{\infty} c_n(x-x_0)^n = c_0 + c_1(x-x_0) + c_2(x-x_0)^2 + \cdots, \qquad (8.118)$$

where $c_0, c_1, c_2, \ldots$ are constants and x_0 is called the point of expansion. Every power series has a radius of convergence defined about the point of expansion. Using the ratio test, the limit

$$\lim_{n\to\infty}\left|\frac{a_{n+1}}{a_n}\right| = \lim_{n\to\infty}\left|\frac{c_{n+1}(x-x_0)^{n+1}}{c_n(x-x_0)^n}\right| = \lim_{n\to\infty}\left|\frac{c_{n+1}}{c_n}\right| |x-x_0|, \qquad (8.119)$$

indicates that the power series converges absolutely for

$$\left[\lim_{n\to\infty}\left|\frac{c_{n+1}}{c_n}\right|\right] |x-x_0| < 1, \qquad (8.120)$$

or for

$$|x-x_0| < R, \qquad (8.121)$$

where R is the **radius of convergence**:

$$R = \frac{1}{\displaystyle\lim_{n\to\infty}\left|\frac{c_{n+1}}{c_n}\right|}. \qquad (8.122)$$

The radius of convergence can be anything including zero and infinity. When $R = 0$, the series converges only at the point x_0. When $R = \infty$, the series converges along the entire x-axis. If two power series $\sum_{n=0}^{\infty} c_n (x - x_0)^n$ and $\sum_{n=0}^{\infty} b_n (x - x_0)^n$ converge to the same function $f(x)$, then the two series are identical, that is, $c_n = b_n$ for all n. In other words, the power series representation of a function is unique. The uniform convergence of power series for the interior of any r satisfying $0 < r < R$, where R is the radius of convergence, follows from the M-test by comparing $\sum_{n=0}^{\infty} c_n (x - x_0)^n$ with $\sum_{n=0}^{\infty} M_n$, $M_n = |c_n| r^n$, where $|c_n (x - x_0)^n| < |c_n| r^n$ for $0 < r < R$.

Example 8.11. *Radius of convergence:* Consider the following power series generated from the geometric series in Example 8.10:

$$
\begin{align*}
\text{(i)} \quad & \frac{1}{(1-x)^2} = \sum_{n=1}^{\infty} n x^{n-1}, \\
\text{(ii)} \quad & \frac{2}{(1-x)^3} = \sum_{n=2}^{\infty} n(n-1) x^{n-2}, \\
\text{(iii)} \quad & \ln \left[\frac{1}{(1-x)} \right] = \sum_{n=0}^{\infty} \frac{x^{n+1}}{(n+1)}.
\end{align*}
\tag{8.123}
$$

We find their radius of convergences by using the ratio test. For the first series, we write the limit

$$
\lim_{n \to \infty} \left| \frac{n+1}{n} \frac{x^n}{x^{n-1}} \right| = |x|,
\tag{8.124}
$$

hence, the series converges for $|x| < 1$ and $R = 1$. For the second series, the ratio test gives

$$
\lim_{n \to \infty} \left| \frac{(n+1)n}{n(n-1)} \frac{x^{n-1}}{x^{n-2}} \right| = |x|,
\tag{8.125}
$$

hence, the radius of convergence is 1 and the series converges for $|x| < 1$. For the last series, we write the limit

$$
\lim_{n \to \infty} \left| \frac{x^{n+2}(n+1)}{(n+2)x^{n+1}} \right| = |x|,
\tag{8.126}
$$

where the radius of convergence is again obtained as 1 and the series converges for $|x| < 1$. In other words, the series generated from the geometric series by successive differentiation and integration have the same radius of convergence. Actually, as the following theorem states, this is in general true for power series:

Theorem 8.10. Within the radius of converge, a power series can be differentiated and integrated as often as one desires. The differentiated or the integrated series also converges uniformly with the same radius of convergence as the original series.

Example 8.12. *Binomial formula:* Using the geometric series:

$$\frac{1}{1-x} = \sum_{n=0}^{\infty} x^n, \quad |x| < 1, \tag{8.127}$$

after k-fold differentiation and using the formula

$$\frac{d^k x^n}{dx^k} = \frac{n! x^{n-k}}{(n-k)!}, \tag{8.128}$$

we obtain

$$\frac{1}{(1-x)^k} = 1 + \frac{kx}{1} + \frac{k(k+1)}{1.2} x^2 + \cdots + \frac{k(k+1)\cdots(k+n-1)}{1.2\ldots n} x^n$$
$$+ \cdots, \quad |x| < 1, \tag{8.129}$$

which is the well-known binomial formula [3] for $(1-x)^{-k}$. In general, the **binomial formula** is given as

$$\boxed{(x+y)^m = \sum_{k=0}^{m} \binom{m}{k} x^m y^{m-k},} \tag{8.130}$$

where $\binom{m}{k}$ are called the **binomial coefficients** defined as

$$\boxed{\binom{m}{k} = \frac{m!}{k!(m-k)!}.} \tag{8.131}$$

If m is negative or noninteger, the upper limit of the series [Eq. 8.130] is infinity. Two special cases,

$$(1+x)^{1/2} = 1 + \frac{1}{2}x - \frac{1}{8}x^2 + \frac{3}{48}x^3 - \cdots \tag{8.132}$$

and

$$(1+x)^{-1/2} = 1 - \frac{1}{2}x + \frac{3}{8}x^2 - \frac{15}{48}x^3 + \cdots, \tag{8.133}$$

are frequently used in obtaining approximate expressions.

8.9   TAYLOR SERIES AND MACLAURIN SERIES

Consider a uniformly convergent power series:

$$f(x) = \sum_{n=0}^{\infty} c_n(x - x_0)^n, \quad x_0 - R < x < x_0 + R, \tag{8.134}$$

with the radius of convergence R. Performing successive differentiations, we write

$$f(x) = c_0 + c_1(x - x_0) + c_2(x - x_0)^2 + c_3(x - x_0)^3 + \cdots,$$
$$f'(x) = c_1 + 2c_2(x - x_0) + 3c_3(x - x_0)^2 + \cdots,$$
$$f''(x) = 2c_2 + 6c_3(x - x_0) + \cdots, \tag{8.135}$$
$$\vdots$$
$$f^{(n)}(x) = n(n - 1)(n - 2) \cdots 3 \cdot 2 \cdot 1 \cdot c_n + \text{positive powers of } (x - x_0).$$

Based on Theorem 8.10, these series converge for $x_0 - R < x < x_0 + R$. Evaluating these derivatives at $x = x_0$, we obtain the coefficients c_n as

$$c_0 = f(x_0),$$
$$c_1 = f'(x_0),$$
$$c_2 = f''(x_0)/2, \tag{8.136}$$
$$\vdots$$
$$c_n = f^{(n)}(x_0)/n!,$$

thus obtaining the **Taylor series** representation of $f(x)$ as

$$\boxed{f(x) = \sum_{n=0}^{\infty} \frac{1}{n!} f^{(n)}(x_0)(x - x_0)^n.} \tag{8.137}$$

When we set $x_0 = 0$, Taylor series reduces to **Maclaurin series**:

$$\boxed{f(x) = \sum_{n=0}^{\infty} \frac{1}{n!} f^{(n)}(0)x^n.} \tag{8.138}$$

8.10   INDETERMINATE FORMS AND SERIES

It is well known that **L'Hôpital's** rule can be used to find the limits of indeterminate expressions like $0/0$ or ∞/∞. However, this method frequently requires multiple differentiations, which makes it rather cumbersome to implement. In such situations, series expansions not only can help us to find the limits of the indeterminate forms but also can allow us to write approximate expressions to any desired level of accuracy. Using the series expansions of elementary functions, we can usually obtain the series expansion of a given expression, which can then be written to any desired level of accuracy.

Example 8.13. *Indeterminate forms:* Consider the limit

$$\lim_{x \to 0} \left[\frac{e^x \sin x}{\cos x - e^x} \right], \tag{8.139}$$

which is indeterminate as $0/0$. Using the Maclaurin series of the elementary functions e^x, $\sin x$, and $\cos x$, we first write the numerator and the denominator in power series as

$$\frac{e^x \sin x}{\cos x - e^x} = \frac{\left[1 + x + \frac{x^2}{2!} + \frac{x^3}{3!} + \cdots \right] \left[x - \frac{x^3}{3!} + \frac{x^5}{5!} + \cdots \right]}{\left[1 - \frac{x^2}{2!} + \frac{x^4}{4!} + \cdots \right] - \left[1 + x + \frac{x^2}{2!} + \frac{x^3}{3!} + \cdots \right]}$$

$$= \frac{x + x^2 + \frac{x^3}{3} - \frac{x^5}{30} - \cdots}{-x - x^2 - \frac{x^3}{6} - \frac{x^5}{120} - \cdots}. \tag{8.140}$$

After a formal division of the numerator with the denominator, we obtain the series

$$\frac{e^x \sin x}{\cos x - e^x} = -1 - \frac{1}{6} x^2 + \frac{1}{6} x^3 + \cdots , \tag{8.141}$$

which yields the limit

$$\lim_{x \to 0} \frac{e^x \sin x}{\cos x - e^x} \to -1. \tag{8.142}$$

This result can also be obtained by the L'Hôpital's rule. However, the series method accomplishes a lot more than this. It not only gives the limit but also allows us to approximate the function in the neighborhood of $x = 0$ to any desired level of accuracy. For example, for small x we can approximate $\frac{e^x \sin x}{\cos x - e^x}$ as

$$\frac{e^x \sin x}{\cos x - e^x} = -1 - \frac{1}{6} x^2 + 0(x^3). \tag{8.143}$$

Example 8.14. *Indeterminate forms:* Let us now consider the function

$$\frac{x^2 + (\cos^2 x - 1)}{x^2[\cos^2 x - 1]}, \tag{8.144}$$

which is $0/0$ in the limit as $x \to 0$. We first write the above expression as

$$\frac{1}{[\cos^2 x - 1]} + \frac{1}{x^2}. \tag{8.145}$$

The first term, which is divergent as $x \to 0$, can be written as

$$\frac{1}{\cos^2 x - 1} = \frac{1}{\left[1 - \frac{x^2}{2!} + \frac{x^4}{4!} + \cdots \right]^2 - 1} \tag{8.146}$$

$$= \frac{1}{-x^2 + \frac{x^4}{3} - \frac{2x^6}{45} + \cdots}, \tag{8.147}$$

which after a formal division gives the series expansion

$$\frac{1}{[\cos^2 x - 1]} = -\frac{1}{x^2} - \frac{1}{3} - \frac{1}{15}x^2 + \cdots . \tag{8.148}$$

Substituting Eq. (8.148) into Eq. (8.145), the divergent piece, $-1/x^2$, is canceled by $1/x^2$ to yield the finite result in the neighborhood of $x = 0$ as

$$\frac{x^2 + (\cos^2 x - 1)}{x^2[\cos^2 x - 1]} = -\frac{1}{3} - \frac{1}{15}x^2 + 0(x^4). \tag{8.149}$$

REFERENCES

1. Kaplan, W., *Advanced Calculus*, Addison-Wesley, Reading, third edition, 1984.

2. Apostol, T.M., *Mathematical Analysis*, Addison-Wesley, Reading, MA, fourth printing, 1971.

3. Dwight, H.B., *Tables of Integrals and Other Mathematical Data*, Prentice Hall, fourth edition, 1961.

PROBLEMS

1. Find the limits of the following sequences:

(i) $\left\{ \dfrac{(3n - 1)^4 + \sqrt{2 + 16n^8}}{2 + n^3 + 7n^4} \right\}$, $\quad n = 1, 2, \ldots$.

(ii) $\left\{ \dfrac{(n!)^2}{(3n)!} \right\}$, $\quad n = 1, 2, \ldots$.

2. Show the divergence of the following series by the nth term test:

$$\text{(i)} \quad \sum_{n=1}^{\infty} \cos \frac{n^2 \pi}{3}.$$

$$\text{(ii)} \quad \sum_{n=1}^{\infty} \frac{2^{2n}}{n^3}.$$

3. Show the divergence of the following series by the comparison test:

$$\text{(i)} \quad \sum_{n=1}^{\infty} \frac{n+3}{n^2 + 2n - 2}.$$

$$\text{(ii)} \quad \sum_{n=1}^{\infty} \frac{1}{\sqrt{n} \ln n}.$$

4. Check the convergence of the following alternating series:

$$\text{(i)} \quad \sum_{n=1}^{\infty} \frac{(-1)^n}{n \ln n}. \qquad \text{(iv)} \quad \sum_{n=1}^{\infty} \frac{(-1)^n n}{10^n}.$$

$$\text{(ii)} \quad \sum_{n=1}^{\infty} \frac{(-1)^n}{\sqrt{n} \ln n}. \qquad \text{(v)} \quad \sum_{n=1}^{\infty} \frac{(-1)^{n+1}}{n^3}.$$

$$\text{(iii)} \quad \sum_{n=1}^{\infty} \frac{(-1)^n n!}{10^n}. \qquad \text{(vi)} \quad \sum_{n=1}^{\infty} \frac{(-1)^{n+1} n^2}{(n + \sqrt{n})}.$$

5. Determine the convergence or the divergence of the following series:

$$\text{(i)} \quad \sum_{n=1}^{\infty} \frac{(4n + 2)n^2}{3n^3 + 1}. \qquad \text{(viii)} \quad \sum_{n=1}^{\infty} \left[\frac{1}{n} - \left(\frac{7}{8} \right)^n \right].$$

$$\text{(ii)} \quad \sum_{n=1}^{\infty} \frac{2n + 5}{n3^n}. \qquad \text{(ix)} \quad \sum_{n=1}^{\infty} \left[\left(\frac{3}{4} \right)^n - \left(\frac{5}{6} \right)^n \right].$$

$$\text{(iii)} \quad \sum_{n=1}^{\infty} \frac{n^2}{n! + 3}. \qquad \text{(x)} \quad \sum_{n=1}^{\infty} \frac{1}{(n + 1)^{3/2}}.$$

$$\text{(iv)} \quad \sum_{n=1}^{\infty} \frac{\cos n\pi}{n + 1}. \qquad \text{(xi)} \quad \sum_{n=1}^{\infty} \frac{n^4 + n^2}{3n^6 + n}.$$

$$\text{(v)} \quad \sum_{n=1}^{\infty} \frac{\ln n}{n + 2 \ln n}. \qquad \text{(xii)} \quad \sum_{n=1}^{\infty} \frac{(n^2 + 1)^n}{n!}.$$

$$\text{(vi)} \quad \sum_{n=2}^{\infty} \frac{1}{n^2 - n}. \qquad \text{(xiii)} \quad \sum_{n=1}^{\infty} \frac{(n^2 + 1)}{n!}.$$

$$\text{(vii)} \quad \sum_{n=1}^{\infty} \frac{n}{2n^3 - 6}. \qquad \text{(xiv)} \quad \sum_{n=1}^{\infty} \frac{n^{10}}{(1.01)^n}.$$

6. Check the convergence of the following series and verify your answers by another method:

$$\text{(i)} \ \sum_{n=1}^{\infty} \frac{1}{n^2}. \qquad \text{(ii)} \ \sum_{n=1}^{\infty} \frac{n}{n^2-3}. \qquad \text{(iii)} \ \sum_{n=1}^{\infty} \frac{1}{n^2+1}.$$

7. Find the range of x for which the following series converge. Check the end points also.

$$\text{(i)} \ \sum_{n=1}^{\infty} \frac{x^n}{2^n n!}. \qquad\qquad \text{(v)} \ \sum_{n=1}^{\infty} \frac{x^n}{n^2+1}.$$

$$\text{(ii)} \ \sum_{n=1}^{\infty} n! x^n. \qquad\qquad \text{(vi)} \ \sum_{n=0}^{\infty} \left(\frac{x}{2}\right)^n.$$

$$\text{(iii)} \ \sum_{n=1}^{\infty} \left(\frac{x^2+1}{2}\right)^n. \qquad \text{(vii)} \ \sum_{n=0}^{\infty} \frac{x^n}{n^2}.$$

$$\text{(iv)} \ \sum_{n=1}^{\infty} \left(\frac{x}{n}\right)^n. \qquad\qquad \text{(viii)} \ \sum_{n=1}^{\infty} \frac{n+\cos^2 nx}{n^3+3n+1}.$$

8. The nth partial sum of a certain series is given as

$$S_n(x) = n^2 x/(1+n^3 x^2).$$

For which one of the intervals, $[-1, 17]$ and $[0.5, 1]$, does this series converges uniformly. Explain.

9. Investigate the uniform convergence of the series with the following nth partial sum:

$$S_n = n^2 x e^{-nx}.$$

10. Check the uniform convergence of

$$f(x) = \sum_{n=1}^{\infty} \frac{nx^2}{n^3+x^2}.$$

11. Verify the following formula for $|x| < 1$:

$$\frac{1}{(1-x)^k} = 1 + \frac{kx}{1} + \frac{k(k+1)}{1.2}x^2 + \cdots + \frac{k(k+1)\cdots(k+n-1)}{1.2\ldots n}x^n + \cdots.$$

12. Using the binomial expansion, show the expansions

$$(1+x)^{1/2} = 1 + \frac{1}{2}x - \frac{1}{8}x^2 + \frac{3}{48}x^3 - \cdots,$$

and

$$(1+x)^{-1/2} = 1 - \frac{1}{2}x + \frac{3}{8}x^2 - \frac{15}{48}x^3 + \cdots.$$

13. Find the Maclaurin expansions of $\sin^2 x$, $\cos^2 x$, and $\sinh x$. Find their radius of convergence and show that they converge uniformly within this radius.

14. Find the Maclaurin expansion of $1/\cos x$ and find its radius of convergence.

15. Find the limit

$$\lim_{x \to 0} \frac{x e^x \sin 2x}{\cos x - 2e^x}.$$

Using the series expansions of elementary functions about $x = 0$, find an approximate expression good to the fourth order in the neighborhood of $x = 0$.

16. Find the interval of convergence of the following series:

(i) $\displaystyle\sum_{n=1}^{\infty} \frac{(n!)^2 x^n}{(2n)!}$.   (iv) $\displaystyle\sum_{n=1}^{\infty} \frac{x^n}{n^2}$.

(ii) $\displaystyle\sum_{n=1}^{\infty} \frac{x^n}{\ln(n+1)}$.   (v) $\displaystyle\sum_{n=1}^{\infty} \frac{\cos nx}{n^3}$.

(iii) $\displaystyle\sum_{n=1}^{\infty} \frac{x^n n^2}{n^2 + 1}$.   (vi) $\displaystyle\sum_{n=1}^{\infty} \frac{x^n}{n!}$.

17. Evaluate the following definite integral to three decimal places:

$$\int_0^1 e^{-x^2}\, dx.$$

18. Expand the following functions:

(i) $f(x) = \frac{1}{3x+2}$,   (ii) $f(x) = \frac{x+2}{(x+3)(x+4)}$,

in Taylor series about $x = 1$ and find their radius of convergence.

19. Expand the following function in Maclaurin series:

$$f(x) = \frac{1}{3x+2}.$$

20. Obtain the Maclaurin series

$$\log\left[\frac{1}{1-x}\right] = x + \frac{x^2}{2} + \frac{x^3}{3} + \cdots , \quad |x| < 1.$$

Show that the series converges for $x = -1$ and hence verify

$$\log 2 = \sum_{n=1}^{\infty} \frac{(-1)^{n+1}}{n}.$$

21. Expand $1/x^2$ in Taylor series about $x = 1$.

22. Find the first three terms of the Maclaurin series of the following functions:

(i) $f(x) = \cos[\ln(x + 1)]$.

(ii) $f(x) = \dfrac{\sin x}{x}$.

(iii) $f(x) = \dfrac{1}{\sqrt{1 + \sin x}}$.

23. Using the binomial formula, write the first two nonzero terms of the series representations of the following expressions about $x = 0$:

(i) $f(x) = \dfrac{(4x - 1)^4 + \sqrt{1 + 4x^8}}{1 + x^3 + 7x^4}$.

(ii) $f(x) = \dfrac{x^2 + 3x^3}{x^3 + 2\sqrt{4 + x^6}}$.

24. Use the L'Hôpital's rule to find the limit

$$\lim_{x \to 0} \frac{\sqrt{x^6 + 4x}}{4x^3 - 3x^2}$$

and then verify your result by finding an appropriate series expansion.

25. Find the limit

$$\lim_{x \to 0} \frac{\sqrt{x^9 + 9x^4}}{5x^3 + 2x^2}$$

by using the L'Hôpital's rule and then verify your result by finding an appropriate series expansion.

26. Evaluate the following limits and check your answers by using Maclaurin series:

(i) $\lim\limits_{x \to 0} \frac{1 - e^x}{x}$.

(ii) $\lim\limits_{x \to 0} \left[\frac{1}{x} - \frac{1}{e^x - 1}\right]$.

CHAPTER 9

COMPLEX NUMBERS AND FUNCTIONS

As Gamow mentioned in his book, *One Two Three ... Infinity: Facts and Speculations of Science*, the sixteenth-century Italian mathematician Cardan is the first *brave* scientist to use the mysterious number called the imaginary i with the property $i^2 = -1$ in writing. Cardan introduced this number to express the roots of cubic and quartic polynomials, albeit with the reservation that it is probably meaningless or fictitious. All imaginary numbers can be written as proportional to the imaginary unit i. It is also possible to define hybrid numbers, $a + ib$, which are known as complex numbers.

Complex analysis is the branch of mathematics that deals with the functions of complex numbers. A lot of the theorems in the real domain can be proven considerably easily with complex analysis. Like control theory and signal analysis, many branches of science and engineering make widespread use of the techniques of complex analysis. As a mathematical tool, complex analysis offers tremendous help in the evaluation of series and improper integrals encountered in physical theories. With the discovery of quantum mechanics, it also became clear that complex numbers are not just convenient computational tools but also have a fundamental bearing on the inner workings of the universe. In this chapter, we introduce the basic elements of complex analysis, that is, complex numbers and their functions.

9.1   THE ALGEBRA OF COMPLEX NUMBERS

A well-known result from mathematical analysis states that a given quadratic equation:

$$ax^2 + bx + c = 0, \quad a, b, c \in \mathbb{R}, \tag{9.1}$$

has always two roots, x_1 and x_2, that can be factored as

$$(x - x_1)(x - x_2) = 0. \tag{9.2}$$

A general expression for the roots of a quadratic equation exists, and it is given by the well-known formula

$$x_{1,2} = \frac{-b \pm \sqrt{b^2 - 4ac}}{2a}. \tag{9.3}$$

When the coefficients satisfy the inequality $b^2 - 4ac \geq 0$, both roots are real. However, when $b^2 - 4ac < 0$, no roots can be found within the set of real numbers. Hence, the number system has to be extended to include a new kind of number, the imaginary i with the property

$$i^2 = -1, \text{ or } i = \sqrt{-1}. \tag{9.4}$$

Now the roots can be expressed in terms of complex numbers as

$$x_{1,2} = \frac{-b \mp i\sqrt{4ac - b^2}}{2a}. \tag{9.5}$$

In general, a **complex number** z is written as

$$\boxed{z = x + iy,} \tag{9.6}$$

where x and y are two real numbers. The real and the imaginary parts of z are written, respectively, as

$$\text{Re } z = x \text{ and Im } z = y. \tag{9.7}$$

Complex numbers are also written as ordered pairs of real numbers:

$$\boxed{z = (x, y).} \tag{9.8}$$

When $y = 0$, that is, Im $z = 0$, we have a real number $z = x$ and when $x = 0$, that is, Re $z = 0$, we have a pure imaginary number $z = iy$. Two **complex numbers**, $z_1 = (x_1, y_1)$ and $z_2 = (x_2, y_2)$, are **equal** if and only if their real and imaginary parts are equal, $x_1 = x_2$, $y_1 = y_2$. **Zero** of the complex number system, $z = 0$, means $x = 0$ and $y = 0$.

Two complex numbers, z_1 and z_2, can be **added** or **subtracted** by adding or subtracting their real and imaginary parts separately:

$$\boxed{z_1 \pm z_2 = (x_1 + iy_1) \pm (x_2 + iy_2) = ((x_1 \pm x_2) + i(y_1 \pm y_2)).} \tag{9.9}$$

Two complex numbers, z_1 and z_2, can be **multiplied** by first writing

$$z_1 z_2 = (x_1 + iy_1)(x_2 + iy_2) \tag{9.10}$$

and then expanding the right-hand side formally as

$$z_1 z_2 = x_1 x_2 + ix_1 y_2 + iy_1 x_2 + i^2 y_1 y_2. \tag{9.11}$$

Using the property $i^2 = -1$, we finally obtain the **product** of two complex numbers as

$$\boxed{z_1 z_2 = (x_1 x_2 - y_1 y_2) + i(x_1 y_2 + y_1 x_2).} \tag{9.12}$$

The **division** of two complex numbers:

$$\frac{z_1}{z_2} = \frac{x_1 + iy_1}{x_2 + iy_2}, \quad z_2 \neq 0, \tag{9.13}$$

can be performed by multiplying and dividing the right-hand side by $(x_2 - iy_2)$:

$$\frac{z_1}{z_2} = \frac{x_1 + iy_1}{x_2 + iy_2} \cdot \frac{x_2 - iy_2}{x_2 - iy_2} = \frac{x_1 x_2 + y_1 y_2 + i(y_1 x_2 - x_1 y_2)}{x_2^2 + y_2^2}, \tag{9.14}$$

$$\boxed{\frac{z_1}{z_2} = \left(\frac{x_1 x_2 + y_1 y_2}{x_2^2 + y_2^2} \right) + i \left(\frac{y_1 x_2 - x_1 y_2}{x_2^2 + y_2^2} \right).} \tag{9.15}$$

Division by zero is not defined. The following properties of **algebraic operations** on complex numbers can be shown to follow from the aforementioned properties:

$$z_1 + z_2 = z_2 + z_1, \tag{9.16}$$

$$z_1 z_2 = z_2 z_1, \tag{9.17}$$

$$z_1 + (z_2 + z_3) = (z_1 + z_2) + z_3, \tag{9.18}$$

$$z_1(z_2 z_3) = (z_1 z_2) z_3, \tag{9.19}$$

$$z_1(z_2 + z_3) = z_1 z_2 + z_1 z_3. \tag{9.20}$$

The **complex conjugate** z^*, or simply the conjugate of a complex number z, is defined as

$$\boxed{z^* = x - iy,}$$
(9.21)

In general, the complex conjugate of a complex quantity is obtained by reversing the sign of the imaginary part or by replacing i with $-i$. **Conjugation** of complex numbers has the following properties:

$$(z_1 + z_2)^* = z_1^* + z_2^*,$$
(9.22)

$$(z_1 z_2)^* = z_1^* z_2^*,$$
(9.23)

$$\left(\frac{z_1}{z_2}\right)^* = \frac{z_1^*}{z_2^*},$$
(9.24)

$$z + z^* = 2\operatorname{Re}z = 2x,$$
(9.25)

$$z - z^* = 2i\operatorname{Im}z = 2iy.$$
(9.26)

The **absolute value**, $|z|$, also called the **modulus**, is the positive real number

$$|z| = |x + iy| = \sqrt{x^2 + y^2}.$$
(9.27)

Modulus has the following properties:

$$|z_1 z_2| = |z_1||z_2|,$$
(9.28)

$$\left|\frac{z_1}{z_2}\right| = \frac{|z_1|}{|z_2|},$$
(9.29)

$$|z| = |z^*|,$$
(9.30)

$$|z|^2 = zz^*.$$
(9.31)

Triangle inequalities:

$$\boxed{|z_1 + z_2| \le |z_1| + |z_2|,}$$
(9.32)

$$\boxed{|z_1 - z_2| \ge ||z_1| - |z_2||,}$$
(9.33)

derive their name from the geometric properties of triangles and they are very useful in applications. The set of all complex numbers with the aforementioned algebraic properties forms a field and is shown as $\mathbb{C}$. Naturally, the set of real numbers, $\mathbb{R}$, is a subfield of $\mathbb{C}$.

A **geometric** representation of complex numbers is possible by introducing the complex z-plane, where the two orthogonal axes, x- and y-axes, represent the real and the imaginary parts of a complex number, respectively (Figure 9.1).

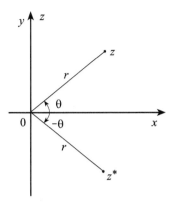

Figure 9.1 Complex z-plane.

From the z-plane it is seen that the modulus of a complex number, $|z| = r$, is equal to the length of the line connecting the point z to the origin O. Using the length of this line, r, and the angle that it makes with the positive x-axis, θ, which is called the **argument** of z and usually written as $\arg z$, we introduce the **polar representation** of complex numbers as

$$z = r(\cos \theta + i \sin \theta). \qquad (9.34)$$

The two representations, $z(x, y)$ and $z(r, \theta)$, are related by the transformation equations

$$x = r \cos \theta,$$
$$y = r \sin \theta, \qquad (9.35)$$

where the inverse transformations are given as

$$r = \sqrt{x^2 + y^2},$$
$$\theta = \tan^{-1}\left(\frac{y}{x}\right). \qquad (9.36)$$

Using the polar representation, we can write the **product** of two complex numbers, $z_1 = r_1(\cos \theta_1 + i \sin \theta_1)$ and $z_2 = r_2(\cos \theta_2 + i \sin \theta_2)$, as

$$z_1 z_2 = r_1 r_2[(\cos \theta_1 \cos \theta_2 - \sin \theta_1 \sin \theta_2) + i(\sin \theta_1 \cos \theta_2 + \cos \theta_1 \sin \theta_2)], \quad (9.37)$$

$$z_1 z_2 = r_1 r_2[\cos(\theta_1 + \theta_2) + i \sin(\theta_1 + \theta_2)]. \qquad (9.38)$$

In other words, the **modulus** of the product of two complex numbers is equal to the product of the moduli of the multiplied numbers:

$$|z_1 z_2| = |z_1||z_2| = r_1 r_2, \qquad (9.39)$$

and the **argument** of the product is equal to the sum of the arguments:

$$\arg z_1 z_2 = \arg z_1 + \arg z_2 = \theta_1 + \theta_2. \tag{9.40}$$

In particular, when a complex number is multiplied with i, its effect on z is to **rotate** it by $\pi/2$:

$$iz = \left(\cos\frac{\pi}{2} + i\sin\frac{\pi}{2}\right) r(\cos\theta + i\sin\theta) = r\left(\cos\left(\theta + \frac{\pi}{2}\right) + i\sin\left(\theta + \frac{\pi}{2}\right)\right). \tag{9.41}$$

Using Eq. (9.38), we can write

$$z_1 z_2 \ldots z_n = r_1 r_2 \ldots r_n[\cos(\theta_1 + \theta_2 + \cdots + \theta_n) + i\sin(\theta_1 + \theta_2 + \cdots + \theta_n)]. \tag{9.42}$$

Consequently, if $z_1 = z_2 = \cdots = z_n = r(\cos\theta + i\sin\theta)$, we obtain

$$\boxed{z^n = r^n[\cos n\theta + i\sin n\theta].} \tag{9.43}$$

When $r = 1$, this becomes the famous **DeMoivre's formula**:

$$\boxed{[\cos\theta + i\sin\theta]^n = \cos n\theta + i\sin n\theta.} \tag{9.44}$$

The **ratio** of two complex numbers can be written in polar form as

$$\frac{z_1}{z_2} = \frac{r_1}{r_2}[\cos(\theta_1 - \theta_2) + i\sin(\theta_1 - \theta_2)], \quad r_2 \neq 0. \tag{9.45}$$

As a special case of this, we can write

$$z^{-1} = r^{-1}[\cos\theta - i\sin\theta], \ r \neq 0, \tag{9.46}$$

which leads to

$$z^{-n} = r^{-n}[\cos n\theta - i\sin n\theta], \ r \neq 0, \ n > 0; \tag{9.47}$$

thus DeMoivre's formula is also valid for negative powers.

9.2 ROOTS OF A COMPLEX NUMBER

Consider the polynomial

$$z^n - z_0 = 0, \quad n = \text{positive integer}, \tag{9.48}$$

which has n roots, $z = z_0^{1/n}$, in the complex z-plane. Using the polar representation of z:

$$z = r(\cos\theta + i\sin\theta), \tag{9.49}$$

and z_0:

$$z_0 = r_0(\cos\theta_0 + i\sin\theta_0), \tag{9.50}$$

we can write Eq. (9.48) as

$$r^n(\cos n\theta + i\sin n\theta) = r_0(\cos\theta_0 + i\sin\theta_0), \tag{9.51}$$

which offers the solutions

$$r = r_0^{1/n}, \quad n\theta \pm 2\pi k = \theta_0, \quad k = 0, 1, 2, \ldots. \tag{9.52}$$

The first equation gives the all equal moduli of the n roots, while the second equation gives their arguments. When k is an integer multiple of n, no new roots emerge. Hence, we obtain the arguments of the n roots as

$$\theta_k = \frac{\theta_0}{n} + 2\pi\left(\frac{k}{n}\right), \quad k = 0, 1, \ldots, (n-1). \tag{9.53}$$

These roots correspond to the n vertices:

$$z = r_0^{1/n}e^{i\theta_k}, \quad \theta_k = \frac{\theta_0}{n} + 2\pi\left(\frac{k}{n}\right), \quad k = 0, 1, \ldots, (n-1), \tag{9.54}$$

of an n-sided polygon inscribed in a circle of radius $r_0^{1/n}$. Arguments of these roots are given as

$$\left(\frac{\theta_0}{n}\right), \left(\frac{\theta_0}{n} + \frac{2\pi.1}{n}\right), \left(\frac{\theta_0}{n} + \frac{2\pi.2}{n}\right), \ldots, \left(\frac{\theta_0}{n} + \frac{2\pi.(n-1)}{n}\right), \tag{9.55}$$

which are separated by

$$\boxed{\Delta\theta_n = \frac{2\pi}{n}.} \tag{9.56}$$

In Figure 9.2, we show the five roots of equation $z^5 - 1 = 0$, where $n = 5$ and $z_0 = 1$. Hence, $r_0 = 1$ and $\theta_0 = 0$. Using Eq. (9.56), this gives the moduli of all the roots as 1 and their arguments, $\arg z_i = \arg z_{i-1} + \frac{2\pi}{5}$, $i = 1, \ldots, 5$, as

$$\arg z_1 = 0,$$

$$\arg z_2 = 0 + \frac{2\pi}{5},$$

$$\arg z_3 = \frac{2\pi}{5} + \frac{2\pi}{5} = \frac{4\pi}{5}, \tag{9.57}$$

$$\arg z_4 = \frac{4\pi}{5} + \frac{2\pi}{5} = \frac{6\pi}{5},$$

$$\arg z_5 = \frac{6\pi}{5} + \frac{2\pi}{5} = \frac{8\pi}{5}.$$

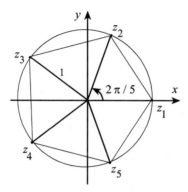

Figure 9.2 Roots of $z^5 - 1 = 0$.

If m and n are positive integers with no common factors, then we can write

$$(z^m)^{1/n} = (z^{1/n})^m = (z)^{m/n} \tag{9.58}$$

$$\boxed{(z^m)^{1/n} = r^{m/n}\left[\cos\frac{m}{n}(\theta + 2\pi k) + i\sin\frac{m}{n}(\theta + 2\pi k)\right],} \tag{9.59}$$

where $k = 0, 1, 2, \ldots, n - 1$.

9.3 INFINITY AND THE EXTENDED COMPLEX PLANE

In many applications, we need to define a number ∞ with the properties

$$z + \infty = \infty + z = \infty, \text{ for all finite } z, \tag{9.60}$$

and

$$z \cdot \infty = \infty \cdot z = \infty, \text{ for all } z \neq 0 \text{ but including } z = \infty. \tag{9.61}$$

The number ∞, which represents infinity, allows us to write

$$\frac{z}{0} = \infty, \ z \neq 0, \tag{9.62}$$

and

$$\frac{z}{\infty} = 0, \ z \neq \infty. \tag{9.63}$$

In the complex z-plane $\mathbb{C}$, there is no point with these properties, thus, we introduce the **extended** complex plane $\widetilde{\mathbb{C}}$, which includes this new point ∞, called the **infinity**:

$$\boxed{\widetilde{\mathbb{C}} = \mathbb{C} + \{\infty\}.} \tag{9.64}$$

A geometric model for the members of the extended complex plane is possible. Consider the three-dimensional unit sphere S:

$$x_1^2 + x_2^2 + x_3^2 = 1. \tag{9.65}$$

For every point on S, except the north pole N at $(0,0,1)$, we associate a complex number

$$z = \frac{x_1 + ix_2}{1 - x_3}. \tag{9.66}$$

This is a one-to-one correspondence with the modulus squared:

$$|z|^2 = \frac{x_1^2 + x_2^2}{(1 - x_3)^2}, \tag{9.67}$$

which, after using Eq. (9.65), becomes

$$|z|^2 = \frac{1 + x_3}{1 - x_3}. \tag{9.68}$$

Solving this equation for x_3 and using Eqs. (9.65) and (9.66) to write x_1 and x_2, we obtain

$$x_1 = \frac{z + z^*}{1 + |z|^2}, \tag{9.69}$$

$$x_2 = \frac{z - z^*}{i(1 + |z|^2)}, \tag{9.70}$$

$$x_3 = \frac{|z|^2 - 1}{|z|^2 + 1}. \tag{9.71}$$

This is a one-to-one correspondence with every point of the z-plane with every point, except $(0,0,1)$, on the surface of the unit sphere. The correspondence with the extended z-plane can be completed by identifying the point $(0,0,1)$ with ∞. Note that from Eq. (9.71) the lower hemisphere, $x_3 < 0$, corresponds to the disc $|z| < 1$, while the upper hemisphere, $x_3 > 0$, corresponds to its outside $|z| > 1$. We identify the z-plane with the $x_1 x_2$-plane, that is, the equatorial plane, and use x_1- and the x_2-axes as the real and the imaginary axes of the z-plane, respectively. In function theory, the unit sphere is called the **Riemann sphere**.

If we write $z = x + iy$ and use Eqs. (9.69)–(9.71), we can establish the ratios

$$\frac{x}{x_1} = \frac{y}{x_2} = \frac{1}{1 - x_3}, \tag{9.72}$$

which with a little help from Figure 9.3 shows that the points z, Z, and N lie on a straight line. In Figure 9.3 and Eq. (9.72), the point N is the north

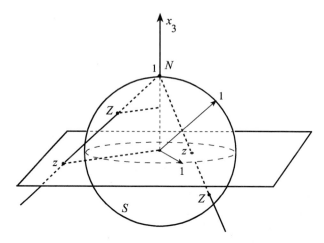

Figure 9.3   Riemann sphere and stereographic projections.

pole $(0,0,1)$ of the Riemann sphere, the point $Z = (x_1, x_2, x_3)$ is the point at which the straight line originating from N pierces the sphere and finally, z is the point where the straight line meets the equatorial plane, which defines the z-plane. This is called the **stereographic projection**.

Geometrically, a stereographic projection maps a straight line in the z-plane into a circle on S, which passes through the pole and vice versa. In general, any circle on the sphere corresponds to a circle or a straight line in the z-plane. Since a circle on the sphere can be defined by the intersection of a plane:

$$ax_1 + bx_2 + cx_3 = d, \quad 0 \le d < 1, \tag{9.73}$$

with the sphere

$$x_1^2 + x_2^2 + x_3^2 = 1, \tag{9.74}$$

using Eqs. (9.69)–(9.71), we can write Eq. (9.73) as

$$a(z + z^*) - bi(z - z^*) + c(|z|^2 - 1) = d(|z|^2 + 1) \tag{9.75}$$

or as

$$(d - c)(x^2 + y^2) - 2ax - 2by + d + c = 0. \tag{9.76}$$

For $d \ne c$ this is the equation of a circle, and it becomes a straight line for $d = c$. Since the transformation is one-to-one, conversely, all circles and straight lines on the z-plane correspond to stereographic projections of circles on the Riemann sphere.

In stereographic projections, there is significant difference between the distances on the Riemann sphere and their projections on the z-plane. Let

(x_1, x_2, x_3) and (x_1', x_2', x_3') be two points on the sphere:

$$x_1^2 + x_2^2 + x_3^2 = 1, \tag{9.77}$$

$$x_1'^2 + x_2'^2 + x_3'^2 = 1, \tag{9.78}$$

where the distance d between these points is given as

$$d^2 = (x_1 - x_1')^2 + (x_2 - x_2')^2 + (x_3 - x_3')^2 \tag{9.79}$$

$$= 2 - 2(x_1 x_1' + x_2 x_2' + x_3 x_3'). \tag{9.80}$$

Using the transformation equations [Eqs. (9.69)–(9.71)], we can write the corresponding distance in the z-plane as

$$d(z, z') = \frac{2|z - z'|}{\sqrt{(1 + |z|^2)(1 + |z'|^2)}}. \tag{9.81}$$

If we take one of the points on the sphere, say z', as the north pole N, then $z' = \infty$, hence, Eq. (9.81) gives

$$d(z, \infty) = \frac{2}{\sqrt{(1 + |z|^2)}}. \tag{9.82}$$

Note that the point $z = x + iy$, where x and/or y are $\pm$ infinity, belongs to the z-plane. Hence, it is not the same point as the ∞ introduced above.

9.4 COMPLEX FUNCTIONS

We can define a real function f as a **mapping** that returns the value $f(x)$ for each point x in its domain of definition:

$$\boxed{f: \ x \to f(x).} \tag{9.83}$$

Graphically, this can be conveniently represented by introducing the rectangular coordinates with two perpendicular axes called the x- and the y-axes. By plotting the value $y = f(x)$ that the function returns along the y-axis directly above the location of x along the x-axis, we obtain a curve as shown in Figure 9.4 called the graph of $f(x)$.

Complex-valued functions are defined similarly as relations that return a complex value for each z in the domain of definition:

$$f: \ z \to f(z). \tag{9.84}$$

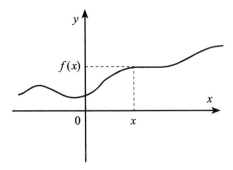

Figure 9.4 Graph of $f(x)$.

Analogous to real functions, we introduce a dependent variable w and write a **complex function** as

$$\boxed{w = f(z).}$$
(9.85)

Since both the dependent and the independent variables have real and imaginary parts:

$$z = x + iy,$$
(9.86)

$$w = u(x, y) + iv(x, y),$$
(9.87)

it is generally simpler to draw w and z on separate planes. Since the function $w = f(z)$ gives the correspondence of the points in the z-plane to the points in the w-plane, it is also called a **mapping** or a **transformation**. This allows us to view complex functions as operations that map curves and regions in their domain of definition to other curves and regions in the w-plane. For example, the function

$$w = \sqrt{x^2 + y^2} + iy$$
(9.88)

maps all points of a circle in the z-plane, $x^2 + y^2 = c^2$, $c \geq 0$, to $u = c$ and $v = y$ in the w-plane. Since the range of y is $-c \leq y \leq c$, the interior of the circle is mapped into the region between the lines $-u \leq v \leq u$ and $u = c$ in the w-plane (Figure 9.5).

The **domain** of definition, D, of $f(z)$, means the set of values that z is allowed to take, while the set of values that the function returns, $w = f(z)$, is called the **range** of w. A function is called **single-valued** in a domain D if it returns a single value w for each z in D. From now on, we use the term function only for the single-valued functions. Multiple-valued functions like $z^{1/2}$ or $\log z$ can be treated as single-valued functions by restricting them to one of their allowed values in a specified domain of the z-plane. Domain of definition of all polynomials:

$$f(z) = a_n z^n + a_{n-1} z^{n-1} + \cdots + a_0,$$
(9.89)

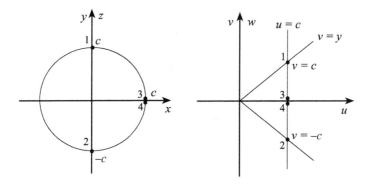

Figure 9.5   The z- and the w-planes for $w = \sqrt{x^2 + y^2} + iy$.

is the entire z-plane, while the function

$$f(z) = \frac{1}{z^2 + 1} \qquad (9.90)$$

is undefined at the points $z = \pm i$.

Each function has a specific real, $u(x, y)$, and an imaginary, $v(x, y)$, part:

$$w = f(z) = u(x, y) + iv(x, y). \qquad (9.91)$$

Consider $f(z) = z^3$. We can write

$$f(z) = z^3 = z^2 z = (x + iy)^2 (x + iy) = [(x^2 - y^2) + i(2xy)](x + iy) \qquad (9.92)$$
$$= [(x^2 - y^2)x - 2xy^2] + i[y(x^2 - y^2) + 2x^2 y], \qquad (9.93)$$

thus obtaining

$$u(x, y) = (x^2 - y^2)x - 2xy^2, \qquad (9.94)$$
$$v(x, y) = y(x^2 - y^2) + 2x^2 y. \qquad (9.95)$$

For $w = \sin z$, the $u(x, y)$ and the $v(x, y)$ functions are simply obtained from the expression $w = \sin(x + iy)$ as $w = \sin x \cosh y + i \cos x \sinh y$.

9.5   LIMITS AND CONTINUITY

Since a complex function can be written as

$$w = u(x, y) + iv(x, y), \qquad (9.96)$$

its limit can be found in terms of the limits of the real functions $u(x, y)$ and $v(x, y)$. Thus, the properties of the limits of complex functions can deduced from the properties of the limits of real functions. Basic results can be summarized in terms of the following theorems:

Theorem 9.1. For a given function:

$$f(z) = u(x, y) + iv(x, y), \tag{9.97}$$

let $z = x + iy$ and $z_0 = x_0 + iy_0$ be two points in the domain of $f(z)$. The limit of $f(z)$ at z_0 exists, that is,

$$\lim_{z \to z_0} f(z) = u_0 + iv_0, \tag{9.98}$$

if and only if

$$\lim_{(x,y) \to (x_0,y_0)} u(x, y) = u_0, \tag{9.99}$$

$$\lim_{(x,y) \to (x_0,y_0)} v(x, y) = v_0. \tag{9.100}$$

Theorem 9.2. If $f_1(z)$ and $f_2(z)$ are two complex functions whose limits exist at z_0:

$$\lim_{z \to z_0} f_1(z) = w_1, \tag{9.101}$$

$$\lim_{z \to z_0} f_2(z) = w_2, \tag{9.102}$$

then the following limits are true:

$$\lim_{z \to z_0} [f_1(z) + f_2(z)] = w_1 + w_2, \tag{9.103}$$

$$\lim_{z \to z_0} [f_1(z) f_2(z)] = w_1 w_2, \tag{9.104}$$

$$\lim_{z \to z_0} \left[\frac{f_1(z)}{f_2(z)} \right] = \frac{w_1}{w_2}, \; w_2 \neq 0. \tag{9.105}$$

The continuity of complex functions can be understood in terms of the continuity of the real functions u and v.

Theorem 9.3. A given function $f(z)$ is continuous at z_0 if and only if all the following three conditions are satisfied:

$$\begin{array}{ll} \text{(i)} & f(z_0) \text{ exists,} \\ \text{(ii)} & \lim_{z \to z_0} f(z) \text{ exists,} \\ \text{(iii)} & \lim_{z \to z_0} f(z) = f(z_0). \end{array} \tag{9.106}$$

This theorem implies that $f(z)$ is continuous if and only if $u(x, y)$ and $v(x, y)$ are continuous.

9.6   DIFFERENTIATION IN THE COMPLEX PLANE

As in real analysis, the **derivative** of a complex function at a point z in its domain of definition is defined as

$$
f'(z) = \frac{df}{dz} = \lim_{\Delta z \to 0} \frac{f(z + \Delta z) - f(z)}{\Delta z}. \tag{9.107}
$$

Nevertheless, there is a fundamental difference between the differentiation of complex and real functions. In the complex z-plane, a given point z can be approached from infinitely many different directions (Figure 9.6). Hence, a meaningful definition of derivative should be independent of the direction of approach. If we approach the point z parallel to the real axis, $\Delta z = \Delta x$, we obtain the derivative

$$
\frac{df}{dz} = \lim_{\Delta x \to 0} \frac{f(z + \Delta x) - f(z)}{\Delta x} \tag{9.108}
$$

$$
= \lim_{\Delta x \to 0} \left[\frac{u(x + \Delta x, y) - u(x, y)}{\Delta x} + i \frac{v(x + \Delta x, y) - v(x, y)}{\Delta x} \right] \tag{9.109}
$$

$$
= \frac{\partial u}{\partial x} + i \frac{\partial v}{\partial x}. \tag{9.110}
$$

On the other hand, if z is approached parallel to the imaginary axis, $\Delta z = i\Delta y$, the derivative becomes

$$
\frac{df}{dz} = \lim_{i\Delta y \to 0} \frac{f(z + i\Delta y) - f(z)}{i\Delta y} \tag{9.111}
$$

$$
= \lim_{i\Delta y \to 0} \left[\frac{u(x, y + i\Delta y) - u(x, y)}{i\Delta y} + i \frac{v(x, y + i\Delta y) - v(x, y)}{i\Delta y} \right] \tag{9.112}
$$

$$
= -i \frac{\partial u}{\partial y} + \frac{\partial v}{\partial y}, \tag{9.113}
$$

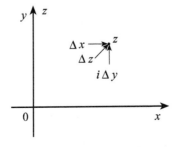

Figure 9.6   Differentiation in the complex plane.

or

$$\frac{df}{dz} = \frac{\partial v}{\partial y} - i\frac{\partial u}{\partial y}. \qquad (9.114)$$

For a meaningful definition of derivative, these two expressions should agree. Thus, giving us the conditions for the existence of derivative at z as

$$\boxed{\frac{\partial u}{\partial x} = \frac{\partial v}{\partial y},} \qquad (9.115)$$

$$\boxed{\frac{\partial v}{\partial x} = -\frac{\partial u}{\partial y}.} \qquad (9.116)$$

These are called the **Cauchy–Riemann conditions.** Note that choosing the direction of approach first along the x- and then along the y-axes is a matter of calculational convenience. A general treatment will also lead to the same conclusion. Cauchy–Riemann conditions shows that the real and the imaginary parts of a differentiable function are related. In summary, the Cauchy–Riemann conditions have to be satisfied for the derivative to exist at a given point. However, as we shall see, in general, they are **not** the **sufficient** conditions.

Example 9.1. *Cauchy–Riemann conditions:* Consider the following simple function:

$$f(z) = z^2, \qquad (9.117)$$

We can find its derivative as the limit

$$\frac{df}{dz} = \lim_{\delta \to 0} \left\{ \frac{(z+\delta)^2 - z^2}{\delta} \right\} = \lim_{\delta \to 0} \left\{ \frac{\delta}{\delta}(2z+\delta) \right\} = 2z. \qquad (9.118)$$

If we write the function $f(z) = z^2$ as

$$f(z) = (x^2 - y^2) + i2xy, \qquad (9.119)$$

we can easily check that the Cauchy–Riemann conditions are satisfied everywhere in the z-plane:

$$\frac{\partial u}{\partial x} = 2x = \frac{\partial v}{\partial y},$$
$$\frac{\partial v}{\partial x} = 2y = -\frac{\partial u}{\partial y}. \qquad (9.120)$$

We now consider the function

$$f(z) = |z|^2 \qquad (9.121)$$

and write the limit

$$\frac{df}{dz} = \lim_{\delta \to 0} \left\{ \frac{1}{\delta} [|z + \delta|^2 - |z|^2] \right\} = \lim_{\delta \to 0} \left\{ \frac{1}{\delta} [(z + \delta)(z^* + \delta^*) - zz^*] \right\} \quad (9.122)$$

$$= \lim_{\delta \to 0} \left\{ z^* + z\frac{\delta^*}{\delta} + \delta^* \right\}. \quad (9.123)$$

At the origin $z = 0$, regardless of the direction of approach, the above limit exists and its value is equal to 0; thus, we can write the derivative $df/dz|_{z=0} = 0$. For the other points, if δ approaches zero along the real axis, $\delta = \epsilon$, we obtain

$$\frac{df}{dz} = \lim_{\varepsilon \to 0} \left\{ z^* + z\frac{\varepsilon}{\varepsilon} + \varepsilon \right\} = z^* + z \quad (9.124)$$

and if δ approaches zero along the imaginary axis, $\delta = i\varepsilon$, we find

$$\frac{df}{dz} = \lim_{i\varepsilon \to 0} \left\{ z^* + z\frac{-i\varepsilon}{i\varepsilon} - i\varepsilon \right\} = z^* - z. \quad (9.125)$$

Hence, except at $z = 0$, the derivative of $f(z) = |z|^2$ does not exist. In fact, the Cauchy–Riemann conditions for $f(z) = |z|^2$ are not satisfied,

$$\frac{\partial u}{\partial x} = 2x \neq \frac{\partial v}{\partial y} = 0, \quad (9.126)$$

$$\frac{\partial v}{\partial x} = 0 \neq -\frac{\partial u}{\partial y} = 2y, \quad (9.127)$$

unless $z = 0$.

Example 9.2. *Cauchy–Riemann conditions:* Consider the function

$$f(z) = \begin{cases} \dfrac{x^3 - y^3}{x^2 + y^2} + i\dfrac{x^3 + y^3}{x^2 + y^2}, & |z| \neq 0, \\ 0, & z = 0. \end{cases} \quad (9.128)$$

At $z = 0$, we can easily check that the Cauchy–Riemann conditions are satisfied:

$$\frac{\partial u(0,0)}{\partial x} = 1 = \frac{\partial v(0,0)}{\partial y}, \quad (9.129)$$

$$\frac{\partial v(0,0)}{\partial x} = 1 = -\frac{\partial u(0,0)}{\partial y}. \quad (9.130)$$

However, if we calculate the derivative using the limits in Eqs. (9.109) and (9.112), we find

$$\frac{df(0)}{dz} = \lim_{\Delta x \to 0} \left[\frac{u(\Delta x, 0) - u(0,0)}{\Delta x} + i\frac{v(\Delta x, 0) - v(0,0)}{\Delta x} \right] \tag{9.131}$$

$$= \lim_{\Delta x \to 0} \left[\frac{(\Delta x)^3}{(\Delta x)^2}\frac{1}{\Delta x} + i\frac{(\Delta x)^3}{(\Delta x)^2}\frac{1}{\Delta x} \right] = 1 + i \tag{9.132}$$

and

$$\frac{df(0)}{dz} = \lim_{i\Delta y \to 0} \left[\frac{u(0, i\Delta y) - u(0,0)}{i\Delta y} + i\frac{v(0, i\Delta y) - v(0,0)}{i\Delta y} \right] \tag{9.133}$$

$$= \lim_{i\Delta y \to 0} \left[\frac{-(i\Delta y)^3}{(i\Delta y)^2}\frac{1}{i\Delta y} + i\frac{(i\Delta y)^3}{(i\Delta y)^2}\frac{1}{i\Delta y} \right] = -1 + i. \tag{9.134}$$

In other words, even though the Cauchy–Riemann conditions are satisfied at $z = 0$, the derivative $f'(0)$ does not exist. That is, the Cauchy–Riemann conditions are **necessary** but **not sufficient** for the existence of derivative.

The following theorem gives the sufficient condition for the existence of derivative at a given point z [1].

Theorem 9.4. If $u(x,y)$ and $v(x,y)$ are real- and single-valued functions with continuous first-order partial derivatives at (x_0, y_0), then the Cauchy–Riemann conditions at (x_0, y_0) imply the existence of $f'(z_0)$.

What happened in Example 9.2 is that in order to satisfy the Cauchy–Riemann conditions at $(0,0)$, all we needed was the existence of the first-order partial derivatives of $u(x,y)$ and $v(x,y)$ at $(0,0)$. However, Theorem 9.4 not only demands the existence of the first partial derivatives of u and v at a given point but also needs their continuity at that point. This means that the first-order partial derivatives should also exist in the neighborhood of a given point for the function to be differentiable.

9.7 ANALYTIC FUNCTIONS

A function is said to be **analytic** at z_0 if its derivative $f'(z)$ exists not only at z_0 but also at every other point in some neighborhood of z_0. Similarly, if a function is analytic at every point of some domain D, then it is called analytic in D. All polynomials [2]:

$$f(z) = a_0 + a_1 z + \cdots + a_n z^n, \tag{9.135}$$

are analytic everywhere in the z-plane. Functions analytic everywhere in the z-plane are called **entire functions**. Since the derivative of

$$f(z) = |z|^2 \tag{9.136}$$

does not exist anywhere except at the origin, it is not analytic anywhere in the z-plane. If a function is analytic at every point in some neighborhood of a point z_0, except the point itself, then the point z_0 is called a **singular point**. For example, the function

$$f(z) = \frac{1}{z-2}; \quad f'(z) = -\frac{1}{(z-2)^2} \tag{9.137}$$

has a singular point at $z = 2$. If two functions are analytic, then their sum and product are also analytic. Their quotient is analytic except at the zeros of the denominator. If we let $f_1(z)$ be analytic in domain D_1 and let $f_2(z)$ be analytic in domain D_2, then the composite function

$$f_1(f_2(z)) \tag{9.138}$$

is also analytic in the domain D_1. For example, since the functions

$$f_1(z) = z^2 + 2 \quad \text{and} \quad f_2(z) = \exp(z) + 1 \tag{9.139}$$

are entire functions, the composite functions

$$f_1(f_2(z)) = (\exp(z) + 1)^2 + 2 \tag{9.140}$$

and

$$f_2(f_1(z)) = \exp(z^2 + 2) + 1 \tag{9.141}$$

are also entire functions.

9.8 HARMONIC FUNCTIONS

For an analytic function, $f(z) = u + iv$, defined in some domain D of the z-plane, the Cauchy–Riemann conditions:

$$\frac{\partial u}{\partial x} = \frac{\partial v}{\partial y}, \tag{9.142}$$

$$\frac{\partial v}{\partial x} = -\frac{\partial u}{\partial y}, \tag{9.143}$$

are satisfied at every point of D. Differentiating the first condition with respect to x and the second condition with respect to y, we get

$$\frac{\partial^2 u}{\partial x^2} = \frac{\partial^2 v}{\partial x \partial y} \quad \text{and} \quad \frac{\partial^2 u}{\partial y^2} = -\frac{\partial^2 v}{\partial y \partial x}. \tag{9.144}$$

For an analytic function, the first-order partial derivatives of u and v are continuous, hence, the mixed derivatives $\partial^2 v/\partial x \partial y$ and $\partial^2 v/\partial y \partial x$ are equal, thus we obtain

$$\boxed{\frac{\partial^2 u}{\partial x^2} + \frac{\partial^2 u}{\partial y^2} = 0.} \tag{9.145}$$

That is, the real part of an analytic function, $u(x, y)$, satisfies the two-dimensional Laplace equation in the domain of definition D. Similarly, differentiating Eq. (9.142) with respect to y and Eq. (9.143) with respect to x and then by adding we obtain

$$\boxed{\frac{\partial^2 v}{\partial x^2} + \frac{\partial^2 v}{\partial y^2} = 0.} \tag{9.146}$$

In other words, the real and the imaginary parts of an analytic function satisfy the two-dimensional Laplace equation. Functions that satisfy the Laplace equation in two dimensions are called **harmonic functions.**They could be used either as the real or the imaginary part of an analytic function. Pairs of harmonic functions (u, v) connected by the Cauchy–Riemann conditions are called the **conjugate harmonic functions.**

Example 9.3. *Conjugate harmonic functions:* Given the real function

$$f(x) = x^3 - 3y^2 x, \tag{9.147}$$

which satisfies the Laplace equation:

$$\frac{\partial^2 f}{\partial x^2} + \frac{\partial^2 f}{\partial y^2} = 6x - 6x = 0. \tag{9.148}$$

Since it is harmonic, we can use it as the real part of an analytic function, $u = x^3 - 3y^2 x$, and find its conjugate pair as follows: We start with the first Cauchy–Riemann condition $\partial u/\partial x = \partial v/\partial y$ to write

$$\frac{\partial v}{\partial y} = 3x^2 - 3y^2, \tag{9.149}$$

which can be integrated immediately as

$$v(x, y) = 3x^2 y - y^3 + \Phi(x), \tag{9.150}$$

where $\Phi(x)$ is an arbitrary function. We now use the second Cauchy–Riemann condition $\partial v/\partial x = -\partial u/\partial y$ to obtain an ordinary differential equation for $\Phi(x)$:

$$6xy + \Phi'(x) = 6yx, \tag{9.151}$$

$$\Phi'(x) = 0, \tag{9.152}$$

the solution of which gives $\Phi(x) = c_0$. Substituting this into Eq. (9.150) yields $v(x, y)$ as

$$v(x, y) = 3x^2 y - y^3 + c_0. \tag{9.153}$$

It can easily be checked that v is also harmonic.

Example 9.4. *C–R conditions in polar coordinates:* In a polar representation, a function can be written as

$$f(z) = u(r, \theta) + iv(r, \theta). \tag{9.154}$$

Using the transformation equations

$$x = r\cos\theta \text{ and } y = r\sin\theta, \tag{9.155}$$

we can write the Cauchy–Riemann conditions as

$$\boxed{\frac{\partial u}{\partial r} = \frac{1}{r}\frac{\partial v}{\partial \theta},} \tag{9.156}$$

$$\boxed{\frac{1}{r}\frac{\partial u}{\partial \theta} = -\frac{\partial v}{\partial r}.} \tag{9.157}$$

Example 9.5. *Derivative in polar representation:* Let us write the derivative of an analytic function $f(z) = u(r, \theta) + iv(r, \theta)$ in polar coordinates as

$$\frac{df}{dz} = \frac{\partial u}{\partial r}\frac{dr}{dz} + \frac{\partial u}{\partial \theta}\frac{d\theta}{dz} + i\frac{\partial v}{\partial r}\frac{dr}{dz} + i\frac{\partial v}{\partial \theta}\frac{d\theta}{dz}. \tag{9.158}$$

Substituting the Cauchy–Riemann conditions [Eqs. (9.156) and (9.157)] in Eq. (9.158) we write

$$\frac{df}{dz} = \frac{\partial u}{\partial r}\frac{dr}{dz} - r\frac{\partial v}{\partial r}\frac{d\theta}{dz} + i\frac{\partial v}{\partial r}\frac{dr}{dz} + ir\frac{\partial u}{\partial r}\frac{d\theta}{dz} \tag{9.159}$$

$$= \frac{\partial u}{\partial r}\left[\frac{dr}{dz} + ir\frac{d\theta}{dz}\right] + i\frac{\partial v}{\partial r}\left[\frac{dr}{dz} + ir\frac{d\theta}{dz}\right]. \tag{9.160}$$

Since $z = re^{i\theta}$, we can write

$$\frac{dz}{dz} = \left[\frac{dr}{dz} e^{i\theta} + ir \frac{d\theta}{dz} e^{i\theta} \right] = 1. \tag{9.161}$$

Hence, the expression inside the square brackets in Eq. (9.160) is

$$\left[\frac{dr}{dz} + ir \frac{d\theta}{dz} \right] = e^{-i\theta}, \tag{9.162}$$

which, when substituted into Eq. (9.160), gives

$$\boxed{\frac{df}{dz} = e^{-i\theta} \left[\frac{\partial u}{\partial r} + i \frac{\partial v}{\partial r} \right].} \tag{9.163}$$

Following similar steps in **rectangular coordinates**, we obtain

$$\boxed{\frac{df}{dz} = \frac{\partial u}{\partial x} + i \frac{\partial v}{\partial x}.} \tag{9.164}$$

9.9 BASIC DIFFERENTIATION FORMULAS

If the derivatives w', w_1', and w_2' exist, then the basic differentiation formulas can be given as
   (I)

$$\frac{dc}{dz} = 0, \ c \in \mathbb{C}, \ \text{ and } \ \frac{dz}{dz} = 1. \tag{9.165}$$

(II)

$$\frac{d(cw)}{dz} = c \frac{dw}{dz}. \tag{9.166}$$

(III)

$$\frac{d}{dz}(w_1 + w_2) = \frac{dw_1}{dz} + \frac{dw_2}{dz}. \tag{9.167}$$

(IV)

$$\frac{d(w_1 w_2)}{dz} = w_1 \frac{dw_2}{dz} + \frac{dw_1}{dz} w_2. \tag{9.168}$$

(V)

$$\frac{d}{dz}\left(\frac{w_1}{w_2} \right) = \frac{w_1' w_2 - w_1 w_2'}{w_2^2}, \ w_2 \neq 0. \tag{9.169}$$

(VI)

$$\frac{d}{dz}[w_1(w_2)] = \frac{dw_1}{dw_2}\left[\frac{dw_2}{dz}\right].$$

(9.170)

(VII)

$$\frac{d}{dz}z^n = nz^{n-1}, \; n > 0,$$

(9.171)

and $z \neq 0$ when $n < 0$ integer.

9.10   ELEMENTARY FUNCTIONS

9.10.1   Polynomials

The simplest analytic function different from a constant is z. Since the product and the sum of analytic functions are also analytic, we conclude that every polynomial of order n:

$$\boxed{P_n(z) = a_0 + a_1 z + \cdots + a_n z^n, a_n \neq 0,}$$

(9.172)

is also an analytic function. All polynomials are also entire functions. The fundamental theorem of algebra states that when n is positive, $P_n(z)$ has at least one root. This simple-sounding theorem, which was the doctoral dissertation of Gauss in 1799, has far-reaching consequences. Assuming that z_1 is a root of P_n, we can reduce its order as

$$P_n(z) = (z - z_1)P_{n-1}(z).$$

(9.173)

Similarly, if z_2 is a root of $P_{n-1}(z)$, we can reduce the order one more time to write

$$P_n(z) = (z - z_1)(z - z_2)P_{n-2}(z).$$

(9.174)

Cascading like this, we eventually reach the bottom of the ladder as

$$P_n(z) = (z - z_1)(z - z_2)\cdots(z - z_n).$$

(9.175)

In other words, a polynomial of order n has n, not necessarily all distinct, roots in the complex plane. Significance of this result becomes clear if we remember how the complex algebra was introduced in the first place. When equations like

$$z^2 + 1 = 0$$

(9.176)

are studied, it is seen that no roots can be found among the set of real numbers. Hence, the number system has to be extended to include the complex numbers. We now see that in general, the set of polynomials with complex coefficients do not have any other roots that are not included in the complex plane $\mathbb{C}$, hence, no further extension of the number system is necessary.

9.10.2 Exponential Function

Let us consider the series expansion of the exponential function with a pure imaginary argument as

$$e^{iy} = \sum_{n=0}^{\infty} \frac{(iy)^n}{n!}. \tag{9.177}$$

We write the even and the odd powers separately:

$$e^{iy} = \sum_{n=0}^{\infty} \frac{i^{2n}y^{2n}}{(2n)!} + \sum_{n=0}^{\infty} \frac{i^{2n+1}y^{2n+1}}{(2n+1)!} = \sum_{n=0}^{\infty} (-1)^n \frac{y^{2n}}{(2n)!} + i \sum_{n=0}^{\infty} (-1)^n \frac{y^{2n+1}}{(2n+1)!}. \tag{9.178}$$

Recognizing the first and the second series as $\cos y$ and $\sin y$, respectively, we obtain

$$\boxed{e^{iy} = \cos y + i \sin y,} \tag{9.179}$$

which is also known as **Euler's formula**. Multiplying this with the real number e^x, we obtain the **exponential function**

$$\boxed{e^z = e^x(\cos y + i \sin y).} \tag{9.180}$$

Since the functions $u = e^x \cos y$ and $v = e^x \sin y$ have continuous first partial derivatives everywhere and satisfy the Cauchy–Riemann conditions, using Theorem 9.4, we conclude that the exponential function is an **entire function**. Using Eq. (9.164):

$$\frac{df}{dz} = \frac{\partial u}{\partial x} + i \frac{\partial v}{\partial x}, \tag{9.181}$$

we obtain the derivative of the exponential function as the usual expression

$$\boxed{\frac{de^z}{dz} = e^z.} \tag{9.182}$$

Using the polar representation in the w-plane, $u = \rho \cos \phi$, $v = \rho \sin \phi$, we write

$$w = e^z = \rho(\cos \phi + i \sin \phi). \tag{9.183}$$

Comparing this with $e^z = e^x(\cos y + i \sin y)$, we obtain $\rho = e^x$ and $\phi = y$, that is, $|e^z| = e^x$ and $\arg e^z = y$. Using the polar representation for two points in the w-plane, $e^{z_1} = \rho_1(\cos \phi_1 + i \sin \phi_1)$ and $e^{z_2} = \rho_2(\cos \phi_2 + i \sin \phi_2)$, we can easily establish the following relations:

$$e^{z_1} e^{z_2} = e^{z_1+z_2}, \tag{9.184}$$

$$\frac{e^{z_1}}{e^{z_2}} = e^{z_1-z_2}, \tag{9.185}$$

$$(e^z)^n = e^{nz}. \tag{9.186}$$

In terms of the exponential function [Eq. (9.179)], the **polar representation** of z:

$$z = r(\cos\theta + i\sin\theta), \qquad (9.187)$$

can be written as

$$\boxed{z = re^{i\theta},} \qquad (9.188)$$

which is quite useful in applications. Another useful property of e^z is that for an integer n we obtain

$$e^{2\pi ni} = (e^{2\pi i})^n = 1, \qquad (9.189)$$

hence, we can write

$$e^{z+2n\pi i} = e^z e^{2n\pi i} = e^z. \qquad (9.190)$$

In other words, e^z is a **periodic function** with the period 2π. Series expansion of e^z is given as

$$\boxed{e^z = 1 + \frac{z}{1!} + \frac{z^2}{2!} + \cdots = \sum_{n=0}^{\infty} \frac{z^n}{n!}.} \qquad (9.191)$$

9.10.3 Trigonometric Functions

Trigonometric functions are defined as

$$\boxed{\cos z = \frac{e^{iz} + e^{-iz}}{2},} \qquad (9.192)$$

$$\boxed{\sin z = \frac{e^{iz} - e^{-iz}}{2i}.} \qquad (9.193)$$

Using the series expansion of e^z [Eq. (9.191)], we can justify these definitions as the usual series expansions:

$$\cos z = 1 - \frac{z^2}{2!} + \frac{z^4}{4!} + \cdots , \quad \sin z = z - \frac{z^3}{3!} + \frac{z^5}{5!} - \cdots . \qquad (9.194)$$

Since e^{iz} and e^{-iz} are entire functions, $\cos z$ and $\sin z$ are also entire functions. Using these series expansions, we obtain the derivatives:

$$\boxed{\frac{d}{dz}\sin z = \cos z, \quad \frac{d}{dz}\cos z = -\sin z.} \qquad (9.195)$$

The other trigonometric functions are defined as

$$\tan z = \frac{\sin z}{\cos z}, \quad \cot z = \frac{\cos z}{\sin z}, \quad \sec z = \frac{1}{\cos z}, \quad \csc z = \frac{1}{\sin z}. \qquad (9.196)$$

The usual **trigonometric identities** are also valid in the complex domain:

$$\sin^2 z + \cos^2 z = 1, \tag{9.197}$$

$$\sin(z_1 \pm z_2) = \sin z_1 \cos z_2 \pm \cos z_1 \sin z_2, \tag{9.198}$$

$$\cos(z_1 \pm z_2) = \cos z_1 \cos z_2 \mp \sin z_1 \sin z_2, \tag{9.199}$$

$$\sin(-z) = -\sin z, \tag{9.200}$$

$$\cos(-z) = \cos z, \tag{9.201}$$

$$\sin\left(\frac{\pi}{2} - z\right) = \cos z, \tag{9.202}$$

$$\sin 2z = 2 \sin z \cos z, \tag{9.203}$$

$$\cos 2z = \cos^2 z - \sin^2 z. \tag{9.204}$$

9.10.4 Hyperbolic Functions

Hyperbolic cosine and sine functions are defined as

$$\boxed{\cosh z = \frac{e^z + e^{-z}}{2},} \tag{9.205}$$

$$\boxed{\sinh z = \frac{e^z - e^{-z}}{2}.} \tag{9.206}$$

Since e^z and e^{-z} are entire functions, $\cosh z$ and $\sinh z$ are also entire functions. The derivatives

$$\frac{d}{dz} \sinh z = \cosh z, \quad \frac{d}{dz} \cosh z = \sinh z \tag{9.207}$$

and some commonly used identities are given as

$$\cosh^2 z - \sinh^2 z = 1, \tag{9.208}$$

$$\sinh(z_1 \pm z_2) = \sinh z_1 \cosh z_2 \pm \cosh z_1 \sinh z_2, \tag{9.209}$$

$$\cosh(z_1 \pm z_2) = \cosh z_1 \cosh z_2 \pm \sinh z_1 \sinh z_2, \tag{9.210}$$

$$\sinh(-z) = -\sinh z, \tag{9.211}$$

$$\cosh(-z) = \cosh z, \tag{9.212}$$

$$\sinh 2z = 2 \sinh z \cosh z. \tag{9.213}$$

Hyperbolic and trigonometric functions can be related through the following formulas:

$$\cos z = \cos(x + iy) = \frac{1}{2}(e^{ix-y} + e^{-ix+y}) \tag{9.214}$$

$$= \frac{1}{2}e^{-y}(\cos x + i \sin x) + \frac{1}{2}e^{y}(\cos x - i \sin x) \tag{9.215}$$

$$= \left(\frac{e^{y} + e^{-y}}{2}\right)\cos x - i\left(\frac{e^{y} - e^{-y}}{2}\right)\sin x \tag{9.216}$$

$$= \cos x \cosh y - i \sin x \sinh y, \tag{9.217}$$

$$\sin z = \sin x \cosh y + i \cos x \sinh y. \tag{9.218}$$

From these formulas, we can deduce the relations

$$\boxed{\sin(iy) = i \sinh y, \ \cos(iy) = \cosh y.} \tag{9.219}$$

9.10.5 Logarithmic Function

Using the polar representation $z = re^{i\theta}$, we can define a **logarithmic function**:

$$w = \log z, \tag{9.220}$$

as

$$\boxed{\log z = \ln r + i\theta, \quad r > 0.} \tag{9.221}$$

Since r is real and positive, an appropriate base for the $\ln r$ can be chosen. Since the points with the same r but with the arguments $\theta \pm 2n\pi$, $n = 0, 1, \ldots$, correspond to the same point in the z-plane, $\log z$ is a **multivalued function**, that is, for a given point in the z-plane, there are infinitely many logarithms, which differ from each other by integral multiples of $2\pi i$:

$$w_n = \log z = \ln|z| + i \arg z, \tag{9.222}$$

$$\boxed{w_n = \ln r + i(\theta \pm 2n\pi), \quad n = 0, 1, \ldots, \quad 0 \le \theta < 2\pi.} \tag{9.223}$$

The value w_0 corresponding to $n = 0$ is called the **principal value**, or the **principal branch** of $\log z$. For $n \ne 0$, w_n gives the nth branch value of $\log z$.

For example, for $z = 5$, $z = -1$, and $z = 1 + i$ we obtain the following logarithms, respectively:

$$w_n = \log 5 = \ln 5 + i \arg 5 \tag{9.224}$$

$$= \ln 5 + i(0 \pm 2n\pi), \tag{9.225}$$

$$w_n = \log(-1) = \ln 1 + i(\pi \pm 2n\pi) \tag{9.226}$$

$$= i(\pi \pm 2n\pi), \tag{}$$

$$w_n = \log(1 + i) = \ln \sqrt{2} + i \left(\frac{\pi}{4} \pm 2n\pi \right) \tag{9.227}$$

$$= \frac{1}{2} \ln 2 + i \left(\frac{\pi}{4} \pm 2n\pi \right). \tag{9.228}$$

For a given value of n, the single-valued function

$$w_n = \log z = \ln|z| + i \arg z \tag{9.229}$$

$$= \ln r + i(\theta \pm 2n\pi),\ 0 \le \theta < 2\pi, \tag{9.230}$$

with the u and the v functions given as

$$u = \ln r, \tag{9.231}$$

$$v = \theta \pm 2n\pi, \tag{9.232}$$

has continuous first-order partial derivatives,

$$\frac{\partial u}{\partial r} = \frac{1}{r},\ \frac{\partial u}{\partial \theta} = 0,\ \frac{\partial v}{\partial r} = 0,\ \frac{\partial v}{\partial \theta} = 1, \tag{9.233}$$

which satisfy the Cauchy–Riemann conditions [Eqs. (9.156) and (9.157)]; hence, Eq. (9.230) defines an analytic function in its domain of definition. Using Eq. (9.163), we can write the **derivative** of $\log z$ as the usual expression:

$$\frac{d}{dz} \log z = e^{-i\theta} \left[\frac{\partial u}{\partial r} + i \frac{\partial v}{\partial r} \right] = \frac{1}{re^{i\theta}}, \tag{9.234}$$

$$\boxed{\frac{d}{dz} \log z = \frac{1}{z}.} \tag{9.235}$$

Using the definition in Eq. (9.230), one can easily show the familiar properties of the $\log z$ function as

$$\log z_1 z_2 = \log z_1 + \log z_2, \tag{9.236}$$

$$\log \frac{z_1}{z_2} = \log z_1 - \log z_2. \tag{9.237}$$

Regardless of which branch is used, we can write the inverse of $w = \log z$ as

$$e^w = e^{\ln z} = e^{(\ln r + i\theta)} = e^{\ln r} e^{i\theta} = re^{i\theta} = z. \tag{9.238}$$

Hence,

$$\boxed{e^{\log z} = z,} \tag{9.239}$$

that is, the exp and the log functions are **inverses** of each other.

9.10.6   Powers of Complex Numbers

Let m be a fixed positive integer. Using Eq. (9.230), we can write

$$m \log z = m \ln r + im(\theta \pm 2n\pi), \quad n = 0, 1, \ldots. \tag{9.240}$$

Using the periodicity of e^z [Eq. (9.190)] and Eq. (9.223), we can also write

$$\log z^m = \log \left[re^{i(\theta \pm 2\pi n)} \right]^m, \quad n = 0, 1, \ldots, \tag{9.241}$$

$$= \ln r^m + im(\theta \pm 2n\pi). \tag{9.242}$$

Comparing Eqs. (9.240) and (9.242), we obtain

$$\boxed{m \log z = \log z^m.} \tag{9.243}$$

Similarly, for a positive integer p, we write

$$\frac{1}{p} \log z = \frac{1}{p} \log[re^{i(\theta \pm 2\pi n)}], \quad n = 0, 1, \ldots, \tag{9.244}$$

$$= \ln r^{1/p} + \frac{i}{p}(\theta \pm 2k\pi), \quad k = 0, 1, \ldots, (p-1). \tag{9.245}$$

We can also write

$$\log z^{1/p} = \log[r^{1/p} e^{(i/p)(\theta \pm 2n\pi)}], \quad n = 0, 1, \ldots, \tag{9.246}$$

$$= \ln r^{1/p} + \frac{i}{p}(\theta \pm 2k\pi), \quad k = 0, 1, \ldots, (p-1). \tag{9.247}$$

Note that due to the periodicity of the exponential function $e^{(i/p)(\theta \pm 2n\pi)}$, no new root emerges when n is an integer multiple of p. Hence, in Eqs. (9.245) and (9.247), we have defined a new integer $k = 0, 1, \ldots, (p-1)$. Comparing Eqs. (9.245) and (9.247) we obtain the familiar expression

$$\log z^{1/p} = \left(\frac{1}{p} \right) \log z. \tag{9.248}$$

In general, we can write

$$\log z^{m/p} = \left(\frac{m}{p}\right) \log z, \tag{9.249}$$

or

$$\boxed{z^{m/p} = e^{(m/p)\log z}.} \tag{9.250}$$

In other words, the p distinct values of $\log z^{m/p}$ give the number $z^{m/p}$. For example, for the principal value of $z^{5/3}$, that is, for $k = 0$, we obtain

$$z^{5/3} = e^{(5/3)\log z} = e^{(5/3)(\ln r + i\theta)} = r^{5/3}e^{i(5/3)\theta}. \tag{9.251}$$

All three of the branches are given as

$$z^{5/3} = r^{5/3}e^{i(5/3)(\theta \pm 2k\pi)}, \quad k = 0, 1, 2. \tag{9.252}$$

We now extend our discussion of powers to cases where the power is complex:

$$w = z^c \text{ or } w = z^{-c}, \tag{9.253}$$

where c is any complex number. For example, for i^{-i} we write

$$i^{-i} = \exp(-i \log i) \tag{9.254}$$

$$= \exp\left\{-i\left[\ln 1 + i\left(\frac{\pi}{2} \pm 2n\pi\right)\right]\right\} \tag{9.255}$$

$$= \exp\left(\frac{\pi}{2} \pm 2n\pi\right), \quad n = 0, 1, \dots. \tag{9.256}$$

Replacing m/p in Eq. (9.250) with c, we write

$$z^c = e^{c \log z}. \tag{9.257}$$

Using the principal value of $\log z$, we can write the derivative

$$\frac{dz^c}{dz} = \left(\frac{c}{z}\right)e^{c \log z} = c\frac{e^{c \log z}}{e^{\log z}} = ce^{(c-1)\log z}. \tag{9.258}$$

The right-hand side is nothing but cz^{c-1}, hence, we obtain the formula

$$\frac{dz^c}{dz} = cz^{c-1}, \tag{9.259}$$

which also allows us to write

$$z^c = e^{c \log z}, \quad z \neq 0. \tag{9.260}$$

Example 9.6. *Complex exponents:* Let us find i^i for the principal branch:

$$i^i = e^{i \log i} = e^{i[\ln 1 + i\pi/2]} = e^{-\pi/2}. \tag{9.261}$$

As another example, we find the principal branch of $(1+i)^i$:

$$(1+i)^i = e^{i \log(1+i)} = e^{i[\ln\sqrt{2} + i\pi/4]} = e^{i\ln\sqrt{2}}e^{-\pi/4} \tag{9.262}$$

$$= 2^{i/2}e^{-\pi/4} = e^{(i/2)\ln 2}e^{-\pi/4}. \tag{9.263}$$

9.10.7 Inverse Trigonometric Functions

Using the definition

$$z = \sin w = \frac{e^{iw} - e^{-iw}}{2i} \tag{9.264}$$

along with

$$(1 - z^2)^{1/2} = \cos w = \frac{(e^{iw} + e^{-iw})}{2}, \tag{9.265}$$

we solve for e^{iw} to obtain

$$e^{iw} = iz + (1 - z^2)^{1/2}, \tag{9.266}$$

which allows us to write

$$w = -i \log[iz + (1 - z^2)^{1/2}]. \tag{9.267}$$

Thus, the inverse sine function is defined as

$$\boxed{\sin^{-1} z = -i \log[iz + (1 - z^2)^{1/2}],} \tag{9.268}$$

which is a multiple-valued function with infinitely many branches. Similarly, one can write the inverses

$$\boxed{\cos^{-1} z = -i \log[z + (z^2 - 1)^{1/2}],} \tag{9.269}$$

$$\boxed{\tan^{-1} z = \frac{i}{2} \log \frac{i + z}{i - z}.} \tag{9.270}$$

REFERENCES

1. Brown, J.W., and R.V. Churchill, *Complex Variables and Applications*, McGraw-Hill, New York, 1995.

2. Bayin, S.S. (2018). *Mathematical Methods in Science and Engineering*, 2e. Hoboken, NJ: Wiley.

PROBLEMS

1. Evaluate the following complex numbers:

   (i) $(\sqrt{3} + i) + i(1 + i\sqrt{3})$,   (iv) $(2, 1)(1, -2)$,

   (ii) $\dfrac{4}{(1 - i)(1 + i)(2 - i)}$,   (v) $(1 - i)^4$,

   (iii) $\dfrac{2 + 3i}{(3 - 2i)(1 + i)}$,   (vi) $\dfrac{(1, 1)}{(2, -1)(1, 3)(2, 2)}$.

2. Evaluate the numbers $z_1 + z_2$, $z_1 - z_2$, and $z_1 z_2$ and show them graphically when

   (i) $z_1 = (1, 1)$, $z_2 = (3, -1)$,   (iii) $z_1 = (1, 3)$, $z_2 = (4, -1)$,

   (ii) $z_1 = (x_1, y_1)$, $z_2 = (x_1, -y_1)$,   (iv) $z_1 = (1 - i)^2$, $z_2 = 1 + 2i$.

3. Prove the commutative law of multiplication:

$$z_1 z_2 = z_2 z_1.$$

4. Prove the following associative laws:

$$z_1 + (z_2 + z_3) = (z_1 + z_2) + z_3,$$
$$(z_1 z_2) z_3 = z_1 (z_2 z_3).$$

5. Prove the distributive law

$$z_1 (z_2 + z_3) = z_1 z_2 + z_1 z_3.$$

6. Find

   (i) $(z^* + 2i)^*$,

   (ii) $(2iz)^*$,

   (iii) $\dfrac{2}{(1 - i)(1 + i)^*}$,

   (iv) $[(1 - i)^4]^*$.

7. Use the polar form to write

   (i) $(z^* + 2i)$,

   (ii) $\dfrac{5}{(1 - i)(1 + i)^*}$,

   (iii) $(iz + 1)^*$,

   (iv) $(1 - i)^*(2 + i)$.

8. Prove the following:

$$(z_1 z_2 z_3)^* = z_1^* z_2^* z_3^*,$$
$$(z^4)^* = (z^*)^4.$$

9. Prove the following:

$$|z_1 z_2| = |z_1||z_2|,$$
$$\left|\frac{z_1}{z_2}\right| = \frac{|z_1|}{|z_2|}.$$

10. Prove and interpret the triangle inequalities:

$$|z_1 + z_2| \le |z_1| + |z_2|,$$
$$|z_1 - z_2| \ge ||z_1| - |z_2||.$$

11. Describe the region of the z-plane defined by

(i) $1 < \operatorname{Im} z < 2$,

(ii) $|z - 1| \ge 2|z + 1|$,

(iii) $|z - 4| > 3$.

12. Show that the following equation describes a circle:

$$\arg\left[\frac{(z - z_1)}{(z_3 - z_1)}\frac{(z_3 - z_2)}{(z - z_2)}\right] = 0.$$

13. Express the following in terms of rectangular and polar coordinates:

$$\left[\frac{(1 - i)}{(1 + i)}\right]^{1+i}.$$

14. Find the roots of $z^4 + i = 0$.

15. If $z^2 = i$, find z.

16. Show that

$$\theta = \left(\frac{1}{2i}\right)\log\left(\frac{z}{z^*}\right).$$

17. Show that

$$\tanh(1 + \pi i) = \frac{e^2 - 1}{e^2 + 1}.$$

18. Find the complex number that is symmetrical to $(1 + 2i)$ with respect to the line $\arg z = \alpha_0$.

19. Find all the values of the following and show them graphically:

$$\text{(i)} \quad z = (3i)^{1/2},$$
$$\text{(ii)} \quad z = (1 + i\sqrt{2})^{3/2},$$
$$\text{(iii)} \quad z = (-1)^{1/3},$$
$$\text{(iv)} \quad z = (1 - i)^{1/3},$$
$$\text{(v)} \quad z = (-8)^{1/3},$$
$$\text{(vi)} \quad z = (1)^{1/4}.$$

20. Derive Eqs. (9.69)–(9.71), which are used in stereographic projections:

$$x_1 = \frac{z + z^*}{1 + |z|^2},$$
$$x_2 = \frac{z - z^*}{i(1 + |z|^2)},$$
$$x_3 = \frac{|z|^2 - 1}{|z|^2 + 1}.$$

21. Show the following ratios [Eq. (9.72)] used in stereographic projections and verify that the points z, Z, and N lie on a straight line:

$$\frac{x}{x_1} = \frac{y}{x_2} = \frac{1}{1 - x_3}.$$

22. Derive Eq. (9.81):

$$d(z, z') = \frac{2|z - z'|}{\sqrt{(1 + |z|^2)(1 + |z'|^2)}},$$

used in stereographic projections to express the distance between two points on the Riemann sphere in terms of the coordinates on the z-plane.

23. Establish the relations

$$e^{z_1} e^{z_2} = e^{z_1 + z_2},$$
$$\frac{e^{z_1}}{e^{z_2}} = e^{z_1 - z_2},$$
$$(e^z)^n = e^{nz}.$$

24. Establish the sum

$$1 + z + z^2 + \cdots + z^n = \frac{1 - z^{n+1}}{1 - z}, \quad z \neq 1,$$

and then show

(i) $1 + \cos\theta + \cos 2\theta + \cdots + \cos n\theta = \cos\left(\dfrac{n\theta}{2}\right)\dfrac{\sin[(n+1)\theta/2]}{\sin(\theta/2)},$

(ii) $\sin\theta + \sin 2\theta + \cdots + \sin n\theta = \sin\left(\dfrac{n\theta}{2}\right)\dfrac{\sin[(n+1)\theta/2]}{\sin(\theta/2)},$

where $0 < \theta < 2\pi$.

25. Show that the following functions are entire:

(i) $f(z) = 2x + y + i(2y - x),$

(ii) $f(z) = -\sin y \cosh x + i \cos y \sinh x,$

(iii) $f(z) = e^{-y}(\cos x + i\sin x),$

(iv) $f(z) = e^z z^2.$

26. Find the singular points of

(i) $\dfrac{z^3 + i}{z^2 - 3z + 3},$

(ii) $\dfrac{3z + 1}{z(z^2 + 2)}.$

Explain why these functions are analytic everywhere except at these points. What are the limits of these functions at the singular points?

27. Show that the following functions are analytic nowhere:

$$f(z) = 2xy + iy,$$

$$f(z) = e^{2y}(\cos x + i\sin 2y).$$

28. Show that for an analytic function, $f(z) = u + iv$, the imaginary part is harmonic, that is,

$$\frac{\partial^2 v}{\partial x^2} + \frac{\partial^2 v}{\partial y^2} = 0.$$

29. Show that

$$f(x,y) = y^2 - x^2 + 2x$$

and

$$f(x,y) = \cosh x \cos y$$

are harmonic functions and find their conjugate harmonic functions.

30. In rectangular coordinates show that the derivative of an analytic function can be written as

$$\frac{df}{dz} = \frac{\partial u}{\partial x} + i\frac{\partial v}{\partial x},$$

or as

$$\frac{df}{dz} = \frac{\partial v}{\partial y} - i\frac{\partial u}{\partial y}.$$

31. Show that

$$e^{1 \pm 3\pi i} = -e,$$
$$e^0 = 1.$$

32. Find all the values such that

$$e^z = -2, \quad e^z = 1 + i\sqrt{2}, \quad e^{(2z+1)} = 1.$$

33. Explain why the following function is entire:

$$f(z) = z^3 + z + e^{z^3} + ze^z.$$

34. Justify the definitions

$$\cos z = \frac{e^{iz} + e^{-iz}}{2}, \quad \sin z = \frac{e^{iz} - e^{-iz}}{2i}$$

and find the inverse functions $\cos^{-1} z$ and $\sin^{-1} z$.

35. Prove the identities

$$\sin(z_1 + z_2) = \sin z_1 \cos z_2 + \cos z_1 \sin z_2,$$
$$\cos(z_1 + z_2) = \cos z_1 \cos z_2 - \sin z_1 \sin z_2.$$

36. Find all the roots of

(i) $\cos z = 2,$

(ii) $\sin z = \cosh 2.$

37. Find the zeros of $\sinh z$ and $\cosh z$.

38. Evaluate

   (i) $(1 \pm i)^i$,

   (ii) $(-2)^{1/\pi}$,

   (iii) $(1 + i\sqrt{3})^{1+i}$.

39. What are the principal values of $(1 - i)^{2i}$, i^{2i}, $(-i)^{2+i}$?

40. In polar coordinates, show that the derivative of an analytic function can be written as

$$\frac{df}{dz} = \frac{1}{r} e^{-i\theta} \left[\frac{\partial v}{\partial \theta} - i \frac{\partial u}{\partial \theta} \right].$$

CHAPTER 10

COMPLEX ANALYSIS

Line integrals, power series, and residues constitute an important part of complex analysis. Complex integration theorems are usually concise but powerful. Many of the properties of analytic functions are quite difficult to prove without the use of these theorems. Complex contour integration also allows us to evaluate various difficult proper or improper integrals encountered in physical theories. Just as in real analysis, in complex integration we distinguish between definite and indefinite integrals. Since differentiation and integration are inverse operations of each other, indefinite integrals can be found by inverting the known differentiation formulas of analytic functions. Definite integrals evaluated over continuous, or at least piecewise continuous, paths are not just restricted to analytic functions and thus can be defined exactly by the same limiting procedure used to define real integrals. Most complex definite integrals can be written in terms of two real integrals. Hence, in their discussion, we heavily rely on the background established in Chapters 1 and 2 on real integrals. One of the most important places, where the theorems of complex integration is put to use is in power series representation of analytic functions. In this regard, Laurent series play an important part in applications, which also allows us to classify singular points.

10.1   CONTOUR INTEGRALS

Each point in the complex plane is represented by two parameters, hence, in contrast to their real counterparts, $\int_{x_1}^{x_2} f(x)\,dx$, complex definite integrals are defined with respect to a **path** or **contour**, C, connecting the upper and the lower bounds of the integral:

$$I = \int_{\substack{z_1 \\ C}}^{z_2} f(z)dz. \tag{10.1}$$

If we write a complex function as $f(z) = u(x, y) + iv(x, y)$, the aforementioned integral can be expressed as the sum of two real integrals:

$$\int_{\substack{z_1 \\ C}}^{z_2} f(z)\,dz = \int_{\substack{(x_1, y_1) \\ C}}^{(x_2, y_2)} [u(x, y) + iv(x, y)][dx + i\,dy] \tag{10.2}$$

$$= \int_{\substack{(x_1, y_1) \\ C}}^{(x_2, y_2)} [u\,dx - v\,dy] + i \int_{\substack{(x_1, y_1) \\ C}}^{(x_2, y_2)} [v\,dx + u\,dy]. \tag{10.3}$$

Furthermore, if the path C is parameterized in terms of a real parameter t:

$$z(t) = x(t) + iy(t), \ t \in [t_1, t_2], \tag{10.4}$$

where the end points t_1 and t_2 are found from

$$z_1 = x(t_1) + iy(t_1) \text{and } z_2 = x(t_2) + iy(t_2), \tag{10.5}$$

the complex integral in Eq. (10.2) can also be written as

$$\int_{\substack{z_1 \\ C}}^{z_2} f(z)\,dz = \int_{\substack{z_1 \\ C}}^{z_2} [u(t) + iv(t)](x'(t) + iy'(t))\,dt \tag{10.6}$$

$$= \int_{\substack{t_1 \\ C}}^{t_2} [ux' - vy']\,dt + i \int_{\substack{t_1 \\ C}}^{t_2} [vx' + y'u]\,dt. \tag{10.7}$$

In the aforementioned equations, we have written $u(x(t), y(t)) = u(t)$, $v(x(t), y(t)) = v(t)$ and $dz = [x'(t) + iy'(t)]\,dt$. Integrals on the right-hand sides of Eqs. (10.3) and (10.7) are real; hence, from the properties of real integrals, we can deduce the following:

$$\int_{z_1 \atop C}^{z_2} f(z)\, dz = -\int_{z_2 \atop C}^{z_1} f(z)\, dz, \tag{10.8}$$

$$\int_{z_1 \atop C}^{z_2} cf(z)\, dz = c\int_{z_1 \atop C}^{z_2} f(z)\, dz, \; c \in \mathbb{C}, \tag{10.9}$$

$$\int_{z_1 \atop C}^{z_2} (f_1(z) + f_2(z))\, dz = \int_{z_1 \atop C}^{z_2} f_1(z)\, dz + \int_{z_1 \atop C}^{z_2} f_2(z)\, dz. \tag{10.10}$$

The two **inequalities**:

$$\left| \int_{z_1 \atop C}^{z_2} f(z)\, dz \right| \leq \int_{z_1 \atop C}^{z_2} |f(z)||dz| \tag{10.11}$$

and

$$\left| \int_{z_1 \atop C}^{z_2} f(z)\, dz \right| \leq M.L, \tag{10.12}$$

where $|f(z)| \leq M$ on C and L is the arclength, are very useful in calculations. When z is a point on C, we can write an infinitesimal arclength as

$$|dz| = |x'(t) + iy'(t)|dt = \sqrt{dx^2 + dy^2}. \tag{10.13}$$

Length of the contour C is now given as

$$\int_C |dz| = L. \tag{10.14}$$

If we parameterize a point on the contour C as $z = z(t)$, we can write

$$I = \int_C f(z)\, dz = \int_C f(z(t)) \frac{dz(t)}{dt}\, dt. \tag{10.15}$$

In another parametric representation of C, where the new parameter τ is related to t by $t = t(\tau)$, we can write the integral [Eq. (10.15)] as

$$I = \int_C f(z(t(\tau))) \left[\frac{dz(t(\tau))}{dt} \frac{dt(\tau)}{d\tau} \right] d\tau, \tag{10.16}$$

which is nothing but

$$I = \int_C f(z(\tau)) \left[\frac{dz(\tau)}{d\tau}\right] d\tau. \tag{10.17}$$

Hence, an important property of the contour integrals is that their value is independent of the parametric representation used.

10.2   TYPES OF CONTOURS

We now introduce the types of contours or paths that are most frequently encountered in the study of complex integrals. A continuous path is defined as the curve

$$x = x(t), \; y = y(t), \; t \in [t_1, t_2], \tag{10.18}$$

where $x(t)$ and $y(t)$ are continuous functions of the real parameter t. If the curve does not intersect itself, that is, when no two distinct values of t in $[t_1, t_2]$ correspond to the same (x, y), we call it a **Jordan arc**. If $x(t_1) = x(t_2)$ and $y(t_1) = y(t_2)$ but no other two distinct values of t correspond to the same point (x, y), we have a **simple closed curve**, which is also called a **Jordan curve**. A piecewise continuous curve like

$$y = t^2, \; x = t, \; t \in [1, 2],$$
$$y = t^3, \; x = t, \; t \in (2, 3], \tag{10.19}$$

is a **Jordan arc**. A circle with the unit radius, $x^2 + y^2 = 1$, which can be expressed in parametric form as $x = \cos t, y = \sin t, t \in [0, 2\pi]$, is a simple closed curve. If the derivatives $x'(t)$ and $y'(t)$ are continuous and do not vanish simultaneously for any value of t, we have a **smooth curve**. For a smooth curve, the **length** exists and is given as

$$\boxed{L = \int_{t_1}^{t_2} \sqrt{x'(t)^2 + y'(t)^2} dt.} \tag{10.20}$$

In general, a contour C is a continuous chain of finite number of smooth curves, $C_1, C_2, \ldots, C_n$. Hence, $C = C_1 + C_2 + \cdots + C_n$. A contour integral over C can now be written as the sum of its parts as

$$\int_C f(z) \, dz = \int_{C_1} f(z) \, dz + \int_{C_2} f(z) \, dz + \cdots + \int_{C_n} f(z) \, dz. \tag{10.21}$$

Contour integrals over closed paths are also written as $\oint_C f(z) \, dz$, where by definition, the **counterclockwise** direction is taken as the **positive** direction.

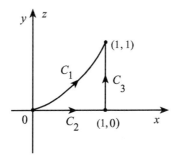

Figure 10.1   Contours for Example 10.1.

Example 10.1. ***Contour integrals:*** Using the contours C_1, C_2, and C_3 shown in Figure 10.1, let us evaluate the following integrals:

$$I_{C_1} = \int_{C_1[y=x^2]} z^2 \, dz, \ I_{C_2} = \int_{C_2[y=0]} z^2 \, dz, \ I_{C_3} = \int_{C_3[x=1]} z^2 \, dz. \qquad (10.22)$$

We first write $I_C = \int_C z^2 \, dz$ as the sum of two real integrals [Eq. (10.3)]:

$$I_C = \int_C [(x^2 - y^2) \, dx - 2xy \, dy] + i \int_C [2xy \, dx + (x^2 - y^2) \, dy]. \qquad (10.23)$$

For the path C_1, we have $y = x^2$ and $dy = 2x \, dx$, hence, the aforementioned integral is evaluated as

$$I_{C_1} = \int_{C_1} [(x^2 - x^4) \, dx - 2x^3 2x \, dx] + i \int_{C_1} [2x^3 \, dx + (x^2 - x^4) 2x \, dx] \qquad (10.24)$$

$$= \int_0^1 (x^2 - 5x^4) \, dx + i \int_0^1 (4x^3 - 2x^5) \, dx = -\frac{2}{3} + i\frac{2}{3}. \qquad (10.25)$$

For the path C_2, we set $y = 0$ and $dy = 0$ to obtain

$$I_{C_2} = \int_{\substack{0 \\ C_2}}^1 [x^2 \, dx] + i[0] = \frac{1}{3}. \qquad (10.26)$$

Finally, for the path C_3, we have $x = 1$ and $dx = 0$, thus obtaining

$$I_{C_3} = -\int_{\substack{0 \\ C_3}}^1 2y \, dy + i \int_{\substack{0 \\ C_3}}^1 (1 - y^2) \, dy = -1 + i\frac{2}{3}. \qquad (10.27)$$

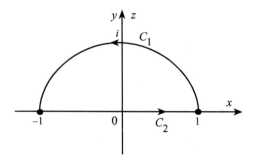

Figure 10.2 Semicircular path.

Example 10.2. *Parametric representation of the contour:* We now consider the semicircular path C_1 in Figure 10.2 for the integral [Eq. (10.3)]:

$$I_{C_1} = \int_{C_1} z^2 \, dz = \int_{C_1} [ux' - vy'] \, dt + i \int_{C_1} [vx' + y'u] \, dt, \qquad (10.28)$$

with $u = x^2 - y^2$, $v = 2xy$. When we use the following parametric form of the path:

$$x(t) = \cos t, \; x'(t) = -\sin t, \qquad (10.29)$$

$$y(t) = \sin t, \; y'(t) = \cos t, \; t \in [0, \pi], \qquad (10.30)$$

Equation (10.28) becomes

$$I_{C_1} = \int_0^\pi [-3\cos^2 t \sin t + \sin^3 t] \, dt + i \int_0^\pi [-3\sin^2 t \cos t + \cos^3 t] \, dt = -\frac{2}{3}. \qquad (10.31)$$

For the path along the real axis, C_2, we can use $x = t$ as a parameter (Figure 10.2):

$$x = t, \; y = 0, \; u = t^2, \; v = 0, \; t \in [-1, 1], \qquad (10.32)$$

hence, the integral $I_{C_2} = \int_{C_2} z^2 \, dz$ becomes

$$I_{C_2} = \int_{-1}^1 t^2 \, dt + i \int_{-1}^1 0 \, dt = \frac{t^3}{3} \Big|_{-1}^1 = \frac{2}{3}. \qquad (10.33)$$

Example 10.3. *Simple closed curves:* For the combined path in Example 10.2, that is, $C = C_1 + C_2$ (Figure 10.2), which is a simple closed curve, the integral $I_C = \oint_C z^2 \, dz$ becomes

$$I_C = \int_{C_1} z^2 \, dz + \int_{C_2} z^2 \, dz = -\frac{2}{3} + \frac{2}{3} = 0. \qquad (10.34)$$

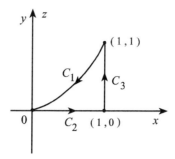

Figure 10.3   Closed simple path.

Similarly, for the closed path C in Figure 10.3, we can use the results obtained in Example 10.1 to write

$$I_C = I_{C_1} + I_{C_2} + I_{C_3} = -\left(-\frac{2}{3} + i\frac{2}{3}\right) + \frac{1}{3} + \left(-1 + i\frac{2}{3}\right) = 0. \qquad (10.35)$$

Note that in the complex plane, geometric interpretation of the integral as the area is no longer true. As we shall see shortly, the fact that we have obtained zero for the integral $\oint z^2 \, dz$ for two very different closed paths is by all means not a coincidence. In the next section, we elaborate this point.

10.3   THE CAUCHY–GOURSAT THEOREM

We have seen that for a closed contour C, the complex contour integral of $f(z)$ can be written in terms of two real integrals [Eq. (10.3)]:

$$\oint_C f(z) \, dz = \oint_C [u \, dx - v \, dy] + i \oint_C [v \, dx + u \, dy]. \qquad (10.36)$$

Let us now look at this integral from the viewpoint of **Green's theorem** (Chapter 2), which states that for two continuous functions, $P(x,y)$ and $Q(x,y)$, defined in a simply connected domain D with continuous first-order partial derivatives within and on a simple closed contour C, we can write the integral

$$\oint_C (P \, dx + Q \, dy) = \int\int_R \left(\frac{\partial Q}{\partial x} - \frac{\partial P}{\partial y}\right) dx \, dy. \qquad (10.37)$$

The positive sense of the contour integral is taken as the counterclockwise direction and R is the region enclosed by the closed contour C. If we apply Green's theorem to the real integrals defining the real and the imaginary parts of the integral in Eq. (10.36), we obtain

$$\oint_C [u \, dx - v \, dy] = -\int\int_R \left(\frac{\partial v}{\partial x} + \frac{\partial u}{\partial y}\right) dx \, dy, \qquad (10.38)$$

$$\oint_C [v \, dx + u \, dy] = \int\int_R \left(\frac{\partial u}{\partial x} - \frac{\partial v}{\partial y}\right) dx \, dy. \qquad (10.39)$$

From the properties of analytic functions [Theorem 9.4], we have seen that a given analytic function:

$$f(z) = u(x, y) + iv(x, y), \tag{10.40}$$

defined in some domain D, has continuous first-order partial derivatives, u_x, u_y, v_x, v_y, and satisfies the **Cauchy–Riemann** conditions:

$$\frac{\partial u}{\partial x} = \frac{\partial v}{\partial y}, \tag{10.41}$$

$$\frac{\partial v}{\partial x} = -\frac{\partial u}{\partial y}. \tag{10.42}$$

Therefore, for an analytic function $f(z)$, the right-hand sides of Eqs. (10.38) and (10.39) are zero. We now state this result as the **Cauchy-Goursat theorem** [1].

Theorem 10.1. *Cauchy-Goursat theorem*: If a function $f(z)$ is analytic within and on a simple closed contour C in a simply connected domain, then

$$\boxed{\oint_C f(z)dz = 0.} \tag{10.43}$$

This is a remarkably simple, but a powerful theorem. For example, to evaluate the integral

$$I = \int_{\substack{z_1 \\ C_1}}^{z_2} f(z)\, dz \tag{10.44}$$

over some complicated path C_1, we first form the closed path: $C_1 + C_2$ (Figure 10.4a). If $f(z)$ is analytic on and within this closed path $C_1 + C_2$, we can use the Cauchy–Goursat theorem to write

$$\oint_{C_1+C_2} f(z)\, dz = 0, \tag{10.45}$$

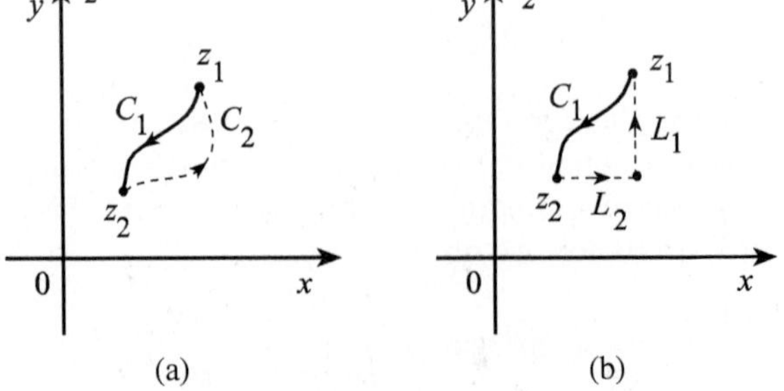

Figure 10.4   We stretch C_2 into $C_2 = L_1 + L_2$.

which allows us to evaluate the desired integral as

$$I = \int_{\substack{z_1 \\ C_1}}^{z_2} f(z)\, dz = -\int_{\substack{z_1 \\ C_2}}^{z_2} f(z)\, dz. \tag{10.46}$$

The general idea is to deform C_2 into a form such that the integral I can be evaluated easily. The Cauchy–Goursat theorem says that we can always do this, granted that $f(z)$ is analytic on and within the closed path $C_1 + C_2$. Figure 10.4b C_2 is composed of two straight-line segments L_1 and L_2. In Example 10.3, for two different closed paths, we have explicitly shown that $\int_C z^2 dz$ is zero. Since z^2 is an entire function, the Cauchy–Goursat theorem says that for any simple closed path the result is zero. Similarly, all polynomials, $P_n(z)$, of order n are entire functions; hence, we can write

$$\oint_C P_n(z)\, dz = 0, \tag{10.47}$$

where C is any simple closed contour.

Example 10.4. *Cauchy–Goursat theorem:* Let us evaluate the integral

$$\int_{\substack{(1,1) \\ C_1}}^{(2,2)} (3z^2 + 1)\, dz \tag{10.48}$$

over any given path C_1 as shown in Figure 10.5a. Since the integrand $f(z) = 3z^2 + 1$ is an entire function, we can form the closed path in Figure 10.5b and use the Cauchy–Goursat theorem to write

$$\oint_{C_1 + L_1 + L_2} (3z^2 + 1)\, dz = 0, \tag{10.49}$$

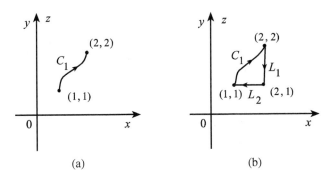

(a)   (b)

Figure 10.5   Definite integrals.

which leads to

$$\int_{C_1} (3z^2 + 1)\, dz = -\int_{L_1} (3z^2 + 1)\, dz - \int_{L_2} (3z^2 + 1)\, dz. \qquad (10.50)$$

From

$$\begin{aligned} f(z) &= 3z^2 + 1 \\ &= [3(x^2 - y^2) + 1] + i(6xy), \end{aligned} \qquad (10.51)$$

we obtain the functions $u = [3(x^2 - y^2) + 1]$ and $v = 6xy$, which are needed in the general formula [Eq. (10.3)]. For L_2, we use the parameterization $x = x$, $y = 1$; hence, we substitute $u(x, 1) = 3x^2 - 2$ and $v(x, 1) = 6x$ into Eq. (10.7) to obtain

$$-\int_{L_2} (3z^2 + 1)\, dz = \int_1^2 (3x^2 - 2)\, dx + i \int_1^2 6x\, dx$$

$$= 5 + 9i. \qquad (10.52)$$

Similarly, for L_1, we use the parameterization $x = 2$, $y = y$; hence, we substitute $u(2, y) = -3y^2 + 13$, $v(2, y) = 12y$ into Eq. (10.7) to get

$$-\int_{L_1} (3z^2 + 1)\, dz = \int_1^2 -12y\, dy + i \int_1^2 (-3y^2 + 13)\, dy \qquad (10.53)$$

$$= -18 + 6i. \qquad (10.54)$$

Finally, substituting Eqs. (10.52) and (10.54) in Eq. (10.50), we obtain

$$\int_{C_1} (3z^2 + 1)\, dz = (5 + 9i) + (-18 + 6i) \qquad (10.55)$$

$$= -13 + 15i. \qquad (10.56)$$

10.4   INDEFINITE INTEGRALS

Let z_0 and z be two points in a simply connected domain D, where $f(z)$ is analytic (Figure 10.6). If C_1 and C_2 are two paths connecting z_0 and z, then by using the Cauchy–Goursat theorem, we can write

$$\int_{C_2} f(z')\, dz' - \int_{C_1} f(z')\, dz' = 0. \qquad (10.57)$$

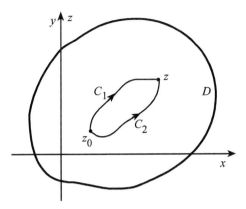

Figure 10.6   Indefinite integrals.

In other words, the integral

$$F(z) = \int_{z_0}^{z} f(z')dz' \qquad C$$

(10.58)

has the same value for all continuous paths (Jordan arcs) connecting the points z_0 and z. In general, we can write

$$\frac{dF}{dz} = f(z),$$

(10.59)

hence, the integral of an analytic function is an analytic function of its upper limit, granted that the path of integration is included in a simply connected domain D, where $f(z)$ is analytic.

Example 10.5. ***Indefinite integrals:*** An indefinite integral of $f(z) = 3z^2 + 1$ exists and is given as

$$\int_{C_1} (3z^2 + 1)\, dz = z^3 + z + c_0,$$

(10.60)

where c_0 is a complex integration constant. Since $(z^3 + z + c_0)$ is an entire function with the derivative $(3z^2 + 1)$, for the integral in Eq. (10.50), we can write

$$\int_{(1,1)}^{(2,2)} (3z^2 + 1)\, dz = (z^3 + z + c_0)|_{(1,1)}^{(2,2)},$$

(10.61)

where C_1 is any continuous path from $(1, 1)$ to $(2, 2)$. Substituting the numbers in the above equation, we naturally obtain the same result in

Eq. (10.56):

$$\int_{C_1} f(z)\, dz = (z(z^2 + 1) + c_0)\big|_{(1,1)}^{(2,2)} = -13 + 15i. \qquad (10.62)$$

10.5   SIMPLY AND MULTIPLY CONNECTED DOMAINS

Simply and multiply connected domains are defined the same way as in real analysis. A **simply connected** domain is an open connected region, where every closed path in this region can be shrunk continuously to a point. An annular region between the two circles (Figure 10.7) with radiuses R_1 and R_2, $R_2 > R_1$, is not simply connected, since the closed path C_0 cannot be shrunk to a point. A region that is not simply connected is called **multiply connected**. The Cauchy–Goursat theorem can be used in multiply connected domains by confining ourselves to a region that is simply connected. In the multiply connected domain shown in Figure 10.7, we have

$$\oint_{C_0} f(z)\, dz \neq 0; \qquad (10.63)$$

however, for C_1 we can write

$$\oint_{C_1} f(z)\, dz = 0, \qquad (10.64)$$

where $f(z)$ is analytic inside the region between the two circles.

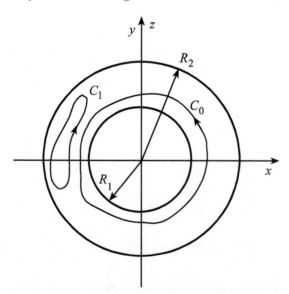

Figure 10.7   Multiply connected domain between two concentric circles.

10.6 THE CAUCHY INTEGRAL FORMULA

The Cauchy–Goursat theorem [Eq. (10.43)] works in simply connected domains, D, where the integrand is analytic within and on the closed contour C included in D. The next theorem is called the Cauchy integral formula. It is about cases where the integrand is of the form

$$F(z) = \frac{f(z)}{(z - z_0)}, \tag{10.65}$$

where z_0 is a point inside C and $f(z)$ is an analytic function within and on C. In other words, the integrand in $\oint_C F(z)\,dz$ has an isolated singular point in C (Figure 10.8).

Theorem 10.2. *Cauchy integral formula:* Let $f(z)$ be analytic at every point within and on a closed contour C in a simply connected domain D. If z_0 is a point inside the region defined by C, then

$$\boxed{f(z_0) = \frac{1}{2\pi i} \oint_{C[\circlearrowleft]} \frac{f(z)\,dz}{(z - z_0)},} \tag{10.66}$$

where $C[\circlearrowleft]$ means the contour C is traced in the counterclockwise direction.

This is another remarkable result from the theory of analytic functions with far-reaching applications in pure and applied mathematics. It basically says that the value of an analytic function, $f(z_0)$, at a point, z_0, inside its domain D of analyticity is determined entirely by the values it takes on a boundary C, which encloses z_0 and which is included in D. The shape of the boundary is not important. Once we decide on a boundary, we have no control over the values that $f(z)$ takes outside the boundary. However, if we change the values that a function takes on a boundary, it will affect the values it takes on the inside. Conversely, if we alter the values of $f(z)$ inside the boundary, a corresponding change has to be implemented on the boundary to preserve the analytic nature of the function.

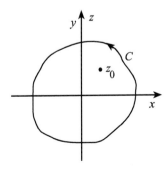

Figure 10.8 Singularity inside the contour.

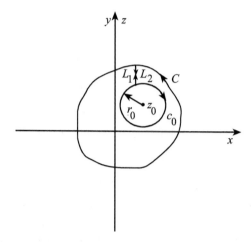

Figure 10.9 Modified path for the Cauchy integral formula.

Proof: To prove this theorem, we modify the path C as shown in Figure 10.9, where we consider the contour c_0 in the limit as its radius goes to zero. Now the integrand, $f(z)/(z - z_0)$, is analytic within and on the combined path $\overline{C}[\circlearrowleft] = L_1[\downarrow] + L_2[\uparrow] + C[\circlearrowleft] + c_0[\circlearrowright]$. By the Cauchy–Goursat theorem, we can write

$$\oint_{C[\circlearrowleft]} + \int_{L_1[\uparrow]} + \int_{L_2[\downarrow]} + \oint_{c_0[\circlearrowright]} \frac{f(z)}{z - z_0}\, dz = 0. \tag{10.67}$$

The two integrals along the straight-line segments cancel each other, thus leaving

$$\oint_{C[\circlearrowleft]} + \oint_{c_0[\circlearrowright]} \frac{f(z)}{z - z_0}\, dz = 0. \tag{10.68}$$

Evaluating both integrals counterclockwise, we write

$$\oint_{C[\circlearrowleft]} \frac{f(z)}{z - z_0}\, dz = \oint_{c_0[\circlearrowleft]} \frac{f(z)}{z - z_0}\, dz. \tag{10.69}$$

We modify the integral on the right-hand side as

$$\oint_{C[\circlearrowleft]} \frac{f(z)}{z - z_0}\, dz = f(z_0)\oint_{c_0[\circlearrowleft]} \frac{1}{z - z_0}\, dz + \oint_{c_0[\circlearrowleft]} \frac{f(z) - f(z_0)}{z - z_0}\, dz = I_1 + I_2. \tag{10.70}$$

For a point on c_0, we can write $z - z_0 = r_0 e^{i\theta}$, $dz = ir_0 e^{i\theta} d\theta$; thus, the first integral, I_1, on the right-hand side of Eq. (10.70) becomes

$$I_1 = f(z_0)\oint_{c_0[\circlearrowleft]} \frac{ir_0 e^{i\theta} d\theta}{r_0 e^{i\theta}} = f(z_0)i\oint_{c_0[\circlearrowleft]} d\theta = f(z_0)2\pi i. \tag{10.71}$$

For the second integral, I_2, when considered in the limit as $r_0 \to 0$, that is, when $z \to z_0$, we can write

$$I_2 = \lim_{r_0 \to 0} \oint_{c_0[\circlearrowleft]} \frac{f(z) - f(z_0)}{z - z_0} \, dz = \oint_{c_0[\circlearrowleft][r_0 \to 0]} \lim_{r_0 \to 0} \left[\frac{f(z) - f(z_0)}{z - z_0} \right] dz. \quad (10.72)$$

The limit

$$\lim_{z \to z_0} \left[\frac{f(z) - f(z_0)}{z - z_0} \right] \quad (10.73)$$

is nothing but the definition of the derivative of $f(z)$ at z_0, that is, $(df/dz)_{z_0}$. Since $f(z)$ is analytic within and on the contour c_0, this derivative exists with a finite modulus $|df(z_0)/dz|$; hence, we can take it outside the integral to write

$$I_2 = \left(\frac{df(z_0)}{dz} \right) \oint_{\circlearrowleft c_0[r_0 \to 0]} 1 \, dz. \quad (10.74)$$

Since 1 is an entire function, using the Cauchy–Goursat theorem, we can write

$$\oint_{c_0[\circlearrowleft][r_0 \to 0]} dz = 0, \quad (10.75)$$

thus obtaining $I_2 = 0$. Substituting Eqs. (10.71) and (10.75) into Eq. (10.70) completes the proof of the Cauchy integral formula.

10.7   DERIVATIVES OF ANALYTIC FUNCTIONS

In the Cauchy-integral formula location of the point z_0 inside the closed contour is entirely arbitrary; hence, we can treat it as a parameter and differentiate Eq. (10.66) with respect to z_0 to write [1, 2]

$$f'(z_0) = \frac{1}{2\pi i} \oint_{C[\circlearrowleft]} \frac{f(z)}{(z - z_0)^2} \, dz. \quad (10.76)$$

Successive differentiation of this formula leads to

$$\boxed{f^{(n)}(z_0) = \frac{n!}{2\pi i} \oint_{C[\circlearrowleft]} \frac{f(z)}{(z - z_0)^{n+1}} dz, \quad n = 1, 2, \ldots .} \quad (10.77)$$

Assuming that this formula is true for any value of n, say $n = k$, one can show that it holds for $n = k + 1$. Based on this formula, we can now present an important result about analytic functions:

Theorem 10.3. If a function is analytic at a given point z_0, then its derivatives of all orders, $f'(z_0), f''(z_0), \ldots$, exist at that point.

In Chapter 9 [Eq. (9.164)], we have shown that the derivative of an analytic function can be written as

$$f'(z) = \frac{\partial u}{\partial x} + i\frac{\partial v}{\partial x},$$

(10.78)

or as

$$f'(z) = \frac{\partial v}{\partial y} - i\frac{\partial u}{\partial y}.$$

(10.79)

Also, Theorem 9.4 says that for a given analytic function, the partial derivatives $u_x, u_y, v_x,$ and v_y exist, and they are continuous functions of x and y. Using Theorem 10.3, we can now conclude that in fact the partial derivatives of all orders of u and v exist and are continuous functions of x and y at each point where $f(z)$ is analytic.

10.8   COMPLEX POWER SERIES

Applications of complex analysis often require manipulations with explicit analytic expressions. To this effect, power series representations of analytic functions are very useful.

10.8.1   Taylor Series with the Remainder

Let $f(z)$ be analytic inside the boundary B and let C be a closed contour inside B (Figure 10.10). Using the Cauchy integral formula [Eq. (10.66)], we can write

$$f(z) = \frac{1}{2\pi i}\oint_{C[\circlearrowleft]} \frac{f(z')\,dz'}{(z'-z)},$$

(10.80)

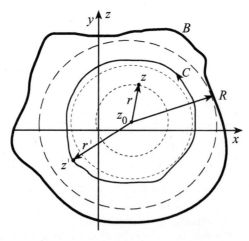

Figure 10.10   Taylor series: $|z - z_0| = r$, $|z' - z_0| = r'$.

where z' is a point on C and z is any point within C. We rewrite the integrand as

$$f(z) = \frac{1}{2\pi i} \oint_{C[\circlearrowleft]} \frac{f(z')\,dz'}{(z'-z_0)-(z-z_0)} = \frac{1}{2\pi i} \oint_{C[\circlearrowleft]} \frac{f(z')\,dz'}{(z'-z_0)\left[1-\frac{(z-z_0)}{(z'-z_0)}\right]}, \qquad (10.81)$$

where z_0 is any point within C that satisfies the inequality

$$|z-z_0| < |z'-z_0|. \qquad (10.82)$$

Note that choosing C as a circle centered at z_0 with any radius greater than r automatically satisfies this inequality. Before we continue with this equation, let us drive a useful formula: We first write the finite sum

$$S = 1 + z + z^2 + \cdots + z^n \qquad (10.83)$$

as

$$S = (1 + z + z^2 + \cdots + z^{n-1}) + z^n, \qquad (10.84)$$

where z is any complex number. We also write S as

$$S - 1 = z + z^2 + \cdots + z^n = z(1 + z + \cdots + z^{n-1}) = z[S - z^n] \qquad (10.85)$$

to obtain

$$S(1-z) = 1 - z^{n+1}, \qquad (10.86)$$

which yields the sum as

$$S = \frac{1}{1-z} - \frac{z^{n+1}}{1-z}. \qquad (10.87)$$

Substituting this in Eq. (10.84), we obtain

$$\boxed{\frac{1}{1-z} = (1 + z + z^2 + \cdots + z^{n-1}) + \frac{z^n}{1-z}.} \qquad (10.88)$$

This is a formula that is quite useful in obtaining complex power series representations.

We now use this result [Eq. (10.88)] to write the quantity inside the square brackets in Eq. (10.81) to obtain

$$f(z) = \frac{1}{2\pi i} \oint_{C[\circlearrowleft]} \frac{f(z')\,dz'}{(z'-z_0)} \left[1 + \frac{(z-z_0)}{(z'-z_0)} + \cdots + \frac{(z-z_0)^{n-1}}{(z'-z_0)^{n-1}} \right.$$

$$\left. + \frac{[(z-z_0)/(z'-z_0)]^n}{\left[1 - \frac{(z-z_0)}{(z'-z_0)}\right]} \right]. \qquad (10.89)$$

Using the derivative formula [Eq. (10.77)]:

$$\frac{1}{2\pi i}\oint_{C[\circlearrowleft]}\frac{f(z')\,dz'}{(z'-z_0)^{k+1}}=\frac{1}{k!}f^{(k)}(z_0),\tag{10.90}$$

we can also write Eq. (10.89) as

$$f(z)=f(z_0)+\frac{1}{1!}f'(z_0)(z-z_0)+\cdots+\frac{1}{(n-1)!}f^{(n-1)}(z_0)(z-z_0)^{n-1}+R_n,$$

$$\tag{10.91}$$

where

$$R_n=\frac{(z-z_0)^n}{2\pi i}\oint_{C[\circlearrowleft]}\frac{f(z')\,dz'}{(z'-z_0)^n(z'-z)}\tag{10.92}$$

is called the **remainder**. Note that the above expression [Eq. (10.91)] is exact, and it is called the **Taylor series** with the remainder term R_n.

Using the triangle inequality,

$$|z'-z|\ge|z'-z_0|-|z-z_0|,\tag{10.93}$$

we can put an upper bound to $|R_n|$ as

$$|R_n|\le\frac{|z-z_0|^n}{2\pi}M\oint_C\frac{|dz'|}{|z'-z_0|^n|z'-z|}\tag{10.94}$$

$$\le\frac{|z-z_0|^n}{2\pi}M\frac{L}{[\min|z'-z_0|]^n\min[|z'-z_0|-|z-z_0|]}\tag{10.95}$$

$$\le\frac{1}{2\pi}M\frac{L}{\min[|z'-z_0|-|z-z_0|]}\left[\frac{|z-z_0|}{\min|z'-z_0|}\right]^n,\tag{10.96}$$

where L is the length of the contour C and M stands for the maximum value that $|f(z)|$ can take on C, and "min" indicates the minimum value of its argument on C. From Eq. (10.82), we have

$$\frac{|z-z_0|}{\min|z'-z_0|}<1,\tag{10.97}$$

which means in the limit as n goes to infinity $|R_n|$ goes to zero. Hence, we obtain the **Taylor series** representation of an analytic function as

$$f(z)=f(z_0)+\frac{1}{1!}f'(z_0)(z-z_0)+\frac{1}{2!}f''(z_0)(z-z_0)^2+\cdots.\tag{10.98}$$

When the function is known to be analytic on and within C, the **convergence** of the above series is assured. The contour integral representation of the coefficients, $f^{(k)}(z_0)/k!$, are given in Eq. (10.90). The **radius of convergence** R is from z_0 to the nearest singular point (Figure 10.10). When $z_0 = 0$, the Taylor series is called the **Maclaurin series**:

$$f(z) = f(0) + \frac{1}{1!}f'(0)z + \frac{1}{2!}f''(0)z^2 + \cdots .$$

(10.99)

Examples of some frequently used power series are

$$e^z = \sum_{n=0}^{\infty} \frac{z^n}{n!}, \qquad\qquad |z| < \infty,$$

$$\sin z = \sum_{n=1}^{\infty} (-1)^{n+1} \frac{z^{2n-1}}{(2n-1)!}, \qquad |z| < \infty,$$

$$\cos z = \sum_{n=0}^{\infty} (-1)^n \frac{z^{2n}}{(2n)!}, \qquad\qquad |z| < \infty,$$

(10.100)

$$\sinh z = \sum_{n=1}^{\infty} \frac{z^{2n-1}}{(2n-1)!}, \qquad\qquad |z| < \infty,$$

$$\cosh z = \sum_{n=0}^{\infty} \frac{z^{2n}}{(2n)!}, \qquad\qquad |z| < \infty,$$

$$\frac{1}{1-z} = \sum_{n=0}^{\infty} z^n, \qquad\qquad |z| < 1.$$

Example 10.6. *Geometric series:* In the following sum [Eq. (10.88)]:

$$\frac{1}{1-z} = (1 + z + z^2 + \cdots + z^{n-1}) + \frac{z^n}{1-z},$$

(10.101)

when $|z| < 1$ we can take the limit $n \to \infty$ to obtain the geometric series

$$\frac{1}{1-z} = \sum_{n=0}^{\infty} z^n, |z| < 1.$$

(10.102)

Example 10.7. *Taylor series of* $1/z$ *about* $z = 1$*:* Consider the function $f(z) = 1/z$, which is analytic everywhere except at $z = 0$. To write its Taylor series representation about the point $z = 1$, we evaluate the derivatives:

$$f^{(0)}(z) = z,$$

$$f^{(1)}(z) = -\frac{1}{z^2},$$

$$\vdots$$

$$f^{(n)}(z) = (-1)^n \frac{n!}{z^{n+1}}, \tag{10.103}$$

at $z = 1$ and obtain the coefficients in Eq. (10.98) as $f^{(n)}(1) = (-1)^n n!$. Hence, the Taylor series of $f(z) = 1/z$ about $z = 1$ becomes

$$\frac{1}{z} = \sum_{n=0}^{\infty} (-1)^n (z-1)^n. \tag{10.104}$$

This series is convergent up to the nearest singular point $z = 0$, hence, we write the radius of convergence as $|z - 1| < 1$.

10.8.2   Laurent Series with the Remainder

Sometimes a function is analytic inside an annular region defined by two boundaries, B_1 and B_2, or has a singular point at z_0 in its domain (Figure 10.11).

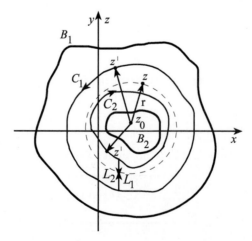

Figure 10.11   Annular region defined by the boundaries, B_1 and B_2, in Laurent series.

In such cases, we can choose the path as shown in Figure 10.11 and use the Cauchy integral theorem to write

$$f(z) = \frac{1}{2\pi i} \oint_{C[\circlearrowleft]} \frac{f(z') \, dz'}{(z' - z)}, \tag{10.105}$$

where z is a point inside the composite path

$$C[\circlearrowleft] = C_1[\circlearrowleft] + C_2[\circlearrowright] + L_1[\downarrow] + L_2[\uparrow]$$

and z' is a point on C. Integrals over the straight-line segments, L_1 and L_2, cancel each other, thus leaving

$$f(z) = \frac{1}{2\pi i} \oint_{C_1[\circlearrowleft]+C_2[\circlearrowright]} \frac{f(z') \, dz'}{(z' - z)} = \frac{1}{2\pi i} \oint_{C_1[\circlearrowleft]} \frac{f(z') \, dz'}{(z' - z)} - \frac{1}{2\pi i} \oint_{C_2[\circlearrowleft]} \frac{f(z') \, dz'}{(z' - z)}, \tag{10.106}$$

where both integrals are now evaluated counterclockwise. We modify the integrands as

$$f(z) = \frac{1}{2\pi i} \oint_{C_1[\circlearrowleft]} \frac{f(z') \, dz'}{(z' - z_0) - (z - z_0)} + \frac{1}{2\pi i} \oint_{C_2[\circlearrowleft]} \frac{f(z') \, dz'}{(z - z_0) - (z' - z_0)} \tag{10.107}$$

$$= \frac{1}{2\pi i} \oint_{C_1[\circlearrowleft]} \frac{f(z') \, dz'}{(z' - z_0) \left[1 - \frac{(z - z_0)}{(z' - z_0)} \right]}$$

$$+ \frac{1}{2\pi i} \oint_{C_2[\circlearrowleft]} \frac{f(z') \, dz'}{(z - z_0) \left[1 - \frac{(z' - z_0)}{(z - z_0)} \right]}, \tag{10.108}$$

where z_0 is any point within the inner boundary B_2. When z' is on C_1, we satisfy the inequality $|z - z_0| < |z' - z_0|$ and when z' is on C_2 we satisfy $|z' - z_0| < |z - z_0|$. Note that choosing C_1 and C_2 as two concentric circles, $|z' - z_0| = r_1$ and $|z' - z_0| = r_2$, respectively, with their radii satisfying $r_1 > r$ and $r_2 < r$ automatically satisfies these inequalities. We now proceed as in the Taylor series derivation and implement Eq. (10.88):

$$\frac{1}{1 - z} = \sum_{k=0}^{n-1} z^k + \frac{z^n}{1 - z}, \tag{10.109}$$

to obtain

$$f(z) = \sum_{k=0}^{n-1} a_k (z - z_0)^k + R_n + \sum_{k=1}^{n} \frac{b_k}{(z - z_0)^k} + Q_n, \tag{10.110}$$

where

$$a_k = \frac{1}{2\pi i} \oint_{C_1[\circlearrowleft]} \frac{f(z')\,dz'}{(z'-z_0)^{k+1}}, \quad k = 0, 1, 2, \ldots, n-1, \tag{10.111}$$

$$b_k = \frac{1}{2\pi i} \oint_{C_2[\circlearrowleft]} \frac{f(z')\,dz'}{(z'-z_0)^{-k+1}}, \quad k = 1, 2, \ldots, n, \tag{10.112}$$

and the remainder terms are written as

$$R_n = \frac{(z-z_0)^n}{2\pi i} \oint_{C_1[\circlearrowleft]} \frac{f(z')\,dz'}{(z'-z_0)^n(z'-z)}, \tag{10.113}$$

$$Q_n = \frac{1}{2\pi i(z-z_0)^n} \oint_{C_2[\circlearrowleft]} \frac{(z'-z_0)^n f(z')}{z-z'}\,dz'. \tag{10.114}$$

The proof that $|R_n|$ approaches to zero as n goes to infinity is exactly the same as in the derivation of the Taylor series. For $|Q_n|$, if we let M be the maximum of $|f(z')|$ on C_2, we can write the inequality

$$|Q_n| \leq \frac{M}{2\pi|z-z_0|^n} \oint_{C_2[\circlearrowleft]} \left| \frac{(z'-z_0)^n}{z-z'}\,dz' \right|. \tag{10.115}$$

Writing the triangle inequality:

$$|z-z'| + |z'-z_0| > |z-z_0|, \tag{10.116}$$

as

$$|z-z'| > |z-z_0| - |z'-z_0|, \tag{10.117}$$

we can also write

$$|Q_n| \leq \frac{M}{2\pi|z-z_0|^n} \frac{\max|z'-z_0|^n}{\min[|z-z_0| - |z'-z_0|]} \oint_{C_2[\circlearrowleft]} |dz'|, \tag{10.118}$$

$$\leq \frac{ML}{2\pi\min[|z-z_0| - |z'-z_0|]} \left[\frac{\max|z'-z_0|}{|z-z_0|} \right]^n, \tag{10.119}$$

where L is the length of the contour C_2 and "min" and "max" stand for the minimum and the maximum values of their arguments on C_2. Since on C_2 we satisfy the inequality

$$\frac{\max|z'-z_0|}{|z-z_0|} < 1, \tag{10.120}$$

as n goes to infinity, $|Q_n|$ goes to zero. Since the function $f(z)$ is analytic inside the annular region defined by the boundaries B_1 and B_2, we can use any closed path encircling B_2 and z within the annular region to evaluate the coefficients

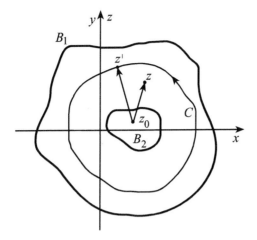

Figure 10.12   Closed contour C in the Laurent theorem.

a_n and b_n without affecting the result. Hence, it is convenient to use the same path C for both coefficients (Figure 10.12). A formal statement of these results is given in the following theorem.

Theorem 10.4. *Laurent series*: Let $f(z)$ be analytic inside the annular region defined by the boundaries B_1 and B_2 and let z_0 be a point inside B_2 (Figure 10.12), then for every point z inside the annular region, $f(z)$ can be represented by the **Laurent series**:

$$f(z) = \sum_{n=0}^{\infty} a_n (z - z_0)^n + \sum_{n=1}^{\infty} \frac{b_n}{(z - z_0)^n}, \tag{10.121}$$

$$a_n = \frac{1}{2\pi i} \oint_{C[\circlearrowleft]} \frac{f(z')\,dz'}{(z' - z_0)^{n+1}}, \quad n = 0, 1, 2, \ldots, \tag{10.122}$$

$$b_n = \frac{1}{2\pi i} \oint_{C[\circlearrowleft]} \frac{f(z')\,dz'}{(z' - z_0)^{-n+1}}, \quad n = 1, 2, \ldots, \tag{10.123}$$

where C is any closed contour inside the annular region encircling B_2 and z_0 (Figure 10.12). If the function $f(z)$ is also analytic inside the central region defined by the boundary B_2, then the coefficients of the negative powers:

$$b_n = \frac{1}{2\pi i} \oint_{C[\circlearrowleft]} f(z')(z' - z_0)^{n-1} dz, \quad n = 1, 2, \ldots, \tag{10.124}$$

all vanish by the Cauchy–Goursat theorem, thus reducing the Laurent series to the Taylor series.

10.9 CONVERGENCE OF POWER SERIES

Concepts of absolute and uniform convergence for series of analytic functions follow from their definitions in real analysis. For the power series

$$\sum_{n=0}^{\infty} a_n (z - z_0)^n, \tag{10.125}$$

we quote the following theorem.

Theorem 10.5. For every power series [Eq. (10.125)], we can find a real number R, $0 \leq R \leq \infty$, called the **radius of convergence** with the properties:

(i) The series converges **absolutely** in $|z - z_0| \leq R$ and **uniformly** for every closed disk $|z - z_0| \leq R' < R$.

(ii) For $|z - z_0| > R$, the series diverges.

(iii) For $|z - z_0| < R$, the sum of the series is an analytic function, hence, its derivative can be obtained by termwise differentiation and the resulting series has the same radius of convergence. Furthermore, if the contour is entirely within the radius of convergence and the sum is a continuous function on C, then the series can be integrated term by term, with the result being equal to the integral of the analytic function that the original series converges to.

Radius of convergence, R, can be found by applying the ratio test as

$$\lim_{n \to \infty} \left| \frac{a_{n+1}(z - z_0)^{n+1}}{a_n(z - z_0)} \right| < 1, \tag{10.126}$$

$$\lim_{n \to \infty} \left| \frac{a_{n+1}}{a_n} \right| |z - z_0| < 1, \tag{10.127}$$

$$|z - z_0| < \lim_{n \to \infty} \left| \frac{a_n}{a_{n+1}} \right|, \tag{10.128}$$

thus

$$\boxed{R = \lim_{n \to \infty} \left| \frac{a_n}{a_{n+1}} \right|.} \tag{10.129}$$

10.10 CLASSIFICATION OF SINGULAR POINTS

Using Laurent series, we can classify the singular points of a function.

Definition 10.1. *Isolated singular point*: If a function is not analytic at z_0 but analytic at every other point in some neighborhood of z_0, then z_0 is called an isolated singular point. For example, $z = 0$ is an isolated singular point of the functions $1/z$ and $1/\sinh z$. The function $1/\sin \pi z$ has infinitely many isolated singular points at $z = 0, \pm 1, \pm 2, \ldots$. However, $z = 0$ is not an isolated singular point of $1/\sin(1/z)$, since every neighborhood of the point $z = 0$ contains other singular points.

Definition 10.2. *Singular point:* In the Laurent series of a function,

$$f(z) = \sum_{n=-\infty}^{\infty} a_n(z - z_0)^n, \tag{10.130}$$

if for $n < -m < 0$, $a_n = 0$ and $a_{-m} \neq 0$, then z_0 is called a **singular point** or **pole** of **order** m.

Definition 10.3. *Essential singular point:* If m is infinity, then z_0 is called an essential singular point. For $\exp(1/z)$, $z = 0$ is an essential singular point.

Definition 10.4. *Simple pole:* If $m = 1$, then z_0 is called a **simple pole**.

Power series representation of an analytic function is **unique**. Once we find a power series that converges to the desired function, we can be sure that it is the power series for that function. In most cases, the needed power series can be constructed without the need for the evaluation of the integrals in Eqs. (10.122) and (10.123) by algebraic manipulations of known series, granted that the series are absolutely and uniformly convergent.

Example 10.8. *Power series representations:* Let us evaluate the power series representation of

$$f(z) = \frac{e^z}{1 - z}. \tag{10.131}$$

We already know the series expansions

$$e^z = 1 + z + \frac{z^2}{2!} + \frac{z^3}{3!} + \cdots, \quad |z| < \infty, \tag{10.132}$$

and

$$\frac{1}{1 - z} = 1 + z + z^2 + \cdots, \quad |z| < 1. \tag{10.133}$$

Hence, we can multiply the two series directly to obtain

$$\frac{e^z}{1 - z} = \left[1 + z + \frac{z^2}{2!} + \frac{z^3}{3!} + \cdots \right] [1 + z + z^2 + z^3 + \cdots] \tag{10.134}$$

$$= 1 + 2z + \frac{5}{2}z^2 + \frac{16}{6}z^3 + \cdots. \tag{10.135}$$

The resulting power series converges in the interval $|z| < 1$.

Example 10.9. *Power series representations:* Power series expansions of $e^{1/z}$ and $\sin(z^2)$ can be obtained by direct substitutions into the expansions [Eq. (10.100)] of e^z and $\sin z$ as

$$e^{1/z} = \sum_{n=0}^{\infty} \frac{1}{n!} \left(\frac{1}{z}\right)^n, \quad z \neq 0, \tag{10.136}$$

$$\sin(z^2) = \sum_{n=1}^{\infty} (-1)^{n+1} \frac{z^{4n-2}}{(2n-1)!}, \quad |z| < \infty. \tag{10.137}$$

The second series converges everywhere, hence, its radius of convergence is infinite, while the first function $e^{1/z}$ has an essential singular point at $z = 0$.

Example 10.10. *Power series representations:* Consider

$$f(z) = \frac{2z - 3}{z^2 - 3z + 2}, \tag{10.138}$$

which can be written as

$$f(z) = \frac{1}{z-1} + \frac{1}{z-2}. \tag{10.139}$$

Hence, $f(z)$ is analytic everywhere except the points $z = 1$ and $z = 2$. If we write $f(z)$ as

$$f(z) = -\frac{1}{2} \left[\frac{1}{1 - z/2}\right] - \frac{1}{1 - z}, \tag{10.140}$$

we can use the geometric series [Eq. (10.102)] to write

$$f(z) = \sum_{n=0}^{\infty} \left[-\frac{1}{2}\left(\frac{z}{2}\right)^n - z^n\right] = \sum_{n=0}^{\infty} [-2^{-n-1} - 1]z^n, \quad |z| < 1. \tag{10.141}$$

This expansion is valid inside the unit circle. We can obtain another expansion if we write $f(z)$ as

$$f(z) = \frac{1}{z}\left(\frac{1}{1 - 1/z}\right) - \frac{1}{2}\frac{1}{1 - z/2} = \sum_{n=0}^{\infty} \left[\frac{1}{z^{n+1}} - \frac{z^n}{2^{n+1}}\right], \quad 1 < |z| < 2, \tag{10.142}$$

which is valid in the annular region between the two circles $|z| = 1$ and $|z| = 2$. We can also write

$$f(z) = \frac{1}{z}\left[\frac{1}{1 - 1/z} + \frac{1}{1 - 2/z}\right] = \sum_{n=0}^{\infty} \frac{1 + 2^n}{z^{n+1}}, \quad |z| > 2, \tag{10.143}$$

which is valid outside the circle $|z| = 2$. Note that the series representations given earlier are all unique in their interval of convergence.

Example 10.11. *Power series representations:* Let us find the power series representation of

$$f(z) = \frac{1}{z^2 \cosh z}. \tag{10.144}$$

Substituting the series expansion of $\cosh z$ [Eq. (10.100)], we write

$$f(z) = \frac{1}{z^2[1 + z^2/2! + z^4/4! + \cdots]} = \frac{1}{z^2 + z^4/2! + z^6/4! + \cdots}. \tag{10.145}$$

Hence, $f(z)$ is analytic everywhere except at $z = 0$. Since the series in the denominator of Eq. (10.145) does not vanish anywhere except at the origin, we can perform a formal division of 1 with the denominator to obtain

$$\frac{1}{z^2 \cosh z} = \frac{1}{z^2} - \frac{1}{2} + \frac{5}{24}z^2 - \cdots, \quad z \neq 0; \tag{10.146}$$

hence, $z = 0$ is a pole of order 2.

10.11   RESIDUE THEOREM

Cauchy integral formula [Eq. (10.66)] deals with cases where the integrand has a simple pole within the closed contour of integration. Armed with the Laurent series representation of functions, we can now tackle contour integrals:

$$\oint_C f(z)\, dz, \tag{10.147}$$

where the integrand has a finite number of isolated singular points of varying orders within the closed contour C (Figure 10.13). We modify the contour as shown in Figure 10.14, where $f(z)$ is analytic in and on the composite path C':

$$C'[\circlearrowleft] = C[\circlearrowleft] + \sum_{j=1}^{n} l_j[\nearrow] + l_j[\swarrow] + \sum_{j=1}^{n} c_i[\circlearrowright]. \tag{10.148}$$

We can now use the Cauchy–Goursat theorem to write

$$\oint_{C'[\circlearrowleft]} f(z)\, dz = 0. \tag{10.149}$$

Integrals over the straight-line segments cancel each other, thus leaving

$$\oint_{C[\circlearrowleft]} f(z)\, dz = \sum_{j=1}^{n} \oint_{c_j[\circlearrowleft]} f(z)\, dz, \tag{10.150}$$

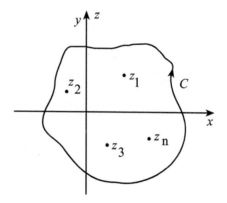

Figure 10.13 Isolated poles.

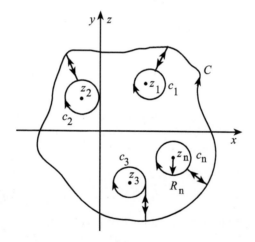

Figure 10.14 Modified path for the residue theorem.

where all integrals are to be evaluated counterclockwise. Using Laurent series expansions [Eq. (10.121)] about the singular points, we write

$$\oint_{C[\circlearrowleft]} f(z)\, dz = \sum_{j=1}^{n} \oint_{c_j[\circlearrowleft]} \left[\sum_{k=0}^{\infty} a_{kj}(z - z_j)^k\, dz + \sum_{k=1}^{\infty} \frac{b_{kj}\, dz}{(z - z_j)^k} \right] \qquad (10.151)$$

$$= \sum_{j=1}^{n} \left[\sum_{k=0}^{\infty} a_{kj} \left(\oint_{c_j[\circlearrowleft]} (z - z_j)^k\, dz \right) + \sum_{k=1}^{\infty} b_{kj} \left(\oint_{c_j[\circlearrowleft]} \frac{dz}{(z - z_j)^k} \right) \right],$$
$$(10.152)$$

where the expansion coefficients a_{kj} and b_{kj} for the jth pole are given in Eqs. (10.122) and (10.123). Since $(z - z_j)^k$ is analytic within and on the contours c_j for all j and k, the first set of integrals vanish:

$$\oint_{c_j[\circlearrowleft]} (z - z_j)^k \, dz = 0, \ j = 1, 2, \ldots, n, \ k = 0, 1, \ldots \ . \tag{10.153}$$

For the second set of integrals, using the parameterization $z - z_j = r_j e^{i\theta}$, we find

$$\oint_{c_j[\circlearrowleft]} \frac{dz}{(z - z_j)^k} = \oint_{c_j[\circlearrowleft]} \frac{r_j i e^{i\theta} \, d\theta}{r_j^k e^{ik\theta}}, \ j = 1, 2, \ldots, n, \ k = 1, \ldots \ , \tag{10.154}$$

$$= \oint_{c_j[\circlearrowleft]} r_j^{(1-k)} e^{i(1-k)\theta} i \, d\theta = \begin{cases} 2\pi i, & k = 1, \\ 0, & k = 2, 3, \ldots . \end{cases} \tag{10.155}$$

In other words,

$$\oint_{C[\circlearrowleft]} f(z) \, dz = 2\pi i \sum_{j=1}^{n} b_{1j}. \tag{10.156}$$

The coefficient of the $1/(z - z_j)$ term, that is, b_{1j}, is called the **residue** of the pole z_j. Hence, the integral $\oint_{C[\circlearrowleft]} f(z) \, dz$ is equal to $2\pi i$ times the sum of the residues of the n isolated poles within the contour C.

This important result is known as the **residue theorem**:

Theorem 10.6. If we let $f(z)$ be an analytic function within and on the closed contour C, except for a finite number of isolated singular points in C, then we obtain

$$\oint_{C[\circlearrowleft]} f(z) dz = 2\pi i \sum_{j=1}^{n} b_{1j}, \tag{10.157}$$

where b_{1j} is the **residue** of the j **th pole**, that is, the coefficient of $1/(z - z_j)$ in the Laurent series expansion of $f(z)$ about z_j. Integral definition of b_{1j} is given as

$$b_{1j} = \frac{1}{2\pi i} \oint_{c_j[\circlearrowleft]} f(z) dz, \ j = 1, 2, \ldots, n. \tag{10.158}$$

Integrals in Eqs. (10.157) and (10.158) are taken in the counterclockwise direction.

Example 10.12. *Residue theorem:* Let us evaluate the integral

$$\oint_{C[\circlearrowleft]} \frac{3z - 1}{z(z - 1)} \, dz, \tag{10.159}$$

where C is the circle about the origin with the radius 2. Since both poles, 0 and 1, are within the contour, we need to find their residues at these points. For the first pole at $z = 0$, we use the expansion

$$\frac{3z-1}{z(z-1)} = \left(3 - \frac{1}{z}\right)\left(\frac{-1}{1-z}\right) = \left(3 - \frac{1}{z}\right)(-)(1 + z + z^2 + \cdots) \quad (10.160)$$

$$= -2 + \frac{1}{z} - 2z - 2z^2 + \cdots, \quad 0 < |z| < 1, \quad (10.161)$$

which yields the residue at $z = 0$ from the coefficient of $1/z$ as $b_1(0) = 1$. For the pole at $z = 1$, we need to expand $1/z$ in powers of $(z-1)$, which is given [Eq. (10.104)] as

$$\frac{1}{z} = 1 - (z-1) + (z-1)^2 - \cdots, \quad |z| < 1. \quad (10.162)$$

Now the series expansion of $\frac{3z-1}{z(z-1)}$ in powers of $(z-1)$ is obtained as

$$\frac{3z-1}{z(z-1)} = \frac{1}{(z-1)}\left[3 - \frac{1}{z}\right] = \frac{1}{(z-1)}[2 + (z-1) - (z-1)^2 - \cdots] \quad (10.163)$$

$$= \frac{2}{(z-1)} + 1 - (z-1) + \cdots, \quad (10.164)$$

which yields the residue of the second pole as $b_1(1) = 2$. Hence, the value of the integral [Eq. (10.159)] is obtained as

$$\oint_C \frac{3z-1}{z(z-1)} dz = 2\pi i[b_1(0) + b_1(1)] = 2\pi i[1 + 2] = 6\pi i. \quad (10.165)$$

Example 10.13. *Finding residues:* To find the residues of unknown functions, we often benefit from the existing series expansions. For example, the residue of $\cos(1/z)$ at $z = 0$ can be read from

$$\cos\frac{1}{z} = 1 - \frac{1}{2!z^2} + \frac{1}{4!z^4} + \cdots \quad (10.166)$$

as $b_1(0) = 0$. Residue of $(\sinh z)/z^2$ can be read from

$$\frac{\sinh z}{z^2} = \frac{1}{z^2}\left[\frac{z}{1!} + \frac{z^3}{3!} + \cdots\right] = \frac{1}{z} + \frac{z}{3!} + \cdots \quad (10.167)$$

as $b_1(0) = 1$.

Example 10.14. *A convenient formula for finding residues:* Since the Laurent expansion of a function that has a pole of order m at z_0 is given as

$$f(z) = \frac{b_{-m}}{(z - z_0)^m} + \cdots + \frac{b_{-1}}{(z - z_0)} + a_0 + a_1(z - z_0) + \cdots,$$

the combination $F(z) = (z - z_0)^m f(z)$ is free of singularities. Hence, the $(m - 1)$-fold differentiation of $F(z)$ gives the residue of $f(z)$ at z_0 as

$$b_1(z_0) = \lim_{z \to z_0} \left\{ \frac{1}{(m - 1)!} \frac{d^{m-1}}{dz^{m-1}} [(z - z_0)^m f(z)] \right\}. \tag{10.168}$$

For example, the function $f(z) = z^4/(z^2 + 1)^4$ has fourth-order poles at $z = \pm i$. Using the above formula, we can find the residues as

$$b_1(i) = \frac{1}{3!} \frac{d^3}{dz^3} \left[\frac{z^4}{(z^2 + i)^4} \right]_{z=i} = -\frac{i}{2^5}, \tag{10.169}$$

$$b_1(-i) = \frac{i}{2^5}. \tag{10.170}$$

Example 10.15. *Residue theorem:* Let us now evaluate the integral

$$I = \oint_{C[\circlearrowleft]} \frac{z^3 + 1}{(z - 1)(z^2 + 4)} \, dz, \tag{10.171}$$

where the integrand has poles at $z = 1$ and $z = \pm 2i$ and C is a circle centered at $z = 3$ with the radius 3. The only pole inside the contour C is $z = 1$. Hence, we calculate its residue as

$$b_1(1) = \lim_{z \to 1} \left[(z - 1) \frac{z^3 + 1}{(z - 1)(z^2 + 4)} \right] = \frac{2}{5}, \tag{10.172}$$

which leads the value of the integral as

$$\oint_{C[\circlearrowleft]} \frac{z^3 + 1}{(z - 1)(z^2 + 4)} \, dz = 2\pi i b_1(1) = \frac{4\pi}{5} i. \tag{10.173}$$

REFERENCES

1. Brown, J.W., and R.V. Churchill, *Complex Variables and Applications*, McGraw-Hill, New York, 1995.

2. Bayin, S.S. (2018). *Mathematical Methods in Science and Engineering*, 2e. Hoboken, NJ: Wiley.

PROBLEMS

1. Evaluate the following integral when C is the straight-line segment from $z = 0$ to $z = 1 + i$:

$$\int_C (y - x + i2x^2)\, dz.$$

2. Evaluate I, where C is two straight-line segments from $z = 0$ to $z = 1 + i$ and from $z = 1 + i$ to $z = 2 + i$:

$$I = \int_C (y - x^2 + i(2x^2 + y))\, dz.$$

3. Evaluate the following integral, where C is the semicircle $z = e^{i\theta}$, $0 \le \theta \le \pi$:

$$\int_C \frac{z+1}{z}\, dz.$$

4. Evaluate I, where C is the square with the vertices at $z = 0$, $z = 1$, $z = 1 + i$, $z = i$:

$$I = \int_C (2z + z^2)\, dz.$$

5. Prove the inequality

$$\left| \int_C f(t)\, dt \right| \le \int_C |f(t)| dt.$$

6. **Bounds for analytic functions**: Let a function $f(z)$ be analytic within and on a circle C, which is centered at z_0 and has the radius R. If $|f(z)| \le M$ on C, then show that the derivatives of $f(z)$ at z_0 satisfy the inequality

$$|f^{(n)}(z_0)| \le \frac{n! M}{R^n}, \quad n = 1, 2, \dots.$$

Using the above inequality, show that the only bounded entire function is the constant function. This result is also known as the **Liouville theorem**.

7. Without integrating show that

$$\left| \int_C \frac{dz}{z^2 + 2} \right| \le \frac{\pi}{2},$$

where C is the arc of the circle with $|z| = 2$ and lying in the first quadrant.

8. Show that

$$\left| \int_C \frac{\log z}{z^2}\, dz \right| \le 2\pi \frac{\pi + \log R}{R}, \quad R > 1,$$

where C is the circle $|z| = R$. What can you say about the value of the integral as $R \to \infty$?

9. For the following integrals, what can you say about their contours?

$$\oint_C \frac{z^2}{z-2} = 0,$$

$$\oint_C z e^{-z} = 0,$$

$$\oint_C z^2 + z + 1 = 0.$$

10. If C is the boundary defined by the rectangle $0 \le x \le 2$, $0 \le y \le 2$ traversed in the counterclockwise direction, show that

$$\oint_{C[\circlearrowleft]} \frac{dz}{z-1-i} = 2\pi i,$$

$$\oint_{C[\circlearrowleft]} \frac{dz}{(z-1-i)^n} = 0, \quad n = 2, 3, \ldots.$$

11. Evaluate the following integrals over the circle $|z| = 2$:

(i) $\oint_{C[\circlearrowleft]} \frac{z^3 + z}{(z-1)^3}\, dz,$

(ii) $\oint_{C[\circlearrowleft]} \frac{2z \sin(\pi/z)}{(z-1)}\, dz.$

12. If $f(z)$ is analytic within and on the closed contour C and z_0 is not on C, then show that

$$\oint_{C[\circlearrowleft]} \frac{f'(z)\, dz}{z - z_0} = \oint_{C[\circlearrowleft]} \frac{f(z)\, dz}{(z - z_0)^2}.$$

13. Find the Taylor series expansions of $\cos z$ and $\sin z$ about $z = \pi/2$ and π.

14. Expand $\sinh z$ about $z = \pi i$ and $z \sinh(z^2)$ about $z = 0$.

15. Find series expansions of $f(z)$:

$$f(z) = \frac{1}{2+z},$$

valid outside the circle $|z| = 2$ and inside the circle $|z| = 1$.

16. Represent the function

$$f(z) = \frac{z}{(z^2 - 3z + 2)}$$

in powers of $(z - 1)$. What is the radius of convergence?

17. Expand

$$f(z) = \frac{2z + 1}{z + 1}$$

in powers of $(z - 1)$ and find the radius of convergence.

18. Expand

$$f(z) = \frac{z - 1}{z^2}$$

in powers of $(z - 1)$ for $|z - 1| > 1$ and for $|z - 1| < 1$.

19. Find the radius of convergence of the following series:

$$\text{(i)} \sum_{n=0}^{\infty} n^3 z^n, \quad \text{(ii)} \sum_{n=0}^{\infty} n! z^n, \quad \text{(iii)} \sum_{n=0}^{\infty} \frac{z^{2n}}{2^n}, \quad \text{(iv)} \sum_{n=0}^{\infty} \frac{z^n}{n!}.$$

20. For what value of z is the series

$$\sum_{n=0}^{\infty} \left(\frac{z}{1 + z} \right)^n$$

convergent? What about the series

$$\sum_{n=0}^{\infty} \frac{z^n}{1 + z^{2n}}.$$

21. Using the geometric series show the following series representations:

$$\frac{1}{(1 - z)^2} = \sum_{n=1}^{\infty} n z^{n-1}, \quad |z| < 1,$$

$$\frac{1}{(1 - z)^3} = \frac{1}{2} \sum_{n=2}^{\infty} n(n - 1) z^{n-2}, \quad |z| < 1.$$

22. Find the following limits:

$$\text{(i)} \lim_{z \to \pi i} \frac{\cosh z}{(z - \pi i)^2},$$

$$\text{(ii)} \lim_{z \to \pi i} \frac{\sinh z}{(z - \pi i)}.$$

23. Classify the singular points of

$$\text{(i)} \ f(z) = \frac{z+2}{z^2+2z},$$

$$\text{(ii)} \ f(z) = \frac{z}{\sin z},$$

$$\text{(iii)} \ f(z) = \frac{z + \exp(z)}{z^4},$$

$$\text{(iv)} \ f(z) = \tanh z,$$

$$\text{(v)} \ f(z) = \frac{\sin z}{z^2} - \frac{1}{z}.$$

24. Evaluate the contour integral

$$I = \oint_{C[\circlearrowleft]} \frac{2z^3+2}{(z-1)(z^2+9)} \, dz,$$

where
(i) C is the circle $|z-2| = 2$,
(ii) C is the circle $|z-2| = 4$.

25. Find the value of

$$I = \oint_{C[\circlearrowleft]} \frac{1}{z^3(z+3)} \, dz,$$

where
(i) C is the circle $|z| = 2$,
(ii) C is the circle $|z+2| = 3$,
(iii) C is the circle $|z+4| = 1$.

26. Using the residue theorem show that

$$\text{(i)} \ \oint_{|z|=2} \frac{3z+1}{z(z-1)^3} dz = 0,$$

$$\text{(ii)} \ \oint_{|z|=1} \cot z \, dz = 2\pi i.$$

CHAPTER 11

ORDINARY DIFFERENTIAL EQUATIONS

Differential equations are proven to be very useful in describing the majority of the physical processes in nature. They are composed of the derivatives of an unknown function, which could depend on several variables. This means that the corresponding differential equation contains partial derivatives of the unknown function. Such equations are called partial differential equations. They are in general quite difficult to solve and deserve separate treatment. On the other hand, in a lot of the physically interesting cases, symmetries of the system allow us to eliminate some of the variables, thus reducing a partial differential equation into an ordinary differential equation. In this chapter, we discuss ordinary differential equations in detail and concentrate on the methods of finding analytic solutions in terms of known functions like polynomials, exponentials, trigonometric functions, etc. We start with a discussion of the first-order differential equations. Since the majority of the differential equations encountered in applications are second-order, we give an extensive treatment of the second-order differential equations and introduce techniques of finding their solutions. The general solution of a differential equation always contains some arbitrary parameters called the integration constants. To facilitate the evaluation of these constants, differential equations have to be supplemented with additional information called the boundary conditions. Boundary conditions play a significant role in the final solution of the problem. We also discuss the conditions for the existence of a unique solution that satisfies the given boundary conditions. We finally discuss the Frobenius method, which can be used to find infinite series solutions when all attempts to find a closed expression for the solution fails. There exists a number of tables for the exactly solvable cases. We highly recommend checking *Ordinary Differential Equations and Their Solutions* by Murphy, before losing all hopes for finding an analytic solution [1].

11.1   BASIC DEFINITIONS FOR ORDINARY DIFFERENTIAL EQUATIONS

Unless otherwise specified, we use y for the **dependent variable** and x for the **independent variable**. Derivatives are written as

$$\frac{dy}{dx} = y', \quad \frac{d^2y}{dx^2} = y'', \quad \frac{d^3y}{dx^3} = y''', \dots, \quad \frac{d^ny}{dx^n} = y^{(n)}. \tag{11.1}$$

The most general differential equation can be given as

$$g(x, y, y', \dots, y^{(n)}) = 0. \tag{11.2}$$

A general solution expressed as

$$F(x, y, C_1, C_2, \dots, C_n) = 0 \tag{11.3}$$

is called an **implicit solution**. For example,

$$2x^2 + 3y^2 + \sin y = 0 \tag{11.4}$$

is an implicit solution of

$$4x + (6y + \cos y)\frac{dy}{dx} = 0. \tag{11.5}$$

An implicit solution can also be given in terms of integrals, which are also called **quadratures**. Solutions that are given as

$$y(x) = G(x, C_1, C_2, \dots, C_n) \tag{11.6}$$

are called **explicit solutions**. For example,

$$y(x) = 2\sin 2x + 3\cos 2x \tag{11.7}$$

is an explicit solution of

$$\frac{d^2y}{dx^2} + 4y = 0. \tag{11.8}$$

Arbitrary constants, $C_1, C_2, \ldots, C_n$, are called the **integration constants**. The order of the highest derivative is also the **order** of the differential equation. The solution of an nth-order differential equation containing n arbitrary constants is called the **general solution**. A solution containing less than the full number of arbitrary constants is called the **special solution** or the **particular solution**. Sometimes a solution that satisfies the differential equation but is not a special case of the general solution can be found. Such a solution is called the **singular solution**. In physical applications, to complete the solution, we also need to determine the arbitrary constants that appear in the solution. In this regard, for a complete specification of a physical problem, an nth-order differential equation has to be complemented with n **initial** or **boundary** conditions. Initial conditions that specify the values of the dependent variable at n different points, $y(x_1) = y_1, \ldots, y(x_n) = y_n$, are called n-**point** initial conditions. When the values of the dependent variable and its $(n-1)$ derivatives at some point x_0 are given, $y_1(x_0) = y_0, y'(x_0) = y'_0, \ldots, y^{(n-1)}(x_0) = y_0^{(n-1)}$, we have **single-point** initial conditions. It is also possible to define **mixed** or more complicated initial conditions. Sometimes as initial conditions we may impose physical principles like causality, where the initial conditions are implicit.

A **linear** differential equation of order n is in general written as

$$\boxed{a_n(x)\frac{d^ny}{dx^n} + a_{n-1}(x)\frac{d^{n-1}y}{dx^{n-1}} + \cdots + a_1(x)\frac{dy}{dx} + a_0(x)y = F(x),} \tag{11.9}$$

where the coefficients, $a_i(x)$, $i = 1, \ldots, n$, and $F(x)$ are only functions of x. If a differential equation is not linear, then it is **nonlinear**. Differential equations with $F(x) = 0$ are called **homogeneous**. If the nonhomogeneous term is different from zero, $F(x) \neq 0$, then the differential equation is called **nonhomogeneous**. Differential equations can also be written as **operator** equations. For example, Eq. (11.9) can be written as

$$\pounds\{y(x)\} = F(x), \tag{11.10}$$

where the differential operator $\pounds$ is defined as

$$\pounds = a_n(x)\frac{d^n}{dx^n} + a_{n-1}(x)\frac{d^{n-1}}{dx^{n-1}} + \cdots + a_1(x)\frac{d}{dx} + a_0(x). \tag{11.11}$$

Linear differential operators satisfy

$$\pounds\{c_1y_1 + c_1y_2\} = c_1\pounds\{y_1\} + c_2\pounds\{y_2\}, \tag{11.12}$$

where c_1 and c_2 are arbitrary constants and y_1 and y_2 are any two solutions of the homogeneous equation:

$$\pounds\{y(x)\} = 0. \tag{11.13}$$

Other definitions will be introduced as we need them.

Example 11.1. *Classification of differential equations:* The following equation:

$$\frac{\partial^2 z(x,y)}{\partial x^2} + \frac{\partial^2 z(x,y)}{\partial y^2} = F(x), \tag{11.14}$$

is a linear, nonhomogeneous, second-order partial differential equation. The equation

$$\frac{\partial^2 z(x,y)}{\partial x^2} + \frac{\partial^2 z(x,y)}{\partial x \partial y} + 2\left(\frac{\partial z(x,y)}{\partial x}\right)^2 = 0, \tag{11.15}$$

is a nonlinear, homogeneous partial differential equation of second-order, while the equation

$$\frac{d^4 y(x)}{dx^4} + 4\frac{d^2 y(x)}{dx^2} + 8y(x) = x^4, \tag{11.16}$$

is a linear, nonhomogeneous, fourth-order ordinary differential equation with constant coefficients. The equation

$$3\frac{d^2 y(x)}{dx^2} + 8y(x)\frac{dy(x)}{dx} = 0, \tag{11.17}$$

is a second-order, homogeneous, and nonlinear ordinary differential equation and

$$(1-x)\frac{d^3 y(x)}{dx^3} + 4x^2\frac{d^2 y(x)}{dx^2} + 8y(x) = 0, \tag{11.18}$$

is a third-order, homogeneous, linear ordinary differential equation with variable coefficients.

11.2 FIRST-ORDER DIFFERENTIAL EQUATIONS

11.2.1 Uniqueness of Solution

The most general first-order differential equation can be given as

$$g(x, y, y') = 0. \tag{11.19}$$

We first investigate equations that can be solved for the first-order derivative as

$$\frac{dy}{dx} = f(x,y). \tag{11.20}$$

Before we introduce methods of finding solutions, we quote a theorem that gives the conditions under which the differential equation [Eq. (11.20)] with the initial condition $y_0 = y(x_0)$, has a unique solution. When all attempts to find an analytic solution fail, it is recommended that the existence and the uniqueness of the solution be checked before embarking on numerical methods.

Theorem 11.1. Consider the initial value problem

$$\boxed{\frac{dy}{dx} = f(x,y), \quad y_0 = y(x_0).} \tag{11.21}$$

If $f(x,y)$ and $\partial f(x,y)/\partial y$ are continuous functions of x and y in the neighborhood of the point (x_0, y_0), then there exists a unique solution $y(x)$ that satisfies the initial condition $y_0 = y(x_0)$.

Example 11.2. *Uniqueness of solutions:* Given the initial value problem

$$\frac{dy}{dx} = 8x^2 - 4xy^3, \quad y(2) = 6. \tag{11.22}$$

Since the functions

$$f(x,y) = 8x^2 - 4xy^3 \quad \text{and} \quad \frac{\partial f}{\partial y} = -12xy^2 \tag{11.23}$$

are continuous in any domain containing the point (2,6), from Theorem 11.1, we can conclude that a unique solution for this initial value problem exists.

Example 11.3. *Uniqueness of solutions:* Consider the initial value problem

$$\frac{dy}{dx} = 6y^{2/3}, \quad y(2) = 0. \tag{11.24}$$

From

$$f(x,y) = 6y^{2/3} \quad \text{and} \quad \frac{\partial f}{\partial y} = \frac{4}{y^{1/3}}, \tag{11.25}$$

we see that $\partial f/\partial y$ is not continuous at $y = 0$. Actually, it does not even exist there. Now the conditions of Theorem 11.1 are not satisfied in the neighborhood of the point $(2, 0)$. Hence, for this initial value problem, we cannot conclude about the existence and the uniqueness of the solution. In fact, this problem has two solutions:

$$y(x) = 0 \quad \text{and} \quad y(x) = (2x - 4)^3. \tag{11.26}$$

11.2.2 Methods of Solution

Under certain conditions, first-order differential equations can immediately be reduced to quadratures.

11.2.3 Dependent Variable Is Missing

When the dependent variable is missing, $f(x, y) = \Phi(x)$, we can write the solution of Eq. (11.20) as the **quadrature**

$$\boxed{y(x) = C + \int \Phi(x)dx.} \qquad (11.27)$$

11.2.4 Independent Variable Is Missing

When the independent variable is missing, $f(x, y) = \Theta(y)$, then the solution is given as the integral

$$\boxed{\int \frac{dy}{\Theta(y)} = x + C.} \qquad (11.28)$$

11.2.5 The Case of Separable $f(x, y)$

If neither variable is missing, but $f(x, y)$ is separable, $f(x, y) = \Phi(x)\Theta(y)$, then the solution can be written as

$$\boxed{\int \frac{dy}{\Theta(y)} = \int \Phi(x)dx + C.} \qquad (11.29)$$

The integrals in Eqs. (11.27) and (11.29) may not always be taken analytically; however, from the standpoint of differential equations, the problem is usually considered as solved once it is reduced to quadratures.

Example 11.4. *Separable equation:* The differential equation

$$\frac{dy}{dx} = \frac{(x-2)y^4}{4x^3(y^2 - 3)} \qquad (11.30)$$

is separable and can be reduced to quadratures:

$$\int \frac{(y^2 - 3)\, dy}{y^4} = \int \frac{(x - 2)\, dx}{4x^3}. \qquad (11.31)$$

The solution is implicit and is found as

$$\frac{1}{y^3} - \frac{1}{y} - \frac{1}{4x^2} + \frac{1}{4x} = C. \qquad (11.32)$$

11.2.6 Homogeneous $f(x, y)$ of Zeroth Degree

Here we use the term "homogeneous" differently. In general, a function satisfying

$$\boxed{f(tx, ty) = t^n \, f(x, y)} \tag{11.33}$$

is called **homogeneous** of **degree** n. Homogeneous equations of the zeroth degree are in general given as

$$y' = f\left(\frac{y}{x}\right), \quad x \neq 0. \tag{11.34}$$

If we use the substitution $y(x) = z(x)x$, we obtain the first-order equation

$$z' = \frac{f(z) - z}{x}, \tag{11.35}$$

which is separable. Solution for $z(x)$ is now written as

$$\int \frac{dz}{f(z) - z} = \int \frac{dx}{x} = \ln x + C. \tag{11.36}$$

11.2.7 Solution When $f(x, y)$ Is a Rational Function

When the differential equation is given as

$$\boxed{y' = \frac{a_1 x + b_1 y + c_1}{a_2 x + b_2 y + c_2}}, \tag{11.37}$$

there are two cases depending on the value of the determinant

$$d = \det\begin{pmatrix} a_1 & b_1 \\ a_2 & b_2 \end{pmatrix}. \tag{11.38}$$

Case I: When $d \neq 0$, that is, when $a_1 b_2 \neq a_2 b_1$, the numerator and the denominator are linearly independent. Hence, we can make the substitutions:

$$x = u + A, \tag{11.39}$$

$$y = v + B, \tag{11.40}$$

so that

$$dx = du, \quad dy = dv, \quad \text{and} \quad dy/dx = dv/du. \tag{11.41}$$

Now Eq. (11.37) becomes

$$\frac{dv}{du} = \frac{a_1 u + b_1 v + a_1 A + b_1 B + c_1}{a_2 u + b_2 v + a_2 A + b_2 B + c_2}. \tag{11.42}$$

Since $d \neq 0$, we can adjust the constants A and B so that

$$a_1 A + b_1 B + c_1 = 0, \qquad (11.43)$$

$$a_2 A + b_2 B + c_2 = 0. \qquad (11.44)$$

This reduces Eq. (11.42) to

$$\frac{dv}{du} = \frac{a_1 u + b_1 v}{a_2 u + b_2 v} = \frac{a_1 + b_1(v/u)}{a_2 + b_2(v/u)}, \qquad (11.45)$$

which is homogeneous of degree zero. Thus, can be solved by the technique introduced in the previous section.

Case II: When the determinant d is zero, that is, when $a_1 b_2 = b_1 a_2$, we can write

$$a_2 x + b_2 y = k(a_1 x + b_1 y), \qquad (11.46)$$

where k is a constant. We now introduce a new unknown $z(x)$:

$$a_1 x + b_1 y = z, \qquad (11.47)$$

so that

$$a_2 x + b_2 y = kz \qquad (11.48)$$

and

$$a_1 + b_1 y' = z'. \qquad (11.49)$$

Now the differential equation for $z(x)$ is separable:

$$z' = a_1 + b_1 \frac{z + c_1}{kz + c_2}, \qquad (11.50)$$

which can be reduced to quadratures immediately.

Example 11.5. *Variable change:* Consider the differential equation

$$y' = \frac{-x + 2y - 4}{2x + y + 3}. \qquad (11.51)$$

Since $d = -5$ [Eq. (11.38)], this is case I. We first write Eqs. (11.43) and (11.44):

$$-A + 2B - 4 = 0, \qquad (11.52)$$

$$2A + B + 3 = 0, \qquad (11.53)$$

and then determine A and B as

$$A = -2, \quad B = 1. \qquad (11.54)$$

Using the substitutions [Eqs. (11.39) and (11.40)]

$$x = u - 2, \tag{11.55}$$

$$y = v + 1, \tag{11.56}$$

we obtain the differential equation [Eq. (11.45)]

$$\frac{dv}{du} = \frac{-u + 2v}{2u + v} = \frac{-1 + 2\dfrac{v}{u}}{2 + \dfrac{v}{u}}. \tag{11.57}$$

Since the right-hand side is homogeneous of zeroth degree, we make another transformation, $z = v/u$, to get

$$\frac{dz}{du} = \left(\frac{dv}{du}\right)\frac{1}{u} - \frac{v}{u^2} = \left[\frac{-1 + 2z}{2 + z} - z\right]\frac{1}{u} = \left[\frac{-1 - z^2}{2 + z}\right]\frac{1}{u}, \tag{11.58}$$

which is separable and can be reduced to quadratures immediately as

$$\int \left[\frac{2 + z}{1 + z^2}\right] dz = -\int \frac{du}{u} = -\ln u + C. \tag{11.59}$$

Evaluating the integral and substituting the original variables, we obtain the solution:

$$2\tan^{-1}\left[\frac{y - 1}{x + 2}\right] + \ln\left[(x + 2)^2 + (y - 1)^2\right]^{1/2} = C. \tag{11.60}$$

11.2.8   Linear Equations of First-order

First-order linear equations in general can be written as

$$\boxed{y' + a(x)y = b(x).} \tag{11.61}$$

For **homogeneous** equations $b(x)$ is zero; hence, the equation becomes separable and the solution can be written as

$$\int \frac{dy}{y} = -\int a(x)\, dx, \tag{11.62}$$

$$\ln y = -\int a(x)\, dx + \ln C, \quad y > 0, \tag{11.63}$$

$$y(x) = Ce^{-\int a(x)\, dx}. \tag{11.64}$$

For the **inhomogeneous** equation, $b(x) \neq 0$, the general solution can be obtained by using the method of **variation of parameters**, where we treat C

in Eq. (11.64) as a function of x. We now differentiate

$$y(x) = C(x)e^{-\int a(x)\,dx} \tag{11.65}$$

as

$$y' = C'e^{-\int a(x)\,dx} - Cae^{-\int a(x)\,dx} \tag{11.66}$$

and substitute into Eq. (11.61) to obtain a differential equation for $C(x)$:

$$C'e^{-\int a(x)\,dx} - Cae^{-\int a(x)\,dx} + aCe^{-\int a(x)\,dx} = b(x), \tag{11.67}$$

$$C' = b(x)e^{\int a(x)\,dx}. \tag{11.68}$$

Solution of Eq. (11.68) gives $C(x)$ as

$$C(x) = \int dx\,\left[b(x)e^{\int a(x)\,dx}\right]. \tag{11.69}$$

Note that we omit the primes in $C(x) = \int^x dx'\,\left[b(x')e^{\int^{x'} a(x'')\,dx''}\right]$. Hence the final solution becomes

$$y(x) = \left[\int dx\,\left(b(x)e^{\int a(x)\,dx}\right)\right]e^{-\int a(x)\,dx} + C_0. \tag{11.70}$$

Example 11.6. *Linear equations:* Consider the linear equation

$$3y' + 6xy = x^3. \tag{11.71}$$

We first solve the corresponding homogeneous equation, $3y' + 6xy = 0$, as

$$\int \frac{dy}{y} = -2\int x\,dx, \tag{11.72}$$

$$\ln(y/C) = -x^2, \quad y/C > 0, \tag{11.73}$$

$$y = Ce^{-x^2}. \tag{11.74}$$

Using the method of variation of parameters, we now substitute $y = C(x)e^{-x^2}$ into Eq. (11.71):

$$C'e^{-x^2} - 2xCe^{-x^2} + 2xCe^{-x^2} = \frac{1}{3}x^3, \tag{11.75}$$

which yields $C(x)$ as

$$C(x) = \frac{1}{3}\int x^3 e^{x^2}\,dx = \frac{1}{6}e^{x^2}(x^2 - 1) + C_0. \tag{11.76}$$

We finally obtain the solution $y = C(x)e^{-x^2}$:

$$y = \frac{1}{6}(x^2 - 1) + C_0 e^{-x^2}. \tag{11.77}$$

11.2.9 Exact Equations

Sometimes the right-hand side of Eq. (11.20) is given as a rational function:

$$\boxed{y' = -\frac{M(x,y)}{N(x,y)}.} \tag{11.78}$$

If we write this as

$$M(x,y) \, dx + N(x,y) \, dy = 0 \tag{11.79}$$

and compare with the total derivative of a function $F(x,y)$:

$$dF(x,y) = \frac{\partial F}{\partial x} \, dx + \frac{\partial F}{\partial y} \, dy = 0, \tag{11.80}$$

we see that finding the solution is equivalent to find a function, $F(x,y) = C$, where C is a constant, $M(x,y)$ and $N(x,y)$ are the partial derivatives:

$$M(x,y) = \frac{\partial F}{\partial x}, \tag{11.81}$$

$$N(x,y) = \frac{\partial F}{\partial y}. \tag{11.82}$$

Since the condition for the existence of $F(x,y)$ is given as

$$\frac{\partial^2 F(x,y)}{\partial x \partial y} = \frac{\partial^2 F(x,y)}{\partial y \partial x}, \tag{11.83}$$

Equation (11.78) can be integrated when

$$\boxed{\frac{\partial M(x,y)}{\partial y} = \frac{\partial N(x,y)}{\partial x}.} \tag{11.84}$$

Starting with Eq. (11.81), we can immediately write the integral

$$F(x,y) = \int M(x,y) \, dx + \Phi(y). \tag{11.85}$$

In order to evaluate $\Phi(y)$, we differentiate Eq. (11.85) with respect to y:

$$\frac{\partial F}{\partial y} = \frac{\partial}{\partial y} \left[\int M(x,y) \, dx + \Phi(y) \right] \tag{11.86}$$

$$= \frac{\partial}{\partial y} \int M(x,y) \, dx + \Phi'(y), \tag{11.87}$$

and use Eq. (11.82) to write

$$\frac{\partial}{\partial y} \int M(x,y) \, dx + \Phi'(y) = N(x,y), \tag{11.88}$$

which gives $\Phi(y)$ as

$$\Phi(y) = \int N(x,y) \, dy - \int \left[\frac{\partial}{\partial y} \int M(x,y) \, dx \right] dy. \tag{11.89}$$

Now the solution [Eq. (11.85)] can be written as

$$F(x,y) = \int M(x,y) dx + \int \left[N(x,y) - \frac{\partial}{\partial y} \int M(x,y) \, dx \right] dy. \tag{11.90}$$

Instead of Eq. (11.81), if we start with

$$\frac{\partial F(x,y)}{\partial y} = N(x,y), \tag{11.91}$$

we obtain an equivalent expression of the solution as

$$F(x,y) = \int N(x,y) dy + \int \left[M(x,y) - \frac{\partial}{\partial x} \int N(x,y) \, dy \right] dx. \tag{11.92}$$

Example 11.7. *Exact equations:* Consider the following exact equation:

$$(x^2 - 2y^2) \, dx + (5 - 4xy) \, dy = 0. \tag{11.93}$$

Using Eq. (11.92), we find the solution as

$$\int N(x,y) \, dy + \int \left[M(x,y) - \frac{\partial}{\partial x} \int N(x,y) \, dy \right] dx = C,$$

$$\int (5 - 4xy) \, dy + \int \left[(x^2 - 2y^2) - \frac{\partial}{\partial x} \int (5 - 4xy) \, dy \right] dx = C,$$

$$5y - \frac{4xy^2}{2} + \int \left[(x^2 - 2y^2) - \frac{\partial}{\partial x} \int (5 - 4xy) \, dy \right] dx = C,$$

$$5y - \frac{4xy^2}{2} + \int \left[(x^2 - 2y^2) - \frac{\partial}{\partial x} \left(5y - \frac{4xy^2}{2} \right) \right] dx = C,$$

$$5y - \frac{4xy^2}{2} + \int [(x^2 - 2y^2) + 2y^2)]dx = C,$$

$$5y - \frac{4xy^2}{2} + \frac{x^3}{3} - 2y^2x + 2y^2x = C,$$

$$5y - 2xy^2 + \frac{x^3}{3} = C. \tag{11.94}$$

It is always a **good idea** to check the solution. Evaluating the total derivative of $F(x, y)$:

$$F(x, y) = 5y - 2xy^2 + \frac{x^3}{3} = C, \tag{11.95}$$

we obtain the original differential equation [Eq. (11.93)]

$$\frac{\partial F}{\partial x} dx + \frac{\partial F}{\partial y} dy = 0,$$

$$(x^2 - 2y^2) dx + (5 - 4xy) dy = 0. \tag{11.96}$$

11.2.10 Integrating Factors

If a given equation:

$$\boxed{M(x, y)dx + N(x, y)dy = 0,} \tag{11.97}$$

is not exact, then

$$\frac{\partial M(x, y)}{\partial y} \neq \frac{\partial N(x, y)}{\partial x}. \tag{11.98}$$

Under most circumstances, it is possible to find an **integrating factor** $I(x, y)$, such that the new functions $I(x, y)M(x, y)$ and $I(x, y)N(x, y)$ in

$$I(x, y)[M(x, y) \, dx + N(x, y) \, dy] = 0, \tag{11.99}$$

satisfy the integrability condition

$$\frac{\partial [I(x, y)M(x, y)]}{\partial y} = \frac{\partial [I(x, y)N(x, y)]}{\partial x}. \tag{11.100}$$

Unfortunately, the **partial differential equation** to be solved for $I(x, y)$:

$$\boxed{M(x, y)\frac{\partial I(x, y)}{\partial y} - N(x, y)\frac{\partial I(x, y)}{\partial x} = I(x, y)\left[\frac{\partial N(x, y)}{\partial x} - \frac{\partial M(x, y)}{\partial y}\right],}$$

$$\tag{11.101}$$

is usually more difficult than the original problem. However, under special circumstances, it is possible to find an integrating factor. For example, when the differential equation is of the form

$$\boxed{\frac{dy}{dx} + P(x)y = Q(x),} \tag{11.102}$$

an integrating factor can be found by a quadrature. We rewrite Eq. (11.102):

$$[P(x)y - Q(x)]dx + dy = 0 \tag{11.103}$$

and identify M and N as

$$M(x,y) = [P(x)y - Q(x)], \tag{11.104}$$

$$N(x,y) = 1. \tag{11.105}$$

Applying the condition for exact differential equations [Eq. (11.84)]:

$$\frac{\partial M(x,y)}{\partial y} = \frac{\partial N(x,y)}{\partial x}, \tag{11.106}$$

$$\frac{\partial [P(x)y - Q(x)]}{\partial y} = \frac{\partial [1]}{\partial x}, \tag{11.107}$$

$$P(x) = 0, \tag{11.108}$$

we see that Eq. (11.102) is not exact unless $P(x)$ is zero. Let us multiply Eq. (11.102) by an integrating factor $I(x)$ that depends only on x as

$$I(x)[P(x)y - Q(x)]dx + I(x)\, dy = 0. \tag{11.109}$$

Now the differential equation that $I(x)$ satisfies becomes

$$\frac{\partial [I(x)(P(x)y - Q(x))]}{\partial y} = \frac{\partial I(x)}{\partial x}, \tag{11.110}$$

$$I(x)P(x) = \frac{dI(x)}{dx}, \tag{11.111}$$

which is easily solved to yield the **integrating factor**

$$\boxed{I(x) = e^{\int P(x)\, dx}.} \tag{11.112}$$

Multiplying Eq. (11.102) with $I(x)$, we write

$$e^{\int P(x)\, dx}\frac{dy}{dx} + e^{\int P(x)\, dx}P(x)y = e^{\int P(x)\, dx}Q(x), \tag{11.113}$$

which can be simplified as

$$\frac{d}{dx}\left[e^{\int P(x)\,dx}y\right] = Q(x)e^{\int P(x)\,dx}. \tag{11.114}$$

This gives the **general solution** of Eq. (11.102) as

$$\boxed{\left(e^{\int P(x)\,dx}\right)y(x) = \int\left[e^{\int P(x)\,dx}Q(x)\right]dx + C.} \tag{11.115}$$

Example 11.8. *Integrating factor:* Consider the following differential equation:

$$\frac{dy}{dx} + \frac{5x+1}{x}y = e^{-5x}. \tag{11.116}$$

Since $P(x) = \frac{5x+1}{x}$, we write the integrating factor as

$$I(x) = e^{\int P(x)\,dx} = \exp\left(\int \frac{5x+1}{x}\,dx\right) = xe^{5x}. \tag{11.117}$$

Multiplying Eq. (11.116) with $I(x)$ and simplifying, we get

$$xe^{5x}\frac{dy}{dx} + e^{5x}(5x+1)y = x, \tag{11.118}$$

$$\frac{d}{dx}[xe^{5x}y] = x, \tag{11.119}$$

$$xe^{5x}y = \frac{x^2}{2} + C, \tag{11.120}$$

thus, the general solution becomes

$$y = \frac{x}{2}e^{-5x} + C\frac{e^{-5x}}{x}. \tag{11.121}$$

If there are two independent integrating factors, I_1 and I_2, that are not constant multiples of each other, then we can write the **general solution** as

$$\boxed{I_1(x,y) = CI_2(x,y).} \tag{11.122}$$

Integrating Factors for Some Terms in Differential Equations:

Term	Integrating Factor
$y - xy'$	$1/x^2,\ 1/y^2,\ 1/xy,\ 1/(x^2 \pm y^2)$
$f(x) + y - xy'$	$1/x^2$
$y + [f(y) - x]y'$	$1/y^2$
$y[1 + f(xy)] - x[1 - f(xy)]y'$	$1/xy$

If an equation is exact and, furthermore, if $M(x, y)$ and $N(x, y)$ are homogeneous of the same degree different from -1, the solution can be written without the need for a quadrature as

$$\boxed{xM(x, y) + yN(x, y) = 0.}$$ (11.123)

First-order differential equations have an **additional advantage**; that is, the roles of dependent and independent variables can be interchanged. Under certain circumstances, this property may come in handy. For example, the equation

$$y^3 \, dx + (5xy - 1) \, dy = 0$$ (11.124)

is clearly not exact. When we write it as

$$\frac{dy}{dx} = \frac{y^3}{(1 - 5xy)},$$ (11.125)

it is nonlinear and not one of the types discussed before. However, if we interchange the roles of x and y:

$$\frac{dx}{dy} = \frac{1 - 5xy}{y^3}$$ (11.126)

and rewrite it as

$$\frac{dx}{dy} + \frac{5x}{y^2} = \frac{1}{y^3},$$ (11.127)

it becomes not only linear in x but also assumes the form

$$\frac{dx}{dy} + P(y)x = Q(y).$$ (11.128)

Of course, it may not always be possible to invert the final solution $x(y)$ as $y(x)$.

11.2.11   Bernoulli Equation

A class of nonlinear first-order differential equation that yields analytic solutions is the Bernoulli equation:

$$\boxed{y' = f(x)y + g(x)y^k.}$$ (11.129)

There are the following special cases:
  (i) When $f(x)$ and $g(x)$ are constants, the Bernoulli equation is separable.
  (ii) When $k = 0$, the equation is linear.
  (iii) When $k = 1$, the equation is separable.

(iv) When k is an arbitrary constant, not necessarily an integer and different from 0 and 1, the solution is given as

$$y^{1-k} = y_1 + y_2, \qquad (11.130)$$

where

$$y_1 = Ce^\phi, \qquad (11.131)$$

$$y_2 = (1-k)e^\phi \int e^{-\phi} g(x)\, dx, \qquad (11.132)$$

$$\phi(x) = (1-k) \int f(x)\, dx. \qquad (11.133)$$

For the proof, make the transformation $y^{1-k} = u$ to get the linear equation

$$u'(x) = (1-k)[g(x) + f(x)u], \qquad (11.134)$$

solution of which leads to the above result [Eqs. (11.130)–(11.133)].

Example 11.9. *Bernoulli equation:* Consider the Bernoulli equation

$$y' = xy + xy^3, \qquad (11.135)$$

where

$$f(x) = x, \quad g(x) = x, \quad k = 3. \qquad (11.136)$$

We first find

$$\phi(x) = (1-3)\int x\, dx = -2\frac{x^2}{2} = -x^2 \qquad (11.137)$$

and then write the solution as

$$\frac{1}{y^2} = Ce^{-x^2} + (1-3)e^{-x^2}\int xe^{x^2} dx = Ce^{-x^2} - 1. \qquad (11.138)$$

11.2.12   Riccati Equation

The generalized **Riccati equation** is defined as

$$\boxed{y'(x) = f(x) + g(x)y + h(x)y^2.} \qquad (11.139)$$

This simple-looking equation is encountered in many different branches of science. Solutions are usually defined in terms of more complicated functions than the elementary transcendental functions. If one or more special solutions can be discovered, then the general solution can be written.

Case I: If a special solution y_1 is discovered, then the transformation

$$y(x) = y_1 + \frac{1}{u} \tag{11.140}$$

changes Eq. (11.139) into a linear form:

$$u' + U(x)u + h(x) = 0, \tag{11.141}$$

where

$$U(x) = g + 2hy_1. \tag{11.142}$$

The general solution can now be written as

$$(y - y_1)\left[C - \int h(x)X(x)\,dx\right] = X(x), \tag{11.143}$$

where

$$X(x) = \exp\left(\int U(x)\,dx\right). \tag{11.144}$$

Case II: Using the transformation

$$y(x) = y_1 + u, \tag{11.145}$$

we can also reduce the Riccati equation into a Bernoulli equation:

$$u'(x) = (g + 2hy_1)u + hu^2, \tag{11.146}$$

the solution of which is given in Eqs. (11.129)–(11.133). For cases where more than one special solution are known and for methods of finding special solutions, we refer the reader to Murphy [1]. Examples of Riccati equations with one special solution can be given as

$$\frac{dy}{dx} = 16x^3(y - 2x)^2 + \frac{y}{x}, \quad y_1 = 2x, \tag{11.147}$$

$$\frac{dy}{dx} = 2(1 - x)y^2 + (2x - 1)y - \frac{x}{2}, \quad y_1 = \frac{1}{2}, \tag{11.148}$$

$$\frac{dy}{dx} = -2y^2 + xy + \frac{1}{2}, \quad y_1 = \frac{x}{2}, \tag{11.149}$$

$$\frac{dy}{dx} = -2xy^2 + x(2x + 1)y - \frac{1}{2}(x^3 + x^2 - 1), \quad y_1 = \frac{x}{2}. \tag{11.150}$$

Sometimes a special solution can be spotted just by inspection. As in the following example, a series substitution with finite number of terms can also be tried.

Example 11.10. *Riccati equation:* Given the Riccati equation

$$y' = x - 2 + \frac{3}{x}y - \frac{1}{x}y^2, \tag{11.151}$$

where

$$f(x) = x - 2, \quad g(x) = \frac{3}{x}, \quad h(x) = -\frac{1}{x}. \tag{11.152}$$

To find a special solution, we can try $y = c_1 x + c_2$ and determine the constants c_1 and c_2 to obtain $y_1 = x$. Making the transformation [Eq. (11.145)]:

$$y = x + u, \tag{11.153}$$

the differential equation to be solved for u becomes one of Bernoulli type:

$$u' = (g + 2hy_1)u + hu^2 = \left(\frac{3}{x} - 2\frac{1}{x}x\right)u - \frac{1}{x}u^2 = \left(\frac{3}{x} - 2\right)u - \frac{1}{x}u^2. \tag{11.154}$$

Solution for $u(x)$ can now be found as

$$\frac{1}{u} = Ce^\phi - e^\phi \int \left(-\frac{1}{x}\right) e^{-\phi} dx, \tag{11.155}$$

where

$$\phi = -\int \left(\frac{3}{x} - 2\right) dx = -3\ln x + 2x. \tag{11.156}$$

Evaluating the integrals, we obtain

$$\frac{1}{u} = C\frac{e^{2x}}{x^3} - \frac{e^{2x}}{x^3} \int x^3 e^{-2x} \left(-\frac{1}{x}\right) dx = C\frac{e^{2x}}{x^3} + \frac{e^{2x}}{x^3} \int x^2 e^{-2x} dx \tag{11.157}$$

$$= \frac{1}{x^3}\left[Ce^{2x} - \frac{x^2}{2} - \frac{x}{2} - \frac{1}{4}\right]. \tag{11.158}$$

Substituting $u(x)$ into Eq. (11.153), we obtain the final solution as

$$y(x) = x + x^3\left[Ce^{2x} - \frac{x^2}{2} - \frac{x}{2} - \frac{1}{4}\right]^{-1}. \tag{11.159}$$

11.2.13 Equations that Cannot Be Solved for y'

First-order equations,

$$g(x, y, y') = 0, \tag{11.160}$$

that cannot be solved for the first derivative may be possible to solve analytically for some specific cases. One such case is **Lagrange's equation:**

$$\boxed{y = \varphi(y')x + \psi(y'),} \tag{11.161}$$

which can always be solved in terms of quadratures. Substitute $y' = p$ in the above equation to write

$$y = \varphi(p)x + \psi(p) \tag{11.162}$$

and then differentiate with respect to x to get

$$p = x\varphi'(p)\frac{dp}{dx} + \varphi(p) + \psi'(p)\frac{dp}{dx}, \tag{11.163}$$

$$p - \varphi = \frac{dp}{dx}(x\varphi' + \psi'). \tag{11.164}$$

We write this as

$$\frac{dx}{dp} = \frac{x\varphi'(p) + \psi'(p)}{p - \varphi(p)} \tag{11.165}$$

$$= \frac{\varphi'(p)}{p - \varphi(p)}x + \frac{\psi'(p)}{p - \varphi(p)}, \quad p \neq \varphi(p), \tag{11.166}$$

which is now a first-order equation, the general solution of which can be given as

$$x = x(p, C). \tag{11.167}$$

Substituting this back into the Lagrange's equation [Eq. (11.162)] gives us a second equation, that is, y in terms of p as

$$y(p, C) = \varphi(p)x(p, C) + \psi(p). \tag{11.168}$$

Eliminating p between Eqs. (11.167) and (11.168) gives us the solution

$$y = y(x, C). \tag{11.169}$$

Example 11.11. *Lagrange's equation:* Consider the equation given as

$$y = 4xy' + y'^2. \tag{11.170}$$

Substituting $y' = p$, we write

$$y = 4px + p^2. \tag{11.171}$$

Differentiating with respect to x gives

$$p = 4p + 4x\frac{dp}{dx} + 2p\frac{dp}{dx},$$

$$-3p = (4x + 2p)\frac{dp}{dx}. \tag{11.172}$$

We now write this as

$$\frac{dx}{dp} = -\frac{4x}{3p} - \frac{2}{3}, \tag{11.173}$$

which is a first-order linear differential equation that can be solved by the substitution $z = x/p$ (Section 11.2.6) as

$$x(p) = p\left[Cp^{-7/3} - \frac{2}{7}\right]. \tag{11.174}$$

Substituting this in Eq. (11.171), we obtain

$$y = 4p.p\left[Cp^{-7/3} - \frac{2}{7}\right] + p^2 = p^2\left[4Cp^{-7/3} - \frac{1}{7}\right]. \tag{11.175}$$

We now have two equations:

$$y(p) = p^2\left[4Cp^{-7/3} - \frac{1}{7}\right], \tag{11.176}$$

$$x(p) = p\left[Cp^{-7/3} - \frac{2}{7}\right], \tag{11.177}$$

which describes the solution in terms of the parameter p.

Example 11.12. *Clairaut equation:* A special case of Lagrange's equation:

$$\boxed{y = xy' + \psi(y'),} \tag{11.178}$$

where $\psi(y')$ is a known function of y', is called the **Clairaut equation**. We first substitute $p \doteq y'$ to write

$$y = xp + \psi(p) \tag{11.179}$$

and then differentiate with respect to x to get

$$\frac{dp}{dx}\left[x + \frac{d\psi}{dp}\right] = 0. \tag{11.180}$$

There are two possibilities: If we take $dp/dx = 0$, the solution becomes

$$p = C. \tag{11.181}$$

Substituting this back into Eq. (11.178) gives the general solution as

$$\boxed{y = Cx + \psi(C),} \tag{11.182}$$

which is a single parameter family of straight lines.

The second alternative is

$$\left[x + \frac{d\psi}{dp}\right] = 0. \tag{11.183}$$

Since ψ is a known function of p, this gives p as a function of x; $p = p(x)$, which when substituted into the Clairaut equation [Eq. (11.179)] gives the **singular solution**

$$\boxed{y = xp(x) + \psi[p(x)].} \tag{11.184}$$

This may be shown to be the envelope of the family of straight lines found before [Eq. (11.182)].

11.3   SECOND-ORDER DIFFERENTIAL EQUATIONS

The most general second-order differential equation can be given as

$$f(x, y, y', y'') = 0, \tag{11.185}$$

where its general solution:

$$F(y, x, C_1, C_2) = 0, \tag{11.186}$$

contains two arbitrary integration constants C_1 and C_2. The general **linear** second-order differential equation can be written as

$$\boxed{A_0(x)y'' + A_1(x)y' + A_2(x)y = f(x).} \tag{11.187}$$

When $f(x)$ is zero, the differential equation is called **homogeneous**, and when $f(x)$ is different from zero, it is called **nonhomogeneous**. The general solution of a second-order linear differential equation can be written as the linear combination of two linearly independent solutions, $y_1(x)$ and $y_2(x)$, as

$$\boxed{y(x) = C_1 y_1 + C_2 y_2,}$$ (11.188)

where C_1 and C_2 are two arbitrary constants. **Linear independence** of two solutions, y_1 and y_2, can be tested by calculating a second-order determinant, $W(y_1, y_2)$, called the **Wronskian**:

$$W(y_1, y_2) = \begin{vmatrix} y_1 & y_2 \\ y_1' & y_2' \end{vmatrix}.$$ (11.189)

Two solutions, y_1 and y_2, are linearly independent if and only if their Wronskian $W(y_1, y_2)$ is nonzero. When the two solutions are linearly dependent, their Wronskian vanishes. Hence, they are related by

$$y_1 y_2' = y_1' y_2,$$ (11.190)

which also means

$$\frac{y_1'}{y_1} = \frac{y_2'}{y_2},$$ (11.191)

$$\ln y_1 = \ln y_2 + \ln C,$$ (11.192)

$$y_1 = C y_2.$$ (11.193)

In other words, one of the solutions is a constant multiple of the other. This also means that for two linearly dependent solutions, we can find two constants, C_1 and C_2, both nonzero, such that

$$C_1 y_1 + C_2 y_2 = 0.$$ (11.194)

11.3.1 The General Case

In the most general case:

$$f(x, y, y', y'') = 0,$$ (11.195)

we can reduce the differential equation to one of first-order when either the dependent or the independent variable is missing.

Independent variable is missing: Using the transformation

$$y' = p, \quad y'' = p\frac{dp}{dy},$$ (11.196)

we can write Eq. (11.195) as

$$f\left(y, p, p\frac{dp}{dy}\right) = 0,$$ (11.197)

which is now a first-order differential equation.

Dependent variable is missing: We now use the substitution

$$y' = p, \quad y'' = p', \quad p' = \frac{dp}{dx} \tag{11.198}$$

and write Eq. (11.195) as

$$f(x, p, p') = 0. \tag{11.199}$$

which is again one of first-order.

Example 11.13. $y'' + f(x)y' + g(y)y'^2 = 0$: This is an exactly solvable case for second-order nonlinear differential equations with variable coefficients. We use the transformation

$$y' = u(x)v(y) \tag{11.200}$$

to write the derivative

$$y'' = u'v + u\frac{dv}{dy}y' = u'v + u^2v\frac{dv}{dy}. \tag{11.201}$$

Substituting Eqs. (11.200) and (11.201) into

$$y'' + f(x)y' + g(y)y'^2 = 0, \tag{11.202}$$

we get

$$u'v + u^2v\frac{dv}{dy} = -fuv - gu^2v^2. \tag{11.203}$$

Rearranging and collecting the x dependence on the left-hand side and the y dependence on the right-hand side, we obtain

$$\frac{u'(x) + f(x)u(x)}{u^2(x)} = -\frac{dv(y)}{dy} - g(y)v(y). \tag{11.204}$$

We can satisfy this equation for all u and v by setting both sides to zero, which gives two first-order equations

$$\frac{u'(x) + f(x)u(x)}{u^2(x)} = 0, \tag{11.205}$$

$$\frac{du}{dx} + f(x)u(x) = 0 \tag{11.206}$$

and

$$\frac{dv(y)}{dy} + g(y)v(y) = 0. \tag{11.207}$$

Both can be integrated to yield

$$u(x) = C_1 e^{-\int f(x)\, dx},\qquad (11.208)$$

$$v(y) = C_2 e^{-\int g(y)\, dy}.\qquad (11.209)$$

Using these in Eq. (11.200) and integrating once more, we obtain the **implicit solution**:

$$\boxed{\int e^{\int g(y)\, dy}\, dy = c_0 \int e^{-\int f(x)\, dx}\, dx + c_1,}\qquad (11.210)$$

where c_0 and c_1 are integration constants.

11.3.2 Linear Homogeneous Equations with Constant Coefficients

In this case, the differential equation is written as

$$\boxed{y'' + a_1 y' + a_2 y = 0,}\qquad (11.211)$$

where a_1 and a_2 are constants. It is clear by inspection that we can satisfy this equation if we find a function whose first and second derivatives are proportional to itself. An obvious candidate with this property is the exponential function

$$y = e^{rx},\qquad (11.212)$$

where r is a constant:

$$y = e^{rx};\quad y' = re^{rx} = ry;\quad y'' = r^2 e^{rx} = r^2 y.\qquad (11.213)$$

Substituting $y = e^{rx}$ into Eq. (11.211), we get a quadratic equation for r called the **characteristic equation:**

$$\boxed{r^2 + a_1 r + a_2 = 0,}\qquad (11.214)$$

which can be factorized to yield the roots as

$$(r - r_1)(r - r_2) = 0,\qquad (11.215)$$

$$r_{1,2} = \frac{-a_1 \pm \sqrt{a_1^2 - 4a_2}}{2} = -\frac{a_1}{2} \pm \left(\frac{a_1^2}{4} - a_2\right)^{1/2}.\qquad (11.216)$$

With r_1 and r_2 as the roots of the characteristic equation, we now have two solutions of the second-order linear differential equation with constant coefficients [Eq. (11.211)] as $e^{r_1 x}$ and $e^{r_2 x}$. Depending on the coefficients a_1 and a_2, we have the following three cases:

Case I: When $\left(\frac{a_1^2}{4} - a_2\right) > 0$, the roots r_1 and r_2 are real and distinct. The corresponding solutions are linearly independent and the **general solution** is given as

$$y = C_1 e^{r_1 x} + C_2 e^{r_2 x}, \tag{11.217}$$

where

$$r_1 = -\frac{a_1}{2} + \left(\frac{a_1^2}{4} - a_2\right)^{1/2}, \quad r_2 = -\frac{a_1}{2} - \left(\frac{a_1^2}{4} - a_2\right)^{1/2}. \tag{11.218}$$

We can also express this solution as

$$y = A_1 e^{-a_1 x/2} \cosh(\beta x + A_2), \tag{11.219}$$

where

$$\beta = \left(\frac{a_1^2}{4} - a_2\right)^{1/2}, \tag{11.220}$$

$$A_1^2 = 4C_1 C_2, \tag{11.221}$$

$$A_2 = \tanh^{-1}\left(\frac{C_1 - C_2}{C_1 + C_2}\right). \tag{11.222}$$

Case II: When $\left(\frac{a_1^2}{4} - a_2\right) = 0$, the roots are real but identical:

$$r = r_1 = r_2 = -\frac{a_1}{2}. \tag{11.223}$$

In this case, we get only one solution. To obtain the second linearly independent solution, we try $y = v(x)e^{rx}$. Substituting its derivatives:

$$y' = v'e^{rx} + rve^{rx}; \quad y'' = v''e^{rx} + rv'e^{rx} + rv'e^{rx} + r^2 ve^{rx}, \tag{11.224}$$

into Eq. (11.211), we obtain

$$v''e^{rx} + rv'e^{rx} + rv'e^{rx} + r^2 ve^{rx} + a_1(v'e^{rx} + rve^{rx}) + a_2 ve^{rx} = 0, \tag{11.225}$$

which after rearranging, becomes

$$[v'' + (r + r + a_1)v' + (r^2 + a_1 r + a_2)v]e^{rx} = 0. \tag{11.226}$$

Since $e^{rx} \neq 0$, we obtain the differential equation to be solved for $v(x)$ as

$$v'' + (2r + a_1)v' + (r^2 + a_1 r + a_2)v = 0. \tag{11.227}$$

Using Eq. (11.223), that is, $r = -\frac{a_1}{2}$, and the relation $a_2 = \frac{a_1^2}{4}$, we write

$$v'' + (-a_1 + a_1)v' + \left(\frac{a_1^2}{4} - \frac{a_1^2}{2} + \frac{a_1^2}{4}\right)v = 0, \tag{11.228}$$

$$v'' = 0. \tag{11.229}$$

This gives v as $v = b_1 + b_2 x$, where b_1 and b_2 are constants. We can now take the second linearly independent solution as $y_2 = xe^{rx}$, which leads to the **general solution:**

$$\boxed{y = C_1 e^{rx} + C_2 x e^{rx}, \quad r = -\frac{a_1}{2}.} \tag{11.230}$$

Note that there is nothing wrong in taking the second linearly independent solution as

$$y_2 = (b_1 + b_2 x)e^{rx}, \tag{11.231}$$

however, y_2 taken like this also contains $y_1 = e^{rx}$.

Case III: When $\left(\frac{a_1^2}{4} - a_2\right) < 0$, the roots r_1 and r_2 are complex, hence, they can be written in terms of two real numbers d_1 and d_2 as

$$r_1 = d_1 + id_2, \quad r_2 = d_1 - id_2, \tag{11.232}$$

where

$$d_1 = -\frac{a_1}{2}, \quad d_2 = \left|\frac{a_1^2}{4} - a_2\right|^{1/2}. \tag{11.233}$$

Now the general solution is given as

$$\boxed{y = e^{d_1 x}[C_1 e^{id_2 x} + C_2 e^{-id_2 x}].} \tag{11.234}$$

Since C_1 and C_2 are arbitrary, we can also take them as complex numbers and write

$$y = e^{d_1 x}[(C_{1R} + iC_{1I})(\cos d_2 x + i \sin d_2 x)$$
$$+ (C_{2R} + iC_{2I})(\cos d_2 x - i \sin d_2 x)] \tag{11.235}$$
$$= e^{d_1 x}\{(C_{1R} + C_{2R})\cos d_2 x + (C_{2I} - C_{1I})\sin d_2 x$$
$$+ i[(C_{1I} + C_{2I})\cos d_2 x + (C_{1R} - C_{2R})\sin d_2 x]\}, \tag{11.236}$$

where C_{iR} and C_{iI}, $i = 1, 2$, are real numbers representing the real and the imaginary parts of C_1 and C_2. Choosing

$$C_{1R} - C_{2R} = 0, \tag{11.237}$$

$$C_{1I} + C_{2I} = 0, \tag{11.238}$$

we can write the solution entirely in terms of real quantities as

$$y(x) = e^{d_1 x}[2C_{1R} \cos d_2 x + 2C_{2I} \sin d_2 x]. \qquad (11.239)$$

This can also be written as

$$\boxed{y(x) = A_0 e^{d_1 x} \cos(d_2 x - A_1),} \qquad (11.240)$$

where we have defined two new real constants A_0 and A_1 as

$$A_0 = 2(C_{1R}^2 + C_{2I}^2)^{1/2}, \qquad (11.241)$$

$$A_1 = \tan^{-1}\frac{C_{2I}}{C_{1R}}. \qquad (11.242)$$

In Eq. (11.240), A_0 is the amplitude and A_1 is the phase of the cosine function:

Example 11.14. ***Harmonic oscillator:**** Small oscillations of a mass m attached to a spring is governed by the equation

$$m\ddot{x} = -kx, \qquad (11.243)$$

where x denotes the displacement from the equilibrium position and k is the spring constant. Usually, we write Eq. (11.243) as

$$\ddot{x} + \omega_0^2 x = 0, \qquad (11.244)$$

where $\omega_0^2 = k/m$ is called the natural frequency of the spring. The general solution of Eq. (11.244) is given as

$$x(t) = a_0 \cos(\omega_0 t - \alpha), \qquad (11.245)$$

where a_0 is the amplitude and α is the phase of the oscillations. Using the initial conditions:

$$x_0 = x(0) = a_0 \cos \alpha, \qquad (11.246)$$

$$v_0 = \dot{x}(0) = a_0 \omega_0 \sin \alpha, \qquad (11.247)$$

we can determine a_0 and α in terms of the initial position and velocity of the mass m as

$$a_0 = \left[\frac{x_0^2 \omega_0^2 + v_0^2}{\omega_0^2}\right]^{1/2}, \qquad (11.248)$$

$$\alpha = \tan^{-1}\left[\frac{v_0}{x_0 \omega_0}\right]. \qquad (11.249)$$

Example 11.15. *Damped harmonic oscillator:* When there is a damping force proportional to velocity, $f_d = -b\dot{x}$, the equation of motion of the harmonic oscillator becomes

$$m\ddot{x} = -b\dot{x} - kx, \tag{11.250}$$

$$\ddot{x} + \frac{b}{m}\dot{x} + \omega_0^2 x = 0, \tag{11.251}$$

where b is the damping factor or the friction coefficient. Roots of the characteristic equation are now given as

$$r_{1,2} = -\frac{b}{2m} \pm \sqrt{\frac{b^2}{4m^2} - \omega_0^2}, \tag{11.252}$$

which we analyze in terms of three separate cases:

Case I: When $b^2/4m^2 > \omega_0^2$, the harmonic oscillator is overdamped and the solution is given entirely in terms of exponentially decaying terms:

$$x(t) = e^{-bt/2m}[C_1 e^{at} + C_2 e^{-at}], \quad a = \sqrt{\frac{b^2}{4m^2} - \omega_0^2}. \tag{11.253}$$

Since the first term blows up at infinity, we set $C_1 = 0$ and write the final solution as

$$x(t) = C_2 e^{-[(b/2m)+a]t}. \tag{11.254}$$

Case II: When $b^2/4m^2 < \omega_0^2$, the harmonic oscillator is underdamped and the roots of the characteristic equation have both real and imaginary parts:

$$r_{1,2} = -\frac{b}{2m} \pm i\omega_1, \quad \omega_1 = \sqrt{\omega_0^2 - \frac{b^2}{4m^2}}. \tag{11.255}$$

Now the general solution is given as

$$x(t) = A_0 e^{-bt/2m} \cos(\omega_1 t - \alpha), \tag{11.256}$$

where the motion is oscillatory, albeit an exponentially decaying amplitude. Using the initial conditions, we can determine A_0 and α as

$$\tan\alpha = \frac{1}{\omega_1}\left(\frac{b}{2m} + \frac{v_0}{x_0}\right), \quad A_0 = \frac{x_0}{\cos\alpha}, \tag{11.257}$$

where $x_0 = x(0)$, $v_0 = \dot{x}(0)$.

Case III: When $b^2/4m^2 = \omega_0^2$, we have the critically damped harmonic oscillator, where the roots of the characteristic equation are equal:

$$r_1 = r_2 = -\frac{b}{2m}. \tag{11.258}$$

The two linearly independent solutions are now taken as $e^{-bt/2m}$ and $te^{-bt/2m}$. Hence, the general solution is written as

$$x(t) = (C_1 + C_2 t)e^{-bt/2m}, \tag{11.259}$$

where the arbitrary constants are given in terms of the initial position x_0 and velocity v_0 as

$$C_1 = x_0, \quad C_2 = v_0 + bx_0/2m. \tag{11.260}$$

11.3.3   Operator Approach

An alternate approach to second-order linear differential equations with constant coefficients is to write the differential equation

$$y'' + a_1 y' + a_2 = 0 \tag{11.261}$$

in terms of a linear differential operator $\mathcal{L}$:

$$\boxed{\mathcal{L} = D^2 + a_1 D + a_2, \quad D = \frac{d}{dx},} \tag{11.262}$$

as

$$\mathcal{L}y(x) = 0. \tag{11.263}$$

We can factorize $\mathcal{L}$ as the product of two linear operators:

$$(D - r_2)(D - r_1)y = 0. \tag{11.264}$$

Since the naming of the roots as r_1 or r_2 is arbitrary, we can also write this as

$$(D - r_1)(D - r_2)y = 0. \tag{11.265}$$

Solutions of the two first-order equations:

$$(D - r_1)y = 0, \tag{11.266}$$

$$(D - r_2)y = 0, \tag{11.267}$$

$$y_1 = C_0 e^{r_1 x}, \quad y_2 = C_1 e^{r_2 x}, \tag{11.268}$$

are also the solutions of Eqs. (11.264) and (11.265), hence, Eq. (11.261). Since the Wronskian of these solutions:

$$W(y_1, y_2) = (r_2 - r_1)e^{(r_1 + r_2)x}, \tag{11.269}$$

is different from zero for **distinct roots**, these solutions are linearly independent and the **general solution** is written as their linear combination:

$$y = C_1 y_1 + C_2 y_2. \tag{11.270}$$

In the case of a **double root**, $r = r_1 = r_2$, Eq. (11.264) or (11.265) becomes

$$(D - r)^2 y = 0, \tag{11.271}$$

$$(D - r)(D - r)y = 0. \tag{11.272}$$

Let us call

$$(D - r)y = u(x) \tag{11.273}$$

so that Eq. (11.272) becomes

$$(D - r)u(x) = 0, \tag{11.274}$$

the solution of which is

$$u = C_0 e^{rx}. \tag{11.275}$$

Substituting this into Eq. (11.273) gives

$$(D - r)y = C_0 e^{rx}, \tag{11.276}$$

which can be solved by using the techniques introduced in Section 11.2.8 to obtain the general solution as

$$y(x) = C_0 e^{rx} + C_1 x e^{rx}. \tag{11.277}$$

As expected, in the case of a double root, the two linearly independent solutions are given as

$$y_1 = e^{rx} \quad \text{and} \quad y_2 = x e^{rx}. \tag{11.278}$$

11.3.4 Linear Homogeneous Equations with Variable Coefficients

We can write the general expression:

$$A_0(x)y'' + A_1(x)y' + A_2(x)y = 0, \tag{11.279}$$

for the second-order linear homogeneous differential equation with variable coefficients as

$$\boxed{y'' + P(x)y' + Q(x)y = 0.} \tag{11.280}$$

When $Q(x)$ is zero, we can substitute

$$v = y' \tag{11.281}$$

and obtain a first-order equation:

$$v' + P(x)v = 0, \tag{11.282}$$

which can be reduced to a quadrature:

$$v = C_1 e^{-\int P(x)\,dx}. \tag{11.283}$$

Using this in Eq. (11.281) and integrating once more, we obtain the solution as

$$\boxed{y = C_1 \int e^{-\int P(x)\,dx}\,dx + C_2.} \tag{11.284}$$

In general, the solution for Eq. (11.280) cannot be given except for certain special cases. However, if a **special solution** $y_1(x)$ is known, then the transformation

$$y = y_1 u(x) \tag{11.285}$$

reduces Eq. (11.280) into the form

$$y_1 u'' + (2y_1' + Py_1)u' = 0, \tag{11.286}$$

where the dependent variable is missing, hence, the solution can be reduced to quadratures with the substitution $w = u'$ as

$$y = y_1 \left[C_1 \int e^{-\int \left[\frac{2y_1' + Py_1}{y_1} \right] dx}\,dx + C_2 \right], \tag{11.287}$$

$$\boxed{y = y_1 \left[C_1 \int \frac{1}{y_1^2} e^{-\int P(x)\,dx}\,dx + C_2 \right].} \tag{11.288}$$

Some suggestions for finding a special solution are [1]:
   (i) If $P(x) + xQ(x) = 0$, then $y_1 = x$.
   (ii) Search for a constant k such that $k^2 + kP + Q = 0$. If you can find such a k, then $y_1 = e^{kx}$.
   (iii) Try a polynomial solution, $y_1 = a_0 + a_1 x + \cdots + a_n x^n$. Sometimes one can solve for the coefficients.
   (iv) Set $x = 0$ or $x = \infty$ in Eq. (11.280). If the result is solvable, then its solution is y_1.

Example 11.16. *When a particular solution is known:* Consider the differential equation

$$y'' + \frac{2}{3x}y' - \frac{1}{9x^{4/3}}y = 0, \tag{11.289}$$

where $P(x) = 2/3x$ and $Q(x) = -1/9x^{4/3}$. A particular solution is given as $y_1 = e^{x^{1/3}}$. We can now use Eq. (11.288) to write the general solution as

$$y = e^{x^{1/3}} \left[C_1 \int e^{-2x^{1/3}} \left[e^{-\int \frac{2}{3x} \, dx} \right] dx + C_2 \right] \tag{11.290}$$

$$= e^{x^{1/3}} \left[C_1 \int e^{-2x^{1/3}} x^{-2/3} \, dx + C_2 \right] \tag{11.291}$$

$$= C_1 e^{-x^{1/3}} + C_2 e^{x^{1/3}}. \tag{11.292}$$

In the last step, we redefined the arbitrary constant C_1.

Example 11.17. Wronskian from the differential equation: An important property of the second-order differential equations is that the Wronskian can be calculated from the differential equation without the knowledge of the solutions. We write the differential equation for two linearly independent solutions, y_1 and y_2, as

$$y_1'' + P(x)y_1' + Q(x)y_1 = 0 \tag{11.293}$$

and

$$y_2'' + P(x)y_2' + Q(x)y_2 = 0. \tag{11.294}$$

Multiply the second equation by y_1 and the first one by y_2:

$$y_1 y_2'' + P(x)y_1 y_2' + Q(x)y_1 y_2 = 0, \tag{11.295}$$

$$y_2 y_1'' + P(x)y_2 y_1' + Q(x)y_2 y_1 = 0, \tag{11.296}$$

and then subtract to write

$$y_1 y_2'' - y_2 y_1'' + P(x)[y_1 y_2' - y_2 y_1'] = 0. \tag{11.297}$$

This is nothing but

$$\frac{dW(x)}{dx} + P(x)W(x) = 0, \tag{11.298}$$

where $W(x) = [y_1 y_2' - y_2 y_1']$. Eq. (11.298) gives the Wronskian, $W(x) = W[y_1(x), y_2(x)]$, in terms of the coefficient $P(x)$ as

$$\boxed{W(x) = W_0 e^{-\int P(x) \, dx}.} \tag{11.299}$$

Note that the solution in Eq. (11.288) can also be written as

$$y(x) = y_1 \left[C_1 \int \frac{W(x)}{y_1^2} \, dx + C_2 \right]. \tag{11.300}$$

11.3.5   Cauchy–Euler Equation

A commonly encountered second-order linear differential equation with variable coefficients is the **Cauchy–Euler equation**:

$$x^2 y'' + a_1 x y' + a_2 y = 0, \tag{11.301}$$

where a_1 and a_2 are constants. Using the transformation

$$y(x) = u(z), \quad z = \ln x \tag{11.302}$$

and the derivatives

$$y' = \frac{u'}{x}, \quad y'' = \frac{1}{x^2}[-u' + u''], \tag{11.303}$$

we can write the Cauchy-Euler equation as

$$u''(z) + (a_1 - 1)u'(z) + a_2 u(z) = 0, \tag{11.304}$$

which has constant coefficients with the characteristic equation

$$r^2 + (a_1 - 1)r + a_2 = 0. \tag{11.305}$$

There are three cases:

Case I: When $(a_1 - 1)^2 > 4a_2$, the general solution for Eq. (11.301) is

$$y = C_1 x^{r_1} + C_2 x^{r_2}, \tag{11.306}$$

where

$$r_{1,2} = \frac{(1 - a_1)}{2} \pm \frac{1}{2}\sqrt{(a_1 - 1)^2 - 4a_2}. \tag{11.307}$$

Case II: When $(a_1 - 1)^2 = 4a_2$, the general solution of Eq. (11.301) becomes

$$y = x^r (C_1 + C_2 \ln x), \quad r = \frac{(1 - a_1)}{2}. \tag{11.308}$$

Case III: When $(a_1 - 1)^2 < 4a_2$, roots of the characteristic equation are complex:

$$r_{1,2} = a \pm ib, \tag{11.309}$$

where a and b are given as

$$a = \frac{(1 - a_1)}{2}, \tag{11.310}$$

$$b = \frac{1}{2}[4a_2 - (a_1 - 1)^2]^{1/2}. \tag{11.311}$$

Now the general solution of Eq. (11.301) can be written as

$$\boxed{y = C_1 x^a \cos[b \ln x - C_2].} \tag{11.312}$$

11.3.6 Exact Equations and Integrating Factors

We now extend the definition of exact equation to second-order differential equations written in the **standard form:**

$$\boxed{A_0(x)y'' + A_1(x)y' + A_2(x)y = 0.} \tag{11.313}$$

For an **exact equation**, we can find two functions as $B_0(x)$ and $B_1(x)$, which allow us to write the above equation in the form

$$\boxed{\frac{d}{dx}[B_0(x)y' + B_1(x)y] = 0,} \tag{11.314}$$

the first integral of which can be written immediately as

$$B_0(x)y' + B_1(x)y = C. \tag{11.315}$$

This is a first-order differential equation and could be approached by the techniques discussed in Section 11.2.

To obtain a test for exact equations, we expand Eq. (11.314):

$$B_0 y'' + (B_0' + B_1)y' + B_1'y = 0; \tag{11.316}$$

then we compare it with Eq. (11.313). Equating the coefficients of y'', y', and y gives the following relations, respectively:

$$B_0 = A_0, \tag{11.317}$$

$$B_0' + B_1 = A_1, \tag{11.318}$$

$$B_1' = A_2. \tag{11.319}$$

Using the first two equations [Eqs. (11.317) and (11.318)], we write $B_1 = A_1 - A_0'$, which after differentiation and substitution of Eq. (11.319) gives the

condition for exactness entirely in terms of the coefficients A_i, $i = 0, 1, 2$ as

$$\boxed{A_1' - A_0'' = A_2.}$$

(11.320)

In the case of first-order differential equations, granted that it has a unique solution, it is always possible to find an integrating factor. For the second-order differential equations, there is no guarantee that an integrating factor exists. If a differential equation fails to satisfy the test for exactness [Eq. (11.320)], we can multiply it with an **integrating factor**, $\mu(x)$, to write

$$\mu(x)[A_0(x)y'' + A_1(x)y' + A_2(x)y] = 0,$$

(11.321)

$$[\mu(x)A_0(x)]y'' + [\mu(x)A_1(x)]y' + [\mu(x)A_2(x)]y = 0.$$

(11.322)

Substituting the coefficients of Eq. (11.322) into Eq. (11.320), we write

$$(\mu A_0)'' - (\mu A_1)' + \mu A_2 = 0,$$

(11.323)

which gives the differential equation to be solved for $\mu(x)$ as

$$\boxed{A_0\mu'' + (2A_0' - A_1)\mu' + (A_0'' - A_1' + A_2)\mu = 0.}$$

(11.324)

This is also a second-order differential equation with variable coefficients, which is not necessarily easier than the original problem.

Example 11.18. *Integrating factor:* Motion of a particle of mass m in uniform gravitational field g is governed by the equation of motion:

$$m\ddot{x} = -mg,$$

(11.325)

where the upward direction is taken as positive. An overhead dot denotes time derivative. Solution is easily obtained as

$$v = \dot{x} = -gt + v_0,$$

(11.326)

$$x = -\frac{1}{2}gt^2 + v_0 t + x_0,$$

(11.327)

where x_0 and v_0 denote the initial position and velocity. In the presence of friction forces proportional to velocity, we write the equation of motion as

$$m\ddot{x} = -b\dot{x} - mg,$$

(11.328)

$$\ddot{x} + \frac{b}{m}\dot{x} = -g,$$

(11.329)

where b is the coefficient of friction. The homogeneous equation:

$$\ddot{x} + \frac{b}{m}\dot{x} = 0, \tag{11.330}$$

is not exact, since the test of exactness [Eq. (11.320)] is indeterminate, that is, $0 - 0 = 0$. Using Eq. (11.324), we write the differential equation to be solved for the integrating factor:

$$\ddot{\mu} - \frac{b}{m}\dot{\mu} = 0, \tag{11.331}$$

the solution of which yields $\mu(t)$ as

$$\mu(t) = e^{bt/m}. \tag{11.332}$$

Multiplying Eq. (11.329) with the integrating factor found gives

$$e^{bt/m}\left[\ddot{x} + \frac{b}{m}\dot{x}\right] = -ge^{bt/m}, \tag{11.333}$$

$$\frac{d}{dt}[\dot{x}e^{bt/m}] = -ge^{bt/m}. \tag{11.334}$$

This can be integrated immediately to yield

$$\int_0^t d\left[\dot{x}e^{bt/m}\right] = -g\int_0^t e^{bt/m}dt, \tag{11.335}$$

$$\dot{x}e^{bt/m} - v_0 = -\frac{gm}{b}\left[e^{bt/m} - 1\right], \tag{11.336}$$

or

$$\dot{x} = \frac{gm}{b}\left[e^{-bt/m}\left(1 + \frac{bv_0}{gm}\right) - 1\right]. \tag{11.337}$$

Note that in the limit as $t \to \infty$, velocity goes to the **terminal velocity**, $v_t = -gm/b$. Eq. (11.337) can be integrated once more to yield

$$x = x_0 + \frac{gm}{b}\left[\frac{m}{b}(1 - e^{-bt/m})\left(1 + \frac{bv_0}{mg}\right) - t\right]. \tag{11.338}$$

In the limit of small friction, this becomes

$$x \simeq \left[x_0 + v_0 t - \frac{1}{2}gt^2\right] + \frac{bg}{6m}t^3 - \frac{bv_0}{2m}t^2 + \cdots. \tag{11.339}$$

11.3.7 Linear Nonhomogeneous Equations

In the presence of a **nonhomogeneous term**, $f(x)$, the **general solution** of the linear equation:

$$A_0(x)y'' + A_1(x)y' + A_2(x)y = f(x), \tag{11.340}$$

is given as

$$\boxed{y = y_c + y_p,} \tag{11.341}$$

where y_c is the solution of the **homogeneous equation**:

$$y_c = C_1 y_1 + C_2 y_2, \tag{11.342}$$

which is also called the **complementary solution**. The **particular solution** y_p is a solution of the nonhomogeneous equation that contains no arbitrary constants. In what follows, we discuss some of the most commonly used techniques to find particular solutions of linear second-order nonhomogeneous differential equations. Even though there is no general technique that works all the time, the techniques we introduce are rather effective when they work.

11.3.8 Variation of Parameters

If we know the **complementary solution** one of the most common techniques used to find a particular solution is the variation of parameters. In this method, we treat the integration constants C_1 and C_2 in the complementary solution as functions of the independent variable x and write Eq. (11.342) as

$$\boxed{y_p = v_1(x)y_1 + v_2(x)y_2.} \tag{11.343}$$

For a given y_1 and y_2, there are many choices for the functions v_1 and v_2 that give the same y_p. Hence, we can use this freedom to simplify equations. We first write the derivative

$$y_p' = v_1' y_1 + v_1 y_1' + v_2' y_2 + v_2 y_2' \tag{11.344}$$

and then use the freedom we have to impose the condition

$$v_1' y_1 + v_2' y_2 = 0, \tag{11.345}$$

so that the first derivative of y_p becomes

$$y_p' = v_1 y_1' + v_2 y_2'. \tag{11.346}$$

Now the second derivative is written as

$$y_p'' = v_1'y_1' + v_1y_1'' + v_2'y_2' + v_2y_2''.$$ (11.347)

Substituting these derivatives into the differential equation [Eq. (11.340)] gives

$$A_0[v_1'y_1' + v_1y_1'' + v_2'y_2' + v_2y_2'']$$
$$+ A_1[v_1y_1' + v_2y_2'] + A_2[v_1y_1 + v_2y_2] = f(x).$$ (11.348)

After rearranging and using the fact that y_1 and y_2 are the solutions of the homogeneous equation, we obtain

$$v_1'y_1' + v_2'y_2' = \frac{f(x)}{A_0(x)}.$$ (11.349)

Along with the condition we have imposed [Eq. (11.345)], this gives two equations:

$$v_1'y_1 + v_2'y_2 = 0,$$ (11.350)

$$v_1'y_1' + v_2'y_2' = \frac{f(x)}{A_0(x)},$$ (11.351)

to be solved for v_1' and v_2'. Determinant of the coefficients is nothing but the Wronskian:

$$W[y_1, y_2] = \begin{vmatrix} y_1 & y_2 \\ y_1' & y_2' \end{vmatrix}.$$ (11.352)

Since y_1 and y_2 are linearly independent solutions of the homogeneous equation, $W[y_1, y_2]$ is different from zero. Hence, we can easily write the solutions for v_1' and v_2' as

$$v_1' = \frac{\begin{vmatrix} 0 & y_2 \\ f/A_0 & y_2' \end{vmatrix}}{\begin{vmatrix} y_1 & y_2 \\ y_1' & y_2' \end{vmatrix}} = -\frac{f(x)y_2(x)}{A_0(x)W[y_1, y_2]},$$ (11.353)

$$v_2' = \frac{\begin{vmatrix} y_1 & 0 \\ y_1' & f/A_0 \end{vmatrix}}{\begin{vmatrix} y_1 & y_2 \\ y_1' & y_2' \end{vmatrix}} = \frac{f(x)y_1(x)}{A_0(x)W[y_1, y_2]}.$$ (11.354)

These can be integrated for v_1 and v_2:

$$v_1 = -\int \frac{f(x)y_2(x)\,dx}{A_0(x)W[y_1, y_2]},$$ (11.355)

$$v_2 = \int \frac{f(x)y_1(x)\,dx}{A_0(x)W[y_1, y_2]},$$ (11.356)

to yield the **particular solution** as

$$\boxed{y_p = v_1(x)y_1 + v_2(x)y_2.} \tag{11.357}$$

11.3.9 Method of Undetermined Coefficients

When this method works, it is usually preferred over the others. It only depends on differentiation and simultaneous solution of linear algebraic equations. Even though its application is usually tedious, it does not require any integrals to be taken, and as in the variation of parameters method, it does not require the knowledge of the solution of the homogeneous equation.

For a general nonhomogeneous term, which can be written as

$$
\begin{aligned}
f(x) = e^{ax}[\cos bx(\alpha_0 + \alpha_1 x + \cdots + \alpha_n x^n) \\
+ \sin bx(\beta_0 + \beta_1 x + \cdots + \beta_n x^n)],
\end{aligned} \tag{11.358}
$$

where a, b, n, α_i, and β_i are constants, we substitute a solution of the form

$$
\begin{aligned}
y_p(x) = x^r e^{ax}[\cos bx(A_{10} + A_{11}x + \cdots + A_{1n}x^n) \\
+ \sin bx(B_{10} + B_{11}x + \cdots + B_{1n}x^n)]
\end{aligned} \tag{11.359}
$$

into the differential equation:

$$A_0(x)y'' + A_1(x)y' + A_2(x)y = f(x). \tag{11.360}$$

The result is $2n$ linear algebraic equations for the coefficients A_{1i}, $i = 0, \ldots, n$, and B_{1i}, $i = 0, \ldots, n$. For the value of r, we take 0, if $a \pm ib$ is not one of the roots of the characteristic equation. In other cases, we take r as the number of roots coinciding with $a \pm ib$, which in no case will be greater than 2 for a second-order differential equation.

Usually, the inhomogeneous term $f(x)$ can be decomposed as $f(x) = f_1(x) + \cdots + f_m(x)$, where each term is of the form in Eq. (11.358). For each term, we repeat the process to obtain the **particular solution** of

$$\boxed{A_0(x)y'' + A_1(x)y' + A_2(x)y = f_1(x) + \cdots + f_m(x),} \tag{11.361}$$

as

$$\boxed{y_p(x) = y_{p1} + y_{p2} + \cdots + y_{pm},} \tag{11.362}$$

where

$$\boxed{A_0(x)y_{pi}'' + A_1(x)y_{pi}' + A_2(x)y_{pi} = f_i(x), \quad i = 0, \ldots, m.} \tag{11.363}$$

Example 11.19. *Damped and driven harmonic oscillator:* The equation of motion is given as

$$\ddot{x} + \frac{b}{m}\dot{x} + \omega_0^2 x = \frac{F_0}{m}\cos\omega t, \tag{11.364}$$

where the nonhomogeneous term is the driving force. Comparing with Eq. (11.358), we set

$$a = 0, \quad b = \omega, \quad \alpha_0 = \frac{F_0}{m}, \quad n = 0, \tag{11.365}$$

hence, write x_p as

$$x_p = t^r[A_{10}\cos\omega t + B_{10}\sin\omega t]. \tag{11.366}$$

We take the value of r as 0, since $i\omega$ does not coincide with any one of the roots of the characteristic equation of [Eq. (11.252)]:

$$\ddot{x} + \frac{b}{m}\dot{x} + \omega_0^2 x = 0. \tag{11.367}$$

Using Eq. (11.366) with $r = 0$ and its derivatives:

$$\dot{x}_p = [-A_{10}\omega\sin\omega t + B_{10}\omega\cos\omega t], \tag{11.368}$$

$$\ddot{x}_p = [-A_{10}\omega^2\cos\omega t - B_{10}\omega^2\sin\omega t], \tag{11.369}$$

in Eq. (11.364), we obtain two linear equations for A_{10} and B_{10}:

$$A_{10}(\omega_0^2 - \omega^2) + B_{10}\left(\frac{b\omega}{m}\right) = \frac{F_0}{m}, \tag{11.370}$$

$$B_{10}(\omega_0^2 - \omega^2) - A_{10}\left(\frac{b\omega}{m}\right) = 0. \tag{11.371}$$

We first square and then add these equations:

$$A_{10}^2 + B_{10}^2 = \frac{(F_0/m)^2}{[(\omega_0^2 - \omega^2)^2 + \left(\frac{b\omega}{m}\right)^2]}. \tag{11.372}$$

Using Eq. (11.371), we can also write the ratio

$$\frac{B_{10}}{A_{10}} = \frac{b\omega/m}{\omega_0^2 - \omega^2}. \tag{11.373}$$

Now the particular solution can be written as

$$x_p(t) = A_0\cos(\omega t + A_1), \tag{11.374}$$

where

$$A_0 = [A_{10}^2 + B_{10}^2]^{1/2}, \quad A_1 = \tan^{-1}\left[\frac{B_{10}}{A_{10}}\right]. \tag{11.375}$$

Finally, the **general solution** becomes

$$x(t) = x_c(t) + x_p(t), \tag{11.376}$$

where x_c is the solution of the homogeneous equation found in Example 11.15. For the underdamped harmonic oscillator, the **complementary solution** is given as [Eq. (11.256)]

$$x_c(t) = C_0 e^{-bt/2m} \cos(\omega_1 t - C_1), \quad \omega_1 = \sqrt{\omega_0^2 - \frac{b^2}{4m^2}}, \tag{11.377}$$

where C_0 and C_1 are the integration constants. Note that for sufficiently large times, $x_c(t)$ dies out leaving only the **particular solution**. It is for this reason that x_c is called the **transient solution** and x_p is called the **steady-state solution**.

Example 11.20. *Method of undetermined coefficients:* Comparing the nonhomogeneous term of the equation

$$y'' - 3y' + 2y = 2xe^x \tag{11.378}$$

with Eq. (11.358) we identify the constants as

$$a = 1, \quad b = 0, \quad \alpha_0 = 0, \quad \alpha_1 = 2, \quad \text{and} \quad n = 1, \tag{11.379}$$

where $a \pm ib$ is just 1. The characteristic equation in this case has the roots $r_1 = 2$ and $r_2 = 1$. Since the second root coincides with 1, we take $r = 1$ and write y_p as

$$y_p = xe^x[A_{10} + A_{11}x]. \tag{11.380}$$

Substituting this into Eq. (11.378) and after rearranging, we get

$$-2A_{11}xe^x + (-A_{10} + 2A_{11})e^x = 2xe^x, \tag{11.381}$$

or

$$-2A_{11}x + (-A_{10} + 2A_{11}) = 2x. \tag{11.382}$$

Equating the equal powers of x, we get two linear equations to be solved for A_{11} and A_{10}:

$$-2A_{11} = 2, \tag{11.383}$$

$$-A_{10} + 2A_{11} = 0, \tag{11.384}$$

which yields

$$A_{11} = -1 \quad \text{and} \quad A_{10} = -2. \tag{11.385}$$

Thus, a particular solution is obtained as

$$y_p = e^x(-2x - x^2). \tag{11.386}$$

It is always a good idea to check.

11.4   LINEAR DIFFERENTIAL EQUATIONS OF HIGHER ORDER

11.4.1   With Constant Coefficients

The homogeneous linear differential equation of **order** n with real constant coefficients:

$$\boxed{a_0 \frac{d^n y}{dx^n} + a_1 \frac{d^{n-1} y}{dx^{n-1}} + \cdots + a_n y = 0,} \tag{11.387}$$

can be solved exactly. If we try a solution of the form $y \propto e^{rx}$, we obtain an nth-order polynomial as the **characteristic equation**:

$$\boxed{a_0 r^n + a_1 r^{n-1} + \cdots + a_n = 0,} \tag{11.388}$$

which has n roots in the complex plane. Depending on the nature of the roots, we have the following cases for the general solution:

Case I: If the characteristic equation has a real root α occurring $k \leq n$ times, then the part of the general solution corresponding to this root is written as

$$[c_0 + c_1 x + c_2 x^2 + \cdots + c_{k-1} x^{k-1}] e^{\alpha x}. \tag{11.389}$$

Case II: If the remaining roots of the characteristic equation are distinct and real, then the general solution is of the form

$$y = [c_0 + c_1 x + c_2 x^2 + \cdots + c_{k-1} x^{k-1}] e^{\alpha x} + c_{k+1} e^{\alpha_{k+1} x} + \cdots + c_n e^{\alpha_n x}. \tag{11.390}$$

Case III: If there are more than one multiple root, then for each multiple root, we add a term similar to Eq. (11.389) for its place in the general solution.

Case IV: If the characteristic equation has complex conjugate roots as $a \pm ib$, then we add

$$y = e^{ax}(c_1 \sin bx + c_2 \cos bx) \tag{11.391}$$

for their place in the general solution.

Case V: If any of the complex conjugate roots are l-fold, then the corresponding part of the general solution is written as

$$y = e^{ax}[(c_0 + c_1 x + \cdots + c_{l-1} x^{l-1}) \sin bx + (c_l + c_{l+1} x + \cdots + c_{2l-1} x^{l-1}) \cos bx]. \tag{11.392}$$

11.4.2   With Variable Coefficients

In general, for nth-order linear homogeneous differential equations with variable or constant coefficients, there are n linearly independent solutions:

$$\{f_1, f_2, \ldots, f_n\}, \tag{11.393}$$

which forms the **fundamental set** of solutions. The general solution is given as their linear combination:

$$y = c_1 f_1 + c_2 f_2 + \cdots + c_n f_n, \tag{11.394}$$

where c_i, $i = 1, \ldots, n$, are arbitrary integration constants. For a linearly independent set $\{f_1, f_2, \ldots, f_n\}$, the Wronskian:

$$W[f_1, f_2, \ldots, f_n \cdot] = \begin{vmatrix} f_1 & f_2 & \cdots & f_n \\ f_1' & f_2' & \cdots & f_n' \\ \vdots & \vdots & \ddots & \vdots \\ f_1^{(n-1)} & f_2^{(n-1)} & \cdots & f_n^{(n-1)} \end{vmatrix}, \tag{11.395}$$

is different from zero. Unlike the case of constant coefficients, for linear differential equations with variable coefficients, no general recipe can be given for the solution. However, if a special solution $v(x)$ is known, then the transformation

$$y(x) = v(x)u(x) \tag{11.396}$$

can be used to reduce the order of the differential equation by 1 to $(n-1)$ in the dependent variable $w = \frac{du}{dx}$.

11.4.3   Nonhomogeneous Equations

If y_{1p} and y_{2p} are the particular solutions of

$$a_0(x)\frac{d^n y}{dx^n} + a_1(x)\frac{d^{n-1} y}{dx^{n-1}} + \cdots + a_n(x)y = f_1(x), \tag{11.397}$$

$$a_0(x)\frac{d^n y}{dx^n} + a_1(x)\frac{d^{n-1} y}{dx^{n-1}} + \cdots + a_n(x)y = f_2(x), \tag{11.398}$$

respectively, then their **linear combination**:

$$\boxed{y = c_1 y_{1p} + c_2 y_{1p},} \tag{11.399}$$

is the **particular solution** of

$$\boxed{a_0(x)\frac{d^n y}{dx^n} + a_1(x)\frac{d^{n-1} y}{dx^{n-1}} + \cdots + a_n(x)y = c_1 f_1(x) + c_2 f_2(x),} \tag{11.400}$$

where c_1 and c_2 are constants. For the nonhomogeneous equation:

$$a_0(x)\frac{d^n y}{dx^n} + a_1(x)\frac{d^{n-1}y}{dx^{n-1}} + \cdots + a_n(x)y = f(x), \tag{11.401}$$

the general solution is the sum of the solution of the homogeneous equation, y_c, which is called the complementary solution, plus a particular solution, y_p, of the nonhomogeneous equation:

$$y = y_c + y_p. \tag{11.402}$$

To find a particular solution, the method of variation of parameters and the method of undetermined coefficients can still be used.

11.5 INITIAL VALUE PROBLEM AND UNIQUENESS OF THE SOLUTION

Consider the initial value problem:

$$y^{(n)} = f(x, y, y', \dots, y^{(n-1)}), \tag{11.403}$$

$$y_0 = y(x_0), \dots, y_0^{(n-1)} = y^{(n-1)}(x_0). \tag{11.404}$$

If the functions

$$f, \quad \frac{\partial f}{\partial y}, \quad \frac{\partial f}{\partial y'}, \dots, \quad \frac{\partial f}{\partial y^{(n-1)}} \tag{11.405}$$

are continuous in the neighborhood of the point $P(x_0, y_0, \dots, y_0^{(n-1)})$ as functions of the $n+1$ variables, then there exists precisely one solution satisfying the initial values

$$y_0 = y(x_0), \dots, \quad y_0^{(n-1)} = y^{(n-1)}(x_0). \tag{11.406}$$

11.6 SERIES SOLUTIONS: FROBENIUS METHOD

So far we have concentrated on analytic techniques, where the solutions of differential equations can be found in terms of elementary functions like exponentials and trigonometric functions. However, for a large class of physically interesting cases, this is not possible. In some cases, when the coefficients of the differential equation satisfies certain conditions, a uniformly convergent series solution may be found.

A second-order linear homogeneous ordinary differential equation with two linearly independent solutions can be written as

$$\boxed{\frac{d^2 y}{dx^2} + P(x)\frac{dy}{dx} + Q(x)y = 0.} \tag{11.407}$$

When x_0 is no worse than a **regular singular point**, that is, if

$$\lim_{x \to x_0} (x - x_0)P(x) \to \text{finite} \tag{11.408}$$

and

$$\lim_{x \to x_0} (x - x_0)^2 Q(x) \to \text{finite}, \tag{11.409}$$

we can seek a series solution of the form

$$\boxed{y(x) = \sum_{k=0}^{\infty} a_k(x - x_0)^{k+\alpha}, \quad a_0 \neq 0.} \tag{11.410}$$

Substituting the series in Eq. (11.410) into the differential equation [Eq. (11.407)], we obtain an equation that in general looks like

$$\sum_{i=0}^{\infty} b_i(a_i, a_{i-1}, a_{i+1}, \alpha)(x - x_0)^{i+\alpha-2} = 0. \tag{11.411}$$

Usually, the first couple of terms go as

$$b_0(a_0, \alpha)(x - x_0)^{\alpha-2} + b_1(a_0, a_1, \alpha)(x - x_0)^{\alpha-1}$$
$$+ b_2(a_2, a_1, a_3, \alpha)(x - x_0)^{\alpha} + \cdots = 0. \tag{11.412}$$

For this equation to be satisfied for all x in our region of interest, all the coefficients, $b_n(a_i, a_{i-1}, a_{i-2}, \alpha)$, must vanish. Setting the coefficient of the lowest power of $(x - x_0)$ with the assumption $a_0 \neq 0$ gives us a quadratic equation for α, which is called the **indicial equation**. For almost all of the physically relevant cases, the indicial equation has two real roots. Setting the coefficients of the remaining powers to zero gives a recursion relation for the remaining coefficients. Depending on the roots of the indicial equation, we have the following possibilities for the two linearly independent solutions of the differential equation [2]:

1. If the two roots differ, $\alpha_1 > \alpha_2$, by a noninteger, then the two linearly independent solutions are given as

$$y_1(x) = |x - x_0|^{\alpha_1} \sum_{k=0}^{\infty} a_k(x - x_0)^k, \quad a_0 \neq 0,$$
$$\tag{11.413}$$
$$y_2(x) = |x - x_0|^{\alpha_2} \sum_{k=0}^{\infty} b_k(x - x_0)^k, \quad b_0 \neq 0.$$

2. If $(\alpha_1 - \alpha_2) = N$, where N is a positive integer, then the two linearly independent solutions are given as

$$y_1(x) = |x - x_0|^{\alpha_1} \sum_{k=0}^{\infty} a_k(x - x_0)^k, \quad a_0 \neq 0,$$

$$y_2(x) = |x - x_0|^{\alpha_2} \sum_{k=0}^{\infty} b_k(x - x_0)^k + Cy_1(x)\ln|x - x_0|, \quad b_0 \neq 0.$$

$$(11.414)$$

The second solution contains a logarithmic singularity, where C is a constant that may or may not be zero. Sometimes α_2 will contain both solutions; hence, it is advisable to start with the smaller root with the hopes that it might provide the general solution.

3. If the indicial equation has a double root, $\alpha_1 = \alpha_2$, then the Frobenius method yields only one series solution. In this case, the two linearly independent solutions can be taken as

$$y(x, \alpha_1) \text{ and } \left.\frac{\partial y(x, \alpha)}{\partial \alpha}\right|_{\alpha=\alpha_1}, \quad (11.415)$$

where the second solution diverges logarithmically as $x \to x_0$. In the presence of a double root, the Frobenius method is usually modified by taking the two linearly independent solutions as

$$y_1(x) = |x - x_0|^{\alpha_1} \sum_{k=0}^{\infty} a_k(x - x_0)^k, \quad a_0 \neq 0,$$

$$(11.416)$$

$$y_2(x) = |x - x_0|^{\alpha_1+1} \sum_{k=0}^{\infty} b_k(x - x_0)^k + y_1(x)\ln|x - x_0|.$$

In all these cases, the general solution is written as

$$y(x) = A_1 y_1(x) + A_2 y_2(x), \quad (11.417)$$

where A_1 and A_2 are the integration constants.

Example 11.21. *A case with distinct roots:* Consider the differential equation

$$x^2 y'' + \left(\frac{x}{2}\right) y' + 2x^2 y = 0. \quad (11.418)$$

Using the Frobenius method, we try a series solution about the regular singular point $x_0 = 0$ as

$$y(x) = |x|^r \sum_{n=0}^{\infty} a_n x^n, \quad a_0 \neq 0. \quad (11.419)$$

Assuming that $x > 0$, we write

$$y(x) = \sum_{n=0}^{\infty} a_n x^{n+r}, \quad (11.420)$$

which gives the derivatives y' and y'':

$$y' = \sum_{n=0}^{\infty} (n+r)a_n x^{n+r-1}, \tag{11.421}$$

$$y'' = \sum_{n=0}^{\infty} (n+r)(n+r-1)a_n x^{n+r-2}. \tag{11.422}$$

Substituting y, y', and y'' into Eq. (11.418) gives

$$\sum_{n=0}^{\infty} (n+r)(n+r-1)a_n x^{n+r} + \frac{1}{2}\sum_{n=0}^{\infty}(n+r)a_n x^{n+r} + 2\sum_{n=0}^{\infty} a_n x^{n+r+2} = 0. \tag{11.423}$$

We express all the series in terms of x^{n+r}:

$$\sum_{n=0}^{\infty} (n+r)(n+r-1)a_n x^{n+r} + \frac{1}{2}\sum_{n=0}^{\infty}(n+r)a_n x^{n+r} + 2\sum_{n=2}^{\infty} a_{n-2} x^{n+r} = 0, \tag{11.424}$$

where we have made the variable change $n+2 \to n'$ in the last series and dropped primes at the end. To start all the series from $n=2$, we write the first two terms of the first two series explicitly:

$$\left[r(r-1) + \frac{r}{2}\right] a_0 x^r + \left[r(r+1) + \frac{(r+1)}{2}\right] a_1 x^{r+1}$$

$$+ \sum_{n=2}^{\infty} \left\{\left[(n+r)(n+r-1) + \frac{1}{2}(n+r)\right] a_n + 2a_{n-2}\right\} x^{n+r} = 0. \tag{11.425}$$

This equation can only be satisfied for all x, if and only if all the coefficients vanish, that is,

$$\left[r\left(r - \frac{1}{2}\right)\right] a_0 = 0, \tag{11.426}$$

$$\left[(r+1)\left(r + \frac{1}{2}\right)\right] a_1 = 0, \tag{11.427}$$

$$\left[(n+r)\left(n+r - \frac{1}{2}\right)\right] a_n + 2a_{n-2} = 0, \quad n \geq 2. \tag{11.428}$$

The coefficient of the first term [Eq. (11.426)] is the indicial equation and with the assumption $a_0 \neq 0$ gives the values of r as

$$r_1 = \frac{1}{2}, \quad r_2 = 0. \tag{11.429}$$

The second equation [Eq. (11.427)] gives $a_1 = 0$ for both values of r and finally, the third equation [Eq. (11.428)] gives the recursion relation:

$$a_n = -\frac{2a_{n-2}}{\left[(n+r)\left(n+r-\frac{1}{2}\right)\right]}, \quad n = 2, 3, \ldots . \tag{11.430}$$

We start with $r_1 = 1/2$, which gives the recursion relation

$$a_n = -\frac{2a_{n-2}}{n(n+\frac{1}{2})}, \quad n = 2, 3, \ldots, \tag{11.431}$$

and hence the coefficients

$$a_2 = -\frac{2a_0}{2(2+\frac{1}{2})} = -\frac{2a_0}{5},$$

$$a_3 = -\frac{2a_1}{3\left(\frac{7}{3}\right)} = 0,$$

$$a_4 = -\frac{2a_2}{4\left(\frac{9}{2}\right)} = \left(\frac{1}{9}\right)\left(\frac{2a_0}{5}\right), \tag{11.432}$$

$$a_5 = 0,$$

$$a_6 = \frac{4a_0}{1755},$$

$$\vdots$$

The first solution is now obtained as

$$y_1 = a_0 x^{\frac{1}{2}}\left[1 - \frac{2x^2}{5} + \frac{2x^4}{45} - \frac{4x^6}{1755} + \cdots\right]. \tag{11.433}$$

Similarly, for the other root, $r_2 = 0$, we obtain the recursion relation

$$a_n = -\frac{2a_{n-2}}{n(n-\frac{1}{2})}, \quad n = 2, 3, \ldots, \tag{11.434}$$

and the coefficients

$$a_1 = 0,$$

$$a_2 = -\frac{2a_0}{3},$$

$$a_3 = 0,$$

$$a_4 = \frac{2a_0}{21}, \tag{11.435}$$

$$a_5 = 0,$$

$$a_6 = -\frac{4a_0}{693},$$

$$\vdots$$

which gives the second solution as

$$y_2 = a_0 \left[1 - \frac{2x^2}{3} + \frac{2x^4}{21} - \frac{4x^6}{693} + \cdots \right]. \tag{11.436}$$

We can now write the general solution as the linear combination

$$y = c_1 x^{\frac{1}{2}} \left[1 - \frac{2x^2}{5} + \frac{2x^4}{45} - \frac{4x^6}{1755} + \cdots \right]$$
$$+ c_2 \left[1 - \frac{2x^2}{3} + \frac{2x^4}{21} - \frac{4x^6}{693} + \cdots \right]. \tag{11.437}$$

Example 11.22. *General expression of the nth term:* In the previous example, we have found the general solution in terms of infinite series. We now carry the solution one step further. That is, we write the general expression for the nth term. The first solution was given in terms of the even powers of x as

$$y_1 = x^{\frac{1}{2}} \left[1 - \frac{2x^2}{5} + \frac{2x^4}{45} - \frac{4x^6}{1755} + \cdots \right] = x^{1/2} \sum_{k=0}^{\infty} a_{2k} x^{2k}. \tag{11.438}$$

Since only the even terms are present, we use the recursion relation [Eq. (11.431)] to write the coefficient of the kth term in the above series as

$$a_{2k} = -\frac{2a_{2(k-1)}}{2k\left(2k+\frac{1}{2}\right)}, \quad k = 1, 2, \ldots . \tag{11.439}$$

We let $k \to k - 1$:

$$a_{2k-2} = -\frac{1}{(k-1)\left(2k-2+\frac{1}{2}\right)} a_{2k-4} \tag{11.440}$$

and use the result back in Eq. (11.439) to write

$$a_{2k} = \frac{1}{k\left(2k+\frac{1}{2}\right)} \frac{1}{(k-1)\left(2k-2+\frac{1}{2}\right)} a_{2k-4}$$
$$= \frac{1}{k(k-1)\left(2k+\frac{1}{2}\right)\left(2k-2+\frac{1}{2}\right)} a_{2(k-2)}. \tag{11.441}$$

We iterate once more. First let $k \to k - 1$ in the above equation:

$$a_{2k-2} = \frac{1}{(k-1)(k-2)\left(2k-2+\frac{1}{2}\right)\left(2k-4+\frac{1}{2}\right)} a_{2(k-3)}, \qquad (11.442)$$

and then substitute the result back into Eq. (11.439) to write

$$a_{2k} = -\frac{1}{k\left(2k+\frac{1}{2}\right)(k-1)(k-2)\left(2k-2+\frac{1}{2}\right)\left(2k-4+\frac{1}{2}\right)} a_{2(k-3)}$$

$$= -\frac{1}{2.2.2.k(k-1)(k-2)\left(k+\frac{1}{4}\right)\left(k-1+\frac{1}{4}\right)\left(k-2+\frac{1}{4}\right)} a_{2(k-3)}$$

$$= -\frac{1}{2^3 k(k-1)(k-2)\left(k+\frac{1}{4}\right)\left(k-1+\frac{1}{4}\right)\left(k-2+\frac{1}{4}\right)} a_{2(k-3)}. \qquad (11.443)$$

After k iterations, we hit a_0. Setting $a_0 = 1$, we obtain

$$a_{2k} = \frac{(-1)^k}{2^k \left[k(k-1)(k-2)\cdots2.1\right]\left(k+\frac{1}{4}\right)\left(k-1+\frac{1}{4}\right)\left(k-2+\frac{1}{4}\right)\cdots\left(1+\frac{1}{4}\right)}. \qquad (11.444)$$

We now use the gamma function:

$$\Gamma(x+1) = x\Gamma(x), \quad x > 0, \quad \Gamma(1) = 1, \qquad (11.445)$$

which allows us to extend the definition of factorial to continuous and fractional integers:

$$\Gamma(n+1) = n\Gamma(n)$$
$$= n(n-1)\Gamma(n-1)$$
$$= n(n-1)(n-2)\Gamma(n-2)$$
$$\vdots$$
$$= n(n-1)(n-2)\cdots(n-k)\Gamma(n-k). \qquad (11.446)$$

This can also be written as

$$n(n-1)(n-2)\cdots(n-k) = \frac{\Gamma(n+1)}{\Gamma(n-k)}, \quad n-k > 0. \qquad (11.447)$$

Using the above formula, we can write

$$\left(k+\frac{1}{4}\right)\left(k-1+\frac{1}{4}\right)\left(k-2+\frac{1}{4}\right)\cdots\left(1+\frac{1}{4}\right) = \frac{\Gamma\left(k+\frac{1}{4}+1\right)}{\Gamma\left(1+\frac{1}{4}\right)}$$

$$= \frac{\Gamma(k+\frac{5}{4})}{\Gamma\left(\frac{5}{4}\right)}. \qquad (11.448)$$

Substituting Eq. (11.448) and $k(k-1)(k-2)\cdots 2\cdot 1 = k!$ into Eq. (11.444), we write a_{2k} as

$$a_{2k} = \frac{(-1)^k \Gamma(5/4)}{2^k k! \Gamma(k+5/4)}, \tag{11.449}$$

which allows us to express the first solution [Eq. (11.438)] in the following compact form:

$$y_1(x) = x^{1/2} \sum_{k=0}^{\infty} \frac{(-1)^k \Gamma(5/4)}{2^k k! \Gamma(k+5/4)} x^{2k}. \tag{11.450}$$

Following similar steps, we can write $y_2(x)$ as

$$y_2(x) = \sum_{k=0}^{\infty} \frac{(-1)^k \Gamma(3/4)}{2^k k! \Gamma(k+3/4)} x^{2k}. \tag{11.451}$$

Example 11.23. *When the roots differ by an integer:* Consider the differential equation

$$x^3 \frac{d^2 y}{dx^2} + 3x^2 \frac{dy}{dx} + \left(x^3 + \frac{3}{4}x\right) y(x) = 0, \quad x \geq 0. \tag{11.452}$$

Since $x_0 = 0$ is a regular singular point, we try a series solution

$$y = \sum_{n=0}^{\infty} a_n x^{n+r}, \quad a_0 \neq 0, \tag{11.453}$$

where the derivatives are given as

$$y' = \sum_{n=0}^{\infty} (n+r) a_n x^{n+r-1}, \tag{11.454}$$

$$y'' = \sum_{n=0}^{\infty} (n+r)(n+r-1) a_n x^{n+r-2}. \tag{11.455}$$

Substituting these into the differential equation [Eq. (11.452)] and rearranging terms as in the previous example, we obtain

$$\left[r(r+2) + \frac{3}{4}\right] a_0 x^{r+1} + \left[(r+1)(r+3) + \frac{3}{4}\right] a_1 x^{r+2}$$

$$+ \sum_{n=2}^{\infty} \left\{\left[(n+r)(n+r+2) + \frac{3}{4}\right] a_n + a_{n-2}\right\} x^{n+r+1} = 0. \tag{11.456}$$

Once again, we set the coefficients of all the powers of x to zero:

$$\left[r(r+2) + \frac{3}{4} \right] a_0 = 0, \quad a_0 \neq 0, \tag{11.457}$$

$$\left[(r+1)(r+3) + \frac{3}{4} \right] a_1 = 0, \tag{11.458}$$

$$\left[(n+r)(n+r+2) + \frac{3}{4} \right] a_n + a_{n-2} = 0, \quad n \geq 2. \tag{11.459}$$

The first equation [Eq. (11.457)] is the indicial equation and with the assumption $a_0 \neq 0$ gives the values of r as

$$r_1 = -\frac{1}{2}, \quad r_2 = -\frac{3}{2}. \tag{11.460}$$

Let us start with the first root, $r_1 = -1/2$. The second equation [Eq. (11.458)] gives

$$\left[\left(-\frac{1}{2} + 1 \right) \left(-\frac{1}{2} + 3 \right) + \frac{3}{4} \right] a_1 = 0, \tag{11.461}$$

$$\left[\frac{8}{4} \right] a_1 = 0, \tag{11.462}$$

$$a_1 = 0. \tag{11.463}$$

The remaining coefficients are obtained from the recursion relation:

$$a_n = -\frac{1}{\left[\left(n - \frac{1}{2} \right) \left(n + \frac{3}{2} \right) + \frac{3}{4} \right]} a_{n-2}, \quad n \geq 2, \tag{11.464}$$

$$= -\frac{4a_{n-2}}{(2n-1)(2n+3)+3}, \quad n \geq 2. \tag{11.465}$$

All the odd terms are zero, and the nonzero terms are obtained as

$$a_2 = -\frac{a_0}{6}, \quad a_4 = \frac{a_0}{120}, \ldots, \quad a_{2n} = (-1)^n \frac{a_0}{(2n+1)!}, \ldots \tag{11.466}$$

Hence, the solution corresponding to $r = -1/2$ is obtained as

$$y_1 = a_0 x^{-1/2} \sum_{n=0}^{\infty} \frac{(-1)^n}{(2n+1)!} x^{2n}. \tag{11.467}$$

We can write this as

$$y_1 = a_0 x^{-3/2} \sum_{n=0}^{\infty} \frac{(-1)^n}{(2n+1)!} x^{2n+1} \tag{11.468}$$

$$= a_0 x^{-3/2} \sin x. \tag{11.469}$$

For the other root, $r_2 = -3/2$, the second equation [Eq. (11.458)] becomes

$$\left[-\frac{1}{2}\left(\frac{3}{2}\right) + \frac{3}{4} \right] a_1 = 0, \tag{11.470}$$

$$0 \cdot a_1 = 0, \tag{11.471}$$

thus $a_1 \neq 0$. The remaining coefficients are given by the recursion relation

$$a_n = -\frac{4a_{n-2}}{(2n-3)(2n+1)+3}, \quad n \geq 2, \tag{11.472}$$

as

$$a_0 \neq 0, \quad a_1 \neq 0, \quad a_2 = -\frac{a_0}{2}, \quad a_3 = -\frac{a_1}{6}, \quad a_4 = \frac{a_0}{24}, \quad a_5 = \frac{a_1}{120}, \ldots \tag{11.473}$$

Now the solution for $r_2 = -3/2$ becomes

$$y = a_0 x^{-3/2}\left(1 - \frac{1}{2}x^2 + \frac{1}{24}x^4 - \cdots \right)$$

$$+ a_1 x^{-3/2}\left(x - \frac{1}{6}x^3 + \frac{1}{120}x^5 - \cdots \right). \tag{11.474}$$

We recognize that the first series in Eq. (11.474) is nothing but $\cos x$ and that the second series is $\sin x$; hence, we write this solution as

$$y = a_0 x^{-3/2} \cos x + a_1 x^{-3/2} \sin x. \tag{11.475}$$

However, this solution may also contain a logarithmic singularity of the form $y_1(x)\ln|x|$:

$$y = a_0 x^{-3/2} \cos x + a_1 x^{-3/2} \sin x + C y_1(x) \ln|x|. \tag{11.476}$$

Substituting this back into Eq. (11.452), we get

$$C(2x^{1/2} \cos x - x^{-1/2} \sin x) = 0. \tag{11.477}$$

The quantity inside the brackets is in general different from zero; hence, we set C to zero. Since Eq. (11.475) also contains the solution obtained for $r_1 = -\frac{1}{2}$, we write the general solution as

$$y = c_0 x^{-3/2} \cos x + c_1 x^{-3/2} \sin x, \tag{11.478}$$

where c_0 and c_1 are arbitrary constants. Notice that in this case the difference between the roots is an integer:

$$r_1 - r_2 = -\frac{1}{2} - \left(-\frac{3}{2}\right) = 1, \tag{11.479}$$

Since it also contains the other solution, starting with the smaller root would have saved us some time. This is not true in general. However, when the roots differ by an integer, it is always advisable to start with the smaller root hoping that it yields both solutions.

11.6.1 Frobenius Method and First-order Equations

It is possible to use Frobenius method for a certain class of first-order differential equations that could be written as

$$y' + p(x)y = 0. \tag{11.480}$$

A singular point of the differential equation, x_0, is now regular, if it satisfies

$$(x - x_0)p(x_0) \to \text{finite.} \tag{11.481}$$

Let us demonstrate how the method works with the following differential equation:

$$xy' + (1 - x)y = 0. \tag{11.482}$$

Obviously, $x_0 = 0$ is a regular singular point. Hence, we substitute

$$y = \sum_{n=0}^{\infty} a_n x^{n+r}, \quad a_0 \neq 0, \tag{11.483}$$

and its derivative:

$$y' = \sum_{n=0}^{\infty} a_n (n + r) x^{n+r-1}, \tag{11.484}$$

into Eq. (11.482) to write

$$\sum_{n=0}^{\infty} a_n (n + r) x^{n+r} + \sum_{n=0}^{\infty} a_n x^{n+r} - \sum_{n=0}^{\infty} a_n x^{n+r+1} = 0. \tag{11.485}$$

Renaming the dummy variable in the first two series as $n \to n' + 1$ and then dropping primes, we obtain

$$\sum_{n=-1}^{\infty} a_{n+1}(n + r + 1)x^{n+r+1} + \sum_{n=-1}^{\infty} a_{n+1} x^{n+r+1} - \sum_{n=0}^{\infty} a_n x^{n+r+1} = 0, \tag{11.486}$$

which after rearranging becomes

$$(r + 1)x^r a_0 + \sum_{n=0}^{\infty} [(n + r + 2)a_{n+1} - a_n]x^{n+r+1} = 0. \tag{11.487}$$

Indicial equation is obtained by setting the coefficient of the term with the lowest power of x to zero:

$$(1 + r)a_0 = 0, \quad a_0 \neq 0, \tag{11.488}$$

which gives $r = -1$. Using this in the recursion relation:

$$a_{n+1} = \frac{a_n}{(n + r + 2)}, \quad n \geq 0, \tag{11.489}$$

we write

$$a_{n+1} = \frac{a_n}{(n + 1)}, \quad n \geq 0, \tag{11.490}$$

and obtain the series solution as

$$y = \frac{a_0}{x}\left(1 + x + \frac{x^2}{2!} + \frac{x^3}{3!} + \cdots + \frac{x^k}{k!} + \cdots\right) = \frac{a_0}{x}e^x. \tag{11.491}$$

This can be justified by direct integration of the differential equation. Applications of the Frobenius method to differential equations of higher than second order can be found in Ince [3].

REFERENCES

1. Murphy, G.M. (1960). *Ordinary Differential Equations and Their Solutions*. Princeton, NJ: Van Nostrand.
2. Ross, S.L. (1984). *Differential Equations*, 3e. New York: Wiley.
3. Ince, E.L. (1958). *Ordinary Differential Equations*. New York: Dover Publications.

PROBLEMS

1. Classify the following differential equations:

(i) $\dfrac{dy}{dx} + x^2 y^2 = 5xe^x.$

(ii) $\dfrac{d^3 y}{dx^3} + x^2 y^2 = 0.$

(iii) $\dfrac{d^4 y}{dx^4} + x\dfrac{d^3 y}{dx^3} - x^2 y = 0.$

(iv) $\dfrac{dy}{dx} + x^2 y = 5.$

(v) $\dfrac{\partial^2 u}{\partial x^2} + \dfrac{\partial^2 u}{\partial y^2} + \dfrac{\partial^2 u}{\partial z^2} = f(x, y, z).$

(vi) $\dfrac{dr}{ds} = \sqrt[3]{1 + \dfrac{d^2 r}{ds^2}}.$

(vii) $\dfrac{\partial^2 u}{\partial x^2} + \dfrac{\partial^2 u}{\partial y \partial x} + \dfrac{\partial^2 u}{\partial z^2} = 0.$

(viii) $\dfrac{d^4 y}{dx^4} + 4\dfrac{d^3 y}{dx^3} - 7y = 0.$

(ix) $\dfrac{d^4 y}{dx^4} + 3\dfrac{d^2 y}{dx^2} - \left(\dfrac{dy}{dx}\right)^2 = 0.$

(x) $x^3 dx + y^2 dy = 0.$

2. Show that the function

$$y(x) = (8x^3 + C)e^{-6x}$$

satisfies the following differential equation:

$$\frac{dy}{dx} = -6y + 24x^2 e^{-6x}.$$

3. Show that the function

$$y(x) = 2 + Ce^{-8x^2}$$

is the solution of the differential equation

$$\frac{dy}{dx} + 16xy = 32x.$$

4. Find the solution of

$$y' = \frac{2x - y + 9}{x - 3y + 2}.$$

5. Given the differential equation

$$\frac{dy}{dx} + 2y = 8xe^{-2x},$$

show that its general solution is given as

$$y = (4x^2 + C)e^{-2x}.$$

6. Show that [Eq. (11.70)]

$$y(x) = \left[\int dx\, b(x)e^{\int a(x)\, dx} \right] e^{-\int a(x)\, dx} + C_0$$

satisfies the differential equation

$$y' + a(x)y = b(x).$$

7. Solve the following initial value problem:

$$\frac{dy}{dx} + 2yx = 16x^3 e^{-x^2}, \quad y(0) = 2.$$

8. Show that the following initial value problems have unique solutions:

$$\text{(i)} \quad \frac{dy}{dx} = 2x^2 \sin y, \qquad y(1) = 2.$$

$$\text{(ii)} \quad \frac{dy}{dx} = \frac{2y^2}{x-2}, \qquad y(1) = 0.$$

$$\text{(iii)} \quad \frac{dy}{dx} + 2y = 8xe^{-2x}, \quad y(0) = 2.$$

9. Which one of the following equations are exact:

$$\text{(i)} \quad (3x + 4y)\, dx + (4x + 4y)\, dy = 0.$$
$$\text{(ii)} \quad (2xy + 2)\, dx + (x^2 + 4y)\, dy = 0.$$
$$\text{(iii)} \quad (y^2 + 1)\cos x\, dx + 2y(\sin x)\, dy = 0.$$
$$\text{(iv)} \quad (3x + 2y)\, dx + (2x + y)\, dy = 0.$$
$$\text{(v)} \quad (4xy - 3)\, dx + (x^2 + 4y)\, dy.$$

10. Solve the following initial value problems:

$$\text{(i)} \quad (8xy - 6)\, dx + (4x^2 + 4y)\, dy = 0, \qquad\qquad y(2) = 2.$$
$$\text{(ii)} \quad (2ye^x + 2e^x + 4y^2)\, dx + 2(e^x + 4xy)\, dy = 0, \quad y(0) = 3.$$
$$\text{(iii)} \quad \frac{dy}{dx} = \frac{2y - 2}{-3 + 2y - 2x}, \qquad\qquad y(-1) = 2.$$

11. Solve the following first-order differential equations:

$$\text{(i)} \quad 16xy\, dx + (4x^2 + 1)\, dy = 0.$$
$$\text{(ii)} \quad x(4y^2 + 1)\, dx + (x^4 + 1)\, dy = 0.$$
$$\text{(iii)} \quad \tan\theta\, dr + 2r\, d\theta = 0.$$
$$\text{(iv)} \quad (x + 2y)\, dx - 2x\, dy = 0.$$
$$\text{(v)} \quad (2x^2 + 2xy + y^2)\, dx + (x^2 + 2xy - y^2)\, dy = 0.$$

12. Find the conditions for which the following equations are exact:

$$\text{(i)} \quad (A_0x + A_1y)\, dx + (B_0x + B_1y)\, dy = 0.$$

$$\text{(ii)} \quad (A_0x^2 + A_1xy + A_2y^2)\, dx + (B_0x^2 + B_1xy + B_2y^2)\, dy = 0.$$

13. Solve the following equations:

$$\text{(i)} \quad y' + (3/x)y = 48x^2.$$
$$\text{(ii)} \quad xy' + [(4x + 1)/(2x + 1)]y = 2x - 1.$$
$$\text{(iii)} \quad y' - (1/x)y = -(1/x^2)y^2.$$
$$\text{(iv)} \quad 2xy' - 4y = 2x^4.$$
$$\text{(v)} \quad 2y' + (8y - 1/y^3)x = 0.$$

14. The generalized Riccati equation is defined as

$$y'(x) = f(x) + g(x)y + h(x)y^2.$$

Using the transformation $y(x) = y_1 + u$, show that the generalized Riccati equation reduces to the Bernoulli equation:

$$u'(x) = (g + 2hy_1)u + hu^2.$$

15. If an equation, $M(x,y) \, dx + N(x,y) \, dy = 0$, is exact and if $M(x,y)$ and $N(x,y)$ are homogeneous of the same degree different from -1, then show that the solution can be written without the need for a quadrature as

$$xM(x,y) + yN(x,y) = 0.$$

16. If there are two independent integrating factors, I_1 and I_2, that are not constant multiples of each other, then show that the general solution of $M(x,y) \, dx + N(x,y) \, dy = 0$ can be written as

$$I_1(x,y) = CI_2(x,y).$$

17. Solve by finding an integrating factor:

    (i) $(10xy + 16y^2 + 1) \, dx + (2x^2 + 8xy) \, dy = 0.$

    (ii) $(8xy^2 + 2y) \, dx + (2y^3 - 2x) \, dy = 0.$

18. Show that the general solution of

$$(D - r)^2 y(x) = 0, \quad D = \frac{d}{dx},$$

is given as $y(x) = c_0 e^{rx} + c_1 x e^{rx}$.

19. Which one of the following sets are linearly independent:

    (i) $\{e^x, e^{2x}, e^{-2x}\}$.

    (ii) $\{x, x^2, x^3\}$.

    (iii) $\{x, 2x, e^{3x}\}$.

    (iv) $\{\sin x, \cos x, \sin 2x\}$.

    (v) $\{e^x, xe^x, x^2 e^x\}$.

20. Given that $y = 2x$ is a solution, find the general solution of

$$\left(x^2 + \frac{1}{4}\right)\frac{d^2 y}{dx^2} - 2x\frac{dy}{dx} + 2y = 0$$

by reduction of order.

21. Given the solution $y = 2x + 1$, solve

$$\left(x + \frac{1}{2}\right)^2 \frac{d^2y}{dx^2} - \left(3x + \frac{3}{2}\right)\frac{dy}{dx} + 3y = 0.$$

22. Find the general solution for

    (i) $y'' + 4y = 0.$

    (ii) $y'' + 2y' + 3y = 0.$

    (iii) $y'' - 2y' + 15y = 0.$

    (iv) $y^{(iv)} + 6y'' + 9y = 0.$

    (v) $y''' - 2y'' + y' - 2y = 0.$

23. Verify that the expression

$$y(x) = y_1(x)\left[C_1 \int \frac{1}{y_1^2} e^{-\int P(x)\,dx}\,dx + C_2\right]$$

satisfies the differential equation

$$y'' + P(x)y' + Q(x)y = 0,$$

where y_1 is a special solution.

24. Use the method of undetermined coefficients to write the general solution of

$$y'' + y = c_0 \sin x + c_1 \cos x.$$

25. Show that a particular solution of the nonhomogeneous equation

$$\frac{d^2y}{dx^2} + n^2y = n^2 f(x)$$

is given as

$$y = n \sin nx \int_0^x f(x) \cos nx \, dx - n \cos nx \int_0^x f(x) \sin nx \, dx.$$

26. Use the method of undetermined coefficients to solve

    (i) $2y'' - 5y' - 3y = x^2 e^x.$

    (ii) $y'' - 2y' - 3y = \sin x.$

    (iii) $y'' - 2y' - 3y = xe^x.$

    (iv) $y'' - 2y' - 3y = 2\sin x + 5xe^x.$

    (v) $y^{(iv)} + y' = x^2 + \sin x + e^x.$

27. Show that the transformation $x = e^t$ reduces the nth-order Cauchy-Euler differential equation,

$$a_0 x^n \frac{d^n y}{dx^n} + a_1 x^{n-1} \frac{d^{n-1} y}{dx^{n-1}} + \cdots + a_n y = F(x),$$

into a linear differential equation with constant coefficients.
Hint: First show for $n = 2$ and then generalize.

28. Find the general solution of

$$x^2 \frac{d^2 y}{dx^2} + x \frac{dy}{dx} + n^2 y = x^m.$$

29. Solve the following Cauchy-Euler equations:

(i) $\quad 2x^2 \dfrac{d^2 y}{dx^2} - 5x \dfrac{dy}{dx} - 3y = 0.$

(ii) $\quad x^2 \dfrac{d^2 y}{dx^2} - 2x \dfrac{dy}{dx} - 3y = 0.$

(iii) $\quad 2x^2 \dfrac{d^2 y}{dx^2} + 3x \dfrac{dy}{dx} + y = 0.$

(iv) $\quad x^2 \dfrac{d^2 y}{dx^2} + x \dfrac{dy}{dx} - 6y = 0.$

(v) $\quad x^3 \dfrac{d^3 y}{dx^3} - x^2 \dfrac{d^2 y}{dx^2} + 2x \dfrac{dy}{dx} - 2y = \ln x.$

(vi) $\quad 2x^2 \dfrac{d^2 y}{dx^2} - 5x \dfrac{dy}{dx} + y = x^3.$

30. Classify the singular points for the following differential equations:

(i) $\quad (x^2 + 3x - 9)y'' + (x + 2)y' - 3x^2 y = 0.$
(ii) $\quad x^3(1 - x)y'' + x^2 \sin x y' - 3xy = 0.$
(iii) $\quad (x^2 - 1)y'' + 2xy' + y = 0.$
(iv) $\quad (x^2 - 4)y'' + (x + 2)y' + 4y = 0.$
(v) $\quad (x^2 - x)y'' + (x - 1)y' - 6y = 0.$
(vi) $\quad x^2 y'' - 2xy' + 2y = 0.$

31. Find series solutions about $x = 0$ and for $x > 0$ for the following differential equations:

(i) $\quad x(x^3 - 1)y'' - (3x^3 - 2)y' + 2x^2 y = 0.$
(ii) $\quad xy'' + 4y' - 4y = 0.$
(iii) $\quad (4x^2 + 1)y'' + 4xy' + 16xy = 0, \ y(0) = 2, \ y'(0) = 6.$
(iv) $\quad 3xy'' + (2 - x)y' - y = 0.$
(v) $\quad xy'' + 4y' - 4xy = 0.$

Discuss the convergence of the series solutions you have found.

32. Find the general expression for the series in Example 11.21:

$$y_2(x) = 1 - \frac{2x^2}{3} + \frac{2x^4}{21} - \frac{4x^6}{693} + \cdots ,$$

where the recursion relation is given as

$$a_n = -\frac{2a_{n-2}}{n\left(n - \frac{1}{2}\right)}, \quad n = 2, 3, \ldots.$$

CHAPTER 12

SECOND-ORDER DIFFERENTIAL EQUATIONS AND SPECIAL FUNCTIONS

In applications, differential equations are usually accompanied by boundary or initial conditions. In the previous chapter, we have concentrated on techniques for finding analytic solutions to ordinary differential equations. We have basically assumed that boundary conditions can in principle be satisfied by a suitable choice of the integration constants in the solution. The general solution of a second-order ordinary differential equation contains two arbitrary constants, which requires two boundary conditions for their determination. The needed information is usually supplied either by giving the value of the solution and its derivative at some point or by giving the value of the solution at two different points. As in chaotic processes, where the system exhibits instabilities with respect to initial conditions, the effect of boundary conditions on the final result can be drastic. In this chapter, we discuss three of the most frequently encountered second-order ordinary differential equations of physics, that is, Legendre, Laguerre, and Hermite equations. We approach these equations from the point of view of the Frobenius method and discuss their solutions in detail. We show that the boundary conditions impose severe restrictions on not just the integration constants but also on the parameters that the differential equation itself includes. Restrictions on such parameters may have rather dramatic effects like the quantization of energy and angular momentum in physical theories.

12.1 LEGENDRE EQUATION

Legendre equation is defined as

$$\boxed{(1-x^2)\frac{d^2y}{dx^2} - 2x\frac{dy}{dx} + ky = 0,} \tag{12.1}$$

where k is a constant parameter with no prior restrictions. In applications, k is related to physical properties like angular momentum, frequency, etc. Range of the independent variable x is $[-1, 1]$.

12.1.1 Series Solution

Legendre equation has two regular singular points at the end points $x = \pm 1$. Since $x = 0$ is a regular point, we use the Frobenius method and try a series solution of the form

$$y(x) = \sum_{n=0}^{\infty} a_n x^{n+s}, \quad a_0 \neq 0. \tag{12.2}$$

Our goal is to find a finite solution in the entire interval $[-1, 1]$. Substituting Eq. (12.2) and its derivatives:

$$y' = \sum_{n=0}^{\infty} a_n(n+s)x^{n+s-1}, \quad y'' = \sum_{n=0}^{\infty} a_n(n+s)(n+s-1)x^{n+s-2}, \tag{12.3}$$

into the Legendre equation, we obtain

$$\sum_{n=0}^{\infty} a_n(n+s)(n+s-1)x^{n+s-2} - \sum_{n=0}^{\infty} a_n(n+s)(n+s-1)x^{n+s}$$

$$-2\sum_{n=0}^{\infty} a_n(n+s)x^{n+s} + k\sum_{n=0}^{\infty} a_n x^{n+s} = 0, \tag{12.4}$$

$$\sum_{n=0}^{\infty} a_n(n+s)(n+s-1)x^{n+s-2}$$

$$+ \sum_{n=0}^{\infty} a_n[-(n+s)(n+s-1) - 2(n+s) + k]x^{n+s} = 0. \tag{12.5}$$

To equate the powers of x, we substitute $n - 2 = n'$ into the first series and drop primes to write

$$\sum_{n=-2}^{\infty} a_{n+2}(n + s + 2)(n + s + 1)x^{n+s}$$

$$+ \sum_{n=0}^{\infty} a_n[-(n + s)(n + s + 1) + k]x^{n+s} = 0. \tag{12.6}$$

Writing the first two terms of the first series explicitly, we get

$$a_0(-2 + s + 2)(-2 + s + 1)x^{s-2} + a_1(-1 + s + 2)(-1 + s + 1)x^{s-1}$$

$$+ \sum_{n=0}^{\infty} \{a_{n+2}(n + s + 2)(n + s + 1) + a_n[-(n + s)(n + s + 1) + k]\}x^{n+s} = 0. \tag{12.7}$$

Since this equation can be satisfied for all x only when the coefficients of all powers of x vanish simultaneously, we write

$$a_0 s(s - 1) = 0, \quad a_0 \neq 0, \tag{12.8}$$

$$a_1 s(s + 1) = 0, \tag{12.9}$$

$$a_{n+2} = -a_n \frac{-(n + s)(n + s + 1) + k}{(n + s + 2)(n + s + 1)}, \quad n = 0, 1, 2, \dots . \tag{12.10}$$

The first equation [Eq. (12.8)] is the **indicial equation** and its roots give the values of s as

$$s_1 = 0 \text{ and } s_2 = 1. \tag{12.11}$$

Starting with the second root, $s = 1$, Eq. (12.9) gives $a_1 = 0$, thus from Eq. (12.10), we obtain the **recursion relation** for the remaining coefficients:

$$\boxed{a_{n+2} = a_n \frac{(n + 1)(n + 2) - k}{(n + 2)(n + 3)}, \quad n = 0, 1, 2, \dots .} \tag{12.12}$$

We can now write the coefficients as

$$a_2 = \frac{1.2 - k}{2.3} a_0, \tag{12.13}$$

$$a_3 = \frac{2.3 - k}{3.4} a_1 = 0, \tag{12.14}$$

$$a_4 = \frac{3.4 - k}{4.5} a_2 = \left[\frac{1.2 - k}{2.3}\right] \left[\frac{3.4 - k}{4.5}\right] a_0, \tag{12.15}$$

$$a_5 = \left[\frac{2.3 - k}{3.4}\right]\left[\frac{4.5 - k}{5.6}\right]a_1 = 0, \tag{12.16}$$

$$\vdots$$

and the series solution becomes

$$y(x) = a_0 x \left[1 + \frac{2 - k}{2.3}x^2 + \left[\frac{1.2 - k}{2.3}\right]\left[\frac{3.4 - k}{4.5}\right]x^4 + \cdots\right]. \tag{12.17}$$

For the time being, we leave this solution aside and continue with the other root, $s = 0$. The recursion relation and the coefficients now become

$$a_{n+2} = a_n \frac{n(n+1) - k}{(n+1)(n+2)}, \quad n = 0, 1, 2, \ldots, \tag{12.18}$$

$$a_0 \neq 0, \tag{12.19}$$

$$a_1 \neq 0, \tag{12.20}$$

$$a_2 = \frac{-k}{1.2}a_0, \tag{12.21}$$

$$a_3 = \frac{1.2 - k}{2.3}a_1, \tag{12.22}$$

$$a_4 = \frac{2.3 - k}{3.4}a_2 = \left[\frac{-k}{1.2}\right]\left[\frac{2.3 - k}{3.4}\right]a_0, \tag{12.23}$$

$$a_5 = \left[\frac{1.2 - k}{2.3}\right]\left[\frac{3.4 - k}{4.5}\right]a_1, \tag{12.24}$$

$$\vdots$$

This gives the series solution

$$y(x) = a_0 \left[1 - \frac{k}{1.2}x^2 - \frac{k(2.3 - k)}{1.2.3.4}x^4 + \cdots\right]$$

$$+ a_1 \left[x + \frac{1.2 - k}{1.2.3}x^3 + \left[\frac{1.2 - k}{1.2.3}\right]\left[\frac{3.4 - k}{4.5}\right]x^5 + \cdots\right] \tag{12.25}$$

$$= a_0 y_1(x) + a_1 y_2(x), \tag{12.26}$$

Note that this solution contains the previous solution [Eq. (12.17)]. Hence, we can take it as the general solution of Legendre equation, where y_1 and y_2 are the two linearly independent solutions and the coefficients a_0 and a_1 are the integration constants. In the Frobenius method, when the roots of the indicial equation differ by an integer, it is always advisable to start with the smaller root with the hopes that it will yield both solutions.

12.1.2   Effect of Boundary Conditions

To check the convergence of these series, we write Eq. (12.25) as

$$y(x) = a_0 \sum c_{2n} x^{2n} + a_1 \sum c_{2n+1} x^{2n+1} \tag{12.27}$$

and consider only the first series with the even powers. Applying the ratio test with the general term $u_{2n} = c_{2n} x^{2n}$ and the recursion relation

$$c_{2n+2} = \frac{2n(2n+1) - k}{(2n+1)(2n+2)} c_{2n}, \quad n = 0, 1, 2, \ldots, \tag{12.28}$$

we obtain

$$\lim_{n \to \infty} \left| \frac{u_{2(n+1)}}{u_{2n}} \right| = \left| \frac{c_{2n+2}}{c_{2n}} \frac{x^{2n+2}}{x^{2n}} \right| = \left| \frac{2n(2n+1) - k}{(2n+1)(2n+2)} x^2 \right| \to |x|^2. \tag{12.29}$$

For convergence, we need this limit to be less than 1. This means that the series converges for the interior of the interval $[-1, 1]$, that is, for $|x| < 1$. For the end points at $x = \pm 1$, the ratio test is inconclusive. We now examine the large n behavior of the series. Since $\lim_{n \to \infty} c_{2n+2}/c_{2n} \to 1$, we can write the high n end of the series:

$$y_1 = \left[1 - \frac{kx^2}{1.2} - \cdots - c_{2n} x^{2n} (1 + x^2 + x^4 + \cdots) \right], \tag{12.30}$$

which diverges at the end points as

$$(1 + x^2 + x^4 + \cdots) = \frac{1}{1 - x^2}. \tag{12.31}$$

The conclusion for the second series with odd powers is exactly the same. Since we want finite solutions everywhere, the divergence at the end points is unacceptable. A finite solution cannot be obtained just by fixing the integration constants. Hence, we turn to the parameter k in our equation. If we restrict the values of k as

$$\boxed{k = l(l+1), \quad l = 0, 1, 2, \ldots,} \tag{12.32}$$

one of the series in Eq. (12.25) terminates after a finite number of terms, while the other series continues to diverge at the end points. However, we still have the coefficients a_0 and a_1 at our disposal [Eq. (12.26)]; hence, we set the coefficient in front of the divergent series to zero and keep only the finite polynomial solution as the meaningful solution in the entire interval $[-1, 1]$. For example, if we take $l = 1$, hence $k = 2$, only the first term survives in the second series, y_2, [Eqs. (12.25) and (12.26)], thus giving

$$y(x) = a_0 \left[1 - \frac{2}{1.2} x^2 - \frac{2(2.3 - 2)}{1.2.3.4} x^4 + \cdots \right] + a_1 x. \tag{12.33}$$

Since the remaining series in Eq. (12.33) diverges at the end points, we set the integration constant a_0 to zero, thus obtaining the finite polynomial solution as

$$y_{l=1}(x) = a_1 x. \qquad (12.34)$$

Similarly, for $l = 2$, hence $k = 6$, Eq. (12.25) becomes

$$y(x) = a_0 \left[1 - \frac{6}{1.2} x^2 \right] + a_1 \left[x + \frac{1.2 - 6}{1.2.3} x^3 + \left[\frac{1.2 - 6}{1.2.3} \right] \left[\frac{3.4 - 6}{4.5} \right] x^5 + \cdots \right],$$

$$(12.35)$$

We now set $a_1 = 0$ to obtain the polynomial solution

$$y_{l=2}(x) = a_0 \left[1 - \frac{6}{1.2} x^2 \right]. \qquad (12.36)$$

In general, the solution is of the form

$$y_l(x) = \begin{cases} a_l x^l + a_{l-2} x^{l-2} + \cdots + a_0, & \text{for even } l, \\[2mm] a_l x^l + a_{l-2} x^{l-2} + \cdots + a_1 x, & \text{for odd } l. \end{cases} \qquad (12.37)$$

12.1.3 Legendre Polynomials

To find a general expression for the coefficients, we substitute $k = l(l+1)$ and write [Eq. (12.18)]:

$$a_n = -a_{n+2} \frac{(n+2)(n+1)}{(l-n)(l+n+1)}, \qquad (12.38)$$

as

$$a_{n-2} = -a_n \frac{n(n-1)}{(l-n+2)(l+n-1)}. \qquad (12.39)$$

Now the coefficients of the decreasing powers of x can be obtained as

$$a_{n-4} = -a_{n-2} \frac{(n-2)(n-3)}{(l-n+4)(l+n-3)}, \qquad (12.40)$$

$$\vdots$$

Starting with the coefficient of the highest power, a_l, we write the subsequent coefficients in the polynomials as

$$a_l, \qquad (12.41)$$

$$a_{l-2} = -a_l \frac{l(l-1)}{2(2l-1)}, \qquad (12.42)$$

$$a_{l-4} = -a_{l-2}\frac{(l-2)(l-3)}{4(2l-3)} = a_l\frac{l(l-1)(l-2)(l-3)}{2.4(2l-1)(2l-3)}, \tag{12.43}$$

$$\vdots$$

Now a trend begins to appear, and after s iterations, we obtain

$$a_{l-2s} = a_l(-1)^s\frac{l(l-1)(l-2)\cdots(l-2s+1)}{2.4\ldots(2s)(2l-1)(2l-3)\cdots(2l-2s+1)}. \tag{12.44}$$

The general expression for the polynomials can be written as

$$y_l(x) = \sum_{s=0}^{[\frac{l}{2}]} a_{l-2s}x^{l-2s}, \tag{12.45}$$

where $[\frac{l}{2}]$ stands for the greatest integer less than or equal to $\frac{l}{2}$. For some l values, $[\frac{l}{2}]$ and the number of terms in y_l are given as

$$
\begin{array}{l c c c c c c c c c}
l & 0 & 1 & 2 & 3 & 4 & 5 & 6 & 7 & 8 \\
[\frac{l}{2}] & 0 & 0 & 1 & 1 & 2 & 2 & 3 & 3 & 4 \;. \\
\text{No. of terms} & 1 & 1 & 2 & 2 & 3 & 3 & 4 & 4 & 5
\end{array} \tag{12.46}
$$

To get a compact expression for a_{2l-s}, we write the numerator of Eq. (12.44) as

$$l(l-1)(l-2)\cdots(l-2s+1)\frac{(l-2s)!}{(l-2s)!} = \frac{l!}{(l-2s)!}. \tag{12.47}$$

Similarly, by multiplying and dividing with the appropriate factors, the denominator is written as

$$2.4\ldots(2s)(2l-1)(2l-3)\cdots(2l-2s+1)$$

$$= \frac{(1.2)(2.2)(2.3)\ldots(2.s)(2l-1)[2l-2](2l-3)[2l-4]\cdots(2l-2s+1)\cdot[2l-2s]!}{[2l-2][2l-4]\cdots[2l-2s+2]\cdot[2l-2s]!} \tag{12.48}$$

$$= 2^s s!\frac{[2l]}{2^s}\frac{(2l-1)[2l-2](2l-3)[2l-4]\cdots(2l-2s+1)\cdot[2l-2s]!\cdot[l-s]!}{[l(l-1)(l-2)\cdots(l-s+1)[l-s]!]\cdot[2l-2s]!} \tag{12.49}$$

$$= \frac{s!(2l)!(l-s)!}{l!(2l-2s)!}. \tag{12.50}$$

Combining Eqs. (12.47) and (12.50) in Eq. (12.44), we obtain

$$a_{l-2s} = a_l(-1)^s\frac{l!}{(l-2s)!}\frac{l!(2l-2s)!}{s!(2l)!(l-s)!} = a_l(-1)^s\frac{(l!)^2(2l-2s)!}{(l-2s)!s!(2l)!(l-s)!}, \tag{12.51}$$

which allows us to write the polynomial solutions as

$$y_l(x) = \sum_{s=0}^{[\frac{l}{2}]} a_l(-1)^s \frac{(l!)^2(2l-2s)!}{(l-2s)!s!(2l)!(l-s)!} x^{l-2s} \tag{12.52}$$

$$= a_l \frac{(l!)^2 2^l}{(2l)!} \sum_{s=0}^{[\frac{l}{2}]} \frac{(-1)^s}{2^l} \frac{(2l-2s)!}{(l-2s)!s!(l-s)!} x^{l-2s}. \tag{12.53}$$

Legendre polynomials are defined by setting

$$a_l = \frac{(2l)!}{(l!)^2 2^l}, \tag{12.54}$$

as

$$\boxed{P_l(x) = \sum_{s=0}^{[\frac{l}{2}]} \frac{(-1)^s}{2^l} \frac{(2l-2s)!}{(l-2s)!s!(l-s)!} x^{l-2s}.} \tag{12.55}$$

These are the finite polynomial solutions of the Legendre equation [Eq. (12.1)] in the entire interval $[-1, 1]$:

Legendre Polynomials

$$P_0(x) = 1,$$
$$P_1(x) = x,$$
$$P_2(x) = (1/2)[3x^2 - 1],$$
$$P_3(x) = (1/2)[5x^3 - 3x], \tag{12.56}$$
$$P_4(x) = (1/8)[35x^4 - 30x^2 + 3],$$
$$P_5(x) = (1/8)[63x^5 - 70x^3 + 15x],$$
$$P_6(x) = (1/16)[231x^6 - 315x^4 + 105x^2 - 5].$$

12.1.4   Rodriguez Formula

Legendre polynomials are also defined by the Rodriguez formula

$$\boxed{P_l(x) = \frac{1}{2^l l!} \frac{d^l}{dx^l}(x^2 - 1)^l.} \tag{12.57}$$

To show its equivalence with the previous formula [Eq. (12.55)], we use the binomial formula and expand $(x^2 - 1)^l$ as

$$(x^2 - 1)^l = \sum_{s=0}^{l} \binom{l}{s}(-1)^s x^{2(l-s)}, \tag{12.58}$$

where the binomial coefficients are defined as

$$\binom{l}{s} = \frac{l!}{s!(l-s)!}. \tag{12.59}$$

We now write Eq. (12.57) as

$$P_l(x) = \frac{1}{2^l l!} \frac{d^l}{dx^l} \sum_{s=0}^{l} (-1)^s \frac{l!}{s!(l-s)!} x^{2(l-s)} \tag{12.60}$$

$$= \frac{1}{2^l l!} \sum_{s=0}^{l} (-1)^s \frac{l!}{s!(l-s)!} \frac{d^l}{dx^l} x^{2(l-s)}. \tag{12.61}$$

When the order of the derivative is greater than the power of x, we get zero:

$$\frac{d^l}{dx^l} x^{2l-2s} = 0, \ l-2s < 0, \ \text{or} \ s > \left[\frac{l}{2}\right]. \tag{12.62}$$

Hence, we can write

$$P_l(x) = \frac{1}{2^l l!} \sum_{s=0}^{\left[\frac{l}{2}\right]} (-1)^s \frac{l!}{s!(l-s)!} \frac{d^l}{dx^l} x^{2(l-s)}. \tag{12.63}$$

Using the useful formula

$$\frac{d^l x^m}{dx^l} = m(m-1)(m-2)\cdots(m-l-1)x^{m-l} = \frac{m!}{(m-l)!} x^{m-l}, \tag{12.64}$$

we obtain

$$P_l(x) = \frac{1}{2^l} \sum_{s=0}^{\left[\frac{l}{2}\right]} (-1)^s \frac{1}{s!(l-s)!} \frac{(2l-2s)!}{(2l-2s-l)!} x^{2l-2s-l} \tag{12.65}$$

$$= \sum_{s=0}^{\left[\frac{l}{2}\right]} \frac{(-1)^s}{2^l} \frac{(2l-2s)!}{s!(l-s)!(l-2s)!} x^{l-2s}, \tag{12.66}$$

which is identical to our previous definition [Eq. (12.55)].

12.1.5 Generating Function

Another useful definition of the Legendre polynomials is the generating function definition:

$$T(x,t) = \frac{1}{(1-2xt+t^2)^{1/2}} = \sum_{l=0}^{\infty} P_l(x)t^l, \quad |t| < 1. \tag{12.67}$$

Equivalence with the previous definitions can be established by using complex contour integrals. Let the roots of $(1 - 2xt + t^2)$ be given as r_1 and r_2 in the complex t-plane. If r is the smaller of $|r_1|$ and $|r_2|$, then the function

$$\frac{1}{(1 - 2xt + t^2)^{1/2}} \tag{12.68}$$

is analytic inside the region $|t| < r$. Hence, we can write the Taylor series

$$T(x,t) = \frac{1}{(1 - 2xt + t^2)^{1/2}} = \sum_{l=0}^{\infty} a_l(x) t^l, \ |t| < r, \tag{12.69}$$

where the coefficients are defined by the contour integral

$$a_l(x) = \frac{1}{2\pi i} \oint_C \frac{(1 - 2xt + t^2)^{-1/2}}{t^{l+1}} \, dt, \tag{12.70}$$

which is evaluated over any closed contour C enclosing $t = 0$ within the region $|t| < r$. We now make the substitution

$$1 - ut = (1 - 2xt + t^2)^{1/2}, \tag{12.71}$$

which is also equal to $t = 2(u - x)/(u^2 - 1)$ and convert the contour integral [Eq. (12.70)] into

$$a_l(x) = \frac{1}{2\pi i} \oint_{C'} \frac{(u^2 - 1)^l \, du}{2^l (u - x)^{l+1}}, \tag{12.72}$$

where C' is now a closed contour (Figure 12.1) in the complex u-plane enclosing the point $u = x$. Note that $t = 0$ corresponds to $u = x$. Using the Cauchy

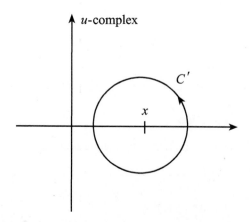

Figure 12.1 Contour C' in the complex u-plane.

integral formula:

$$\frac{d^n f(z_0)}{dz^n} = \frac{n!}{2\pi i}\oint_{C'} \frac{f(z)\,dz}{(z-z_0)^{n+1}}, \tag{12.73}$$

where $f(z)$ is analytic within and on C', with the replacements $z = u$, $z_0 = x$, $n = l$ and

$$f(u) = \frac{(u^2-1)^l}{2^l}, \tag{12.74}$$

we obtain

$$a_l(x) = \frac{1}{2^l l!}\frac{d^l(u^2-1)^l}{du^l}\bigg|_{u=x}. \tag{12.75}$$

Substituting $a_l(x)$ into Eq. (12.69):

$$T(x,t) = \frac{1}{(1-2xt+t^2)^{1/2}} = \sum_{l=0}^{\infty}\left[\frac{1}{2^l l!}\frac{d^l}{dx^l}(x^2-1)^l\right]t^l = \sum_{l=0}^{\infty}P_l(x)t^l, \quad |t| < 1, \tag{12.76}$$

we obtain the Rodriguez formula:

$$P_l(x) = \frac{1}{2^l l!}\frac{d^l}{dx^l}(x^2-1)^l. \tag{12.77}$$

12.1.6   Special Values

Special values of the Legendre polynomials are given as

$$\boxed{\begin{array}{ll} P_l(1) = 1, & P_{2l}(0) = (-1)^l\dfrac{(2l)!}{2^{2l}(l!)^2}, \\ P_l(-1) = (-1)^l, & P_{2l+1}(0) = 0. \end{array}} \tag{12.78}$$

We can prove the special values at the end points, $x = \mp 1$, by using the generating function [Eq. (12.67)] and the geometric series:

$$T(\mp 1, t) = \frac{1}{(1\pm 2t+t^2)^{1/2}} = \sum_{l=0}^{\infty}P_l(\mp 1)t^l, \tag{12.79}$$

$$\frac{1}{(1\pm t)} = \sum_{l=0}^{\infty}P_l(\mp 1)t^l, \tag{12.80}$$

$$\sum_{l=0}^{\infty}(\mp 1)^l t^l = \sum_{l=0}^{\infty}P_l(\mp 1)t^l, \tag{12.81}$$

thus obtaining $P_l(\mp 1) = (\mp 1)^l$. For the special values at the origin, we write

$$T(0, t) = \frac{1}{(1+t^2)^{1/2}} = \sum_{l=0}^{\infty}t^l P_l(0), \tag{12.82}$$

and use the binomial expansion to write the left-hand side as

$$(1+t^2)^{-1/2} = 1 + \left(-\frac{1}{2}\right)t^2 + \cdots + \frac{\left(-\frac{1}{2}\right)\left(-\frac{3}{2}\right)\cdots\left(-\frac{2l-1}{2}\right)}{l!}(t^2)^l + \cdots \quad (12.83)$$

$$= \sum_{l=0}^{\infty} (\mp 1)^l \frac{1.3.5\ldots(2l-1)t^{2l}}{2^l l!} \quad (12.84)$$

$$= \sum_{l=0}^{\infty} (\mp 1)^l \frac{1.2.3\ldots(2l-1)(2l)t^{2l}}{2^l l! 2.4.6\ldots(2l-2)(2l)} \quad (12.85)$$

$$= \sum_{l=0}^{\infty} (\mp 1)^l \frac{(2l)!t^{2l}}{2^l l! 2^l l!} \quad (12.86)$$

$$= \sum_{l=0}^{\infty} (\mp 1)^l \frac{(2l)!t^{2l}}{2^{2l}(l!)^2}. \quad (12.87)$$

Finally, comparing with the right-hand side of Eq. (12.82), we get the desired result.

12.1.7   Recursion Relations

Differentiating the generating function with respect to x, we write

$$\frac{\partial T(x,t)}{\partial x} = \frac{\partial}{\partial x}\left[\frac{1}{(1-2xt+t^2)^{1/2}}\right] = \frac{\partial}{\partial x}\sum_{l=0}^{\infty} P_l(x)t^l, \quad (12.88)$$

$$\frac{t}{(1-2xt+t^2)^{3/2}} = \sum_{l=0}^{\infty} P_l'(x)t^l, \quad (12.89)$$

$$\frac{t}{(1-2xt+t^2)^{1/2}} = \sum_{l=0}^{\infty} P_l'(x)(1-2xt+t^2)t^l, \quad (12.90)$$

$$\sum_{l=0}^{\infty} P_l(x)t^{l+1} = \sum_{l=0}^{\infty} P_l'(x)t^l - 2x\sum_{l=0}^{\infty} P_l'(x)t^{l+1} + \sum_{l=0}^{\infty} P_l'(x)t^{l+2}. \quad (12.91)$$

In the first sum on the right-hand side, we make the change $l = l' + 1$, and in last sum, we let $l = l'' - 1$ and then drop primes to write

$$\sum_{l=0}^{\infty} P_l(x)t^{l+1} = \sum_{l=-1}^{\infty} P_{l+1}'(x)t^{l+1} - 2x\sum_{l=0}^{\infty} P_l'(x)t^{l+1} + \sum_{l=1}^{\infty} P_{l-1}'(x)t^{l+1}. \quad (12.92)$$

Since

$$P_0' = 0, \quad (12.93)$$

we can start all the sums from $l = 0$ to write

$$\sum_{l=0}^{\infty} [P_l(x) - P'_{l+1}(x) + 2xP'_l(x) - P'_{l-1}(x)]t^{l+1} = 0. \qquad (12.94)$$

This equation can be satisfied for all t only when the coefficients of all the powers of t vanish simultaneously, thus yielding the recursion relation

$$\boxed{P_l(x) = P'_{l+1}(x) - 2xP'_l(x) + P'_{l-1}(x).} \qquad (12.95)$$

A second relation can be obtained by differentiating the generating function with respect to t as

$$\boxed{(2l + 1)xP_l(x) = (l + 1)P_{l+1}(x) + lP_{l-1}(x).} \qquad (12.96)$$

These two basic recursion relations are very useful in calculations, and others can be generated by using these.

12.1.8   Orthogonality

We write the Legendre equation [Eq. (12.1)] for the indices l and m as

$$\frac{d}{dx}\left[(1 - x^2)\frac{dP_l}{dx}\right] + l(l + 1)P_l = 0, \qquad (12.97)$$

$$\frac{d}{dx}\left[(1 - x^2)\frac{dP_m}{dx}\right] + m(m + 1)P_m = 0. \qquad (12.98)$$

We now multiply the first equation by P_m and then subtract the second equation multiplied by P_l, and integrate the result over the entire range to write

$$\int_{-1}^{1} P_m \frac{d}{dx}\left[(1 - x^2)\frac{dP_l}{dx}\right] dx - \int_{-1}^{1} P_l \frac{d}{dx}\left[(1 - x^2)\frac{dP_m}{dx}\right] dx$$

$$+ [l(l + 1) - m(m + 1)] \int_{-1}^{1} P_l(x)P_m(x)\, dx = 0. \qquad (12.99)$$

Integrating the first two integrals by parts gives

$$P_m(1 - x^2)\frac{dP_l}{dx}\bigg|_{-1}^{1} - \int_{-1}^{1} \frac{dP_m}{dx}(1 - x^2)\frac{dP_l}{dx}\, dx$$

$$- P_l(1 - x^2)\frac{dP_m}{dx}\bigg|_{-1}^{1} + \int_{-1}^{1} \frac{dP_l}{dx}(1 - x^2)\frac{dP_m}{dx}\, dx$$

$$+ [l(l + 1) - m(m + 1)] \int_{-1}^{1} P_l(x)P_m(x)\, dx = 0. \qquad (12.100)$$

Since the surface terms vanish, we are left with the equation

$$-\int_{-1}^{1} \frac{dP_m}{dx}(1-x^2)\frac{dP_l}{dx}\,dx + \int_{-1}^{1} \frac{dP_l}{dx}(1-x^2)\frac{dP_m}{dx}\,dx$$

$$+ [l(l+1) - m(m+1)]\int_{-1}^{1} P_l(x)P_m(x)\,dx = 0, \tag{12.101}$$

which is

$$[l(l+1) - m(m+1)]\int_{-1}^{1} P_l(x)P_m(x)\,dx = 0. \tag{12.102}$$

We now have two possibilities:

$$\boxed{\begin{aligned} l \neq m; \quad & \int_{-1}^{1} P_l(x)P_m(x)\,dx = 0, \\ l = m; \quad & \int_{-1}^{1} [P_l(x)]^2\,dx = N_l. \end{aligned}} \tag{12.103}$$

This is called the orthogonality relation of Legendre polynomials. To evaluate the **normalization constant** N_l, we use the Rodriguez formula [Eq. (12.57)] to write

$$N_l = \int_{-1}^{1} [P_l(x)]^2\,dx \tag{12.104}$$

$$= \left(\frac{1}{2^l l!}\right)^2 \int_{-1}^{1} \left[\frac{d^l}{dx^l}(x^2-1)^l\right]\left[\frac{d^l}{dx^l}(x^2-1)^l\right]\,dx. \tag{12.105}$$

l-fold integration by parts gives

$$N_l = \frac{1}{2^{2l}(l!)^2}\left[(-1)^l \int_{-1}^{1} (x^2-1)^l \frac{d^{2l}}{dx^{2l}}(x^2-1)^l\,dx\right]. \tag{12.106}$$

Since the highest power in the expansion of $(x^2-1)^l$ is x^{2l}, we use the formula

$$\frac{d^{2l}}{dx^{2l}}(x^2-1)^l = (2l)! \tag{12.107}$$

to write

$$N_l = \frac{(-1)^l(2l)!}{2^{2l}(l!)^2}\int_{-1}^{1} (x^2-1)^l\,dx. \tag{12.108}$$

Using trigonometric substitution and contour integrals, the aforementioned integral can be evaluated as [1]

$$\int_{-1}^{1} (x^2-1)^l\,dx = \frac{(-1)^l 2^{2l+1}(l!)^2}{(2l+1)!}, \tag{12.109}$$

thus yielding the normalization constant N_l:

$$N_l = \frac{(-1)^l (2l)!}{2^{2l}(l!)^2} \cdot \frac{(-1)^l 2^{2l+1}(l!)^2}{(2l+1)!} \tag{12.110}$$

$$\boxed{N_l = \frac{2}{2l+1}.} \tag{12.111}$$

12.1.9 Legendre Series

If a real function $F(x)$ is continuous apart from a finite number of discontinuities in the interval $[-1,1]$, that is $F(x)$ is at least piecewise continuous and the integral $\int_{-1}^{1} [F(x)]^2 \, dx$ is finite, then the Legendre series $\sum_{l=0}^{\infty} c_l P_l(x)$ converges to $F(x)$:

$$\boxed{F(x) = \sum_{l=0}^{\infty} c_l P_l(x),} \tag{12.112}$$

when x is not a point of discontinuity. Expansion coefficients c_l are calculated using the orthogonality relation:

$$\int_{-1}^{1} P_l(x) P_m(x) \, dx = [2/(2m+1)]\delta_{lm}, \tag{12.113}$$

as

$$\boxed{c_l = \frac{2l+1}{2} \int_{-1}^{1} F(x) P_l(x) dx.} \tag{12.114}$$

At the point of discontinuity, the series converges to the mean [2]:

$$\boxed{\frac{1}{2}[F(x^+) + F(x^-)].} \tag{12.115}$$

Example 12.1. *Generating function:* A practical use for the generating function can be given in electromagnetic theory. Electrostatic potential $\Psi(r, \theta, \phi)$ of a point charge Q displaced from the origin along the z-axis by a distance a is given as (Figure 12.2)

$$\Psi(r, \theta, \phi) = \frac{Q}{R} = \frac{Q}{\sqrt{r^2 + a^2 - 2ar \cos\theta}} \tag{12.116}$$

$$= \frac{Q}{r} \frac{1}{\sqrt{1 + \dfrac{a^2}{r^2} - \dfrac{2a}{r} \cos\theta}}. \tag{12.117}$$

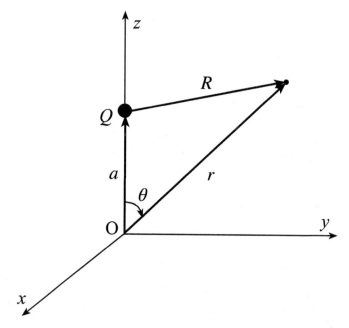

Figure 12.2   Electrostatic potential of a point charge.

If we substitute $t = a/r$ and $t = \cos\theta$, electrostatic potential becomes

$$\Psi(r,\theta,\phi) = \frac{Q}{r}\frac{1}{\sqrt{1 - 2xt + t^2}}, \quad |t| < 1. \tag{12.118}$$

Using the generating function of Legendre polynomials [Eq. (12.67)], we can express $\Psi(r,\theta,\phi)$ in terms of the Legendre polynomials as

$$\Psi(r,\theta,\phi) = \frac{Q}{r}\sum_{l=0}^{\infty} P_l(\cos\theta)\left(\frac{a}{r}\right)^l. \tag{12.119}$$

Example 12.2. *Another derivation for N_l:* The normalization constant can also be obtained from the generating function. We multiply two generating functions and integrate between -1 and 1 to write

$$\int_{-1}^{1} T(x,t)T(x,t)dx = \int_{-1}^{1} \frac{1}{\sqrt{1 - 2xt + t^2}}\frac{1}{\sqrt{1 - 2xt + t^2}}\,dx \tag{12.120}$$

$$= -\frac{1}{2t}\ln(1 - 2xt + t^2)\Big|_{-1}^{1} \tag{12.121}$$

$$= -\frac{1}{t}\ln(1 - t) + \frac{1}{t}\ln(1 + t). \tag{12.122}$$

This is also equal to

$$\sum_{l=0}^{\infty}\sum_{k=0}^{\infty}\left[\int_{-1}^{1}dx\,P_l P_k\right]t^{l+k} = -\frac{1}{t}\ln(1-t) + \frac{1}{t}\ln(1+t). \qquad (12.123)$$

Expanding the right-hand side in powers of t and using the orthogonality relation [Eq. (12.103)], we write

$$\sum_{l=0}^{\infty}\left[\int_{-1}^{1}dx\,P_l^2\right]t^{2l} = \sum_{l'=0}^{\infty}\frac{t^{l'}}{(l'+1)}(1+(-1)^{l'}), \qquad (12.124)$$

$$\sum_{l=0}^{\infty}N_l t^{2l} = \sum_{l=0}^{\infty}\frac{2t^{2l}}{(2l+1)}, \qquad (12.125)$$

thus obtaining the normalization constant as

$$N_l = \int_{-1}^{1}dx\,P_l^2(x) = \frac{2}{2l+1}. \qquad (12.126)$$

Example 12.3. *Evaluation of the integral* $I_{kl} = \int_{-1}^{1}x^k P_l(x)\,dx$, $k \le l$: This is a useful integral in many applications with Legendre polynomials. We use the Rodriguez formula to write I_{kl} as

$$I_{kl} = \frac{1}{2^l l!}\int_{-1}^{1}x^k\frac{d^l}{dx^l}(x^2-1)^l\,dx, \quad k \le l. \qquad (12.127)$$

Integration by parts gives

$$I_{kl} = \frac{1}{2^l l!}\left[x^k\frac{d^{l-1}(x^2-1)^l}{dx^{l-1}}\bigg|_{-1}^{1} - \int_{-1}^{1}kx^{k-1}\frac{d^{l-1}}{dx^{l-1}}(x^2-1)^l\,dx\right], \qquad (12.128)$$

where the surface term vanishes. A second integration by parts gives

$$I_{kl} = \frac{1}{2^l l!}\left[-kx^{k-1}\frac{d^{l-2}(x^2-1)^l}{dx^{l-2}}\bigg|_{-1}^{1}\right.$$
$$\left. + (-1)^2\int_{-1}^{1}k(k-1)x^{k-2}\frac{d^{l-2}}{dx^{l-2}}(x^2-1)^l\,dx\right], \qquad (12.129)$$

where the surface term vanishes again and a trend appears. After k-fold integration by parts, all the surface terms vanish, thus giving

$$I_{kl} = \frac{(-1)^k k!}{2^l l!}\int_{-1}^{1}\frac{d^{l-k}}{dx^{l-k}}(x^2-1)^l\,dx, \quad k \le l. \qquad (12.130)$$

For $k < l$, the integral, I_{kl}, vanishes. This can be seen by writing I_{kl} as

$$I_{kl} = \frac{(-1)^k k!}{2^l l!} \int_{-1}^{1} \frac{d}{dx} \left[\frac{d^{l-k-1}}{dx^{l-k-1}} (x^2 - 1)^l \right] dx, \qquad (12.131)$$

which becomes

$$I_{kl} = \frac{(-1)^k k!}{2^l l!} \int_{-1}^{1} d \left[\frac{d^{l-k-1}}{dx^{l-k-1}} (x^2 - 1)^l \right] = \frac{(-1)^k k!}{2^l l!} \left[\frac{d^{l-k-1}}{dx^{l-k-1}} (x^2 - 1)^l \right]_{-1}^{1}. \qquad (12.132)$$

Since $k < l$ and both k and l are integers, the highest k value without k being equal to l is $l - 1$, which makes the integral zero:

$$I_{(l-1)l} = \frac{(-1)^{l-1}(l-1)!}{2^l l!} [(x^2 - 1)^l]_{-1}^{1} = 0. \qquad (12.133)$$

For the lower values of k, the derivative in Eq. (12.132) always contains positive powers of $(x^2 - 1)$, which vanishes when the end points are substituted. When $k = l$, no derivative is left in Eq. (12.130); hence, the integral reduces to

$$I_{ll} = \frac{(-1)^l l!}{2^l l!} \int_{-1}^{1} (x^2 - 1)^l dx. \qquad (12.134)$$

This is the same integral in Eq. (12.109). Substituting its value, we obtain

$$I_{ll} = \frac{2^{l+1} (l!)^2}{(2l + 1)!}. \qquad (12.135)$$

Now the complete result is given as

$$\int_{-1}^{1} x^k P_l(x) dx = \begin{cases} 0, & k < l, \\ \dfrac{2^{l+1} (l!)^2}{(2l + 1)!}, & k = l. \end{cases} \qquad (12.136)$$

12.2 HERMITE EQUATION

12.2.1 Series Solution

Hermite equation is defined as

$$h''(x) - 2xh'(x) + (\epsilon - 1)h(x) = 0, \quad x \in [-\infty, \infty], \qquad (12.137)$$

where ϵ is a real continuous parameter, which is unrestricted at this point. Hermite equation is primarily encountered in the study of quantum mechanical harmonic oscillator, where ϵ stands for the energy of the oscillator. We look for a finite solution in the entire interval $[-\infty, \infty]$. Since $x = 0$ is a regular point, we use the Frobenius method and try a series solution of the form

$$h(x) = \sum_{k=0}^{\infty} a_k x^{k+s}, \quad a_0 \neq 0. \tag{12.138}$$

Substituting $h(x)$ and its derivatives:

$$h'(x) = \sum_{k=0}^{\infty} a_k(k+s)x^{k+s-1}, \quad h''(x) = \sum_{k=0}^{\infty} a_k(k+s)(k+s-1)x^{k+s-2},$$

$$\tag{12.139}$$

into Eq. (12.137), we obtain

$$\sum_{k=0}^{\infty} a_k(k+s)(k+s-1)x^{k+s-2} - 2\sum_{k=0}^{\infty} a_k(k+s)x^{k+s}$$

$$+ (\epsilon - 1)\sum_{k=0}^{\infty} a_k x^{k+s} = 0. \tag{12.140}$$

To equate the powers of x, we let $k - 2 = k'$ in the first series and then drop primes:

$$\sum_{k=-2}^{\infty} a_{k+2}(k+s+1)(k+s+2)x^{k+s} - 2\sum_{k=0}^{\infty} a_k(k+s)x^{k+s}$$

$$+ (\epsilon - 1)\sum_{k=0}^{\infty} a_k x^{k+s} = 0. \tag{12.141}$$

If we write the first two terms of the first series explicitly, we can write the remaining terms under the same sum as

$$s(s-1)a_0 x^{s-2} + s(s+1)a_1 x^{s-1}$$

$$+ \sum_{k=0}^{\infty} \{a_{k+2}(k+s+1)(k+s+2) - a_k[2(k+s) - (\epsilon - 1)]\}x^{k+s} = 0.$$

$$\tag{12.142}$$

Setting the coefficients of all the powers of x to zero, we obtain

$$s(s-1)a_0 = 0, \quad a_0 \neq 0, \tag{12.143}$$

$$s(s+1)a_1 = 0, \tag{12.144}$$

$$a_{k+2} = a_k \frac{2(k+s) - \epsilon + 1}{(k+s+1)(k+s+2)}, \quad k = 0, 1, \ldots . \tag{12.145}$$

The first equation is the **indicial equation**, the roots of which give the values of s as 0 and 1. Since their difference is an integer, guided by our past experience, we start with the smaller root, hoping that it will directly yield the general solution. Now the recursion relation becomes

$$\boxed{a_{k+2} = a_k \frac{2k - \epsilon + 1}{(k+1)(k+2)}, \quad k = 0, 1, 2, \ldots,} \tag{12.146}$$

which yields the coefficients:

$$a_0 \neq 0,$$
$$a_1 \neq 0,$$
$$a_2 = \frac{1 - \epsilon}{1.2} a_0,$$
$$a_3 = \frac{3 - \epsilon}{2.3} a_1, \tag{12.147}$$
$$a_4 = \frac{5 - \epsilon}{3.4} a_2 = \frac{1 - \epsilon}{1.2} \cdot \frac{5 - \epsilon}{3.4} a_0,$$
$$a_5 = \frac{7 - \epsilon}{4.5} a_3 = \frac{3 - \epsilon}{2.3} \cdot \frac{7 - \epsilon}{4.5} a_1,$$
$$\vdots$$

and the following series solution:

$$h(x) = a_0 \left[1 + \frac{1 - \epsilon}{1.2} x^2 + \frac{1 - \epsilon}{1.2} \cdot \frac{5 - \epsilon}{3.4} x^4 + \cdots \right]$$
$$+ a_1 \left[x + \frac{3 - \epsilon}{2.3} x^3 + \frac{3 - \epsilon}{2.3} \cdot \frac{7 - \epsilon}{4.5} x^5 + \cdots \right]. \tag{12.148}$$

As we hoped, this solution contains two linearly independent solutions:

$$h_1(x) = 1 + \frac{1 - \epsilon}{1.2} x^2 + \frac{1 - \epsilon}{1.2} \cdot \frac{5 - \epsilon}{3.4} x^4 + \cdots , \tag{12.149}$$
$$h_2(x) = x + \frac{3 - \epsilon}{2.3} x^3 + \frac{3 - \epsilon}{2.3} \cdot \frac{7 - \epsilon}{4.5} x^5 + \cdots . \tag{12.150}$$

Since a_0 and a_1 are completely arbitrary, from now on we treat them as the integration constants and take the general solution as

$$h(x) = a_0 h_1(x) + a_2 h_2(x). \tag{12.151}$$

To check the convergence of these series, we write them as

$$h_1 = \sum_{n=0}^{\infty} a_{2n} x^{2n} \text{ and } h_2 = \sum_{n=0}^{\infty} a_{2n+1} x^{2n+1} \tag{12.152}$$

and concentrate on the first one with the even powers:

$$h_1(x) = a_0[1 + a_2 x^2 + a_4 x^4 + \cdots]. \tag{12.153}$$

From the recursion relation [Eq. (12.146)], the ratio of the coefficients of two consecutive terms is given as

$$\frac{a_{2n+2}}{a_{2n}} = \frac{4n - \epsilon + 1}{(2n+1)(2n+2)}. \tag{12.154}$$

For sufficiently large n, this has the limit

$$\lim_{n \to \infty} \frac{a_{2n+2}}{a_{2n}} \to \frac{1}{n}, \tag{12.155}$$

hence we can write the series

$$h_1(x) = a_0 \left[1 + a_2 x^2 + \cdots + a_{2n} x^{2n} \left(1 + \frac{x^2}{1!} + \frac{x^4}{2!} + \cdots \right) \right], \tag{12.156}$$

where the series in parenthesis is nothing but

$$e^{x^2} = \sum_{j=0}^{\infty} \frac{x^{2j}}{j!} = \left[1 + \frac{x^2}{1!} + \frac{x^4}{2!} + \cdots \right]. \tag{12.157}$$

In other words, $h_1(x)$ diverges as e^{x^2} when $x \to \pm\infty$. This conclusion is also valid for the second series $h_2(x)$ with the odd powers of x. Our aim is to find solutions that are finite everywhere in the interval $[-\infty, \infty]$. This cannot be accomplished by just adjusting the integration constants a_0 and a_1. Hence, we turn to the parameter ϵ in the differential equation and restrict it to the following values:

$$\epsilon = 2n + 1, \ n = 0, 1, 2, \ldots, \tag{12.158}$$

$$\begin{array}{lccccc} n = & 0 & 1 & 2 & 3 & 4 & \cdots, \\ \epsilon = & 1 & 3 & 5 & 7 & 9 & \cdots. \end{array} \tag{12.159}$$

This allows us to terminate one of the series in Eq. (12.148) after a finite number of terms. For the even values of n, the general solution now looks like

$$h_n(x) = a_0[1 + a_2 x^2 + \cdots a_n x^n] + a_1[x + a_3 x^3 + \cdots], \ n = \text{even}. \tag{12.160}$$

Since the second series diverges as $x \to \pm\infty$, we set $a_1 = 0$ and keep only the polynomial solutions. Similarly, for the odd values of n, we have

$$h_n(x) = a_0[1 + a_2 x^2 + \cdots] + a_1[x + a_3 x^3 + \cdots + a_n x^n], \quad n = \text{odd.} \quad (12.161)$$

We now set $a_0 = 0$ and keep only the polynomial solutions with the odd powers of x. In summary, we obtain the following finite polynomial solutions:

$$\begin{aligned}
n &= 0, \quad h_0(x) = a_0, \\
n &= 1, \quad h_1(x) = a_1 x, \\
n &= 2, \quad h_2(x) = a_0(1 - 2x^2), \\
n &= 3, \quad h_3(x) = a_1(x - \tfrac{2}{3}x^3),
\end{aligned} \qquad (12.162)$$

$$\vdots \qquad \vdots$$

12.2.2   Hermite Polynomials

To find a general expression for the polynomial solutions of the Hermite equation, we write the recursion relation [Eq. (12.146)] for the coefficients as

$$a_{k-2} = -a_k \frac{k(k-1)}{2(n-k+2)}, \quad k = 2, 3, \ldots. \qquad (12.163)$$

Starting with a_n, that is, the coefficient of the highest power of x, we write the coefficients of the decreasing powers:

$$a_{n-2} = -a_n \frac{n(n-1)}{2(n-n+2)} = -a_n \frac{n(n-1)}{2.2}, \qquad (12.164)$$

$$a_{n-4} = -a_{n-2} \frac{(n-2)(n-3)}{2.4} = (-)(-)a_n \frac{n(n-1)(n-2)(n-3)}{2.2(2.4)}, \qquad (12.165)$$

which after j iterations gives

$$a_{n-2j} = a_n \frac{(-1)^j n(n-1)(n-2)\cdots(n-2j+1)}{2^j 2.4\ldots(2j)} \qquad (12.166)$$

$$= a_n \frac{(-1)^j n!}{(n-2j)! 2^j 2^j j!}. \qquad (12.167)$$

We can now write $h_n(x)$ as

$$h_n(x) = \sum_{j=0}^{\left[\frac{n}{2}\right]} \frac{a_n (-1)^j n!}{(n-2j)! 2^{2j} j!} x^{n-2j}, \qquad (12.168)$$

where $[\frac{n}{2}]$ stands for the greatest integer less than or equal to $\frac{n}{2}$. If we take $a_n = 2^n$, the resulting polynomials are called the **Hermite polynomials**:

$$H_n(x) = \sum_{j=0}^{[\frac{n}{2}]} \frac{(-1)^j n! 2^{n-2j}}{(n-2j)! j!} x^{n-2j}. \tag{12.169}$$

Hermite Polynomials

$$
\begin{aligned}
H_0(x) &= 1, \\
H_1(x) &= 2x, \\
H_2(x) &= 4x^2 - 2, \\
H_3(x) &= 8x^3 - 12x, \\
H_4(x) &= 16x^4 - 48x^2 + 12,
\end{aligned}
\tag{12.170}
$$

$$\vdots$$

12.2.3 Contour Integral Representation

Hermite polynomials can also be defined by the complex contour integral

$$H_n(x) = \frac{(-1)^n n!}{2\pi i} \oint_C \frac{e^{x^2 - z^2}\, dz}{(z-x)^{n+1}}, \tag{12.171}$$

where C is any closed contour enclosing the point x. We can evaluate this integral by using the residue theorem as

$$\oint_C \frac{e^{x^2 - z^2}\, dz}{(z-x)^{n+1}} = 2\pi i \; \text{Residue} \left\{ \frac{e^{x^2 - z^2}}{(z-x)^{n+1}} \right\}_{z=x}. \tag{12.172}$$

Using the expansion $e^z = \sum_m z^m / m!$, we write

$$\frac{e^{x^2 - z^2}}{(z-x)^{n+1}} = \frac{e^{-(z+x)(z-x)}}{(z-x)^{n+1}} = \sum_{m=0}^{\infty} \frac{(-1)^m (z+x)^m (z-x)^m}{m!(z-x)^{n+1}}. \tag{12.173}$$

Substitution of the binomial expansion:

$$(z+x)^m = [2x + (z-x)]^m = \sum_{l=0}^{m} \frac{m!}{l!(m-l)!} (2x)^{m-l} (z-x)^l, \tag{12.174}$$

gives

$$\frac{e^{x^2 - z^2}}{(z-x)^{n+1}} = \sum_{m=0}^{\infty} \sum_{l=0}^{m} \frac{(-1)^m m!}{l!(m-l)!} (2x)^{m-l} (z-x)^l \frac{(z-x)^m}{m!(z-x)^{n+1}} \tag{12.175}$$

$$= \sum_{m=0}^{\infty} \sum_{l=0}^{m} \frac{(-1)^m 2^{m-l}}{l!(m-l)!} x^{m-l} (z-x)^{m+l-n-1}. \tag{12.176}$$

Since the desired residue is the coefficient of the $(z - x)^{-1}$ term, in the afore-mentioned double sum [Eq. (12.176)], we set

$$m + l - n - 1 = -1, \tag{12.177}$$

$$m + l - n = 0, \tag{12.178}$$

$$m = n - l, \tag{12.179}$$

which gives the residue as

$$\text{Residue}\left\{ \frac{e^{x^2 - z^2}}{(z - x)^{n+1}} \right\}_{z=x} = \sum_{l=0}^{\left[\frac{n}{2}\right]} \frac{(-1)^{n-l} 2^{n-2l}}{l!(n - 2l)!} x^{n-2l}. \tag{12.180}$$

This yields the value of the integral in Eq. (12.172) as

$$\oint_C \frac{e^{x^2 - z^2} dz}{(z - x)^{n+1}} = 2\pi i \sum_{l=0}^{\left[\frac{n}{2}\right]} \frac{(-1)^{n-l} 2^{n-2l}}{l!(n - 2l)!} x^{n-2l}. \tag{12.181}$$

Hence, Hermite polynomials [Eq. (12.171)] become

$$H_n(x) = \sum_{l=0}^{\left[\frac{n}{2}\right]} \frac{(-1)^l n! 2^{n-2l}}{l!(n - 2l)!} x^{n-2l}, \tag{12.182}$$

which is identical to our previous definition [Eq. (12.169)].

12.2.4   Rodriguez Formula

We first write the contour integral definition [Eq. (12.171)]:

$$H_n(x) = \frac{(-1)^n n!}{2\pi i} e^{x^2} \oint_C \frac{e^{-z^2} dz}{(z - x)^{n+1}} \tag{12.183}$$

and the derivative formula:

$$\frac{d^n}{dz^n} f(z_0) = \frac{n!}{2\pi i} \oint_C \frac{f(z) dz}{(z - z_0)^{n+1}}, \tag{12.184}$$

where $f(z)$ is analytic within and on the contour C, which encloses the point z_0. Making the replacements $f(z) = e^{-z^2}$ and $z_0 = x$ in Eq. (12.184):

$$\oint_C \frac{e^{-z^2} dz}{(z - x)^{n+1}} = \frac{2\pi i}{n!} \frac{d^n e^{-z^2}}{dz^n}\bigg|_{z=x}, \tag{12.185}$$

and using this in Eq. (12.183), we obtain the **Rodriguez formula:**

$$
H_n(x) = (-1)^n e^{x^2} \frac{d^n e^{-x^2}}{dx^n}, \quad n = 0, 1, 2, \ldots \tag{12.186}
$$

12.2.5   Generating Function

Starting with the definition

$$
H_n(x) = \sum_{j=0}^{\left[\frac{n}{2}\right]} \frac{(-1)^j n! 2^{n-2j}}{(n-2j)! j!} x^{n-2j}, \tag{12.187}
$$

we aim to construct a series of the form

$$
\sum_{n=0}^{\infty} \frac{H_n(x) t^n}{n!}. \tag{12.188}
$$

Using Eq. (12.187), we write

$$
\sum_{n=0}^{\infty} \frac{H_n(x) t^n}{n!} = \sum_{n=0}^{\infty} \sum_{j=0}^{\left[\frac{n}{2}\right]} \frac{(-1)^j n! 2^{n-2j}}{(n-2j)! j! n!} x^{n-2j} t^n \tag{12.189}
$$

and make the dummy variable change $n \to n' + j$. After dropping primes, we obtain

$$
\sum_{n=0}^{\infty} \frac{H_n(x) t^n}{n!} = \sum_{n=0}^{\infty} \sum_{j=0}^{n} \frac{(-1)^j 2^{n-j}}{(n-j)! j!} x^{n-j} t^{n+j} \tag{12.190}
$$

$$
= \sum_{n=0}^{\infty} \sum_{j=0}^{n} \frac{(-1)^j 2^{n-j} n!}{(n-j)! j! n!} x^{n-j} t^{n+j}, \tag{12.191}
$$

where the second equation is multiplied and divided by $n!$. We rearrange Eq. (12.191) as

$$
\sum_{n=0}^{\infty} \frac{H_n(x) t^n}{n!} = \sum_{n=0}^{\infty} \frac{t^n}{n!} \sum_{j=0}^{n} \frac{n!}{(n-j)! j!} (2x)^{n-j} (-t)^j. \tag{12.192}
$$

The second series on the right-hand side is the binomial expansion:

$$
\sum_{j=0}^{n} \frac{n!}{(n-j)! j!} (2x)^{n-j} (-t)^j = (2x - t)^n, \tag{12.193}
$$

hence Eq. (12.192) becomes

$$\sum_{n=0}^{\infty} \frac{H_n(x)t^n}{n!} = \sum_{n=0}^{\infty} \frac{t^n(2x-t)^n}{n!}. \tag{12.194}$$

Furthermore, the series on the right-hand side is nothing but the exponential function with the argument $t(2x - t)$:

$$\sum_{n=0}^{\infty} \frac{t^n(2x-t)^n}{n!} = e^{t(2x-t)}, \tag{12.195}$$

thus giving us the **generating function** $T(x,t)$ of $H_n(x)$ as

$$\boxed{T(x,t) = e^{t(2x-t)} = \sum_{n=0}^{\infty} \frac{H_n(x)t^n}{n!}.} \tag{12.196}$$

12.2.6 Special Values

Using the generating function, we can find the special values at the origin:

$$T(0,t) = e^{-t^2} = \sum_{n=0}^{\infty} \frac{H_n(0)t^n}{n!}, \tag{12.197}$$

$$\sum_{n=0}^{\infty} (-1)^n \frac{t^{2n}}{n!} = \sum_{n=0}^{\infty} \frac{H_n(0)t^n}{n!}, \tag{12.198}$$

which gives

$$\boxed{H_{2n}(0) = (-1)^n \frac{(2n)!}{n!},} \tag{12.199}$$

$$\boxed{H_{2n+1}(0) = 0.} \tag{12.200}$$

12.2.7 Recursion Relations

By using the generating function, we can drive the two basic recursion relations of the Hermite polynomials. Differentiating $T(x,t)$ with respect to x, we write

$$\frac{\partial T(x,t)}{\partial x} = 2te^{2xt-t^2} = \sum_{n=0}^{\infty} \frac{H_n'(x)}{n!} t^n. \tag{12.201}$$

Substituting the definition of the generating function to the left-hand side, we rewrite this as

$$2t \sum_{n=0}^{\infty} \frac{H_n(x)t^n}{n!} = \sum_{n=0}^{\infty} \frac{H_n'(x)}{n!} t^n, \tag{12.202}$$

$$2\sum_{n=0}^{\infty}\frac{H_n(x)t^{n+1}}{n!}=\sum_{n=0}^{\infty}\frac{H_n'(x)}{n!}t^n. \tag{12.203}$$

Making a dummy variable change in the first series, $n \to n' - 1$, and dropping primes, we write

$$2\sum_{n=1}^{\infty}\frac{H_{n-1}(x)t^n}{(n-1)!}=\sum_{n=0}^{\infty}\frac{H_n'(x)}{n!}t^n, \tag{12.204}$$

$$\sum_{n=1}^{\infty}[2nH_{n-1}(x)-H_n'(x)]\frac{t^n}{n!}=0, \tag{12.205}$$

which gives the first recursion relation as

$$\boxed{2nH_{n-1}(x)=H_n'(x), \quad n=1,2,\ldots .} \tag{12.206}$$

Note that $H_0' = 0$. For the second recursion relation, we differentiate $T(x,t)$ with respect to t and write

$$\frac{\partial T(x,t)}{\partial t}=(2x-2t)e^{2xt-t^2}=\sum_{n=0}^{\infty}\frac{H_n(x)}{n!}nt^{n-1}, \tag{12.207}$$

$$2(x-t)e^{2xt-t^2}=\sum_{n=1}^{\infty}\frac{H_n(x)}{(n-1)!}t^{n-1}, \tag{12.208}$$

$$2(x-t)\sum_{n=0}^{\infty}\frac{H_n(x)t^n}{n!}=\sum_{n=1}^{\infty}\frac{H_n(x)}{(n-1)!}t^{n-1}, \tag{12.209}$$

$$2x\sum_{n=0}^{\infty}\frac{H_n(x)t^n}{n!}-2\sum_{n=0}^{\infty}\frac{H_n(x)t^{n+1}}{n!}=\sum_{n=1}^{\infty}\frac{H_n(x)}{(n-1)!}t^{n-1}. \tag{12.210}$$

To equate the powers of t, we let $n \to n' - 1$ in the second series and $n \to n'' + 1$ in the third series and then drop primes to write

$$\sum_{n=0}^{\infty}[2xH_n-2nH_{n-1}-H_{n+1}]\frac{t^n}{n!}=0. \tag{12.211}$$

Setting the coefficients of all the powers of t to zero gives us the second recursion relation:

$$\boxed{2xH_n-2nH_{n-1}=H_{n+1}.} \tag{12.212}$$

12.2.8   Orthogonality

We write the Hermite equation [Eq. (12.137)] as

$$H_n'' - 2xH_n' = -2nH_n, \tag{12.213}$$

where we have substituted $2n = \epsilon - 1$. The left-hand side can be made exact (Chapter 11) by multiplying the Hermite equation with e^{-x^2}:

$$e^{-x^2} H_n'' - 2xe^{-x^2} H_n' = -2ne^{-x^2} H_n, \tag{12.214}$$

which can now be written as

$$\frac{d}{dx}\left[e^{-x^2}\frac{dH_n}{dx}\right] = -2nH_n(x)e^{-x^2}. \tag{12.215}$$

We now multiply both sides by $H_m(x)$ and integrate over $[-\infty, \infty]$ to write

$$\int_{-\infty}^{\infty} H_m \frac{d}{dx}\left[e^{-x^2}\frac{dH_n}{dx}\right] dx = -2n \int_{-\infty}^{\infty} H_m(x)H_n(x)e^{-x^2} dx, \tag{12.216}$$

which, after integration by parts, becomes

$$H_m' H_n' e^{-x^2}\Big|_{-\infty}^{\infty} - \int_{-\infty}^{\infty} H_m' \, H_n' e^{-x^2} dx = -2n \int_{-\infty}^{\infty} H_m(x)H_n(x)e^{-x^2} dx. \tag{12.217}$$

Since the surface term vanishes, we have

$$\int_{-\infty}^{\infty} H_m' \, H_n' e^{-x^2} dx = 2n \int_{-\infty}^{\infty} H_m(x)H_n(x)e^{-x^2} dx. \tag{12.218}$$

Interchanging n and m gives another equation:

$$\int_{-\infty}^{\infty} H_m' \, H_n' e^{-x^2} dx = 2m \int_{-\infty}^{\infty} H_m(x)H_n(x)e^{-x^2} dx, \tag{12.219}$$

which we subtract from the original equation [Eq. (12.218)] to get

$$2(m-n) \int_{-\infty}^{\infty} H_m(x)H_n(x)e^{-x^2} dx = 0. \tag{12.220}$$

We have two cases:

$$\boxed{\begin{array}{ll} m \neq n; & \int_{-\infty}^{\infty} H_m(x)H_n(x)e^{-x^2} dx = 0, \\[2ex] m = n; & \int_{-\infty}^{\infty} H_n^2(x)e^{-x^2} dx = N_n, \end{array}} \tag{12.221}$$

which shows that the Hermite polynomials are orthogonal with respect to the **weight factor** e^{-x^2}.

To complete the orthogonality relation, we need to calculate the normalization constant. Replace n by $n-1$ in the recursion relation [Eq. (12.212)] to write

$$2xH_{n-1} - 2(n-1)H_{n-2} = H_n. \tag{12.222}$$

Multiply this by H_n:

$$2xH_nH_{n-1} - 2nH_nH_{n-2} + 2H_nH_{n-2} = H_n^2. \tag{12.223}$$

We also multiply the recursion relation (12.212) by H_{n-1} to obtain a second equation:

$$2xH_{n-1}H_n - 2nH_{n-1}^2 = H_{n-1}H_{n+1}. \tag{12.224}$$

Subtracting Eq. (12.224) from (12.223) gives

$$2xH_nH_{n-1} - 2nH_nH_{n-2} + 2H_nH_{n-2} - 2xH_{n-1}H_n + 2nH_{n-1}^2$$
$$= H_n^2 - H_{n-1}H_{n+1}, \tag{12.225}$$

or

$$-2nH_nH_{n-2} + 2H_nH_{n-2} + 2nH_{n-1}^2 + H_{n-1}H_{n+1} = H_n^2. \tag{12.226}$$

After multiplying by the weight factor e^{-x^2} and integrating over $[-\infty, \infty]$, the aforementioned equation becomes

$$-2n\int_{-\infty}^{\infty} dx e^{-x^2} H_nH_{n-2} + 2\int_{-\infty}^{\infty} dx e^{-x^2} H_nH_{n-2}$$
$$+ 2n\int_{-\infty}^{\infty} dx e^{-x^2} H_{n-1}^2 + \int_{-\infty}^{\infty} dx e^{-x^2} H_{n-1}H_{n+1} = \int_{-\infty}^{\infty} dx e^{-x^2} H_n^2. \tag{12.227}$$

Using the orthogonality relation [Eq. (12.221)], this simplifies to

$$2n\int_{-\infty}^{\infty} dx e^{-x^2} H_{n-1}^2 = \int_{-\infty}^{\infty} dx e^{-x^2} H_n^2, \tag{12.228}$$

$$2nN_{n-1} = N_n, \quad n = 1, 2, 3, \ldots . \tag{12.229}$$

Starting with N_n, we iterate this formula to write

$$N_n = 2nN_{n-1}$$
$$= 2n2(n-1)N_{n-2}$$
$$= 2n2(n-1)2(n-2)N_{n-3}$$

$$\vdots$$

$$= 2^{j+1}n(n-1)\cdots(n-j)N_{n-j-1}. \tag{12.230}$$

We continue until we hit $j = n - 1$, thus

$$N_n = 2^n n! N_0. \tag{12.231}$$

We evaluate N_0 using $H_0 = 1$ as

$$N_0 = \int_{-\infty}^{\infty} e^{-x^2} H_0^2(x) \, dx = \int_{-\infty}^{\infty} e^{-x^2} \, dx = \sqrt{\pi}, \tag{12.232}$$

which yields N_n as

$$\boxed{N_n = 2^n n! \sqrt{\pi}, \quad n = 0, 1, 2, \dots .} \tag{12.233}$$

12.2.9   Series Expansions in Hermite Polynomials

A series expansion for any sufficiently smooth function, $f(x)$, defined in the infinite interval $[-\infty, \infty]$ can be given as

$$\boxed{f(x) = \sum_{n=0}^{\infty} c_n H_n(x).} \tag{12.234}$$

Using the orthogonality relation:

$$\int_{-\infty}^{\infty} e^{-x^2} H_m(x) H_n(x) \, dx = 2^n n! \sqrt{\pi} \delta_{mn}, \tag{12.235}$$

we can evaluate the expansion coefficients c_n as

$$\boxed{c_n = \frac{1}{2^n n! \sqrt{\pi}} \int_{-\infty}^{\infty} e^{-x^2} f(x) H_n(x) dx, \quad n = 0, 1, 2, \dots .} \tag{12.236}$$

Convergence of this series is assured, granted that the real function $f(x)$ defined in the infinite interval $(-\infty, \infty)$ is piecewise smooth in every subinterval $[-a, a]$ and the integral

$$\int_{-\infty}^{\infty} e^{-x^2} f^2(x) \, dx \tag{12.237}$$

is finite. At the points of discontinuity, the series converges to

$$\boxed{\frac{1}{2} [f(x^+) + f(x^-)].} \tag{12.238}$$

A proof of this theorem can be found in [2].

Example 12.4. *Expansion of* $f(x) = e^{ax}$, a *is a constant:* Since $f(x)$ is a sufficiently smooth function, we can write the convergent series:

$$e^{ax} = \sum_{n=0}^{\infty} c_n H_n(x), \tag{12.239}$$

where the coefficients are given as

$$c_n = \frac{1}{2^n n! \sqrt{\pi}} \int_{-\infty}^{\infty} e^{-x^2 + ax} H_n(x) \, dx. \tag{12.240}$$

Using the Rodriguez formula [Eq. (12.186)], we can evaluate the integral to obtain c_n:

$$c_n = \frac{(-1)^n}{2^n n! \sqrt{\pi}} \int_{-\infty}^{\infty} e^{-x^2 + ax} e^{x^2} \frac{d^n (e^{-x^2})}{dx^n} \, dx \tag{12.241}$$

$$= \frac{(-1)^n}{2^n n! \sqrt{\pi}} \int_{-\infty}^{\infty} e^{ax} \frac{d^n (e^{-x^2})}{dx^n} \, dx \tag{12.242}$$

$$= \frac{a^n}{2^n n! \sqrt{\pi}} \int_{-\infty}^{\infty} e^{ax - x^2} \, dx \tag{12.243}$$

$$= \frac{a^n}{2^n n!} e^{a^2/4}. \tag{12.244}$$

We have used n-fold integration by parts in Eq. (12.242) and completed the square in the last integral. Now, the final result can be expressed as

$$e^{ax} = e^{a^2/4} \sum_{n=0}^{\infty} \frac{a^n}{2^n n!} H_n(x). \tag{12.245}$$

12.3 LAGUERRE EQUATION

Laguerre equation is defined as

$$\boxed{x \frac{d^2 y}{dx^2} + (1 - x) \frac{dy}{dx} + ny = 0, \quad x \in [0, \infty],} \tag{12.246}$$

where n is a real continuous parameter. It is usually encountered in the study of single electron atoms in quantum mechanics, where the free parameter n is related to the energy of the atom.

12.3.1   Series Solution

Since $x = 0$ is a regular singular point, we can use the Frobenius method and attempt a series solution of the form

$$y(x, s) = \sum_{r=0}^{\infty} a_r x^{r+s}, \quad a_0 \neq 0, \tag{12.247}$$

with the derivatives

$$y'(x, s) = \sum_{r=0}^{\infty} a_r(r + s)x^{r+s-1}, \quad y''(x, s) = \sum_{r=0}^{\infty} a_r(r + s)(r + s - 1)x^{r+s-2}. \tag{12.248}$$

Substituting these into the Laguerre equation, we get

$$\sum_{r=0}^{\infty} a_r(r + s)(r + s - 1)x^{r+s-1} + \sum_{r=0}^{\infty} a_r(r + s)x^{r+s-1}$$

$$- \sum_{r=0}^{\infty} a_r(r + s)x^{r+s} + n\sum_{r=0}^{\infty} a_r x^{r+s} = 0, \tag{12.249}$$

$$\sum_{r=0}^{\infty} a_r(r + s)^2 x^{r+s-1} - \sum_{r=0}^{\infty} a_r(r + s - n)x^{r+s} = 0. \tag{12.250}$$

In the first series, we let $r - 1 = r'$ and drop primes to write

$$\sum_{r=-1}^{\infty} a_{r+1}(r + s + 1)^2 x^{r+s} - \sum_{r=0}^{\infty} a_r(r + s - n)x^{r+s} = 0, \tag{12.251}$$

$$a_0 s^2 x^{s-1} + \sum_{r=0}^{\infty} [a_{r+1}(r + s + 1)^2 - a_r(r + s - n)]x^{r+s} = 0. \tag{12.252}$$

Equating all the coefficients of the equal powers of x to zero gives us

$$a_0 s^2 = 0, \quad a_0 \neq 0, \tag{12.253}$$

$$a_{r+1} = a_r \frac{(r + s - n)}{(r + s + 1)^2}, \quad r = 0, 1, \ldots . \tag{12.254}$$

In this case, the indicial equation [Eq.(12.253)] has a double root at $s = 0$, hence the **recursion relation** becomes

$$\boxed{a_{r+1} = -a_r \frac{n - r}{(r + 1)^2}, \quad r = 0, 1, \ldots,} \tag{12.255}$$

which leads to the series solution

$$y(x) = a_0 \left[1 - \frac{n}{1!}x + \frac{n(n-1)}{(2!)^2}x^2 + \cdots \right.$$

$$\left. +(-1)^r \frac{n(n-1)\cdots(n-r+1)}{(r!)^2}x^r + \cdots \right]. \qquad (12.256)$$

This can be written as

$$y(x) = a_0 \left[\sum_{r=0}^{\infty} (-1)^r \frac{n(n-1)\cdots(n-r+1)}{(r!)^2}x^r \right]. \qquad (12.257)$$

Laguerre equation also has a second linearly independent solution. However, it diverges logarithmically as $x \to 0$, hence we set its coefficient in the general solution to zero and continue with the series solution given previously.

12.3.2 Laguerre Polynomials

As we add more and more terms, this series behaves as

$$y(x) = a_0 \left[1 - \frac{n}{1!}x + \cdots + a_r x^r \left(1 + \frac{x}{1!} + \frac{x^2}{2!} + \cdots \right) \right], \qquad (12.258)$$

which diverges as e^x as $x \to \infty$. For a finite solution, everywhere in the interval $[0, \infty]$, we have no choice but to restrict n to integer values, the effect of which is to terminate the series [Eq. (12.256)] after a finite number of terms, thus leading to the polynomial solutions:

$$y_n(x) = a_0 \left[\sum_{r=0}^{n} (-1)^r \frac{n(n-1)\cdots(n-r+1)}{(r!)^2}x^r \right] \qquad (12.259)$$

$$= a_0 \left[\sum_{r=0}^{n} (-1)^r \frac{n!}{(r!)^2(n-r)!}x^r \right]. \qquad (12.260)$$

Polynomials defined as

$$L_n(x) = \sum_{r=0}^{n} (-1)^r \frac{n!}{(r!)^2(n-r)!}x^r \qquad (12.261)$$

are called the **Laguerre polynomials** and constitute the everywhere finite solutions of the **Laguerre equation**:

$$x\frac{d^2 L_n(x)}{dx^2} + (1-x)\frac{dL_n(x)}{dx} + nL_n(x) = 0, \quad x \in [0, \infty]. \qquad (12.262)$$

12.3.3   Contour Integral Representation

Laguerre polynomials are also defined by the complex contour integral

$$L_n(x) = \frac{1}{2\pi i} \oint_C \frac{z^n e^{x-z} dz}{(z-x)^{n+1}}, \qquad (12.263)$$

where the contour C is any closed path enclosing the point $z = x$. To show the equivalence of the two definitions, we evaluate the contour integral by using the residue theorem:

$$L_n(x) = \frac{1}{2\pi i} 2\pi i \operatorname{Residue}\left\{ \frac{z^n e^{x-z}}{(z-x)^{n+1}} \right\}_{z=x}. \qquad (12.264)$$

To find the residue, we use the expansions

$$z^n = (z + x - x)^n = [(z - x) + x]^n = \sum_{l=0}^{n} \frac{n!}{l!(n-l)!}(z-x)^l x^{n-l} \qquad (12.265)$$

and

$$e^{x-z} = e^{-(z-x)} = \sum_{m=0}^{\infty} (-1)^m \frac{(z-x)^m}{m!}. \qquad (12.266)$$

Using these, we can write the integrand of the contour integral [Eq. (12.263)] as

$$\frac{z^n e^{x-z}}{(z-x)^{n+1}} = \sum_{l=0}^{n} \frac{n!}{l!(n-l)!} \frac{(z-x)^l x^{n-l}}{(z-x)^{n+1}} \sum_{m=0}^{\infty} (-1)^m \frac{(z-x)^m}{m!} \qquad (12.267)$$

$$= \sum_{m=0}^{\infty} \sum_{l=0}^{n} \frac{(-1)^m n!}{m! l! (n-l)!}(z-x)^{l-n-1+m} x^{n-l}. \qquad (12.268)$$

For the residue, we need the coefficient of the $(z-x)^{-1}$ term, that is, the term with

$$l - n - 1 + m = -1, \qquad (12.269)$$

$$l = n - m. \qquad (12.270)$$

Therefore, the residue is obtained as

$$\operatorname{Residue}\left\{ \frac{z^n e^{x-z}}{(z-x)^{n+1}} \right\}_{z=x} = \sum_{m=0}^{n} \frac{(-1)^m n! x^m}{m!(n-m)! m!}. \qquad (12.271)$$

Substituting into Eq. (12.264), we obtain

$$L_n(x) = \sum_{m=0}^{n} \frac{(-1)^m n! x^m}{(n-m)!(m!)^2}, \qquad (12.272)$$

which agrees with our previous definition [Eq. (12.261)].

12.3.4   Rodriguez Formula

Using the contour integral representation [Eq. (12.263)] and the Cauchy derivative formula:

$$\frac{2\pi i}{n!} f^{(n)}(z_0) = \oint \frac{f(z)dz}{(z - z_0)^{n+1}},$$

(12.273)

we can write the Rodriguez formula of Laguerre polynomials as

$$\boxed{L_n(x) = \frac{e^x}{n!} \frac{d^n(x^n e^{-x})}{dx^n}.}$$

(12.274)

12.3.5   Generating Function

To obtain the generating function $T(x, t)$, we multiply $L_n(x)$ [Eq. (12.272)] by t^n and sum over n to write

$$T(x, t) = \sum_{n=0}^{\infty} L_n(x)t^n = \sum_{n=0}^{\infty} \sum_{r=0}^{n} \frac{(-1)^r n! x^r t^n}{(n - r)!(r!)^2}.$$

(12.275)

We introduce a new dummy variable s as $n = r + s$ to write

$$T(x, t) = \sum_{r=0}^{\infty} \sum_{s=0}^{\infty} \frac{(-1)^r (r + s)! x^r t^{r+s}}{(r!)^2 s!}.$$

(12.276)

Note that both sums now run from zero to infinity. We rearrange this as

$$T(x, t) = \sum_{r=0}^{\infty} \frac{(-1)^r x^r t^r}{r!} \sum_{s=0}^{\infty} \frac{(r + s)! t^s}{(r!)s!}.$$

(12.277)

If we note that the second sum is nothing but [3]

$$\sum_{s=0}^{\infty} \frac{(r + s)! t^s}{(r!)s!} = \frac{1}{(1 - t)^{r+1}},$$

(12.278)

we can rewrite Eq. (12.277) as

$$T(x, t) = \frac{1}{(1 - t)} \sum_{r=0}^{\infty} \frac{[-xt/(1 - t)]^r}{r!}.$$

(12.279)

Finally, using $e^x = \sum_{r=0}^{\infty} x^r/r!$, we obtain the generating function definition of $L_n(x)$ as

$$\boxed{T(x, t) = \frac{1}{(1 - t)} \exp\left[\frac{-xt}{(1 - t)}\right] = \sum_{n=0}^{\infty} L_n(x)t^n.}$$

(12.280)

12.3.6 Special Values and Recursion Relations

Using the generating function and the geometric series, $1/(1-t) = \sum_{n=0}^{\infty} t^n$, we easily obtain the special value

$$\boxed{L_n(0) = 1.}$$ (12.281)

From the Laguerre equation [Eq. (12.262)], we also get by inspection

$$\boxed{L'_n(0) = -n.}$$ (12.282)

Differentiating the generating function with respect to t gives us the first recursion relation:

$$\boxed{(n+1)L_{n+1}(x) = (2n+1-x)L_n(x) - nL_{n-1}(x).}$$ (12.283)

Similarly, differentiating with respect to x, we obtain the second recursion relation:

$$\boxed{L'_{n+1}(x) - L'_n(x) = -L_n(x).}$$ (12.284)

Using the first recursion relation [Eq. (12.283)], the second recursion relation can also be written as

$$\boxed{xL'_n(x) = nL_n(x) - nL_{n-1}(x).}$$ (12.285)

12.3.7 Orthogonality

If we multiply the Laguerre equation [Eq. (12.262)] by e^{-x}:

$$e^{-x}x\frac{d^2L_n}{dx^2} + (1-x)e^{-x}\frac{dL_n}{dx} = -ne^{-x}L_n(x),$$ (12.286)

the left-hand side becomes exact, thus can be written as (see Chapter 11)

$$\frac{d}{dx}\left[xe^{-x}\frac{dL_n}{dx}\right] = -ne^{-x}L_n.$$ (12.287)

We first multiply both sides by $L_m(x)$ and then integrate by parts:

$$\int_0^\infty L_m \frac{d}{dx}\left[xe^{-x}\frac{dL_n}{dx}\right]dx = -n\int_0^\infty e^{-x}L_nL_m\,dx,$$ (12.288)

$$L_mxe^{-x}L'_n\big|_0^\infty - \int_0^\infty \frac{dL_m}{dx}\left[xe^{-x}\frac{dL_n}{dx}\right]dx = -n\int_0^\infty e^{-x}L_nL_m\,dx,$$ (12.289)

$$-\int_0^\infty \frac{dL_m}{dx}\left[xe^{-x}\frac{dL_n}{dx}\right]dx = -n\int_0^\infty e^{-x}L_nL_m\,dx.$$ (12.290)

Interchanging m and n, we obtain another equation:

$$-\int_0^\infty \frac{dL_n}{dx}\left[xe^{-x}\frac{dL_m}{dx}\right]dx = -m\int_0^\infty e^{-x}L_nL_m\,dx, \qquad (12.291)$$

which, when subtracted from Eq. (12.290) gives

$$(m-n)\int_0^\infty e^{-x}L_nL_m\,dx = 0. \qquad (12.292)$$

This gives the orthogonality relation as

$$\boxed{\int_0^\infty e^{-x}L_nL_m\,dx = N_n\delta_{nm}.} \qquad (12.293)$$

Using the generating function [Eq. (12.280)], the normalization constant N_n can be obtained as 1.

12.3.8   Series Expansions in Laguerre Polynomials

Like the previous special functions we have studied, any sufficiently smooth real function in the interval $[0,\infty]$ can be expanded in terms of Laguerre polynomials as

$$\boxed{f(x) = \sum_{n=0}^\infty c_n L_n(x),} \qquad (12.294)$$

where the coefficients c_n are found by using the orthogonality relation:

$$\boxed{c_n = \int_0^\infty e^{-x}f(x)L_n(x)dx.} \qquad (12.295)$$

Convergence of this series to $f(x)$ is guaranteed when the real function $f(x)$ is piecewise smooth in every subinterval, $[x_1,x_2]$, where $0 < x_1 < x_2 < \infty$, of $[0,\infty)$ and x is not a point of discontinuity, and the integral

$$\int_0^\infty e^{-x}[f(x)]^2dx \qquad (12.296)$$

is finite. At the points of discontinuity, the series converges to [2]

$$\boxed{\frac{1}{2}[f(x^+) + f(x^-)].} \qquad (12.297)$$

Example 12.5. ***Laguerre series of*** e^{-ax}***:*** This function satisfies the conditions stated in Section 12.3.8 for $a > 0$; hence we can write the series

$$e^{-ax} = \sum_{n=0}^{\infty} c_n L_n(x), \qquad (12.298)$$

where the expansion coefficients are obtained as

$$c_n = \int_0^{\infty} e^{-(a+1)x} L_n(x) \, dx \qquad (12.299)$$

$$= \frac{1}{n!} \int_0^{\infty} e^{-ax} \frac{d^n}{dx^n}(e^{-x} x^n) \, dx \qquad (12.300)$$

$$= \frac{a^n}{n!} \int_0^{\infty} e^{-(a+1)x} x^n \, dx \qquad (12.301)$$

$$= \frac{a^n}{(a+1)^{n+1}}, \quad n = 0, 1, \dots . \qquad (12.302)$$

REFERENCES

1. Bayin, S.S. (2018). *Mathematical Methods in Science and Engineering*, 2e. Wiley.

2. Lebedev, N.N. (1965). *Special Functions and Their Applications*. Englewood Cliffs, NJ: Prentice-Hall.

3. Dwight, H.B. (1961). *Tables of Integrals and Other Mathematical Data*, 4e. New York: Macmillan.

PROBLEMS

1. Find Legendre series expansion of the step function:

$$f(x) = \begin{cases} 0, & -1 \le x < a, \\ 1, & a < x \le 1. \end{cases}$$

Discuss the behavior of the series you found at $x = a$.

Hint: Use the asymptotic form of the Legendre series given as

$$P_l(x) \approx \sqrt{\frac{2}{\pi l \sin \theta}} \sin\left[\left(l + \frac{1}{2}\right)\theta + \frac{\pi}{4}\right], \, l \to \infty, \, \varepsilon \le \theta \le \pi - \varepsilon,$$

where ε is any positive number [1, 2].

2. Show the parity relation of Legendre polynomials:

$$P_l(-x) = (-1)^l P_l(x).$$

3. Using the basic recursion relations [Eqs. (12.95) and (12.96)], derive

(i) $P'_{l+1}(x) = (l+1)P_l(x) + xP'_l(x)$.

(ii) $P'_{l-1}(x) = -lP_l(x) + xP'_l(x)$.

(iii) $(1-x^2)P'_l(x) = lP_{l-1}(x) - lxP_l(x)$.

(iv) $(1-x^2)P'_l(x) = (l+1)xP_l(x) - (l+1)P_{l+1}(x)$.

4. Show the relation

$$\frac{1-t^2}{(1-2xt+t^2)^{3/2}} = \sum_{l=0}^{\infty}(2l+1)P_l(x)t^l.$$

5. Show that Legendre expansion of the Dirac-delta function is

$$\delta(x) = \sum_{l=0}^{\infty}(-1)^l \frac{(4l+1)(2l)!}{2^{2l+1}(l!)^2}P_{2l}(x).$$

6. Show that Hermite polynomials satisfy the parity relation

$$H_n(x) = (-1)^n H_n(-x).$$

7. (i) Show that Hermite polynomials can also be defined as

$$H_n(x) = \frac{n!}{2\pi i}\oint_C t^{-n-1}e^{-t^2+2tx}\,dt.$$

(ii) Define your contour and evaluate the integral to justify your result.

8. Show the integral

$$\int_{-\infty}^{\infty} x^2 e^{-x^2}[H_n(x)]^2\,dx = \pi^{1/2}2^n n!\left(n+\frac{1}{2}\right).$$

9. Show that

$$\int_{-\infty}^{\infty} x^2 e^{-x^2}H_n(x)H_m(x)\,dx = 2^{n-1}\pi^{1/2}(2n+1)n!\delta_{nm}$$

$$+ 2^n \pi^{1/2}(n+2)!\delta_{n+2,m} + 2^{n-2}\pi^{1/2}n!\delta_{n-2,m}.$$

10. Show the Laguerre expansion

$$x^m = \sum_{n=0}^{m} c_n L_n(x), \quad m = 0,1,2,\ldots,$$

where

$$c_n = (-1)^n \frac{(m!)^2}{n!(m-n)!}.$$

11. Using the generating function definition of Laguerre polynomials, show that the normalization constant N_n is 1, where

$$\int_0^\infty e^{-x} L_n L_m \, dx = N_n \delta_{nm}.$$

12. Prove the basic recursion relations of the Laguerre polynomials:

$$(n+1)L_{n+1}(x) = (2n+1-x)L_n(x) - nL_{n-1}(x),$$

$$L'_{n+1}(x) - L'_n(x) = -L_n(x).$$

13. Using the basic recursion relations obtained in Problem 12 derive

$$xL'_n(x) = nL_n(x) - nL_{n-1}(x).$$

CHAPTER 13

BESSEL'S EQUATION AND BESSEL FUNCTIONS

Bessel functions are among the most frequently encountered special functions in physics and engineering. They are very useful in quantum mechanics in WKB approximations. Since they are usually encountered in solving potential problems with cylindrical boundaries, they are also called cylinder functions. Bessel functions are used even in abstract number theory and mathematical analysis. Like the other special functions, they form a complete and an orthogonal set. Therefore, any sufficiently smooth function can be expanded in terms of Bessel functions. However, their orthogonality is not with respect to their order but with respect to a parameter in their argument, which usually assumes the values of the infinitely many roots of the Bessel function. In this chapter, we introduce the basic Bessel functions and their properties. We also discuss the modified Bessel functions and the spherical Bessel functions. There exists a wealth of literature on special functions. Like the classic treatise by Watson, some of them are solely devoted to Bessel functions and their applications [1].

Essentials of Mathematical Methods in Science and Engineering, Second Edition. Selçuk Ş. Bayın. **629**
© 2020 John Wiley & Sons, Inc. Published 2020 by John Wiley & Sons, Inc.

13.1   BESSEL'S EQUATION AND ITS SERIES SOLUTION

Bessel's equation is defined as

$$\boxed{x^2\frac{d^2y_m}{dx^2} + x\frac{dy_m}{dx} + (x^2 - m^2)y_m = 0, \ x \geq 0,} \tag{13.1}$$

where the range of the independent variable could be taken as the entire real axis or even the entire complex plane. At this point, we restrict m to positive and real values. Since $x = 0$ is a regular singular point, we can try a series solution of the form

$$y_m(x) = \sum_{k=0}^{\infty} c_k x^{k+r}, \quad c_0 \neq 0, \tag{13.2}$$

with the derivatives

$$y'_m = \sum_{k=0}^{\infty} c_k(k+r)x^{k+r-1}, \quad y''_m = \sum_{k=0}^{\infty} c_k(k+r)(k+r-1)x^{k+r-2}. \tag{13.3}$$

Substituting these into the Bessel's equation, we write

$$\sum_{k=0}^{\infty} c_k(k+r)(k+r-1)x^{k+r} + \sum_{k=0}^{\infty} c_k(k+r)x^{k+r} + \sum_{k=0}^{\infty} c_k x^{k+r+2}$$

$$- m^2\sum_{k=0}^{\infty} c_k x^{k+r} = 0, \tag{13.4}$$

which can be arranged as

$$\sum_{k=0}^{\infty}[(k+r)(k+r-1) + (k+r) - m^2]c_k x^{k+r} + \sum_{k=0}^{\infty} c_k x^{k+r+2} = 0. \tag{13.5}$$

We now let $k + 2 = k'$ in the second sum and drop primes to write

$$\sum_{k=0}^{\infty}[(k+r)(k+r-1) + (k+r) - m^2]c_k x^{k+r} + \sum_{k=2}^{\infty} c_{k-2} x^{k+r} = 0. \tag{13.6}$$

Writing the first two terms of the first series explicitly, we can have both sums starting from $k = 2$, thus

$$(r^2 - m^2)c_0 x^r + [(r+1)^2 - m^2]c_1 x^{r+1} + \sum_{k=2}^{\infty}[((k+r)^2 - m^2)c_k + c_{k-2}]x^{k+r} = 0. \tag{13.7}$$

Equating coefficients of the equal powers of x to zero, we obtain

$$(r^2 - m^2)c_0 = 0, \quad c_0 \neq 0, \tag{13.8}$$

$$[(r+1)^2 - m^2]c_1 = 0, \tag{13.9}$$

$$[(k+r)^2 - m^2]c_k + c_{k-2} = 0, \quad k = 2, 3, \ldots . \tag{13.10}$$

The first equation is the **indicial equation**, the solution of which gives the values of r as $r = \pm m$, $m > 0$. For the time being, we take $r = m$, hence the second equation [Eq. (13.9)] gives

$$[(m+1)^2 - m^2]c_1 = 0, \tag{13.11}$$

$$(2m+1)c_1 = 0, \tag{13.12}$$

which determines c_1 as zero. Finally, the third equation [Eq. (13.10)] gives the **recursion relation** for the remaining coefficients as

$$\boxed{c_k = -\frac{c_{k-2}}{[(k+r)^2 - m^2]}, \quad k = 2, 3, \ldots .} \tag{13.13}$$

Now, with $r = m$, all the nonzero coefficients become

$$c_0 \neq 0, \tag{13.14}$$

$$c_2 = \frac{c_0}{(m+2)^2 - m^2}, \tag{13.15}$$

$$c_4 = \frac{c_0}{[(m+2)^2 - m^2][(m+4)^2 - m^2]}, \tag{13.16}$$

$$\vdots$$

A similar procedure gives the series solution for $r = -m$; hence for $r = m$, we can write

$$y_m = c_0 x^m[1 + c_2 x^2 + c_4 x^4 + \cdots] = \sum_{k=0}^{\infty} c_{2k} x^{2k+m}, \quad m > 0. \tag{13.17}$$

To check convergence, we write the general term as $u_k = c_{2k}x^{2k+m}$. From the limit

$$\lim_{k \to \infty} \left| \frac{u_k}{u_{k-1}} \right| = \lim_{k \to \infty} \left| \frac{c_{2k} x^{2k+m}}{c_{2(k-1)} x^{2(k-1)+m}} \right| \tag{13.18}$$

$$= \lim_{k \to \infty} \left| \frac{c_{2k}}{c_{2(k-1)}} x^2 \right| \tag{13.19}$$

$$= \lim_{k \to \infty} \left| \frac{x^2}{(2k+m)^2 - m^2} \right| \tag{13.20}$$

$$= \lim_{k \to \infty} \left| \frac{x^2}{4k^2} \right| \to 0 < 1, \tag{13.21}$$

we conclude that the series converges on the real axis for $r = +m$. A parallel argument leads to the same conclusion for $r = -m$.

To obtain a general expression for the kth term of this series, we use the recursion relation [Eq. (13.13)] with $r = m$ to write

$$c_{2k} = -\frac{c_{2(k-1)}}{2^2 k(k+m)}, \tag{13.22}$$

$$c_{2(k-1)} = -\frac{c_{2(k-2)}}{2^2 (k-1)(k+m-1)}, \tag{13.23}$$

$$c_{2(k-2)} = -\frac{c_{2(k-3)}}{2^2 (k-2)(k+m-2)}, \tag{13.24}$$

$$\vdots$$

Substitution of Eq. (13.23) into Eq. (13.22) gives

$$c_{2k} = (-1)^2 \frac{c_{2(k-2)}}{2^2 2^2 k(k-1)(k+m)(k+m-1)}. \tag{13.25}$$

Next we substitute Eq. (13.24) into the aforementioned equation:

$$c_{2k} = (-1)^3 \frac{c_{2(k-3)}}{2^2 2^2 2^2 k(k-1)(k-2)(k+m)(k+m-1)(k+m-2)}, \tag{13.26}$$

where a trend begins to appear. After s iterations, we obtain

$$c_{2k} = (-1)^s \frac{c_{2(k-s)}}{2^{2s} k(k-1)\cdots(k-s+1)(k+m)(k+m-1)\cdots(k+m-s+1)} \tag{13.27}$$

and after k iterations, we hit c_0:

$$c_{2k} = (-1)^k \frac{c_0}{2^{2k} k(k-1)\cdots 2 \cdot 1 (k+m)(k+m-1)\cdots(m+1)}. \tag{13.28}$$

Dividing and multiplying by $m!$, we can write this as

$$c_{2k} = (-1)^k \frac{c_0 m!}{2^{2k} k!(k+m)!}. \tag{13.29}$$

Now the series solution [Eq. (13.17)] becomes

$$y_m(x) = c_0 m! x^m \sum_{k=0}^{\infty} (-1)^k \frac{(x/2)^{2k}}{k!(k+m)!}. \tag{13.30}$$

Bessel function of order m is defined by setting

$$c_0 2^m m! = 1 \tag{13.31}$$

as

$$\boxed{J_m(x) = \left(\frac{x}{2}\right)^m \sum_{k=0}^{\infty} (-1)^k \frac{(x/2)^{2k}}{k!(k+m)!}.} \tag{13.32}$$

$J_m(x)$ are also called the Bessel function of the **first kind**. Series expressions of the first two Bessel functions are

$$J_0(x) = \left[1 - \frac{(x/2)^2}{(1!)^2} + \frac{(x/2)^4}{(2!)^2} - \frac{(x/2)^6}{(3!)^2} + \cdots \right], \tag{13.33}$$

$$J_1(x) = \frac{x}{2} \left[1 - \frac{(x/2)^2}{1!2!} + \frac{(x/2)^4}{2!3!} - \frac{(x/2)^6}{3!4!} + \cdots \right]. \tag{13.34}$$

Bessel functions of higher order can be expressed in terms of J_0 and J_1: We first multiply Eq. (13.32) by x^m and then differentiate with respect to x to write

$$\frac{d}{dx}[x^m J_m] = \frac{d}{dx} \sum_{k=0}^{\infty} (-1)^k \frac{(x/2)^{2k+2m}}{2^{-m}k!(k+m)!} = \frac{d}{dx} \sum_{k=0}^{\infty} (-1)^k \frac{x^{2k+2m}}{2^{2k+m}k!(k+m)!} \tag{13.35}$$

$$= \sum_{k=0}^{\infty} (-1)^k \frac{(2k+2m)x^{2k+2m-1}}{2^{2k+m}k!(k+m)!} \tag{13.36}$$

$$= x^m \left[\left(\frac{x}{2}\right)^{m-1} \sum_{k=0}^{\infty} (-1)^k \frac{(x/2)^{2k}}{k!(k+m-1)!}\right] = x^m J_{m-1}. \tag{13.37}$$

Writing the left-hand side explicitly:

$$m x^{m-1} J_m + x^m J_m' = x^m J_{m-1}, \tag{13.38}$$

we obtain

$$\boxed{\frac{m}{x} J_m + J_m' = J_{m-1}.} \tag{13.39}$$

Similarly, multiplying Eq. (13.32) by x^{-m} and then differentiating with respect to x, we obtain another relation:

$$-\frac{m}{x}J_m + J'_m = -J_{m+1}. \tag{13.40}$$

Subtracting Eq. (13.40) from Eq. (13.39) gives

$$\frac{2m}{x}J_m = J_{m-1} + J_{m+1}, \quad m = 1, 2, \ldots, \tag{13.41}$$

while their sum yields

$$2J'_m = J_{m-1} - J_{m+1}, \quad m = 1, 2, \ldots . \tag{13.42}$$

Repeated applications of these formulas allow us to find J_m and J'_m in terms of J_0 and J_1. For $m = 0$, Eq. (13.42) is replaced by

$$J'_0 = -J_1. \tag{13.43}$$

13.1.1   Bessel Functions $J_{\pm m}(x)$, $N_m(x)$, and $H_m^{(1,2)}(x)$

The series solution of the Bessel's equation can be extended to all positive values of m through the use of the **gamma function** as (see Problem 6.)

$$J_m(x) = \sum_{k=0}^{\infty} \frac{(-1)^k}{k!\Gamma(m+k+1)} \left(\frac{x}{2}\right)^{m+2k}, \tag{13.44}$$

which is called the Bessel function of the first kind of order m. A second solution can be written as

$$J_{-m}(x) = \sum_{k=0}^{\infty} \frac{(-1)^k}{k!\Gamma(-m+k+1)} \left(\frac{x}{2}\right)^{-m+2k}. \tag{13.45}$$

However, the second solution is independent of the first solution only for the noninteger values of m. For the integer values, $m = n$, we multiply J_{-n} by $(-1)^{-n}$ and write

$$(-1)^{-n}J_{-n}(x) = \left(\frac{x}{2}\right)^{-n}\left(\frac{x}{2}\right)^{n}\left(\frac{x}{2}\right)^{-n}\sum_{k=n}^{\infty}(-1)^{k-n}\frac{(x/2)^{2k}}{k!(k-n)!} \tag{13.46}$$

$$= \left(\frac{x}{2}\right)^{n}\sum_{k=n}^{\infty}(-1)^{k-n}\frac{(x/2)^{2(k-n)}}{k!(k-n)!}. \tag{13.47}$$

We now let $k - n \to k'$ and then drop primes to see that J_{-n} and J_n are proportional to each other as

$$J_{-n}(x) = (-1)^n J_n(x).$$

(13.48)

When m takes integer values, the second and linearly independent solution can be taken as

$$N_m(x) = \frac{\cos m\pi J_m(x) - J_{-m}(x)}{\sin m\pi},$$

(13.49)

which is called the **Neumann function** or the Bessel function of the **second kind**. Note that $N_m(x)$ and $J_m(x)$ are linearly independent even for the non-integer values of m. Hence, it is common practice to take $N_m(x)$ and $J_m(x)$ as the two linearly independent solutions for all m (Figure 13.1).

Since N_m is indeterminate for the integer values of m, we can find the series expression of N_m by considering the limit $m \to n$ (integer) and the L'Hôpital's rule:

$$N_n(x) = \lim_{m \to n} N_m(x) = \lim_{m \to n} \frac{\partial_m[\cos m\pi J_m(x) - J_{-m}(x)]}{\partial_m[\sin m\pi]}$$

(13.50)

$$= \frac{1}{\pi} \lim_{m \to n} \left[\frac{\partial J_m}{\partial m} - (-1)^m \frac{\partial J_{-m}}{\partial m} \right].$$

(13.51)

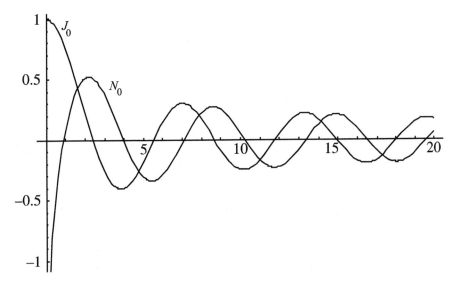

Figure 13.1   Bessel functions, J_0 and N_0.

We first evaluate the derivative $\partial_m J_m$ as

$$\frac{\partial J_m}{\partial m} = J_m(x)\ln\left(\frac{x}{2}\right) - \left(\frac{x}{2}\right)^m \sum_{k=0}^{\infty}(-1)^k \frac{\Psi(m+k+1)}{k!\Gamma(m+k+1)}\left(\frac{x}{2}\right)^{2k}, \qquad (13.52)$$

where we used the definition

$$\Psi(x) = \frac{\Gamma'(x)}{\Gamma(x)}. \qquad (13.53)$$

Some of the properties of $\Psi(x)$ are $\Psi(1) = -\gamma$, where $\gamma = 0.57721566$ is the **Euler constant** and

$$\Psi(m+1) = -\gamma + 1 + \frac{1}{2} + \cdots + \frac{1}{m}, \quad m = 1, 2, 3, \ldots . \qquad (13.54)$$

We can now write the limit

$$\lim_{m\to n}\frac{\partial J_m}{\partial m} = J_n(x)\ln\left(\frac{x}{2}\right) - \left(\frac{x}{2}\right)^n \sum_{k=0}^{\infty}(-1)^k \frac{\Psi(n+k+1)}{k!\Gamma(n+k+1)}\left(\frac{x}{2}\right)^{2k} \qquad (13.55)$$

$$= \sum_{k=0}^{\infty}(-1)^k \frac{(x/2)^{n+2k}}{k!(n+k)!}\left[\ln\frac{x}{2} - \Psi(k+n+1)\right]. \qquad (13.56)$$

Similarly, we write the derivative

$$\frac{\partial J_{-m}}{\partial m} = \sum_{k=0}^{\infty}(-1)^k \frac{(x/2)^{-m+2k}}{k!\Gamma(k-m+1)}\left[-\ln\frac{x}{2} + \Psi(k-m+1)\right]. \qquad (13.57)$$

In the limit as $m \to n$ (integer), for the first n terms, $k = 0, 1, \ldots, n-1$, $\Gamma(k-m+1)$ and $\Psi(k-m+1)$ are infinite since their arguments are zero or negative integer (Problem 6):

$$\lim_{m\to n}\begin{cases}\Gamma(k-m+1) &\to \infty, \\[2mm] \Psi(k-m+1) &\to \infty.\end{cases}, \quad k = 0, 1, \ldots, (n-1), \qquad (13.58)$$

However, using the well-known results from the theory of gamma functions:

$$\Gamma(x)\Gamma(1-x) = \frac{\pi}{\sin \pi x}, \qquad (13.59)$$

$$\Psi(1-x) - \Psi(x) = \pi \cot \pi x, \qquad (13.60)$$

their ratio is constant:

$$\lim_{m \to n} \frac{\Psi(k-m+1)}{\Gamma(k-m+1)} = \lim_{m \to n} \left[\Gamma(m-k) \sin \pi(m-k) \frac{\Psi(m-k) + \pi \cot \pi(m-k)}{\pi} \right]$$

(13.61)

$$= (-1)^{n-k}(n-k-1)!, \ k = 0, 1, \ldots, (n-1). \qquad (13.62)$$

We can now write

$$\lim_{m \to n} \frac{\partial J_{-m}}{\partial m} = (-1)^n \sum_{k=0}^{n-1} \frac{(n-k-1)!}{k!} \left(\frac{x}{2}\right)^{2k-n}$$

$$+ (-1)^n \sum_{p=0}^{\infty} (-1)^p \frac{(x/2)^{n+2p}}{p!(n+p)!} \left[-\ln \frac{x}{2} + \Psi(p+1) \right], \qquad (13.63)$$

where we have defined a new dummy index in the second sum as

$$p = k - n. \qquad (13.64)$$

We now substitute the limits in Eqs. (13.56) and (13.63) into Eq. (13.51) to find the series expression of $N_n(x)$ as

$$N_n(x) = -\frac{1}{\pi} \sum_{k=0}^{n-1} \frac{(n-k-1)!}{k!} \left(\frac{x}{2}\right)^{2k-n}$$

$$+ \frac{1}{\pi} \sum_{k=0}^{\infty} (-1)^k \frac{(x/2)^{n+2k}}{k!(n+k)!} \left[2 \ln \frac{x}{2} - \Psi(k+1) - \Psi(k+n+1) \right],$$

(13.65)

where $n = 0, 1, \ldots$, and the first sum should be set to zero for $n = 0$. We can also write this as

$$N_n(x) = \frac{2}{\pi} J_n(x) \ln \frac{x}{2} - \frac{1}{\pi} \sum_{k=0}^{n-1} \frac{(n-k-1)!}{k!} \left(\frac{x}{2}\right)^{2k-n}$$

$$- \frac{1}{\pi} \sum_{k=0}^{\infty} (-1)^k \frac{(x/2)^{n+2k}}{k!(n+k)!} [\Psi(k+1) + \Psi(k+n+1)]. \qquad (13.66)$$

Other linearly independent solutions of the Bessel's equation are given as the **Hankel functions**, $H_m^{(1)}(x)$, $H_m^{(2)}(x)$, which are defined as

$$\boxed{H_m^{(1)}(x) = J_m(x) + iN_m(x)} \qquad (13.67)$$

and

$$H_m^{(2)}(x) = J_m(x) - iN_m(x).$$

(13.68)

Hankel functions are also called the Bessel functions of the **third kind**. The motivation for introducing $H_m^{(1,2)}(x)$ is that they have very simple asymptotic expressions for large x, which makes them very useful in applications. In the limit as $x \to 0$, the Bessel function $J_m(x)$ is finite for $m \geq 0$ and behaves as

$$\lim_{x \to 0} J_m(x) \to \frac{1}{\Gamma(m+1)} \left(\frac{x}{2}\right)^m.$$

(13.69)

All the other functions diverge as

$$\lim_{x \to 0} N_m(x) \to \begin{cases} \frac{2}{\pi}\left[\ln\left(\frac{x}{2}\right) + \gamma\right], & m = 0, \\ -\frac{\Gamma(m)}{\pi}\left(\frac{2}{x}\right)^m, & m \neq 0, \end{cases}$$

(13.70)

$$\lim_{x \to 0} H_0^{(1)}(x) \to 1 - i\frac{2}{\pi}\ln\left(\frac{2}{x}\right),$$

(13.71)

$$\lim_{x \to 0} H_0^{(2)}(x) \to 1 + i\frac{2}{\pi}\ln\left(\frac{2}{x}\right),$$

(13.72)

$$\lim_{x \to 0} H_m^{(1)}(x) \to -i\frac{\Gamma(m)}{\pi}\left(\frac{2}{x}\right)^m, \quad m \neq 0,$$

(13.73)

$$\lim_{x \to 0} H_m^{(2)}(x) \to i\frac{\Gamma(m)}{\pi}\left(\frac{2}{x}\right)^m, \quad m \neq 0,$$

(13.74)

In the limit as $x \to \infty$, the Bessel functions, $J_m(x)$, $N_m(x)$, $H_m^{(1)}(x)$, and $H_m^{(2)}(x)$, with $m > 0$, behave as

$$\lim_{x \to \infty} J_m(x) \to \sqrt{\frac{2}{\pi x}} \cos\left(x - \frac{m\pi}{2} - \frac{\pi}{4}\right),$$

(13.75)

$$\lim_{x \to \infty} N_m(x) \to \sqrt{\frac{2}{\pi x}} \sin\left(x - \frac{m\pi}{2} - \frac{\pi}{4}\right).$$

(13.76)

$$\lim_{x \to \infty} H_m^{(1)}(x) \to \sqrt{\frac{2}{\pi x}} \exp\left[i\left(x - \frac{m\pi}{2} - \frac{\pi}{4}\right)\right],$$

(13.77)

$$\lim_{x \to \infty} H_m^{(2)}(x) \to \sqrt{\frac{2}{\pi x}} \exp\left[-i\left(x - \frac{m\pi}{2} - \frac{\pi}{4}\right)\right].$$

(13.78)

We remind the reader that $m \geq 0$ in these equations.

13.1.2   Recursion Relations

Using the series definition of Bessel functions, we have obtained the following recursion relations [Eqs. (13.41) and (13.42)]:

$$J_{m-1}(x) + J_{m+1}(x) = \frac{2m}{x} J_m(x), \quad m = 1, 2, \ldots, \tag{13.79}$$

$$J_{m-1}(x) - J_{m+1}(x) = 2J'_m(x), \quad m = 1, 2, \ldots. \tag{13.80}$$

First by adding and then by subtracting these equations, we obtain two more relations:

$$J_{m-1}(x) = \frac{m}{x} J_m(x) + J'_m(x) = x^{-m} \frac{d}{dx} [x^m J_m(x)], \tag{13.81}$$

$$J_{m+1}(x) = \frac{m}{x} J_m(x) - J'_m(x) = -x^m \frac{d}{dx} [x^{-m} J_m(x)]. \tag{13.82}$$

Other Bessel functions, N_n, $H_n^{(1)}$, and $H_n^{(2)}$, satisfy the same recursion relations.

13.1.3   Generating Function

Similar to the other special functions, we look for a generating function, $g(x, t)$, which produces Bessel functions as

$$g(x, t) = \sum_{m=-\infty}^{\infty} t^m J_m(x). \tag{13.83}$$

Multiply the relation

$$J_{m-1} + J_{m+1} = \frac{2m}{x} J_m \tag{13.84}$$

with t^m and sum over m:

$$\sum_{m=-\infty}^{\infty} t^m J_{m-1} + \sum_{m=-\infty}^{\infty} t^m J_{m+1} = \frac{2}{x} \sum_{m=-\infty}^{\infty} m t^m J_m, \tag{13.85}$$

$$t \sum_{m=-\infty}^{\infty} t^{m-1} J_{m-1} + \frac{1}{t} \sum_{m=-\infty}^{\infty} t^{m+1} J_{m+1} = \frac{2}{x} \sum_{m=-\infty}^{\infty} m t^m J_m. \tag{13.86}$$

We first substitute $m - 1 = m'$ into the first sum and $m + 1 = m''$ into the second and then drop primes to get

$$t \sum_{m=-\infty}^{\infty} t^m J_m + \frac{1}{t} \sum_{m=-\infty}^{\infty} t^m J_m = \frac{2t}{x} \frac{\partial g}{\partial t}, \tag{13.87}$$

$$tg + \frac{1}{t}g = \frac{2t}{x}\frac{\partial g}{\partial t}, \tag{13.88}$$

$$\left(t + \frac{1}{t}\right)g = \frac{2t}{x}\frac{\partial g}{\partial t}. \tag{13.89}$$

This can be reduced to the quadrature

$$\int \frac{dg}{g} = \frac{x}{2}\int\left(1 + \frac{1}{t^2}\right)dt, \tag{13.90}$$

to yield the generating function $g(x,t)$ as

$$g(x,t) = \phi(x)\exp\left[\frac{x}{2}\left(t - \frac{1}{t}\right)\right], \tag{13.91}$$

where $\phi(x)$ is a function to be determined. We now write the aforementioned equation as

$$g(x,t) = \phi(x)e^{xt/2}e^{-x/2t} \tag{13.92}$$

and substitute series expressions for the exponentials to get

$$g(x,t) = \phi(x)\left[\sum_{n=0}^{\infty}\left(\frac{xt}{2}\right)^n\frac{1}{n!}\right]\left[\sum_{m=0}^{\infty}\left(\frac{-x}{2t}\right)^m\frac{1}{m!}\right] \tag{13.93}$$

$$= \phi(x)\sum_{n=0}^{\infty}\sum_{m=0}^{\infty}(-1)^m\frac{x^{n+m}t^{n-m}}{2^{n+m}n!m!}. \tag{13.94}$$

Since $g(x,t) = \sum_{m=-\infty}^{\infty}t^m J_m$, we can write

$$\phi(x)\sum_{n=0}^{\infty}\sum_{m=0}^{\infty}(-1)^m\frac{x^{n+m}t^{n-m}}{2^{n+m}n!m!} = \sum_{m'=-\infty}^{\infty}t^{m'}J_{m'} \tag{13.95}$$

$$= \cdots + \frac{1}{t}J_{-1} + J_0 + tJ_1 + \cdots. \tag{13.96}$$

To extract $\phi(x)$, it is sufficient to look at the simplest term, that is, the coefficient of t^0. In the double sum on the left-hand side, this means taking only the terms with $m = n$. Hence,

$$\phi(x)\left[\sum_{m}(-1)^m\frac{x^{2m}}{2^{2m}(m!)^2}\right] = J_0. \tag{13.97}$$

The quantity inside the square brackets is nothing but J_0, thus determining $\phi(x)$ as 1. We can now write the generating function as

$$\boxed{\exp\left[\left(\frac{x}{2}\right)\left(t - \frac{1}{t}\right)\right] = \sum_{m=-\infty}^{\infty}t^m J_m.} \tag{13.98}$$

13.1.4 Integral Definitions

Since the generating function definition works only for integer orders, integral definitions are developed for arbitrary orders. Among the most commonly used integral definitions, we have

$$J_m(x) = \frac{1}{\pi} \int_0^\pi \cos[m\varphi - x\sin\varphi]d\varphi - \frac{\sin m\pi}{\pi} \int_0^\infty e^{-x\sinh\theta - m\theta}d\theta, \quad (13.99)$$

and

$$J_m(x) = \frac{(x/2)^m}{\sqrt{\pi}\Gamma(m + \frac{1}{2})} \int_{-1}^1 dt(1 - t^2)^{m-\frac{1}{2}} \cos xt, \quad m > -\frac{1}{2}. \quad (13.100)$$

For the integer values of m, the second term in the first definition [Eq. (13.99)] vanishes, thus leaving

$$J_m(x) = \frac{1}{\pi} \int_0^\pi \cos[m\varphi - x\sin\varphi]d\varphi, \quad m = 0, \pm 1, \pm 2, \ldots, \quad (13.101)$$

which can be proven by using the generating function (Problem 14). We prove the second definition [Eq. (13.100)], which is due to Poisson, by using the integral representation of the gamma function (Problem 6):

$$\frac{1}{\Gamma(k + m + 1)} = \frac{1}{\Gamma(k + \frac{1}{2})\Gamma(m + \frac{1}{2})} \int_{-1}^1 t^{2k}(1 - t^2)^{m-1/2}dt, \quad m > -1/2, \quad (13.102)$$

which, when substituted into the definition of the Bessel function [Eq. (13.44)]:

$$J_m(x) = \sum_{k=0}^\infty \frac{(-1)^k}{\Gamma(k + 1)\Gamma(m + k + 1)} \left(\frac{x}{2}\right)^{m+2k}, \quad (13.103)$$

gives

$$J_m(x) = \sum_{k=0}^\infty \frac{(-1)^k(x/2)^{m+2k}}{\Gamma(k + 1)\Gamma(k + \frac{1}{2})\Gamma(m + \frac{1}{2})} \int_{-1}^1 t^{2k}(1 - t^2)^{m-1/2}dt. \quad (13.104)$$

Since the series converges absolutely, we can interchange the summation and the integral signs to write

$$J_m(x) = \frac{(x/2)^m}{\Gamma(m + \frac{1}{2})} \int_{-1}^1 dt(1 - t^2)^{m-1/2} \left[\sum_{k=0}^\infty \frac{(-1)^k(xt/2)^{2k}}{\Gamma(k + 1)\Gamma(k + \frac{1}{2})}\right]. \quad (13.105)$$

Finally, using the so-called **duplication formula** of the gamma functions:

$$2^{2k}\Gamma(k+1)\Gamma\left(k+\frac{1}{2}\right) = \Gamma(1/2)\Gamma(2k+1) \tag{13.106}$$

$$= \Gamma(1/2)(2k)!, \tag{13.107}$$

where $\Gamma(1/2) = \sqrt{\pi}$, we obtain the desired integral representation:

$$J_m(x) = \frac{(x/2)^m}{\Gamma(m+\frac{1}{2})} \int_{-1}^{1} dt(1-t^2)^{m-1/2} \sum_{k=0}^{\infty} \frac{(-1)^k(xt)^{2k}}{\Gamma(1/2)(2k)!} \tag{13.108}$$

$$= \frac{(x/2)^m}{\Gamma(1/2)\Gamma(m+\frac{1}{2})} \int_{-1}^{1} dt(1-t^2)^{m-1/2} \cos xt. \tag{13.109}$$

13.1.5   Linear Independence of Bessel Functions

Two functions, u_1 and u_2, are linearly independent if and only if their Wronskian:

$$W[u_1, u_2] = \begin{vmatrix} u_1 & u_2 \\ u_1' & u_2' \end{vmatrix}, \tag{13.110}$$

does not vanish identically. Let u_1 and u_2 be two solutions of Bessel's equation:

$$\frac{d(xu_1')}{dx} + \left(x - \frac{m^2}{x}\right)u_1 = 0, \tag{13.111}$$

$$\frac{d(xu_2')}{dx} + \left(x - \frac{m^2}{x}\right)u_2 = 0. \tag{13.112}$$

We first multiply Eq. (13.111) by u_2 and Eq. (13.112) by u_1 and then subtract to write

$$u_1\frac{d}{dx}(xu_2') - u_2\frac{d}{dx}(xu_1') = 0, \tag{13.113}$$

$$\frac{d}{dx}[x(u_1u_2' - u_2u_1')] = 0, \tag{13.114}$$

$$\frac{d}{dx}[xW(x)] = 0. \tag{13.115}$$

Hence, the Wronskian of two solutions of the Bessel's equation is

$$W[u_1, u_2] = \frac{C}{x}, \tag{13.116}$$

where C is a constant. If we take

$$u_1 = J_m \text{ and } u_2 = J_{-m}, \tag{13.117}$$

C can be calculated by the limit

$$C = \lim_{x \to 0} xW[J_m, J_{-m}], \tag{13.118}$$

where m is a noninteger. Using the asymptotic expansion of $J_m(x)$ [Eq. (13.69)] and the following properties of the gamma function:

$$\Gamma(x+1) = x\Gamma(x),$$
$$\Gamma(x)\Gamma(1-x) = \frac{\pi}{\sin \pi x}, \tag{13.119}$$

we can determine C as

$$C = \lim_{x \to 0} \frac{-2m}{\Gamma(1+m)\Gamma(1-m)}[1 + 0(x^2)] = -\frac{2\sin m\pi}{\pi},$$

thus obtaining

$$W[J_m, J_{-m}] = -\frac{2\sin m\pi}{\pi x}. \tag{13.120}$$

Similarly, we can show the following **Wronskians**:

$$\boxed{W[J_m, H_m^{(2)}] = -\frac{2i}{\pi x},} \tag{13.121}$$

$$\boxed{W[H_m^{(1)}, H_m^{(2)}] = -\frac{4i}{\pi x},} \tag{13.122}$$

$$\boxed{W[J_m, N_m] = \frac{2}{\pi x}.} \tag{13.123}$$

This establishes the linear independence of the functions J_m, N_m, $H_m^{(1)}$, and $H_m^{(2)}$. Hence, the solution of the Bessel's equation:

$$x^2\frac{d^2y_m}{dx^2} + x\frac{dy_m}{dx} + (x^2 - m^2)y_m = 0, \tag{13.124}$$

can be written in any one of the following ways:

$$y_m(x) = a_1 J_m + a_2 N_m \tag{13.125}$$
$$= b_1 J_m + b_2 H_m^{(1)} \tag{13.126}$$
$$= c_1 J_m + c_2 H_m^{(2)} \tag{13.127}$$
$$= d_1 H_m^{(1)} + d_2 H_m^{(1)}, \tag{13.128}$$

where $a_1, a_2, \ldots, d_2$ are constants to be determined from the initial conditions.

13.1.6 Modified Bessel Functions $I_m(x)$ and $K_m(x)$

If we take the arguments of the Bessel functions $J_m(x)$ and $H_m^{(1)}(x)$ as imaginary, we obtain the modified Bessel functions (Figures 13.2 and 13.3):

$$I_m(x) = \frac{J_m(ix)}{i^m} \qquad (13.129)$$

and

$$K_m(x) = \frac{\pi i^{m+1}}{2} H_m^{(1)}(ix). \qquad (13.130)$$

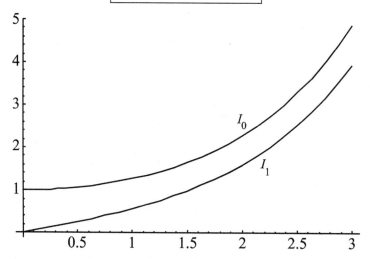

Figure 13.2 Modified Bessel functions I_0 and I_1.

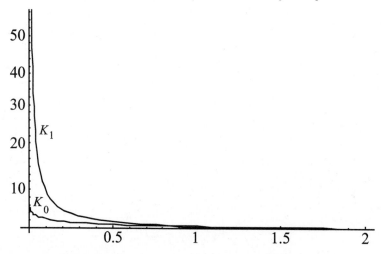

Figure 13.3 Modified Bessel functions K_0 and K_1.

These functions are linearly independent solutions of the differential equation

$$\frac{d^2y}{dx^2} + \frac{1}{x}\frac{dy}{dx} - \left(1 + \frac{m^2}{x^2}\right)y(x) = 0. \tag{13.131}$$

Their $x \to 0$ and $x \to \infty$ limits are given as (real $m \geq 0$)

$$\lim_{x\to 0} I_m(x) \to \frac{(x/2)^m}{\Gamma(m+1)}, \tag{13.132}$$

$$\lim_{x\to 0} K_m(x) \to \begin{cases} -\left[\ln\left(\frac{x}{2}\right) + \gamma\right], & m = 0, \\ \frac{\Gamma(m)}{2}\left(\frac{2}{x}\right)^m, & m \neq 0 \end{cases} \tag{13.133}$$

and

$$\lim_{x\to\infty} I_m(x) \to \frac{1}{\sqrt{2\pi x}}e^x\left[1 + 0\left(\frac{1}{x}\right)\right], \tag{13.134}$$

$$\lim_{x\to\infty} K_m(x) \to \sqrt{\frac{\pi}{2x}}e^{-x}\left[1 + 0\left(\frac{1}{x}\right)\right]. \tag{13.135}$$

13.1.7 Spherical Bessel Functions $j_l(x)$, $n_l(x)$, and $h_l^{(1,2)}(x)$

Bessel functions with half-integer orders often appear in applications with the factor $\sqrt{\pi/2x}$, so that they are given a special name. Spherical Bessel functions, $j_l(x)$, $n_l(x)$, and $h_l^{(1,2)}(x)$, are defined as (Figures 13.4 and 13.5)

$$j_l(x) = \sqrt{\frac{\pi}{2x}}J_{l+\frac{1}{2}}(x), \tag{13.136}$$

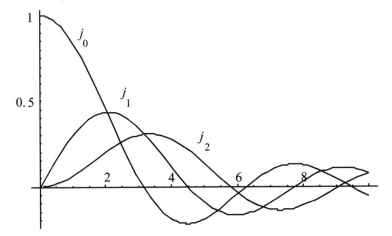

Figure 13.4 Spherical Bessel functions j_0, j_1, and j_2.

$$n_l(x) = \sqrt{\frac{\pi}{2x}} N_{l+\frac{1}{2}}(x), \tag{13.137}$$

$$h_l^{(1,2)}(x) = \left(\frac{\pi}{2x}\right)^{1/2} \left[J_{l+\frac{1}{2}}(x) \pm iN_{l+\frac{1}{2}}(x)\right], \tag{13.138}$$

where $l = 0, \pm 1, \pm 2, \ldots$. Bessel functions with half-integer indices, $J_{l+\frac{1}{2}}(x)$ and $N_{l+\frac{1}{2}}(x)$, satisfy the differential equation

$$\frac{d^2y_l(x)}{dx^2} + \frac{1}{x}\frac{dy_l(x)}{dx} + \left[1 - \frac{\left(l+\frac{1}{2}\right)^2}{x^2}\right] y_l(x) = 0, \tag{13.139}$$

while the spherical Bessel functions, $j_l(x)$, $n_l(x)$, and $h_l^{(1,2)}(x)$ satisfy

$$\frac{d^2y_l(x)}{dx^2} + \frac{2}{x}\frac{dy_l(x)}{dx} + \left[1 - \frac{l(l+1)}{x^2}\right] y_l(x) = 0. \tag{13.140}$$

Series expansions of the spherical Bessel functions are given as

$$j_l(x) = 2^l x^l \sum_{n=0}^{\infty} \frac{(-1)^n (l+n)!}{n!(2n+2l+1)!} x^{2n}, \tag{13.141}$$

$$n_l(x) = \frac{(-1)^{l+1}}{2^l x^{l+1}} \sum_{n=0}^{\infty} \frac{(-1)^n (n-l)!}{n!(2n-2l)!} x^{2n}, \tag{13.142}$$

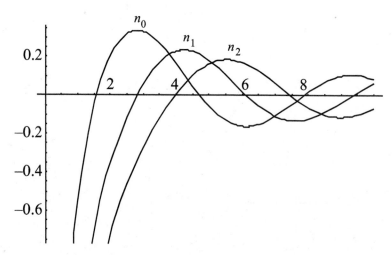

Figure 13.5   Spherical Bessel functions n_0, n_1, and n_2.

where the first equation can be derived using Eq. (13.44), the duplication formula [Eq. (13.107)]: $2^{2l+2n}(l+n)!\Gamma(l+n+1/2) = \sqrt{\pi}(2l+2n)!$ and $\Gamma(n+1) = n\Gamma(n)$. Spherical Bessel functions can also be defined as

$$j_l(x) = (-x)^l \left(\frac{1}{x}\frac{d}{dx}\right)^l \left(\frac{\sin x}{x}\right), \qquad (13.143)$$

$$n_l(x) = (-x)^l \left(\frac{1}{x}\frac{d}{dx}\right)^l \left(-\frac{\cos x}{x}\right). \qquad (13.144)$$

The first two spherical Bessel functions are

$$j_0(x) = \frac{\sin x}{x}, \qquad n_0(x) = -\frac{\cos x}{x},$$

$$j_1(x) = \frac{\sin x}{x^2} - \frac{\cos x}{x}, \qquad n_1(x) = -\frac{\cos x}{x^2} - \frac{\sin x}{x} \qquad (13.145)$$

and

$$h_0^{(1)}(x) = -i\frac{e^{ix}}{x}, \qquad h_0^{(2)}(x) = i\frac{e^{-ix}}{x},$$

$$h_1^{(1)}(x) = e^{ix}\left[-\frac{1}{x} - \frac{i}{x^2}\right], \qquad h_1^{(2)}(x) = e^{-ix}\left[\frac{i}{x^2} - \frac{1}{x}\right]. \qquad (13.146)$$

Using the recursion relations given in Eqs. (13.81) and (13.82), one can easily obtain the following **recursion relations:**

$$\frac{d}{dx}[x^{l+1}y_l(x)] = x^{l+1}y_{n-1}(x), \qquad (13.147)$$

$$\frac{d}{dx}[x^{-l}y_l(x)] = -x^{-l}y_{n+1}(x), \qquad (13.148)$$

where y_l stands for any of the spherical Bessel functions, j_l, n_l, or $h_l^{(1,2)}$. **Asymptotic forms** of the spherical Bessel functions are given as

$$j_l(x) \rightarrow \frac{x^l}{(2l+1)!!}\left(1 - \frac{x^2}{2(2l+3)} + \cdots\right), \qquad x \ll 1, \qquad (13.149)$$

$$n_l(x) \rightarrow -\frac{(2l-1)!!}{x^{l+1}}\left(1 - \frac{x^2}{2(1-2l)} + \cdots\right), \qquad x \ll 1, \qquad (13.150)$$

$$j_l(x) \rightarrow \frac{1}{x}\sin\left(x - \frac{l\pi}{2}\right), \qquad x \gg 1, \qquad (13.151)$$

$$n_l(x) \rightarrow -\frac{1}{x}\cos\left(x - \frac{l\pi}{2}\right), \qquad x \gg 1, \qquad (13.152)$$

where the **double factorial** is defined as $(2l + 1)!! = (2l + 1)(2l − 1)$ $(2l − 3) \cdots 5 \cdot 3 \cdot 1 = (2l + 1)!/2^l l!$.

13.2 ORTHOGONALITY AND THE ROOTS OF BESSEL FUNCTIONS

Bessel functions also satisfy an orthogonality relation. Unlike the Legendre, Hermite, or Laguerre polynomials, their orthogonality is not with respect to their order but with respect to a parameter in their argument. We now write the Bessel equation with respect to a new argument, kx, and replace x with kx in Eq. (13.1) to write

$$\frac{d^2 J_m(kx)}{dx^2} + \frac{1}{x}\frac{dJ_m(kx)}{dx} + \left(k^2 - \frac{m^2}{x^2}\right) J_m(kx) = 0, \quad (13.153)$$

where k is a new parameter. For another parameter l, we write

$$\frac{d^2 J_m(lx)}{dx^2} + \frac{1}{x}\frac{dJ_m(lx)}{dx} + \left(l^2 - \frac{m^2}{x^2}\right) J_m(lx) = 0. \quad (13.154)$$

Multiply Eq. (13.153) by $xJ_m(lx)$ and Eq. (13.154) by $xJ_m(kx)$ and then subtract to obtain

$$\frac{d}{dx}[x(J_m(kx)J'_m(lx) - J'_m(kx)J_m(lx))] = (k^2 - l^2)xJ_m(kx)J_m(lx). \quad (13.155)$$

In physical applications, the range is usually $[0, a]$, hence we integrate both sides from zero to a to write

$$(k^2 - l^2)\int_0^a xJ_m(kx)J_m(lx)\, dx = a[J_m(ka)J'_m(la) - J'_m(ka)J_m(la)] \quad (13.156)$$

and substitute the definitions

$$k = \frac{x_{mi}}{a}, \ l = \frac{x_{mj}}{a}, \quad (13.157)$$

which gives

$$(x_{mi}^2 - x_{mj}^2)\left(\frac{1}{a^2}\right)\int_0^a xJ_m\left(\frac{x_{mi}}{a}x\right) J_m\left(\frac{x_{mj}}{a}x\right) dx$$

$$= a\left[J_m\left(\frac{x_{mi}}{a}a\right) J'_m\left(\frac{x_{mj}}{a}a\right) - J'_m\left(\frac{x_{mi}}{a}a\right) J_m\left(\frac{x_{mj}}{a}a\right)\right]. \quad (13.158)$$

In most physical applications, the boundary conditions imposed on the system are such that the right-hand side vanishes:

$$J_m\left(\frac{x_{mi}}{a}a\right) J'_m\left(\frac{x_{mj}}{a}a\right) = J'_m\left(\frac{x_{mi}}{a}a\right) J_m\left(\frac{x_{mj}}{a}a\right). \quad (13.159)$$

Along with the aforementioned condition, when $x_{mi} \neq x_{mj}$, Eq. (13.158) implies

$$\int_0^a x J_m\left(\frac{x_{mi}}{a}x\right) J_m\left(\frac{x_{mj}}{a}x\right) dx = 0. \tag{13.160}$$

To complete the orthogonality relation, we have to evaluate the integral

$$I_{ij} = \int_0^a x J_m\left(\frac{x_{mi}}{a}x\right) J_m\left(\frac{x_{mj}}{a}x\right) dx \tag{13.161}$$

for $x_{mi} = x_{mj}$, that is $i = j$. Calling

$$\frac{x_{mi}}{a}x = t, \tag{13.162}$$

we write $I_{ii} = I$ as

$$I = \int_0^a x J_m^2\left(\frac{x_{mi}}{a}x\right) dx = \left(\frac{a}{x_{mi}}\right)^2 \int_0^{x_{mi}} t J_m^2(t)\, dt. \tag{13.163}$$

When we integrate Eq. (13.163) by parts, $\int u\, dv = uv - \int v\, du$, with the definitions $u = J_m^2(t)$ and $dv = t\, dt$, we obtain

$$\left(\frac{x_{mi}}{a}\right)^2 I = \frac{1}{2}t^2 J_m^2(t)\Big|_0^{x_{mi}} - \int_0^{x_{mi}} J_m(t) J_m'(t) t^2\, dt. \tag{13.164}$$

Using the Bessel's equation [Eq. (13.1)]:

$$t^2 J_m(t) = m^2 J_m(t) - t J_m'(t) - t^2 J''(t), \tag{13.165}$$

we now write Eq. (13.164) as

$$\left(\frac{x_{mi}}{a}\right)^2 I = \frac{1}{2}t^2 J_m^2(t)\Big|_0^{x_{mi}} - \int_0^{x_{mi}} J_m'[m^2 J_m - t J_m'^2 - t^2 J_m'']\, dt, \tag{13.166}$$

which can be simplified:

$$\left(\frac{x_{mi}}{a}\right)^2 I = \frac{1}{2}t^2 J_m^2(t)\Big|_0^{x_{mi}} - m^2\int_0^{x_{mi}} J_m J_m'\, dt + \frac{1}{2}\int_0^{x_{mi}} \frac{d}{dt}[t^2 J_m'^2]\, dt \tag{13.167}$$

$$= \frac{1}{2}t^2 J_m^2(t)\Big|_0^{x_{mi}} - m^2\frac{1}{2}\int_0^{x_{mi}} \frac{d}{dt}[J_m^2]\, dt + \frac{1}{2}\int_0^{x_{mi}} \frac{d}{dt}[t^2 J_m'^2]\, dt \tag{13.168}$$

$$= \left[\frac{1}{2}t^2 J_m^2(t) - \frac{1}{2}m^2 J_m^2(t) + \frac{1}{2}t^2 J_m'^2(t)\right]_0^{x_{mi}}. \tag{13.169}$$

For $m > -1$, the right-hand side vanishes for $x = 0$. For $x = a$, we usually impose the boundary condition

$$J_m \left(\frac{x_{mi}}{a} x \right) \bigg|_{x=a} = 0,$$

or

$$J_m(x_{mi}) = 0. \tag{13.170}$$

In other words, x_{mi} are the roots of $J_m(x)$.

From the asymptotic form [Eq. (13.75)] of the Bessel function, it is clear that it has infinitely many distinct positive roots (Figure 13.6).

$$J_m(x_{mi}) = 0, \quad i = 1, 2, 3, \ldots, \tag{13.171}$$

where x_{mi} stands for the ith root of the mth-order Bessel function. When m takes integer values, the first three roots are given as

$$
\begin{array}{llll}
m = 0 & x_{01} = 2.405 & x_{02} = 5.520 & x_{03} = 8.654 & \cdots \\
m = 1 & x_{11} = 3.832 & x_{12} = 7.016 & x_{13} = 10.173 & \cdots \\
m = 2 & x_{21} = 5.136 & x_{22} = 8.417 & x_{23} = 11.620 & \cdots
\end{array}
\tag{13.172}
$$

Higher-order roots are approximately given by the formula

$$\boxed{x_{mi} \simeq \pi i + \left(m - \frac{1}{2} \right) \frac{\pi}{2}, \quad i = 1, 2, \ldots .} \tag{13.173}$$

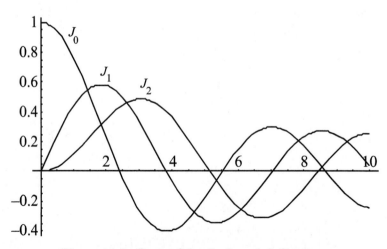

Figure 13.6 Roots of the J_0, J_1, and J_2 functions.

We now write Eq. (13.169) as

$$\left(\frac{x_{mi}}{a}\right)^2 I = \frac{1}{2}x_{mi}^2 J_m'^2(x_{mi}), \tag{13.174}$$

which, after using with Eq. (13.163), gives

$$\int_0^a x J_m^2\left(\frac{x_{mi}}{a}x\right) dx = \frac{1}{2}a^2 J_m'^2(x_{mi}). \tag{13.175}$$

Using the recursion relation [Eq. (13.82)]

$$J_{m+1} = \frac{m}{x}J_m - J_m' \tag{13.176}$$

and the boundary condition $J_m(x_{mi}) = 0$, we can also write this as

$$\int_0^a x J_m^2\left(\frac{x_{mi}}{a}x\right) dx = \frac{1}{2}a^2 J_{m+1}^2(x_{mi}). \tag{13.177}$$

The **orthogonality relation** of Bessel functions can now be given as

$$\boxed{\int_0^a x J_m\left(\frac{x_{mi}}{a}x\right) J_m\left(\frac{x_{mj}}{a}x\right) dx = \frac{a^2}{2}J_{m+1}^2(x_{mi})\delta_{ij}, \quad m \geq -1.} \tag{13.178}$$

Since Bessel functions form a **complete set**, any sufficiently smooth function, $f(x)$, in the interval $x \in [0, a]$ can be expanded as

$$\boxed{f(x) = \sum_{i=1}^{\infty} A_{mi} J_m\left(\frac{x_{mi}}{a}x\right), \quad m \geq -1,} \tag{13.179}$$

where the expansion coefficients, A_{mi}, are found from

$$\boxed{A_{mi} = \frac{2}{a^2 J_{m+1}^2(x_{mi})}\int_0^a x f(x) J_m\left(\frac{x_{mi}}{a}x\right) dx.} \tag{13.180}$$

Series expansions [Eq. (13.179)] with the coefficients calculated using Eq. (13.180) are called the **Fourier–Bessel series**.

13.2.1   Expansion Theorem

If a given real function, $f(x)$, is piecewise continuous in the interval $(0, a)$ and of bounded variation in every subinterval $[a_1, a_2]$, where $0 < a_1 < a_2 < a$, and if the integral

$$\int_0^a \sqrt{x} |f(x)| dx \tag{13.181}$$

is finite, then the Fourier-Bessel series converges to $f(x)$ at every point of continuity. At points of discontinuity, the series converges to the mean of the right and left limits at that point:

$$\boxed{\frac{1}{2}[f(x^+) + f(x^-)].} \tag{13.182}$$

For the proof and the definition of bounded variation, we refer to [2, p. 355] and [1, p. 591]. However, bounded variation basically means that the displacement in the y direction, δy, as we move along the graph of the function is finite.

13.2.2   Boundary Conditions for the Bessel Functions

For the roots given in Eq. (13.172), we have used the **Dirichlet** boundary condition, that is,

$$J_m(ka) = 0, \tag{13.183}$$

which gives us the roots

$$k = \frac{x_{mi}}{a}. \tag{13.184}$$

Now the set of functions

$$\left\{ J_m\left(\frac{x_{mi}}{a} x\right) \right\}, \quad i = 1, 2, \ldots, \quad m \geq 0, \tag{13.185}$$

form a complete and an orthogonal set with respect to the index i in the interval $[0, a]$. The same conclusion holds for the **Neumann** boundary condition defined as

$$\left. \frac{dJ_m(x_{mi}x/a)}{dx} \right|_{x=a} = 0. \tag{13.186}$$

Similarly, the general boundary condition is written as

$$\left[A_0 \frac{dJ_m\left(x_{mi}\frac{x}{a}\right)}{dx} + B_0 J_m\left(x_{mi}\frac{x}{a}\right) \right]_{x=a} = 0. \tag{13.187}$$

In the case of **general** boundary conditions, using Eqs. (13.160) and (13.169), the orthogonality condition becomes

$$
\int_0^a x J_m \left(\frac{x_{mi}}{a} x\right) J_m \left(\frac{x_{mj}}{a} x\right) dx
$$

$$
= \begin{cases} 0, & x_{mi} \neq x_{mj}, \\ \left(\frac{a^2}{2}\right) \left[J'^2(x_{mi}) + \left(1 - \frac{m^2}{x_{mi}^2}\right) J^2(x_{mi}) \right], & x_{mi} = x_{mj}. \end{cases}
\tag{13.188}
$$

Now the expansion coefficients in the series [Eq. (13.179)]

$$
f(x) = \sum_{i=1}^{\infty} A_{mi} J_m \left(\frac{x_{mi}}{a} x\right), \quad m \geq -1,
\tag{13.189}
$$

become

$$
A_{mi} = \frac{2 \int_0^a x f(x) J_m \left(x_{mi} \frac{x}{a}\right) dx}{a^2 \left[J'^2(x_{mi}) + \left(1 - \frac{m^2}{x_{mi}^2}\right) J^2(x_{mi}), \right]} .
\tag{13.190}
$$

In terms of Bessel function $J_n(kx)$, Neumann and the general boundary conditions are written, respectively, as

$$
k \frac{dJ_n(x)}{dx} \bigg|_{x=ka} = 0
\tag{13.191}
$$

and

$$
\left[A_0 J_n(x) + B_0 k \frac{dJ_n(x)}{dx} \right]_{x=ka} = 0,
\tag{13.192}
$$

For the Neumann boundary condition [Eq. (13.186)], there exist infinitely many roots, which can be read from existing tables. For the general boundary condition, roots depend on the values that A_0 and B_0 take. Thus, each case must be handled separately by numerical analysis. From all three types of boundary conditions, we obtain a complete and an orthogonal set as

$$
\left\{ J_n \left(x_{nl} \frac{r}{a}\right) \right\}, \quad l = 1, 2, 3, \ldots .
\tag{13.193}
$$

Example 13.1. *Evaluation of* $\int_0^\infty e^{-kx} J_0(lx)$, $k, l > 0$: To evaluate this integral, we make use of the integral representation in Eq. (13.100). Replacing J_0 with its integral representation, we obtain

$$\int_0^\infty e^{-kx} J_0(lx)\, dx = \int_0^\infty dx\, e^{-kx} \frac{2}{\pi} \int_0^{\pi/2} \cos[lx \sin \varphi] d\varphi \qquad (13.194)$$

$$= \frac{2}{\pi} \int_0^{\pi/2} d\varphi \int_0^\infty dx\, e^{-kx} \cos[lx \sin \varphi] \qquad (13.195)$$

$$= \frac{2}{\pi} \int_0^{\pi/2} \frac{k\, d\varphi}{k^2 + l^2 \sin^2 \varphi} \qquad (13.196)$$

$$= \frac{1}{\sqrt{k^2 + l^2}}, \quad k, l > 0. \qquad (13.197)$$

Since the integral $\int_0^\infty e^{-kx} J_0(lx)\, dx$ is convergent, we have interchanged the order of the φ and x integrals in Eq. (13.195).

Example 13.2. *Evaluate* $\int_0^\infty e^{-k^2 x^2} J_m(lx) x^{m+1}\, dx$: This is also called the **Weber integral**, where $k, l > 0$ and $m > -1$. We use the series representation of J_m [Eq. (13.44)] to write

$$\int_0^\infty e^{-k^2 x^2} J_m(lx) x^{m+1}\, dx$$

$$= \int_0^\infty dx\, e^{-k^2 x^2} x^{m+1} \sum_{n=0}^\infty \frac{(-1)^n}{n! \Gamma(m+n+1)} \left(\frac{lx}{2}\right)^{m+2n} \qquad (13.198)$$

$$= \sum_{n=0}^\infty \frac{(-1)^n}{n! \Gamma(m+n+1)} \left(\frac{l}{2}\right)^{m+2n} \int_0^\infty dx\, e^{-k^2 x^2} x^{2m+2n+1} \qquad (13.199)$$

$$= \sum_{n=0}^\infty \frac{(-1)^n}{n! \Gamma(m+n+1)} \left(\frac{l}{2}\right)^{m+2n} \frac{1}{2k^{2m+2n+2}} \int_0^\infty dt\, e^{-t} t^{m+n} \qquad (13.200)$$

$$= \frac{l^m}{(2k^2)^{m+1}} \sum_{n=0}^\infty \frac{(-l^2/4k^2)^n}{n!} \qquad (13.201)$$

$$= \frac{l^m}{(2k^2)^{m+1}} e^{-l^2/4k^2}, \quad k, l > 0 \text{ and } m > -1. \qquad (13.202)$$

Since the sum converges absolutely, we have interchanged the summation and the integration signs and defined a new variable, $t = k^2 x^2$, in Eq. (13.199).

Example 13.3. *Crane problem:* We now consider small oscillations of a mass raised (or lowered) by a crane with uniform velocity. Equation of motion is given by

$$\frac{d}{dt}(ml^2 \dot{\theta}) + mgl \sin \theta = 0, \qquad (13.203)$$

where l is the length of the cable and m is the mass raised. For small oscillations, we can take $\sin \theta \simeq \theta$. For a crane operator changing the length of the cable with uniform velocity, $\frac{dl}{dt} = v_0$, we write the equation of motion as

$$l\ddot{\theta} + 2v_0\dot{\theta} + g\theta = 0. \tag{13.204}$$

We now switch to l as our independent variable. Using the derivatives

$$\dot{\theta} = \frac{d\theta}{dl}\frac{dl}{dt} = v_0\frac{d\theta}{dl}, \tag{13.205}$$

$$\ddot{\theta} = v_0\frac{d^2\theta}{dl^2}\frac{dl}{dt} = v_0^2\frac{d^2\theta}{dl^2}, \tag{13.206}$$

we can write the equation of motion in terms of l as

$$\frac{d^2\theta}{dl^2} + \frac{2}{l}\frac{d\theta}{dl} + \frac{g}{lv_0^2}\theta(l) = 0. \tag{13.207}$$

In applications, we usually encounter differential equations of the form

$$y'' + \frac{(1-2a)}{x}y' + \left[(bcx^{c-1})^2 + \frac{a^2 - p^2c^2}{x^2}\right]y(x) = 0, \tag{13.208}$$

solutions of which can be expressed in terms of Bessel functions as

$$y(x) = x^a[A_0 J_p(bx^c) + A_1 N_p(bx^c)]. \tag{13.209}$$

Applying to our case, we identify

$$1 - 2a = 2, \tag{13.210}$$

$$a^2 - p^2c^2 = 0, \tag{13.211}$$

$$c = \frac{1}{2}, \tag{13.212}$$

$$b^2c^2 = \frac{g}{v_0^2}, \tag{13.213}$$

which gives

$$a = -\frac{1}{2}, \ b = \frac{2\sqrt{g}}{v_0}, \ c = \frac{1}{2}, \ p = 1. \tag{13.214}$$

We can now write the general solution of Eq. (13.207) as

$$\theta(l) = \frac{1}{\sqrt{l}}\left[A_0 J_1\left(\frac{2\sqrt{gl}}{v_0}\right) + A_1 N_1\left(\frac{2\sqrt{gl}}{v_0}\right)\right]. \tag{13.215}$$

Time-dependent solution, $\theta(t)$, is obtained with the substitution $l(t) = l_0 + v_0 t$.

REFERENCES

1. Watson, G.N. (1962). *A Treatise on the Theory of Bessel Functions*, 2e. London: Cambridge University Press.

2. Titchmarsh, E.C. (1939). *The Theory of Functions*. New York: Oxford University Press.

PROBLEMS

1. Derive the following recursion relations:

$$\text{(i)} \quad J_{m-1}(x) + J_{m+1}(x) = \frac{2m}{x} J_m(x), \quad m = 1, 2, \ldots,$$

$$\text{(ii)} \quad J_{m-1}(x) - J_{m+1}(x) = 2J'_m(x), \quad m = 1, 2, \ldots.$$

Use the first equation to express a Bessel function of arbitrary order, $m = 0, 1, 2, \ldots$, in terms of $J_0(x)$ and $J_1(x)$. Also show that for $m = 0$ the second equation is replaced by $J'_0(x) = -J_1(x)$.

2. Derive Eq. (13.52):

$$\frac{\partial J_m}{\partial m} = J_m(x) \ln \left(\frac{x}{2}\right) - \left(\frac{x}{2}\right)^m \sum_{k=0}^{\infty} (-1)^k \frac{\Psi(m+k+1)}{k! \Gamma(m+k+1)} \left(\frac{x}{2}\right)^{2k}$$

and Eq. (13.57):

$$\frac{\partial J_{-m}}{\partial m} = \sum_{k=0}^{\infty} (-1)^k \frac{(x/2)^{-m+2k}}{k! \Gamma(k-m+1)} \left[-\ln \frac{x}{2} + \Psi(k-m+1)\right].$$

3. Verify the following Wronskians:

$$W[J_m, H_m^{(2)}] = -\frac{2i}{\pi x},$$

$$W[H_m^{(1)}, H_m^{(2)}] = -\frac{4i}{\pi x},$$

$$W[J_m, N_m] = \frac{2}{\pi x}.$$

4. Find the constant C in the Wronskian

$$W[I_m(x), K_m(x)] = \frac{C}{x}.$$

5. Verify the Wronskian

$$W[J_m, J_{-m}] = -\frac{2 \sin m\pi}{\pi x}.$$

6. **Gamma function:** To extend the definition of factorial to noninteger values, we can use the integral

$$\Gamma(x) = \int_0^\infty t^{x-1} e^{-t}\, dt, \quad x > 0,$$

where $\Gamma(x)$ is called the gamma function. Using integration by parts, show that for $x \geq 1$

$$\Gamma(x+1) = x\Gamma(x).$$

Use the integral definition to establish that $\Gamma(1) = 1$ and then show that when n is a positive integer $\Gamma(n+1) = n!$. Because of the divergence at $x = 0$, the integral definition does not work for $x \leq -1$. However, definition of the gamma function can be extended to negative values of x by writing the aforementioned formula as

$$\Gamma(x) = \frac{1}{x}\Gamma(x+1),$$

provided that for negative integers, we define

$$\frac{1}{\Gamma(-n)} = 0, \ n \text{ is integer.}$$

Using these, first show that

$$\Gamma(-1/2) = \sqrt{\pi},$$

and then, find the value of $\Gamma(-3/2)$. Also evaluate the following values of the gamma function:

$$\Gamma(0), \ \Gamma(1/2), \ \Gamma(-3/2), \ \Gamma(3/2).$$

7. Evaluate the integral

$$\int_0^\infty \frac{x^{m+1} J_m(bx)}{(x^2 + a^2)^{n+1}}\, dx, \quad a, b > 0, \quad 2n + \frac{3}{2} > m > -1.$$

Hint: Use the substitution

$$\frac{1}{(x^2 + a^2)^{n+1}} = \frac{1}{\Gamma(n+1)} \int_0^\infty e^{-(x^2 + a^2)t} t^n\, dt, \quad n > -1.$$

8. For the integer values of n, prove that

$$N_{-n}(x) = (-1)^n N_n(x).$$

9. Use the generating function:

$$\exp\left[\frac{x}{2}\left(t - \frac{1}{t}\right)\right] = \sum_{n=-\infty}^{\infty} J_n(x)t^n,$$

to prove the relations

$$J_n(-x) = (-1)^n J_n(x)$$

and

$$J_n(x + y) = \sum_{r=-\infty}^{\infty} J_r(x)J_{n-r}(y),$$

which is also known as the **addition formula** of Bessel functions.

10. Prove the formula

$$J_n(x) = (-1)^n x^n \left(\frac{1}{x}\frac{d}{dx}\right)^n J_0(x)$$

by induction, that is, assume it to be true for $n = N$ and then show for $N + 1$.

11. Derive the formula

$$e^{iz\cos\theta} = \sum_{m=-\infty}^{\infty} i^m J_m(z)e^{im\theta},$$

where z is a point in the complex plane. This is called the **Jacobi–Anger expansion** and gives a plane wave in terms of cylindrical waves.

12. In Fraunhofer diffraction, from a circular aperture, we encounter the integral

$$I \sim \int_0^a r\,dr \int_0^{2\pi} d\theta\, e^{ibr\cos\theta},$$

where a and b are constants depending on the physical parameters of the problem. Using the integral definition:

$$J_m(x) = \frac{1}{\pi}\int_0^\pi \cos[m\varphi - x\sin\varphi]d\varphi, \quad m = 0, \pm 1, \pm 2, \ldots,$$

first show that

$$I \sim 2\pi \int_0^a J_0(br)r\,dr$$

and then integrate to find

$$I = (2\pi a/b)J_1(ab).$$

Hint: Using the recursion relation

$$\frac{m}{x} J_m + J'_m = J_{m-1},$$

first prove that

$$\frac{d}{dx}[x^m J_m(x)] = x^m J_{m-1}(x).$$

Also note that a similar equation:

$$\frac{d}{dx}[x^{-m} J_m(x)] = -x^{-m} J_{m+1}(x),$$

can be obtained via

$$-\frac{m}{x} J_m + J'_m = -J_{m+1}.$$

13. Prove the following integral definition of the Bessel functions:

$$J_m(x) = \frac{(\frac{1}{2}x)^m}{\sqrt{\pi}\Gamma(m + \frac{1}{2})} \int_{-1}^{1} (1 - t^2)^{m-\frac{1}{2}} e^{\pm ixt} \, dt, \quad m > -\frac{1}{2}.$$

14. Using the generating function definition of Bessel functions, show the integral representation

$$J_m(x) = \frac{1}{\pi} \int_0^{\pi} \cos[m\varphi - x \sin \varphi] d\varphi, \quad m = 0, \pm 1, \pm 2, \ldots .$$

15. Prove

$$\int_0^{\infty} J_n(bx) \, dx = \frac{1}{b},$$

where n is positive integer.

16. Show that

$$\int_0^{\infty} \frac{J_n(x)}{x} \, dx = \frac{1}{n}.$$

17. Show that the spherical Bessel functions satisfy the differential equation

$$\frac{d^2 y_l(x)}{dx^2} + \frac{2}{x} \frac{dy_l(x)}{dx} + \left[1 - \frac{l(l+1)}{x^2}\right] y_l(x) = 0.$$

18. Spherical Bessel functions, $j_l(x)$ and $n_l(x)$, can also be defined as

$$j_l(x) = (-x)^l \left(\frac{1}{x} \frac{d}{dx}\right)^l \frac{\sin x}{x},$$

$$n_l(x) = (-x)^l \left(\frac{1}{x} \frac{d}{dx}\right)^l \left(-\frac{\cos x}{x}\right).$$

Prove these formulas by induction.

19. Using series representations of the Bessel functions J_m and N_m, show that the series representations of the spherical Bessel functions j_l and n_l are given as

$$j_l(x) = 2^l x^l \sum_{n=0}^{\infty} \frac{(-1)^n (l+n)!}{n!(2n+2l+1)!} x^{2n},$$

$$n_l(x) = \frac{(-1)^{l+1}}{2^l x^{l+1}} \sum_{n=0}^{\infty} \frac{(-1)^n (n-l)!}{n!(2n-2l)!} x^{2n}.$$

Hint: Use the duplication formula:

$$2^{2k} \Gamma(k+1) \Gamma(k+1/2) = \Gamma(1/2) \Gamma(2k+1) = \sqrt{\pi} (2k)!.$$

20. Using the recursion relations given in Eqs. (13.81) and (13.82):

$$J_{m-1}(x) = x^{-m} \frac{d}{dx} [x^m J_m(x)],$$

$$J_{m+1}(x) = -x^m \frac{d}{dx} [x^{-m} J_m(x)],$$

show that spherical Bessel functions satisfy the following recursion relations:

$$\frac{d}{dx} [x^{l+1} y_l(x)] = x^{l+1} y_{l-1}(x),$$

$$\frac{d}{dx} [x^{-l} y_l(x)] = -x^{-l} y_{l+1}(x),$$

where y_l stands for anyone of the spherical Bessel functions, j_l, n_l, or $h_l^{(1,2)}$.

21. Show that the solution of the general equation

$$y'' + \frac{(1-2a)}{x} y' + \left[(bcx^{c-1})^2 + \frac{a^2 - p^2 c^2}{x^2} \right] y(x) = 0,$$

can be expressed in terms of Bessel functions as

$$y(x) = x^a [A_0 J_p(bx^c) + A_1 N_p(bx^c)].$$

CHAPTER 14

PARTIAL DIFFERENTIAL EQUATIONS AND SEPARATION OF VARIABLES

Majority of the differential equations of physics and engineering are partial differential equations. Laplace equation

$$\vec{\nabla}^2 \Psi(\vec{r}) = 0, \tag{14.1}$$

which plays a central role in potential theory, is encountered in electrostatics, magnetostatics, and stationary flow problems. Diffusion and flow or transfer problems are commonly described by the equation

$$\vec{\nabla}^2 \Psi(\vec{r}, t) - \frac{1}{\alpha^2} \frac{\partial \Psi(\vec{r}, t)}{\partial t} = 0, \tag{14.2}$$

where α is a physical constant depending on the characteristics of the environment. The wave equation

$$\vec{\nabla}^2 \Psi(\vec{r}, t) - \frac{1}{v^2} \frac{\partial^2 \Psi(\vec{r}, t)}{\partial t^2} = 0, \tag{14.3}$$

where v stands for the wave velocity, is used to study wave phenomena in many different branches of science and engineering. Helmholtz equation

$$\overrightarrow{\nabla}^2 \Psi(\overrightarrow{r}) + k_0^2 \Psi(\overrightarrow{r}) = 0, \tag{14.4}$$

is encountered in the study of waves and oscillations. Nonhomogeneous versions of these equations, where the right-hand side is nonzero, are also frequently encountered. In general, the nonhomogeneous term represents sources, sinks, or interactions that may be present. In quantum mechanics, the time-independent Schrödinger equation is written as

$$-\frac{\hbar^2}{2m} \overrightarrow{\nabla}^2 \Psi(\overrightarrow{r}) + V(\overrightarrow{r})\Psi(\overrightarrow{r}) = E\Psi(\overrightarrow{r}), \tag{14.5}$$

while the time-dependent Schrödinger equation is given as

$$-\frac{\hbar^2}{2m} \overrightarrow{\nabla}^2 \Psi(\overrightarrow{r},t) + V(\overrightarrow{r})\Psi(\overrightarrow{r},t) = i\hbar\frac{\partial \Psi(\overrightarrow{r},t)}{\partial t}. \tag{14.6}$$

Partial differential equations are in general more difficult to solve. Integral transforms and Green's functions are among the most commonly used techniques to find analytic solutions [1]. However, in many of the interesting cases, it is possible to convert a partial differential equation into a set of ordinary differential equations by the method of **separation of variables.**

Most of the frequently encountered partial differential equations of physics and engineering can be written as a special case of the general equation:

$$\boxed{\overrightarrow{\nabla}^2 \Psi(\overrightarrow{r},t) + \kappa\Psi(\overrightarrow{r},t) = a\frac{\partial^2 \Psi(\overrightarrow{r},t)}{\partial t^2} + b\frac{\partial \Psi(\overrightarrow{r},t)}{\partial t},} \tag{14.7}$$

where a and b are usually constants but κ could be a function of $\overrightarrow{r}$. In this chapter, we discuss treatment of this general equation by the method of separation of variables in Cartesian, spherical, and cylindrical coordinates. Our results can be adopted to specific cases by an appropriate choice of the parameters κ, a, and b.

14.1   SEPARATION OF VARIABLES IN CARTESIAN COORDINATES

In Cartesian coordinates, we start by separating the time variable in Eq. (14.7) by the substitution

$$\Psi(\overrightarrow{r},t) = F(\overrightarrow{r})T(t) \tag{14.8}$$

and write

$$T(t)\overrightarrow{\nabla}^2 F(\overrightarrow{r}) + \kappa F(\overrightarrow{r})T(t) = F(\overrightarrow{r})\left[a\frac{d^2T}{dt^2} + b\frac{dT}{dt}\right], \tag{14.9}$$

where we take κ as a constant. Dividing both sides by $F(\vec{r})T(t)$ gives

$$\frac{1}{F(\vec{r})}[\vec{\nabla}^2 F(\vec{r}) + \kappa F(\vec{r})] = \frac{1}{T(t)}\left[a\frac{d^2T}{dt^2} + b\frac{dT}{dt}\right], \qquad (14.10)$$

where the left-hand side is only a function of $\vec{r}$ and the right-hand side is only a function of t. Since $\vec{r}$ and t are independent variables, the only way this equation can be true for all $\vec{r}$ and t is when both sides are equal to the same constant. Calling this constant $-k^2$, we obtain two equations:

$$\frac{1}{T(t)}\left[a\frac{d^2T}{dt^2} + b\frac{dT}{dt}\right] = -k^2, \qquad (14.11)$$

and

$$\frac{1}{F(\vec{r})}[\vec{\nabla}^2 F(\vec{r}) + \kappa F(\vec{r})] = -k^2. \qquad (14.12)$$

The choice of a minus sign in front of k^2 is arbitrary. In some problems, boundary conditions may require a plus sign if we want to keep k as a real parameter. In Cartesian coordinates, the second equation is written as

$$\frac{\partial^2 F(\vec{r})}{\partial x^2} + \frac{\partial^2 F(\vec{r})}{\partial y^2} + \frac{\partial^2 F(\vec{r})}{\partial z^2} + (\kappa + k^2)F(\vec{r}) = 0. \qquad (14.13)$$

We now separate the x variable by the substitution

$$F(\vec{r}) = X(x)G(y, z) \qquad (14.14)$$

and write

$$G(y, z)\frac{d^2X(x)}{dx^2} + X(x)\left[\frac{\partial^2 G(y, z)}{\partial y^2} + \frac{\partial^2 G(y, z)}{\partial z^2} + (\kappa + k^2)G(y, z)\right] = 0, \qquad (14.15)$$

which after division by $X(x)G(y, z)$ becomes

$$-\frac{1}{X(x)}\frac{d^2X(x)}{dx^2} = \frac{1}{G(y, z)}\left[\frac{\partial^2 G(y, z)}{\partial y^2} + \frac{\partial^2 G(y, z)}{\partial z^2} + (\kappa + k^2)G(y, z)\right]. \qquad (14.16)$$

Similarly, the only way this equality can hold for all x and (y, z) is when both sides are equal to the same constant, k_x^2, which gives the equations

$$\frac{1}{X(x)}\frac{d^2X(x)}{dx^2} + k_x^2 = 0 \qquad (14.17)$$

and

$$\frac{\partial^2 G(y, z)}{\partial y^2} + \frac{\partial^2 G(y, z)}{\partial z^2} + (\kappa + k^2 - k_x^2)G(y, z) = 0. \qquad (14.18)$$

Finally, we separate the last equation by the substitution

$$G(y, z) = Y(y)Z(z), \tag{14.19}$$

which gives

$$Z(z)\frac{d^2Y(y)}{dy^2} + Y(y)\left[\frac{d^2Z(z)}{dz^2} + (\kappa + k^2 - k_x^2)Z(z)\right] = 0. \tag{14.20}$$

Dividing by $Y(y)Z(z)$, we write

$$-\frac{1}{Y(y)}\frac{d^2Y(y)}{dy^2} = \frac{1}{Z(z)}\left[\frac{d^2Z(z)}{dz^2} + (\kappa + k^2 - k_x^2)Z(z)\right]; \tag{14.21}$$

and by using the same argument used for the other variables, we set both sides equal to a constant, k_y^2, to obtain

$$-\frac{1}{Y(y)}\frac{d^2Y(y)}{dy^2} = k_y^2, \tag{14.22}$$

$$\frac{d^2Z(z)}{dz^2} + (\kappa + k^2 - k_x^2 - k_y^2)Z(z) = 0. \tag{14.23}$$

In summary, after separating the variables, we have reduced the partial differential equation [Eq. (14.7)] to four ordinary differential equations:

$$\boxed{a\frac{d^2T}{dt^2} + b\frac{dT}{dt} + k^2T(t) = 0,} \tag{14.24}$$

$$\boxed{\frac{d^2X(x)}{dx^2} + k_x^2X(x) = 0,} \tag{14.25}$$

$$\boxed{\frac{d^2Y(y)}{dy^2} + k_y^2Y(y) = 0,} \tag{14.26}$$

$$\boxed{\frac{d^2Z(z)}{dz^2} + (\kappa + k^2 - k_x^2 - k_y^2)Z(z) = 0.} \tag{14.27}$$

During this process, three constants, k, k_x, k_y, which are called the **separation constants**, have entered into our equations. The final solution is now written as

$$\boxed{\Psi(\vec{r}, t) = T(t)X(x)Y(y)Z(z).} \tag{14.28}$$

14.1.1   Wave Equation

One of the most frequently encountered partial differential equations of physics and engineering is the wave equation:

$$\vec{\nabla}^2 \Psi(\vec{r}, t) - \frac{1}{v^2} \frac{\partial^2 \Psi(\vec{r}, t)}{\partial t^2} = 0. \qquad (14.29)$$

For its separable solutions, we set

$$a = \frac{1}{v^2}, \; b = 0, \; \kappa = 0, \qquad (14.30)$$

where v is the wave speed. Introducing

$$\omega = kv, \; k^2 = k_x^2 + k_y^2 + k_z^2, \qquad (14.31)$$

which stands for the angular frequency, we find the equations to be solved in Cartesian coordinates as

$$\frac{d^2 T}{dt^2} + \omega^2 T(t) = 0, \qquad (14.32)$$

$$\frac{d^2 X(x)}{dx^2} + k_x^2 X(x) = 0, \qquad (14.33)$$

$$\frac{d^2 Y(y)}{dy^2} + k_y^2 Y(y) = 0, \qquad (14.34)$$

$$\frac{d^2 Z(z)}{dz^2} + k_z^2 Z(z) = 0. \qquad (14.35)$$

All these equations are of the same type. If we concentrate on the first equation, the two linearly independent solutions are $\cos \omega t$ and $\sin \omega t$. Hence, the general solution can be written as

$$T(t) = a_0 \cos \omega t + a_1 \sin \omega t,$$

or as

$$T(t) = A \cos(\omega t + \delta),$$

where (a_0, a_1) and (A, δ) are arbitrary constants to be determined from the boundary conditions. In anticipation of applications to quantum mechanics, one can also take the two linearly independent solutions as $e^{\pm i\omega t}$. Now the solutions of Eqs. (14.32)– (14.35) can be conveniently combined to write

$$\Psi(\vec{r}, t) = c_0 e^{i(\vec{k} \cdot \vec{r} - \omega t)} + c_1 e^{-i(\vec{k} \cdot \vec{r} + \omega t)}, \qquad (14.36)$$

where

$$\vec{k} = (k_x, k_y, k_z),$$ (14.37)

$$\vec{r} = (x, y, z).$$ (14.38)

These are called the plane wave solutions of the wave equation, and $\Psi(\vec{r}, t)$ corresponds to the superposition of two plane waves moving in opposite directions.

14.1.2 Laplace Equation

In Eq. (14.7) if we set $\kappa = 0$ and assume no time dependence, we obtain the Laplace equation:

$$\boxed{\vec{\nabla}^2 \Psi(\vec{r}) = 0.}$$ (14.39)

Since there is no time dependence, in Eq. (14.11) we also set $k = 0$ and $T(t) = 1$, thus obtaining the equations to be solved for a separable solution [Eq. (14.28)] as

$$\frac{d^2 X(x)}{dx^2} + k_x^2 X(x) = 0,$$ (14.40)

$$\frac{d^2 Y(y)}{dy^2} + k_y^2 Y(y) = 0,$$ (14.41)

$$\frac{d^2 Z(z)}{dz^2} - (k_x^2 + k_y^2) Z(z) = 0,$$ (14.42)

Depending on the boundary conditions, solutions of these equations are given in terms of trigonometric or hyperbolic functions.

Example 14.1. *Laplace equation inside a rectangular region:* If a problem has translational symmetry along one of the Cartesian axes, say the z-axis, then the solution is independent of z. Hence, we solve the Laplace equation in two dimensions:

$$\frac{\partial^2 \Psi(x, y)}{\partial x^2} + \frac{\partial^2 \Psi(x, y)}{\partial y^2} = 0,$$ (14.43)

where the solution consists of a family of curves in the xy-plane. Solutions in three dimensions are obtained by extending these curves along the z direction to form a family of surfaces. Consider a rectangular region (Figure 14.1) defined by

$$x \in [0, a], \ y \in [0, b],$$ (14.44)

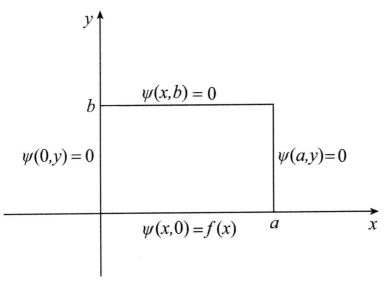

Figure 14.1 Boundary conditions for the Laplace equation inside the rectangular region in Example 14.1.

and the boundary conditions given as

$$\Psi(x,0) = f(x), \tag{14.45}$$

$$\Psi(x,b) = 0, \tag{14.46}$$

$$\Psi(0,y) = 0, \tag{14.47}$$

$$\Psi(a,y) = 0. \tag{14.48}$$

In the general equation [Eq. (14.7)], we set

$$a = b = \kappa = 0. \tag{14.49}$$

No time dependence gives $k = 0$ and $T(t) = 1$. Since there is no z dependence, Eq. (14.27) also gives $Z(z) = 1$, and $k_x^2 + k_y^2 = 0$; hence we define

$$\lambda^2 = k_x^2 = -k_y^2, \tag{14.50}$$

which gives the equations to be solved for a separable solution as

$$\frac{d^2 X(x)}{dx^2} + \lambda^2 X(x) = 0, \tag{14.51}$$

$$\frac{d^2 Y(y)}{dy^2} - \lambda^2 Y(y) = 0. \tag{14.52}$$

Solutions of these equations can be written immediately as

$$X(x) = a_0 \sin \lambda x + a_1 \cos \lambda x, \tag{14.53}$$

$$Y(y) = b_0 \sinh \lambda y + b_1 \cosh \lambda y. \tag{14.54}$$

Imposing the third boundary condition [Eq. (14.47)], we set $a_1 = 0$, which yields

$$X(x) = a_0 \sin \lambda x. \tag{14.55}$$

Using the last condition [Eq. (14.48)], we find the allowed values of λ:

$$\lambda_n = \frac{n\pi}{a}, \quad n = 1, 2, \ldots, \tag{14.56}$$

which gives

$$X_n(x) = a_0 \sin \frac{n\pi x}{a}, \quad n = 1, 2, \ldots, \tag{14.57}$$

$$Y_n(y) = b_0 \sinh \frac{n\pi y}{a} + b_1 \cosh \frac{n\pi y}{a}. \tag{14.58}$$

Hence, the separable solution, $\Psi_n(x, y) = X_n(x)Y_n(y)$, of the Laplace equation becomes

$$\Psi_n(x, y) = a_0 \left[\sin \frac{n\pi x}{a} \right] \left[b_0 \sinh \frac{n\pi y}{a} + b_1 \cosh \frac{n\pi y}{a} \right]. \tag{14.59}$$

Without any loss of generality, we can also write this as

$$\Psi_n(x, y) = A \left[\sin \frac{n\pi x}{a} \right] \left[B \sinh \frac{n\pi y}{a} + \cosh \frac{n\pi y}{a} \right]. \tag{14.60}$$

We now impose the second condition [Eq. (14.46)] to write

$$\Psi_n(x, b) = A \left[\sin \frac{n\pi x}{a} \right] \left[B \sinh \frac{n\pi b}{a} + \cosh \frac{n\pi b}{a} \right] = 0 \tag{14.61}$$

and obtain B as

$$B = -\frac{\cosh \dfrac{n\pi b}{a}}{\sinh \dfrac{n\pi b}{a}}. \tag{14.62}$$

Substituting this back into Eq. (14.60), we write

$$\Psi_n(x, y) = A \left[\sin \frac{n\pi x}{a} \right] \left[\frac{\sinh \dfrac{n\pi}{a}(b - y)}{\sinh \dfrac{n\pi}{a} b} \right]. \tag{14.63}$$

So far, we have satisfied all the boundary conditions except the first one, that is, Eq. (14.45). However, the solution set

$$\left\{ X_n(x) = a_0 \sin \frac{n\pi x}{a} \right\}, \quad n = 1, 2, \ldots, \tag{14.64}$$

like the special functions we have seen in Chapters 12 and 13, forms a complete and an orthogonal set satisfying the **orthogonality relation**

$$\int_0^a \left[\sin \frac{m\pi x}{a} \right] \left[\sin \frac{n\pi x}{a} \right] dx = \left(\frac{a}{2} \right) \delta_{mn}. \tag{14.65}$$

Using these base solutions, we can express a **general solution** as the infinite series

$$\Psi(x, y) = \sum_{n=1}^{\infty} c_n \Psi_n(x, y) = \sum_{n=1}^{\infty} c_n \left[\sin \frac{n\pi x}{a} \right] \left[\frac{\sinh \frac{n\pi}{a}(b - y)}{\sinh \frac{n\pi}{a} b} \right]. \tag{14.66}$$

Note that the set $\{\Psi_n(x, y)\}$, $n = 1, 2, \ldots$, is also complete and orthogonal. At this point, we suffice by saying that this series converges to $\Psi(x, y)$ for any continuous and sufficiently smooth function. Since the aforementioned series is basically a Fourier series, we will be more specific about what is meant from sufficiently smooth when we introduce the (trigonometric) Fourier series in Chapter 15. We now impose the final boundary condition [Eq. (14.45)] to write

$$\Psi(x, 0) = f(x) = \sum_{n=1}^{\infty} c_n \sin \frac{n\pi x}{a}. \tag{14.67}$$

To find the expansion coefficients, we multiply the aforementioned equation by $\sin \frac{m\pi}{a} x$ and integrate over $[0, a]$ and then use the orthogonality relation [Eq. (14.65)] to obtain

$$c_n = \left(\frac{2}{a} \right) \int_0^a f(x) \sin \frac{n\pi x}{a} \, dx. \tag{14.68}$$

Since each term in the series (14.66) satisfies the homogeneous boundary conditions [Eqs. (14.46)–(14.48)], so does the sum.

Now, let us consider a different set of boundary conditions (Figure 14.2):

$$\Psi(x, 0) = 0, \tag{14.69}$$

$$\Psi(x, b) = 0, \tag{14.70}$$

$$\Psi(0, y) = F(y), \tag{14.71}$$

$$\Psi(a, y) = 0. \tag{14.72}$$

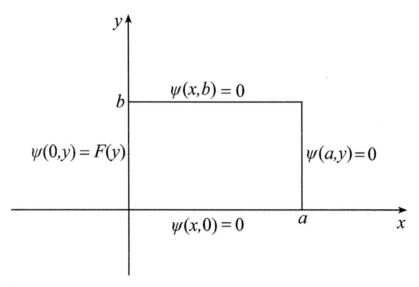

Figure 14.2   A different choice for the boundary conditions in Example 14.1.

In this case, the solution can be found by following similar steps:

$$\Psi(x, y) = \sum_{n=1}^{\infty} C_n \left[\sin \frac{n\pi y}{b}\right] \left[\frac{\sinh \frac{n\pi}{b}(a - x)}{\sinh \frac{n\pi}{b} a}\right], \tag{14.73}$$

where

$$C_n = \left(\frac{2}{b}\right) \int_0^b F(y) \sin \frac{n\pi y}{b} \, dy. \tag{14.74}$$

Note that in this case, the boundary conditions forces us to take $\lambda^2 = -k_x^2 = k_y^2$ [Eq. (14.50)]. Solution for the more general boundary conditions (Figure 14.3):

$$\Psi(x, 0) = f(x), \tag{14.75}$$

$$\Psi(x, b) = 0, \tag{14.76}$$

$$\Psi(0, y) = F(y), \tag{14.77}$$

$$\Psi(a, y) = 0, \tag{14.78}$$

can now be written as the linear combination of the solutions given in Eqs. (14.66) and (14.73) as

$$\Psi(x, y) = \sum_{n=1}^{\infty} c_n \sin \frac{n\pi x}{a} \left[\frac{\sinh \frac{n\pi}{a}(b - y)}{\sinh \frac{n\pi}{a} b}\right] + \sum_{n=1}^{\infty} C_n \sin \frac{n\pi y}{b} \left[\frac{\sinh \frac{n\pi}{b}(a - x)}{\sinh \frac{n\pi}{b} a}\right], \tag{14.79}$$

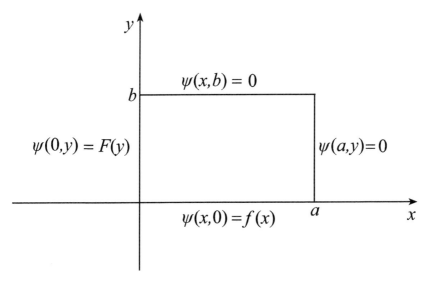

Figure 14.3 For more general boundary conditions.

where the coefficients are found as in Eqs. (14.68) and (14.74). Similarly, when all the boundary conditions are not homogeneous, the general solution is written as a superposition of all four cases.

14.1.3 Diffusion and Heat Flow Equations

For the heat flow and diffusion problems, we need to solve the equation

$$\overrightarrow{\nabla}^2 \Psi(\overrightarrow{r}, t) - \frac{1}{\alpha^2} \frac{\partial \Psi(\overrightarrow{r}, t)}{\partial t} = 0, \tag{14.80}$$

which can be obtained from the general equation [Eq. (14.7)] by setting

$$\kappa = 0, \ a = 0, \quad \text{and} \quad b = \frac{1}{\alpha^2}. \tag{14.81}$$

Now the equations to be solved for a separable solution [Eq. (14.28)] becomes

$$\frac{1}{\alpha^2} \frac{dT}{dt} + k^2 T(t) = 0, \tag{14.82}$$

$$\frac{d^2 X(x)}{dx^2} + k_x^2 X(x) = 0, \tag{14.83}$$

$$\frac{d^2 Y(y)}{dy^2} + k_y^2 Y(y) = 0, \tag{14.84}$$

$$\frac{d^2 Z(z)}{dz^2} + k_z^2 Z(z) = 0. \tag{14.85}$$

In the last equation, we have substituted

$$k^2 - k_x^2 - k_y^2 = k_z^2. \tag{14.86}$$

Solution of the first equation gives the time dependence as

$$\boxed{T(t) = T_0 e^{-k^2 \alpha^2 t},} \tag{14.87}$$

while the remaining equations have the solutions

$$X(x) = a_0 \cos k_x x + a_1 \sin k_x x, \tag{14.88}$$

$$Y(y) = b_0 \cos k_y y + b_1 \sin k_y y, \tag{14.89}$$

$$Z(z) = c_0 \cos k_z z + c_1 \sin k_z z. \tag{14.90}$$

Example 14.2. *Heat transfer equation in a rectangular region:* First consider the one-dimensional problem

$$\frac{d^2 \Psi(x,t)}{dx^2} = \frac{1}{\alpha^2} \frac{\partial \Psi(x,t)}{\partial t}, \ x \in [0,a], \ t > 0, \tag{14.91}$$

with the boundary conditions

$$\Psi(0,t) = 0, \ \Psi(a,t) = 0, \ \Psi(x,0) = f(x). \tag{14.92}$$

Using the time dependence given in Eq. (14.87) and the Fourier series method of Example 14.1, we can write the general solution as

$$\Psi(x,t) = \sum_{n=1}^{\infty} a_n \left(\sin \frac{n\pi x}{a} \right) e^{-(n^2 \pi^2 \alpha^2 / a^2)t}, \tag{14.93}$$

where the boundary conditions determined k as $k = n\pi/a$, $n = 1, 2, \ldots$, and the expansion coefficients are given as

$$a_n = \left(\frac{2}{a} \right) \int_0^a f(x) \sin \frac{n\pi x}{a} \, dx. \tag{14.94}$$

For the heat transfer equation in two dimensions,

$$\frac{d^2 \Psi(x,y,t)}{dx^2} + \frac{d^2 \Psi(x,y,t)}{dy^2} = \frac{1}{\alpha^2} \frac{\partial \Psi(x,y,t)}{\partial t}, \tag{14.95}$$

the solution over the rectangular region $x \in [0, a]$, $y \in [0, b]$, $t > 0$, satisfying the boundary conditions

$$\Psi(x, y, 0) = f(x, y), \tag{14.96}$$

$$\Psi(x, 0, t) = 0, \tag{14.97}$$

$$\Psi(x, b, t) = 0, \tag{14.98}$$

$$\Psi(0, y, t) = 0, \tag{14.99}$$

$$\Psi(a, y, t) = 0, \tag{14.100}$$

can be written as the series

$$\Psi(x, y, t) = \sum_{m,n=1}^{\infty} A_{mn} \left[\sin \frac{m\pi x}{a}\right] \left[\sin \frac{n\pi y}{b}\right] e^{-\left[\frac{m^2\pi^2}{a^2} + \frac{n^2\pi^2}{a^2}\right]\alpha^2 t}. \tag{14.101}$$

The expansion coefficients are now obtained from the integral

$$A_{mn} = \left(\frac{4}{ab}\right) \int_0^b \int_0^a f(x, y) \left[\sin \frac{m\pi x}{a}\right] \left[\sin \frac{n\pi y}{b}\right] dx dy. \tag{14.102}$$

14.2 SEPARATION OF VARIABLES IN SPHERICAL COORDINATES

In spherical coordinates, Eq. (14.7) is written as

$$\frac{1}{r^2}\frac{\partial}{\partial r}\left[r^2\frac{\partial}{\partial r}\Psi(\overrightarrow{r}, t)\right] + \frac{1}{r^2 \sin\theta}\frac{\partial}{\partial \theta}\left[\sin\theta\frac{\partial}{\partial \theta}\Psi(\overrightarrow{r}, t)\right]$$

$$+ \frac{1}{r^2\sin^2\theta}\frac{\partial^2}{\partial \phi^2}\Psi(\overrightarrow{r}, t) + \kappa\Psi(\overrightarrow{r}, t) = a\frac{\partial^2\Psi(\overrightarrow{r}, t)}{\partial t^2} + b\frac{\partial\Psi(\overrightarrow{r}, t)}{\partial t}, \tag{14.103}$$

where the ranges of the independent variables are $r \in [0, \infty]$, $\theta \in [0, \pi]$, and $\phi \in [0, 2\pi]$. We first substitute a solution of the form $\Psi(\overrightarrow{r}, t) = F(\overrightarrow{r})T(t)$ and write Eq. (14.103) as

$$T(t)\left\{\frac{1}{r^2}\frac{\partial}{\partial r}\left[r^2\frac{\partial}{\partial r}F(\overrightarrow{r})\right] + \frac{1}{r^2 \sin\theta}\frac{\partial}{\partial \theta}\left[\sin\theta\frac{\partial}{\partial \theta}F(\overrightarrow{r})\right]\right.$$

$$\left. + \frac{1}{r^2\sin^2\theta}\frac{\partial^2}{\partial \phi^2}F(\overrightarrow{r}) + \kappa F(\overrightarrow{r})\right\} = \left[a\frac{\partial^2 T(t)}{\partial t^2} + b\frac{\partial T(t)}{\partial t}\right] F(\overrightarrow{r}). \tag{14.104}$$

Dividing the above equation by $T(t)F(\overrightarrow{r})$ and collecting the position dependence on the left-hand side and the time dependence on the right-hand side, we obtain

$$
\frac{1}{F(\overrightarrow{r})}\left\{\frac{1}{r^2}\frac{\partial}{\partial r}\left[r^2\frac{\partial}{\partial r}F(\overrightarrow{r})\right] + \frac{1}{r^2\sin\theta}\frac{\partial}{\partial\theta}\left[\sin\theta\frac{\partial}{\partial\theta}F(\overrightarrow{r})\right]\right.
$$
$$
\left. + \frac{1}{r^2\sin^2\theta}\frac{\partial^2}{\partial\phi^2}F(\overrightarrow{r}) + \kappa F(\overrightarrow{r})\right\} = \left[a\frac{d^2T(t)}{dt^2} + b\frac{dT(t)}{dt}\right]\frac{1}{T(t)}. \qquad (14.105)
$$

Since $\overrightarrow{r}$ and t are independent variables, the only way to satisfy this equation for all $\overrightarrow{r}$ and t is to set both sides equal to the same constant, say $-k^2$. Hence, we obtain the following two equations:

$$
\frac{1}{r^2}\frac{\partial}{\partial r}\left[r^2\frac{\partial}{\partial r}F(\overrightarrow{r})\right] + \frac{1}{r^2\sin\theta}\frac{\partial}{\partial\theta}\left[\sin\theta\frac{\partial}{\partial\theta}F(\overrightarrow{r})\right]
$$
$$
+ \frac{1}{r^2\sin^2\theta}\frac{\partial^2}{\partial\phi^2}F(\overrightarrow{r}) + \kappa F(\overrightarrow{r}) = -k^2 F(\overrightarrow{r}), \qquad (14.106)
$$
$$
a\frac{d^2T(t)}{dt^2} + b\frac{dT(t)}{dt} = -k^2 T(t), \qquad (14.107)
$$

where the equation for $T(t)$ is an ordinary differential equation. We continue separating variables by using the substitution $F(\overrightarrow{r}) = R(r)Y(\theta,\phi)$, and write Eq. (14.106) as

$$
Y(\theta,\phi)\left[\frac{1}{r^2}\frac{d}{dr}\left(r^2\frac{d}{dr}R(r)\right) + (\kappa + k^2)R(r)\right]
$$
$$
= -\frac{R(r)}{r^2}\left[\frac{1}{\sin\theta}\frac{\partial}{\partial\theta}\left(\sin\theta\frac{\partial}{\partial\theta}Y(\theta,\phi)\right) + \frac{1}{\sin^2\theta}\frac{\partial^2 Y(\theta,\phi)}{\partial\phi^2}\right]. \qquad (14.108)
$$

Multiplying both sides by $r^2/R(r)Y(\theta,\phi)$, we obtain

$$
\frac{1}{R(r)}\frac{d}{dr}\left[r^2\frac{d}{dr}R(r)\right] + (\kappa + k^2)r^2
$$
$$
= -\frac{1}{Y(\theta,\phi)}\left[\frac{1}{\sin\theta}\frac{\partial}{\partial\theta}\left(\sin\theta\frac{\partial}{\partial\theta}Y(\theta,\phi)\right) + \frac{1}{\sin^2\theta}\frac{\partial^2 Y(\theta,\phi)}{\partial\phi^2}\right]. \qquad (14.109)
$$

Since r and (θ,ϕ) are independent variables, this equation can only be satisfied for all r and (θ,ϕ) when both sides of the equation are equal to the same constant. We call this constant λ and write

$$
\frac{d}{dr}\left(r^2\frac{d}{dr}R(r)\right) + [(\kappa + k^2)r^2 - \lambda]R(r) = 0 \qquad (14.110)
$$

and

$$
\frac{1}{\sin\theta}\frac{\partial}{\partial\theta}\left[\sin\theta\frac{\partial Y(\theta,\phi)}{\partial\theta}\right] + \frac{1}{\sin^2\theta}\frac{\partial^2 Y(\theta,\phi)}{\partial\phi^2} + \lambda Y(\theta,\phi) = 0. \qquad (14.111)
$$

Equation (14.110) for $R(r)$ is now an ordinary differential equation. We finally separate the θ and ϕ variables in $Y(\theta, \phi)$ as $Y(\theta, \phi) = \Theta(\theta)\Phi(\phi)$ and write

$$\frac{\Phi(\phi)}{\sin\theta}\frac{d}{d\theta}\left[\sin\theta\frac{d}{d\theta}\Theta(\theta)\right] + \lambda\Theta(\theta)\Phi(\phi) = -\frac{\Theta(\theta)}{\sin^2\theta}\frac{\partial^2\Phi(\phi)}{\partial\phi^2}. \tag{14.112}$$

Multiplying both sides by $\sin^2\theta/\Theta(\theta)\Phi(\phi)$ and calling the new separation constant m^2, we write

$$\frac{\sin\theta}{\Theta(\theta)}\frac{d}{d\theta}\left[\sin\theta\frac{d\Theta}{d\theta}\right] + \lambda\sin^2\theta = -\frac{1}{\Phi(\phi)}\frac{d^2\Phi(\phi)}{d\phi^2} = m^2. \tag{14.113}$$

We now obtain the differential equations to be solved for $\Theta(\theta)$ and $\Phi(\phi)$ as

$$\sin\theta\frac{d}{d\theta}\left[\sin\theta\frac{d\Theta}{d\theta}\right] + [\lambda\sin^2\theta - m^2]\Theta(\theta) = 0 \tag{14.114}$$

and

$$\frac{d^2\Phi(\phi)}{d\phi^2} + m^2\Phi(\phi) = 0. \tag{14.115}$$

In summary, via the method of separation of variables, in spherical coordinates we have reduced the partial differential equation:

$$\boxed{\vec{\nabla}^2\Psi(\vec{r}, t) + \kappa\Psi(\vec{r}, t) = a\frac{\partial^2\Psi(\vec{r}, t)}{\partial t^2} + b\frac{\partial\Psi(\vec{r}, t)}{\partial t},} \tag{14.116}$$

to four ordinary differential equations:

$$\boxed{a\frac{d^2T_k(t)}{dt^2} + b\frac{dT_k(t)}{dt} + k^2T_k(t) = 0,} \tag{14.117}$$

$$\boxed{\frac{d}{dr}\left(r^2\frac{d}{dr}R_{k\lambda}(r)\right) + [(\kappa + k^2)r^2 - \lambda]R_{k\lambda}(r) = 0,} \tag{14.118}$$

$$\boxed{\sin\theta\frac{d}{d\theta}\left[\sin\theta\frac{d\Theta_{\lambda m}}{d\theta}\right] + [\lambda\sin^2\theta - m^2]\Theta_{\lambda m}(\theta) = 0,} \tag{14.119}$$

$$\boxed{\frac{d^2\Phi_m(\phi)}{d\phi^2} + m^2\Phi_m(\phi) = 0,} \tag{14.120}$$

which have to be solved simultaneously with the appropriate boundary conditions to yield the final solution as

$$\boxed{\Psi_{k\lambda m}(\vec{r}, t) = T_k(t)R_{k\lambda}(r)\Theta_{\lambda m}(\theta)\Phi_m(\phi).} \tag{14.121}$$

During this process, three separation constants, k, λ, and m, indicated as subscripts, have entered into our equations. For the **time-independent** cases we set $a = b = k = 0$ and $T(t) = 1$. For problems with **azimuthal symmetry**, where there is no ϕ dependence, we take $m = 0$, $\Phi(\phi) = 1$. For the ϕ dependent solutions, we impose the periodic boundary condition:

$$\Phi_m(\phi + 2\pi) = \Phi_m(\phi), \tag{14.122}$$

to write the general solution of Eq. (14.120) as

$$\Phi_m(\phi) = a_0 \cos m\phi + a_1 \sin m\phi, \ m = 0, 1, 2, \ldots. \tag{14.123}$$

Note that with applications to **quantum mechanics** in mind, the general solution can also be written as

$$\boxed{\Phi_m(\phi) = a_0 e^{im\phi} + a_1 e^{-im\phi}.} \tag{14.124}$$

Since the set

$$\{\cos m\phi, \ \sin m\phi\ \}, \ m = 0, 1, 2, \ldots, \tag{14.125}$$

is complete and orthogonal, an arbitrary solution satisfying the periodic boundary conditions can be expanded as

$$\Phi(\phi) = \sum_{m=0}^{\infty} A_m \cos m\phi + B_m \sin m\phi. \tag{14.126}$$

This is basically the **trigonometric Fourier series**. We postpone a formal treatment of Fourier series to Chapter 15 and continue with Eq. (14.119). Defining a new independent variable as $x = \cos\theta$, $x \in [-1, 1]$, we write Eq. (14.119) as

$$\boxed{(1 - x^2)\frac{d^2\Theta_{\lambda m}(x)}{dx^2} - 2x\frac{d\Theta_{\lambda m}(x)}{dx} + \left[\lambda - \frac{m^2}{(1 - x^2)}\right]\Theta_{\lambda m}(x) = 0.} \tag{14.127}$$

For $m = 0$, this reduces to the **Legendre equation**. If we impose the boundary condition that $\Theta_{\lambda 0}(x)$ be finite over the entire interval including the end points, the separation constant λ has to be restricted to integer values:

$$\lambda = l(l + 1), \ l = 0, 1, 2, \ldots. \tag{14.128}$$

Thus, the finite solutions of Eq. (14.127) become the **Legendre polynomials** (Chapter 13):

$$\boxed{\Theta_{l0}(x) = P_l(x).} \tag{14.129}$$

Since the Legendre polynomials form a complete and an orthogonal set, a general solution can be expressed in terms of the Legendre series as

$$\Theta(x) = \sum_{l=0}^{\infty} C_l P_l(x).$$ (14.130)

For the cases with $m \neq 0$, Eq. (14.127) is called the **associated Legendre equation**, polynomial solutions of which are given as

$$\Theta_{lm}(x) = P_l^m(x) = (1 - x^2)^{m/2} \frac{d^m P_l(x)}{dx^m}, \ l = 0, 1, \ldots$$ (14.131)

For a solution with general angular dependence, $Y(\theta, \phi)$, we expand in terms of the combined complete and orthogonal set

$$u_{lm} = P_l^m(\cos\theta)[A_{lm} \cos m\phi + B_{lm} \sin m\phi], \quad l = 0, 1, \ldots, \quad m = 0, 1, \ldots, l.$$

A particular complete and orthogonal set constructed by using the θ and ϕ solutions is called the **spherical harmonics**. A detailed treatment of the associated Legendre polynomials and spherical harmonics is given in Bayin [1], hence we suffice by saying that the set $\{P_l^m(x)\}$, $l = 0, 1, \ldots$, is also complete and orthogonal. A general solution of Eq. (14.127) can now be written as the series

$$\Theta(\theta) = \sum_{l=0}^{\infty} C_l P_l^m(\cos\theta).$$ (14.132)

So far, nothing has been said about the parameters a, b, and κ; hence the solutions found for the ϕ and θ dependences, $\Phi(\phi)$ and $\Theta(x)$, are usable for a large class of cases. To proceed with the remaining equations that determine the t and the r dependences [Eqs. (14.117) and (14.118)], we have to specify the values of the parameters a, b, and κ, where there are a number of cases that are predominantly encountered in applications.

14.2.1 Laplace Equation

For the Laplace equation:

$$\vec{\nabla}^2 \Psi(\vec{r}) = 0,$$ (14.133)

in Eq. (14.7) we set $\kappa = a = b = 0$. Since there is no time dependence, k is also zero; hence, the radial equation [Eq. (14.118)] along with Eq. (14.128) becomes

$$\frac{d}{dr}\left[r^2 \frac{dR}{dr}\right] - l(l+1)R(r) = 0,$$ (14.134)

or

$$r^2 \frac{d^2 R}{dr^2} + 2r \frac{dR}{dr} - l(l+1)R(r) = 0. \tag{14.135}$$

This is nothing but the **Cauchy–Euler** equation (Chapter 11), the general solution of which can be written as

$$R_l(r) = C_0 r^l + C_1 \frac{1}{r^{l+1}}. \tag{14.136}$$

We can now write the general solution of the Laplace equation in spherical coordinates as

$$\Psi(r, \theta, \phi) = \sum_{l=0}^{\infty} \sum_{m=0}^{l} \left[A_{lm} r^l + B_{lm} \frac{1}{r^{l+1}} \right] P_l^m(\cos\theta)[a_{lm} \cos m\phi + b_{lm} \sin m\phi],$$

$$\tag{14.137}$$

where A_{lm}, B_{lm}, a_{lm}, and b_{lm} are the expansion coefficients to be determined from the boundary conditions. In problems with **azimuthal** or **axial symmetry**, the solution does not depend on the variable ϕ, hence we set $m = 0$, thus obtaining the series solution as

$$\Psi(r, \theta) = \sum_{l=0}^{\infty} \left[A_l r^l + B_l \frac{1}{r^{l+1}} \right] P_l(\cos\theta). \tag{14.138}$$

14.2.2   Boundary Conditions for a Spherical Boundary

Boundary conditions for $\Psi(r, \theta)$ on a spherical boundary with radius a is usually given as one of the following three types:

I. Dirichlet: The Dirichlet boundary condition is defined by specifying the value of $\Psi(r, \theta)$ on the boundary $r = a$ as

$$\Psi(a, \theta) = f(\theta). \tag{14.139}$$

II. Neumann: When the derivative is specified:

$$\left. \frac{d}{dr} \Psi(r, \theta) \right|_{r=a} = g(\theta), \tag{14.140}$$

we have the Neumann boundary condition.

III. General: When the boundary condition is given as

$$\left(\Psi(r, \theta) + d_0 \frac{d}{dr} \Psi(r, \theta) \right)_{r=a} = h(\theta), \tag{14.141}$$

where d_0 could be a function of θ, it is called the general boundary condition.

For finite solutions inside a sphere, we set $B_l = 0$ in Eq. (14.138) and take the solution as

$$\Psi(r, \theta) = \sum_{l=0}^{\infty} A_l r^l P_l(\cos\theta). \tag{14.142}$$

For the Dirichlet condition:

$$\Psi(a, \theta) = f(\theta) = \sum_{l=0}^{\infty} A_l a^l P_l(\cos\theta), \tag{14.143}$$

the remaining coefficients, A_l, can be evaluated by using the orthogonality relation [Eqs. (12.103) and (12.111)] of the Legendre polynomials as

$$A_l = \frac{1}{a^l}\left(l + \frac{1}{2}\right) \int_0^{\pi} f(\theta) P_l(\cos\theta) \sin\theta \, d\theta. \tag{14.144}$$

Outside the spherical boundary and for finite solutions at infinity, we set $A_l = 0$ and take the solution as

$$\Psi(r, \theta) = \sum_{l=0}^{\infty} B_l \frac{1}{r^{l+1}} P_l(\cos\theta). \tag{14.145}$$

Now the expansion coefficients are found from

$$B_l = a^{l+1}\left(l + \frac{1}{2}\right) \int_0^{\pi} f(\theta) P_l(\cos\theta) \sin\theta \, d\theta. \tag{14.146}$$

For a domain bounded by two concentric circles with radii a and b, both A_l and B_l in Eq. (14.138) are nonzero. For the Dirichlet conditions $\Psi(a, \theta) = f_1(\theta)$ and $\Psi(b, \theta) = f_2(\theta)$, we now write

$$\Psi(a, \theta) = f_1(\theta) = \sum_{l=0}^{\infty}\left[A_l a^l + B_l \frac{1}{a^{l+1}}\right] P_l(\cos\theta), \tag{14.147}$$

$$\Psi(b, \theta) = f_2(\theta) = \sum_{l=0}^{\infty}\left[A_l b^l + B_l \frac{1}{b^{l+1}}\right] P_l(\cos\theta). \tag{14.148}$$

Using the orthogonality relation of the Legendre polynomials, we obtain two linear equations:

$$A_l a^l + B_l \frac{1}{a^{l+1}} = \left(l + \frac{1}{2}\right) \int_0^{\pi} f_1(\theta) P_l(\cos\theta) \sin\theta \, d\theta, \tag{14.149}$$

$$A_l b^l + B_l \frac{1}{b^{l+1}} = \left(l + \frac{1}{2}\right) \int_0^{\pi} f_2(\theta) P_l(\cos\theta) \sin\theta \, d\theta, \tag{14.150}$$

which can be solved for A_l and B_l.

Solutions satisfying the Neumann boundary condition [Eq. (14.140)] or the general boundary conditions [Eq. (14.141)] are obtained similarly. For more general cases involving both angular variables θ and ϕ, the general solution [Eq. (14.137)] is given in terms of the associated Legendre polynomials. This time, the Dirichlet condition for a spherical boundary is given as $\Psi(a, \theta, \phi) = f(\theta, \phi)$, and the coefficients A_{lm} and B_{lm} in Eq. (14.137) are evaluated by using the orthogonality relation of the new basis functions:

$$u_{lm}(\theta, \phi) = P_l^m(\cos\theta)[a_1 \cos m\phi + a_2 \sin m\phi]. \qquad (14.151)$$

Example 14.3. *Potential of a point charge inside a sphere:* Consider a hollow conducting sphere of radius a held at zero potential. We place a charge q at point A along the z-axis at r' as shown in Figure 14.4. Due to the linearity of the Laplace equation, we can write the potential $\Omega(\overrightarrow{r})$ at a point inside the conductor as the sum of the potential of the point charge and the potential $\Psi(\overrightarrow{r})$ due to the induced charge on the conductor as

$$\Omega(\overrightarrow{r}) = \frac{q}{|\overrightarrow{r} - \overrightarrow{r}'|} + \Psi(\overrightarrow{r}), \qquad (14.152)$$

where

$$|\overrightarrow{r} - \overrightarrow{r}'| = \sqrt{r^2 + r'^2 - 2rr'\cos\theta}. \qquad (14.153)$$

Due to axial symmetry $\Omega(\overrightarrow{r})$ has no ϕ dependence, hence we write it as $\Omega(r, \theta)$. Since $\Omega(r, \theta)$ must vanish on the surface of the sphere, we have a Dirichlet problem. The boundary condition we have to satisfy is now written as

$$-\frac{q}{\sqrt{a^2 + r'^2 - 2ar'\cos\theta}} = \Psi(a, \theta). \qquad (14.154)$$

Using the generating function definition of the Legendre polynomials:

$$T(x, t) = \frac{1}{(1 - 2xt + t^2)^{1/2}} = \sum_{l=0}^{\infty} P_l(x)t^l, \quad |t| < 1, \qquad (14.155)$$

we can write the left-hand side of Eq. (14.154) as

$$-\frac{q}{a}\sum_{l=0}^{\infty}\left(\frac{r'}{a}\right)^l P_l(\cos\theta) = \Psi(a, \theta). \qquad (14.156)$$

Since $\frac{r'}{a} < 1$, the previously mentioned series is uniformly convergent. Using the Legendre expansion of $\Psi(r, \theta)$:

$$\Psi(r, \theta) = \sum_{l=0}^{\infty} A_l r^l P_l(\cos\theta), \qquad (14.157)$$

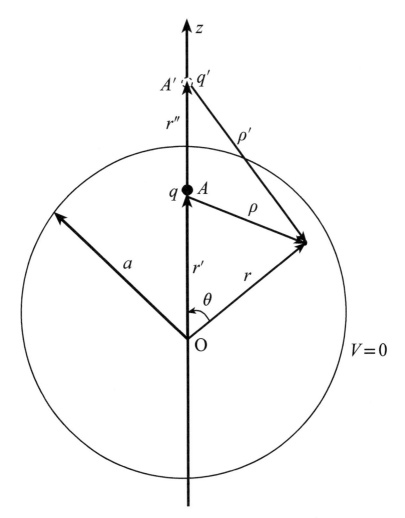

Figure 14.4 Point charge inside a grounded conducting sphere.

we can also write the Dirichlet condition [Eq. (14.156)] as

$$\Psi(a, \theta) = \sum_{l=0}^{\infty} A_l a^l P_l(\cos \theta). \tag{14.158}$$

Comparing the two expressions for $\Psi(a, \theta)$ [Eqs. (14.156) and (14.158)], we obtain the expansion coefficients:

$$A_l = -\frac{q}{a} \left(\frac{r'}{a^2} \right)^l, \tag{14.159}$$

which allows us to write $\Psi(r, \theta)$ [Eq. (14.157)] as

$$\Psi(r, \theta) = -\frac{q}{a} \sum_{l=0}^{\infty} \left(\frac{rr'}{a^2} \right)^l P_l(\cos\theta). \tag{14.160}$$

Using the generating function definition of P_l [Eq. (14.155)], we rewrite $\Psi(r, \theta)$ as

$$\Psi(r, \theta) = -\frac{q}{a} \frac{1}{\sqrt{1 - 2\left(\dfrac{r'r}{a^2} \right)\cos\theta + \left(\dfrac{rr'}{a^2} \right)^2}}. \tag{14.161}$$

Now the potential at $\overrightarrow{r}$ becomes

$$\Omega(\overrightarrow{r}) = \frac{q}{|\overrightarrow{r} - \overrightarrow{r'}|} + \Psi(r, \theta)$$

$$= \frac{q}{|\overrightarrow{r} - \overrightarrow{r'}|} - \frac{q}{a} \frac{1}{\sqrt{1 - 2\left(\dfrac{r'r}{a^2} \right)\cos\theta + \left(\dfrac{rr'}{a^2} \right)^2}}. \tag{14.162}$$

We rearrange this as

$$\Omega(\overrightarrow{r}) = \frac{q}{|\overrightarrow{r} - \overrightarrow{r'}|} + \frac{-q\dfrac{a}{r'}}{\sqrt{r^2 + \dfrac{a^4}{r'^2} - 2\dfrac{a^2 r}{r'}\cos\theta}}. \tag{14.163}$$

If we call $\rho = |\overrightarrow{r} - \overrightarrow{r'}|$ and introduce q', r'', and ρ' such that

$$q' = -q\frac{a}{r'}, \quad r' = \frac{a^2}{r'}, \quad \rho' = \sqrt{r^2 + r''^2 - 2r''r\cos\theta}, \tag{14.164}$$

we can also write $\Omega(\overrightarrow{r})$ as

$$\Omega(\overrightarrow{r}) = \frac{q}{\rho} + \frac{q'}{\rho'}. \tag{14.165}$$

Note that this is the result that one would get by using the **image method**, where an image charge q' is located along the z axis at A' at a distance of r'' from the origin (Figure 14.4).

14.2.3   Helmholtz Equation

In our general equation [Eq. (14.7)], we set $a = b = 0$, $\kappa = k_0^2$, to obtain the Helmholtz equation:

$$\boxed{\overrightarrow{\nabla}^2 \Psi(\overrightarrow{r}) + k_0^2 \Psi(\overrightarrow{r}) = 0.} \tag{14.166}$$

Since there is no time dependence in the separated equations [Eqs. (14.117)–(14.120)], we also set $k = 0$ and $T(t) = 1$. The radial part of the Helmholtz equation [Eq. (14.118)] becomes

$$\frac{d}{dr}\left[r^2\frac{dR}{dr}\right] + [k_0^2 r^2 - l(l+1)]R(r) = 0, \tag{14.167}$$

$$\frac{d^2R}{dr^2} + \frac{2}{r}\frac{dR}{dr} + \left[k_0^2 - \frac{l(l+1)}{r^2}\right]R(r) = 0, \tag{14.168}$$

where we have substituted $\lambda = l(l+1)$. The general solution of Eq. (14.168) is given in terms of the spherical Bessel functions as

$$R_l(r) = c_0 j_l(k_0 r) + c_1 n_l(k_0 r). \tag{14.169}$$

The general solution of the Helmholtz equation in spherical coordinates can now be written as the series

$$\Psi(\overrightarrow{r}, t) = \sum_{l=0}^{\infty}\sum_{m=0}^{l}[A_{lm}j_l(k_0 r) + B_{lm}n_l(k_0 r)]P_l^m(\cos\theta)\cos(m\phi + \delta_{lm}),$$

$$\tag{14.170}$$

where the coefficients A_{lm}, B_{lm}, and δ_{lm} are to be determined from the boundary conditions. Including the important problem of diffraction of electromagnetic waves from the earth's surface, many of the problems of mathematical physics can be expressed as such superpositions. In problems involving steady-state oscillations, time dependence is described by $e^{i\omega t}$ or $e^{-i\omega t}$, and in Eq. (14.170) $n_l, (k_0 r)$ is replaced by $h_l^{(2)}(k_0 r)$ or $h_l^{(1)}(k_0 r)$.

14.2.4 Wave Equation

In Eq. (14.7), we set $\kappa = 0$, $b = 0$, and $a = 1/v^2$ and obtain the wave equation

$$\overrightarrow{\nabla}^2\Psi(\overrightarrow{r}, t) = \frac{1}{v^2}\Psi(\overrightarrow{r}, t), \tag{14.171}$$

where v stands for the wave velocity. The time-dependent part of the solution, $T(t)$, is now determined by the equation [Eq. (14.117)]

$$\frac{d^2T}{dt^2} = -k^2 v^2 T(t), \tag{14.172}$$

the solution of which can be written in the following equivalent ways:

$$T(t) = a_0 \cos\omega t + a_1 \sin\omega t, \tag{14.173}$$

$$T(t) = A_0 \cos(\omega t + A_1), \quad \omega = kv, \tag{14.174}$$

where a_0, a_1, A_0, and A_1 are the integration constants. The radial equation [Eq. (14.118)] with $\lambda = l(l+1)$ is now written as

$$\frac{d^2R}{dr^2} + \frac{2}{r}\frac{dR}{dr} + \left[k^2 - \frac{l(l+1)}{r^2}\right]R(r) = 0, \tag{14.175}$$

where the solution is given in terms of the spherical Bessel functions as

$$R_l(r) = c_0 j_l(kr) + c_1 n_l(kr). \tag{14.176}$$

We can now write the general solution of the wave equation [Eq. (14.171)] as

$$\Psi(\overrightarrow{r},t) = \sum_{l=0}^{\infty}\sum_{m=0}^{l}[A_{lm}j_l(kr)+B_{lm}n_l(kr)]P_l^m(\cos\theta)\cos(m\phi+\delta_{lm})\cos(\omega t+A_l),$$

$$\tag{14.177}$$

where the coefficients A_{lm}, B_{lm}, δ_{lm}, and A_l are to be determined from the boundary conditions.

14.2.5   Diffusion and Heat Flow Equations

In Eq. (14.7), we set $\kappa = 0$, $b \neq 0$, and $a = 0$ to obtain

$$\overrightarrow{\nabla}^2\Psi(\overrightarrow{r},t) = b\frac{\partial\Psi(\overrightarrow{r},t)}{\partial t}, \tag{14.178}$$

which is the governing equation for diffusion or heat flow phenomenon. Since $k^2 \neq 0$, using Eq. (14.117), we write the differential equation to be solved for $T(t)$ as

$$b\frac{dT(t)}{dt} + k^2 T(t) = 0, \tag{14.179}$$

which gives the time dependence as

$$T(t) = Ce^{-k^2t/b}, \tag{14.180}$$

where C is an integration constant to be determined from the initial conditions. Radial dependence with $\lambda = l(l+1)$ is determined by Eq. (14.118):

$$\frac{d^2R}{dr^2} + \frac{2}{r}\frac{dR}{dr} + \left[k^2 - \frac{l(l+1)}{r^2}\right]R(r) = 0, \tag{14.181}$$

the solutions of which are given in terms of the spherical Bessel functions as

$$R_l(r) = A_0 j_l(kr) + B_0 n_l(kr). \tag{14.182}$$

Now the general solution of the diffusion equation can be written as

$$\Psi(\overrightarrow{r},t) = \sum_{l=0}^{\infty}\sum_{m=0}^{l}[A_{lm}j_l(kr) + B_{lm}n_l(kr)]P_l^m(\cos\theta)\cos(m\phi + \delta_{lm})e^{-k^2t/b},$$

(14.183)

where the coefficients A_{lm}, B_{lm}, and δ_{lm} are to be determined from the boundary conditions.

14.2.6 Time-Independent Schrödinger Equation

For a particle of mass m moving under the influence of a central potential $V(r)$, the time-independent Schrödinger equation is written as

$$-\frac{\hbar^2}{2m}\overrightarrow{\nabla}^2\Psi(\overrightarrow{r}) + V(r)\Psi(\overrightarrow{r}) = E\Psi(\overrightarrow{r}),$$

(14.184)

where E stands for the energy of the system. To compare with Eq. (14.7), we rewrite this as

$$\overrightarrow{\nabla}^2\Psi(\overrightarrow{r}) + \left[\frac{2mE}{\hbar^2} - \frac{2mV(r)}{\hbar^2}\right]\Psi(\overrightarrow{r}) = 0.$$

(14.185)

In the general equation [Eq. (14.7)], we now set $a = b = 0$. Using Eq. (14.117), we also set $k = 0$ and $T(t) = 1$. Note that κ is now a function of r:

$$\kappa(r) = \frac{2m}{\hbar^2}[E - V(r)],$$

(14.186)

hence the radial equation [Eq. (14.118)] becomes

$$\frac{d}{dr}\left(r^2\frac{dR}{dr}\right) + \left(\frac{2m}{\hbar^2}[E - V(r)] - l(l+1)\right)R(r) = 0.$$

(14.187)

For the Coulomb potential, solutions of this equation are given in terms of the **associated Laguerre polynomials**, which are closely related to the Laguerre polynomials [1].

14.2.7 Time-Dependent Schrödinger Equation

For the central force problems, the time-dependent Schrödinger equation is

$$-\frac{\hbar^2}{2m}\overrightarrow{\nabla}^2\Psi(\overrightarrow{r},t) + V(r)\Psi(\overrightarrow{r},t) = i\hbar\frac{\partial\Psi(\overrightarrow{r},t)}{\partial t},$$

(14.188)

which can be rewritten as

$$\vec{\nabla}^2 \Psi(\vec{r}, t) - \frac{2mV(r)}{\hbar^2} \Psi(\vec{r}, t) = -i\frac{2m}{\hbar}\frac{\partial \Psi(\vec{r}, t)}{\partial t}. \tag{14.189}$$

We now have

$$\kappa = -\frac{2mV(r)}{\hbar^2}, \quad a = 0, \quad b = -\frac{2mi}{\hbar}. \tag{14.190}$$

The time-dependent part of the solution satisfies

$$-\frac{2mi}{\hbar}\frac{dT}{dt} + k^2 T = 0, \tag{14.191}$$

which has the solution

$$T(t) = T_0 e^{-ik^2 \hbar t/2m}. \tag{14.192}$$

We relate the separation constant k^2 with the energy E as $k^2 = 2mE/\hbar$, hence $T(t)$ is written as

$$\boxed{T(t) = T_0 e^{-iEt/\hbar}.} \tag{14.193}$$

The radial part of the Schrödinger equation [Eq. (14.118)] is now given as

$$\boxed{\frac{d}{dr}\left(r^2\frac{dR}{dr}\right) + \left(\frac{2m}{\hbar^2}[E - V(r)] - l(l+1)\right)R(r) = 0,} \tag{14.194}$$

where $l = 0, 1, \ldots$. Solutions of Eq. (14.194) are given in terms of the **associated Laguerre polynomials**. Angular part of the solution, $\Psi(\vec{r}, t)$, comes from Eqs. (14.119) and (14.120), which can be expressed in terms of the **spherical harmonics** $Y_{lm}(\theta, \phi)$.

14.3 SEPARATION OF VARIABLES IN CYLINDRICAL COORDINATES

In cylindrical coordinates, we can write the general equation:

$$\vec{\nabla}^2 \Psi(\vec{r}, t) + \kappa\Psi(\vec{r}, t) = a\frac{\partial^2 \Psi(\vec{r}, t)}{\partial t^2} + b\frac{\partial \Psi(\vec{r}, t)}{\partial t}, \tag{14.195}$$

as

$$\frac{1}{r}\frac{\partial}{\partial r}\left[r\frac{\partial \Psi(\vec{r}, t)}{\partial r}\right] + \frac{1}{r^2}\frac{\partial^2 \Psi(\vec{r}, t)}{\partial \phi^2} + \frac{\partial^2 \Psi(\vec{r}, t)}{\partial z^2} + \kappa\Psi(\vec{r}, t)$$
$$= a\frac{\partial^2 \Psi(\vec{r}, t)}{\partial t^2} + b\frac{\partial \Psi(\vec{r}, t)}{\partial t}. \tag{14.196}$$

Separating the time variable, $\Psi(\overrightarrow{r}, t) = F(\overrightarrow{r})T(t)$, we write

$$
T(t)\left[\frac{1}{r}\frac{\partial}{\partial r}\left(r\frac{\partial F(\overrightarrow{r})}{\partial r}\right) + \frac{1}{r^2}\frac{\partial^2 F(\overrightarrow{r})}{\partial \phi^2} + \frac{\partial^2 F(\overrightarrow{r})}{\partial z^2} + \kappa F(\overrightarrow{r})\right]
$$
$$
= F(\overrightarrow{r})\left[a\frac{d^2 T(t)}{dt^2} + b\frac{dT(t)}{dt}\right], \tag{14.197}
$$

Dividing by $F(\overrightarrow{r})T(t)$ gives the separated equation:

$$
\frac{1}{F(\overrightarrow{r})}\left[\frac{1}{r}\frac{\partial}{\partial r}\left(r\frac{\partial F(\overrightarrow{r})}{\partial r}\right) + \frac{1}{r^2}\frac{\partial^2 F(\overrightarrow{r})}{\partial \phi^2} + \frac{\partial^2 F(\overrightarrow{r})}{\partial z^2} + \kappa F(\overrightarrow{r})\right]
$$
$$
= \frac{1}{T(t)}\left[a\frac{d^2 T(t)}{dt^2} + b\frac{dT(t)}{dt}\right]. \tag{14.198}
$$

Setting both sides equal to the same constant, $-\chi^2$, we obtain

$$
a\frac{d^2 T(t)}{dt^2} + b\frac{dT(t)}{dt} + \chi^2 T(t) = 0, \tag{14.199}
$$
$$
\frac{1}{r}\frac{\partial}{\partial r}\left[r\frac{\partial F(\overrightarrow{r})}{\partial r}\right] + \frac{1}{r^2}\frac{\partial^2 F(\overrightarrow{r})}{\partial \phi^2} + \frac{\partial^2 F(\overrightarrow{r})}{\partial z^2} + (\kappa + \chi^2)F(\overrightarrow{r}) = 0. \tag{14.200}
$$

We now separate the z variable by the substitution $F(\overrightarrow{r}) = G(r, \phi)Z(z)$ to write

$$
Z(z)\left[\frac{1}{r}\frac{\partial}{\partial r}\left(r\frac{\partial G(r, \phi)}{\partial r}\right) + \frac{1}{r^2}\frac{\partial^2 G(r, \phi)}{\partial \phi^2} + \chi^2 G(r, \phi)\right]
$$
$$
= -G(r, \phi)\left[\kappa Z(z) + \frac{d^2 Z(z)}{dz^2}\right]. \tag{14.201}
$$

Dividing by $G(r, \phi)Z(z)$ and setting both sides equal to the same constant, $-\lambda^2$, we get

$$
\frac{d^2 Z(z)}{dz^2} + (\kappa - \lambda^2)Z(z) = 0, \tag{14.202}
$$
$$
\frac{1}{r}\frac{\partial}{\partial r}\left[r\frac{\partial G(r, \phi)}{\partial r}\right] + \frac{1}{r^2}\frac{\partial^2 G(r, \phi)}{\partial \phi^2} + (\chi^2 + \lambda^2)G(r, \phi) = 0. \tag{14.203}
$$

Finally, we substitute $G(r, \phi) = R(r)\Phi(\phi)$ to write the second equation as

$$
\Phi(\phi)\left[\frac{1}{r}\frac{\partial}{\partial r}\left(r\frac{\partial R(r)}{\partial r}\right) + (\chi^2 + \lambda^2)R(r)\right] = -\frac{R(r)}{r^2}\left[\frac{d^2 \Phi(\phi)}{d\phi^2}\right]. \tag{14.204}
$$

Dividing by $R(r)\Phi(\phi)/r^2$ and setting both sides to the constant μ^2 gives us the last two equations:

$$\frac{1}{r}\frac{d}{dr}\left(r\frac{dR(r)}{dr}\right) + \left(\chi^2 + \lambda^2 - \frac{\mu^2}{r^2}\right)R(r) = 0, \tag{14.205}$$

$$\frac{d^2\Phi(\phi)}{d\phi^2} + \mu^2\Phi(\phi) = 0. \tag{14.206}$$

In summary, we have reduced the partial differential equation [Eq. (14.195)] in cylindrical coordinates to the following ordinary differential equations:

$$\boxed{a\frac{d^2T(t)}{dt^2} + b\frac{dT(t)}{dt} + \chi^2 T(t) = 0,} \tag{14.207}$$

$$\boxed{\frac{1}{r}\frac{d}{dr}\left(r\frac{dR(r)}{dr}\right) + \left(\chi^2 + \lambda^2 - \frac{\mu^2}{r^2}\right)R(r) = 0,} \tag{14.208}$$

$$\boxed{\frac{d^2\Phi(\phi)}{d\phi^2} + \mu^2\Phi(\phi) = 0,} \tag{14.209}$$

$$\boxed{\frac{d^2Z(z)}{dz^2} - (\lambda^2 - \kappa)Z(z) = 0.} \tag{14.210}$$

Combining the solutions of these equations with the appropriate boundary conditions, we write the general solution of Eq. (14.195) as

$$\boxed{\Psi(\overrightarrow{r},t) = T(t)R(r)\Phi(\phi)Z(z).} \tag{14.211}$$

When there is no time dependence, we set in Eqs. (14.207)–(14.210)

$$a = 0, \ b = 0, \ \chi = 0, \ T(t) = 1. \tag{14.212}$$

For azimuthal symmetry, there is no ϕ dependence, hence we set

$$\mu = 0, \ \Phi(\phi) = 1. \tag{14.213}$$

14.3.1   Laplace Equation

When we set $\kappa = 0$, $a = 0$, and $b = 0$, Eq. (14.195) becomes the Laplace equation:

$$\boxed{\overrightarrow{\nabla}^2\Psi(\overrightarrow{r}) = 0.} \tag{14.214}$$

For a time-independent separable solution, $\Psi(\overrightarrow{r}, t) = R(r)\Phi(\phi)Z(z)$, we also set $\chi = 0$ and $T = 1$ in Eq. (14.207); hence the equations to be solved become

$$\frac{1}{r}\frac{d}{dr}\left(r\frac{dR(r)}{dr}\right) + \left(\lambda^2 - \frac{\mu^2}{r^2}\right)R(r) = 0, \tag{14.215}$$

$$\frac{d^2\Phi(\phi)}{d\phi^2} + \mu^2\Phi(\phi) = 0, \tag{14.216}$$

$$\frac{d^2Z(z)}{dz^2} - \lambda^2 Z(z) = 0. \tag{14.217}$$

Solutions can be written, respectively, as

$$\boxed{R(r) = a_0 J_\mu(\lambda r) + a_1 N_\mu(\lambda r),} \tag{14.218}$$

$$\boxed{\Phi(\phi) = b_0 \cos\mu\phi + b_1 \sin\mu\phi,} \tag{14.219}$$

$$\boxed{Z(z) = c_0 \cosh\lambda z + c_1 \sinh\lambda z.} \tag{14.220}$$

14.3.2 Helmholtz Equation

In Eq. (14.195), we set $\kappa = k^2$, $a = 0$, and $b = 0$ to obtain the Helmholtz equation:

$$\boxed{\overrightarrow{\nabla}^2\Psi(\overrightarrow{r}) + k^2\Psi(\overrightarrow{r}) = 0,} \tag{14.221}$$

which when a separable solutions of the form $\Psi(\overrightarrow{r}) = R(r)\Phi(\phi)Z(z)$ is substituted, leads to the following differential equations:

$$\frac{1}{r}\frac{d}{dr}\left[r\frac{dR(r)}{dr}\right] + (\lambda^2 - \frac{\mu^2}{r^2})R(r) = 0, \tag{14.222}$$

$$\frac{d^2\Phi(\phi)}{d\phi^2} + \mu^2\Phi(\phi) = 0, \tag{14.223}$$

$$\frac{d^2Z(z)}{dz^2} - (\lambda^2 - k^2)Z(z) = 0. \tag{14.224}$$

In terms of the separation constants, the solution is now written as

$$\boxed{\Psi_{\lambda\mu}(\overrightarrow{r}) = R_{\lambda\mu}(r)\Phi_\mu(\phi)Z_\lambda(z).} \tag{14.225}$$

Note that in Eq. (14.207), we have set $\chi = 0$ and $T(t) = 1$ for no time dependence. We can now write the solution of the radial equation:

$$\boxed{R_{\lambda\mu}(r) = a_0 J_\mu(\lambda r) + a_1 N_\mu(\lambda r),} \tag{14.226}$$

and the solution of Eq. (14.223) gives the ϕ dependence:

$$\Phi_\mu(\phi) = b_0 \cos \mu\phi + b_1 \sin \mu\phi. \tag{14.227}$$

Finally, for the solution of the z equation [Eq. (14.224)], we define $\lambda^2 - k^2 = k_0^2$ to write the choices

$$Z_\lambda(z) = c_0 \left\{ \begin{array}{c} \cos k_0 z \\ \cosh k_0 z \end{array} \right\} + c_1 \left\{ \begin{array}{c} \sin k_0 z \\ \sinh k_0 z \end{array} \right\} for \left\{ \begin{array}{c} \lambda^2 - k^2 \le 0 \\ \lambda^2 - k^2 > 0 \end{array} \right\}. \tag{14.228}$$

14.3.3   Wave Equation

For the choices $\kappa = 0$, $b = 0$, $a = 1/v^2$, Eq. (14.195) becomes the wave equation:

$$\vec{\nabla}^2 \Psi(\vec{r}, t) = \frac{1}{v^2} \frac{\partial^2 \Psi(\vec{r}, t)}{\partial t^2}, \tag{14.229}$$

where v stands for the wave velocity. For a separable solution,

$$\Psi(\vec{r}, t) = T(t) R(r) \Phi(\phi) Z(z), \tag{14.230}$$

equations to be solved become [Eqs. (14.207)–(14.210)]

$$\frac{1}{v^2} \frac{d^2 T(t)}{dt^2} + \chi^2 T(t) = 0, \tag{14.231}$$

$$\frac{1}{r} \frac{d}{dr} \left[r \frac{dR(r)}{dr} \right] + (\chi^2 + \lambda^2 - \frac{\mu^2}{r^2}) R(r) = 0, \tag{14.232}$$

$$\frac{d^2 \Phi(\phi)}{d\phi^2} + \mu^2 \Phi(\phi) = 0, \tag{14.233}$$

$$\frac{d^2 Z(z)}{dz^2} - \lambda^2 Z(z) = 0, \tag{14.234}$$

where χ, λ, and μ are the separation constants. Solution of the time-dependent equation gives

$$T(t) = a_0 \cos \omega t + a_1 \sin \omega t, \tag{14.235}$$

where we have defined $\omega = v\chi$. Defining a new parameter as $\chi^2 + \lambda^2 = m^2$, solution of the radial equation is immediately written in terms of Bessel functions:

$$R(r) = a_0 J_\mu(mr) + a_1 N_\mu(mr). \tag{14.236}$$

The solution of Eq. (14.233) is

$$\Phi(\phi) = b_0 \cos \mu\phi + b_1 \sin \mu\phi \qquad (14.237)$$

and for the solution of the z equation, we write

$$Z(z) = c_0 \cosh \lambda z + c_1 \sinh \lambda z. \qquad (14.238)$$

14.3.4   Diffusion and Heat Flow Equations

For the diffusion and the heat flow equations, in Eq. (14.195), we set $\kappa = 0$, $a = 0$, $b \neq 0$ to obtain

$$\overrightarrow{\nabla}^2 \Psi(\overrightarrow{r}, t) = b \frac{\partial \Psi(\overrightarrow{r}, t)}{\partial t}. \qquad (14.239)$$

Now for the separated solution, $\Psi(\overrightarrow{r}, t) = T(t)R(r)\Phi(\phi)Z(z)$, we have to solve the following differential equations:

$$b \frac{dT(t)}{dt} + \chi^2 T(t) = 0, \qquad (14.240)$$

$$\frac{1}{r}\frac{d}{dr}\left(r\frac{dR(r)}{dr}\right) + (\chi^2 + \lambda^2 - \frac{\mu^2}{r^2})R(r) = 0, \qquad (14.241)$$

$$\frac{d^2\Phi(\phi)}{d\phi^2} + \mu^2\Phi(\phi) = 0, \qquad (14.242)$$

$$\frac{d^2Z(z)}{dz^2} - \lambda^2 Z(z) = 0. \qquad (14.243)$$

The time-dependent part can be solved immediately to yield

$$T(t) = T_0 e^{-\chi^2 t/b}, \qquad (14.244)$$

while the remaining equations have the solutions

$$R(r) = a_0 J_\mu(mr) + a_1 N_\mu(mr), \qquad (14.245)$$

$$\Phi(\phi) = b_0 \cos \mu\phi + b_1 \sin \mu\phi, \qquad (14.246)$$

$$Z(z) = c_0 \cosh \lambda z + c_1 \sinh \lambda z, \qquad (14.247)$$

where $m^2 = \chi^2 + \lambda^2$.

Example 14.4. *Dirichlet problem for the Laplace equation:* Consider the Laplace equation, $\vec{\nabla}^2 \Psi(r, \phi, z) = 0$, with the following Dirichlet conditions for a cylindrical domain:

$$\Psi(a, \phi, z) = 0, \tag{14.248}$$

$$\Psi(r, \phi, 0) = f_0(r), \tag{14.249}$$

$$\Psi(r, \phi, l) = f_1(r). \tag{14.250}$$

This could be a problem where we find the temperature distribution inside a cylinder with the temperature distributions at its top and bottom surfaces are given as shown in Figure 14.5, and the side surface is held at 0 temperature. Since the boundary conditions are independent of ϕ, we search for axially symmetric separable solutions of the form

$$\Psi(r, \phi, z) = R(r)Z(z).$$

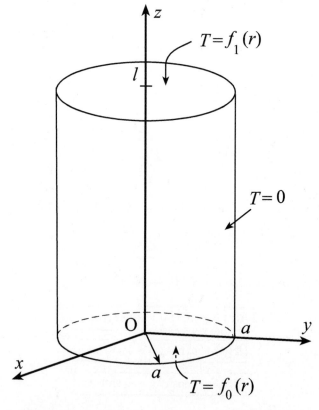

Figure 14.5 Boundary conditions for Example 14.4.

Setting $\mu = 0$, $\Phi(\phi) = 1$ for no ϕ dependence, we use Eqs. (14.218) and (14.220) to write $R(r)$ and $Z(z)$ as

$$R(r) = a_0 J_0(\lambda r) + a_1 N_0(\lambda r), \tag{14.251}$$

$$Z(z) = c_0 \cosh \lambda z + c_1 \sinh \lambda z. \tag{14.252}$$

Since $N_\mu(x) \to \infty$ when $r \to 0$, for physically meaningful solutions that are finite along the z-axis, we set $a_1 = 0$ in Eq. (14.251). Using the first boundary condition [Eq. (14.248)], we write

$$J_0(\lambda a) = 0 \tag{14.253}$$

and obtain the admissible values of λ as

$$\lambda_n = \frac{x_{0n}}{a}, \quad n = 1, 2, \ldots, \tag{14.254}$$

where x_{0n} are the zeros of the Bessel function $J_0(x)$. Now a general solution can be written in terms of the complete and orthogonal set:

$$\left\{ \Psi_n(r, z) = \left[c_{0n} \cosh \left(\frac{x_{0n} z}{a} \right) + c_{1n} \sinh \left(\frac{x_{0n} z}{a} \right) \right] J_0 \left(\frac{x_{0n} r}{a} \right) \right\}, \tag{14.255}$$

$n = 1, 2, \ldots$, as

$$\Psi(r, z) = \sum_{n=1}^{\infty} \left[A_n \cosh \left(\frac{x_{0n} z}{a} \right) + B_n \sinh \left(\frac{x_{0n} z}{a} \right) \right] J_0 \left(\frac{x_{0n} r}{a} \right). \tag{14.256}$$

Using the remaining boundary conditions [Eqs. (14.249) and (14.250)], we also write

$$f_0(r) = \Psi(r, 0) = \sum_{n=1}^{\infty} A_n J_0 \left(\frac{x_{0n}}{a} r \right), \tag{14.257}$$

$$f_1(r) = \Psi(r, l) = \sum_{n=1}^{\infty} \left[A_n \cosh \left(\frac{x_{0n}}{a} l \right) + B_n \sinh \left(\frac{x_{0n}}{a} l \right) \right] J_0 \left(\frac{x_{0n}}{a} r \right). \tag{14.258}$$

Using the orthogonality relation

$$\int_0^a r J_0 \left(\frac{x_{0m}}{a} r \right) J_0 \left(\frac{x_{0n}}{a} r \right) dr = \frac{a^2}{2} [J_1(x_{0n})]^2 \delta_{mn}, \tag{14.259}$$

we can evaluate the expansion coefficients A_n and B_n as

$$A_n = \frac{2 \int_0^a r f_0(r) J_0 \left(\frac{x_{0n}}{a} r \right) dr}{a^2 [J_1(x_{0n})]^2}, \tag{14.260}$$

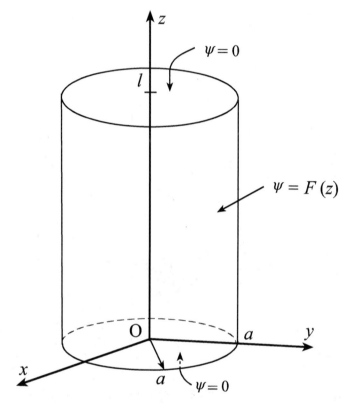

Figure 14.6 Boundary conditions for Example 14.5.

$$B_n = \frac{2\left[\int_0^a r f_1(r) J_0\left(\frac{x_{0n}r}{a}\right) dr - \cosh\left(\frac{x_{0n}l}{a}\right)\int_0^a r f_0(r) J_0\left(\frac{x_{0n}r}{a}\right) dr\right]}{a^2[J_1(x_{0n})]^2 \sinh\left(\frac{x_{0n}l}{a}\right)}.$$

$$(14.261)$$

Example 14.5. *Another boundary condition for the Laplace equation:* We now solve the Laplace equation with the following boundary conditions (Figure 14.6):

$$\Psi(a, z) = F(z), \qquad (14.262)$$

$$\Psi(r, 0) = 0, \qquad (14.263)$$

$$\Psi(r, l) = 0. \qquad (14.264)$$

Because of axial symmetry, we again take $\mu = 0$, $\Phi(\phi) = 1$ in Eq. (14.216). To satisfy the second condition, we set $c_0 = 0$ in Eq. (14.220):

$$Z(z) = c_0 \cosh \lambda z + c_1 \sinh \lambda z, \qquad (14.265)$$

to write

$$Z(z) = c_1 \sinh \lambda z. \tag{14.266}$$

If we write sinh as sin with an imaginary argument

$$\sinh \lambda z = -i \sin i\lambda z, \tag{14.267}$$

and use the third condition [Eq. (14.264)], the allowed values of λ are found as

$$\lambda_n = \frac{n\pi i}{l}, \quad n = 1, 2, \ldots. \tag{14.268}$$

Now the solutions can be written in terms of the modified Bessel functions as

$$R(r) = a_0 J_0 \left(\frac{n\pi i}{l} r \right) + a_1 H_0^{(1)} \left(\frac{n\pi i}{l} r \right)$$

$$= a_0 I_0 \left(\frac{n\pi}{l} r \right) + a_1 \left(\frac{2}{\pi i} \right) K_0 \left(\frac{n\pi}{l} r \right), \tag{14.269}$$

$$Z(z) = c_1 \sin \left(\frac{n\pi}{l} z \right).$$

Since

$$K_0 \left(\frac{n\pi}{l} r \right) \to \infty \text{ as } r \to 0,$$

we also set $a_1 = 0$, thus obtaining the complete and orthogonal set

$$\left\{ \Psi_n(r, z) = a_0 I_0 \left(\frac{n\pi}{l} r \right) \sin \left(\frac{n\pi}{l} z \right) \right\}, \quad n = 1, 2, \ldots. \tag{14.270}$$

We can now write a general solution as the series

$$\Psi(r, z) = \sum_{n=1}^{\infty} A_n I_0 \left(\frac{n\pi}{l} r \right) \sin \left(\frac{n\pi}{l} z \right). \tag{14.271}$$

Using the orthogonality relation

$$\int_0^l \sin \left(\frac{m\pi}{l} z \right) \sin \left(\frac{n\pi}{l} z \right) dz = \frac{l}{2} \delta_{mn}, \tag{14.272}$$

we find the expansion coefficients as

$$A_n = \left(\frac{2}{l} \right) \left[I_0 \left(\frac{n\pi}{l} a \right) \right]^{-1} \int_0^l F(z) \sin \left(\frac{n\pi}{l} z \right) dz. \tag{14.273}$$

Example 14.6. *Periodic boundary conditions:* Consider the Laplace equation in cylindrical coordinates with the following periodic boundary conditions:

$$\Psi(r, 0, z) = \Psi(r, 2\pi, z), \tag{14.274}$$

$$\frac{\partial \Psi(r, \phi, z)}{\partial \phi}\bigg|_{\phi=0} = \frac{\partial \Psi(r, \phi, z)}{\partial \phi}\bigg|_{\phi=2\pi} \tag{14.275}$$

and

$$\Psi(a, \phi, z) = 0. \tag{14.276}$$

Using the first two conditions [Eqs. (14.274) and (14.275)] with Eq. (14.219)

$$\Phi(\phi) = b_0 \cos \mu\phi + b_1 \sin \mu\phi,$$

which we write as

$$\Phi(\phi) = \delta_0 \cos(m\phi + \delta_m), \tag{14.277}$$

where δ_0 and δ_m are constants, we obtain the allowed values of m as

$$\mu = m = 0, 1, 2, \dots. \tag{14.278}$$

For finite solutions along the z-axis, we set $a_1 = 0$ in the radial solution [Eq. (14.218)] to write

$$R(r) = a_0 J_m(\lambda r). \tag{14.279}$$

Imposing the final boundary condition [Eq. (14.276)]

$$J_m(\lambda a) = 0, \tag{14.280}$$

we obtain the admissible values of λ as

$$\lambda_n = \frac{x_{mn}}{a}, \quad n = 1, 2, \dots, \tag{14.281}$$

where x_{mn} are the roots of $J_m(x)$. Finally, for the z-dependent solution, we use Eq. (14.220) and take the basis functions as the set

$$\left\{ \Psi_{mn} = J_m\left(\frac{x_{mn}r}{a}\right) [\cos(m\phi + \delta_m)] \left[c_0 \cosh \frac{x_{mn}z}{a} + c_1 \sinh \frac{x_{mn}z}{a} \right] \right\}, \tag{14.282}$$

where $m = 0, 1, \ldots$ and $n = 1, 2, \ldots$. We can now write a general solution as the series

$$\Psi(r, \phi, z)$$
$$= \sum_m \sum_n J_m \left(\frac{x_{mn} r}{a}\right) \cos(m\phi + \delta_m) \left[A_{mn} \cosh \frac{x_{mn} z}{a} + B_{mn} \sinh \frac{x_{mn} z}{a}\right].$$
$$(14.283)$$

Example 14.7. *Cooling of a long circular cylinder:* Consider a long circular cylinder with radius a, initially heated to a uniform temperature T_1, while its surface is maintained at a constant temperature T_0. Assume the length to be so large that the z-dependence of the temperature can be ignored. Since we basically have a two-dimensional problem, using the cylindrical coordinates, we write the heat transfer equation [Eq. (14.239)] as

$$b \frac{\partial \Psi(r, t)}{\partial t} = \frac{\partial^2 \Psi(r, t)}{\partial r^2} + \frac{1}{r} \frac{\partial \Psi(r, t)}{\partial r}, \qquad (14.284)$$

where b is a constant depending on the physical parameters of the system. We take the boundary condition at the surface as

$$\Psi(a, t) = T_0, \quad 0 < t < \infty, \qquad (14.285)$$

while the initial condition at $t = 0$ is

$$\Psi(r, 0) = T_1, \quad 0 \leq r < a. \qquad (14.286)$$

We can work with the homogeneous boundary condition by defining a new dependent variable as

$$\Omega(r, t) = \Psi(r, t) - T_0, \qquad (14.287)$$

where $\Omega(r, t)$ satisfies the differential equation

$$b \frac{\partial \Omega(r, t)}{\partial t} = \frac{\partial^2 \Omega(r, t)}{\partial r^2} + \frac{1}{r} \frac{\partial \Omega(r, t)}{\partial r} \qquad (14.288)$$

with the following boundary conditions:

$$\Omega(a, t) = 0, \ 0 < t < \infty, \qquad (14.289)$$

$$\Omega(r, 0) = T_1 - T_0, \ 0 \leq r < a. \qquad (14.290)$$

We need a finite solution for $0 \leq r < a$ and one that satisfies $\Omega(r, t) \to 0$ as $t \to \infty$. Substituting a separable solution of the form

$\Omega(r,t) = R(r)T(t)$, we obtain the differential equations to be solved for $R(r)$ and $T(t)$ [Eqs. (14.240)–(14.243)] as

$$b\frac{dT}{dt} + \chi^2 T = 0, \qquad (14.291)$$

$$\frac{1}{r}\frac{d}{dr}\left[r\frac{dR}{dr}\right] + \chi^2 R = 0, \qquad (14.292)$$

where χ is the separation constant. Note that Eqs. (14.291) and (14.292) can also be obtained by first choosing $a = 0$ and $\kappa = 0$ in Eq. (14.195) and then from Eqs. (14.207)–(14.210) with the choices $\mu = 0$, $\Phi(\phi) = 1$, and $\lambda = 0$, $Z(z) = 1$. Solution of the time-dependent equation [Eq. (14.291)] can be written immediately:

$$T(t) = Ce^{-\chi^2 t/b}, \qquad (14.293)$$

while the solution [Eq. (14.245)] of the radial equation [Eq. (14.292)] is

$$R(r) = a_0 J_0(\chi r) + a_1 N_0(\chi r). \qquad (14.294)$$

Since $N_0(\chi r)$ diverges as $r \to 0$, we set a_1 to 0, thus obtaining

$$\Omega(r,t) = a_0 J_0(\chi r)e^{-\chi^2 t/b}. \qquad (14.295)$$

To satisfy the condition in Eq. (14.289), we write

$$J_0(\chi a) = 0, \qquad (14.296)$$

which gives the allowed values of χ as the zeros of $J_0(x)$:

$$\chi_n = \frac{x_{0n}}{a}, \quad n = 1, 2, \ldots. \qquad (14.297)$$

Now the solution becomes

$$\Omega_n(r,t) = A_n J_0\left(\frac{x_{0n}}{a}r\right)e^{-x_{0n}^2 t/ab}. \qquad (14.298)$$

Since these solutions form a complete and orthogonal set, we can write a general solution as the series

$$\Omega(r,t) = \sum_{n=1}^{\infty} C_n J_0\left(\frac{x_{0n}}{a}r\right)e^{-x_{0n}^2 t/ab}. \qquad (14.299)$$

Since $\Omega_n(r,t)$ satisfies all the conditions except Eq. (14.290), their linear combination will also satisfy the same conditions. To satisfy the remaining condition [Eq. (14.290)], we write

$$\Omega_n(r,0) = T_1 - T_0 = \sum_{n=1}^{\infty} C_n J_0\left(\frac{x_{0n}}{a}r\right). \tag{14.300}$$

We now use the orthogonality relation:

$$\int_0^a rJ_0\left(\frac{x_{0m}}{a}r\right)J_0\left(\frac{x_{0n}}{a}r\right)dr = \frac{a^2}{2}[J_1(x_{0n})]^2\delta_{mn}, \tag{14.301}$$

along with the recursion relation

$$x^m J_{m-1}(x) = \frac{d}{dx}[x^m J_m(x)] \tag{14.302}$$

and the special value $J_1(0) = 0$, to write the expansion coefficients as

$$C_n = \frac{2(T_1 - T_0)}{x_{0n}^2 J_1(x_{0n})}. \tag{14.303}$$

Example 14.8. *Symmetric   vibrations   of   a   circular   drum-head:* Consider a circular membrane fixed at its rim and oscillating freely. For oscillations symmetric about the origin, we have only r dependence. Hence, we write the wave equation [Eq. (14.229)] in cylindrical coordinates as

$$\frac{\partial^2 \Psi(r,t)}{\partial t^2} = v^2\left[\frac{\partial^2 \Psi(r,t)}{\partial r^2} + \frac{1}{r}\frac{\partial \Psi(r,t)}{\partial r}\right], \tag{14.304}$$

where $\Psi(r,t)$ represents the vertical displacement of the membrane from its equilibrium position. For a separable solution, $\Psi(r,t) = R(r)T(t)$, the corresponding equations to be solved are

$$\frac{1}{v^2}\frac{d^2T}{dt^2} + \chi^2 T = 0, \tag{14.305}$$

$$\frac{d^2R}{dr^2} + \frac{1}{r}\frac{dR}{dr} + \chi^2 R = 0, \tag{14.306}$$

where χ is the separation constant. These equations can again be obtained from our basic equations [Eq. (14.195) and Eqs. (14.231)–(14.234)] with the substitutions $\kappa = 0$, $b = 0$, $a = 1/v^2$ and $\lambda = 0$, $Z(z) = 1$, $\mu = 0$, $\Phi(\phi) = 1$. The time-dependent equation can be solved immediately as

$$T(t) = \delta_0 \cos(\omega t + \delta_1), \tag{14.307}$$

where we have defined $\omega^2 = \chi^2 v^2$. The general solution of Eq. (14.306) is

$$R(r) = a_0 J_0 \left(\frac{\omega}{v} r\right) + a_1 N_0 \left(\frac{\omega}{v} r\right). \tag{14.308}$$

We again set $a_1 = 0$ for regular solutions at the origin, which leads to

$$\Psi(r,t) = A_0 J_0 \left(\frac{\omega}{v} r\right) \cos(\omega t + \delta_1). \tag{14.309}$$

Since the membrane is fixed at its rim, we write

$$\Psi(a,t) = J_0 \left(\frac{\omega}{v} r\right) = 0. \tag{14.310}$$

This gives the allowed frequencies as

$$\omega_n = x_{0n} \frac{v}{a}, \quad n = 1, 2, \ldots, \tag{14.311}$$

where x_{0n} are the zeros of $J_0(x)$. We now have the complete set of orthogonal functions

$$\Psi_n(r,t) = J_0 \left(\frac{x_{0n}}{a} r\right) \cos \left(x_{0n} \frac{v}{a} t + \delta_n\right), \quad n = 1, 2, \ldots, \tag{14.312}$$

which can be used to write a general solution as

$$\Psi(r,t) = \sum_{n=1}^{\infty} A_n J_0 \left(\frac{x_{0n}}{a} r\right) \cos \left(x_{0n} \frac{v}{a} t + \delta_n\right). \tag{14.313}$$

Expansion coefficients A_n and the phases δ_n come from the initial conditions:

$$\Psi(r,0) = f(r), \tag{14.314}$$

$$\frac{\partial \Psi}{\partial t}\bigg|_{t=0} = F(r), \tag{14.315}$$

as

$$\delta_n = \tan^{-1} \frac{Y_{0n}}{X_{0n}}, \tag{14.316}$$

$$A_n = \sqrt{X_{0n}^2 + Y_{0n}^2}, \tag{14.317}$$

where

$$X_{0n} = \frac{2}{a^2 J_1^2(x_{0n})} \int_0^a f(r) r J_0 \left(\frac{x_{0n}}{a} r\right) dr, \tag{14.318}$$

$$Y_{0n} = \frac{-2}{x_{0n} v a J_1^2(x_{0n})} \int_0^a F(r) r J_0 \left(\frac{x_{0n}}{a} r\right) dr. \tag{14.319}$$

REFERENCES

1. Bayin, S.S. (2018). *Mathematical Methods in Science and Engineering*, 2e. Wiley.

2. Lebedev, N.N. (1965). *Special Functions and Their Applications*. Englewood Cliffs, NJ: Prentice-Hall.

PROBLEMS

1. Solve the two-dimensional Laplace equation:

$$\frac{\partial^2 \Psi(x,y)}{\partial x^2} + \frac{\partial^2 \Psi(x,y)}{\partial y^2} = 0,$$

inside a rectangular region, $x \in [0, a]$ and $y \in [0, b]$, in Cartesian coordinates with the following boundary conditions:

$$\Psi(x,0) = 0,$$
$$\Psi(x,b) = f_0,$$
$$\Psi(0,y) = 0,$$
$$\Psi(a,y) = 0.$$

2. Solve the Laplace equation:

$$\frac{\partial^2 \Psi(x,y)}{\partial x^2} + \frac{\partial^2 \Psi(x,y)}{\partial y^2} = 0,$$

inside a rectangular region, $x \in [0, a]$ and $y \in [0, b]$, in Cartesian coordinates with the following boundary conditions:

$$\Psi(x,0) = 0,$$
$$\Psi(x,b) = f_0,$$
$$\Psi(0,y) = f_1,$$
$$\Psi(a,y) = 0,$$

where f_0 and f_1 are constants (Figure 14.7).

3. In Example 14.1, show that the Laplace equation with the boundary conditions

$$\Psi(x,0) = 0,$$
$$\Psi(x,b) = 0,$$
$$\Psi(0,y) = F(y),$$
$$\Psi(a,y) = 0,$$

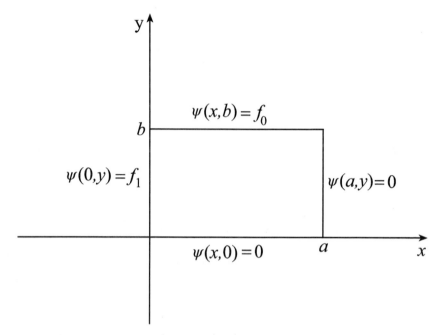

Figure 14.7   Boundary conditions for Problem 2.

leads to the solution

$$\Psi(x,y) = \sum_{n=1}^{\infty} C_n \left[\sin \frac{n\pi y}{b}\right] \left[\frac{\sinh \frac{n\pi}{b}(a-x)}{\sinh \frac{n\pi}{b}a}\right],$$

where

$$C_n = \frac{2}{b} \int_0^b F(y) \sin\left(\frac{n\pi y}{b}\right) dy.$$

4. Under what conditions will the Helmholtz equation:

$$\vec{\nabla}^2 \Psi(\vec{r}) + k^2(\vec{r})\Psi(\vec{r}) = 0,$$

be separable in Cartesian coordinates.

5. Toroidal coordinates (α, β, ϕ) are defined as

$$x = \frac{c \sinh\alpha \cos\phi}{\cosh\alpha - \cos\beta}, \ y = \frac{c \sinh\alpha \sin\phi}{\cosh\alpha - \cos\beta}, \ z = \frac{c \sin\beta}{\cosh\alpha - \cos\beta},$$

where $\alpha \in [0, \infty)$, $\beta \in (-\pi, \pi]$, $\phi \in (-\pi, \pi]$ and the scale factor c is positive definite. Toroidal coordinates are useful in solving problems with a torous as the bounding surface or domains bounded by two intersecting spheres (see Lebedev for a discussion of various coordinate systems available [2]).

(i) Show that the Laplace equation:

$$\overrightarrow{\nabla}^2 \Psi(\alpha, \beta, \phi) = 0,$$

in toroidal coordinates is given as

$$\frac{\partial}{\partial\alpha}\left[\frac{\sinh\alpha}{\cosh\alpha - \cos\beta}\frac{\partial\Psi}{\partial\alpha}\right] + \frac{\partial}{\partial\beta}\left[\frac{\sinh\alpha}{\cosh\alpha - \cos\beta}\frac{\partial\Psi}{\partial\beta}\right]$$

$$+ \frac{1}{(\cosh\alpha - \cos\beta)\sinh\alpha}\frac{\partial^2\Psi}{\partial\phi^2} = 0.$$

(ii) Show that as it stands this equation is not separable.

(iii) However, show that with the substitution

$$\Psi(\alpha, \beta, \phi) = (2\cosh\alpha - 2\cos\beta)^{1/2}\Omega(\alpha, \beta, \phi),$$

the resulting equation is separable as

$$\Omega(\alpha, \beta, \phi) = A(\alpha)B(\beta)C(\phi)$$

and find the corresponding ordinary differential equations for $A(\alpha)$, $B(\beta)$, and $C(\phi)$.

6. Using your result in Problem 5, find separable solutions of the heat flow equation in toroidal coordinates.

7. Consider a cylinder of length l and radius a whose ends are kept at temperature zero. Find the steady-state distribution of temperature inside the sphere when the rest of the surface is maintained at temperature T_0.

8. Find the electrostatic potential inside a closed cylindrical conductor of length l and radius a, with the bottom and the lateral surfaces held at potential V and the top surface held at zero potential. The top surface is separated by a thin insulator from the rest of the cylinder.

9. Show that the stationary distribution of temperature in the upper half-space, $z > 0$, satisfying the boundary condition

$$T(x, y, 0) = F(r) = \begin{cases} T_0, & r < a, \\ 0, & r > a, \end{cases}$$

is given as

$$T(r, z) = T_0 a \int_0^\infty e^{-\lambda z} J_0(\lambda r) J_1(\lambda a) d\lambda.$$

Hint: Use the relation

$$\frac{1}{r}\delta(r - r') = \int_0^\infty \lambda J_m(\lambda r) J_m(\lambda r')d\lambda.$$

Can you derive it?

10. Consider a grounded conducting sphere of radius a and a point charge Q outside the conductor at a distance d from the center. Find the potential outside the conductor. Discuss your result.

CHAPTER 15

FOURIER SERIES

In 1807, Fourier announced in a seminal paper that a large class of functions can be written as linear combinations of sines and cosines. Today, infinite series representation of functions in terms of sinusoidal functions is called the Fourier series, which has become an indispensable tool in signal analysis. Spectroscopy is the branch of science that deals with the analysis of a given signal in terms of its components. Image processing and data compression are among other important areas of application for the Fourier series.

15.1 ORTHOGONAL SYSTEMS OF FUNCTIONS

After the introduction of Fourier series, it became clear that they are only a part of a much more general branch of mathematics called the theory of **orthogonal functions**. Legendre polynomials, Hermite polynomials, and Bessel functions are among the other commonly used orthogonal function sets. Certain features of this theory are incredibly similar to geometric vectors, where in n dimensions a given vector can be written as a linear combination of n linearly independent

Essentials of Mathematical Methods in Science and Engineering, Second Edition. Selçuk Ş. Bayın.
© 2020 John Wiley & Sons, Inc. Published 2020 by John Wiley & Sons, Inc.

basis vectors. In the theory of orthogonal functions, we can express almost any *arbitrary* function as the linear combination of a set of basis functions. Many of the tools used in the study of ordinary vectors have counterparts in the theory of orthogonal functions [1]. Among the most important ones is the definition of **inner product**, which is the analog of the scalar or the dot product for ordinary vectors.

Definition 15.1. If f and g are two complex-valued functions, both (Riemann) integrable in the interval $[a, b]$, their **inner product** is defined as the integral

$$(f, g) = \int_a^b f^*(x)g(x)dx. \tag{15.1}$$

For real-valued functions, the complex conjugate becomes superfluous. From this definition, it follows that the inner product satisfies the properties

$$(f, g) = (g, f)^*, \tag{15.2}$$

$$(cf, g) = c^*(f, g), \tag{15.3}$$

$$(f, cg) = c(f, g), \tag{15.4}$$

$$(f_1 + f_2, g) = (f_1, g) + (f_2, g), \tag{15.5}$$

$$(f, g_1 + g_2) = (f, g_1) + (f, g_2), \tag{15.6}$$

where c is a complex number. The nonnegative number, $(f, f)^{1/2}$, is called the **norm** of f. It is usually denoted by $\|f\|$, and it is the analog of the magnitude of a vector. The following inequalities follow directly from the properties of inner product:

Cauchy–Schwarz inequality:

$$|(f, g)| \leq \|f\|\|g\|. \tag{15.7}$$

Minkowski inequality:

$$\|f + g\| \leq \|f\| + \|g\|, \tag{15.8}$$

which is the analog of the triangle inequality.

Definition 15.2. Let $S = \{u_0, u_1, \ldots\}$ be a set of integrable functions in the interval $[a, b]$. If

$$(u_m, u_n) = 0 \quad \text{for all } m \neq n, \tag{15.9}$$

then the set S is called **orthogonal**. Furthermore, when the norm of each element of S is normalized to unity, $\|u_n\| = 1$, we have

$$(u_m, u_n) = \delta_{mn} \tag{15.10}$$

and the set S is called **orthonormal**. We have seen that Legendre polynomials, Hermite polynomials, Laguerre polynomials, and Bessel functions form orthogonal sets. As the reader can verify, the set

$$\left\{ u_n(x) = \frac{e^{inx}}{\sqrt{2\pi}} \right\}, \quad n = 0, 1, \ldots, \quad x \in [0, 2\pi] \tag{15.11}$$

is one of the most important orthonormal sets in applications.

Linear independence of function sets is defined similar to ordinary vectors. A set of functions,

$$S = \{u_0, u_1, \ldots, u_n\}, \tag{15.12}$$

is called **linearly independent** in the interval $[a, b]$ if the equation

$$c_0 u_0 + c_1 u_1 + \cdots + c_n u_n = 0, \tag{15.13}$$

where $c_0, c_1, \ldots, c_n$ are in general complex numbers, cannot be satisfied unless all c_i are zero:

$$c_0 = c_1 = \cdots = c_n = 0. \tag{15.14}$$

An **infinite set** $S = \{u_0, u_1, \ldots\}$ is called **linearly independent** in $[a, b]$, if every finite subset of S is linearly independent. It is clear that every orthogonal/orthonormal set is linearly independent. All the similarities with vectors suggest that an arbitrary function may be expressed as the linear combination of the elements of an orthonormal set. An expansion of this sort will naturally look like

$$\boxed{f(x) = \sum_{n=0}^{\infty} c_n u_n(x), \quad x \in [a, b].} \tag{15.15}$$

There are basically two important questions to be addressed: first, how to find the expansion coefficients, and second, will the series $\sum_{n=0}^{\infty} c_n u_n$ converge to $f(x)$? Finding the coefficients, at least formally, is possible. Using the orthonormality relation, $\int u_m^* u_n \, dx = \delta_{mn}$, we can find c_n as

$$(u_n, f) = \sum_{m=0}^{\infty} c_m (u_n, u_m) = \sum_{m=0}^{\infty} c_m \delta_{nm} = c_n. \tag{15.16}$$

In other words, the coefficients of expansion are found as

$$\boxed{c_n = \int_a^b u_n^* f \, dx.} \tag{15.17}$$

As far as the **convergence** of the series $\sum_{n=0}^{\infty} c_n u_n$ is concerned, the following questions come to mind: Does it converge uniformly? Does it converge only at certain points of the interval $[a, b]$? Does it converge pointwise, that is, for all the points of the interval $[a, b]$? Uniform convergence implies the integrability of $f(x)$ and justifies the steps leading to Eq. (15.17). In other words, when the series $\sum_{n=0}^{\infty} c_n u_n$ converges uniformly to a function $f(x)$, then that function is integrable, and the coefficients c_n are found as in Eq. (15.17). Note that uniform convergence implies pointwise convergence but not vice versa.

For an orthonormal set $S = \{u_0, u_1, \ldots\}$ defined over the interval $[a, b]$, an integrable function can be written as the series

$$f(x) = \sum_{n=0}^{\infty} c_n u_n(x), \quad c_n = (u_n, f), \tag{15.18}$$

which is called the **generalized Fourier series** of $f(x)$ with respect to the set S. A general discussion of the convergence of these series is beyond the scope of this book, but for majority of the physically interesting problems, they converge [2]. However, for the (trigonometric) Fourier series, we address this question in detail. In the meantime, to obtain further justification of the series representation in Eq. (15.18), consider the partial sum

$$S_n(x) = \sum_{k=0}^{n} c_k u_k(x), \quad n = 0, 1, 2, \ldots, \tag{15.19}$$

and the finite sum

$$t_n(x) = \sum_{k=0}^{n} b_k u_k(x), \quad n = 0, 1, 2, \ldots, \tag{15.20}$$

where b_k are arbitrary complex numbers. We now write the expression

$$\int_a^b |f - t_n|^2 dx = (f - t_n, f - t_n) \tag{15.21}$$

$$= (f, f) - (f, t_n) - (t_n, f) + (t_n, t_n). \tag{15.22}$$

Using the properties of inner product, along with the orthogonality relation [Eq. (15.10)], we can write

$$(t_n, t_n) = \left(\sum_{k=0}^{n} b_k u_k(x), \sum_{l=0}^{n} b_l u_l(x) \right) \tag{15.23}$$

$$= \sum_{k=0}^{n} \sum_{l=0}^{n} b_k^* b_l (u_k, u_l) = \sum_{k=0}^{n} |b_k|^2, \tag{15.24}$$

$$(f, t_n) = \left(f, \sum_{k=0}^{n} b_k u_k(x) \right) \tag{15.25}$$

$$= \sum_{k=0}^{n} b_k(f, u_k(x)) = \sum_{k=0}^{n} b_k c_k^* \tag{15.26}$$

and

$$(t_n, f) = (f, t_n)^* = \sum_{k=0}^{n} b_k^* c_k. \tag{15.27}$$

Using Eqs. (15.24), (15.26), and (15.27), Eq. (15.22) can be written as

$$\int_a^b |f - t_n|^2 dx = \|f\|^2 + \sum_{k=0}^{n} |b_k|^2 - \sum_{k=0}^{n} b_k c_k^* - \sum_{k=0}^{n} b_k^* c_k \tag{15.28}$$

$$= \|f\|^2 - \sum_{k=0}^{n} |c_k|^2 + \sum_{k=0}^{n} (b_k - c_k)(b_k^* - c_k^*) \tag{15.29}$$

$$= \|f\|^2 - \sum_{k=0}^{n} |c_k|^2 + \sum_{k=0}^{n} |b_k - c_k|^2. \tag{15.30}$$

Since the right-hand side [Eq. (15.30)] is smallest when $b_k = c_k$, we can also write

$$\int_a^b |f - S_n|^2 dx \leq \int_a^b |f - t_n|^2 dx. \tag{15.31}$$

Significance of these results becomes clear if we notice that each linear combination, t_n, can be thought of as an approximation of the function $f(x)$. The integral on the left-hand side of Eq. (15.31) can now be interpreted as the **mean square error** of this approximation. The inequality in Eq. (15.31) states that among all possible approximations, $t_n = \sum_{k=0}^{n} b_k u_k$, of $f(x)$, the nth partial sum of the generalized Fourier series represents the best approximation in terms of the mean square error.

Using Eq. (15.30), we can obtain two more useful expressions. If we set $c_k = b_k$ and notice that the right-hand side is always positive, we can write

$$\boxed{\int_a^b |f|^2 dx \geq \sum_{k=0}^{\infty} |c_k|^2,} \tag{15.32}$$

which is called the **Bessel's inequality**. With the substitution $c_k = b_k$, Eq. (15.30) also implies

$$\|f - S_n\|^2 = \int_a^b |f - S_n|^2 dx \tag{15.33}$$

$$= \int_a^b |f|^2 dx - \sum_{k=0}^n |c_k|^2. \tag{15.34}$$

Since as $n \to \infty$ we have $\|f - S_n\| \to 0$, we obtain

$$\boxed{\sum_{k=0}^\infty |c_k|^2 = \int_a^b |f|^2 dx,} \tag{15.35}$$

which is known as the **Parseval's formula**. We conclude this section with the definition of **completeness**:

Definition 15.3. Given an orthonormal set of integrable functions defined over the interval $[a, b]$:

$$S = \{u_0, u_1, \ldots\}. \tag{15.36}$$

In the expansion

$$f(x) = \sum_{n=0}^\infty c_n u_n(x), \tag{15.37}$$

if the limit

$$\lim_{n \to \infty} \left\| \sum_{k=0}^n c_k u_k - f \right\| \to 0 \tag{15.38}$$

is true for every integrable f, then the set S is called **complete**, and we say the series $\sum_{n=0}^\infty c_n u_n(x)$ **converges in the mean** to $f(x)$. Convergence in the mean is not as strong as uniform or pointwise convergence, but for most practical purposes, it is sufficient.

From Bessel's inequality [Eq. (15.32)] it is seen that for **absolutely integrable** functions, that is, when the integral $\int_a^b |f| dx$ exists, the series $\sum_{k=0}^\infty |c_k|^2$ also converges, hence the nth term of this series necessarily goes to 0 as $n \to \infty$. In particular, if we use the complete set

$$S = \{u_n = e^{inx}\}, \quad n = 0, 1, 2, \ldots, \quad x \in [0, 2\pi], \tag{15.39}$$

we obtain the limit

$$\lim_{n \to \infty} \int_0^{2\pi} f(x) e^{-inx} dx = 0, \tag{15.40}$$

which can be used to deduce the following useful formulas:

$$\lim_{n \to \infty} \int_0^{2\pi} f(x) \cos nx \, dx = 0, \tag{15.41}$$

$$\lim_{n \to \infty} \int_0^{2\pi} f(x) \sin nx \, dx = 0. \tag{15.42}$$

This result can be generalized as the **Riemann–Lebesgue lemma** [3, p. 471] for absolutely integrable functions on $[a, b]$ as

$$\lim_{\alpha \to \infty} \int_a^b f(x) \sin(\alpha x + \beta) dx = 0, \tag{15.43}$$

which holds for all real α and β, and where the lower limit a could be $-\infty$ and the upper limit b could be ∞.

15.2 FOURIER SERIES

The term Fourier series usually refers to series expansions in terms of the **orthogonal system**

$$S = \{u_0, u_{2n-1}, u_{2n}\}, \quad n = 1, 2, \ldots, \ x \in [-\pi, \pi], \tag{15.44}$$

where

$$u_0 = \frac{1}{\sqrt{2\pi}}, \quad u_{2n-1} = \frac{\cos nx}{\sqrt{\pi}}, \quad u_{2n} = \frac{\sin nx}{\sqrt{\pi}}, \tag{15.45}$$

which satisfy the orthogonality relations

$$\frac{1}{\pi} \int_{-\pi}^{\pi} \cos nx \cos mx \, dx = \delta_{nm}, \tag{15.46}$$

$$\frac{1}{\pi} \int_{-\pi}^{\pi} \sin nx \sin mx \, dx = \delta_{nm}, \tag{15.47}$$

$$\frac{1}{\pi} \int_{-\pi}^{\pi} \sin nx \cos mx \, dx = 0. \tag{15.48}$$

Since the basis functions are periodic with the period 2π, their linear combinations are also periodic with the same period. We shall see that every periodic function satisfying certain smoothness conditions can be expressed as the **Fourier series**

$$f(x) = \frac{a_0}{2} + \sum_{n=1}^{\infty} [a_n \cos nx + b_n \sin nx]. \tag{15.49}$$

Using the orthogonality relations, we can evaluate the **expansion coefficients** as

$$a_n = \frac{1}{\pi} \int_{-\pi}^{\pi} f(t) \cos nt \, dt, \quad n = 0, 1, 2, \ldots, \tag{15.50}$$

$$b_n = \frac{1}{\pi} \int_{-\pi}^{\pi} f(t) \sin nt \, dt, \quad n = 1, 2, \ldots. \tag{15.51}$$

Substituting the coefficients [Eqs. (15.50) and (15.51)] back into Eq. (15.49) and using trigonometric identities, we can write the Fourier series in more **compact form** as

$$f(x) = \frac{1}{2\pi} \int_{-\pi}^{\pi} f(t) dt + \frac{1}{\pi} \sum_{n=1}^{\infty} \int_{-\pi}^{\pi} f(t) \cos n(x - t) dt. \tag{15.52}$$

Note that the first term $a_0/2$ is the mean value of $f(x)$ in the interval $[-\pi, \pi]$.

15.3   EXPONENTIAL FORM OF THE FOURIER SERIES

Using the relations

$$\cos nx = \frac{1}{2}(e^{inx} + e^{-inx}),$$

$$\sin nx = \frac{1}{2i}(e^{inx} - e^{-inx}), \tag{15.53}$$

we can express the Fourier series in exponential form as

$$f(x) = \frac{c_0}{2} + \sum_{n=1}^{\infty} [c_n e^{inx} + c_n^* e^{-inx}], \tag{15.54}$$

where

$$c_n = \frac{1}{2}(a_n - ib_n), \quad c_n^* = \frac{1}{2}(a_n + ib_n), \quad c_0 = a_0. \tag{15.55}$$

Since $c_n^* = c_{-n}$, we can also write this in compact form as

$$f(x) = \sum_{n=-\infty}^{\infty} c_n e^{inx}. \tag{15.56}$$

15.4   CONVERGENCE OF FOURIER SERIES

We now turn to the convergence problem of Fourier series. The whole theory can be built on two fundamental formulas. Let us now write the partial sum

$$S_n(x) = \frac{a_0}{2} + \sum_{k=1}^{n}[a_k \cos kx + b_k \sin kx], \quad n = 1, 2, \ldots \quad . \tag{15.57}$$

Substituting the definitions of the coefficients a_k and b_k [Eqs. (15.50) and (15.51)] into the aforementioned equation, we obtain

$$S_n(x) = \frac{1}{\pi} \int_{-\pi}^{\pi} f(t) \left\{ \frac{1}{2} + \sum_{k=1}^{n} (\cos kt \cos kx + \sin kt \sin kx) \right\} dt \tag{15.58}$$

$$= \frac{1}{\pi} \int_{-\pi}^{\pi} f(t) \left\{ \frac{1}{2} + \sum_{k=1}^{n} \cos k(t - x) \right\} dt \tag{15.59}$$

$$= \frac{1}{\pi} \int_{-\pi}^{\pi} f(t) D_n(t - x) \, dt, \tag{15.60}$$

where we have introduced the function

$$D_n(t) = \frac{1}{2} + \sum_{k=1}^{n} \cos kt. \tag{15.61}$$

Since both $f(t)$ and $D_n(t)$ are periodic functions with the period 2π, after a variable change, $t - x \to t$, we can write

$$S_n(x) = \frac{1}{\pi} \int_{-\pi-x}^{\pi-x} f(x + t) D_n(t) \, dt = \frac{1}{\pi} \int_{-\pi}^{\pi} f(x + t) D_n(t) \, dt. \tag{15.62}$$

Using the fact that $D_n(-t) = D_n(t)$, this can also be written as

$$S_n(x) = \frac{2}{\pi} \int_{0}^{\pi} \frac{f(x + t) + f(x - t)}{2} D_n(t) \, dt \tag{15.63}$$

$$= \frac{4}{\pi} \int_{0}^{\pi/2} \frac{f(x + 2t) + f(x - 2t)}{2} D_n(2t) \, dt. \tag{15.64}$$

Using the trigonometric relation (see Problem 6):

$$\sum_{k=1}^{n} \cos kt = \left(\sin \frac{nt}{2} \cos \frac{(n+1)t}{2} \right) / \sin \frac{t}{2} \tag{15.65}$$

$$= -\frac{1}{2} + \frac{1}{2} \left(\sin \frac{(2n+1)t}{2} \right) / \sin \frac{t}{2}, \tag{15.66}$$

we can also write $D_n(2t)$ as

$$D_n(2t) = \begin{cases} \frac{\sin(2n+1)t}{2\sin t}, & t \neq m\pi, m \text{ is an integer,} \\ \left(n + \frac{1}{2}\right), & t = m\pi, m \text{ is an integer,} \end{cases} \tag{15.67}$$

thus obtaining the partial sum as

$$S_n(x) = \frac{2}{\pi} \int_0^{\pi/2} \frac{f(x+2t) + f(x-2t)}{2} \left[\frac{\sin(2n+1)t}{\sin t}\right] dt, \tag{15.68}$$

which is called the **integral representation** of Fourier series. It basically says that the Fourier series written for $f(x)$ **converges** at the point x, if and only if the following limit exists:

$$\lim_{n \to \infty} \frac{2}{\pi} \int_0^{\pi/2} \frac{f(x+2t) + f(x-2t)}{2} \left[\frac{\sin(2n+1)t}{\sin t}\right] dt. \tag{15.69}$$

In the case that this limit exists, it is equal to the sum of the Fourier series.

We now write the **Dirichlet integral**:

$$\lim_{n \to \infty} \frac{2}{\pi} \int_0^{\delta} g(t) \frac{\sin nt}{t} dt = g(0^+), \tag{15.70}$$

which Jordan has shown to be true when $g(t)$ is of bounded variation. That is, when we move along the x-axis, the change in $g(x)$ is finite [3, p. 473]. Basically, the Dirichlet integral says that the value of the integral depends entirely on the local behavior of the function $g(t)$ near 0. Using the Riemann–Lebesgue lemma, granted that $g(t)/t$ is absolutely integrable in the interval $[\varepsilon, \delta]$, $0 < \varepsilon < \delta$, we can replace t in the denominator of the Dirichlet integral with another function, like $\sin t$, that has the same behavior near 0 without affecting the result. In the light of these, since the function

$$F(t) = \begin{cases} \frac{1}{t} - \frac{1}{\sin t}, & 0 < t \leq \frac{\pi}{2}, \\ 0, & t = 0 \end{cases} \tag{15.71}$$

is continuous at $t = 0$, we can write the integral

$$\lim_{n \to \infty} \frac{2}{\pi} \int_0^{\pi/2} \left(\frac{1}{t} - \frac{1}{\sin t}\right) \left[\frac{f(x+2t) + f(x-2t)}{2}\right] \sin(2n+1)t \; dt = 0. \tag{15.72}$$

Now, the convergence problem of the Fourier series reduces to finding the conditions on $f(x)$, which guarantee the **existence** of the limit [Eq. (15.69)]:

$$\lim_{n \to \infty} \frac{2}{\pi} \int_0^{\pi/2} \left[\frac{f(x+2t) + f(x-2t)}{2}\right] \frac{\sin(2n+1)t}{t} dt. \tag{15.73}$$

Employing the Riemann–Lebesgue lemma one more time, we can replace the upper limit of this integral with δ, where δ is any positive number less than $\pi/2$. This follows from the Riemann–Lebesque lemma, which assures that the following limit is true:

$$\lim_{n\to\infty} \frac{2}{\pi} \int_\delta^{\pi/2} \left[\frac{f(x+2t) + f(x-2t)}{2} \right] \frac{\sin(2n+1)t}{t}\, dt \to 0. \tag{15.74}$$

We can now quote the Riemann localization theorem:

Theorem 15.1. Riemann localization theorem: Assume that $f(x)$ is absolutely integrable in the interval $[0, 2\pi]$ and periodic with the period 2π. Then, the Fourier series produced by $f(x)$ converges for a given x, if and only if the following limit exists:

$$\boxed{\lim_{n\to\infty} \frac{2}{\pi} \int_0^\delta \left[\frac{f(x+2t) + f(x-2t)}{2} \right] \frac{\sin(2n+1)t}{t} dt,} \tag{15.75}$$

where $\delta < \frac{\pi}{2}$ is a positive number. When the limit exists, it is equal to the sum of the Fourier series produced by $f(x)$.

Importance of this result lies in the fact that the convergence of a Fourier series at a given point is determined by the behavior of $f(x)$ in the neighborhood of that point. This is surprising, since the coefficients [Eqs. (15.50) and (15.51)] are determined through integrals over the entire interval.

15.5 SUFFICIENT CONDITIONS FOR CONVERGENCE

We now present a theorem due to Jordan, which gives the sufficient conditions for the convergence of a Fourier series at a point [3, p. 478].

Theorem 15.2. Let $f(x)$ be absolutely integrable in the interval $(0, 2\pi)$ with the period 2π and consider the interval $[x-\delta, x+\delta]$ centered at x in which $f(x)$ is of bounded variation. Then, the Fourier series generated by $f(x)$ converges for this value of x to the sum

$$\boxed{\frac{f(x^+) + f(x^-)}{2}.} \tag{15.76}$$

Furthermore, if $f(x)$ is continuous at x, then the series converges to $f(x)$.

Proof of this theorem is based on showing that the limit in the Riemann localization theorem

$$\lim_{n\to\infty} \frac{2}{\pi} \int_0^\delta g(t) \frac{\sin(2n+1)t}{t} dt, \tag{15.77}$$

exists for

$$g(t) = \frac{f(x+2t) + f(x-2t)}{2}, \quad t \in [0, \delta], \quad \delta < \pi/2, \tag{15.78}$$

and equals $g(0^+)$.

This theorem is about the convergence of Fourier series at a given point. However, it says nothing about **uniform convergence**. For this, we present the so called **Fundamental theorem**. We first define the concepts of piecewise continuous, smooth, and very smooth functions. A function defined in the closed interval $[a, b]$ is **piecewise continuous** if the interval can be subdivided into a finite number of subintervals, where in each of these intervals, the function is continuous and has finite limits at both ends of the interval. Furthermore, if the function $f(x)$ coincides with a continuous function $f_i(x)$ in the ith subinterval and if $f_i(x)$ has continuous first derivatives, then we say that $f(x)$ is **piecewise smooth**. If, in addition, the function $f_i(x)$ has continuous second derivatives, we say $f(x)$ is **piecewise very smooth**.

15.6   THE FUNDAMENTAL THEOREM

Theorem 15.3. Let $f(x)$ be a piecewise very smooth function in the interval $[-\pi, \pi]$ with the period 2π, then the Fourier series

$$\boxed{f(x) \sim \frac{a_0}{2} + \sum_{n=1}^{\infty}(a_n \cos nx + b_n \sin nx),} \tag{15.79}$$

where the coefficients are given as

$$a_n = \frac{1}{\pi} \int_{-\pi}^{\pi} f(x) \cos nx \, dx, \quad n = 0, 1, \ldots, \tag{15.80}$$

$$b_n = \frac{1}{\pi} \int_{-\pi}^{\pi} f(x) \sin nx \, dx, \quad n = 1, 2, \ldots \tag{15.81}$$

converges uniformly to $f(x)$ in every closed interval where $f(x)$ is continuous. At each point of **discontinuity**, x_1, inside the interval $[-\pi, \pi]$, Fourier series converges to

$$\boxed{\frac{1}{2}\left[\lim_{x \to x_1^-} f(x) + \lim_{x \to x_1^+} f(x) \right],} \tag{15.82}$$

and at the end points $x = \pm\pi$ to

$$\frac{1}{2}\left[\lim_{x \to \pm\pi^-} f(x) + \lim_{x \to \pm\pi^+} f(x) \right]. \tag{15.83}$$

For most practical situations, the requirement can be weakened from **very smooth** to **smooth**. For the proof of this theorem, we refer the reader to Kaplan [2, p. 490]. We remind the reader that all that is required for the convergence of a Fourier series is the piecewise continuity of the first and the second derivatives of the function. This result is remarkable in itself, since for the convergence of Taylor series, derivatives of all orders have to exist and the remainder term has to go to zero.

15.7   UNIQUENESS OF FOURIER SERIES

Theorem 15.4. Let $f(x)$ and $g(x)$ be two piecewise continuous functions in the interval $[-\pi, \pi]$ with the same Fourier coefficients, that is

$$\frac{1}{\pi} \int_{-\pi}^{\pi} f(x) \cos nx \, dx = \frac{1}{\pi} \int_{-\pi}^{\pi} g(x) \cos nx \, dx, \tag{15.84}$$

$$\frac{1}{\pi} \int_{-\pi}^{\pi} f(x) \sin nx \, dx = \frac{1}{\pi} \int_{-\pi}^{\pi} g(x) \sin nx \, dx. \tag{15.85}$$

Then $f(x) = g(x)$, except perhaps at the points of discontinuity.

Proof of the uniqueness theorem follows at once, if we define a new piecewise continuous function as

$$h(x) = f(x) - g(x) \tag{15.86}$$

and write the Fourier expansion of $h(x)$ and use Eqs. (15.84) and (15.85).

15.8   EXAMPLES OF FOURIER SERIES

15.8.1   Square Wave

A square wave is defined as the periodic extension of the function (Figure 15.1)

$$f(x) = \begin{cases} -1, & -\pi \leq x < 0, \\ +1 & 0 \leq x \leq \pi. \end{cases} \tag{15.87}$$

We first evaluate the expansion coefficients as

$$a_n = -\frac{1}{\pi} \int_{-\pi}^{0} \cos nx \, dx + \frac{1}{\pi} \int_{0}^{\pi} \cos nx \, dx = 0, \quad n = 0, 1, \ldots, \tag{15.88}$$

$$b_n = -\frac{1}{\pi} \int_{-\pi}^{0} \sin nx \, dx + \frac{1}{\pi} \int_{0}^{\pi} \sin nx \, dx = \begin{cases} 0, & n = 2, 4, \ldots, \\ \dfrac{4}{n\pi}, & n = 1, 3, \ldots, \end{cases} \tag{15.89}$$

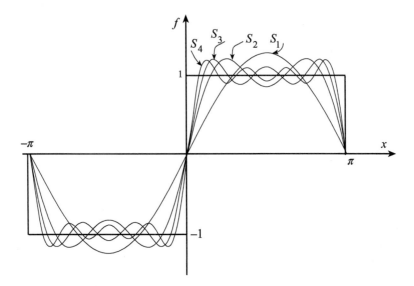

Figure 15.1 Square wave and partial sums.

which gives the Fourier series

$$f(x) = \frac{4}{\pi} \sum_{n=1}^{\infty} \frac{\sin(2n-1)x}{(2n-1)} \tag{15.90}$$

$$= \frac{4}{\pi} \sin x + \frac{4}{3\pi} \sin 3x + \frac{4}{5\pi} \sin 5x + \cdots. \tag{15.91}$$

We plot the first four partial sums,

$$S_1 = \frac{4}{\pi} \sin x, \tag{15.92}$$

$$S_2 = S_1 + \frac{4}{3\pi} \sin 3x, \tag{15.93}$$

$$S_3 = S_2 + \frac{4}{5\pi} \sin 5x, \tag{15.94}$$

$$S_4 = S_3 + \frac{4}{7\pi} \sin 7x, \tag{15.95}$$

in Figure 15.1, which clearly demonstrates the convergence of the Fourier series. Note that at the points of discontinuity, the series converges to the average

$$f(0) = \frac{1}{2}[f(0^+) + f(0^-)] = 0. \tag{15.96}$$

Convergence near the discontinuity is rather poor. In fact, near the discontinuity, Fourier series overshoots the value of the function at that point. This is a general feature of all discontinuous functions, which is called the **Gibbs phenomenon**. The first term $a_0/2$ is zero, which represents the mean value of $f(x)$ over the interval $(-\pi, \pi)$.

15.8.2 Triangular Wave

Let us now consider the triangular wave with the period 2π, which is continuous everywhere (Figure 15.2):

$$f(x) = \begin{cases} \dfrac{\pi}{2} + x, & -\pi \leq x \leq 0, \\[3mm] \dfrac{\pi}{2} - x, & 0 \leq x \leq \pi. \end{cases} \tag{15.97}$$

We find

$$a_n = \frac{1}{\pi}\int_{-\pi}^{0}\left(\frac{\pi}{2}+x\right)\cos nx\ dx + \frac{1}{\pi}\int_{0}^{\pi}\left(\frac{\pi}{2}-x\right)\cos nx\ dx \tag{15.98}$$

$$= \frac{1}{\pi}\left[\left(\frac{\pi}{2}+x\right)\frac{\sin nx}{n}\bigg|_{-\pi}^{0} - \frac{1}{n}\int_{-\pi}^{0}\sin nx\ dx + \cdots\right] \tag{15.99}$$

$$= \frac{1}{\pi}\left[\frac{1}{n^2}(1-\cos n\pi) + \frac{1}{n^2}(1-\cos n\pi)\right] \tag{15.100}$$

$$= \frac{2}{\pi n^2}(1-\cos n\pi), \quad n = 1, 2, \ldots . \tag{15.101}$$

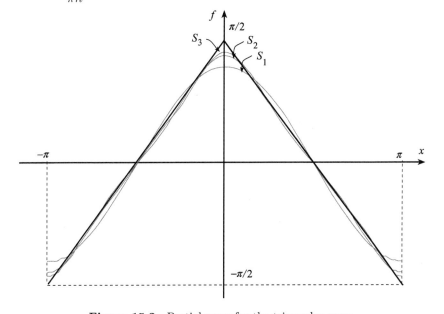

Figure 15.2   Partial sums for the triangular wave.

The coefficient a_0 needs to be calculated separately as

$$a_0 = \frac{1}{\pi} \int_{-\pi}^{0} \left(\frac{\pi}{2} + x\right) dx + \frac{1}{\pi} \int_{0}^{\pi} \left(\frac{\pi}{2} - x\right) dx = 0. \tag{15.102}$$

Similarly, we evaluate the other coefficients:

$$b_n = \frac{1}{\pi} \int_{-\pi}^{0} \left(\frac{\pi}{2} + x\right) \sin nx \ dx + \frac{1}{\pi} \int_{0}^{\pi} \left(\frac{\pi}{2} - x\right) \sin nx \ dx \tag{15.103}$$

$$= 0, \quad n = 1, 2, \ldots. \tag{15.104}$$

Hence, the Fourier expansion of the triangular wave becomes

$$f(x) = \frac{4}{\pi} \cos x + \frac{4}{9\pi} \cos 3x + \frac{4}{25\pi} \cos 5x + \cdots \tag{15.105}$$

$$= \frac{4}{\pi} \sum_{n=1}^{\infty} \frac{\cos(2n-1)x}{(2n-1)^2}. \tag{15.106}$$

The first three partial sums are plotted in Figure 15.2. For the triangular wave, the function is continuous everywhere, but the first derivative has a jump discontinuity at the vertices of the triangle, where the convergence is weaker.

15.8.3   Periodic Extension

Each term of the Fourier series is periodic with the period 2π, hence the sum is also periodic with the same period. In this regard, Fourier series cannot be used to represent a function for all x if it is not periodic. On the other hand, if a Fourier series converges to a function in the interval $[-\pi, \pi]$, then the series represents the periodic extension of $f(x)$ with the period 2π. Consider the function shown in Figure 15.3:

$$f(x) = \begin{cases} 0, & -\pi < x \le 0, \\ x, & 0 \le x < \pi, \end{cases} \tag{15.107}$$

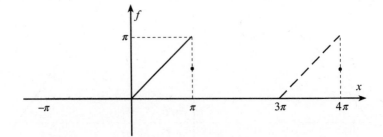

Figure 15.3   Periodic extension.

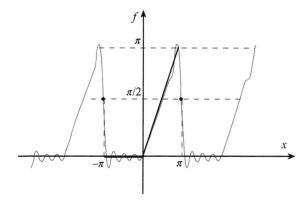

Figure 15.4 Fourier series representation of $f(x)$ [Eq. (15.107)] by S_7.

where the dashed lines represent its periodic extension.

Fourier coefficients of $f(x)$ are found as

$$a_0 = \frac{1}{\pi} \int_{-\pi}^{\pi} f(x) \; dx = \frac{1}{\pi} \left(\int_{-\pi}^{\pi} 0 \; dx + \int_0^{\pi} x \; dx \right) = \frac{\pi}{2}, \tag{15.108}$$

$$a_n = \frac{1}{\pi} \int_0^{\pi} x \cos nx \; dx = \frac{\cos nx + nx \sin nx}{\pi n^2} \Big|_0^{\pi} = \frac{(-1)^n - 1}{\pi n^2}, \tag{15.109}$$

$$b_n = \frac{1}{\pi} \int_0^{\pi} x \sin nx \; dx = \frac{\sin nx - nx \cos nx}{\pi n^2} \Big|_0^{\pi} = -\frac{(-1)^n}{n}, \tag{15.110}$$

where $n = 1, 2, \ldots$. Thus, the Fourier series representation of $f(x)$ in the interval $(-\pi, \pi)$ is

$$f(x) = \frac{\pi}{4} + \sum_{n=1}^{\infty} \left[\frac{(-1)^n - 1}{\pi n^2} \cos nx - \frac{(-1)^n}{n} \sin nx \right]. \tag{15.111}$$

Fourier extension of $f(x)$ converges at the points of discontinuity to (Figure 15.4)

$$\frac{1}{2}[f(\pi^-) + f(\pi^+)] = \frac{\pi}{2}. \tag{15.112}$$

15.9 FOURIER SINE AND COSINE SERIES

So far, we have considered periodic functions with the period 2π or functions defined in the interval $[-\pi, \pi]$. If the function has the period 2π, then any interval $[c, c + 2\pi]$ can be used. For such an interval, the **Fourier series** becomes

$$f(x) = \frac{a_0}{2} + \sum_{n=1}^{\infty} a_n \cos nx + b_n \sin nx, \tag{15.113}$$

$$a_n = \frac{1}{\pi} \int_c^{c+2\pi} f(x) \cos nx \ dx, \tag{15.114}$$

$$b_n = \frac{1}{\pi} \int_c^{c+2\pi} f(x) \sin nx \ dx. \tag{15.115}$$

Using $[-\pi, \pi]$ as the fundamental interval has certain advantages. If the function is even, that is, $f(-x) = f(x)$, all b_n are zero, thus reducing the Fourier series to the **Fourier cosine series**:

$$f(x) = \frac{a_0}{2} + \sum_{n=1}^{\infty} a_n \cos nx, \tag{15.116}$$

$$a_n = \frac{2}{\pi} \int_0^{\pi} f(x) \cos nx \ dx. \tag{15.117}$$

Similarly, for an odd function where $f(-x) = -f(x)$, all a_n are zero, thus giving the **Fourier sine series**:

$$f(x) = \sum_{n=1}^{\infty} b_n \sin nx, \tag{15.118}$$

$$b_n = \frac{2}{\pi} \int_0^{\pi} f(x) \sin nx \ dx. \tag{15.119}$$

15.10 CHANGE OF INTERVAL

If the function $f(x)$ has the period $2L$, that is, $f(x + 2L) = f(x)$, the Fourier series is written as

$$f(x) = \frac{a_0}{2} + \sum_{n=1}^{\infty} a_n \cos \left(\frac{n\pi}{L}x\right) + b_n \sin \left(\frac{n\pi}{L}x\right), \tag{15.120}$$

$$a_n = \frac{1}{L} \int_{-L}^{L} f(x) \cos \left(\frac{n\pi}{L}x\right) dx, \tag{15.121}$$

$$b_n = \frac{1}{L} \int_{-L}^{L} f(x) \sin \left(\frac{n\pi}{L}x\right) dx. \tag{15.122}$$

In this case, the Fourier cosine series is given as

$$f(x) = \frac{a_0}{2} + \sum_{n=1}^{\infty} a_n \cos \left(\frac{n\pi}{L}x\right), \quad x \in [0, L], \tag{15.123}$$

$$a_n = \frac{2}{L} \int_0^{L} f(x) \cos \left(\frac{n\pi}{L}x\right) dx, \tag{15.124}$$

and the Fourier sine series become

$$f(x) = \sum_{n=1}^{\infty} b_n \sin\left(\frac{n\pi}{L}x\right), \tag{15.125}$$

$$b_n = \frac{2}{L} \int_0^L f(x) \sin\left(\frac{n\pi}{L}x\right) dx. \tag{15.126}$$

Example 15.1. *Power in periodic signals:* For periodic signals, the **total power** is given by the expression

$$\boxed{P_{tot} = \frac{1}{T} \int_{-T/2}^{T/2} f^2(x) dx.} \tag{15.127}$$

Substituting the Fourier expansion of $f(x)$, we write

$$P_{tot} = \frac{1}{2L} \int_{-L}^{L} \left[\frac{a_0}{2} + \sum_{n=1}^{\infty} a_n \cos\left(\frac{n\pi}{L}x\right) + b_n \sin\left(\frac{n\pi}{L}x\right) \right]^2 dx \tag{15.128}$$

$$= \frac{a_0^2}{4} + \frac{1}{2L} \int_{-L}^{L} \sum_{n=1}^{\infty} \sum_{m=1}^{\infty} [a_n a_m \cos(n\omega_0 x) \cos(m\omega_0 x)$$

$$+ 2a_n b_m \cos(n\omega_0 x) \sin(m\omega_0 x) + b_n b_m \sin(n\omega_0 x) \sin(m\omega_0 x)] dx, \tag{15.129}$$

where we have substituted $\omega_0 = \pi/L$. Using the orthogonality relations [Eqs. (15.46)−(15.48)], we finally find

$$\boxed{P_{tot} = \frac{a_0^2}{4} + \sum_{n=1}^{\infty} \frac{(a_n^2 + b_n)^2}{2}.} \tag{15.130}$$

15.11 INTEGRATION AND DIFFERENTIATION OF FOURIER SERIES

The term by term integration of Fourier series,

$$f(x) = \frac{a_0}{2} + \sum_{n=1}^{\infty} [a_n \cos nx + b_n \sin nx], \tag{15.131}$$

yields

$$\int_{x_1}^{x_2} dx f(x) = \frac{a_0(x_2 - x_1)}{2} + \sum_{n=1}^{\infty} \left[\frac{a_n}{n}(-1)\sin nx + \frac{b_n}{n}\cos nx \right]_{x_1}^{x_2}. \tag{15.132}$$

Hence, the effect of integration is to divide each coefficient by n. If the original series converges, then the integrated series will also converge. In fact, the integrated series will converge more rapidly than the original series. On the other hand, differentiation multiplies each coefficient by n, hence convergence is not guaranteed.

REFERENCES

1. Bayin, S.S. (2018). *Mathematical Methods in Science and Engineering*, 2e. Wiley.

2. Kaplan, W. (1984). *Advanced Calculus*, 3e. Reading, MA: Addison-Wesley.

3. Apostol, T.M. (1971). *Mathematical Analysis*. Reading, MA: Addison-Wesley, fourth printing.

PROBLEMS

1. Using properties of the inner product, prove the following inequalities:

$$|(f,g)| \leq \|f\| \, \|g\|$$

and

$$\|f + g\| \leq \|f\| + \|g\|.$$

2. Show that the following set of functions is orthonormal:

$$\left\{ u_n(x) = \frac{e^{inx}}{\sqrt{2\pi}} \right\}, \quad n = 0, 2, \ldots, \quad x \in [0, 2\pi].$$

3. Show the following orthogonality relations of the Fourier series:

$$\frac{1}{\pi} \int_{-\pi}^{\pi} \cos nx \cos mx \; dx = \delta_{nm},$$

$$\frac{1}{\pi} \int_{-\pi}^{\pi} \sin nx \sin mx \; dx = \delta_{nm},$$

$$\frac{1}{\pi} \int_{-\pi}^{\pi} \sin nx \cos mx \; dx = 0.$$

4. Show that the Fourier series can be written as

$$f(x) = \frac{1}{2\pi} \int_{-\pi}^{\pi} f(t) \; dt + \frac{1}{\pi} \sum_{n=1}^{\infty} \int_{-\pi}^{\pi} f(t) \cos n(x - t) \, dt, \quad n = 0, 1, 2, \ldots \; .$$

5. **Dirichlet integral** is given as

$$\lim_{n \to \infty} \frac{2}{\pi} \int_{0}^{\delta} g(t) \frac{\sin nt}{t} \; dt = g(0^+),$$

which says that the value of the integral depends entirely on the local behavior of the function $g(t)$ near 0. Show that Riemann–Lebesgue lemma allows us to replace t with another function, say $\sin t$, that has the same behavior near 0 without affecting the result.

6. After proving the following trigonometric relations:

$$\sum_{k=1}^{n} \cos kt = \left(\sin \frac{nt}{2} \cos \frac{(n+1)t}{2} \right) / \sin \frac{t}{2}$$

$$= -\frac{1}{2} + \frac{1}{2} \left(\sin \frac{(2n+1)t}{2} \right) / \sin \frac{t}{2}$$

and

$$\sum_{k=1}^{n} \sin kt = \left(\sin \frac{nt}{2} \sin \frac{(n+1)t}{2} \right) / \sin \frac{t}{2},$$

show that the function $D_n(t)$ [Eq. (15.61)]:

$$D_n(t) = \frac{1}{2} + \sum_{k=1}^{n} \cos kt,$$

can be written as

$$D_n(2t) = \begin{cases} \dfrac{\sin(2n+1)t}{2\sin t}, & t \neq m\pi, m \text{ is an integer,} \\ \left(n + \frac{1}{2}\right), & t = m\pi, m \text{ is an integer.} \end{cases}$$

Hint: First show that

$$\sum_{k=1}^{n} e^{ikt} = e^{it} \frac{1 - e^{int}}{1 - e^{it}} = \frac{\sin(nt/2)}{\sin(t/2)} e^{i(n+1)t/2}$$

and then take the real and imaginary parts of this equation, which gives two useful relations, where the real part is the desired equation.

7. Expand $f(x) = 3x + 2$ in Fourier series for the following intervals:
   (i) $(-\pi, \pi)$.
   (ii) $(0, 2\pi)$.
   (iii) In Fourier cosine series for $[0, \pi]$.
   (iv) In Fourier sine series for $(0, \pi)$.
   (v) In Fourier series $(0, \pi)$.
   Graph the first three partial sums of the series you have obtained.

8. Expand $f(x) = x^2$ in Fourier series for the interval $(0, 2\pi)$ and plot your result.

9. Expand $\sin x$ in Fourier cosine series for $[0, \pi]$.

10. Expand $\sin x$ in Fourier series for $[0, 2\pi]$.

11. Find the Fourier cosine series of $f(x) = x$ in the interval $[0, \pi]$.

12. Find the Fourier series of the saw tooth wave:

$$f(x) = \begin{cases} x, & 0 \le x < \pi, \\ x - 2\pi, & \pi \le x \le 2\pi, \end{cases}$$

and plot the first three partial sums.

13. Fourier analyze the asymmetric square wave

$$f(x) = \begin{cases} d, & 0 < x < \pi, \\ 0, & \pi < x < 2\pi. \end{cases}$$

14. Show that

$$\delta(x - t) = \frac{1}{2\pi} + \frac{1}{\pi} \sum_{n=1}^{\infty} \cos n(x - t).$$

15. Show that the Fourier expansion of

$$f(x) = \begin{cases} -(1/2)(\pi + x), & -\pi \le x < 0, \\ +(1/2)(\pi - x), & 0 < x \le \pi, \end{cases}$$

is given as

$$f(x) = \sum_{n=1}^{\infty} \frac{\sin nx}{n}.$$

16. Show that the Fourier expansion of

$$f(x) = \begin{cases} x, & 0 \le x \le \pi, \\ -x, & -\pi \le x \le 0, \end{cases}$$

is given as

$$f(x) = \frac{\pi}{2} - \frac{4}{\pi} \sum_{n=0}^{\infty} \frac{\cos(2n + 1)x}{(2n + 1)^2}.$$

17. Show that

$$|\sin x| = \frac{8}{\pi} \sum_{n=1}^{\infty} \frac{\sin^2 nx}{4n^2 - 1}.$$

CHAPTER 16

FOURIER AND LAPLACE TRANSFORMS

Using Fourier series, periodic signals can be expressed as infinite sums of sines and cosines. For a general nonperiodic signal, we use the Fourier transforms. It turns out that most of the signals encountered in applications can be broken down into linear combinations of sines and cosines. This process is called spectral analysis. In this chapter, Fourier transforms are introduced along with the basics of signal analysis. Apart from signal analysis, Fourier transforms are also very useful in quantum mechanics and in the study of scattering phenomenon. Another widely used integral transform is the Laplace transform, which will be introduced with its basic properties and applications to differential equations and to transfer functions.

16.1 TYPES OF SIGNALS

Signals are essentially modulated forms of energy. They exist in many types and shapes. Gravitational waves emitted by neutron star mergers and the electromagnetic pulses emitted from a neutron star are natural signals. In quantum mechanics, particles can be viewed as signals in terms of probability waves. In technology, electromagnetic waves and sound waves are two predominant sources of signals. They are basically used to transmit and receive information. Signals exist in two basic forms as **analog** and **digital**. Analog signals are given as continuous functions of time. Digital signals are obtained from analog signals either by **sampling** or **binary coding**.

In sampling (Figure 16.1), we extract information from an analog signal with a certain sampling frequency $f_s = 1/\Delta t$. Given a sampled signal, the main problem is to generate the underlying analog signal. In general, there exist infinitely many possibilities. For the sampled signal on the top in Figure 16.2, any one of the functions:

$$\cos 5t, \quad \cos 10t, \quad \cos 25t, \ldots \tag{16.1}$$

with the frequencies $5/2\pi$, $10/2\pi$, $25/2\pi, \ldots$, could be the underlying signal. To eliminate the ambiguity, we can either increase the sampling frequency or

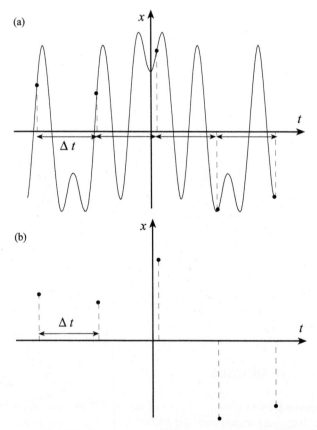

Figure 16.1 Analog signal (a) and the signal (b) with the sampling frequency $1/\Delta t$.

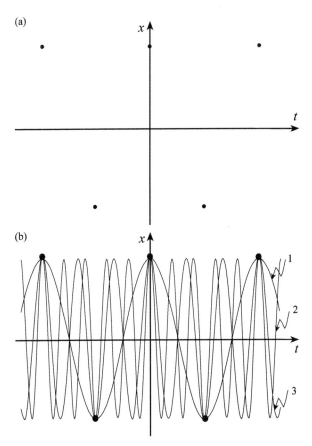

Figure 16.2 Ambiguity (b) in the sampled signal (a): $1 = \cos 5t$, $2 = \cos 10t$, $3 = \cos 25t$.

obtain some prior information about the underlying signal. For example, we may determine that the frequency cannot be greater than $(8/2\pi)\mathrm{Hz}$, which in this case implies that the underlying signal is $\cos 5t$. In fact, a theorem by Shannon, also called the **sampling theorem**, says that in order to recover the original signal unambiguously, the sampling frequency has to be at least twice the highest frequency present in the sampled signal. This minimum sampling frequency is also called the **Nyquist sampling frequency** [1]. In characterizing signals, we often refer to **amplitude** and **phase spectrums**. For the signal

$$x(t) = \cos(5t + 1) + 2\cos(10t - 2) - 3\cos(25t + 2), \qquad (16.2)$$

the amplitude and phase spectrums are obtained by plotting the amplitudes and the phases of the individual cosine waves that make up the signal as functions of frequency. In Figure 16.3, on the top we show the underlying analog signal [Eq. (16.2)] and on the bottom the corresponding amplitude

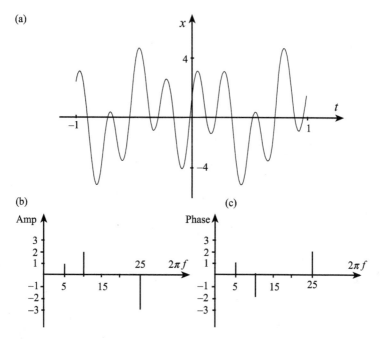

Figure 16.3 Spectral analysis of the analog signal on the top, $x(t) = \cos(5t + 1) + 2\cos(10t - 2) - 3\cos(25t + 2)$ and its amplitude (b) and phase (c) spectrums.

and phase spectrums. The amplitude and phase spectrums of a signal completely characterize the signal, and together they are called the **frequency spectrum** [2].

16.2 SPECTRAL ANALYSIS AND FOURIER TRANSFORMS

For a given analog signal (Figure 16.3a), it is important to determine the amplitude, frequency, and the phases of the individual cosine waves that make up the signal. For this purpose, we introduce the **correlation function** or the **correlation coefficient** denoted by r. The correlation coefficient between two samples:

$$x = \{x_i\} \quad \text{and} \quad y = \{y_i\}, \quad i = 1, 2, \ldots, N, \tag{16.3}$$

is defined as

$$r = \frac{\langle xy \rangle - \langle x \rangle \langle y \rangle}{\sigma_x \sigma_y}, \tag{16.4}$$

where σ_x and σ_y are the standard deviations of the samples, and $\langle \rangle$ denotes the mean value of the corresponding quantity. When the two samples are identical, $x \equiv y$, their correlation coefficient is 1. When there is no systematic relation between the two samples, their correlation coefficient is 0. Hence, the closer r

is to unity, the stronger the relation between the two samples. We usually take y as a cosine or a sine wave with unit amplitude, where the standard deviation is fixed and the mean is 0, hence we define the correlation coefficient as

$$R_0 = \langle xy \rangle = \frac{1}{N} \sum_{i=0}^{N-1} x_i y_i, \tag{16.5}$$

which is also called the **modified correlation coefficient**. For continuous signals, the correlation coefficient is defined as the integral

$$R_0 = \int_{-T/2}^{T/2} x(t)y(t)dt, \tag{16.6}$$

where T is the **duration** of the signal or the **observation time**.

16.3 CORRELATION WITH COSINES AND SINES

Let the **signal** be a single cosine function:

$$x(t) = A\cos(2\pi f_0 t + \theta_0), \tag{16.7}$$

where A stands for the amplitude, f_0 is the frequency measured in Hz, and θ_0 is the phase. To find the modified correlation coefficient R_0, we use a cosine function with zero phase:

$$y(t) = \cos(2\pi f t), \tag{16.8}$$

where f is a *test* frequency, to write

$$R_0 = \int_{-T/2}^{T/2} A\cos(2\pi f_0 t + \theta_0)\cos(2\pi f t)dt. \tag{16.9}$$

Using the trigonometric identity

$$\cos A \cos B = \frac{1}{2}\cos(A+B) + \frac{1}{2}\cos(A-B), \tag{16.10}$$

we write R_0 as

$$R_0 = \frac{A}{2}\int_{-T/2}^{T/2} \cos(2\pi(f_0+f)t + \theta_0)\ dt$$

$$+ \frac{A}{2}\int_{-T/2}^{T/2} \cos(2\pi(f_0-f)t + \theta_0)\ dt. \tag{16.11}$$

This is easily integrated to yield

$$
R_0 = \frac{A}{4\pi(f_0 + f)} \left[\sin\left(2\pi(f_0 + f)\frac{T}{2} + \theta_0 \right) + \sin\left(2\pi(f_0 + f)\frac{T}{2} - \theta_0 \right) \right]
$$
$$
+ \frac{A}{4\pi(f_0 - f)} \left[\sin\left(2\pi(f_0 - f)\frac{T}{2} + \theta_0 \right) + \sin\left(2\pi(f_0 - f)\frac{T}{2} - \theta_0 \right) \right].
$$
$$(16.12)$$

Using the relation

$$
\sin(A \pm B) = \sin A \cos B \pm \cos A \sin B, \qquad (16.13)
$$

Equation (16.12) can be simplified to

$$
R_0 = \frac{A}{2\pi(f + f_0)} \sin\left(2\pi(f + f_0)\frac{T}{2} \right) \cos\theta_0
$$
$$
+ \frac{A}{2\pi(f - f_0)} \sin\left(2\pi(f - f_0)\frac{T}{2} \right) \cos\theta_0. \qquad (16.14)
$$

In Figure 16.4, we plot R_0 for

$$
x(t) = 15\cos(2\pi 4 t + 1.5) \qquad (16.15)
$$

with $T = 1$. There are two notable peaks, one of which is at $f = -4$ and the other at $f = 4$. The first peak at the negative frequency is meaningless; however, the other peak gives the frequency of the signal.

The next problem is to find the amplitude and the phase of the given signal. For this, we look at R_0 in Eq. (16.14) more carefully. Near the peak frequency, $f = f_0$, the second term dominates the first; thus the peak value R_0^{max} can be written as

$$
R_0^{\text{max}} = \left(\frac{AT}{2} \right) \left[\lim_{f \to f_0} \frac{\sin\left(2\pi(f - f_0)\frac{T}{2} \right)}{2\pi(f - f_0)\frac{T}{2}} \right] \cos\theta_0, \qquad (16.16)
$$

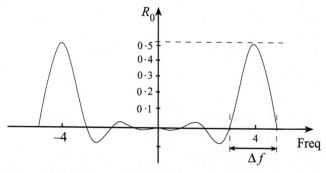

Figure 16.4   Modified correlation function R_0 for $x(t) = 15\cos(2\pi 4 t + 1.5)$ and $T = 1$.

where we have multiplied and divided by $T/2$. Using the limit

$$\lim_{f \to f_0} \frac{\sin\left(2\pi(f - f_0)\frac{T}{2}\right)}{2\pi(f - f_0)\frac{T}{2}} \to 1, \tag{16.17}$$

we obtain

$$R_0^{\max} = \left(\frac{AT}{2}\right)\cos\theta_0. \tag{16.18}$$

In other words, the peak value of the correlation function gives us a relation between the amplitude A and the phase θ_0 of the signal. To find A and θ_0, we need another relation. For this we now correlate the signal with a sine function:

$$y(t) = \sin(2\pi f t), \tag{16.19}$$

and write

$$S_0 = \int_{-T/2}^{T/2} A\cos(2\pi f_0 t + \theta_0)\sin(2\pi f t)dt. \tag{16.20}$$

Following similar steps, we can evaluate this integral to obtain

$$S_0 = \frac{A}{2\pi(f + f_0)}\sin\left(2\pi(f_0 + f)\frac{T}{2}\right)\sin\theta_0$$
$$- \frac{A}{2\pi(f - f_0)}\sin\left(2\pi(f - f_0)\frac{T}{2}\right)\sin\theta_0. \tag{16.21}$$

For the signal $x(t) = 15\cos(2\pi 4t + 1.5)$, S_0 peaks again at $f = 4$ but with a negative value (Figure 16.5). Following similar steps that lead to $R_0^{\max}$, we can

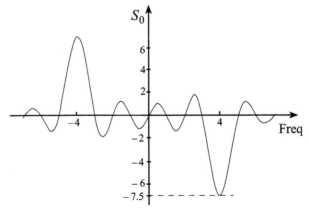

Figure 16.5   Correlation function S_0 for $x(t) = 15\cos(2\pi 4t + 1.5)$ and $T = 1$.

write the peak value of S_0 as

$$S_0^{\max} = -\frac{AT}{2}\sin\theta_0, \tag{16.22}$$

which gives another relation between A and θ_0. We can now use Eqs. (16.18) and (16.22) to evaluate the amplitude and the phase of the underlying signal as

$$A = \frac{2}{T}\sqrt{(R_0^{\max})^2 + (S_0^{\max})^2}, \tag{16.23}$$

$$\theta_0 = \tan^{-1}\left[-\frac{S_0^{\max}}{R_0^{\max}}\right]. \tag{16.24}$$

For the signal $x(t) = 15\cos(2\pi 4t + 1.5)$, we read the values of $R_0^{\max}$ and $S_0^{\max}$ from the Figures 16.4 and 16.5 as

$$R_0^{\max} = 0.52, \quad S_0^{\max} = -7.5, \tag{16.25}$$

which yield the amplitude and the phase as

$$A = 15 \quad \text{and} \quad \theta_0 = 1.5, \tag{16.26}$$

as expected.

Note that the **modified correlation functions** R_0 and S_0 [Eqs. (16.9) and (16.20)] are **linear integral operators**. Hence for signals composed of several cosine waves, the corresponding correlation functions will be the sum of the correlation functions of the individual terms. For example, the correlation function R_0 of

$$x(t) = 40\cos(2\pi 2t + 1.5) + 15\cos(2\pi 4t + 1) \tag{16.27}$$

has two peaks at $f_1 = 2$ and $f_2 = 4$ (Figure 16.6).

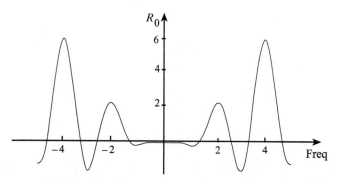

Figure 16.6 R_0 for $x(t) = 40\cos(2\pi 2t + 1.5) + 15\cos(2\pi 4t + 1)$ for $T = 1.5$.

16.4   CORRELATION FUNCTIONS AND FOURIER TRANSFORMS

Using complex notation, we introduce the modified complex correlation function $X(f)$ as

$$X(f) = R_0 - iS_0 \tag{16.28}$$

$$= \int_{-T/2}^{T/2} A\cos(2\pi f_0 t + \theta_0)\cos(2\pi f t)dt$$

$$-i\int_{-T/2}^{T/2} A\cos(2\pi f_0 t + \theta_0)\sin(2\pi f t)dt, \tag{16.29}$$

which for a general signal, $x(t)$, can be written as

$$X(f) = \int_{-T/2}^{T/2} x(t)[\cos(2\pi f t) - i\sin(2\pi f t)]dt \tag{16.30}$$

$$= \int_{-T/2}^{T/2} x(t)e^{-2\pi i f t}dt. \tag{16.31}$$

If the signal has infinite duration or it is observed for an infinite amount of time, we let $T \to \infty$, thus obtaining the expression

$$\boxed{X(f) = \int_{-\infty}^{\infty} x(t)e^{-2\pi i f t}dt,} \tag{16.32}$$

which is called the **Fourier transform** of the signal $x(t)$.

The **amplitude spectrum** of the signal, $A(f)$, is now the modulus of $X(f)$:

$$\boxed{A(f) = |X(f)| = (R_0^2 + S_0^2)^{1/2}.} \tag{16.33}$$

The **phase spectrum**, $\theta(f)$, is the argument of $X(f)$, that is,

$$\boxed{\theta(f) = \arg X(f) = \tan^{-1}\left(-\frac{S_0}{R_0}\right).} \tag{16.34}$$

Note that the phase spectrum is meaningful only when the amplitude spectrum is nonzero.

16.5   INVERSE FOURIER TRANSFORM

We have seen that **Fourier transform** of a signal is defined as

$$X(f) = \int_{-\infty}^{\infty} x(t)e^{-2\pi i f t} \, dt. \tag{16.35}$$

Multiplying both sides by $e^{i2\pi f t}$ and integrating over f gives

$$\int_{-\infty}^{\infty} X(f)e^{i2\pi f t} \, df = \int_{-\infty}^{\infty} df \, e^{i2\pi f t} \int_{-\infty}^{\infty} x(t')e^{-i2\pi f t'} \, dt' \tag{16.36}$$

$$= \int_{-\infty}^{\infty} x(t') \left[\int_{-\infty}^{\infty} e^{-i2\pi f(t'-t)} \, df \right] dt'. \tag{16.37}$$

Using one of the representations of the Dirac-delta function:

$$\delta(t' - t) = \int_{-\infty}^{\infty} e^{-i2\pi f(t'-t)} \, df, \tag{16.38}$$

we can write Eq. (16.37) as

$$\int_{-\infty}^{\infty} X(f)e^{i2\pi f t} \, df = \int_{-\infty}^{\infty} x(t')\delta(t' - t) \, dt' = x(t). \tag{16.39}$$

Thus, the **inverse Fourier transform** is defined as

$$x(t) = \int_{-\infty}^{\infty} X(f)e^{i2\pi f t} df. \tag{16.40}$$

16.6   FREQUENCY SPECTRUMS

In Figure 16.4, we have seen that the modified correlation function R_0 for $x(t) = 15\cos(2\pi 4t + 1.5)$ is peaked at $f = 4$. Let us now consider the amplitude spectrum $A(f)$ for this signal [Eq. (16.33)]. The plot in Figure 16.7a gives $A(f)$ for $T = 1$. Ideally speaking, if the integration was carried out for an infinite amount of time, the amplitude spectrum would look like the plot in Figure 16.7b, that is, an infinitely long spike at $f = 4$, which is a Dirac-delta function. Similarly, when T goes to infinity, the plot of R_0 in Figure 16.4 will consist of two infinitely long spikes at $f = -4$ and $f = 4$. To see what is going on, let us consider the cosine signal

$$x(t) = A\cos(2\pi f_0 t), \tag{16.41}$$

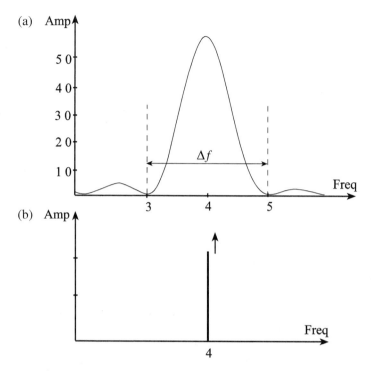

Figure 16.7 Amplitude spectrum of $x(t) = 15\cos(2\pi 4t + 1.5)$; (a) for $T = 1$ and (b) for as T goes to infinity.

where we have set the phase to zero for simplicity. The modified correlation function R_0 for this signal is written as

$$R_0 = \frac{A}{2\pi(f + f_0)} \sin\left[2\pi(f + f_0)\frac{T}{2}\right] + \frac{A}{2\pi(f - f_0)} \sin\left[2\pi(f - f_0)\frac{T}{2}\right]. \quad (16.42)$$

Near the peak, R_0 is predominantly determined by the second term. Hence, near f_0, we can write

$$R_0 \simeq \frac{A}{2\pi(f - f_0)} \sin\left[2\pi(f - f_0)\frac{T}{2}\right], \quad (16.43)$$

which has the peak value

$$R_0^{\max} = \frac{AT}{2}. \quad (16.44)$$

The first two zeros on either side of the peak are symmetrically situated about f_0, which are determined by the zeros of the sine function in Eq. (16.43):

$$2\pi(f - f_0)\frac{T}{2} = \pi \quad \text{and} \quad 2\pi(f - f_0)\frac{T}{2} = -\pi, \quad (16.45)$$

which give the frequencies

$$f_1 = f_0 + \frac{1}{T}, \tag{16.46}$$

$$f_2 = f_0 - \frac{1}{T}. \tag{16.47}$$

In other words, the width of the peak is determined by

$$\boxed{\Delta f = f_1 - f_2 = \frac{2}{T}.} \tag{16.48}$$

In Figure 16.7a, Δf is shown for $T = 1$. The longer the **duration** of the signal or the **observation time**, the sharper and taller the peak will be. In the limit as $T \to \infty$, it will be an infinitely long spike (Figure 16.7b).

16.7   DIRAC-DELTA FUNCTION

One of the most frequently used functions of mathematical physics is the **Dirac-delta function** or the **impulse function**. There are several ways of defining a Dirac-delta function. For our purposes, the following equivalent forms are very useful:

$$\delta(x - x') = \frac{1}{2\pi} \int_{-\infty}^{+\infty} dk e^{ik(x - x')} \tag{16.49}$$

$$= \frac{1}{2\pi} \frac{\sin k(x - x')}{(x - x')} \Big|_{-\infty}^{+\infty} \tag{16.50}$$

$$= \frac{1}{\pi} \lim_{k \to \infty} \left[\frac{\sin k(x - x')}{(x - x')} \right]. \tag{16.51}$$

When the argument of the Dirac-delta function goes to zero, we obtain infinity:

$$\lim_{x \to x'} \delta(x - x') \to \lim_{k \to \infty} \frac{k}{\pi} \to \infty. \tag{16.52}$$

Area under a Dirac-delta function can be found by using the well-known definite integral

$$\int_{-\infty}^{+\infty} \frac{\sin(ky)}{y} \, dy = \pi \tag{16.53}$$

as

$$\boxed{\int_{-\infty}^{+\infty} \delta(x - x') dx = 1.} \tag{16.54}$$

Dirac-delta function allows us to pick the value of a function at the point that makes the argument of the Dirac-delta function zero:

$$\int_{-\infty}^{+\infty} f(x)\delta(x - x')dx = f(x').$$

(16.55)

This is also called the **sampling property**. To visualize a Dirac-delta function $\delta(x - x')$, consider a rectangular region of unit area centered at $x = x'$ with the base length ΔL and height $\delta(0)$. A Dirac-delta function described by Eqs. (16.54) and (16.55) is obtained in the limit as $\Delta L \to 0$, while the area of the rectangle is preserved as 1. In this limit, we obtain an infinitely long spike, that is, a function that is zero everywhere but assumes the value infinity at $x = x'$. However, the area under this function is 1 (Problems 11 and 12). Comparing the correlation function R_0 [Eq. (16.43)] with the Dirac-delta function [Eq. (16.51)], it is seen that as $T \to \infty$, peaks become infinitely long spikes. The same argument goes for the peaks in S_0 and the amplitude spectrum.

Properties of Dirac-delta function
$\int_{-\infty}^{\infty} f(x)\delta^{(n)}(x - a)dx = (-1)^n f^{(n)}(a),$
$\delta(-x) = \delta(x),$
$\delta(ax) = \frac{1}{|a|}\delta(x), a \neq 0,$
$\delta(x - a) = \theta'(x - a),$

where $\theta(x - a)$ is the unit step function: 0 for $x < a$ and 1 for $x > a$.

16.8   A CASE WITH TWO COSINES

Let us now consider the amplitude spectrum for a signal composed of two cosines:

$$x(t) = 40 \cos(2\pi 2t + 1.5) + 15 \cos(2\pi 4t + 1).$$

(16.56)

The amplitude spectrum for $T = 0.5$ is shown in Figure 16.8. Instead of the expected two peaks, we have a single peak. What happened is that the integration time is too short for us to resolve the individual peaks. That is, the two peaks are so wide that they overlap. In Figure 16.9, we increase the integration time to $T = 2.5$. Now the two peaks can be seen clearly. Note that the first peak at $f = 2$ is off the scale.

The combined signal from a close binary pulsar should in principle have two peaks corresponding to the rotational frequencies of the individual pulsars. Signal from a single pulsar normally has a single peak corresponding to the rotation frequency of the neutron star. After timing the radio pulses for sometime, in 1974, Hulse and Taylor discovered that the pulsar PSR B1913+16, previously thought to be a single pulsar, has another periodic variation in its signal, thus indicating that it is a binary pulsar. In this case, the pulses from

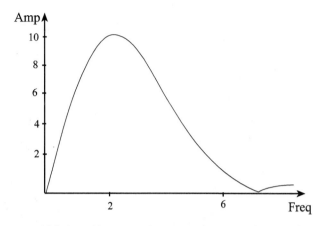

Figure 16.8 $A(f)$ for $x(t) = 40\cos(2\pi 2t + 1.5) + 15\cos(2\pi 4t + 1)$ for $T = 0.5$.

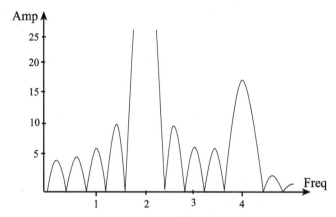

Figure 16.9 $A(f)$ for $x(t) = 40\cos(2\pi 2t + 1.5) + 15\cos(2\pi 4t + 1)$ for $T = 2.5$.

the companion pulsar were never observed, but Hulse and Taylor noticed that there is a systematic change in the arrival time of the pulses, which could only mean that it must be a binary system. After evaluating the orbital parameters, they concluded the orbit decays due to gravitational radiation as predicted by Einstein's theory of gravitation. This constituted the first indirect verification of the existence of gravitational waves and was awarded the 1993 Nobel Prize in Physics [3].

16.9   GENERAL FOURIER TRANSFORMS AND THEIR PROPERTIES

In general, the Fourier transform of a function, $x(\beta)$, in terms of a parameter α is defined as

$$X(\alpha) = \int_{-\infty}^{\infty} x(\beta)e^{-i\alpha\beta} \, d\beta. \tag{16.57}$$

The inverse Fourier transform is written as

$$x(\beta) = \int_{-\infty}^{\infty} X(\alpha)e^{i\alpha\beta}\,d\alpha, \tag{16.58}$$

where α and β are two conjugate variables related by the Fourier transform. Symbolically, Fourier transforms are shown as

$$X(\alpha) = \mathcal{F}\{x(\beta)\} \quad \text{and} \quad x(\beta) = \mathcal{F}^{-1}\{X(\alpha)\}. \tag{16.59}$$

For the existence of a Fourier transform, the integral $\int_{-\infty}^{\infty} x(\beta)e^{-i\alpha\beta}d\beta$ has to exist. Since $|e^{-i\alpha\beta}| = 1$, we can write the inequality

$$\left| \int_{-\infty}^{\infty} x(\beta)e^{-i\alpha\beta}\,d\beta \right| \leq \int_{-\infty}^{\infty} |x(\beta)|d\beta. \tag{16.60}$$

Hence, the Fourier transform $X(\alpha)$ exists if the function $x(\beta)$ is **absolutely integrable**, that is, if the integral $\int_{-\infty}^{\infty} |x(\beta)|d\beta$ is finite.

The Fourier transform has a number of very useful properties that make it exceptionally versatile. First of all, $\mathcal{F}$ and $\mathcal{F}^{-1}$ are **linear operators**. If

$$X_1(\alpha) = \mathcal{F}\{x_1(\beta)\} \quad \text{and} \quad X_2(\alpha) = \mathcal{F}\{x_2(\beta)\},$$

then

$$\mathcal{F}\{c_1 x_1(\beta) + c_2 x_2(\beta)\} = c_1 X_1(\alpha) + c_2 X_2(\alpha), \tag{16.61}$$

where c_1 and c_2 are constants. Similar equations for the inverse operator $\mathcal{F}^{-1}$ hold. Next, when $x(\beta)$ has continuous derivatives through order n, we can write

$$\mathcal{F}\left\{ \frac{d^n x(\beta)}{d\beta^n} \right\} = (i\alpha)^n \mathcal{F}\{x(\beta)\}, \quad n = 1, 2, \ldots, \tag{16.62}$$

$$\mathcal{F}\left\{ \frac{d^n x(\beta)}{d\beta^n} \right\} = (i\alpha)^n X(\alpha), \tag{16.63}$$

which makes Fourier transforms extremely useful for applications to differential equations. Also, if c is a real constant, then we can write

$$\mathcal{F}\{x(\beta - c)\} = e^{-ic\alpha} X(\alpha), \tag{16.64}$$

$$\mathcal{F}\{e^{ic\beta} x(\beta)\} = X(\alpha - c). \tag{16.65}$$

Using two functions, $x(\beta)$ and $y(\beta)$, we can form their **convolution**, $x * y = h(\beta)$, to obtain another function $h(\beta)$ as

$$h(\beta) = x * y = \int_{-\infty}^{+\infty} x(t)y(\beta - t)dt. \tag{16.66}$$

Using the definition of Fourier transforms, it can be shown that

$$x * y = \mathcal{F}^{-1}\{ X(\alpha)Y(\alpha)\}. \tag{16.67}$$

In other words, the inverse Fourier transform of the product of two Fourier transforms is the convolution of the original functions.

Let $X(\alpha)$ and $x(\beta)$ be two functions related by the Fourier transform equation [Eq. (16.57)]. Taking the complex conjugate of $X(\alpha)$, we write

$$X^*(\alpha) = \int_{-\infty}^{\infty} x^*(\beta) \exp[i\alpha\beta]d\beta. \tag{16.68}$$

Integrating $|X(\alpha)|^2$ over all space and using the definition of Dirac-delta function, we can write

$$\int_{-\infty}^{+\infty} |X(\alpha)|^2 d\alpha = \int_{-\infty}^{\infty} d\alpha \left[\int_{-\infty}^{+\infty} x^*(\beta) \exp[i\alpha\beta]d\beta \right]$$
$$\times \left[\int_{-\infty}^{+\infty} x(\beta') \exp[-i\alpha\beta']d\beta' \right] \tag{16.69}$$
$$= \int_{-\infty}^{+\infty} \int_{-\infty}^{+\infty} x^*(\beta)x(\beta') \left[\int_{-\infty}^{+\infty} \exp[-i\alpha(\beta' - \beta)]d\alpha \right] d\beta'd\beta \tag{16.70}$$
$$= \int_{-\infty}^{+\infty} \int_{-\infty}^{+\infty} x^*(\beta)x(\beta')\delta(\beta' - \beta)d\beta'd\beta \tag{16.71}$$
$$= \int_{-\infty}^{+\infty} x^*(\beta) \left[\int_{-\infty}^{+\infty} x(\beta')\delta(\beta' - \beta)d\beta' \right] d\beta. \tag{16.72}$$

Using the defining property of Dirac-delta function [Eq. (16.55)], this gives us an important identity:

$$\int_{-\infty}^{+\infty} |X(\alpha)|^2 d\alpha = \int_{-\infty}^{+\infty} |x(\beta)|^2 d\beta, \tag{16.73}$$

which is known as the **Parseval's theorem**. One of the main applications of Fourier transforms is to quantum mechanics, where the state functions in position space, $\Psi(x)$, are related to their counterparts in momentum space, $\Phi(p)$, via the Fourier transform

$$\Phi(p) = \int_{-\infty}^{\infty} \Psi(x)e^{-i\hbar px}\, dx, \tag{16.74}$$

where $\hbar$ is the Planck's constant. Further properties of Fourier transforms can be found in Bayin [4].

16.10 BASIC DEFINITION OF LAPLACE TRANSFORM

We define Laplace transform of a function, $x(t)$, as

$$\boxed{X(s) = \pounds\{x(t)\} = \int_{0}^{\infty} x(t)e^{-st}dt,} \tag{16.75}$$

where the operator $\pounds$ indicates the Laplace transform of its argument. Laplace transforms are linear transformations, which satisfy

$$\boxed{\pounds\{x(t) + y(t)\} = \pounds\{x(t)\} + \pounds\{y(t)\},} \tag{16.76}$$

$$\boxed{\pounds\{ax(t)\} = a\pounds\{x(t)\},} \tag{16.77}$$

where a is a constant. Because of their widespread use, extensive tables of Laplace transforms are available. The following list of elementary Laplace transforms can be verified by direct evaluation of the integral in Eq. (16.75):

Elementary Laplace Transforms

	$x(t)$	$X(s)$
(1)	1	$1/s$, $s > 0$
(2)	t^n	$n!/s^{n+1}$, $s > 0$, $n > -1$
(3)	e^{at}	$1/(s - a)$, $s > a$
(4)	$\sin bt$	$b/(s^2 + b^2)$, $s > 0$
(5)	$\cos bt$	$s/(s^2 + b^2)$, $s > 0$
(6)	$t^n e^{-at}$	$n!/(s + a)^{n+1}$, $n > -1$, $s + a > 0$
(7)	$\sinh bt$	$b/(s^2 - b^2)$, $s > b$
(8)	$\cosh bt$	$s/(s^2 - b^2)$, $s > b$
(9)	$t \sin bt$	$2bs/(s^2 + b^2)^2$, $s > 0$
(10)	$t \cos bt$	$(s^2 - b^2)/(s^2 + b^2)^2$, $s > 0$
(11)	$\delta(t - a)$	e^{-as}, $a \geq 0$

Inverse Laplace transform is shown as $\pounds^{-1}$, and it is also a linear operator:

$$\pounds^{-1}\{X(s) + Y(s)\} = \pounds^{-1}\{X(s)\} + \pounds^{-1}\{Y(s)\}, \tag{16.78}$$

$$\pounds^{-1}\{aX(s)\} = a\pounds^{-1}\{X(s)\}, \quad a \text{ is a constant.} \tag{16.79}$$

The aforementioned table can also be used to write inverse Laplace transforms. For example, using the first entry, we can write the inverse transform

$$\pounds^{-1}\left\{\frac{1}{s}\right\} = 1. \tag{16.80}$$

Two useful properties of Laplace transforms are given as

$$\pounds\{x(t-c)\} = e^{-cs}X(s), \ c > 0, \ t > c \tag{16.81}$$

and

$$\pounds\{e^{bt}x(t)\} = X(s-b), \tag{16.82}$$

where more such relations can be found in Bayin [4]. The **convolution** $x * y$ is defined as

$$x * y = \int_0^t x(t')y(t-t')\, dt'. \tag{16.83}$$

It can be shown that the convolution of two functions, $x(t)$ and $y(t)$, is the inverse Laplace transform of the product of their Laplace transforms:

$$x * y = \pounds^{-1}\{X(s)Y(s)\}. \tag{16.84}$$

In most cases, by using the aforementioned properties along with the linearity of the Laplace transforms, the needed inverse can be generated from a list of elementary transforms.

Example 16.1. *Inverse Laplace transforms:* Let us find the inverses of the following Laplace transforms:

$$X_1(s) = \frac{s}{(s+1)(s+3)}, \tag{16.85}$$

$$X_2(s) = \frac{1}{s^2 + 2s + 3}, \tag{16.86}$$

$$X_3(s) = \frac{se^{-cs}}{s^2 - 1}, \quad c > 0. \tag{16.87}$$

Using partial fractions [4], we can write $X_1(s)$ as

$$\frac{s}{(s+1)(s+3)} = -\frac{1}{2}\left(\frac{1}{s+1}\right) + \frac{3}{2}\left(\frac{1}{s+3}\right). \tag{16.88}$$

Using the linearity of $\pounds^{-1}$ and the third entry in the table, we obtain

$$x_1(t) = \pounds^{-1}\left\{-\frac{1}{2}\left(\frac{1}{s+1}\right) + \frac{3}{2}\left(\frac{1}{s+3}\right)\right\} \tag{16.89}$$

$$= -\frac{1}{2}\pounds^{-1}\left\{\frac{1}{s+1}\right\} + \frac{3}{2}\pounds^{-1}\left\{\frac{1}{s+3}\right\} = -\frac{1}{2}e^{-t} + \frac{3}{2}e^{-3t}. \tag{16.90}$$

For the second inverse, we complete the square to write

$$\pounds^{-1}\{X_2(s)\} = \pounds^{-1}\left\{\frac{1}{s^2 + 2s + 3}\right\} \tag{16.91}$$

$$= \pounds^{-1}\left\{\frac{1}{s^2 + 2s + 1 + 2}\right\} = \pounds^{-1}\left\{\frac{1}{(s+1)^2 + 2}\right\}. \tag{16.92}$$

We now use the fourth entry in the table:

$$\pounds^{-1}\left\{\frac{1}{s^2 + 2}\right\} = \frac{\sin\sqrt{2}t}{\sqrt{2}}, \tag{16.93}$$

and employ the inverse of the property in Eq. (16.82):

$$\pounds^{-1}\{X(s-b)\} = e^{bt}x(t), \tag{16.94}$$

to obtain

$$\pounds^{-1}\{X_2(s)\} = e^{-t}\left(\frac{\sin\sqrt{2}t}{\sqrt{2}}\right). \tag{16.95}$$

For the third inverse, we use the property in Eq. (16.81) to write the inverse:

$$\pounds^{-1}\{e^{-cs}X(s)\} = x(t-c), \quad c > 0, \quad t > c, \tag{16.96}$$

along with

$$\pounds^{-1}\left\{\frac{s}{s^2 - 1}\right\} = \cosh t, \tag{16.97}$$

thus obtaining the desired inverse as

$$\pounds^{-1}\{X_3(t)\} = \cosh(t-c). \tag{16.98}$$

16.11   DIFFERENTIAL EQUATIONS AND LAPLACE TRANSFORMS

An important application of the Laplace transforms is to ordinary differential equations with constant coefficients. We first write the Laplace transform of a derivative:

$$\pounds\left\{\frac{dx}{dt}\right\} = \int_0^\infty \left(\frac{dx}{dt}\right) e^{-st}\, dt, \tag{16.99}$$

which, after integration by parts, becomes

$$\pounds\left\{\frac{dx}{dt}\right\} = x(t)e^{-st}|_0^\infty + s\int_0^\infty x(t)e^{-st}\, dt. \tag{16.100}$$

Assuming that $s > 0$ and $\lim_{t\to\infty} x(t)e^{-st} \to 0$ is true, we obtain

$$\pounds\left\{\frac{dx}{dt}\right\} = sX(s) + x(0^+), \tag{16.101}$$

where $x(0^+)$ means the origin is approached from the positive t-axis. Similarly, we have

$$\pounds\left\{\frac{d^2x}{dt^2}\right\} = x'(t)e^{-st}|_0^\infty + s\pounds\left\{\frac{dx}{dt}\right\} \tag{16.102}$$

$$= s^2X(s) - sX(0^+) - X'(0^+), \tag{16.103}$$

where we have assumed that all the surface terms vanish in the limit as $t \to \infty$ and $s > 0$. Under similar conditions, for the nth derivative, we can write

$$\boxed{\pounds\{x^{(n)}(t)\} = s^n X(s) - s^{n-1}x(0^+) - s^{n-2}x'(0^+) - \cdots - x^{(n-1)}(0^+).} \tag{16.104}$$

Example 16.2. *Solution of differential equations:* Consider the following nonhomogeneous ordinary differential equation with constant coefficients:

$$\frac{d^2x}{dt^2} + 2\frac{dx}{dt} + 4x(t) = \sin 2t, \tag{16.105}$$

along with the initial conditions

$$x(0) = 0 \quad \text{and} \quad x'(0) = 0. \tag{16.106}$$

Assuming that Laplace transform of the solution, $X(s)$, exists and using the fact that $\pounds$ is a linear operator, we take the Laplace transform of the differential equation [Eq. (16.105)] to write

$$\pounds\left\{\frac{d^2x}{dt^2} + 2\frac{dx}{dt} + 4x(t)\right\} = \pounds\{\sin 2t\}, \tag{16.107}$$

$$\pounds\left\{\frac{d^2x}{dt^2}\right\} + 2\pounds\left\{\frac{dx}{dt}\right\} + 4\pounds\{x(t)\} = \pounds\{\sin 2t\}, \tag{16.108}$$

$$[s^2X(s) - sx(0) - x'(0)] + 2[sX(s) - x(0)] + 4X(s) = \frac{2}{s^2 + 4}. \tag{16.109}$$

Imposing the boundary conditions [Eq. (16.106)], we obtain the Laplace transform of the solution as

$$s^2X(s) + 2sX(s) + 4X(s) = \frac{2}{s^2 + 4}, \tag{16.110}$$

$$X(s) = \frac{2}{(s^2 + 2s + 4)(s^2 + 4)}. \tag{16.111}$$

To find the solution, we now have to find the inverse transform

$$x(t) = \pounds^{-1}\left\{\frac{2}{(s^2 + 2s + 4)(s^2 + 4)}\right\}. \tag{16.112}$$

Using partial fractions, we can write this as

$$x(t) = \pounds^{-1}\left\{\frac{-(1/4)s}{(s^2 + 4)} + \frac{(1/2) + (1/4)s}{(s^2 + 2s + 4)}\right\} \tag{16.113}$$

$$= -\frac{1}{4}\pounds^{-1}\left\{\frac{s}{(s^2 + 4)}\right\} + \frac{1}{4}\pounds^{-1}\left\{\frac{1}{(s^2 + 2s + 4)}\right\}$$

$$+ \frac{1}{4}\pounds^{-1}\left\{\frac{s + 1}{(s^2 + 2s + 4)}\right\} \tag{16.114}$$

$$= -\frac{1}{4}\pounds^{-1}\left\{\frac{s}{(s^2 + 4)}\right\} + \frac{1}{4}\pounds^{-1}\left\{\frac{1}{(s + 1)^2 + 3}\right\}$$

$$+ \frac{1}{4}\pounds^{-1}\left\{\frac{s + 1}{(s + 1)^2 + 3}\right\}. \tag{16.115}$$

Using the fourth and the fifth entries in the table and Eq. (16.82), we obtain the solution as

$$x(t) = -\frac{1}{4}\cos 2t + \frac{e^{-t}}{4}\left(\frac{\sin\sqrt{3}t}{\sqrt{3}} + \cos\sqrt{3}t\right). \tag{16.116}$$

16.12   TRANSFER FUNCTIONS AND SIGNAL PROCESSORS

There are extensive applications of Laplace transforms to signal processing, control theory, and communications. Here, we consider only some of the basic applications, which require the introduction of the **transfer function**. We now introduce a **signal processor** as a general device, which for a given input signal $u(t)$ produces an output signal $x(t)$. For electromagnetic signals, the internal structure of a signal processor is composed of electronic circuits. The effect of the device on the input signal can be represented by a differential operator, which we take to be a linear ordinary differential operator with constant coefficients, say

$$\boxed{\mathbf{O} = a\frac{d}{dt} + 1, \quad a > 0.}$$
(16.117)

The role of $\mathbf{O}$ is to relate the **input** signal $u(t)$ to the **output** signal $x(t)$ as

$$\mathbf{O}x(t) = u(t),$$
(16.118)

$$a\frac{dx(t)}{dt} + x(t) = u(t).$$
(16.119)

Taking the Laplace transform of this equation, we obtain

$$asX(s) - ax(0) + X(s) = U(s).$$
(16.120)

Since there is no signal out when there is no signal in, we take the initial conditions as

$$x(0) = 0 \quad \text{when} \quad u(0) = 0.$$
(16.121)

Hence, we write

$$(as + 1)X(s) = U(s).$$
(16.122)

The **transfer function** $G(s)$ is defined as

$$\boxed{G(s) = \frac{X(s)}{U(s)} = \frac{1}{as + 1}}$$
(16.123)

A general linear signal processor, $G(s)$, allows us to obtain the Laplace transform, $X(s)$, of the output signal from the Laplace transform, $U(s)$, of the input signal as

$$\boxed{X(s) = G(s)U(s).}$$
(16.124)

A single component signal processor can be shown as in Figure 16.10. For the signal processor represented by

$$G(s) = \frac{1}{1 + as},$$
(16.125)

Figure 16.10 A single signal processor.

consider a sinusoidal input as

$$u(t) = \sin \omega t. \tag{16.126}$$

Since the Laplace transform of the input $u(t)$ is

$$U(s) = \frac{\omega}{s^2 + \omega^2}, \tag{16.127}$$

Equation (16.124) gives the Laplace transform of the output signal as

$$X(s) = \frac{\omega}{(s^2 + \omega^2)(1 + as)}. \tag{16.128}$$

Using partial fractions, we can write the inverse transform as

$$x(t) = \pounds^{-1}\{X(s)\} \tag{16.129}$$

$$= \left(\frac{\omega}{1 + \omega^2 a^2}\right) \pounds^{-1}\left\{\frac{1}{s^2 + \omega^2}\right\} - \left(\frac{\omega a}{1 + \omega^2 a^2}\right) \pounds^{-1}\left\{\frac{s}{s^2 + \omega^2}\right\}$$

$$+ \left(\frac{\omega a^2}{1 + \omega^2 a^2}\right) \pounds^{-1}\left\{\frac{1}{1 + as}\right\}, \tag{16.130}$$

which yields

$$x(t) = \left[\frac{1}{1 + \omega^2 a^2}\right] \sin \omega t - \left[\frac{\omega a}{1 + \omega^2 a^2}\right] \cos \omega t + \left[\frac{\omega a}{1 + \omega^2 a^2}\right] e^{-t/a}. \tag{16.131}$$

The last term is called the **transient signal**, which dies out for large times, hence leaving the **stationary signal** as

$$x(t) = \left[\frac{1}{1 + \omega^2 a^2}\right] \sin \omega t - \left[\frac{\omega a}{1 + \omega^2 a^2}\right] \cos \omega t. \tag{16.132}$$

This can also be written as

$$x(t) = \left[\frac{1}{1 + \omega^2 a^2}\right]^{1/2} \sin(\omega t - \delta), \tag{16.133}$$

where $\delta = \tan^{-1} a\omega$.

In summary, for the processor represented by the differential operator

$$\mathbf{O} = a\frac{d}{dx} + 1, \quad a > 0,$$

(16.134)

when the input signal is a sine wave with zero phase, unit amplitude, and angular frequency ω, the output signal is again a sine wave with the same angular frequency ω but with the amplitude $(1 + \omega^2 a^2)^{-1/2}$ and the phase $\delta = \tan^{-1} a\omega$, both of which now depend on ω.

16.13   CONNECTION OF SIGNAL PROCESSORS

In practice, we may need to connect several signal processors to obtain the desired effect. For example, if we connect two signal processors in series (Figure 16.11), thus feeding the output of the first processor into the second one as the input:

$$X_1(s) = G_1(s)U_1(s),$$

(16.135)

$$X_2(s) = G_2(s)X_1(s),$$

(16.136)

we obtain

$$X_2(s) = G_2(s)G_1(s)U_1(s).$$

(16.137)

In other words, the effective transfer function of the two processors, G_1 and G_2, connected in **series** become their product:

$$\boxed{G(s) = G_2(s)G_1(s).}$$

(16.138)

On the other hand, if we connect two processors in parallel (Figure 16.12), thus feeding the same input into both processors:

$$X_1(s) = G_1(s)U(s),$$

(16.139)

$$X_2(s) = G_2(s)U(s),$$

(16.140)

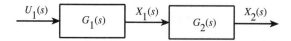

Figure 16.11   Series connection of signal processors.

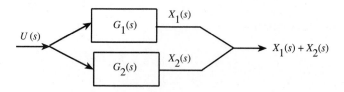

Figure 16.12   Parallel connection of signal processors.

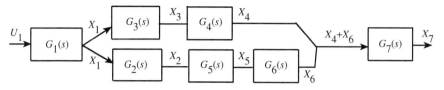

Figure 16.13 $G = G_1(G_3G_4 + G_2G_5G_6)G_7$.

along with combining their outputs:

$$X_2(s) + X_1(s) = [G_2(s) + G_1(s)]U(s), \tag{16.141}$$

we obtain the effective transfer function as their sum:

$$\boxed{G(s) = G_2(s) + G_1(s).} \tag{16.142}$$

For the combination in Figure 16.13, the effective transfer function is given as

$$G = G_1(G_3G_4 + G_2G_5G_6)G_7. \tag{16.143}$$

Example 16.3. *Signal processors in series:* Consider two linear signal processors represented by the following differential equations:

$$\dot{x}(t) + x(t) = u(t) \tag{16.144}$$

with the initial condition $x(0) = 0$ and

$$\ddot{x}(t) + 2\dot{x}(t) + 4x = u(t) \tag{16.145}$$

with the initial conditions

$$x(0) = \dot{x}(0) = 0. \tag{16.146}$$

The individual transfer functions are given, respectively, as

$$G_1(s) = \frac{1}{1+s}, \tag{16.147}$$

$$G_2(s) = \frac{1}{s^2 + 2s + 4}. \tag{16.148}$$

Connecting them in series, their effective transfer function becomes

$$G(s) = G_2(s)G_1(s) \tag{16.149}$$

$$= \frac{1}{(s^2 + 2s + 4)(1 + s)}. \tag{16.150}$$

For an input signal represented by the sine function:

$$u(t) = \sin t, \tag{16.151}$$

we write the Laplace transform of the input as

$$\pounds\{u(t)\} = U(s) = \frac{1}{1 + s^2}. \tag{16.152}$$

We can now write the Laplace transform of the output signal as

$$X_2(s) = G_2(s)G_1(s)U(s) \tag{16.153}$$

$$= \left[\frac{1}{(s^2 + 2s + 4)(1 + s)} \right] \frac{1}{(s^2 + 1)}. \tag{16.154}$$

Using partial fractions, this can be written as

$$X_2(s) = \frac{14/3(26) + 2s/3(26)}{s^2 + 2s + 4} + \frac{13/3(26)}{1 + s} + \frac{1/26 - 5s/26}{s^2 + 1}. \tag{16.155}$$

We rewrite this so that the inverse transform can be found easily:

$$X_2(s) = \frac{14}{3(26)} \left(\frac{1}{(s+1)^2 + 3} \right) + \frac{2}{3(26)} \left(\frac{(s+1)}{(s+1)^2 + 3} \right)$$

$$- \frac{2}{3(26)} \left(\frac{1}{(s+1)^2 + 3} \right) + \frac{13}{3(26)} \left(\frac{1}{1+s} \right)$$

$$+ \frac{1}{26} \left(\frac{1}{1+s^2} \right) - \frac{5}{26} \left(\frac{s}{1+s^2} \right). \tag{16.156}$$

This is the Laplace transform of the output signal. Using our table of Laplace transforms and Eq. (16.82), we take its inverse to find the physical signal as

$$\pounds^{-1}\{X_2(s)\} = x(t) = e^{-t} \left[\frac{12}{3(26)\sqrt{3}} \sin \sqrt{3}t - \frac{2}{3(26)} \cos \sqrt{3}t + \frac{13}{3(26)} \right]$$

$$+ \sqrt{\frac{1}{26}} \sin(t + \delta), \tag{16.157}$$

where $\delta = -\tan^{-1} 5$. Note that the output of the first processor G_1 is

$$x_1(t) = \frac{1}{2}[e^{-t} + \sin t - \cos t],$$

which satisfies the initial conditions in Eq. (16.146).

REFERENCES

1. Kusse, B.R. and Westwig, E.A. (2006). *Mathematical Physics: Applied Mathematics for Scientists and Engineers*, 2e. Weinheim: Wiley-VCH.

2. Woolfson, M.M. and Woolfson, M.S. (2007). *Mathematics for Physics*. Oxford: Oxford University Press.

3. Weisberg, J.M., Nice, D.J., and Taylor, J.H. (2010). Timing measurements of the relativistic binary pulsar PSR B1913+16. *Astrophys. J.* 722: 1030.

4. Bayin, S.S. (2018). *Mathematical Methods in Science and Engineering*, 2e. Wiley.

PROBLEMS

1. The correlation coefficient is defined as

$$r = \frac{\langle xy \rangle - \langle x \rangle \langle y \rangle}{\sigma_x \sigma_y},$$

where σ_x and σ_y are the standard deviations of the samples. Show that doubling all the values of x does not change the value of r.

2. Show that the correlation function R_0 [Eq. (16.9)] can be written as

$$R_0 = \frac{A}{2\pi(f + f_0)} \sin\left(2\pi(f + f_0)\frac{T}{2}\right) \cos\theta_0$$
$$+ \frac{A}{2\pi(f - f_0)} \sin\left(2\pi(f - f_0)\frac{T}{2}\right) \cos\theta_0.$$

3. Show that the correlation function S_0 [Eq. (16.20)] is given as

$$S_0 = \frac{A}{2\pi(f + f_0)} \sin\left(2\pi(f_0 + f)\frac{T}{2}\right) \sin\theta_0$$
$$- \frac{A}{2\pi(f - f_0)} \sin\left(2\pi(f - f_0)\frac{T}{2}\right) \sin\theta_0.$$

4. Show that

$$S_0^{\max} = -\frac{AT \sin\theta_0}{2}.$$

5. Show that the amplitude spectrum is an even function and the phase spectrum is an odd function, that is

$$A(f) = A(-f),$$
$$\theta(f) = -\theta(-f).$$

6. Find the Fourier transform of a Dirac-delta function.

7. Dirac-delta functions are very useful in representing discontinuities in physical theories. Using the Dirac-delta function, express the three dimensional density of an infinitesimally thin shell of mass M. Verify your answer by taking a volume integral.

8. Find the Fourier transform of a Gaussian:

$$f(x) = \frac{\alpha}{\sqrt{\pi}} \exp[-\alpha^2 x^2].$$

Also show that

$$\int_{-\infty}^{\infty} |f(x)|^2 dx = \int_{-\infty}^{\infty} |F(s)| ds = 1,$$

where

$$F(s) = \int_{-\infty}^{\infty} f(x) e^{-isx} dx.$$

9. If $X(s)$ is the Fourier transform of $x(t)$, show that the Fourier transform of its derivative dx/dt is given as

$$X'(s) = (i2\pi f) X(s),$$

granted that $x(t) \to 0$ as $t \to \pm\infty$. Under similar conditions, generalize this to derivatives of arbitrary order.

10. Given the signal

$$f(t) = \begin{cases} 4e^{-t}, & t \geq 0, \\ 0, & t < 0, \end{cases}$$

find its Fourier transform.

11. A normalized Gaussian pulse has the form

$$x(t) = \frac{\alpha}{\sqrt{\pi}} e^{-\alpha^2 t^2}.$$

Show that the area under this pulse is independent of α and is equal to one.

12. With the aid of a plot program, see how the Gaussian

$$x(t) = \frac{\alpha}{\sqrt{\pi}} e^{-\alpha^2 t^2}$$

scales with α and how the Dirac-delta function is obtained in the limit as $\alpha \to \infty$. Also show that $\alpha/\sqrt{\pi}$ is the peak value of the Gaussian.

13. Find the Fourier transform of

$$f(t) = t^3, \quad 0 < t < 1.$$

14. Given the exponentially decaying signal,

$$f(t) = \begin{cases} 0, & t < 0, \\ e^{-at}, & t > 0, \end{cases}$$

find its Fourier transform.

15. Show that the Fourier transform of a damped sinosoidal signal,

$$f(t) = \begin{cases} 0, & t < 0, \\ e^{-at}\sin\omega t, & t > 0, \end{cases}$$

is

$$F(\omega) = -\frac{\omega_0/\sqrt{2\pi}}{(\omega - \omega_1)(\omega - \omega_2)},$$

where $\omega_1 = \omega + ia$ and $\omega_2 = \omega - ia$.

16. For the signal

$$x(t) = 40\cos(2\pi 2t + 1.5) + 15\cos(2\pi 4t + 1)$$

plot

$$R_0, \quad S_0, \quad A(f), \quad \theta(f)$$

for various values of T. Using these plots, generate the analog signal given previously.

Hint: The plot program you use may show one of the peaks off scale as in Figure 16.9. In that case, you can read the peak value by restricting the plot to the neighborhood of the critical frequency you are investigating. You can also check the maximum values you read from the plots by direct substitution.

17. Find the Laplace transforms of
    (i) $e^{-t}\sin t$,
    (ii) $\cosh t \ \sinh t$,
    (iii) $\sin(t - 2)$,
    (iv) t^n,
    (v) $t^3 e^t$.

18. Find the inverse Laplace transforms of
    (i) $1/s(s + 1)$,
    (ii) $1/(s + 1)(s + 2)'s + 3)$,

(iii) $s^2/(s^4 - a^4)$,

(iv) $s^2/(s^2 - 1)^2$.

19. Using Laplace transforms, solve the following differential equations:

(i) $x''(t) - 4x' + 4x = \cos t$, $\quad x(0) = 1 \quad x'(0) = 1$,

(ii) $x''(t) - 2x' + x = \sin t$, $\quad x(0) = 1$, $\quad x'(0) = 1$.

20. Action of a signal processor is described by the differential equation

$$2x'(t) + x(t) = u(t).$$

Find the transfer function for the following inputs:

(i) $u(t) = e^{-t} \cos t$,

(ii) $u(t) = e^{-t} \sin t$.

21. Two signal processors are described by the differential equations

$$2x_1' + x_1(t) = u_1(t)$$

and

$$x_2' + x_2(t) = u_2(t).$$

(i) Find the transfer functions for the cases where they are connected in series.

(ii) Find the transfer functions for the cases where they are connected in parallel.

(iii) Find their outputs when the input is a step function:

$$u(t) = \begin{cases} 0, & t \leq 0, \\ 1, & t > 0. \end{cases}$$

22. Make a diagram representing the following combination of signal processors:

$$G(s) = G_1(s)\{G_2(s) + G_3(s) + G_4(s)\}G_5(s)G_6(s).$$

23. Using Laplace transforms, solve the following differential equation:

$$x''(t) - 2x'(t) + x(t) = \sin t - \cos t, \quad x(0) = 1, \quad x'(0) = 1.$$

CHAPTER 17

CALCULUS OF VARIATIONS

The extremum point of a surface, $z = z(x, y)$, is defined as the point where the first differential, $\Delta^{(1)} z(x, y)$, vanishes:

$$\Delta^{(1)} z(x, y) = \frac{\partial z}{\partial x}\, dx + \frac{\partial z}{\partial y}\, dy = 0. \tag{17.1}$$

Since dx and dy are independent infinitesimal displacements, Eq. (17.1) can only be satisfied when the coefficients of dx and dy vanish simultaneously:

$$\frac{\partial z}{\partial x} = \frac{\partial z}{\partial y} = 0. \tag{17.2}$$

A point that satisfies these conditions is called the **stationary point** or the **extremum point** of the surface. To determine whether an extremum

Essentials of Mathematical Methods in Science and Engineering, Second Edition. Selçuk Ş. Bayın.

corresponds to a maximum or a minimum, one has to check the second differential, $\Delta^{(2)} z(x, y)$, which involves second-order partial derivatives of $z(x, y)$. For a function of n independent variables

$$f = f(x_1, x_2, \ldots, x_n), \tag{17.3}$$

stationary points are determined by n independent conditions:

$$\frac{\partial z}{\partial x_1} = \frac{\partial z}{\partial x_2} = \cdots = \frac{\partial z}{\partial x_n} = 0. \tag{17.4}$$

The simultaneous solution of these n conditions gives the coordinates of the stationary points, which, when substituted into $f(x_1, x_2, \ldots, x_n)$, gives the **stationary values**, that is, the local maximum or the minimum values of $f(x_1, x_2, \ldots, x_n)$. All this, which sounds familiar from Chapter 2, is actually a simple application of variational analysis to functions of numbers. In general, the calculus of variations is the branch of mathematics that investigates the stationary values of a **generalized function**, defined in terms of some **generalized variables**. These variables could be numbers, functions, tensors, etc. In this regard, calculus of variations has found a wide range of applications in science and engineering. In this chapter, we introduce the basic elements of this interesting branch of mathematics.

17.1   A SIMPLE CASE

A particular type of variational problem, where the unknown is a function that appears under an integral sign, has proven to be quite useful. For example, if we want to find the shortest path between two points, we have to extremize the line integral that gives the distance, d, as a function of the path, $y(x)$, as

$$d[y(x)] = \int_1^2 ds. \tag{17.5}$$

In the aforementioned equation, $y(x)$ is any path connecting the points 1 and 2 and ds is the infinitesimal arclength. In flat space, ds is given by the Pythagoras theorem:

$$ds = \sqrt{1 + \left(\frac{dy}{dx}\right)^2}\, dx, \tag{17.6}$$

hence, the variational problem to be solved is to find the path that minimizes $d[y(x)]$:

$$d[y(x)] = \int_1^2 \sqrt{1 + \left(\frac{dy}{dx}\right)^2}\, dx. \tag{17.7}$$

Since distance depends on the path taken, $d[y(x)]$ is a function of a function. Hence, it is called a **functional**. This is a special case of the general variational problem:

$$J[y(x)] = \int_1^2 f(y, y', x) \, dx, \quad y' = \frac{dy}{dx}, \tag{17.8}$$

where $J[y(x)]$ is the functional to be extremized, $y(x)$ is the desired function with the independent variable x, and $f(y, y', x)$ is a known function of y, y', and x.

17.2   VARIATIONAL ANALYSIS

We first introduce the basic type of variational problem, where we look for a function or a path that extremizes the functional

$$\boxed{J[y(x)] = \int_1^2 F(y, y', x) dx} \tag{17.9}$$

and satisfies certain boundary conditions. To find the stationary function, we need to analyze how $J[y(x)]$ changes when the path is varied. Similar to finding the extremum of a function of numbers, we look for a $y(x)$, which in its neighborhood, the change in $J[y(x)]$ vanishes to first order. We parameterize all allowed paths connecting the points x_1 and x_2 in terms of a continuous and differentiable parameter ε as $y(x, \varepsilon)$, where $y(x, 0)$ is the desired stationary path. It is important to keep in mind that x is not varied, hence it is not a function of ε. Substituting $y(x, \varepsilon)$ into Eq. (17.9), we obtain J as a function of ε as

$$J(\varepsilon) = \int_1^2 F(y(x, \varepsilon), y'(x, \varepsilon), x) \, dx, \tag{17.10}$$

which can be expanded in powers of ε:

$$J(\varepsilon) = J(0) + \frac{dJ}{d\varepsilon}\bigg|_{\varepsilon=0} \varepsilon + \frac{1}{2!} \frac{d^2J}{d\varepsilon^2}\bigg|_{\varepsilon=0} \varepsilon^2 + \cdots. \tag{17.11}$$

The extremum of $J(\varepsilon)$ can now be found as in ordinary calculus by imposing the condition $\frac{dJ}{d\varepsilon}\big|_{\varepsilon=0} = 0$. Differentiating $J(\varepsilon)$ with respect to ε, we get

$$\frac{dJ}{d\varepsilon} = \frac{d}{d\varepsilon} \int_{x_1}^{x_2} F(y(x, \varepsilon), y'(x, \varepsilon), x) \, dx \tag{17.12}$$

$$= \int_{x_1}^{x_2} \frac{d}{d\varepsilon} F(y(x, \varepsilon), y'(x, \varepsilon), x) \, dx \tag{17.13}$$

$$= \int_{x_1}^{x_2} \left[\frac{\partial F}{\partial y} \frac{dy}{d\varepsilon} + \frac{\partial F}{\partial y'} \frac{dy'}{d\varepsilon} \right] dx, \tag{17.14}$$

where in writing Eq. (17.13), we assumed that integration with respect to x and differentiation with respect to ε can be interchanged. Also assuming that

$$\frac{d}{d\varepsilon}\left(\frac{dy}{dx}\right) = \frac{d}{dx}\left(\frac{dy}{d\varepsilon}\right),\tag{17.15}$$

we can write the last integral as

$$\frac{dJ}{d\varepsilon} = \int_{x_1}^{x_2}\left[\frac{\partial F}{\partial y}\frac{dy}{d\varepsilon} + \frac{\partial F}{\partial y'}\frac{d}{dx}\left(\frac{dy}{d\varepsilon}\right)\right] dx.\tag{17.16}$$

Integrating the second term on the right-hand side by parts gives

$$\frac{dJ}{d\varepsilon} = \frac{\partial F}{\partial y'}\frac{dy}{d\varepsilon}\Big|_{x_1}^{x_2} + \int_{x_1}^{x_2}\left[\frac{\partial F}{\partial y} - \frac{d}{dx}\left(\frac{\partial F}{\partial y'}\right)\right]\left(\frac{dy}{d\varepsilon}\right) dx.\tag{17.17}$$

For the stationary path, we set $\frac{dJ}{d\varepsilon} = 0$ at $\varepsilon = 0$ to get

$$\frac{dJ}{d\varepsilon}\Big|_{\varepsilon=0} = \left(\frac{\partial F}{\partial y'}\frac{dy}{d\varepsilon}\right)_{\varepsilon=0}\Big|_{x_1}^{x_2} + \int_{x_1}^{x_2}\left[\frac{\partial F}{\partial y} - \frac{d}{dx}\left(\frac{\partial F}{\partial y'}\right)\right]_{\varepsilon=0}\left(\frac{dy}{d\varepsilon}\right)_{\varepsilon=0} dx = 0.\tag{17.18}$$

To see how $y(x, \varepsilon)$ behaves in the neighborhood of the desired path, we expand in powers of ε:

$$y(x, \varepsilon) = y(x, 0) + \left(\frac{dy}{d\varepsilon}\right)_{\varepsilon=0}\varepsilon + \frac{1}{2!}\left(\frac{d^2y}{d\varepsilon^2}\right)_{\varepsilon=0}\varepsilon^2 + \cdots,\tag{17.19}$$

which to first order in ε can be written as

$$y(x, \varepsilon) = y(x, 0) + \left(\frac{dy}{d\varepsilon}\right)_{\varepsilon=0}\varepsilon.\tag{17.20}$$

Note that since x is not varied, we can write $\partial y(x, \varepsilon)/\partial\varepsilon = dy(x, \varepsilon)/d\varepsilon$ and $\partial y(x, \varepsilon)/\partial x = dy(x, \varepsilon)/dx$. Since $(dy/d\varepsilon)_{\varepsilon=0}$ is a function of x only, we introduce the continuous and differentiable function $\eta(x)$,

$$\eta(x) = \left(\frac{dy}{d\varepsilon}\right)_{\varepsilon=0},\tag{17.21}$$

which reduces Eq. (17.18) to

$$\frac{dJ}{d\varepsilon}\Big|_{\varepsilon=0} = \frac{\partial F}{\partial y'}\eta(x)\Big|_{x_1}^{x_2} + \int_{x_1}^{x_2}\left[\frac{\partial F}{\partial y} - \frac{d}{dx}\left(\frac{\partial F}{\partial y'}\right)\right]\eta(x)\, dx = 0.\tag{17.22}$$

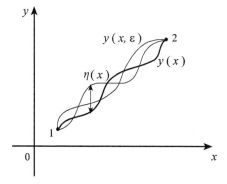

Figure 17.1   Stationary path $y(x)$, the varied path $y(x, \varepsilon)$, and the variation $\eta(x)$.

From Eq. (17.20), it is seen that the variation δy, that is, the infinitesimal difference between the varied path and the desired path is given as

$$\delta y = y(x, \varepsilon) - y(x, 0) = \eta(x)\varepsilon. \tag{17.23}$$

In other words, ε is a measure of the smallness of the variation, and $\eta(x)$ gives how the variation changes with position between the end points (Figure 17.1). Since Eq. (17.22) has to be satisfied by the desired path for all allowed variations, we have dropped the subscript $\varepsilon = 0$. There are now two cases that need to be discussed separately:

17.2.1   Case I: The Desired Function is Prescribed at the End Points

In this case, the variation at the end points is zero:

$$\eta(x_1) = \eta(x_2) = 0; \tag{17.24}$$

hence the surface term in Eq. (17.22) vanishes, that is,

$$\left(\frac{\partial F}{\partial y'}\right) \eta(x) \Big|_{x_1}^{x_2} = 0, \tag{17.25}$$

thus leaving

$$\int_{x_1}^{x_2} \left[\frac{\partial F}{\partial y} - \frac{d}{dx}\left(\frac{\partial F}{\partial y'}\right)\right] \eta(x)\, dx = 0. \tag{17.26}$$

For all allowed variations, $\eta(x)$, Eq. (17.26) can only be satisfied if the quantity inside the square brackets vanishes:

$$\boxed{\frac{\partial F}{\partial y} - \frac{d}{dx}\left(\frac{\partial F}{\partial y'}\right) = 0.} \tag{17.27}$$

This is called the **Euler equation**. In summary, when the variation at the end points is zero, to find the stationary function, we have to solve the Euler equation, which is a second-order differential equation. Of course, at the end, we have to evaluate the integration constants by imposing the boundary conditions, which are the prescribed values of the function at the end points:

$$\boxed{y(x_1) = y_1 \ \text{ and } \ y(x_2) = y_2.}$$
(17.28)

17.2.2 Case II: Natural Boundary Conditions

In cases where the variations at the end points do not vanish, that is, when $y(x)$ does not have prescribed values at the end points, the integral in Eq. (17.22) is still zero for the stationary function (path):

$$\int_{x_1}^{x_2} \left[\frac{\partial F}{\partial y} - \frac{d}{dx} \left(\frac{\partial F}{\partial y'} \right) \right] \eta(x) \, dx = 0;$$
(17.29)

but in order to satisfy Eq. (17.22), we also need to have

$$\left[\frac{\partial F}{\partial y'} \eta(x) \right]_{x_2} = \left[\frac{\partial F}{\partial y'} \eta(x) \right]_{x_1},$$
(17.30)

$$\left(\frac{\partial F}{\partial y'} \right)_{x_2} \eta(x_2) = \left(\frac{\partial F}{\partial y'} \right)_{x_1} \eta(x_1).$$
(17.31)

Since the variations $\eta(x_1)$ and $\eta(x_1)$ are completely arbitrary, the only way to satisfy Eq. (17.31) is to have

$$\left(\frac{\partial F}{\partial y'} \right)_{x_2} = 0 \ \text{ and } \ \left(\frac{\partial F}{\partial y'} \right)_{x_1} = 0.$$
(17.32)

These conditions are usually called the **natural boundary conditions**. If $y(x)$ is predetermined at one of the end points, say $x = x_1$, then $\eta(x_1)$ is zero but $\eta(x_2)$ is arbitrary. Hence, we impose the boundary conditions

$$y(x_1) = y_1 \ \text{ and } \ \left(\frac{\partial F}{\partial y'} \right)_{x_2} = 0.$$
(17.33)

Similarly, if $y(x)$ is predetermined at $x = x_2$ but free at the other end, then we use

$$\left(\frac{\partial F}{\partial y'} \right)_{x_1} = 0 \ \text{ and } \ y(x_2) = y_2.$$
(17.34)

17.3 ALTERNATE FORM OF EULER EQUATION

In the Euler equation:

$$\frac{d}{dx}\left(\frac{\partial F}{\partial y'}\right) - \frac{\partial F}{\partial y} = 0, \tag{17.35}$$

if we write the total derivative in the first term explicitly:

$$\frac{d}{dx}\left(\frac{\partial F}{\partial y'}\right) = \frac{\partial}{\partial x}\left(\frac{\partial F}{\partial y'}\right) + \frac{\partial}{\partial y}\left(\frac{\partial F}{\partial y'}\right)\frac{dy}{dx} + \frac{\partial}{\partial y'}\left(\frac{\partial F}{\partial y'}\right)\frac{dy'}{dx}, \tag{17.36}$$

we obtain

$$\frac{\partial^2 F}{\partial y'^2}\left(\frac{dy'}{dx}\right) + \frac{\partial^2 F}{\partial y \partial y'}\left(\frac{dy}{dx}\right) + \frac{\partial^2 F}{\partial x \partial y'} - \frac{\partial F}{\partial y} = 0. \tag{17.37}$$

It is useful to note that this equation is equivalent to

$$\frac{1}{y'}\left[\frac{d}{dx}\left(F - \frac{\partial F}{\partial y'}\frac{dy}{dx}\right) - \frac{\partial F}{\partial x}\right] = 0, \tag{17.38}$$

which gives the alternate form of the Euler equation as

$$\boxed{\frac{d}{dx}\left(F - \frac{\partial F}{\partial y'}\frac{dy}{dx}\right) = \frac{\partial F}{\partial x}.} \tag{17.39}$$

This form is particularly useful when $F(x, y, y')$ does not depend on x explicitly, which in such a case allows us to write the **first integral** immediately as

$$\boxed{F(y, y') - \frac{\partial F}{\partial y'}y' = C,} \tag{17.40}$$

where C is constant. This is now a first-order differential equation to be solved for the stationary function.

Example 17.1. *Shortest path between two points:* We have seen that finding the shortest path between two points on a plane is basically a variation problem, where

$$F(x, y, y') = (1 + y'^2)^{1/2}. \tag{17.41}$$

Since F does not depend on x explicitly, it is advantageous to use the second form of the Euler equation [Eq. (17.40)]. We first write the needed derivative:

$$\frac{\partial F}{\partial y'} = \frac{1}{2}(1 + y'^2)^{-1/2}2y' = (1 + y'^2)^{-1/2}y', \tag{17.42}$$

and then obtain the Euler equation to be solved for $y(x)$ as

$$(1 + y'^2)^{1/2} - y'^2(1 + y'^2)^{-1/2} = C. \tag{17.43}$$

This has the solution

$$y(x) = c_0 x + c_1, \tag{17.44}$$

which is the equation of a straight line.

Example 17.2. *A case with different boundary conditions:* Let us now consider the functional

$$J[y(x)] = \int_0^l (Ty'^2 - \rho\omega^2 y^2)\, dx, \tag{17.45}$$

where T, ρ, and ω are positive constants. The Euler equation is now written as

$$\frac{d}{dx}\left(T\frac{dy}{dx}\right) + \rho\omega^2 y = 0, \tag{17.46}$$

where the general solution can be immediately written as

$$y(x) = c_1 \cos \beta x + c_2 \sin \beta x, \quad \beta^2 = \frac{\rho\omega^2}{T}. \tag{17.47}$$

We first consider the boundary conditions where the values of $y(x)$ are prescribed at both end points as

$$y(0) = 0 \ \text{ and } \ y(l) = 1, \tag{17.48}$$

which determines the integration constants as

$$c_1 = 0 \ \text{ and } \ c_2 = \frac{1}{\sin \beta l}, \tag{17.49}$$

thus giving the stationary function

$$y(x) = \frac{\sin \beta x}{\sin \beta l}, \quad \beta l \neq 0, \pi, 2\pi, \ldots. \tag{17.50}$$

Next, we consider the case where $y(x)$ is prescribed at one of the end points, $x = l$, as $y(l) = 1$ but left free at the other end point. We now use Eq. (17.34) with $F = Ty'^2 - \rho\omega^2 y^2$ to write the second condition to be imposed as

$$\left(\frac{\partial F}{\partial y'}\right)_{x=0} = 2Ty'(0) = 0 \tag{17.51}$$

to obtain the solution

$$y(x) = \frac{\cos \beta x}{\cos \beta l}, \quad \beta l \neq \frac{\pi}{2}, \frac{3\pi}{2}, \ldots . \tag{17.52}$$

Finally, when neither $y(0)$ nor $y(l)$ are prescribed, we impose the natural boundary conditions [Eq. (17.32)]:

$$\left(\frac{\partial F}{\partial y'}\right)_{x=0} = 0; \quad 2Ty'(0) = 0, \tag{17.53}$$

$$\left(\frac{\partial F}{\partial y'}\right)_{x=l} = 0; \quad 2Ty'(l) = 0, \tag{17.54}$$

which yields the stationary function as

$$y(x) = \begin{cases} 0, & \beta l \neq \pi, 2\pi, \ldots, \\ c_1 \cos \beta x, & \beta l = \pi, 2\pi, \ldots, \end{cases} \tag{17.55}$$

where c_1 is an arbitrary constant.

17.4 VARIATIONAL NOTATION

So far, we have confined our discussion of variational analysis to functionals of the form

$$J[y(x)] = \int_{x_1}^{x_2} F(x, y, y') \, dx. \tag{17.56}$$

In order to be able to discuss more general cases with several dependent and several independent variables and functionals with higher-order derivatives, we introduce the variational differential "δ". Compared to "d", which indicates an infinitesimal change in a quantity between two infinitesimally close points: $dx, df(x)$, etc., the variational differential δ denotes an infinitesimal change in a quantity between two neighboring paths: δy, $\delta F(x, y, y')$, etc. We can write $\delta F(x, y, y')$ as

$$\delta F = F(x, y + \varepsilon \eta(x), y' + \varepsilon \eta'(x)) - F(x, y, y'). \tag{17.57}$$

Expanding the first term on the right-hand side in the neighborhood of (x, y, y') and keeping only the first-order terms in ε, we obtain

$$\delta F = \frac{\partial F}{\partial y} \varepsilon \eta(x) + \frac{\partial F}{\partial y'} \varepsilon \eta'(x), \tag{17.58}$$

which, after substituting $\delta y = \varepsilon \eta(x)$ [Eq. (17.23)], becomes

$$\delta F(x, y, y') = \frac{\partial F}{\partial y} \delta y + \frac{\partial F}{\partial y'} \delta y'. \tag{17.59}$$

Note that this expression is consistent with the assumption $\delta x = 0$. We now consider the variational integral

$$\delta J = \delta \int_{x_1}^{x_2} F(x, y, y') \, dx. \tag{17.60}$$

Since the variation of the independent variable, δx, and the variation of the end points, δx_1 and δx_2, are zero, we can interchange the variation and the integration signs to write

$$\delta J = \int_{x_1}^{x_2} \delta F(x, y, y') \, dx, \tag{17.61}$$

hence

$$\delta J = \int_{x_1}^{x_2} \left[\frac{\partial F}{\partial y} \delta y + \frac{\partial F}{\partial y'} \delta y' \right] dx \tag{17.62}$$

$$= \int_{x_1}^{x_2} \left[\frac{\partial F}{\partial y} \delta y + \frac{\partial F}{\partial y'} \delta \left(\frac{dy}{dx} \right) \right] dx. \tag{17.63}$$

Since the derivative with respect to an independent variable and the variation sign can be interchanged:

$$\delta \left[\frac{dy}{dx} \right] = \frac{d}{dx} [\delta y], \tag{17.64}$$

we can also write

$$\delta J = \int_{x_1}^{x_2} \left[\frac{\partial F}{\partial y} \delta y + \frac{\partial F}{\partial y'} \frac{d}{dx} (\delta y) \right] dx. \tag{17.65}$$

After integrating the second term on the right-hand side by parts, we obtain

$$\delta J = \frac{\partial F}{\partial y'} \delta y \Big|_{x_1}^{x_2} + \int_{x_1}^{x_2} \left[\frac{\partial F}{\partial y} - \frac{d}{dx} \left(\frac{\partial F}{\partial y'} \right) \right] \delta y \, dx. \tag{17.66}$$

This is the analog of the first differential $\Delta^{(1)} z$; hence, we also write it as $\delta^{(1)} F(x, y, y')$. For the stationary functions, we set the first variation to zero, thus obtaining

$$\delta^{(1)} J = \frac{\partial F}{\partial y'} \delta y \Big|_{x_1}^{x_2} + \int_{x_1}^{x_2} \left[\frac{\partial F}{\partial y} - \frac{d}{dx} \left(\frac{\partial F}{\partial y'} \right) \right] \delta y \, dx = 0. \tag{17.67}$$

Since the aforementioned equation has to be satisfied for any arbitrary variation δy, we reach the same conclusions introduced in Section 17.2 as Cases I and II. If we notice that $\delta y = \varepsilon \eta(x)$, which reduces Eq. (17.67) to Eq. (17.22), the connection between the two approaches becomes clear. It can be easily verified that the variation of sums, products, ratios, powers, etc., are completely analogous to the corresponding rules of differentiation:

$$\delta F = \frac{\partial F}{\partial y}\,\delta y + \frac{\partial F}{\partial y'}\,\delta y', \tag{17.68}$$

$$\delta(F_1 + F_2) = \delta F_1 + \delta F_2, \tag{17.69}$$

$$\delta(F_1 F_2) = F_1\,\delta F_2 + F_2\,\delta F_1, \tag{17.70}$$

$$\delta\left(\frac{F_1}{F_2}\right) = \frac{F_2\,\delta F_1 - F_1\,\delta F_2}{F_2^2}, \tag{17.71}$$

$$\delta F^n = nF^{n-1}\,\delta F. \tag{17.72}$$

So far, nothing has been said about the stationary function being a maximum or a minimum. As in ordinary extremum problems, the answer comes from a detailed analysis of the second variation $\delta^{(2)}J$, which is obtained by keeping the second-order terms in Eq. (17.11), where $y(x,\varepsilon)$ to second order is now taken as

$$y(x,\varepsilon) = y(x,0) + \eta(x)\varepsilon + \frac{1}{2}\eta'(x)\varepsilon^2. \tag{17.73}$$

However, in most of the physically interesting cases, this analysis, which could be quite cumbersome, can be evaded. It is usually intuitively clear that the particular stationary function at hand is the needed maximum or minimum. In other cases, where we may have more than one stationary function, we can pick the one that actually maximizes or minimizes the given functional by direct substitution. It is also possible that in some cases, we are only interested in the extremum, regardless of it being a maximum or a minimum.

17.5 A MORE GENERAL CASE

Consider the functional

$$J = \int\!\int_R F(x, y, u, v, u_x, u_y, v_x, v_y)dx\,dy, \ u_x = \frac{\partial}{\partial x}, \ \text{etc.}, \tag{17.74}$$

where there are two independent variables, x and y, and two dependent variables, $u(x,y)$ and $v(x,y)$. Using variational notation, we write the first variation $\delta^{(1)}J$ and set it to zero for stationary functions:

$$\delta^{(1)}J = \int\!\int_R \left[\left(\frac{\partial F}{\partial u}\,\delta u + \frac{\partial F}{\partial u_x}\,\delta u_x + \frac{\partial F}{\partial u_y}\,\delta u_y\right)\right.$$
$$\left. + \left(\frac{\partial F}{\partial v}\,\delta v + \frac{\partial F}{\partial v_x}\,\delta v_x + \frac{\partial F}{\partial v_y}\,\delta v_y\right)\right]dx\,dy = 0. \tag{17.75}$$

Here we assume that δu and δv are continuous and differentiable functions over the region R. We now consider a typical term, which involves variation of a derivative:

$$\int\int_R \frac{\partial F}{\partial u_x} \delta u_x dx \, dy, \tag{17.76}$$

and write

$$\int\int_R \frac{\partial F}{\partial u_x} \delta\left(\frac{\partial u}{\partial x}\right) dx \, dy = \int\int_R \frac{\partial F}{\partial u_x} \frac{\partial(\delta u)}{\partial x} dx \, dy \tag{17.77}$$

$$= \int\int_R \frac{\partial}{\partial x}\left(\frac{\partial F}{\partial u_x}\delta u\right) dx \, dy - \int\int_R \frac{\partial}{\partial x}\left(\frac{\partial F}{\partial u_x}\right)\delta u \, dx \, dy. \tag{17.78}$$

Using the two-dimensional version of the divergence theorem [Eq. (2.225)], we can write the following useful formulas:

$$\int\int_R \frac{\partial \phi}{\partial x} dx \, dy = \oint_C \phi \cos\theta \, ds, \tag{17.79}$$

$$-\int\int_R \frac{\partial \phi}{\partial y} dx \, dy = \oint_C \phi \sin\theta \, ds, \tag{17.80}$$

where θ is the angle between the positive x-axis and the outward normal to the boundary C of the region R and ds is the infinitesimal arclength along $d\vec{s}$. Using Eq. (17.79), we can transform the first integral on the right-hand side of Eq. (17.78) to write

$$\int\int_R \frac{\partial F}{\partial u_x} \delta\left(\frac{\partial u}{\partial x}\right) dx \, dy = \oint_C \frac{\partial F}{\partial u_x} \cos\theta\delta u \, ds - \int\int_R \frac{\partial}{\partial x}\left(\frac{\partial F}{\partial u_x}\right)\delta u \, dx \, dy. \tag{17.81}$$

Treating other such terms in Eq. (17.75) with the variation of a derivative, we can write the equation to be satisfied by the stationary functions as

$$\delta^{(1)}J = \oint_C \left[\left(\frac{\partial F}{\partial u_x}\cos\theta - \frac{\partial F}{\partial u_y}\sin\theta\right)\delta u + \left(\frac{\partial F}{\partial v_x}\cos\theta - \frac{\partial F}{\partial v_y}\sin\theta\right)\delta v\right] ds$$

$$+ \int\int_R \left\{\left[\frac{\partial F}{\partial u} - \frac{\partial}{\partial x}\left(\frac{\partial F}{\partial u_x}\right) - \frac{\partial}{\partial y}\left(\frac{\partial F}{\partial u_y}\right)\right]\delta u\right.$$

$$\left.+ \left[\frac{\partial F}{\partial v} - \frac{\partial}{\partial x}\left(\frac{\partial F}{\partial v_x}\right) - \frac{\partial}{\partial y}\left(\frac{\partial F}{\partial v_y}\right)\right]\delta v\right\} dx \, dy = 0. \tag{17.82}$$

If the values of the functions u and v are prescribed on the boundary, then the first integral over the boundary C vanishes. Since the variations δu and δv are arbitrary, the only way to satisfy Eq. (17.82) is to have

$$\frac{\partial F}{\partial u} - \frac{\partial}{\partial x}\left(\frac{\partial F}{\partial u_x}\right) - \frac{\partial}{\partial y}\left(\frac{\partial F}{\partial u_y}\right) = 0, \tag{17.83}$$

$$\frac{\partial F}{\partial v} - \frac{\partial}{\partial x}\left(\frac{\partial F}{\partial v_x}\right) - \frac{\partial}{\partial y}\left(\frac{\partial F}{\partial v_y}\right) = 0. \tag{17.84}$$

These are now the **Euler equations** to be solved simultaneously for the stationary functions $u(x,y)$ and $v(x,y)$.

For this case, the **natural boundary conditions** are given as

$$\frac{\partial F}{\partial u_x}\cos\theta - \frac{\partial F}{\partial u_y}\sin\theta = 0 \text{ on } C \tag{17.85}$$

when $u(x,y)$ is not prescribed on C and

$$\frac{\partial F}{\partial v_x}\cos\theta - \frac{\partial F}{\partial v_y}\sin\theta = 0 \text{ on } C \tag{17.86}$$

when $v(x,y)$ is not prescribed on C. Generalization to cases with m-dependent and n-independent variables is obvious [1].

Example 17.3. Minimal surfaces: Consider a smooth surface $z = z(x,y)$ bounded by a simple closed curve C (Figure 17.2), whose surface area can be written as (Chapters 2 and 3)

$$S = \int\int_{R_{xy}} \sqrt{1 + \left(\frac{\partial z}{\partial x}\right)^2 + \left(\frac{\partial z}{\partial y}\right)^2}\, dx\, dy, \tag{17.87}$$

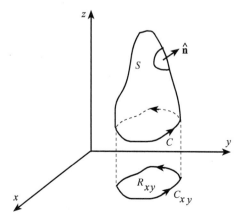

Figure 17.2 Minimal surfaces.

where R_{xy} is the region in the xy-plane bounded by the projection C_{xy} onto the xy-plane. To find the minimal surface with the minimal area, we have to extremize S with

$$F = \sqrt{1 + \left(\frac{\partial z}{\partial x}\right)^2 + \left(\frac{\partial z}{\partial y}\right)^2}, \tag{17.88}$$

where we have one dependent, z, and two independent, x and y, variables. Using Eq. (17.83) for the only Euler equation we have, we write

$$\frac{\partial F}{\partial z} - \frac{\partial}{\partial x}\left(\frac{\partial F}{\partial z_x}\right) - \frac{\partial}{\partial y}\left(\frac{\partial F}{\partial z_y}\right) = 0, \tag{17.89}$$

$$\frac{\partial}{\partial x}\left[\frac{z_x}{(1 + z_x^2 + z_y^2)^{1/2}}\right] + \frac{\partial}{\partial y}\left[\frac{z_y}{(1 + z_x^2 + z_y^2)^{1/2}}\right] = 0, \tag{17.90}$$

which can be simplified as

$$(1 + z_y^2)z_{xx} - 2z_x z_y z_{xy} + (1 + z_x^2)z_{yy} = 0. \tag{17.91}$$

Example 17.4. *Laplace equation:* Let us find the Euler equation for the following variational problem:

$$\delta \int \int \int_R (\phi_x^2 + \phi_y^2 + \phi_z^2)\, dx\, dy\, dz, \tag{17.92}$$

where the only dependent variable $\phi(x, y, z)$ takes prescribed values on the boundary of R. Using $F = (\phi_x^2 + \phi_y^2 + \phi_z^2)$ and the generalization of Eq. (17.83) to three independent variables as

$$\frac{\partial F}{\partial \phi} - \frac{\partial}{\partial x}\left(\frac{\partial F}{\partial \phi_x}\right) - \frac{\partial}{\partial y}\left(\frac{\partial F}{\partial \phi_y}\right) - \frac{\partial}{\partial z}\left(\frac{\partial F}{\partial \phi_z}\right) = 0, \tag{17.93}$$

we obtain

$$\frac{\partial^2 \phi}{\partial x^2} + \frac{\partial^2 \phi}{\partial y^2} + \frac{\partial^2 \phi}{\partial z^2} = 0, \tag{17.94}$$

which is the Laplace equation.

Example 17.5. *The inverse problem:* Sometimes we are interested in writing a variational integral whose Euler equation is a given differential equation. For example, consider the differential equation

$$\frac{d}{dx}\left(F\frac{dy}{dx}\right) + \rho\omega^2 y + p = 0, \tag{17.95}$$

which is the governing equation for the small displacements of a rotating string of length l, where $y(x)$ represents the displacement from the axis of rotation, $F(x)$ is the tension, $\rho(x)$ is the linear mass density, $\omega(x)$ is the angular velocity of the string, and $p(x)$ is the intensity of the load distribution. To find the corresponding variational problem, we multiply the aforementioned equation with $\delta y(x)$ and integrate over $(0, l)$:

$$\int_0^l \frac{d}{dx}\left(F\frac{dy}{dx}\right)\delta y\ dx + \int_0^l \rho\omega^2 y\delta y\ dx + \int_0^l p\delta y\ dx = 0. \qquad (17.96)$$

The second and third integrals are nothing but

$$\int_0^l \rho\omega^2 y\delta y\ dx + \int_0^l p\delta y\ dx = \delta \int_0^l \frac{1}{2}\rho\omega^2 y\ dx + \delta \int_0^l py\ dx. \qquad (17.97)$$

Using integration by parts, the first term in Eq. (17.96) can be written as

$$\int_0^l \frac{d}{dx}\left(F\frac{dy}{dx}\right)\delta y\ dx = \left[F\frac{dy}{dx}\delta y\right]_0^l - \int_0^l \left(F\frac{dy}{dx}\right)\delta\left(\frac{dy}{dx}\right)dx$$

$$= \left[F\frac{dy}{dx}\delta y\right]_0^l - \delta \int_0^l \frac{1}{2}F\left(\frac{dy}{dx}\right)^2 dx. \qquad (17.98)$$

Substituting Eqs. (17.97) and (17.98) into Eq. (17.96), we obtain

$$\left[F\frac{dy}{dx}\delta y\right]_0^l + \delta \int_0^l \left[\frac{1}{2}\rho\omega^2 y^2 + py - \frac{1}{2}F\left(\frac{dy}{dx}\right)^2\right]dx = 0. \qquad (17.99)$$

If the boundary conditions imposed are such that the surface term vanishes:

$$\left[F\frac{dy}{dx}\delta y\right]_0^l = 0, \qquad (17.100)$$

that is, when $y(x)$ is prescribed at both the end points, which implies

$$\delta y(0) = \delta y(l) = 0, \qquad (17.101)$$

or when the natural boundary conditions are imposed:

$$\left(F\frac{dy}{dx}\delta y\right)_l = \left(F\frac{dy}{dx}\delta y\right)_0, \qquad (17.102)$$

the variational integral that produces Eq. (17.95) as its Euler equation becomes

$$\delta \int_0^l \left[\frac{1}{2}\rho\omega^2 y^2 + py - \frac{1}{2}F\left(\frac{dy}{dx}\right)^2\right]dx = 0. \qquad (17.103)$$

17.6 HAMILTON'S PRINCIPLE

In the previous example, we considered the inverse problem, where we started with a differential equation and obtained a variational integral whose Euler equation is the given differential equation. Let us now apply this to classical mechanics. In Newton's theory, a single particle of mass m, moving under the influence of a force field $\overrightarrow{f}$, follows a path that satisfies the equation of motion

$$m\frac{d^2\overrightarrow{r}}{dt^2} - \overrightarrow{f} = 0, \quad \overrightarrow{r}(t_1) = \overrightarrow{r}_1, \quad \overrightarrow{r}(t_2) = \overrightarrow{r}_2, \tag{17.104}$$

where $\overrightarrow{r}_1$ and $\overrightarrow{r}_2$ are the initial and the final positions of the particle. Let us now consider variations $\delta\overrightarrow{r}$ about the actual path $\overrightarrow{r}(t)$. Since the end points are prescribed, we take the variations at the end points as zero:

$$\delta\overrightarrow{r}(t_1) = \delta\overrightarrow{r}(t_2) = 0. \tag{17.105}$$

At any intermediate time, the varied path is given by $\overrightarrow{r}(t) + \delta\overrightarrow{r}(t)$. We now take the scalar product of the equation of motion [Eq. (17.104)] with $\delta\overrightarrow{r}(t)$ and integrate over time from t_1 to t_2 :

$$\int_{t_1}^{t_2} \left(m\frac{d^2\overrightarrow{r}}{dt^2} \cdot \delta\overrightarrow{r} - \overrightarrow{f} \cdot \delta\overrightarrow{r} \right) dt = 0. \tag{17.106}$$

Integrating by parts and interchanging $\frac{d}{dt}$ and δ, the first term becomes

$$\int_{t_1}^{t_2} m\frac{d}{dt}\left[\frac{d\overrightarrow{r}}{dt} \right] \cdot \delta\overrightarrow{r}\, dt = m\left\{ \left[\frac{d\overrightarrow{r}}{dt} \cdot \delta\overrightarrow{r} \right]_{t_1}^{t_2} - \int_{t_1}^{t_2} \frac{d\overrightarrow{r}}{dt} \cdot \delta\left(\frac{d\overrightarrow{r}}{dt} \right) dt \right\}. \tag{17.107}$$

Since variation at the end points is zero, the surface term vanishes. Utilizing the relation

$$\frac{d\overrightarrow{r}}{dt} \cdot \delta\left(\frac{d\overrightarrow{r}}{dt} \right) = \frac{1}{2}\delta\left(\frac{d\overrightarrow{r}}{dt} \right)^2, \tag{17.108}$$

we can write

$$\int_{t_1}^{t_2} m\frac{d^2\overrightarrow{r}}{dt^2} \cdot \delta\overrightarrow{r}\, dt = -\frac{m}{2}\int_{t_1}^{t_2} \delta\left(\frac{d\overrightarrow{r}}{dt} \right)^2 dt = -\delta\int_{t_1}^{t_2} \frac{1}{2}m\left(\frac{d\overrightarrow{r}}{dt} \right)^2 dt \tag{17.109}$$

$$= -\delta\int_{t_1}^{t_2} T\, dt, \tag{17.110}$$

where $T = \frac{1}{2}m(d\overrightarrow{r}/dt)^2$ is the well-known expression for the kinetic energy, and we have interchanged the variation and the integral signs. In the light of these, Eq. (17.106) becomes

$$\int_{t_1}^{t_2} (\delta T + \overrightarrow{f} \cdot \delta\overrightarrow{r})\, dt = 0. \tag{17.111}$$

Furthermore, if the particle is moving under the influence of a conservative force field, we can introduce a potential energy function $V(\overrightarrow{r})$ such that

$$\overrightarrow{f} = -\frac{dV}{d\overrightarrow{r}}. \tag{17.112}$$

We can now write $\overrightarrow{f} \cdot \delta \overrightarrow{r}$ as

$$\overrightarrow{f} \cdot \delta \overrightarrow{r} = -\frac{dV}{d\overrightarrow{r}} \cdot \delta \overrightarrow{r} = -\delta V \tag{17.113}$$

which gives the final form of Eq. (17.111) as

$$\boxed{\delta \int_{t_1}^{t_2} (T - V)\, dt = 0.} \tag{17.114}$$

This is called **Hamilton's principle**. It basically says that the path that a particle follows is the one that makes the integral $\int_{t_1}^{t_2} (T - V)\, dt$ stationary, that is, an extremum. In most cases, the extremum is a minimum. The difference $(T - V)$ is called the **Lagrangian, L,** and the **functional**

$$\boxed{J[\overrightarrow{r}(t)] = \int_{t_1}^{t_2} L(\overrightarrow{r}, \dot{\overrightarrow{r}}, t)\, dt, \quad L = T - V,} \tag{17.115}$$

is called the **action**. Hamilton's principle can be extended to systems with N particles by writing Eq. (17.111) for a single particle and then by summing over all particles within the system:

$$\int_{t_1}^{t_2} dt \sum_{k=1}^{N} \left(\delta \left[\frac{1}{2} m_k \left(\frac{d\overrightarrow{r_k}}{dt} \right)^2 \right] - \overrightarrow{f}_k \cdot \delta \overrightarrow{r}_k \right) = 0, \tag{17.116}$$

where $\overrightarrow{f}_k$ is the net force acting on the kth particle [1–3].

17.7 LAGRANGE'S EQUATIONS OF MOTION

In a dynamical system with n degrees of freedom, it is usually possible to find n independent parameters called the **generalized coordinates** that uniquely describe the state of the system. For example, for a pendulum consisting of a mass m suspended by a string of fixed length l, the position of the mass can be completely specified by giving the angle that the string makes with the vertical (Figure 17.3). If we were to use Cartesian coordinates of m, the x and y coordinates will not be independent, since they are related by the constraining equation

$$x^2 + y^2 = l^2, \tag{17.117}$$

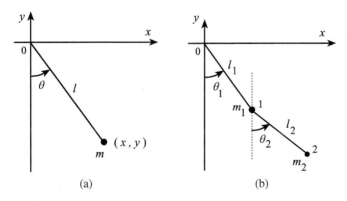

Figure 17.3   Generalized coordinates for the simple pendulum (a) and for the double pendulum (b).

which indicates that the length of the string is fixed. Similarly, for a double pendulum, the two angles, θ_1 and θ_2, are sufficient to completely specify the state of the system at any given time (Figure 17.3). In this case, the four Cartesian coordinates, (x_1, y_1) for m_1 and (x_2, y_2) for m_2, are related by the two constraining equations:

$$x_1^2 + y_1^2 = l_1^2, \tag{17.118}$$

$$(x_2 - x_1)^2 + (y_2 - y_1)^2 = l_2^2. \tag{17.119}$$

In general, the total kinetic energy of a system will be a function of the n generalized coordinates, $q_1, q_2, \ldots, q_n$, and the n generalized velocities, $\dot{q}_1, \dot{q}_2, \ldots, \dot{q}_n$. In terms of generalized coordinates, the $\vec{f} \cdot \delta \vec{r}$ term in Eq. (17.111) is replaced by

$$Q_1 \delta q_1 + Q_2 \delta q_2 + \cdots + Q_n \delta q_n, \tag{17.120}$$

where Q_i are the components of the generalized force. For conservative systems, we can define a generalized potential energy function, which depends only on the generalized coordinates as

$$V(q_1, q_2, \ldots, q_n), \tag{17.121}$$

where the generalized forces are given as

$$-\frac{\partial V}{\partial q_i} = Q_i. \tag{17.122}$$

In terms of generalized coordinates, Hamilton's principle:

$$\delta \int_{t_1}^{t_2} (T - V)\, dt = 0, \tag{17.123}$$

determines the dynamics of an n-dimensional system by the n Euler equations:

$$\frac{d}{dt}\left[\frac{\partial(T-V)}{\partial\dot{q}_i}\right] - \frac{\partial(T-V)}{\partial q_i} = 0, \quad i = 1, 2, \ldots, n, \tag{17.124}$$

which are now called **Lagrange's equations of motion**. In addition to the conservative forces, if the system also has nonconservative forces, $Q'_1, Q'_2, \ldots, Q'_n$, Lagrange's equations become

$$\frac{d}{dt}\left[\frac{\partial(T-V)}{\partial\dot{q}_i}\right] - \frac{\partial(T-V)}{\partial q_i} = Q'_i, \quad i = 1, 2, \ldots, n. \tag{17.125}$$

In general, Lagrangian formulation of mechanics is given in terms of **Hamilton's principle** as

$$\boxed{\delta \int_{t_1}^{t_2} L(q_1, q_2, \ldots, q_n, \dot{q}_1, \dot{q}_2, \ldots, \dot{q}_n, t)dt = 0,} \tag{17.126}$$

where $q_1, q_2, \ldots, q_n$ are the n independent generalized coordinates. The corresponding n Lagrange's equations of motion:

$$\boxed{\frac{d}{dt}\left(\frac{\partial L}{\partial \dot{q}_i}\right) - \frac{\partial L}{\partial q_i} = 0, \quad i = 1, 2, \ldots, n,} \tag{17.127}$$

have to be solved simultaneously with the appropriate boundary conditions. As we have seen, for **conservative systems**, the Lagrangian can be written as

$$\boxed{L = T - V.} \tag{17.128}$$

Example 17.6. *Simple pendulum:* For a simple pendulum (Figure 17.3), the kinetic energy is written as

$$T = \frac{1}{2}m(l\dot{\theta})^2. \tag{17.129}$$

In the absence of friction, we can write the gravitational potential energy with respect to the equilibrium position as

$$V = mgl(1 - \cos\theta), \tag{17.130}$$

which leads to the Lagrangian

$$L = \frac{1}{2}m(l\dot{\theta})^2 - mgl(1 - \cos\theta). \tag{17.131}$$

The corresponding Lagrange's equation for θ is now written as

$$\frac{d}{d\theta}(ml^2\dot{\theta}) - 0 + mgl\sin\theta = 0, \tag{17.132}$$

$$\ddot{\theta} + \frac{g}{l}\sin\theta = 0, \tag{17.133}$$

which is the well-known result for the equation of motion of a simple pendulum. For small oscillations, $\sin\theta \simeq \theta$, the solution is given as $\theta(t) = a_0\cos(\omega t + a_1)$, where $\omega^2 = g/l$ and a_0 and a_1 are the integration constants to be determined from initial conditions. Solutions for finite oscillations are given in terms of elliptic functions [2].

Example 17.7. *Double pendulum:* For the double pendulum in Figure 17.3, we write the Cartesian coordinates of the mass m_1 as

$$x_1 = l_1\sin\theta_1, \; y_1 = -l_1\cos\theta_1 \tag{17.134}$$

and for m_2 as

$$x_2 = l_1\sin\theta_1 + l_2\sin\theta_2, \; y_2 = -l_1\cos\theta_1 - l_2\cos\theta_2. \tag{17.135}$$

The total kinetic energy of the system is now written as

$$T = \frac{1}{2}m_1(\dot{x}_1^2 + \dot{y}_1^2) + \frac{1}{2}m_2(\dot{x}_2^2 + \dot{y}_2^2) \tag{17.136}$$

$$= \frac{1}{2}(m_1 + m_2)l_1^2\dot{\theta}_1^2 + m_2l_1l_2\dot{\theta}_1\dot{\theta}_2\cos(\theta_1 - \theta_2) + \frac{1}{2}m_2l_2^2\dot{\theta}_2^2, \tag{17.137}$$

while the total potential energy is

$$V = m_1gy_1 + m_2gy_2 \tag{17.138}$$

$$= -(m_1 + m_2)gl_1\cos\theta_1 - m_2gl_2\cos\theta_2. \tag{17.139}$$

Using Eqs. (17.137) and (17.139), we can write the Lagrangian, $L = T - V$, and obtain Lagrange's equation of motion for θ_1 as

$$\frac{d}{dt}\left[\frac{\partial L}{\partial\dot{\theta}_1}\right] - \frac{\partial L}{\partial\theta_1} = 0, \tag{17.140}$$

$$(m_1 + m_2)l_1\ddot{\theta}_1 + m_2l_2\left[\ddot{\theta}_2\cos(\theta_1 - \theta_2) + \dot{\theta}_2^2\sin(\theta_1 - \theta_2)\right]$$

$$+ (m_1 + m_2)g\sin\theta_1 = 0 \tag{17.141}$$

and for θ_2 as

$$\frac{d}{dt}\left[\frac{\partial L}{\partial \dot{\theta}_2}\right] - \frac{\partial L}{\partial \theta_2} = 0, \qquad (17.142)$$

$$l_1\ddot{\theta}_1\cos(\theta_1 - \theta_2) + l_2\ddot{\theta}_2 - l_1\dot{\theta}_1^2\sin(\theta_1 - \theta_2) + g\sin\theta_2 = 0. \qquad (17.143)$$

We have now two coupled differential equations [Eqs. (17.141) and (17.143)] to be solved for $\theta_1(t)$ and $\theta_2(t)$.

17.8   DEFINITION OF LAGRANGIAN

We have seen that for one-dimensional conservative systems in Cartesian coordinates, Hamilton's principle is written as

$$\delta I = \delta \int_{t_1}^{t_2} L(x, \dot{x}, t) \, dt \qquad (17.144)$$

$$= \int_{t_1}^{t_2} \delta L(x, \dot{x}, t) \, dt = 0, \qquad (17.145)$$

where the Lagrangian is defined as $L(x, \dot{x}, t) = T - V$. Let us now consider the following Lagrangian:

$$L' = \frac{1}{2}m\dot{x}^2 + k\dot{x}xt,$$

which does not look like anything familiar. However, if we write the corresponding Lagrange's equation of motion:

$$\frac{d}{dt}\left[\frac{\partial L'}{\partial \dot{x}}\right] - \frac{\partial L'}{\partial x} = 0, \qquad (17.146)$$

$$\frac{d}{dt}(m\dot{x} + kxt) - k\dot{x}t = 0, \qquad (17.147)$$

$$m\ddot{x} + k\dot{x}t + kx - k\dot{x}t = 0, \qquad (17.148)$$

$$m\ddot{x} + kx = 0, \qquad (17.149)$$

we see that it is nothing but the harmonic oscillator equation of motion for a mass m attached to a spring with the spring constant k. This equation of motion can also be obtained from the well-known Lagrangian

$$L = \frac{1}{2}m\dot{x}^2 - \frac{1}{2}kx^2, \qquad (17.150)$$

where $V = \frac{1}{2}kx^2$ is the familiar expression for the potential energy stored in the spring. If we look a little carefully, we notice that the two Lagrangians are related by the equation

$$L = L' - k\dot{x}xt - \frac{1}{2}kx^2$$

$$= L' + \frac{d}{dt}\left[-\frac{1}{2}kx^2t\right], \tag{17.151}$$

where they differ by a total time derivative. In fact, from Hamilton's principle,

$$\delta I = \int_{t_1}^{t_2} \delta L \, dt = 0, \tag{17.152}$$

it is seen that the definition of Lagrangian is arbitrary up to a total time derivative. That is, the Lagrangians $L(q, \dot{q}, t)$ and $L'(q, \dot{q}, t)$ that differ by a total time derivative:

$$\boxed{L'(q, \dot{q}, t) = L(q, \dot{q}, t) + \frac{d}{dt}F(q, \dot{q}, t),} \tag{17.153}$$

have the same equation of motion. The proof follows directly from Hamilton's principle:

$$\delta I = \delta \int_{t_1}^{t_2} L' \, dt \tag{17.154}$$

$$= \delta \int_{t_1}^{t_2} \left(L + \frac{d}{dt}F\right) dt, \tag{17.155}$$

$$= \delta \int_{t_1}^{t_2} L \, dt + \delta \int_{t_1}^{t_2} \frac{dF}{dt} \, dt, \tag{17.156}$$

$$= \delta \int_{t_1}^{t_2} L \, dt + \delta F(t_2) - \delta F(t_1). \tag{17.157}$$

Since the variation at the end points is zero, the last two terms, $\delta F(t_2)$ and $\delta F(t_1)$, vanish, thus yielding

$$\boxed{\delta \int_{t_1}^{t_2} L' \, dt = \delta \int_{t_1}^{t_2} L \, dt,} \tag{17.158}$$

which means that the Lagrangians L and L' have the same equation of motion. For an n-dimensional system, we replace q by $q_1, q_2, \ldots, q_n$, and $\dot{q}$ by $\dot{q}_1, \dot{q}_2, \ldots, \dot{q}_n$, but the conclusion remains the same.

In this regard, in writing a Lagrangian we are not restricted with the form $T - V$. In **special relativity**, the Lagrangian for a freely moving particle is given as

$$L = m_0^2 c^2 \sqrt{1 - \frac{v^2}{c^2}},$$

(17.159)

which is not the relativistic kinetic energy of the particle. In general relativity, **Einstein's field equations** are obtained from the variational integral

$$\delta \int_{t_1}^{t_2} L dt = \delta \int_{t_1}^{t_2} \left[\int \int \int_V R \sqrt{-g} d^3 x \right] dt = 0,$$

(17.160)

where R is the curvature scalar and g is the determinant of the metric tensor. In **electromagnetic theory**, Hamilton's principle is written as

$$\delta \int_{t_1}^{t_2} L dt = \delta \int_{t_1}^{t_2} \left[\int \int \int_V \left(-\frac{1}{16\pi} F_{\alpha\beta} F^{\alpha\beta} - \frac{1}{c} J_\alpha A^\alpha \right) d^3 x \right] dt = 0,$$

(17.161)

where $F^{\alpha\beta}$ is the field strength tensor and J^α and A^α are the four current and four potential, respectively. Lagrangian formulation of mechanics makes applications of Newton's theory to many-body problems and to continuous systems possible. It is also an indispensable tool in classical and quantum field theories.

17.9   PRESENCE OF CONSTRAINTS IN DYNAMICAL SYSTEMS

For a system with n degrees of freedom, one can in principle find n independent generalized coordinates to write Hamilton's principle as

$$\delta J(q_i, \dot{q}_i, t) = \delta \int_{t_1}^{t_2} L(q_i, \dot{q}_i, t) \, dt = 0.$$

(17.162)

Using the condition that the variation at the end points vanish:

$$\delta q_i(t_1) = \delta q_i(t_2) = 0, \quad i = 1, 2, \ldots, n,$$

(17.163)

we write

$$\delta J = \int_{t_1}^{t_2} \sum_{i=1}^{n} \left[\frac{\partial L}{\partial q_i} \delta q_i + \frac{\partial L}{\partial \dot{q}_i} \delta \dot{q}_i \right] dt = 0$$

(17.164)

$$= \int_{t_1}^{t_2} \sum_{i=1}^{n} \left[\frac{d}{dt} \left(\frac{\partial L}{\partial \dot{q}_i} \right) - \frac{\partial L}{\partial q_i} \right] \delta q_i \, dt = 0.$$

(17.165)

Since δq_i are independent and arbitrary, the only way to satisfy Eq. (17.165) is to have the coefficients of all δq_i vanish simultaneously:

$$\frac{d}{dt}\left(\frac{\partial L}{\partial \dot{q}_i}\right) - \frac{\partial L}{\partial q_i} = 0, \quad i = 1, 2, \ldots, n, \qquad (17.166)$$

thus obtaining n Lagrange's equations to be solved for the n independent generalized coordinates $q_i(t)$.

Sometimes, we may have k **constraints** imposed on the system:

$$\phi_1(q_1, q_2, \ldots, q_n) = 0,$$

$$\phi_2(q_1, q_2, \ldots, q_n) = 0,$$

$$\vdots$$

$$\phi_k(q_1, q_2, \ldots, q_n) = 0. \qquad (17.167)$$

In this case, the n δq_is in Eq. (17.165) are no longer independent, hence we cannot set their coefficients to zero. One way to face this situation is to find a reduced set of $(n - k)$ generalized coordinates that are all independent. However, this may not always be possible or convenient. In such cases, a practical way out is to use the **Lagrange undetermined multipliers**. Using the constraining equations [Eq. (17.167)], we write the following variational conditions:

$$\frac{\partial \phi_1}{\partial q_1}\delta q_1 + \frac{\partial \phi_1}{\partial q_2}\delta q_2 + \cdots + \frac{\partial \phi_1}{\partial q_n}\delta q_n = 0,$$

$$\frac{\partial \phi_2}{\partial q_1}\delta q_1 + \frac{\partial \phi_2}{\partial q_2}\delta q_2 + \cdots + \frac{\partial \phi_2}{\partial q_n}\delta q_n = 0,$$

$$\vdots$$

$$\frac{\partial \phi_k}{\partial q_1}\delta q_1 + \frac{\partial \phi_k}{\partial q_2}\delta q_2 + \cdots + \frac{\partial \phi_k}{\partial q_n}\delta q_n = 0. \qquad (17.168)$$

In short, these k equations can be written as

$$\sum_{i=1}^{n} \frac{\partial \phi_l}{\partial q_i}\delta q_i, \quad l = 1, 2, \ldots, k. \qquad (17.169)$$

We now multiply each one of these equations with a Lagrange undetermined multiplier λ_l, to write

$$\lambda_l \sum_{i=1}^{n} \frac{\partial \phi_l}{\partial q_i}\delta q_i = 0, \quad l = 1, 2, \ldots, k, \qquad (17.170)$$

and then add them to have

$$\sum_{l=1}^{k} \lambda_l \sum_{i=1}^{n} \frac{\partial \phi_l}{\partial q_i} \delta q_i = 0. \tag{17.171}$$

Integrate this with respect to time from t_1 to t_2:

$$\int_{t_1}^{t_2} \sum_{l=1}^{k} \lambda_l \sum_{i=1}^{n} \frac{\partial \phi_l}{\partial q_i} \delta q_i \, dt = 0, \tag{17.172}$$

and then add to Eq. (17.165) to write

$$\delta J = \int_{t_1}^{t_2} \sum_{i=1}^{n} \left[\frac{d}{dt} \left(\frac{\partial L}{\partial \dot{q}_i} \right) - \frac{\partial L}{\partial q_i} + \sum_{l=1}^{k} \lambda_l \frac{\partial \phi_l}{\partial q_i} \right] \delta q_i \, dt = 0. \tag{17.173}$$

In this equation, δq_i are still not independent. However, we have the k Lagrange undetermined multipliers, λ_l, at our disposal to set the quantities inside the square brackets to zero:

$$\frac{d}{dt} \left(\frac{\partial L}{\partial \dot{q}_i} \right) - \frac{\partial L}{\partial q_i} + \sum_{l=1}^{k} \lambda_l \frac{\partial \phi_l}{\partial q_i} = 0, \quad i = 1, \ldots, n, \tag{17.174}$$

thus obtaining the new Lagrange's equations. We now have $n + k$ unknowns, that is, the n q_is and the k Lagrange undetermined multipliers. We can solve for these $n + k$ unknowns by using n Lagrange's equations and the k constraining equations. If we write Lagrange's equations [Eq. (17.174)] as

$$\boxed{\frac{d}{dt} \left(\frac{\partial L}{\partial \dot{q}_i} \right) - \frac{\partial L}{\partial q_i} = -\sum_{l=1}^{k} \lambda_l \frac{\partial \phi_l}{\partial q_i}, \quad i = 1, \ldots, n,} \tag{17.175}$$

and compare with Eq. (17.125), we see that the right-hand side

$$\boxed{Q_i' = -\sum_{l=1}^{k} \lambda_l \frac{\partial \phi_l}{\partial q_i},}$$

behaves like the ith component of a generalized force. In fact, it represents the **constraining force** on the system [3].

Example 17.8. *Constraining force:* Consider the simple pulley shown in Figure 17.4, which has only one degree of freedom. We solve the problem with two generalized coordinates, q_1 and q_2, and one constraining equation

$$q_1 + q_2 = l, \tag{17.176}$$

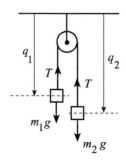

Figure 17.4 Constraining force.

that is

$$\phi(q_1, q_2) = q_1 + q_2 - l = 0, \tag{17.177}$$

which indicates that the string has fixed length. We write the Lagrangian as

$$L = \frac{1}{2} m_1 \dot{q}_1^2 + \frac{1}{2} m_2 \dot{q}_2^2 + m_1 q_1 g + m_2 q_2 g.$$

Now Lagrange's equation of motion for q_1 becomes

$$\frac{d}{dt}\left(\frac{\partial L}{\partial \dot{q}_1}\right) - \frac{\partial L}{\partial q_1} = -\lambda \frac{\partial \phi}{\partial q_1}, \tag{17.178}$$

$$\frac{d}{dt}(m_1 \dot{q}_1) - m_1 g = -\lambda, \tag{17.179}$$

$$m_1 \ddot{q}_1 - m_1 g = -\lambda \tag{17.180}$$

and for q_2, we obtain

$$\frac{d}{dt}\left(\frac{\partial L}{\partial \dot{q}_2}\right) - \frac{\partial L}{\partial q_2} = -\lambda \frac{\partial \phi}{\partial q_2}, \tag{17.181}$$

$$\frac{d}{dt}(m_2 \dot{q}_2) - m_2 g = -\lambda, \tag{17.182}$$

$$m_2 \ddot{q}_2 - m_2 g = -\lambda. \tag{17.183}$$

All together, we have the following three equations to be solved for q_1, q_2, and λ :

$$m_1 \ddot{q}_1 - m_1 g = -\lambda, \tag{17.184}$$

$$m_2 \ddot{q}_2 - m_2 g = -\lambda, \tag{17.185}$$

$$q_1 + q_2 - l = 0. \tag{17.186}$$

Eliminating λ from the first two equations and using $\ddot{q}_1 = -\ddot{q}_2$ from the third gives the accelerations:

$$\ddot{q}_1 = -\ddot{q}_2 = \left(\frac{m_1 - m_2}{m_1 + m_2}\right)g. \tag{17.187}$$

Using this in Eq. (17.184) or (17.185), we obtain λ as

$$-\lambda = -\frac{2m_1 m_2}{m_1 + m_2}g. \tag{17.188}$$

Note that in this case $-\lambda$

$$Q_1' = -\lambda\frac{\partial\phi}{\partial q_1} = -\lambda, \tag{17.189}$$

$$Q_2' = -\lambda\frac{\partial\phi}{\partial q_1} = -\lambda, \tag{17.190}$$

corresponds to the tension T in the rope, which is the constraining force.

17.10 CONSERVATION LAWS

Given the Lagrange equations:

$$\frac{d}{dt}\left(\frac{\partial L}{\partial \dot{q}_i}\right) - \frac{\partial L}{\partial q_i} = 0, \quad i = 1, 2, \ldots, n, \tag{17.191}$$

if the Lagrangian

$$L(q_1, q_2, \ldots, q_{i-1}, -, q_{i+1}, \ldots, q_n, \dot{q}_1, \dot{q}_2, \ldots, \dot{q}_{i-1}, -, \dot{q}_{i+1}, \ldots, \dot{q}_n, t), \tag{17.192}$$

does not depend on the ith generalized coordinate explicitly, then $\partial L/\partial \dot{q}_i$ corresponds to a conserved quantity:

$$\frac{\partial L}{\partial \dot{q}_i} = C, \tag{17.193}$$

where C is a constant. In Cartesian coordinates, one-dimensional motion of a free particle is governed by the Lagrangian

$$L = \frac{1}{2}m\dot{x}^2, \tag{17.194}$$

hence the conserved quantity:

$$\frac{\partial L}{\partial \dot{x}} = m\dot{x} = C, \tag{17.195}$$

corresponds to the linear momentum of the particle. In general,

$$p_i = \partial L / \partial \dot{q}_i, \quad i = 1, 2, \ldots, n, \tag{17.196}$$

corresponds to the ith **generalized momentum** of the system (Problem 17.16). Additional properties and examples of variational analysis, along with its connections to optimum control theory can be found in Bayin [4].

REFERENCES

1. Hildebrand, F.B. (1992). *Methods of Applied Mathematics*, 2nd reprint edition. New York: Dover Publications.

2. Bradbury, T.C. (1968). *Theoretical Mechanics*. New York: Wiley.

3. Goldstein, H., Poole, C., and Safko, J. (2002). *Classical Mechanics*, 3e. San Francisco, CA: Addison-Wesley.

4. Bayin, S.S. (2018). *Mathematical Methods in Science and Engineering*, 2e. Wiley.

PROBLEMS

1. Determine the dimensions of a rectangular parallelepiped, which has sides parallel to the coordinate planes with the maximum volume and that is inscribed in the ellipsoid

$$\frac{x^2}{a^2} + \frac{y^2}{b^2} + \frac{z^2}{c^2} = 1.$$

2. Find the lengths of the principal semiaxes of the ellipse

$$C_1 x^2 + 2C_2 xy + C_3 y^2 = 1, \quad C_1 C_3 > C_2^2.$$

3. Consider solids of revolution generated by all parabolas passing through $(0,0)$ and $(2,1)$ and rotated about the x-axis. Find the one with the least volume between $x = 0$ and $x = 1$.
   Hint: Take $y = x + C_0 x(2 - x)$ and determine C_0.

4. In the derivation of the alternate form of the Euler equation, verify the equivalence of the following relations:

$$\frac{\partial^2 F}{\partial y'^2} \left(\frac{dy'}{dx} \right) + \frac{\partial^2 F}{\partial y \partial y'} \left(\frac{dy}{dx} \right) + \frac{\partial^2 F}{\partial x \partial y'} - \frac{\partial F}{\partial y} = 0,$$

$$\frac{1}{y'} \left[\frac{d}{dx} \left(F - \frac{\partial F}{\partial y'} \frac{dy}{dx} \right) - \frac{\partial F}{\partial x} \right] = 0.$$

5. Write the Euler equations for the following functionals:

(i)
$$F = 2y'^2 + xyy' - y^2,$$

(ii)
$$F = y'^2 + c \sin y,$$

(iii)
$$F = x^3 y'^2 - x^2 y^2 + 2yy',$$

(iv)
$$F = x^2 y'^2 + 2yy' + y^3.$$

6. A geodesic is a curve on a surface which gives the shortest distance between two points. On a plane, geodesics are straight lines. Find the geodesics for the following surfaces:

(i) Right circular cylinder.

(ii) Right circular cone.

(iii) Sphere.

7. Determine the stationary functions of the functional

$$J = \int_0^1 (C_0 y'^2 + C_1 yy' + C_2 y') \, dx$$

for the following boundary conditions:

(i) The end conditions $y(0) = 0$ and $y(1) = 1$ are satisfied.

(ii) Only the condition $y(0) = 0$ is prescribed.

(iii) Only the condition $y(1) = 1$ is prescribed.

(iv) No end conditions are prescribed.

8. **The brachistochrone problem**: Find the shape of the curve joining two points, along which a particle initially at rest falls freely under the influence of gravity from the higher point to the lower point in the least amount of time.

9. Find the Euler equation for the problem

$$\delta \int_{x_1}^{x_2} F(x, y, y', y'') \, dx = 0$$

and discuss the associated boundary conditions.

10. Derive the Euler equation for the problem

$$\delta \int \int_R F(x, y, u, u_x, u_y) \, dx \, dy = 0,$$

subject to the condition that $u(x, y)$ is prescribed on the closed boundary of R.

11. Derive the Euler equation for the problem

$$\delta \int_{x_1}^{x_2} \int_{y_1}^{y_2} F(x, y, u, u_x, u_y, u_{xx}, u_{xy}, u_{yy}) \, dx \, dy = 0.$$

What are the associated natural boundary conditions in this case.

12. Write Hamilton's principle for a particle of mass m moving vertically under the action of uniform gravity and a resistive force proportional to the displacement from an equilibrium position. Write Lagrange's equation of motion for this particle.

13. Write Hamilton's principle for a particle of mass m moving vertically under the action of uniform gravity and a drag force proportional to its velocity.

14. Write Lagrange's equations of motion for a triple pendulum consisting of equal masses, m, connected with inextensible strings of equal length l. Use the angles θ_1, θ_2, and θ_3 that each pendulum makes with the vertical as the generalized coordinates. For small displacements from equilibrium show that the Lagrangian reduces to

$$L = \frac{ml^2}{2}(3\dot{\theta}_1^2 + 2\dot{\theta}_2^2 + \dot{\theta}_3^2 + 4\dot{\theta}_1\dot{\theta}_2 + 2\dot{\theta}_2\dot{\theta}_3 + 2\dot{\theta}_1\dot{\theta}_3)$$
$$- \frac{mgl}{2}(3\theta_1^2 + 2\theta_2^2 + \theta_3^2).$$

15. Small deflections of a rotating shaft of length l, subjected to an axial end load of P and transverse load of intensity $p(x)$ is described by the differential equation

$$\frac{d^2}{dx^2}\left(EI\frac{d^2y}{dx^2}\right) + P\frac{d^2y}{dx^2} - \rho\omega^2 y - p(x) = 0,$$

where EI is the bending stiffness of the shaft, ρ is the density, and ω is the angular frequency of rotation. Show that this differential equation can be obtained from the variational principle

$$\delta \int_0^l \left[\frac{1}{2}EIy''^2 - \frac{1}{2}Py'^2 - \frac{1}{2}\rho\omega^2 y^2 - py\right] dx = 0.$$

What boundary conditions did you impose? For other examples from the theory of elasticity, see Hildebrand [1].

16. A pendulum that is not restricted to oscillate on a plane is called a **spherical pendulum**. Using spherical coordinates, obtain the equations of motion corresponding to r, θ, and ϕ:

$$mg\cos\theta - T = -m(s\dot{\theta}^2 + s\sin^2\theta\dot{\phi}^2),$$

$$-mg\sin\theta = m(s\ddot{\theta} - s\sin\theta\cos\theta\dot{\phi}^2),$$

$$0 = \frac{1}{s\sin\theta}\frac{d}{dt}(ms^2\sin^2\theta\dot{\phi}),$$

where s is the length of the pendulum and T is the tension in the rope. Since there are basically two independent coordinates, θ and ϕ, show that the equations of motion can be written as

$$\ddot{\theta} - \dot{\phi}^2\sin\theta\cos\theta + \frac{g}{s}\sin\theta = 0 \quad \text{and} \quad ms^2\sin^2\theta\dot{\phi} = l,$$

where l is a constant. Show that the constant l is actually the x_3 component of the angular momentum:

$$l_3 = m(x_1\dot{x}_2 - \dot{x}_1 x_2).$$

17. When we introduced the natural boundary conditions, we used Eq. (17.22):

$$\frac{dJ}{d\varepsilon}\bigg|_{\varepsilon=0} = \frac{\partial F}{\partial y'}\eta(x)\bigg|_{x_1}^{x_2} + \int_{x_1}^{x_2}\left[\frac{\partial F}{\partial y} - \frac{d}{dx}\left(\frac{\partial F}{\partial y'}\right)\right]\eta(x)\,dx = 0,$$

where for stationary paths, we set

$$\frac{\partial F}{\partial y'}\eta(x_2) - \frac{\partial F}{\partial y'}\eta(x_1) = 0$$

and

$$\int_{x_1}^{x_2}\left[\frac{\partial F}{\partial y} - \frac{d}{dx}\left(\frac{\partial F}{\partial y'}\right)\right]\eta(x)\,dx = 0.$$

Explain.

18. Write the components of the generalized momentum for the following problems:

(i) plane pendulum,

(ii) spherical pendulum,

(iii) motion of earth around the sun

and discuss whether they are conserved or not.

CHAPTER 18

PROBABILITY THEORY AND DISTRIBUTIONS

Probability theory is the science of random events. It has long been known that there are definite regularities among large numbers of random events. In ordinary scientific parlance, certain initial conditions, which can be rather complicated, lead to certain events. For example, if we know the initial position and velocity of a planet, we can be certain of its position and velocity at a later time. In fact, one of the early successes of Newton's theory was its prediction of the solar and lunar eclipses for decades and centuries ahead of time. An event that is definitely going to happen when certain conditions are met is called **certain**. If there is no set of conditions that could make an event happen, then that event is called **impossible**. If under certain conditions an event may or may not happen, we call it **random**. From here, it is clear that the certainty, impossibility, and randomness of an event depends on the set of existing conditions. Randomness could result from a number of reasons. Some of these are the presence of large numbers of interacting parts, insufficient knowledge about the initial conditions, properties of the system, and also the environment. Probability is also a word commonly used in everyday language.

Using the available information, we often base our decisions on how probable or improbable we think certain chains of events are going to unravel. The severity of the consequences of our decisions can vary greatly. For example, deciding not to take an umbrella, thinking that it will probably not rain, may just result in getting wet. However, in other situations, the consequences could be much more dire. They could even jeopardize many lives including ours, corporate welfare, and in some cases, even our national security. Probability not only depends on the information available but also has strong correlations with the intended purpose and expectations of the observer. In this regard, many different approaches to probability have been developed, which can be grouped under three categories:

(1) Definitions of mathematical probability based on the knowledge and the priorities of the observer.

(2) Definitions that reduce to a reproducible concept of probability for the same initial conditions for all observers. This is also called the **classical definition** of probability. The most primitive concept of this definition is **equal probability** for each possible event or outcome.

(3) Statistical definitions, which follow from the frequency of the occurrence of an event among a large number of trials with the same initial conditions.

The concept of probability in quantum mechanics does not fit into any one of the aforementioned categories. We discuss quantum probability along with (1) in Chapter 19 within the context of information theory. In this chapter, we concentrate on (2) and (3), which are related.

18.1   INTRODUCTION TO PROBABILITY THEORY

Origin of the probability theory can be traced as far back as to the communications between Pascal (1623–1662) and Fermat (1601–1665). An extensive discussion of the history of probability can be found in Todhunter [1]. Modern probability theory is basically founded by Kolmogorov (1903–1987).

18.1.1   Fundamental Concepts

We define the **sample space** of an experiment as the set S of all possible outcomes $A, B, \ldots$:

$$S = \{A, B, \ldots\}. \tag{18.1}$$

If we roll a die, the sample space is

$$S = \{1, 2, 3, 4, 5, 6\}, \tag{18.2}$$

where each term corresponds to the number of dots showing on each side. We obviously exclude the events where the die breaks or stands on edge. These events could be rendered as practically impossible by adjusting the conditions. For a single toss of a coin, the sample space is going to be composed of two elements.

$$S = \{\text{head, tail}\}. \tag{18.3}$$

An **event** E is defined as a set of points chosen from the sample space. For example,

$$E = \{1, 3, 5\} \tag{18.4}$$

corresponds to the case where the die comes up with an odd number, that is a 1, 3, or 5. An event may be a single element, such as

$$E = \{4\}, \tag{18.5}$$

where the die comes up 4. Events that are single elements are called **simple** or **elementary events**. Events that are not single elements are called **compound elements**.

18.1.2   Basic Axioms of Probability

We say that S is a **probability space** if for every event in S we could find a number $P(E)$ with

(1) $P(E) \geqslant 0$,
(2) $P(S) = 1$,
(3) If E_1 and E_2 are two **mutually exclusive** events in S, that is, their intersection is a null set, $E_1 \cap E_2 = \emptyset$, then the probability of their union is the sum of their probabilities:

$$P(E_1 \cup E_2) = P(E_1) + P(E_2). \tag{18.6}$$

Any function $P(E)$ satisfying the properties (1)–(3) defined on the events of S is called the **probability** of E. These axioms are due to Kolmogorov [2]. They are sufficient for sample spaces with finite number of elements. For sample spaces with infinite number of events, they have to be modified.

18.1.3   Basic Theorems of Probability

Based on these axioms, we now list the basic theorems of probability theory [3].

Theorem 18.1. If E_1 is a subset of E_2, that is, $E_1 \subset E_2$, then

$$P(E_1) \leqslant P(E_2). \tag{18.7}$$

Theorem 18.2. For any event E in S,

$$P(E) \leqslant 1. \tag{18.8}$$

Theorem 18.3. For the complementary set E^c of E, we have

$$P(E^c) = 1 - P(E),\qquad(18.9)$$

where $E^c + E = S$, also shown as $E^c \cup E = S$.

Theorem 18.4. Probability of no event happening is zero:

$$P(\emptyset) = 0,\qquad(18.10)$$

where $\emptyset$ denotes the null set.

Theorem 18.5. For any two events in S, we can write

$$P(E_1 \cup E_2) = P(E_1) + P(E_2) - P(E_1 \cap E_2).\qquad(18.11)$$

Theorem 18.6. If $E_1, E_2, \ldots, E_m$ are **mutually exclusive** events with $E_i \cap E_j = \emptyset$, $1 \leqslant i$, $j \leqslant m$ and $i \neq j$, then we have what is also called the **addition theorem**:

$$\boxed{P(E_1 \cup E_2 \cup \cdots \cup E_m) = \sum_{i=1}^{m} P(E_i).}\qquad(18.12)$$

So far, we said nothing about how to find $P(E)$ explicitly. When S is finite with elementary events $E_1, E_2, \ldots, E_m$, any choice of positive numbers $P_1, P_2, \ldots, P_m$ satisfying

$$\sum_{i=1}^{m} P_i = 1\qquad(18.13)$$

will satisfy the three axioms. In some situations, symmetry of the problem allows us to assign equal probability to each elementary event. For example, when a die is manufactured properly, there is no reason to favor one face over the other. Hence, we assign equal probability to each face as

$$P_1 = P_2 = \cdots = P_6 = \frac{1}{6}.\qquad(18.14)$$

To clarify some of these points, let us choose the following events for a die rolling experiment:

$$E_1 = \{1, 3, 5\},\quad E_2 = \{1, 5, 6\},\quad \text{and}\quad E_3 = \{3\},\qquad(18.15)$$

which give

$$P(E_1) = \frac{1}{6} + \frac{1}{6} + \frac{1}{6} = \frac{1}{2},$$

$$P(E_2) = \frac{1}{6} + \frac{1}{6} + \frac{1}{6} = \frac{1}{2},\qquad(18.16)$$

$$P(E_3) = \frac{1}{6}.$$

Note that in agreement with Theorem 18.1, we have

$$P(E_3) < P(E_1). \tag{18.17}$$

We can also use

$$E_1^c = \{2, 4, 6\}, \quad E_3^c = \{1, 2, 4, 5, 6\}, \quad E_1 \cup E_2 = \{1, 3, 5, 6\}, \tag{18.18}$$

and

$$E_1 \cap E_2 = \{1, 5\} \tag{18.19}$$

to write

$$P(E_1^c) = 1 - P(E_1) = 1 - \frac{1}{2} = \frac{1}{2}, \tag{18.20}$$

$$P(E_3^c) = 1 - P(E_3) = 1 - \frac{1}{6} = \frac{5}{6}, \tag{18.21}$$

which are in agreement with Theorem 18.3. Also in conformity with Theorem 18.5, we have

$$P(E_1 \cup E_2) = P(E_1) + P(E_2) - P(E_1 \cap E_2) \tag{18.22}$$

$$= \frac{1}{2} + \frac{1}{2} - \frac{2}{6} = \frac{2}{3}. \tag{18.23}$$

For any finite sample space with N elements:

$$S = \{E_1, E_2, \ldots, E_N\}, \tag{18.24}$$

for an event

$$E = E_1 \cup E_2 \cup E_3 \cup \cdots \cup E_m, \quad m \leqslant N, \tag{18.25}$$

where $E_1, E_2, \ldots, E_m$ are elementary events with $E_i \cap E_j = \emptyset$, $1 \leqslant i, j \leqslant m$, $i \neq j$, we can write

$$P(E) = \sum_{i=1}^{m} P(E_i). \tag{18.26}$$

If the problem is symmetric, so that we can assign equal probability, $1/N$, to each elementary event, then the probability of E becomes

$$P(E) = \frac{m}{N}, \tag{18.27}$$

where m is the number of elements of S in E. In such cases, finding $P(E)$ reduces to simply counting the number of elements of S in E. We come back to these points after we introduce permutations and combinations.

18.1.4 Statistical Definition of Probability

In the previous section (18.1.3), using the symmetries of the system, we have assigned, a priori, equal probability to each elementary event. In the case of a *perfect* die, for each face, this gives a probability of 1/6 and for a coin toss, it assigns equal probability of 1/2 to the two possibilities: heads or tails. An experimental justification of these probabilities can be obtained by repeating die rolls or coin tosses sufficiently many times and by recording the frequency of occurrence of each elementary event. The catch here is, How identical can we make the conditions and how many times is sufficient? Obviously, in each roll or toss, the conditions are slightly different. The twist of our wrist, the positioning of our fingers, etc., all contribute to a different velocity, position, and orientation of the die or the coin at the instant it leaves our hand. We also have to consider variation of the conditions where the die or the coin lands. However, unless we intentionally control our wrist movements and the initial positioning of the die or the coin to change the odds, for a large number of rolls or tosses we expect these variations, which are not necessarily small, to be random, and to cancel each other. Hence, we can define the **statistical probability** of an event E as the frequency of occurrences:

$$P(E) = \lim_{n \to \infty} \frac{\text{number of occurrences of } E}{n}. \tag{18.28}$$

Obviously, for this definition to work, the limit must exist. It turns out that for situations where it is possible to define a classical probability, fluctuation of the frequency occurs about the probability of the event, and the magnitude of fluctuations die out as the number of tries increases. There are plenty of data to verify this fact. In the case of a coin toss experiment, the probability for heads or tails quickly converges to 1/2. In the case of a *loaded die*, frequencies may yield the probabilities:

$$P(1) = \frac{1}{2}, \quad P(2) = \frac{1}{8}, \quad P(3) = \frac{1}{4}, \quad P(4) = \frac{1}{16}, \quad P(5) = \frac{1}{32}, \quad P(6) = \frac{1}{32},$$

which also satisfy the axioms of probability. In fact, one of the ways to find out that a die is loaded or manufactured improperly is to determine its probability distribution.

When it comes to scientific and technological applications, classical definition of probability usually runs into serious difficulties. First of all, it is generally difficult to isolate the equiprobable elements of the sample space. In some cases, the sample space could be infinite with infinite number of possible outcomes or the possible outcomes could be distributed continuously, thus making it difficult, if not always impossible, to enumerate. In order to circumvent some of these difficulties, the **statistical probability** concept comes in very handy.

18.1.5   Conditional Probability and Multiplication Theorem

Let us now consider an experiment where two dice are rolled, the possibilities are given as follows:

(1,1)	(1,2)	(1,3)	(1,4)	(1,5)	(1,6)
(2,1)	(2,2)	(2,3)	(2,4)	(2,5)	(2,6)
(3,1)	(3,2)	(3,3)	(3,4)	(3,5)	(3,6)
(4,1)	(4,2)	(4,3)	(4,4)	(4,5)	(4,6)
(5,1)	(5,2)	(5,3)	(5,4)	(5,5)	(5,6)
(6,1)	(6,2)	(6,3)	(6,4)	(6,5)	(6,6)

The first number in parentheses gives the outcome of the first die, and the second number is the outcome of the second die. If we are interested in the outcomes $\{A\}$, where the sum is 6, obviously there are 5 desired results out of a total of 36 possibilities. Thus, the conditional probability is

$$P(A) = \frac{5}{36}. \tag{18.29}$$

We now look for the probability that the sum 6 comes (event A) if it is known that the sum is an even number (event B). In this case, the sample space contains only 18 elements. Since we look for the event A after the event B has been realized, the probability is given as

$$P(A/B) = \frac{5}{18}. \tag{18.30}$$

$P(A/B)$, which is also shown as $P(A \mid B)$, is called the **conditional probability** of A. It is the probability of A occurring after B has occurred.

Let us now generalize this to a case where $\{C_1, C_2, \ldots, C_n\}$ is the set of uniquely possible, mutually exclusive, and equiprobable events. Among this set, let

$m \leqslant n$ denote the number of events acceptable by A,

$k \leqslant n$ denote the number of events acceptable by B,

r denote the number of events acceptable by both A and B.

We show the events acceptable by both A and B as AB or $A \cap B$. Obviously, $r \leqslant k$ and $r \leqslant m$. This means that the probability of A happening after B has happened is

$$P(A/B) = \frac{r}{k} = \frac{r/n}{k/n} = \frac{P(AB)}{P(B)}. \tag{18.31}$$

Similarly,

$$P(B/A) = \frac{r}{m} = \frac{r/n}{m/n} = \frac{P(BA)}{P(A)} = \frac{P(AB)}{P(A)}. \tag{18.32}$$

Note that if $P(B)$ is an impossible event, that is, $P(B) = 0$, then Eq. (18.31) becomes meaningless. Similarly, Eq. (18.32) is meaningless when $P(A) = 0$. Equations (18.31) and (18.32), which are equivalent, represent the **multiplication theorem**, and we write them as

$$P(AB) = P(A)P(B/A) = P(B)P(A/B). \qquad (18.33)$$

For **independent events**, that is, the occurrence of A, or B, is independent of B, or A, occurring, then the multiplication theorem takes on a simple form:

$$P(AB) = P(A)P(B). \qquad (18.34)$$

For example, in an experiment we first roll a die, event A, and then toss a coin, event B. Clearly, the two events are independent. The probability of getting the number 5 in the die roll and a head in the coin toss is

$$P(AB) = \frac{1}{6} \cdot \frac{1}{2} = \frac{1}{12}. \qquad (18.35)$$

18.1.6   Bayes' Theorem

Let us now consider a tetrahedron with its faces colored as the first face is red, A, the second face is green, B, the third face is blue, C, and finally the fourth face is in all three colors, ABC. In a roll of the tetrahedron, the color red has the probability

$$P(A) = \frac{1}{4} + \frac{1}{4} = \frac{1}{2}. \qquad (18.36)$$

This follows from the fact that the color red shows in two of the four faces. Similarly, we can write the following probabilities:

$$P(B) = P(C) = \frac{1}{2},$$

$$P(A/B) = P(B/C) = \frac{1}{2},$$

$$P(C/A) = P(B/A) = \frac{1}{2},$$

$$P(C/B) = P(A/C) = \frac{1}{2}. \qquad (18.37)$$

This means that the events A, B, and C are **pairwise independent**. However, if it is known that B and C has occurred, then we can be certain that A has also occurred, that is,

$$P(A/BC) = 1, \qquad (18.38)$$

which means that events A, B, and C are **collectively dependent**.

Let us now consider an event B that can occur together with one and only one of the n **mutually exclusive** events:

$$\{A_1, A_2, \ldots, A_n\}. \tag{18.39}$$

Since BA_i and BA_j with $i \neq j$ are also **mutually exclusive** events, we can use the **addition theorem** [Eq. (18.12)] of probabilities to write

$$P(B) = \sum_{i=1}^{n} P(BA_i). \tag{18.40}$$

Using the **multiplication theorem** [Eq. (18.33)], this becomes

$$P(B) = \sum_{i=1}^{n} P(A_i)P(B/A_i), \tag{18.41}$$

which is also called the **total probability**.

We now drive an important formula called **Bayes' formula**. It is required to find the probability of event A_i provided that event B has already occurred. Using the multiplication theorem [Eq. (18.33)], we can write

$$P(A_iB) = P(B)P(A_i/B) = P(A_i)P(B/A_i), \tag{18.42}$$

which gives

$$P(A_i/B) = \frac{P(A_i)P(B/A_i)}{P(B)}. \tag{18.43}$$

Using the formula of total probability [Eq. (18.41)], this gives Bayes' formula:

$$P(A_i/B) = \frac{P(A_i)P(B/A_i)}{\sum_{j=1}^{n} P(A_j)P(B/A_j)}, \tag{18.44}$$

which gives the probability of event A_i provided that B has occurred first. Bayes' formula has interesting applications in decision theory and is also used extensively for data analysis in physical sciences [4, 5].

Example 18.1. *Colored balls in six bags:* Three bags have composition A_1 with 2 white and 1 black ball each, one bag has composition A_2 with 9 black balls, and the remaining 2 bags have composition A_3 with 3 white balls and 1 black ball each. We select a bag randomly and draw one ball from it. What is the probability that this ball is white? Call this event B. Since the ball could come from any one of the six bags with compositions

A_1, A_2, and A_3, we can write

$$B = A_1B + A_2B + A_3B. \tag{18.45}$$

Using the formula of total probability [Eq. (18.41)], we write

$$P(B) = P(A_1)P(B/A_1) + P(A_2)P(B/A_2) + P(A_3)P(B/A_3), \tag{18.46}$$

where

$$P(A_1) = \frac{3}{6}, \quad P(A_2) = \frac{1}{6}, \quad P(A_3) = \frac{2}{6}, \tag{18.47}$$

$$P(B/A_1) = \frac{2}{3}, \quad P(B/A_2) = 0, \quad P(B/A_3) = \frac{3}{4}, \tag{18.48}$$

to obtain

$$P(B) = \frac{3}{6} \cdot \frac{2}{3} + \frac{1}{6} \cdot 0 + \frac{2}{6} \cdot \frac{3}{4} = \frac{7}{12}. \tag{18.49}$$

Example 18.2. *Given six identical bags:* The bags have the following contents:

3 bags with contents A_1 composed of 2 white and 3 black balls each,
2 bags with contents A_2 composed of 1 white and 4 black balls each,
1 bag with contents A_3 composed of 4 white and 1 black balls each.

We pick a ball from a randomly selected bag, which turns out to be white. Call this event B. We now want to find, after the ball is picked, the probability that the ball was taken from the bag of the third composition. We have the following probabilities:

$$P(A_1) = \frac{3}{6}, \quad P(A_2) = \frac{2}{6}, \quad P(A_3) = \frac{1}{6}, \tag{18.50}$$

$$P(B/A_1) = \frac{2}{5}, \quad P(B/A_2) = \frac{1}{5}, \quad P(B/A_3) = \frac{4}{5}. \tag{18.51}$$

Using the Bayes' formula [Eq. (18.44)], we obtain

$$P(A_3/B) = \frac{P(A_3)P(B/A_3)}{P(A_1)P(B/A_1) + P(A_2)P(B/A_2) + P(A_3)P(B/A_3)} \tag{18.52}$$

$$= \frac{(1/6)(4/5)}{(3/6)(2/5) + (2/6)(1/5) + (1/6)(4/5)} = \frac{1}{3}. \tag{18.53}$$

18.1.7 Geometric Probability and Buffon's Needle Problem

We have mentioned that the classical definition of probability is insufficient when we have infinite sample spaces. It also fails when the possible outcomes of an experiment are distributed continuously. Consider the following general problem: On a plane, we have a region R and in it another region r. We want to define the probability of a *thrown* point landing in the region r. Another way to pose this problem is: What is the probability of a point, coordinates of which are picked randomly, falling into the region r. Guided by our intuition, we can define this probability as

$$\boxed{P = \frac{\text{area of } r}{\text{area of } R}}, \tag{18.54}$$

which satisfies the basic three axioms. We can generalize this formula as

$$\boxed{P = \frac{\text{measure of } r}{\text{measure of } R}}, \tag{18.55}$$

where the measure stands for length, area, volume, etc.

Example 18.3. *Buffon's needle problem:* We partition a plane by two parallel lines separated by a distance of $2a$. A needle of length $2l$ is thrown randomly onto this plane ($l < a$). We want to find the probability of the needle intersecting one of the lines. We show the distance from the center of the needle to the closest line with x and the angle of the needle with θ (Figure 18.1). Configuration of the needle is completely specified by x and θ. For the needle to cross one of the lines, it is necessary and sufficient that

$$x \leqslant l \sin\theta \tag{18.56}$$

be satisfied. Now the probability is the ratio of the region under the curve $x = l\sin\theta$ to the area of the rectangle $\int_0^\pi l\sin\theta d\theta$ in Figure 18.2:

$$P = \frac{\int_0^\pi l\sin\theta\, d\theta}{a\pi} = \frac{2l}{a\pi}. \tag{18.57}$$

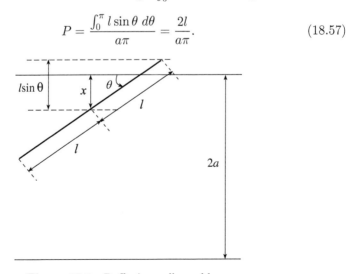

Figure 18.1 Buffon's needle problem.

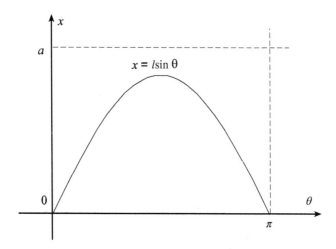

Figure 18.2 Area in the Buffon's needle problem.

Historically, the Buffon's needle problem was the starting point in solving certain problems in the theory of gunfire with varying shell sizes. It has also been used for purposes of estimating the approximate value of π. For more on the geometric definition of probability and its limitations, we refer to Gnedenko [6].

18.2 PERMUTATIONS AND COMBINATIONS

We mentioned that in symmetric situations, where the sample space is finite and each event is equally probable, assigning probabilities reduces to a simple counting process. To introduce the basic ideas, we use a bag containing a number of balls numbered as $1, 2, \ldots, N$. As we shall see, the bag and the balls could actually stand for many things in scientific and technological applications.

18.2.1 The Case of Distinguishable Balls with Replacement

We draw a ball from the bag, record the number, and then throw the ball back into the bag. Repeating this process k times, we form a k-tuple of numbers $(x_1, x_2, \ldots, x_i, \ldots, x_k)$, where x_i denotes the number of the ith draw. Let S be the totality of such k-tuples. In the first draw, we could get any one of the N balls; hence, there are N possible and equiprobable outcomes. In the second draw, since the ball is thrown back into the bag or replaced with an identical ball, we again have N possible outcomes. Altogether, this gives N^2 possible outcomes for the first two draws. For k draws, naturally the sample space contains N^k k-tuples as possible outcomes.

18.2.2 The Case of Distinguishable Balls Without Replacement

We now repeat the same process, but this time do not replace the balls. In the first draw, there are N independent possibilities. For the second draw, there are $N - 1$ balls left in the bag, hence only $N - 1$ possible outcomes. For the rth draw, $r \leqslant k$, there will be $N - r + 1$ balls left, thus giving only $N - r + 1$ possibilities. For k draws, the sample space will contain $N^{(k)}$ elements:

$$N^{(k)} = N(N-1)(N-2)\cdots(N-k+1), \tag{18.58}$$

which can also be written as

$$N^{(k)} = \frac{N(N-1)\cdots(N-k+1)[(N-k)(N-k-1)\cdots 2 \cdot 1]}{[(N-k)(N-k-1)\cdots 2 \cdot 1]} = \frac{N!}{(N-k)!}. \tag{18.59}$$

Permutation is a selection of objects with a definite order. N distinguishable objects distributed into k numbered spaces has $N^{(k)}$ distinct possibilities, which is also written as $_N P_k$. When $k = N$, we have $N!$ possibilities, thus we can write

$$N^{(N)} = \frac{N!}{0!} = N!. \tag{18.60}$$

This is also taken as the definition of $0!$ as $0! = 1$.

Example 18.4. *A coin is tossed 6 times:* If we identify heads with 1 and tails with 2, this is identical to the bag problem with replacement, where $N = 2$ and $k = 6$. Thus, there are $2^6 = 64$ possibilities for the 6-tuple numbers. The possibility of any one of them coming, say $E = (1, 2, 1, 1, 1, 2)$, is $P(E) = 1/64$.

Example 18.5. *A die is rolled five times:* This is identical to the bag problem with replacement, where $N = 6$ and $k = 5$. We now have $6^5 = 7776$ possible outcomes.

Example 18.6. *Five cards selected from a deck of 52 playing cards:* If we assume that the order in which the cards are selected is irrelevant, then this is equivalent to the bag problem without replacement. Now, $N = 52$ and $k = 5$, which gives

$$52^{(5)} = \frac{52!}{(52-5)!} = 311{,}875{,}200 \tag{18.61}$$

possible outcomes.

Example 18.7. *Friends to visit:* Let us say that we arrived at our home town and have 5 friends to visit. There are $5! = 120$ different orders that we can do this. If we have time for only three visits, then there are $5^{(3)} = 60$ different ways.

Example 18.8. *Number of different numbers:* How many different numbers can we make from the digits $1, 2, 3, 4$? If we use two digits and if repeats are permitted, there are $4^2 = 16$ possibilities. If we do not allow repeats, then there are only $4^{(2)} = 12$ possibilities.

18.2.3 The Case of Indistinguishable Balls

Let us now consider N balls, where not all of them are different. Let there be n_1 balls of one kind, n_2 balls of the second kind, $\ldots, n_k$ balls of the kth kind such that $n_1 + n_2 + \cdots + n_k = N$. We also assume that balls of the same kind are indistinguishable. A natural question to ask is, In how many distinct ways, $N^{(n_1 n_2 \cdots n_k)}$, also written as $_N P_{n_1 n_2 \cdots n_k}$, can we arrange these balls? When all the balls are distinct, we have $N!$ possibilities but n_1 balls of the first kind are indistinguishable. Thus, $n_1!$ of these possibilities, that is, the permutations of the n_1 balls among themselves, lead to identical configurations. Hence, for distinct arrangements, we have to divide $N!$ by $n_1!$. Arguing the same way for the other kinds, we obtain

$$_N P_{n_1 n_2 \cdots n_k} = N^{(n_1 n_2 \cdots n_k)} = \frac{N!}{n_1! n_2! \cdots n_k!}. \tag{18.62}$$

Permutation is an outcome with a particular ordering. For example, 1234 is a different permutation of 4231. In many situations, we are interested in selection of objects with no regard to their order. We call such arrangements **combinations** and show them as $_n C_r$, which means the number of ways r objects can be selected out of n objects with no attention paid to their order. Since the order of the remaining $n - r$ objects is also irrelevant, among the $_n P_r = n!/r!$ permutations, there are $(n - r)!$ that give the same combination, thus

$$_n C_r = \frac{_n P_r}{(n - r)!} = \frac{n!}{(n - r)! r!}. \tag{18.63}$$

Combinations are often shown as

$$_n C_r = \binom{n}{r}. \tag{18.64}$$

It is easy to show that

$$\binom{n}{r} = \binom{n}{n - r}. \tag{18.65}$$

Example 18.9. *Number of poker hands:* In a poker hand, there are 5 cards from an ordinary deck of 52 cards. These 52 cards can be arranged in 52! different ways. Since the order of the 5 cards in a player's hand

and the order of the remaining 47 cards do not matter, the number of possible poker hands is

$$\binom{52}{5} = \frac{52!}{(5!)(47!)} = 2\,598\,960. \tag{18.66}$$

18.2.4 Binomial and Multinomial Coefficients

Since $_nC_r$ also appear in the **binomial expansion**:

$$(x+y)^n = \sum_{j=0}^{n} \binom{n}{j} x^j y^{n-j}, \tag{18.67}$$

they are also called the **binomial coefficients**. Similarly, the multinomial expansion is given as

$$(x_1 + x_2 + \cdots + x_r)^n = \sum_{k_1,k_2,\ldots,k_r \geqslant 0} \binom{n}{k_1,k_2,\ldots,k_{r-1}} x_1^{k_1} \cdots x_r^{k_r}, \tag{18.68}$$

where the sum is over all nonnegative integer r-tuples $(k_1, k_2, \ldots, k_r)$ with their sum $k_1 + k_2 + \cdots + k_r = n$. The coefficients defined as

$$\binom{n}{k_1,k_2,\ldots,k_{r-1}} = \frac{n!}{k_1! k_2! \cdots k_{r-1}!} \tag{18.69}$$

are called the **multinomial coefficients**. Some useful properties of the binomial coefficients are as follows:

$$\sum_{j=0}^{n} \binom{n}{j} = 2^n, \tag{18.70}$$

$$\sum_{j=0}^{n} (-1)^j \binom{n}{j} = 0, \tag{18.71}$$

$$\binom{n+1}{r} = \binom{n}{r} + \binom{n}{r-1}, \tag{18.72}$$

$$\sum_{j=0}^{N} \binom{N}{j} \binom{M}{n-j} = \binom{N+M}{n}, \tag{18.73}$$

$$\sum_{k_1,k_2,\ldots,k_r \geqslant 0} \binom{n}{k_1,k_2,\ldots,k_{r-1}} = r^n. \tag{18.74}$$

18.3 APPLICATIONS TO STATISTICAL MECHANICS

An important application of the probability concepts discussed so far comes from statistical mechanics. For most practical applications, it is sufficient to consider gases or solids as collection of independent particles, which move freely except for the brief moments during collisions. In a solid, we can consider atoms vibrating freely essentially independent of each other. According to quantum mechanics, such quasi-independent particles can only have certain discrete energies given by the energy eigenvalues ε_1, $\varepsilon_2, \dots$. Specific values of these energies depend on the details of the system. At a given moment and at a certain temperature, the state of a system can be described by giving the

number of particles with energy ε_1,
number of particles with energy ε_2,

$$\vdots$$

Our basic goal in statistical mechanics is to find how these particles are distributed among these energy levels subject to the conditions

$$\sum_i n_i = N, \tag{18.75}$$

$$\sum_i n_i \varepsilon_i = U, \tag{18.76}$$

where N is the total number of particles and U is the internal energy of the system. To clarify some of these points, consider a simple model with 3 atoms, a, b, and c [7]. Let the available energies be 0, ε, 2ε, and 3ε. Let us also assume that the internal energy of the system is 3ε. Among the three atoms, this energy could be distributed as

a	ε	3ε	0	0	2ε	2ε	ε	ε	0	0
b	ε	0	3ε	0	ε	0	2ε	0	2ε	ε
c	ε	0	0	3ε	0	ε	0	2ε	ε	2ε

It is seen that altogether, there are 10 possible configurations or **complexions**, also called the **microstates**, in which the 3ε amount of energy can be distributed among the three atoms. Since atoms interact, no matter how briefly, through collisions, the system fluctuates between these possible complexions.

We now introduce the fundamental assumption of statistical mechanics by postulating, a priori, that all possible complexions are equally probable. If we look at the complexions, a little more carefully, we see that they can be grouped into three states (S_1, S_2, S_3) with respect to the **occupancy numbers**

$(n_0, n_\varepsilon, n_{2\varepsilon}, n_{3\varepsilon})$ as

	S_1	S_2	S_3
n_0	0	2	1
n_ε	3	0	1
$n_{2\varepsilon}$	0	0	1
$n_{3\varepsilon}$	0	1	0

.

Note that only 1 complexion corresponds to state S_1, 3 complexions to S_2, and 6 complexions to state S_3. Since all the complexions are equiprobable, probabilities of finding the system in the states S_1, S_2, and S_3 are $\frac{1}{10}$, $\frac{3}{10}$, and $\frac{6}{10}$, respectively. This means that if we make sufficiently many observations, 6 out of 10 times the system will be seen in state S_3, 3 out of 10 times it will be in state S_2, and only 1 out of 10 times it will be in state S_1. In terms of a time parameter, given sufficient time, the system can be seen in all three states. However, this simple 3-atom model will spend most of its time in state S_3, which can be considered as its **equilibrium state**.

18.3.1 Boltzmann Distribution for Solids

We now extend this simple model to a solid with energy eigenvalues $\varepsilon_1, \varepsilon_2, \ldots$. A particular state can be specified by giving the occupancy numbers $n_1, n_2, \ldots$ of the energy levels. We now have a problem of N distinguishable atoms distributed into boxes labeled $\varepsilon_1, \varepsilon_2, \ldots$, so that there are n_1 atoms in box 1, n_2 atoms in box 2, etc. Atoms are considered as **distinguishable** in the sense that we can identify them in terms of their locations in the lattice. Since how atoms are distributed in each box or the energy level is irrelevant, the number of complexions corresponding to a particular state is given as

$$\boxed{W = \frac{N!}{n_1! n_2! \cdots}.} \tag{18.77}$$

The most probable state is naturally the one with the maximum number of complexions subject to the following constraints:

$$N = \sum_i n_i, \tag{18.78}$$

$$U = \sum_i n_i \varepsilon_i. \tag{18.79}$$

Mathematically, this problem is solved by finding the occupancy numbers that make W a maximum. For reasons to be clear shortly, we maximize $\ln W$

and write

$$\delta \ln W = \frac{\partial(\ln W)}{\partial n_1}\delta n_1 + \frac{\partial(\ln W)}{\partial n_2}\delta n_2 + \cdots \tag{18.80}$$

$$= \sum_i \frac{\partial(\ln W)}{\partial n_i}\delta n_i. \tag{18.81}$$

The maximum number of complexions satisfy the condition

$$\delta \ln W = 0, \tag{18.82}$$

subject to the constraints

$$\sum_i \delta n_i = 0, \tag{18.83}$$

$$\sum_i \delta n_i \varepsilon_i = 0. \tag{18.84}$$

We now introduce two Lagrange undetermined multipliers as α and β. Multiplying Eq. (18.83) by α and Eq. (18.84) by β and then adding to Eq. (18.81) gives

$$\sum_i \left(\frac{\partial(\ln W)}{\partial n_i} + \alpha + \beta \varepsilon_i \right) \delta n_i = 0. \tag{18.85}$$

With the introduction of the Lagrange undetermined multipliers, we can treat all δn_i in Eq. (18.85) as independent and set their coefficients to zero:

$$\frac{\partial(\ln W)}{\partial n_i} + \alpha + \beta \varepsilon_i = 0. \tag{18.86}$$

We now turn to Eq. (18.77) and write $\ln W$ as

$$\ln W = \ln \frac{N!}{n_1!n_2!\cdots} = \ln N! - \sum_i \ln(n_i!). \tag{18.87}$$

Using the **Stirling approximation** for the factorial of a large number:

$$\ln n_i! \simeq n_i \ln n_i - n_i, \tag{18.88}$$

this can also be written as

$$\ln W = \ln N! - \sum_i [n_i \ln(n_i) - n_i]. \tag{18.89}$$

After differentiation, we obtain

$$\frac{\partial(\ln W)}{\partial n_i} = -\frac{\partial}{\partial n_i}(n_i \ln n_i - n_i) = -\ln n_i. \tag{18.90}$$

Substituting this into Eq. (18.86), we write

$$-\ln n_i + \alpha + \beta \varepsilon_i = 0 \tag{18.91}$$

to obtain

$$\boxed{n_i = A e^{\beta \varepsilon_i},} \tag{18.92}$$

where we have called $A = e^{\alpha}$. This is the well-known **Boltzmann formula**. As we shall see shortly, β is given as

$$\boxed{\beta = -\frac{1}{kT},} \tag{18.93}$$

where k is the Boltzmann constant and T is the temperature and A is determined from the condition which gives the total number of particles as $N = \sum_i n_i$.

18.3.2 Boltzmann Distribution for Gases

Compared to solids, the case of gases have basically two differences. First of all, atoms are now free to move within the entire volume of the system. Hence, they are not localized. Second, the distribution of energy eigenvalues is practically continuous. There are many more energy levels than the number of atoms; hence, the occupancy number of each level is usually either 0 or 1. Mostly 0 and almost never greater than 1. For example, in 1 cc of helium gas at 1 atm and 290 K, there are approximately 10^6 times more levels than atoms. Since atoms cannot be localized, we treat them as **indistinguishable** particles. For the second difference, we group neighboring energy levels in bundles so that a complexion is now described by saying

n_1 particles in the first bundle of g_1 levels with energy ε_1,
n_2 particles in the second bundle of g_2 levels with energy ε_2,

$$\vdots$$

The choice of g_k is quite arbitrary, except that n_k has to be large enough to warrant usage of the Stirling approximation of factorial. Also, g_k must be large but not too large so that each bundle can be approximated by the average energy ε_k.

As before, the most probable values of n_k are the ones corresponding to the maximum number of complexions. However, W is now more complicated. We first concentrate on the kth bundle, where there are g_k levels available for n_k particles. For the first particle, naturally all g_k levels are available. For the second particle, there will be $(g_k - 1)$ levels left. If we keep going on like this, we find

$$g_k(g_k - 1) \cdots (g_k - n_k + 1) \tag{18.94}$$

different possibilities. Since $g_k \gg n_k$, we can write this as $g_k^{n_k}$. Within the kth bundle, it does not matter how we order the n_k particles, thus we divide $g_k^{n_k}$ with $n_k!$. This gives the number of distinct complexions for the kth bundle as

$$\frac{g_k^{n_k}}{n_k!}. \tag{18.95}$$

Similar expressions for all the other bundles can be written. Hence, the total number of complexions become

$$W = \prod_k \frac{g_k^{n_k}}{n_k!}. \tag{18.96}$$

This has to be maximized subject to the conditions

$$N = \sum_k n_k, \tag{18.97}$$

$$U = \sum_k n_k \varepsilon_k. \tag{18.98}$$

Proceeding as for the gases and introducing two Lagrange undetermined multipliers, α and β, we write the variation of $\ln W$ as

$$\delta \ln W = \sum_k \left(\ln \frac{g_k}{n_k} + \alpha + \beta \varepsilon_k \right) \delta n_k = 0. \tag{18.99}$$

This gives the number of complexions as

$$n_k = A g_k e^{\beta \varepsilon_k}, \tag{18.100}$$

where A comes from $N = \sum_k n_k$, and β is again equal to $-1/kT$.

18.3.3 Bose–Einstein Distribution for Perfect Gases

We now remove the restriction on n_k. For the Bose–Einstein distribution, there is no restriction on the number of particles that one can put in each level. We first consider the number of different ways that we can distribute n_k particles over the g_k levels of the kth bundle. This is equivalent to finding the number of different ways that one can arrange N indistinguishable particles in g_k boxes. Let us consider a specific case with 2 balls, $n_k = 0, 1, 2$, and three boxes, $g_k = 3$. The six distinct possibilities, which can be described by the formula

$$6 = \frac{[2 + (3 - 1)]!}{2!(3 - 1)!}, \tag{18.101}$$

are shown below:

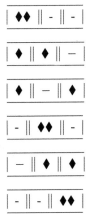

This can be understood by the fact that for three boxes, there are two partitions, shown by the double lines, which is one less than the number of boxes. The numerator in Eq. (18.101), $[2 + (3 - 1)]!$, gives the number of permutations of the number of balls plus the number of partitions. However, the permutations of the balls, $2!$, and the permutations of the partitions, $(3-1)!$, among themselves do not lead to any new configurations, which explains the denominator. Since the number of partitions is always one less than the number of boxes, this formula can be generalized as

$$\frac{(n_k + g_k - 1)!}{n_k!(g_k - 1)!}. \tag{18.102}$$

For the whole system, this gives W as

$$\boxed{W = \prod_k \frac{(n_k + g_k - 1)!}{n_k!(g_k - 1)!}.} \tag{18.103}$$

Proceeding as in the previous cases, that is, by introducing two Lagrange undetermined multipliers, α and β, for the two constraints, $N = \sum_k n_k$ and $U = \sum_k n_k \varepsilon_k$, respectively and then maximizing $\ln W$, we obtain the **Bose–Einstein distribution** as

$$\boxed{n_k = \frac{g_k}{e^{-\alpha - \beta \varepsilon_k} - 1}.} \tag{18.104}$$

One can again show that $\beta = -1/kT$. Notice that for high temperatures, where the -1 in the denominator is negligible, Bose–Einstein distribution reduces to the Boltzmann distribution [Eq. (18.100)].

Bose–Einstein Condensation:
Using the method of **ensembles**, one can show that the distribution is also written as

$$n_i = \frac{g_i}{e^{-\alpha + \varepsilon_i/kT} - 1}, \tag{18.105}$$

where n_i is now the average number of particles in the ith level, not the group of levels, with the energy ε_i. For the lowest level, $i = 1$, this becomes

$$n_1 = \frac{g_1}{e^{-\alpha + \varepsilon_1/kT} - 1}, \tag{18.106}$$

which means that we can populate the lowest level by as many particles as we desire by making α very close to ε_1/kT. This phenomenon with very interesting applications is called the **Bose –Einstein condensation.**

18.3.4 Fermi–Dirac Distribution

In the case of Fermi–Dirac distribution, the derivation of n_i proceeds exactly the same way as in the Bose–Einstein distribution. However, with the exception that due to **Pauli exclusion principle**, each level can only be occupied by only one particle. For the first particle, there are g_k levels available, which leaves only $(g_k - 1)$ levels for the second particle and so on, thus giving the number of arrangements as

$$g_k(g_k - 1)\cdots(g_k - n_k + 1) = \frac{g_k!}{(g_k - n_k)!}. \tag{18.107}$$

Since the particles are **indistinguishable**, $n_k!$ arrangements among themselves have no significance, which for the kth bundle gives the number of possible arrangements as

$$\frac{g_k!}{n_k!(g_k - n_k)!}. \tag{18.108}$$

For the whole system, this gives

$$W = \prod_k \frac{g_k!}{n_k!(g_k - n_k)!}. \tag{18.109}$$

Using the method of Lagrange undetermined multipliers and the constraints $N = \sum_k n_k$ and $U = \sum_k n_k \varepsilon_k$, one obtains the Fermi–Dirac distribution function as

$$n_k = \frac{g_k}{e^{-\alpha + \varepsilon_k/kT} + 1}. \tag{18.110}$$

We have again written $\beta = -1/kT$ and α is to be determined from the condition $N = \sum_k n_k$. With respect to the Bose–Einstein distribution [Eq. (18.104)], the change in sign in the denominator is crucial. It is the source of the enormous pressures that hold up white dwarfs and neutron stars.

18.4   STATISTICAL MECHANICS AND THERMODYNAMICS

All the distribution functions considered so far contained two arbitrary constants, α and β, which were introduced as Lagrange undetermined multipliers. In order to be able to determine the values of these constants, we have to make contact with thermodynamics. In other words, we have to establish the relation between the microscopic properties like the occupation numbers, energy levels, number of complexions, etc., and the macroscopic properties like the density (ρ), pressure (P), temperature (T), and entropy (S).

18.4.1   Probability and Entropy

We know that in reaching equilibrium, isolated systems acquire their most probable state, that is, the state with the most number of complexions. This is analogous to the second law of thermodynamics, which says that isolated systems seek their maximum entropy state. In this regard, it is natural to expect a connection between the number of complexions W and the thermodynamic entropy S. To find this connection, let us bring two thermodynamic systems, A and B, with their respective entropies, S_A and S_B, in thermal contact with each other. The total entropy of the system is

$$\boxed{S = S_A + S_B.} \tag{18.111}$$

If W_A and W_B are their respective number of complexions, also called microstates, the total number of complexions is

$$\boxed{W = W_A \cdot W_B.} \tag{18.112}$$

If we call the desired relation $S = f(W)$, Eq. (18.111) means that

$$f(W_A) + f(W_B) = f(W_A W_B). \tag{18.113}$$

Differentiating with respect to W_B gives us

$$f'(W_B) = W_A f'(W_A W_B). \tag{18.114}$$

Differentiating once more but this time with respect to W_A, we get

$$0 = f'(W_A W_B) + W_A W_B f''(W_A W_B), \tag{18.115}$$

$$f'(W) = -W f''(W), \tag{18.116}$$

$$\frac{f''(W)}{f'(W)} = -\frac{1}{W}. \tag{18.117}$$

The first integral of this gives

$$\ln f'(W) = -\ln W + \text{constant}, \tag{18.118}$$

or

$$\frac{df(W)}{dW} = \frac{k}{W}.$$

(18.119)

Integrating once more, we obtain $f(W) = k \ln W +$ constant, where k is some constant to be determined. We can now write the relation between the entropy and the number of complexions as

$$S = k \ln W + S_0.$$

(18.120)

If we define the entropy of a completely ordered state, that is, $W = 1$ as 0, we obtain the final expression for the relation between the thermodynamic entropy and the number of complexions W as

$$\boxed{S = k \ln W.}$$

(18.121)

18.4.2 Derivation of β

Consider two systems, one containing N and the other N' particles, brought into thermal contact with each other. State of the first system can be described by giving the occupation numbers as

n_1 particles in the energy states ε_1
n_2 particles in the energy states ε_2
$$\vdots$$

Similarly, the second system can be described by giving the occupation numbers as

n_1' particles in the energy states ε_1'
n_2' particles in the energy states ε_2'
$$\vdots$$

Now, the total number of complexions for the combined system is

$$W = W_1 \cdot W_2 = \frac{N!}{n_1! n_2! \cdots} \cdot \frac{N'!}{n_1'! n_2'! \cdots}.$$

(18.122)

When both systems reach thermal equilibrium, their occupation numbers become such that $\ln W$ is a maximum subject to the following conditions:

$$N = \sum_i n_i,$$

(18.123)

$$N' = \sum_i n_i',$$

(18.124)

$$U = \sum_i (n_i \varepsilon_i + n_i' \varepsilon_i'),$$

(18.125)

where N, N', and the total energy U are constants. Introducing the Lagrange undetermined multipliers α, α', and β, we write

$$\alpha \sum_i \delta n_i = 0, \tag{18.126}$$

$$\alpha' \sum_i \delta n_i' = 0, \tag{18.127}$$

$$\beta \sum_i \delta n_i \varepsilon_i + \delta n_i' \varepsilon_i' = 0. \tag{18.128}$$

Proceeding as in the previous cases, we now write $\delta \ln W = 0$ and employ the Stirling's formula [Eq. (18.88)] to obtain

$$\sum_i (-\ln n_i + \alpha + \beta \varepsilon_i)\delta n_i + \sum_j (-\ln n_j' + \alpha' + \beta \varepsilon_j')\delta n_j' = 0. \tag{18.129}$$

For this to be true for all δn_i and $\delta n_j'$, we have to have

$$\begin{aligned}
n_i &= A e^{\beta \varepsilon_i}, \\
n_j' &= A' e^{\beta \varepsilon_j'},
\end{aligned} \tag{18.130}$$

where $A = e^{\alpha}$ and $A' = e^{\alpha'}$. In other words, β is the same for two systems in thermal equilibrium.

To find an explicit expression for β, we slowly add dQ amount of heat into a system in equilibrium. During this process, which is taking place reversibly at constant temperature T, the allowed energy values remain the same but the occupation numbers, n_i, of each level change such that W is still a maximum after the heat is added. Using Eq. (18.81):

$$\delta \ln W = \sum_i \frac{\partial(\ln W)}{\partial n_i} \delta n_i \tag{18.131}$$

and [Eq. (18.85)]:

$$\sum_i \left(\frac{\partial(\ln W)}{\partial n_i} + \alpha + \beta \varepsilon_i \right) \delta n_i = 0, \tag{18.132}$$

we can write the change in $\ln W$ as

$$\delta \ln W = -\sum_i (\alpha + \beta \varepsilon_i)\delta n_i = -\alpha \sum_i \delta n_i - \beta \sum_i \varepsilon_i \delta n_i. \tag{18.133}$$

During this process, the total number of particles does not change, hence

$$\sum_i \delta n_i = 0. \tag{18.134}$$

Since the heat added to the system can be written as

$$dQ = \sum_i \delta n_i \varepsilon_i, \qquad (18.135)$$

Eq. (18.133) becomes

$$\beta = -\frac{d \ln W}{dQ}. \qquad (18.136)$$

Using the definition of entropy, $S = k \ln W$, obtained in Eq. (18.121), Eq. (18.136) can also be written as

$$\beta = -\frac{1}{k}\frac{dS}{dQ}. \qquad (18.137)$$

In thermodynamics, in any reversible heat exchange taking place at constant temperature T, dQ is related to the change in entropy as

$$dQ = T dS. \qquad (18.138)$$

Comparing Eqs. (18.137) and (18.138), we obtain the desired relation as

$$\boxed{\beta = -\frac{1}{kT},} \qquad (18.139)$$

where k can also be identified as the Boltzmann constant by further comparisons with thermodynamics.

18.5   RANDOM VARIABLES AND DISTRIBUTIONS

The concept of random variable is one of the most important elements of the probability theory. The number of rain drops impinging on a selected area is a random variable, which depends on a number of random factors. The number of passengers arriving at a subway station at certain times of the day is also a random variable. Velocities of gas molecules take on different values depending on the random collisions with the other molecules. In the case of electronic noise, voltages and currents change from observation to observation in a random way. All these examples show that random variables are encountered in many different branches of science and technology.

Despite the diversity of these examples, mathematical description is similar. Under random effects, each of these variables is capable of taking a variety of values. It is imperative that we know the range of values that a random variable can take. However, this is not sufficient. We also need to know the frequencies with which a random variable assumes these values. Since random variables could be continuous or discrete, we need a unified formalism to study their behavior. Hence, we introduce the **distribution function** of probabilities of the random variable X as

$$F_X(x) = P(X \leqslant x). \tag{18.140}$$

From now on, we show random variables with the uppercase Latin letters, $X, Y, \ldots$, and the values that they can take with the lowercase Latin letters, $x, y, \ldots$.

Before we introduce what exactly $F_X(x)$ means, consider a time function $X(t)$ shown as in Figure 18.3. The independent variable t usually stands for time, but it could also be considered as any parameter that changes continuously. We now define the **distribution function $F_X(x)$** as

$$F_X(x) = \lim_{T \to \infty} \frac{1}{2T} \int_{-T}^{T} C_x[X(t)]dt, \tag{18.141}$$

where C_x is defined as

$$C_x[X] = \begin{cases} 1, & X \leqslant x, \\ 0, & X > x. \end{cases} \tag{18.142}$$

The role of C_x can be understood from Figure 18.4, where the integral in Eq. (18.141) is evaluated over the total duration of time during which $X(x)$ is less than or equal to x in the interval $[-T, T]$. The interval over which $X \leqslant x$ is indicated by thick lines. Thus,

$$\frac{1}{2T} \int_{-T}^{T} C_x[X(t)] \, dt \tag{18.143}$$

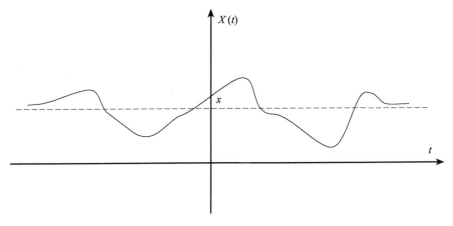

Figure 18.3 Time function $X(t)$.

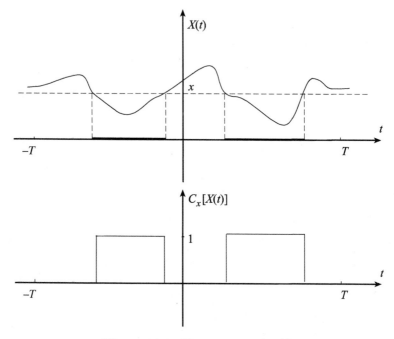

Figure 18.4 Time average of $X(t)$.

is the time average evaluated over the fraction of the time that $X(t)$ is less than or equal to x. Note that the distribution $F_X(x)$ gives not a single time average but an infinite number of time averages, that is, one for each x.

Example 18.10. *Arcsine distribution:* Let us now consider the function

$$X(t) = \sin \omega t. \tag{18.144}$$

During the period $[0, \frac{2\pi}{\omega}]$, $X(t)$ is less than x in the intervals indicated by thick lines in Figure 18.5. When $x > 1$, $X(t)$ is always less than x, hence

$$F_X(x) = 1, \quad x > 1. \tag{18.145}$$

When $x < -1$, $X(t)$ is always greater than x, hence

$$F_X(x) = 0, \quad x < -1. \tag{18.146}$$

For the regions indicated in Figure 18.5, we write

$$\int_0^{2\pi/\omega} C_x[X(t)]dt = \left[\frac{1}{\omega}\sin^{-1}x - 0\right] + \left[\frac{2\pi}{\omega} - \frac{1}{\omega}(\pi - \sin^{-1}x)\right] \tag{18.147}$$

$$= \frac{1}{\omega}[\pi + 2\sin^{-1}x], \tag{18.148}$$

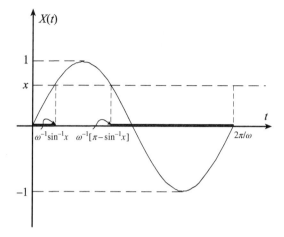

Figure 18.5 Arcsine distribution.

thus obtaining the arcsine distribution as

$$
F_X(x) = \begin{cases}
1, & x > 1, \\
\dfrac{1}{2} + \dfrac{1}{\pi}\sin^{-1}x, & |x| \leqslant 1, \\
0, & x < -1.
\end{cases} \tag{18.149}
$$

Arcsine distribution is used in **communication problems**, where an interfering signal with an unknown constant sinusoid of unknown phase may be thought to hide the desired signal.

18.6 DISTRIBUTION FUNCTIONS AND PROBABILITY

In the aforementioned example, evaluation of the distribution function was simple, since the time function $X(t)$ was given in terms of a simple mathematical expression. In most practical situations due to the random nature of the conditions, $X(t)$ cannot be known ahead of time. However, the distribution function may still be determined by some other means. The point that needs to be emphasized is that random processes in nature are usually defined in terms of certain averages like distribution functions.

If we remember the geometric definition of probability given in Section 18.1.7, the fraction of time that $X(x) \leqslant x$ is actually the probability of the event $\{X(t) \leqslant x\}$ happening. In the same token, we call the fraction of time that $x_1 < X(t) \leqslant x_2$ the probability of the event $\{x_1 < X(t) \leqslant x_2\}$ and show it as $P\{x_1 < X(t) \leqslant x_2\}$. From this, it follows that

$$P\{X(t) \leqslant x\} = F_X(x). \tag{18.150}$$

Since

$$C_b(x) - C_a(x) = \begin{cases} 1, & a < x \leqslant b, \\ 0, & \text{otherwise,} \end{cases} \tag{18.151}$$

we can write

$$P\{x_1 < X(t) \leqslant x_2\} = \lim_{T \to \infty} \frac{1}{2T} \int_{-\infty}^{\infty} \{C_{x_2}[X(t)] - C_{x_1}[X(t)]\} \, dt \tag{18.152}$$

$$= F_X(x_2) - F_X(x_1). \tag{18.153}$$

Thus, the probability $P\{x_1 < X(t) \leqslant x_2\}$ is expressed in terms of the distribution function $F_X(x)$. This argument can be extended to nonoverlapping intervals $[x_1, x_2], [x_3, x_4], \ldots$ as

$$P\{x_1 < X(t) \leqslant x_2, x_3 < X(t) \leqslant x_4, \cdots \}$$
$$= [F_X(x_2) - F_X(x_1)] + [F_X(x_4) - F_X(x_3)] + \cdots. \tag{18.154}$$

From the definition of the distribution function, also called the **cumulative distribution** function, one can easily check that the following conditions are satisfied:
  (i) $0 \leqslant F_X(x) \leqslant 1$,
  (ii) $\lim_{x \to -\infty} F_X(x) = 0$ and $\lim_{x \to \infty} F_X(x) = 1$,
  (iii) $F_X(x) \leqslant F_X(x')$, if and only if $x \leqslant x'$.
  This means that $F_X(x)$ satisfies all the basic axioms of probability given in Section 18.1.2. It can be proven that a real valued function of a real variable which satisfies the aforementioned conditions is a distribution function. In other words, it is possible to construct at least one time function $X(t)$, the distribution function of which coincides with the given function. This result removes any doubts about the existence of the limit

$$F_X(x) = \lim_{T \to \infty} \frac{1}{2T} \int_{-T}^{T} C_x[X(t)] \, dt. \tag{18.155}$$

Note that condition (iii) implies that wherever the derivative of $F_X(x)$ exists, it is always positive. At the points of discontinuity, we can use the Dirac-delta function to represent the derivative of $F_X(x)$. This is usually sufficient to cover the large majority of the physically meaningful cases. With this understanding, we can write all distributions as integrals:

$$\boxed{F_X(x) = \int_{-\infty}^{x} p_X(x')dx',} \tag{18.156}$$

where

$$p_X(x) = \frac{dF_X(x)}{dx} \geqslant 0.$$

(18.157)

The converse of this statement is that if $p_X(x)$ is any nonnegative integrable function,

$$\int_{-\infty}^{\infty} p_X(x')dx' = 1,$$

(18.158)

then $F_X(x)$ defined in Eq. (18.156) satisfies the conditions (i)–(iii). The function $p_X(x)$ obtained from $F_X(x)$ is called the **probability density**. This name is justified if we write the event $\{X(t) \text{ in } D\}$, where D represents some set of possible outcomes over the real axis, as

$$P\{X(t) \text{ in } D\} = \int_D p_X(x') \, dx',$$

(18.159)

or as

$$P\{x \leqslant X(t) \leqslant x + dx\} = p_X(x) \, dx.$$

(18.160)

18.7 EXAMPLES OF CONTINUOUS DISTRIBUTIONS

In this section, we introduce some of the most commonly encountered continuous distribution functions.

18.7.1 Uniform Distribution

Probability density for the uniform distribution is given as

$$p_X(x) = \begin{cases} \dfrac{1}{2a}, & |x| \leqslant a, \\[2mm] 0, & |x| > a, \end{cases}$$

(18.161)

where a is a positive number. The distribution function $F_X(x)$ is easily obtained from Eq. (18.156) as (Figure 18.6)

$$F_X(x) = \begin{cases} 0, & x < -a, \\[2mm] \dfrac{(x+a)}{2a}, & |x| \leqslant a, \\[2mm] 1, & x > a. \end{cases}$$

(18.162)

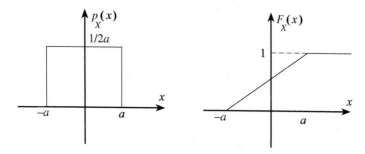

Figure 18.6 The uniform distribution.

18.7.2 Gaussian or Normal Distribution

The bell-shaped **Gauss distribution** is defined by the **probability density**

$$p_X(x) = \frac{1}{\sqrt{2\pi}\sigma}e^{-(1/2\sigma^2)(x-m)^2}, \quad \sigma > 0, \quad -\infty < x < \infty, \tag{18.163}$$

and the **distribution function** (Figure 18.7)

$$F_X(x) = \Phi\left(\frac{x-m}{\sigma}\right), \tag{18.164}$$

where

$$\Phi(x) = \frac{1}{\sqrt{2\pi}}\int_{-\infty}^{x} e^{-\xi^2/2}d\xi. \tag{18.165}$$

Gaussian distribution is extremely useful in many different branches of science and technology. In the limit as $\sigma \to 0$, $p_X(x)$ becomes one of the most commonly

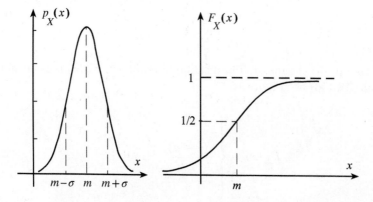

Figure 18.7 The Gauss or the normal distribution.

used representations of the Dirac-delta function. In this sense, the Dirac-delta function is a probability density.

18.7.3 Gamma Distribution

The Gamma distribution is defined by the probability density

$$p_X(x) = \begin{cases} \dfrac{\alpha^n e^{-\alpha x} x^{n-1}}{\Gamma(n)}, & x > 0, \\ 0, & x \leqslant 0, \end{cases} \tag{18.166}$$

where $\alpha, n > 0$. It is clear that $p_X(x) \geqslant 0$ and $\int_{-\infty}^{\infty} p_X(x)\, dx = 1$.

There are two cases that deserves mentioning:

(i) The case where $n = 1$ is called the **exponential distribution**:

$$p_X(x) = \begin{cases} \alpha e^{-\alpha x}, & x > 0, \\ 0, & x \leqslant 0. \end{cases} \tag{18.167}$$

(ii) The case where $\alpha = 1/2$, $n = m/2$, where m is a positive integer, is called the χ^2 **distribution** (Chi-square), with m degrees of freedom:

$$p_X(x) = \begin{cases} \dfrac{e^{-x/2} x^{(m/2)-1}}{2^{m/2}\Gamma(m/2)}, & x > 0, \\ 0, & x \leqslant 0. \end{cases} \tag{18.168}$$

In general, the integral

$$F_X(x) = \int_{-\infty}^{x} p_X(x')\, dx' \tag{18.169}$$

cannot be evaluated analytically for the χ^2 distribution; however, there exists extensive tables for the values of $F_X(x)$. The χ^2 distribution is extremely useful in checking the fit of an experimental data to a theoretical one.

18.8 DISCRETE PROBABILITY DISTRIBUTIONS

When X is a discrete random variable, the distribution function $F_X(x)$ becomes a step function. Hence, it can be specified by giving a sequence of numbers, $x_1, x_2, \ldots$, and the sequence of probabilities, $p_X(x_1), p_X(x_2), \ldots$, satisfying the following conditions:

$$p_X(x_i) > 0, \quad i = 1, 2, \ldots, \tag{18.170}$$

$$\sum_{i=0}^{\infty} p_X(x_i) = 1. \tag{18.171}$$

Now, the distribution function, $F_X(x)$, is given as

$$F_X(x) = \sum_{x_i \leqslant x} p_X(x_i). \tag{18.172}$$

Some of the commonly encountered discrete distributions are given in the following:

18.8.1 Uniform Distribution

Given a bag with N balls numbered as $1, 2, \ldots, N$. Let X be the number of the ball drawn. When one ball is drawn at random, the probability of any outcome $x = 1, 2, \ldots, N$ is

$$p_X(x) = \frac{1}{N}. \tag{18.173}$$

The (cumulative) distribution function is given in terms of the step function as

$$F_X(x) = \frac{1}{N} \sum_{i=1}^{N} \theta(x - i), \quad 1 \leqslant x \leqslant N, \tag{18.174}$$

where i is an integer and

$$\theta(x - i) = \begin{cases} 1, & x \geqslant i, \\ 0, & x < i. \end{cases} \tag{18.175}$$

18.8.2 Binomial Distribution

Let us consider a set of n independent experiments, where each experiment has two possible outcomes as 0 and 1. If Y_i is the outcome of the ith experiment, then

$$X = \sum_{i=1}^{n} Y_i \tag{18.176}$$

is the sum of the outcomes with the result 1. We also let the probability of 1 occurring in any one of the events as

$$P\{Y_i = 1\} = p, \tag{18.177}$$

where $i = 1, 2, \ldots, n$ and $0 < p < 1$. Now, for any event, that is, the set of n experiments with the results

$$Y_1 = y_1, \quad Y_2 = y_2, \ldots, Y_n = y_n, \tag{18.178}$$

we have the probability

$$P\{Y_1 = y_1, \quad Y_2 = y_2, \ldots, Y_n = y_n\} = cp^x(1 - p)^{n-x}, \tag{18.179}$$

where y_i takes the values 0 or 1 and x is the number of 1's obtained. The number c is the number of rearrangements of the symbols 0 and 1 that does not change the number of 1's. Hence, the probability of obtaining x number of 1's becomes

$$P\{X = x\} = \binom{n}{x} p^x (1 - p)^{n-x}. \tag{18.180}$$

The reason for multiplying with

$$\binom{n}{x} = \frac{n!}{x!(n-x)!} \tag{18.181}$$

is that the permutations of 0's and the permutations of 1's do not change the number of 1's.

We now define the **binomial probability density** function as

$$p_X(x) = \binom{n}{x} p^x (1 - p)^{n-x}, \quad x = 0, 1, \ldots, n. \tag{18.182}$$

Notice that

$$p_X(x) > 0, \quad x = 0, 1, \ldots, n, \tag{18.183}$$

and

$$\sum_{x=0}^{n} p_X(x) = 1. \tag{18.184}$$

This is clearly seen by using the binomial expansion [Eq. (18.67)] as

$$\sum_{x=0}^{n} p_X(x) = \sum_{x=0}^{n} \binom{n}{x} p^x (1 - p)^{n-x} = (p + 1 - p)^n = 1. \tag{18.185}$$

Finally, we write the **binomial distribution** function as

$$F_X(x) = \sum_{x \leqslant n} \binom{n}{x} p^x (1 - p)^{n-x}, \quad x = 0, 1, \ldots, n. \tag{18.186}$$

Example 18.11. *A coin tossed 6 times:* We now have $p = \frac{1}{2}$, $n = 6$, and X stands for the total number of heads. Using Eqs. (18.182) and (18.186), we can summarize the results in terms of the following table:

$X(x)$	0	1	2	3	4	5	6
$p_X(x)$	$\frac{1}{64}$	$\frac{6}{64}$	$\frac{15}{64}$	$\frac{20}{64}$	$\frac{15}{64}$	$\frac{6}{64}$	$\frac{1}{64}$
$F_X(x)$	$\frac{1}{64}$	$\frac{7}{64}$	$\frac{22}{64}$	$\frac{42}{64}$	$\frac{57}{64}$	$\frac{63}{64}$	$\frac{64}{64}$

where the histograms of $p_X(x)$ and $F_X(x)$ are given as in Figure 18.8.

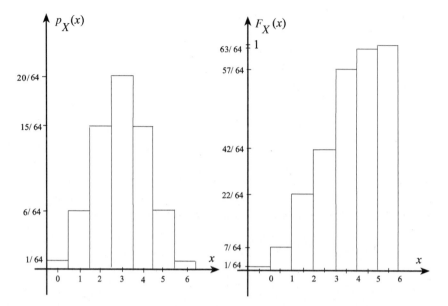

Figure 18.8   Plots of $p_X(x)$ and $F_X(x)$ for a coin tossed 6 times.

18.8.3   Poisson Distribution

Poisson probability density is given by

$$p_X(x) = \frac{e^{-\lambda}\lambda^x}{x!}, \quad \lambda > 0, \quad x = 0, 1, 2, \dots . \tag{18.187}$$

It is obvious that $p_X(x)$ is positive for every nonnegative integer and

$$\sum_{x=0}^{\infty} p_X(x) = e^{-\lambda} \sum_{x=0}^{\infty} \frac{\lambda^x}{x!} = e^{-\lambda}e^{\lambda} = 1. \tag{18.188}$$

Poisson distribution function is now written as

$$F_X(x) = \sum_{k=0}^{x} \frac{e^{-\lambda}\lambda^k}{k!}, \quad x = 0, 1, 2, \dots . \tag{18.189}$$

Poisson distribution expresses the probability of a number of events occurring in a given period of time with a known average rate and which are independent of the time since the last event. Other distributions and examples can be found in Harris [3].

18.9   FUNDAMENTAL THEOREM OF AVERAGES

We now state an important theorem, which says that all time averages of the form

$$\lim_{T \to \infty} \frac{1}{2T} \int_{-T}^{T} \phi[X(t)] \, dt, \tag{18.190}$$

where the limit exists and ϕ is a real valued function of a real variable, can be calculated by using the probability density $p_X(x)$ as

$$\lim_{T \to \infty} \frac{1}{2T} \int_{-T}^{T} \phi[X(t)] \, dt = \int_{-\infty}^{\infty} \phi(x) p_X(x) \, dx \tag{18.191}$$

or by using the distribution function $F_X(x)$ as

$$\lim_{T \to \infty} \frac{1}{2T} \int_{-T}^{T} \phi[X(t)] \, dt = \int_{-\infty}^{\infty} \phi(x) \, dF_X(x). \tag{18.192}$$

This theorem is also called the **quasi-ergodic theorem**. We shall not discuss the proof, since it is far too technical for our purposes. However, a plausibility argument can be found in Margenau and Murphy [8] by Hofstetter. Significance of this theorem is in the fact that it allows us to calculate time averages in terms of distributions or probability densities, which are easier to obtain and to work with. The quantity on the right-hand side is called the **expected value** of $\phi(x)$ and it is usually shown as

$$\boxed{\int_{-\infty}^{\infty} \phi(x) p_X(x) dx = \langle \phi \rangle.} \tag{18.193}$$

Of course, the time function $X(t)$ is still a quantity of prime importance to the theory. Proving the equivalence of relations like Eqs. (18.191) and (18.192), which establishes the connections between the expected values calculated by using the distribution functions and the time averages, is done in a series of philosophically involved theorems usually referred to as the **law of large numbers** or the **ergodic theorems**. Note that the expected value is basically a weighted average with the weights determined by $p_X(x)$.

In the case of **discrete random** variables, since

$$p_X(x_i) = P\{X = x_i\}, \tag{18.194}$$

the **expected value** is written as

$$\boxed{\langle g(X) \rangle = \sum_{i=1}^{\infty} g(x_i) p_X(x_i).} \tag{18.195}$$

Basic rules for manipulations with expected values are given as

$$\langle a \rangle = a, \tag{18.196}$$

$$\langle ag_1 + bg_2 \rangle = a\langle g_1 \rangle + b\langle g_2 \rangle, \tag{18.197}$$

$$\langle g \rangle \leqslant \langle |g| \rangle, \tag{18.198}$$

where a and b are constants and g, g_1, and g_2 are functions.

18.10   MOMENTS OF DISTRIBUTION FUNCTIONS

We mentioned that the majority of the time averages can be calculated by using distribution functions. However, in most cases by concentrating on a few but more easily measured parameters of the distribution function, such as the **moments** of the distribution functions, one can avoid the complications that measuring or specifying a complete distribution function involves.

The nth **moment** of a distribution function F_X is defined by the equation

$$\alpha_n = \langle X^n \rangle = \int_{-\infty}^{\infty} x^n p_X(x) dx, \tag{18.199}$$

where α_n is called the nth **moment** of the random variable X. First-order moment α_1, which gives the **expected value** or the **most probable value** of X is also called the **mean** and is shown as m or m_X. Interpretation of the higher moments becomes easier if we define the nth **central moment** μ_n as

$$\mu_n = \langle (X - m)^n \rangle = \int_{-\infty}^{\infty} (x - m)^n p_X(x) dx. \tag{18.200}$$

The **second central moment** μ_2 is usually called the **variance** of the distribution, and it is shown as σ^2 or σ_X^2. Variance can often be calculated by the formula

$$\sigma^2 = \langle (X - m)^2 \rangle = \langle X^2 \rangle - 2m\langle X \rangle + m^2 = \alpha_2 - m^2. \tag{18.201}$$

The square root of variation is called the **standard deviation**, and it is a measure of the spread about the mean value. Another important concept is the **median** m_e:

$$\int_{-\infty}^{m_e} p_X(x) dx = \frac{1}{2}, \tag{18.202}$$

which gives the midpoint of the distribution. In the case of Gaussian or the uniform distribution, the median coincides with the mean. However, as in the case with binomial distribution with $p \neq 1/2$, this is not always the case. In the case of discrete random variables, $X = x_i^{(k)}$, where $x^{(k)} = \frac{x!}{(x-k)!}$, we refer to $\langle X \rangle = \alpha_{[k]}$ as the kth **factorial moment**.

18.10.1   Moments of the Gaussian Distribution

The **mean**, α_1, of the Gaussian distribution [Eq. (18.163)] is written as

$$\alpha_1 = \int_{-\infty}^{\infty} x \frac{1}{\sqrt{2\pi}\sigma} e^{-(1/2\sigma^2)(x-m)^2} \, dx. \tag{18.203}$$

After a suitable change of variable, $\xi = (x - m)/\sigma$, and evaluating the integrals, we obtain

$$\alpha_1 = \frac{1}{\sqrt{2\pi}} \int_{-\infty}^{\infty} (\sigma\xi + m) e^{-\xi^2/2} \, d\xi \tag{18.204}$$

$$= \frac{\sigma}{\sqrt{2\pi}} \int_{-\infty}^{\infty} \xi e^{-\xi^2/2} \, d\xi + \frac{m}{\sqrt{2\pi}} \int_{-\infty}^{\infty} e^{-\xi^2/2} \, d\xi \tag{18.205}$$

$$= 0 + m \tag{18.206}$$

$$= m. \tag{18.207}$$

With the same variable change, the **variance**, μ_2, is obtained as

$$\mu_2 = \int_{-\infty}^{\infty} (x - m)^2 \frac{1}{\sqrt{2\pi}\sigma} e^{-(1/2\sigma^2)(x-m)^2} \, dx \tag{18.208}$$

$$= \frac{\sigma^2}{\sqrt{2\pi}} \int_{-\infty}^{\infty} \xi^2 e^{-\xi^2/2} \, d\xi \tag{18.209}$$

$$= \sigma^2. \tag{18.210}$$

Central moments of Gaussian distribution are given as

$$\boxed{\mu_{2n} = \frac{(2n)!}{2^n n!} \sigma^{2n}, \quad \mu_{2n+1} = 0, \quad n = 0, 1, 2, \ldots .} \tag{18.211}$$

18.10.2   Moments of the Binomial Distribution

The mean, α_1, of the binomial distribution [Eq. (18.182)] is written as

$$\alpha_1 = \sum_{x=0}^{n} x \binom{n}{x} p^x (1 - p)^{n-x} \tag{18.212}$$

$$= \sum_{x=0}^{n} \frac{xn!}{(n-x)!x!} p^x (1 - p)^{n-x} \tag{18.213}$$

$$= \sum_{x=1}^{n} \frac{n!}{(n-x)!(x-1)!} p^x (1 - p)^{n-x} \tag{18.214}$$

$$= n \sum_{x=1}^{n} \frac{(n-1)!}{(n-x)!(x-1)!} p^x (1-p)^{n-x} \tag{18.215}$$

$$= n \sum_{x=1}^{n} \frac{(n-1)!}{[(n-1)-(x-1)]!(x-1)!} p^x (1-p)^{n-x} \tag{18.216}$$

$$= n \sum_{x=1}^{n} \binom{n-1}{x-1} p^x (1-p)^{n-x}. \tag{18.217}$$

We now make the substitutions $x - 1 = y$ and $n - 1 = m'$ to write Eq. (18.217) as

$$\alpha_1 = np \sum_{y=0}^{m'} \binom{m'}{y} p^y (1-p)^{m'-y}. \tag{18.218}$$

Since the sum is equal to 1 [Eq. (18.185)], we obtain the **mean** as

$$\boxed{(m =) \; \alpha_1 = np.} \tag{18.219}$$

To find the variance, σ^2, we use Eq. (18.201):

$$(\sigma^2 =) \; \mu_2 = \alpha_2 - m^2. \tag{18.220}$$

To evaluate α_2, we first write $\langle X^2 \rangle$ as $\langle X^2 \rangle - \langle X \rangle + \langle X \rangle$, and then proceed as follows:

$$\alpha_2 = \langle X^2 \rangle = \langle X(X-1) \rangle + \langle X \rangle \tag{18.221}$$

$$= \sum_{x=2}^{n} x(x-1) \binom{n}{x} p^x (1-p)^{n-x} + m \tag{18.222}$$

$$= \sum_{x=2}^{n} \frac{n!}{(x-2)!(n-x)!} p^x (1-p)^{n-x} + m \tag{18.223}$$

$$= n(n-1)p^2 \sum_{x=2}^{n} \binom{n-2}{x-2} p^{x-2}(1-p)^{n-x} + m. \tag{18.224}$$

Using the substitutions $y = x - 2$ and $m' = n - 2$, we write this as

$$\alpha_2 = n(n-1)p^2 \sum_{y=0}^{m'} \binom{m'}{y} p^y (1-p)^{m'-y} + m. \tag{18.225}$$

Summation is again 1, thus yielding

$$\boxed{\alpha_2 = n(n-1)p^2 + m.} \tag{18.226}$$

Using this in Eq. (18.220) and the expression for the mean in Eq. (18.219), we finally obtain the **variance**, $\sigma^2 = \alpha_2 - m^2$, as

$$\boxed{\sigma^2 = np(1 - p).}$$

(18.227)

18.10.3   Moments of the Poisson Distribution

We write the **factorial moments** of the Poisson distribution as

$$\alpha_{[k]} = \sum_{x=0}^{\infty} x^{(k)} \frac{e^{-\lambda} \lambda^x}{x!}, \quad \text{where } x^{(k)} = \frac{x!}{(x - k)!}.$$

(18.228)

Thus,

$$\alpha_{[k]} = \sum_{x=k}^{\infty} \frac{e^{-\lambda} \lambda^x}{(x - k)!}$$

(18.229)

$$= \sum_{y=0}^{\infty} \frac{e^{-\lambda} \lambda^{y+k}}{y!}$$

(18.230)

$$= \lambda^k e^{-\lambda} \sum_{y=0}^{\infty} \frac{\lambda^y}{y!}$$

(18.231)

$$= \lambda^k,$$

(18.232)

where we have defined a new variable as $x - k = y$. Since $x^{(1)} = x$, we write the **mean** as

$$(m = \alpha_1 =) \ \alpha_{[1]} = \lambda.$$

(18.233)

Since $x^{(2)} = x(x - 1)$, we can write $\alpha_2 = \alpha_{[2]} + \alpha_{[1]}$; hence, the **variance** σ^2 is obtained as

$$\sigma^2 = \alpha_2 - m^2$$

(18.234)

$$= \alpha_{[2]} + \alpha_{[1]} - m^2$$

(18.235)

$$= \lambda^2 + \lambda - \lambda^2$$

(18.236)

$$= \lambda.$$

(18.237)

Using the Dirac-delta function, we can also write the Poisson probability density [Eq. (18.187)] as a continuous function of x:

$$p_X(x) = \sum_{k=0}^{\infty} \frac{\lambda^x}{\Gamma(x + 1)} e^{-\lambda} \delta(x - k),$$

(18.238)

and find the mean m as the integral

$$m = \langle X \rangle = \int_{-\infty}^{\infty} x \left[\sum_{k=0}^{\infty} \frac{\lambda^x}{\Gamma(x+1)} e^{-\lambda} \delta(x-k) \right] dx \tag{18.239}$$

$$= \sum_{k=0}^{\infty} e^{-\lambda} \left[\int_{-\infty}^{\infty} x \frac{\lambda^x}{\Gamma(x+1)} \delta(x-k) dx \right] \tag{18.240}$$

$$= \sum_{k=0}^{\infty} e^{-\lambda} k \frac{\lambda^k}{k!} \tag{18.241}$$

$$= \lambda e^{-\lambda} \sum_{k=1}^{\infty} \frac{\lambda^{k-1}}{(k-1)!} \tag{18.242}$$

$$= \lambda e^{-\lambda} e^{\lambda} \tag{18.243}$$

$$= \lambda. \tag{18.244}$$

Note that the gamma function $\Gamma(x+1)$ is equal to $k!$ when $x = k$, integer.
Similarly, we evaluate the second moment α_2 as

$$\alpha_2 = \langle X^2 \rangle = \int_{-\infty}^{\infty} x^2 \left[\sum_{k=0}^{\infty} \frac{\lambda^x}{\Gamma(x+1)} e^{-\lambda} \delta(x-k) \right] dx \tag{18.245}$$

$$= \sum_{k=0}^{\infty} e^{-\lambda} \left[\int_{-\infty}^{\infty} x^2 \frac{\lambda^x}{\Gamma(x+1)} \delta(x-k) dx \right] \tag{18.246}$$

$$= \sum_{k=0}^{\infty} e^{-\lambda} k^2 \frac{\lambda^k}{k!} \tag{18.247}$$

$$= \lambda e^{-\lambda} \sum_{k=1}^{\infty} \frac{\lambda^{k-1} k}{(k-1)!} \tag{18.248}$$

$$= \lambda e^{-\lambda} \sum_{k=0}^{\infty} \frac{\lambda^k (k+1)}{k!} \tag{18.249}$$

$$= \lambda e^{-\lambda} \left[\sum_{k=0}^{\infty} \frac{\lambda^k k}{k!} + \sum_{k=0}^{\infty} \frac{\lambda^k}{k!} \right] \tag{18.250}$$

$$= \lambda e^{-\lambda} [\lambda e^{\lambda} + e^{\lambda}] \tag{18.251}$$

$$= \lambda^2 + \lambda. \tag{18.252}$$

Using definition of the variance, $\sigma^2 = \alpha_2 - m^2$, we finally obtain $\sigma^2 = \lambda$.

18.11 CHEBYSHEV'S THEOREM

To demonstrate how well σ and σ^2 represent the spread in the probability density $p_X(x)$, we prove a useful theorem which says that if m and σ are the mean and the standard deviation of a random variable X, then for any positive constant k, the probability that X will take on a value within k standard deviations of the mean is at least $1 - \frac{1}{k^2}$, that is,

$$p_X(|x - m| - k\sigma) \geqslant 1 - \frac{1}{k^2}, \quad \sigma \neq 0.$$

(18.253)

Proof:

Using the definition of variance, we write

$$\sigma^2 = \int_{-\infty}^{\infty} (x - m)^2 p_X(x) \, dx$$

(18.254)

and divide the region of integration as

$$\sigma^2 = \int_{-\infty}^{m-k\sigma} (x - m)^2 p_X(x) \, dx + \int_{m-k\sigma}^{m+k\sigma} (x - m)^2 p_X(x) \, dx$$

$$+ \int_{m+k\sigma}^{\infty} (x - m)^2 p_X(x) \, dx.$$

(18.255)

Since $p_X(x) \geqslant 0$, the second integral is always $\geqslant 0$, thus we can also write

$$\sigma^2 \geqslant \int_{-\infty}^{m-k\sigma} (x - m)^2 p_X(x) \, dx + \int_{m+k\sigma}^{\infty} (x - m)^2 p_X(x) \, dx.$$

(18.256)

In the intervals $x \leqslant m - k\sigma$ and $x \geqslant m + k\sigma$, the inequality

$$(x - m)^2 \geqslant k^2 \sigma^2$$

(18.257)

is true; hence, Eq. (18.256) becomes

$$\sigma^2 \geqslant \int_{-\infty}^{m-k\sigma} k^2 \sigma^2 p_X(x) \, dx + \int_{m+k\sigma}^{\infty} k^2 \sigma^2 p_X(x) \, dx$$

(18.258)

$$\geqslant k^2 \sigma^2 \left[\int_{-\infty}^{m-k\sigma} p_X(x) \, dx + \int_{m+k\sigma}^{\infty} p_X(x) \, dx \right],$$

(18.259)

or

$$\frac{1}{k^2} \geqslant \int_{-\infty}^{m-k\sigma} p_X(x) \, dx + \int_{m+k\sigma}^{\infty} p_X(x) \, dx.$$

(18.260)

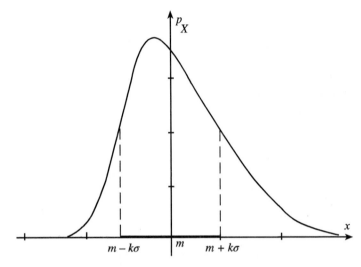

Figure 18.9   Chebyshev theorem.

The sum of the integrals on the right-hand side is the probability that X will take a value in the regions $x \leqslant m - k\sigma$ and $x \geqslant m + k\sigma$ (Figure 18.9), which means

$$p_X(|x - m| \geqslant k\sigma) \leqslant \frac{1}{k^2}, \quad \sigma \neq 0. \tag{18.261}$$

Since $\int_{-\infty}^{\infty} p_X(x)\, dx = 1$, we can write

$$p_X(|x - m| < k\sigma) \geqslant 1 - \frac{1}{k^2}, \tag{18.262}$$

thereby proving the theorem. The proof for discrete distributions can be given by similar arguments. Note that the Chebyshev's theorem gives a lower bound to the probability of observing a random variable within k standard deviations of the mean. The actual probability is usually higher.

18.12   LAW OF LARGE NUMBERS

One of the fundamental concepts of the probability theory is the law of large numbers. It describes how the mean of a randomly selected sample from a large population is likely to be close to the mean of the entire population. In other words, for an event of probability p, the frequency of occurrence, that is, the ratio of the number of times that event has actually occurred to the total number of observations, approaches p as the number of observations become arbitrarily large. For example, the average mass of 20 oranges randomly picked from a box of 100 oranges is probably closer to the actual average than the average mass calculated by just 10 oranges. Similarly, the average calculated by randomly picking 95 oranges will be much closer to the actual average found by using the entire sample. Even though this sounds as if we are not saying much, the law of large numbers allows us to give precise measurements of the likelihood that an estimate is close to the right or the correct value. This allows us to make forecasts or predictions that otherwise would not be possible. A formal proof of the law of large numbers is rather technical for our purposes [6, 9]. However, to demonstrate how the law of large numbers work, consider a random variable X having a **binomial distribution** [Eq. (18.186)] with the parameters n and p. We now define a new random variable:

$$Y = \frac{X}{n}, \tag{18.263}$$

where Y is the ratio of success in n tries. For example, X could be the number of times a certain face of a die comes in n number of rolls. Using the results for X, that is, np for the mean and $np(1-p)$ for the variation [Eqs. (18.219) and (18.227)], we can write the mean m_Y and the variance σ_Y^2 of Y as

$$m_Y = p, \quad \sigma_Y^2 = \frac{p(1-p)}{n}. \tag{18.264}$$

Using the Chebyshev's theorem with $k = c_0/\sigma_Y$, where c_0 is a positive constant, we can conclude that the probability of obtaining a particular value for the ratio Y within the range $m_Y - c_0$ and $m_Y + c_0$, that is, between

$$p - c_0 \quad \text{and} \quad p + c_0 \tag{18.265}$$

in n trials is at least

$$\boxed{1 - \frac{p(1-p)}{nc_0^2}.} \tag{18.266}$$

The actual frequency p is observed only when the number of tries goes to infinity, which in this case [Eq. (18.266)] approaches 1. This result is called the **law of large numbers**. For a coin toss experiment, the actual frequency is $p = 0.5$; hence, our chance of observing the frequency of heads or tails in the interval $(0.5 - c_0, 0.5 + c_0)$ is at least $\left(1 - \frac{0.25}{nc_0^2}\right)$. Since the probability in Eq. (18.266) is always a number < 1, this means that n and c_0 have to satisfy the inequality $n > 0.25/c_0^2$.

REFERENCES

1. Todhunter, I. (1949). *A History of the Theory of Probability From the Time of Pascal to Laplace*. New York: Chelsea Publishing Company.

2. Kolmogorov, A.N. (1950). *Foundations of the Theory of Probability*. New York: Chelsea Publishing Company.

3. Harris, B. (1966). *Theory of Probability*. Reading, MA: Addison-Wesley.

4. Bather, J.A. (2000). *Decision Theory, An Introduction to Programming and Sequential Decisions*. Chichester: Wiley.

5. Sivia, D.S. and Skilling, J. (2006). *Data Analysis: A Bayesian Tutorial*, 2e. New York: Oxford University Press.

6. Gnedenko, B.V. (1973). *The Theory of Probability*. Moscow: MIR Publishers, second printing.

7. Wilks, J. (1961). *The Third Law of Thermodynamics*. London: Oxford University Press.

8. Margenau, H. and Murphy, G.M. (eds) (1964). *The Mathematics of Physics and Chemistry*. Princeton, NJ: Van Nostrand.

9. Grimmett, G.R. and Stirzaker, D.R. (2001). *Probability and Random Processes*, 3e. Clarendon: Oxford University Press.

PROBLEMS

1. A coin is tossed four times. Write two different sample spaces in line with the following criteria:

   (i) Only the odd number of tails is of interest.

   (ii) Outcome of each individual toss is of interest.

2. Write the sample spaces when

   (i) a die rolled 3 times,

   (ii) 3 dice rolled simultaneously,

   (iii) 4 balls selected without replacement from a bag containing 7 black and 3 white balls.

3. If a pair of dice are rolled, what is the probability of the sum being 8?

4. When a tetrahedron with its sides colored as the first face is red, A, the second face is green, B, the third face is blue, C, and the fourth face is in all three colors, ABC, justify the following probabilities by writing the appropriate sample spaces:

$$P(A) = P(B) = P(C) = \frac{1}{2},$$

$$P(A/B) = P(B/C) = \frac{1}{2},$$

$$P(C/A) = P(B/A) = \frac{1}{2},$$

$$P(C/B) = P(A/C) = \frac{1}{2},$$

$$P(A/BC) = P(B/AC) = P(C/AB) = 1.$$

5. Two individuals decide to toss a coin n times. One bets on heads and the other on tails. What is the probability of them coming even after n tosses?

6. Given 6 bags, where 3 of them have the composition A_1 with 4 white and 2 black balls each, 1 bag has the composition A_2 with 6 black balls, and the remaining 2 bags have the composition A_3 with 4 white balls and 2 black balls each. We select a bag randomly and draw one ball from it. What is the probability that this ball is black?

7. Given 8 identical bags with the following contents: 4 bags with the contents A_1 composed of 3 white and 3 black balls each, 3 bags with the contents A_2 composed of 1 white and 2 black balls each, and 1 bag with the contents A_3 composed of 3 white and 1 black ball. We pick a ball from a randomly selected bag. It turns out to be black; call this event B. Find, after the ball is picked, its probability of coming from the second bag.

8. In a coin toss experiment, you know that the first 4 tosses came heads. What would you bet on for the next toss? What would you bet on if you did not know the result of the previous tosses.

9. In a die roll experiment, two dice are rolled. What is the probability of the sum being 2, 5, 9? What is the probability of these sums coming in succession?

10. Four coins are tossed.

    (i) What is the probability of the third coin coming heads?

    (ii) What is the probability of having an even number of tails?

11. A bag contains 20 black and 30 white balls. What is the probability of drawing a black and a white ball in succession?

12. Five cards are selected from an ordinary deck of 52 playing cards. Find the probability of having two cards of the same face value and three cards of different face value from the remaining 12 face values.

13. Given 5 bags: Two of them have the composition A_1 with two white and one black ball each, one bag has the composition A_2 with 10 black balls, and the remaining two bags have the composition A_3 with three white balls and one black ball each.
    A ball is drawn from a randomly selected bag.

    (i) Find the probability of this ball being white.

    (ii) Find the probability of this ball being black.

14. How many different six letter "words" can you make using the English alphabet? Count every combination as a word. How many words if consonants and vowels must alternate?

15. What is the number of ways of distributing k indistinguishable objects in N boxes with no empty box.

16. In how many ways can you distribute k indistinguishable balls into N boxes so that there is at most one ball in each box? Obviously, $k \leqslant N$.

17. How many distinguishable rearrangements of the letters in the word "distinguishable."

18. Prove the binomial theorem:

$$(x + y)^n = \sum_{k=0}^{n} \binom{n}{k} x^k y^{n-k},$$

by induction.

19. Show the following:

(i)

$$\sum_{k=0}^{n} k \binom{n}{k} = n2^{n-1}, \quad n \geqslant 0, \quad \text{integer.}$$

(ii)

$$\sum_{k=0}^{n} (-1)^k k \binom{n}{k} = 0, \quad n \geqslant 0, \quad \text{integer.}$$

(iii)

$$\sum_{k=0}^{n} \binom{n}{k}^2 = \binom{2n}{n}, \quad n \geqslant 0, \quad \text{integer.}$$

(iv)

$$\sum_{k=0}^{m} (-1)^k \binom{n-k}{m-k} \binom{n}{k} = 0, \quad m, n \text{ are integers with } n \geqslant m \geqslant 0.$$

20. Prove the following properties of the binomial coefficients:

(i)

$$\sum_{j=0}^{n} \binom{n}{j} = 2^n,$$

(ii)

$$\sum_{j=0}^{n} (-1)^j \binom{n}{j} = 0,$$

(iii)

$$\binom{n+1}{r} = \binom{n}{r} + \binom{n}{r-1},$$

(iv)

$$\sum_{j=0}^{n} \binom{n}{j} \binom{m}{n-j} = \binom{n+m}{n},$$

(v)

$$\sum_{k_1,k_2,\ldots,k_r \geqslant 0} \binom{n}{k_1, k_2, \ldots, k_{r-1}} = r^n, \quad \sum_{i=1}^{r} k_i = n.$$

21. In a roll of three dice, what is the chance of getting for the sum either a 5, 9, or a 10.

22. Ten black and 10 white balls to be ordered along a line.

   (i) What is the probability that black and white balls alternate?

   (ii) If they are positioned along a circle, what is the probability that they alternate?

23. A bag contains three balls; red, white, and green. Five balls are drawn with replacement. What is the probability of green balls appearing three times?

24. Starting with [Eq. (18.103)]

$$W = \prod_k \frac{(n_k + g_k - 1)!}{n_k!(g_k - 1)!},$$

give a complete derivation of the Bose–Einstein distribution:

$$n_k = \frac{g_k}{e^{-\alpha + \varepsilon_k/kT} - 1}.$$

Also show that Eq. (18.96):

$$W = \prod_k \frac{g_k^{n_k}}{n_k!},$$

used in the derivation of the Boltzmann distribution for gasses can be obtained from

$$W = \prod_k \frac{(n_k + g_k - 1)!}{n_k!(g_k - 1)!}$$

with the appropriate assumption.

25. Complete the intermediate steps of the derivation of the Fermi–Dirac distribution:

$$n_k = \frac{g_k}{e^{-\alpha + \varepsilon_k/kT} + 1}.$$

26. **Hypergeometric distribution**: Consider a shipment of N objects, in which M are defective. Show that the probability density function:

$$p_X(x) = P\{X = x\} = \frac{\binom{M}{x}\binom{N-M}{n-x}}{\binom{N}{n}}, \quad x = 0, 1, 2, \ldots, n,$$

where X is the number of defective items, gives the probability of finding x number of defective items in a random selection of n objects without replacement from the original shipment. Check that $p_X(x) \geqslant 0$ and $\sum_x p_X(x) = 1$. As a specific illustration make a table of $p_X(x)$ vs. x for $N = 15$, $M = 5$, $n = 5$.

27. Find the probability density of the **Cauchy distribution**:

$$F_X(x) = \frac{1}{\pi} \left[\arctan x + \frac{\pi}{2} \right].$$

Verify that your result is indeed a probability density function.

28. Show that

$$f(x) = \begin{cases} x, & 0 < x \leqslant 1, \\ 2 - x, & 1 < x < 2, \\ 0, & \text{otherwise}, \end{cases}$$

is a probability density. Find the corresponding distribution function, $F_X(x)$, which is called the **double triangle distribution**. Plot both $f(x)$ and $F_X(x)$.

29. A die is rolled until a 5 appears. If X is the number of rolls needed, find $P\{X \leqslant x\}$.

30. Two dice are rolled. If X is the sum of the number of spots on the faces showing, find $P\{X \leqslant x\}$.

31. On the average, 10 % of the goods produced in a factory are defective. In a package of 100 items, what is the probability of 3 or more of them being defective.

32. A book of 400 pages has 150 misprints. What is the probability of having 2 misprints in a randomly chosen page? What is the probability of having 2 or more misprints?

33. A fair die is rolled until every face has appeared twice. If X is the number of rolls, what is the probability density of X?

34. Show that when X has binomial distribution, then

$$P\{X = \text{even}\} = \frac{1}{2}[1 + (q - p)^n].$$

Also show that when X has the Poisson distribution, then

$$P\{X = \text{even}\} = \frac{1}{2}[1 + e^{-2\lambda}].$$

35. Show that the following are probability density functions:
(i)

$$\sqrt{\frac{2}{\pi}} e^{-x^2/2}, \quad x > 0.$$

(ii) **Rayleigh distribution:**

$$\beta^2 x e^{-\beta^2 x^2/2}, \quad \beta > 0, \quad x > 0.$$

36. Find the factorial moments of the **hypergeometric distribution** described by the probability density

$$p_X(x) = \frac{\binom{M}{x}\binom{N-M}{n-x}}{\binom{N}{n}}, \quad \text{where } x = 0, 1, 2, \ldots, n.$$

37. Find the factorial moments of the **Polya distribution** described by the following probability density function:

$$p_X(x) = \frac{\binom{\alpha + x - 1}{x}\binom{\beta + N - x - 1}{N - x}}{\binom{\alpha + \beta + N - 1}{N}},$$

    where $x = 0, 1, \ldots, N$, α, $\beta > 0$. Using this, write the mean and the variance.

38. Given the distribution function

$$F_X(x) = 2^{2/3}\pi^{-1/2}x e^{-x^4/2}, \quad x > 0,$$

    find the mean and the variance.

39. Complete the details of the integrals involved in the derivation of the moments of the Gaussian distribution.

40. Given the following probability density of X:

$$f(x) = \begin{cases} c_0 x^4(1-x)^4, & 0 < x < 1, \\ 0, & \text{elsewhere.} \end{cases}$$

    (i) Find the normalization constant c_0.

    (ii) Find the mean m and the variance σ^2.

    (iii) What is the probability that X will take a value within two standard deviations of the mean?

    (iv) Compare your result with the Chebyshev's theorem.

41. Find a bound on the probability that a random variable is within 4 standard deviations of its mean.

42. Using the Gaussian probability distribution, find the probability of an event to be found within 3 standard deviations of its mean and compare your result with Chebyshev's theorem.

43. Show that the mean, m, and the variation, σ^2, of the **hypergeometric distribution**:

$$p_X(x) = \frac{\dbinom{M}{x}\dbinom{N-M}{n-x}}{\dbinom{N}{n}},$$

where

$$x = 0, 1, 2, \ldots, n, \quad x \leqslant M \text{ and } n - x \leqslant N - M,$$

are given as

$$m = \frac{nM}{N}, \quad \sigma^2 = \frac{nM(N-M)(N-n)}{N^2(N-1)}.$$

44. Using a suitable coin toss simulator, which can be found on the internet, demonstrate the law of large numbers.

CHAPTER 19

INFORMATION THEORY

Information is a frequently used word in everyday language, and almost everybody has some notion of what it stands for. Today, as a technical term, information has also entered into many different branches of science and engineering. It is reasonable to start the beginning of modern information theories with the discovery of communication devices and computers. Before we introduce the mathematical theory of information, let us start by sorting out some of the complex and interrelated concepts that this seemingly simple word implies.

For a scientific treatment of information, it is essential that we define a meaningful measure for its amount. Let us start with an idealized but illustrative example. Consider large numbers of aeroplanes and cars of various kinds and models, all separated into small parts and put into separate boxes. Also assume that these parts are separated in such a way that by just looking at any one of them, you cannot tell whether it came from an aeroplane or a car. These boxes are then stored randomly in a warehouse with codes (x, y, z) printed on them indicating their locations in the warehouse. These parts are also separated in

Essentials of Mathematical Methods in Science and Engineering, Second Edition. Selçuk Ş. Bayın.
© 2020 John Wiley & Sons, Inc. Published 2020 by John Wiley & Sons, Inc.

such a way that they can be brought together with only a few relatively simple instructions and tools like screwdrivers, wrenches, etc. Using the parts in the warehouse, in principle, it is now possible for anybody who can operate a few simple tools to build a car or an aeroplane. Instructions will always be simple like: Take the parts in the boxes at (x_1, y_1, z_1) and (x_2, y_2, z_2), bring their indicated sides together, and then connect them with a certain size screw, bolt, etc. At each step of the way, the assembler does not have to know what is being built, nor will there be any decisions to make. Obviously, building a full-size luxury car is going to require a lot more instructions than for a subcompact car. Similarly, a Boeing 747 will require much more information than either one of them. This shows that it should be possible to define a measure for the amountof information, at least for comparable tasks. In the mathematical theory of information, we use **bit** as the unit of information, where 1 bit is the information content of the answer to a binary question like the right or left? top or bottom? etc.

This example also contains several other important elements of the information theory such as the data, information, knowledge, and goal, which will be discussed shortly. In the aforementioned example, data consist of the locations of the needed parts, information is the set of instructions consisting of the sequence in which the parts are to be put together and how, knowledge is the ability to understand and to follow the instructions. Finally, the goal is the product. Adding purposeto these, we can draw the block diagram in Figure 19.1 for a basic information processing mechanism, where information is included as a part of knowledge. In the aforementioned example, purpose lies within whoever supplies the information to the assembler. Only the person who supplies the information knows what the final product looks like and why it is being built.

An important aspect of information is its **value** or **relevance**, which is subjective and naturally very hard to quantify. Sometimes 1b of information like the answer to a simple binary question as go or no go, buy or sell, push or don't push may have life or death consequences to a person, to a company, or even to an entire nation. Information that makes one person respond as "great" may only get "so what" from another. Yet another person may respond as "what else is new?" In the last case, can we still consider it as information to that person?

In summary, information depends not only on the receiver but also on where and when it is received. Being too late, too little, or out of place may easily render otherwise valuable information worthless. Besides, information not only depends on the receiver but also on the sender and how reliable the sender is. In many instances, the sender may be trying to manipulate the receiver's decision processes by deliberately giving carefully selected wrong or incomplete

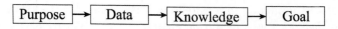

Figure 19.1   Basic information processing mechanism.

information. When the sender is not reliable, the receiver may be perplexed even by the correct information, hence lose precious time in reaching a critical decision. These become significant tools for manipulating one's opponent, especially when important decisions have to be made over a very short period of time.

Another important aspect of information is how it is transmitted. This process always involves some kind of **coding** and then decoding device so that the sender and the receiver can understand each other. In the above example, instructions can be transmitted via written documents or e-mail. The **security** of the channels used for transmission is also another major concern and brings a third party into the system. Coding theory is primarily interested in developing a spy-proof means of communications.

We also need ways to store information for future uses. In such cases, problems regarding the efficient and safe means of information **storage** and its rapid and reliable retrieval are among the major technological challenges. Sometimes, the information we have may involve gaps or parts that have changed or parts that cannot be read for a number of reasons. A major problem of information science is the need to develop methods of recovering the missing or degraded information.

So far, the information we talked about has a sender and a receiver with some communication device in between. Sometimes, the information we require is in the environment, which could be social, political, electronic, or physical. In such cases we need mechanisms to extract the needed information from the environment quickly and correctly and possibly without affecting the environment.

From the above introduction, one may get the impression that information is just a bag of worms that will defy mathematical treatment for years to come. In part this is true. However, beginning with the classical paper of Shannon in 1948 [1, 2], mathematical theory of information has become a serious interdisciplinary science at the crossroads of mathematics, physics, statistics, psychology, computer science, and electrical engineering. Its impact on the development of CDs, DVDs, smart phones, internet, search engines, and the study of linguistics and of human perception has been extremely important.

Even the part of information that has defied proper mathematical treatment is being carefully studied and extensively used in political science and economics for manipulating target audiences at rational levels through the controlled flow of information over various direct or indirect channels without the subject noticing. Also, certain channels, when they cannot be closed or censored, can be rendered useless by filling them with irrelevant information. This is essentially lowering the signal-to-noise ratio of that channel. Such tactics with military and political applications are usually discussed within the context of **perception management**, where deceit and disinformation are important tools. Today, perception management is widely used by companies against consumers, various organizations against the public, or political parties and governments against the voters. Perception management is usually used for good causes like improving company or product image, or raising public awareness on an important issue. However, since the target audiences can loose or have deformations in their sense of reality and ethical values, it could also have rather dire consequences threatening national security. In this regard, it is also a part of modern information warfare, which should be combatted seriously with matching wits. Note that disinformation should not be confused with **misinformation**, which is not intentional.

19.1   ELEMENTS OF INFORMATION PROCESSING MECHANISMS

We have mentioned that the basic elements of information processing can be shown by the block diagram in Figure 19.1. We will now discuss what each term stands for in greater detail and add more structure to this diagram. In general, the subject of an information processing mechanism could be a factory, a person, a team of individuals, a company, a social network, or even an entire nation. All these subjects could be considered as having a collective intelligence and memory – in short, a "mind." For each one of these subjects – depending on their purpose – data, information, knowledge, and goal have different meanings. The effect of the information processing mechanism on the subject is to change it from one state of mind, $|\Psi_1\rangle$, to another, $|\Psi_2\rangle$. The state of mind function is going to depend on a number of internal parameters, which at this point is not important for us. For an individual, $|\Psi\rangle$ represents a certain pattern of neural networks in that person's brain.

Let us now consider a typical scientific endeavor, which is basically an information processing mechanism that can be generalized to any other situation relatively easily (Figure 19.2). A scientist starts the process with a definite state of mind, $|\Psi_1\rangle$, with an **idea** or **incentive**, which helps to define the **purpose** of the entire process. The next step is probably one of the most critical steps, that is, the **initiative**. What starts the entire process is the initiative, not the existence of purpose or incentive. Life is full of examples where two individuals who had the same idea/purpose and the knowledge, but somehow only one of them got the whole process started and finalized with a definite result. The next step involves **data**. It usually involves raw numbers consisting of the outputs of some measuring instruments in terms of currents, voltages, times, etc. After the data are reduced, we obtain what we call **information**. They are now luminosities, electric fields, distances, temperatures, pressures, etc. This information can now be fed into the next box, which involves **knowledge**. In scientific endeavors, knowledge is a model or a theory like Maxwell's theory. In this case, information is the **initial conditions** needed by the model, which will produce **new information** that needs to be checked by new experiments and data. If the **new data** support the new information, we will have gained a better understanding of the phenomenon that we are investigating. Or, this may also lead to new questions and thus to new incentives for new scientific endeavors, which will take us back to the beginning of this process to start all

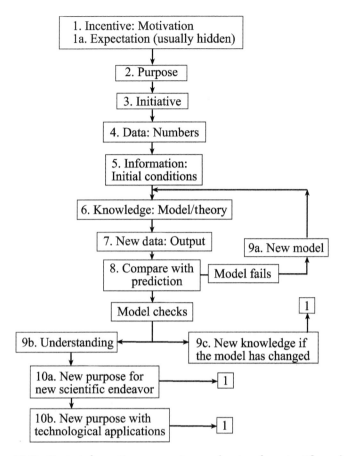

Figure 19.2   Basic information processing mechanism for scientific endeavor.

over again. Our improved understanding may also bring in sight new technological applications, which could start a new round of information processing.
If the new data do not support the new information, we have three options:

(i) We may go back and modify our model and resume the process.
(ii) If we trust our model, we may recheck our data and information.
(iii) If nothing works, we may go back to the beginning and redefine our
      purpose.

Another important aspect of this entire process, which is usually ignored,
is its time scale. One of the things that affects the time scale is the subject's
expectation, which is what the subject expects or hopes will happen when
the goal is realized. Expectation could be quite different from purpose, and it
is usually neither openly declared nor clearly formulated by the subject. When
expectations are deliberately hidden, they are called **ulterior motives**, which
could be an important part of the strategy. **Anticipation** is another thing
that affects the time scale, which involves projections about what may happen

along the way. This not only affects the speed of the entire process by making it possible to skip some of the time-consuming intermediate steps but also helps to avoid potential hazards and traps that may halt or delay the process.

At any point along the way, targets of opportunity may pop up, the merits of which may outweigh the potential gains of the ongoing process. This may make it necessary to abandon or postpone the existing process and to start a new one, which may not even be remotely connected to the old one. Among all these steps, **incentive**, **expectation**, and **understanding** are the ones that most critically rely on the presence of a **human brain**. Understanding naturally occurs when the subjects "mind" reaches a new state, $|\Psi_2\rangle$. The entire process can be represented in terms an operator $G(2,1)$, which represents all the intermediate steps as

$$|\Psi_2\rangle = G(2,1)|\Psi_1\rangle. \tag{19.1}$$

In real-life situations, there are usually several ongoing processes collaborating or competing with each other that complicates the structure of $|\Psi_1\rangle$, $|\Psi_2\rangle$, and $G(2,1)$. For another person – say an entrepreneur – data, information, knowledge, etc., mean different things but the aforementioned process basically remains intact.

19.2 CLASSICAL INFORMATION THEORY

So far, we have discussed the conceptual side of information. Classical theory of information quantifies certain aspects of information but does not concern itself with by whom and why it was sent and why one should have it. It is also not interested in its potential effects on both the sender and the receiver. Shannon's classical information theory basically concerns itself with computers, communication systems, and control mechanisms. It treats information content of a message independent of its meaning and purpose and defines a **measure** for its **amount**, thus allowing one to study its degradation during transmission, storage, and retrieval. This theory works with qualitative expressions concerning the possible outcomes of an experiment and the amount of information to be gained when one of the possibilities has actually occurred. In this sense, it is closely related to the **probability theory**. This is also the reason why the term **statistical information** is often used in classical information theory. To emphasize the statistical side of information, it is frequently related to dice rolls, coin tosses, or pinball machines. We prefer to work with a different model with properties that will be useful to us when we discuss other aspects of information and its relation to decision theory. Our system is composed of a car with a driver and a road map, where at each junction the road bifurcates. The road could have as many junctions as one desires and naturally a corresponding number of potential destinations. We can also have an **observer** who can find out to which destination the car has arrived by checking each terminal. If the driver flips a coin at each junction to choose which way to go, we basically have the Roederer's pinball machine [3], where the car is replaced by a ball and the gravity does the driving. Junctions are the pins of the pinball machine that deflect the balls to the right or the left with a definite probability, usually equally, and the terminals are now the bins of the pinball machine.

For the time being, let us continue with the pinball machine with one pin and two bins as shown in Figure 19.3. Using binary notation, we show the state of absence of a ball in one of the bins, left or right, as $|0\rangle$ and its presence as $|1\rangle$. Note that observing only one of the bins is sufficient. In classical information theory, if the ball is not in one of the bins, we can be certain that it is in the other bin. If the machine is operated N times, the probability of observing $|0\rangle$ is $p_0 = N_0/N$, where N_0 is the number of occurrences of $|0\rangle$ in N runs. Similarly, the probability of observing the state $|1\rangle$ is $p_1 = N_1/N$, where N_1 is the number of occurrences of the state $|1\rangle$. Naturally, $N = N_0 + N_1$ and $p_0 + p_1 = 1$. In the symmetric situation, $p_0 = p_1 = 0.5$, by checking only one of the bins, an observer **gains** on the average 1b of information. That is the amount of information equivalent to the answer of a yes or no question. In the classical theory, the ball being in the right or the left bin has nothing to do with the observer checking the bins. Even if the observer is not interested in checking the bins, the event has occurred, and the result does not change. Let us now consider the case where the observer has adjusted the pin so that the ball always falls into the left bin. In this case, $p_0 = 0$ and $p_1 = 1$. Since the observer is certain that the ball is in the left bin, by checking the bins, left or the right, there will be no information gain. In other words, the observer already knows the outcome, and no matter how many times the pinball machine is run, the result will not change. This is a case where the observer may respond as "What else is new?" If the pin is adjusted so that p_1 is very close to zero, but not zero, then among N observations, the observer will find the ball in the left

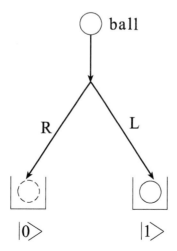

Figure 19.3   Pinball machine with one pin.

bin only in very very few of the cases. Now the response is more likely to be "Wow!"

The classical information theory also defines an objective **information value** – as Roederer calls it, the "Wow" factor. It is also called the **novelty value** or the **information content**, which is basically a measure of how surprising or how rare the result is to the observer. Information value, I, will naturally be a function of how probable that event is among all the other possible outcomes, that is $I(p)$. To find a suitable expression, let us call the information value of observing the state $|0\rangle$ as I_0 and $|1\rangle$ as I_1. For the case where we adjusted the pin so that the ball always falls into the same bin, that is $p_i = 1$ ($i = 0$ or 1), we define the information value as $I_i = 0$. This is so, since the result is certain and no information is to be gained by an observation. We also define $I_i \to \infty$ for the cases where $p_i \to 0$. For independent events, $a, b, \ldots$, we expect the information value to be additive, that is $I = I_a + I_b + \cdots$. From the probability theory, we know that for independent events with probabilities $p_a, p_b, \ldots$, the total probability is given as the product $p = p_a p_b \ldots$, hence we need a function satisfying the relation

$$I(p_a p_b \ldots) = I_a(p_a) + I_b(p_b) + \cdots = I(p_a) + I(p_b) + \cdots. \tag{19.2}$$

We also require the information value to satisfy the inequality $I(p_i) > I(p_j)$, when $p_i > p_j$. A logarithmic function of the form

$$I(p) = -C \log p, \ C = \text{const.} \tag{19.3}$$

satisfies all these conditions. Since $p \leq 1$, the negative sign is needed to ensure $I > 0$. To determine the constant C, we use the symmetric case of the pinball machine, where the average information gain is 1b. Setting $I(0.5)$ to 1:

$$I(0.5) = -C \log(0.5) = 1, \tag{19.4}$$

we find $C = 1/\log 2$. Thus, the **information value** for the **pinball machine** can be written as

$$I_i = -\frac{1}{\log 2} \log p_i = -\log_2 p_i, \ i = 0 \text{ or } 1, \tag{19.5}$$

where $\log_2$ is the logarithm with respect to base 2.

19.2.1   Prior Uncertainty and Entropy of Information

Since Shannon and Weaver [2] were interested only in the information content of the message received with respect to the set of all potentially possible messages that could be received, they called I the **information content**. One of the most important quantities introduced by the Shannon's theory answers the question: Given the probabilities of each alternative, how much information on the average can we expect to gain when one of the possibilities is realized beforehand? In other words, what is the **prior uncertainty** of the outcome? For this, Shannon introduced the entropy of the source of information or in short the **entropy of information**, H, as [1]

$$H = p_0 I_0 + p_1 I_1 = -p_0 \log_2 p_0 - p_1 \log_2 p_2. \qquad (19.6)$$

Note that H is basically the **average information value** or the **expected information gain**. For the pinball machine, if the probability of finding the ball in the left bin is p, then the probability of not finding it is $(1 - p)$, which gives the H function as

$$H = -p \log_2 p - (1 - p)\log_2(1 - p). \qquad (19.7)$$

For a pinball machine with two possible outcomes, the entropy of information, H, is shown in Figure 19.4. For the symmetric case, $p = 0.5$, the expected gain of information is 1b. When $p = 0$ or $p = 1$, H is zero since there is no prior uncertainty and we already know the result.

We can extend our road map or the pinball machine to cases with more than two junctions or pins. At each junction or pin, there will be two possibilities. In general, we can write H as

$$H = -\sum_{i=0}^{N-1} p_i \log_2 p_i, \qquad (19.8)$$

where N is the number of possible final outcomes and $\sum_i p_i = 1$. When all the possible outcomes are equiprobable, H has an absolute maximum

$$H = \log_2 N. \qquad (19.9)$$

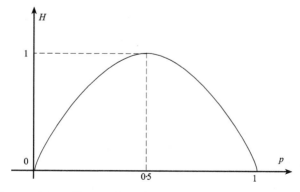

Figure 19.4 Shannon's H function for the pinball machine.

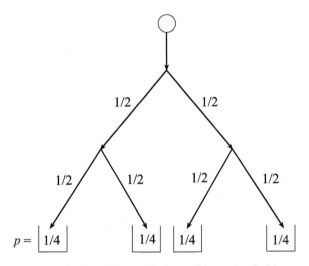

Figure 19.5   Pinball machine with 4 possible equiprobable outcomes.

From the aforementioned definition [Eq. (19.8)], it is clear that H is zero only when all p_i except one is zero. Since $\sum_i p_i = 1$, the nonzero p_i is equal to 1. Note that when $p_i = 0$, we define $p_i \log_2 p_i$ as 0. In the case of four equiprobable possible outcomes (Figure 19.5), probability of finding the ball in one of the bins in the final level is 1/4. In this case, $H = 2b$.

Example 19.1. *Pinball machine:* In the binary pinball machine, if the pin is adjusted so that $p_0 = 2/3$ and $p_1 = 1/3$, the expected information gain, H, is

$$H = \frac{2}{3} \log_2 \frac{3}{2} + \frac{1}{3} \log_2 3 = 0.92b. \tag{19.10}$$

Example 19.2. *Expected information gain:* With the prior probabilities

$$0.3, \ 0.2, \ 0.2, \ 0.1, \ 0.2, \tag{19.11}$$

H is found as

$$H = - \ [0.3 \log_2 0.3 + 3(0.2) \log_2 0.2 + 0.1 \log_2 0.1] = 2.25b. \tag{19.12}$$

Similarly, for a single roll with an unbiased die, which can be viewed as a pinball machine with a single pin, where the ball has six all equiprobable paths to go, the value of H is obtained as

$$H = -6 \left(\frac{1}{6}\right) \log_2 \frac{1}{6} = 2.58b. \tag{19.13}$$

Example 19.3. *Scale of H:* We have the freedom to choose the basis of the logarithm function in the definition of H. In general, we can use $\log_r$, where r is the number of possible outcomes. In the most frequently used binary case, r is 2. However, a change of basis from r to s is always possible. Taking the logarithm of $x = r^{\log_r x}$ to the base s, we write

$$\log_s x = [\log_s r] \log_r x, \quad x > 0, \tag{19.14}$$

and obtain the scale relation between H_s and H_r as

$$H_s = [\log_s r] H_r. \tag{19.15}$$

We can now use logarithms with respect to base 10 to find H_2 as

$$H_2 = \left[\frac{1}{\log_{10} 2}\right] H_{10} = 3.32 H_{10}. \tag{19.16}$$

In what follows, unless otherwise specified, we use the binary basis. We also use the notation where log means the logarithm with respect to base 10 and ln means the logarithm with respect to base e as usual.

19.2.2 Joint and Conditional Entropies of Information

The Case of Joint Events

Consider two chance events, A and B, with m and n possibilities, respectively. Let $p(i,j)$ be the **joint probability** of the ith and jth possibilities occurring for A and B, respectively. For the joint event, we write the entropy of information as

$$H(A, B) = -\sum_i^m \sum_j^n p(i,j) \log_2 p(i,j). \tag{19.17}$$

For the individual events, A and B, we have, respectively,

$$H(A) = -\sum_i^m \left[\sum_j^n p(i,j) \log_2 \sum_{j'}^n p(i,j')\right], \tag{19.18}$$

$$H(B) = -\sum_j^n \left[\sum_i^m p(i,j) \log_2 \sum_{i'}^m p(i',j)\right]. \tag{19.19}$$

Comparing

$$H(A) + H(B) = -\sum_i^m \sum_j^n p(i,j) \left(\log_2 \left[\sum_{j'}^n p(i,j') \right] + \log_2 \left[\sum_{i'}^m p(i',j) \right] \right)$$

with Eq. (19.17) it is easily seen that the inequality

$$\boxed{H(A,B) \leq H(A) + H(B)} \tag{19.20}$$

holds. Equality is true for independent events, where

$$p(i,j) = p(i)p(j). \tag{19.21}$$

Equation (19.20) says that the uncertainty or the average information value of a joint event is always less than the sum for the individual events.

The Case of Conditional Probabilities

We write the conditional probability [Eq. (18.32)] of B assuming the value j after A has assumed the value i as

$$p(j/i) = \frac{p(i,j)}{\sum_{j'}^n p(i,j')}, \tag{19.22}$$

where the denominator is the total probability of A assuming the value i that is acceptable by B. The conditional entropy of information of B, that is, $H(B/A)$, is now defined as the average of the conditional entropy:

$$\left[\sum_j^n p(j/i) \log_2 p\,(j/i) \right], \tag{19.23}$$

as

$$H(B/A) = -\sum_i^m \left(\sum_{j'}^n p(i,j') \right) \left[\sum_j^n p(j/i) \log_2 p\,(j/i) \right] \tag{19.24}$$

$$= -\sum_i^m \sum_j^n \left(\sum_{j'}^n p(i,j') \right) p(j/i) \log_2 p\,(j/i) \tag{19.25}$$

$$= -\sum_i^m \sum_j^n p(i,j) \log_2 p\,(j/i), \tag{19.26}$$

where in the last step we have substituted $p(i,j)$ for $\left[\sum_{j'}^n p(i,j') \right] p(j/i)$ [Eq. (19.22)]. The quantity $H(B/A)$ is a measure of the average uncertainty in B when A has occurred. In other words, it is the expected average information gain by observing B after A has happened. Substituting the value of $p(j/i)$

[Eq. (19.22)] and using Eq. (19.18), and after rearranging, we obtain

$$H(B/A) = -\sum_i^m \sum_j^n p(i,j) \log_2 p(i,j) + \sum_i^m \left\{ \left[\sum_j^n p(i,j) \right] \log_2 \sum_{j'}^n p(i,j') \right\}$$

(19.27)

$$= H(A,B) - H(A).$$

(19.28)

Thus,

$$\boxed{H(A,B) = H(A) + H(B/A).}$$

(19.29)

Using Eqs. (19.20) and (19.29), we can write the inequality

$$H(A) + H(B) \geq H(A,B) = H(A) + H(B/A),$$

(19.30)

hence

$$\boxed{H(B) \geq H(B/A).}$$

(19.31)

In other words, the average information value to be gained by observing B after A has been realized can never be greater than the average information value of the event B alone.

Entropy of Information for Continuous Distributions

For discrete set of probabilities, $p_1, p_1, \ldots, p_n$, the entropy of information was defined as

$$H = -\sum_i^n p_i \log_2 p_i.$$

(19.32)

For continuous distribution of probabilities, H is defined as

$$\boxed{H = -\int_{-\infty}^{\infty} p(x) \log_2 p(x) dx.}$$

(19.33)

For probabilities with two arguments, we can write $H(x,y)$ as

$$H(x,y) = -\int \int p(x,y) \log_2 p(x,y) \, dx \, dy.$$

(19.34)

Now the conditional probabilities become

$$H(y/x) = -\int \int p(x,y) \log_2 \frac{p(x,y)}{p(x)} \, dx \, dy,$$

(19.35)

$$H(x/y) = -\int \int p(x,y) \log_2 \frac{p(x,y)}{p(y)} \, dx \, dy.$$

(19.36)

Example 19.4. H for the Gaussian distribution: For the Gaussian probability distribution:

$$p(x) = \frac{1}{\sqrt{2\pi}\sigma} e^{-(x^2/2\sigma^2)}, \tag{19.37}$$

we can find H as

$$H(x) = -\int_{-\infty}^{\infty} p(x) \log_2 p(x) \, dx \tag{19.38}$$

$$= \left(\frac{1}{\ln 2}\right) \left[\int_{-\infty}^{\infty} p(x) \ln \sqrt{2\pi}\sigma \, dx + \int_{-\infty}^{\infty} p(x) \frac{x^2}{2\sigma^2} \, dx\right] \tag{19.39}$$

$$= \left(\frac{1}{\ln 2}\right) \left(\ln \sqrt{2\pi}\sigma + \frac{\sigma^2}{2\sigma^2}\right) = \left(\frac{1}{\ln 2}\right) \left(\ln \sqrt{2\pi}\sigma + \ln \sqrt{e}\right) \tag{19.40}$$

$$= \left(\frac{1}{\ln 2}\right) \left(\ln \sqrt{2\pi e}\sigma\right). \tag{19.41}$$

19.2.3 Decision Theory

We now go back to our car and driver model. For the sake of argument, we use three junctions or "decision" points shown as 1, 2, and 3 (Figure 19.6). At each junction, the road bifurcates and the driver has to decide which way to go. This allows us to add purpose, decisions, and strategy into our model, where the driver has a specific target, say, a friend in terminal 3. When $p_i = 1/8$, $i = 1, \ldots, 8$, we have the maximum entropy of information of this system, which is $3b$:

$$H = -\sum_{i=1}^{8} p_i \log_2 p_i = \sum_{i=1}^{8} p_i I_i = 3b. \tag{19.42}$$

If the driver is not given any instructions and flips, a fair coin at each junction to decide which way to go, the probability of reaching terminal 3 in a given try is $1/8$. Reaching terminal 3 will naturally require numerous tries involving backtracking and recalling past decisions and then reversing to the other alternative. On the average, only one out of eight tries the driver will be successful. Once the terminal 3 is reached, the driver will have acquired information with the information value $I_3 = 3b$:

$$I_3 = -\log_2 \frac{1}{8} \tag{19.43}$$

$$= -\frac{1}{\log 2} \log 0.125 \tag{19.44}$$

$$= 3b. \tag{19.45}$$

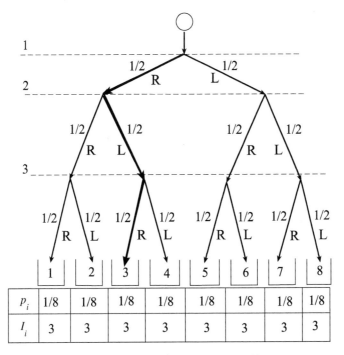

	1	2	3	4	5	6	7	8
p_i	1/8	1/8	1/8	1/8	1/8	1/8	1/8	1/8
I_i	3	3	3	3	3	3	3	3

Figure 19.6   Car and driver with 8 possible targets.

If the driver is given the sequence of instructions about which way to turn at the junctions, like *right–left–right*, the information content of which is $3b$, then he or she can reach the desired terminal with certainty. However, the information expected to be gained by the driver when he or she gets there is now zero. The information given by the driver's friend has removed all the prior uncertainty.

Now, consider a case where some of the roads are blocked as shown in Figure 19.7. Now, the H value of the system is

$$H = -\sum_{i=1}^{8} p_i \log_2 p_i = \sum_{i=1}^{8} p_i I_i \tag{19.46}$$

$$= -[3(0.25)\log_2 0.25 + 2(0.125)\log_2 0.125] \tag{19.47}$$

$$= -\left[\frac{1}{\log 2}\right][3(0.25)\log 0.25 + 2(0.125)\log 0.125] \tag{19.48}$$

$$= 2.25b. \tag{19.49}$$

Since the prior uncertainty has decreased, H is naturally less than $3b$. However, the driver still has to make three binary decisions. Hence, the needed information to reach the terminal 3 is still worth $3b$. This is also equal to the information (novelty) value I_3 of terminal 3. The difference originates from

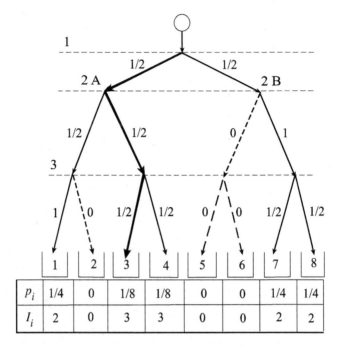

Figure 19.7   Car and driver with some of the roads blocked.

the fact that in Shannon's theory, H is defined independent of purpose. It is basically the average information expected to be gained by an observer from the entire system. It doesn't matter which terminal is intended and why. On the other hand, if the driver aims to reach terminal 7, then there are only two binary decisions to be made, *left–right*; hence the amount of information needed is 2b. Note that at junction 2B (Figure 19.7), the right road is blocked. Hence, there is no need for a decision by the driver. The information value I_7 of terminal 7, is also 2b.

19.2.4   Decision Theory and Game Theory

To demonstrate the basic elements of the decision theory and its connections with the game theory, we consider a case where a company has to decide whether it should expand its business or not. In this case, each terminal of the decision tree has a prize/profit or penalty/loss waiting for our subject. Advisors tell that if the company expands (E), the electronic goods department (EGD), in recession (R) it will lose $50 000. However, if the economy remains good (EG), the electronic goods department will make $120 000. On the other hand, if the company expands the household items department (HHI), and gets caught in recession, it will lose $30 000 but if the economy remains good, it will make $80 000. Finally, if the company does not expand (DNE), in recession, it will make $2000; and if the economy remains good, it will make $30 000.

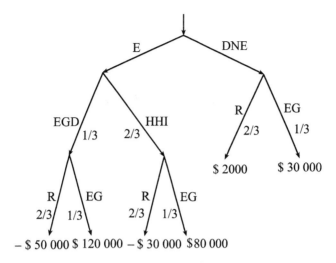

Figure 19.8   Decision tree for the merchant.

Advisors also assess that the probabilities for the economy going into recession and remaining good are 2/3 and 1/3, respectively. They also think that if the company expands, the probabilities for the electronic goods and household items departments outperforming each other are 1/3 and 2/3, respectively. What should this company do?

We can draw the decision tree shown in Figure 19.8. We can calculate the company's expected losses or gains for the next fiscal year as follows:

If the company expands the electronic goods department, the expected gain is

$$-(50\ 000)\frac{1}{3} \times \frac{2}{3} + (120\ 000)\frac{1}{3} \times \frac{1}{3} = 2222. \tag{19.50}$$

If the company expands the household items department, the expected gain is

$$-(30\ 000)\left(\frac{2}{3}\right) \times \left(\frac{2}{3}\right) + (80\ 000)\left(\frac{2}{3}\right) \times \left(\frac{1}{3}\right) = 4444 \tag{19.51}$$

If the company does not expand, the expected gain is

$$2000\left(\frac{2}{3}\right) + 30\ 000\left(\frac{1}{3}\right) = 11\ 333. \tag{19.52}$$

Since the company's expected profit in the last option, $11 333, is greater than the previous two cases, it should delay expanding capacity for another year. This method is called **Bayes' criteria** and works when we can identify the decision points and assess the probabilities.

If the company has no idea about the probabilities and if the owner is a pessimist afraid of losing money, then the company should decide to wait for another year and avoid the risk of losing \$50 000. This is called the **minimax criteria**. That is, you minimize the maximum expected loss. Minimax criteria is among the many that can be used in such situations. In this example, the company appears as if it is playing a game with the economy. The company has four moves: expand or wait and if it decides to expand, expand the electronic goods department or the household items department. The company also gets to make the first move. On the other hand, economy has two moves, go into recession or remain good, and it does not care about what the company has decided.

For a game with two players, A and B, each having two moves, a_1, a_2 and b_1, b_2, respectively, we can write the following **payoff matrix**, which is also called the **normal form** representation:

$$\text{Player } A$$

		a_1	a_2
Player B	b_1	$L(a_1, b_1)$	$L(a_2, b_1)$
	b_2	$L(a_1, b_2)$	$L(a_2, b_2)$

(19.53)

In this representation, $L(a_i, b_j)$ is the loss function for the player A, when A chooses strategy a_i and B chooses strategy b_j, where $i, j = 1, 2$. For the player A, the loss function $L(a_i, b_j)$ is positive for losses and negative for gains and vice versa for the player B. Since whatever one player wins, the other player loses, such games are called **zero-sum** games. In other words, there is no cut for the house, and no capital is neither generated nor lost. Zero-sum games could be symmetric or asymmetric under the change of the identities of the players. That is, in general, $L(a_i, b_j) \neq -L(b_j, a_i)$. Depending on which player makes the first move, the above payoff matrix can also be shown with the decision trees in Figure 19.9. These are called **extensive form** representations. In extensive

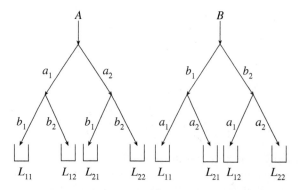

Figure 19.9   Decision trees for the players A and B, where $L_{ij} = L(a_i, b_j)$, $i, j = 1, 2$.

form games, players act sequentially and they are aware of the earlier moves made. There are also games where players act without knowing what their opponent has decided. Games where players act simultaneously are essentially games of this type. Games where the players act with incomplete information about their opponents moves have also been designed. In such games, normal form representation is usually preferred over the extensive form.

Let us now consider the game depicted in Figure 19.10, where the player A makes the first move at the first decision point, where it has two alternatives, a_1 and a_2. At the second decision point, player B gets to make its move with four choices: b_1, b_2 when A decides a_1 and b_3, b_4 when A decides a_2. We now introduce two random variables, x and y, that can only take the values 0 and 1 at the decision points 1 and 2, respectively. The decision function $d_1(x)$ at point 1 is defined as

$$d_1(x) = \begin{cases} a_1, & x = 0, \\ a_2, & x = 1 \end{cases} \qquad (19.54)$$

Similarly, the decision functions for the player B are defined as

$$D_1(y) = \begin{cases} b_1, & y = 0, \\ b_2, & y = 1, \end{cases} \qquad D_2(y) = \begin{cases} b_3, & y = 0, \\ b_4, & y = 1. \end{cases} \qquad (19.55)$$

In this game, the random variables can be tied to the actions of a third element. For example, in designing a strategy for a political confrontation, one often has

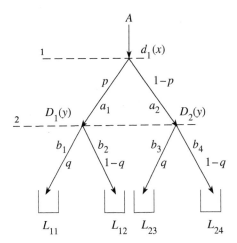

Figure 19.10 Statistical game, where $L_{ij} = L(a_i, b_j)$, $j = 1, 2$ when $i = 1$ and $j = 3, 4$ when $i = 2$.

to factor in the potential responses of other countries. Let the player A assign the probabilities p and q for the potential actions of this third element, which affects the decisions of both A and B (Figure 19.10). To compare the merits of all these decisions for A, we also define the **expected risk/loss function** $R(a_i, b_j)$ as

$$R(a_i, b_j) = E\{L(d_1(x), D_{j(x)}(y))\}, \ j(0) = 1 \text{ and } j(1) = 2, \qquad (19.56)$$

where the expected value E has to be taken with respect to the random variables x and y as

$$R(a_1, b_1) = pqL(a_1, b_1),$$
$$R(a_1, b_2) = p(1 - q)L(a_1, b_2),$$
$$R(a_2, b_3) = (1 - p)qL(a_2, b_3),$$
$$R(a_2, b_4) = (1 - p)(1 - q)L(a_2, b_4).$$

Player A can now write the payoff matrix

$$\begin{bmatrix} R(a_1, b_1) & R(a_2, b_3) \\ R(a_1, b_2) & R(a_2, b_4) \end{bmatrix}$$

and minimize the expected maximum losses or risks by using the minimax criteria.

Example 19.5. *Normal form games and payoff matrices:* In decision theory, given the alternatives, it is important to make an informed choice. Depending on the situation, costs or gains could stand for many different things, and the payoff matrices can be made as complex as one desires. Consider the following payoff matrix:

Player A

		a_1	a_2
Player B	b_1	7	-4
	b_2	3	5

(19.57)

The corresponding decision trees are given in Figure 19.11.

When no probabilities can be assigned, using the first decision tree and the minimax criteria, player A chooses strategy a_2 to avoid the risk of losing 7 points in case everything goes bad (remember that plus sign is for losses for player A).

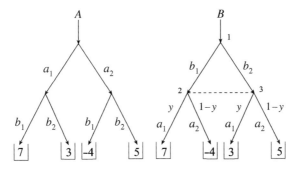

Figure 19.11 Decision trees for the players A and B in Example 19.5.

Example 19.6. *Another game and strategy for A:* Let us consider player A in the second decision tree (Figure 19.11). Now, B makes the first move and A has to decide without knowing what B has decided. Since A acts without seeing the opponent's move, we connected points 2 and 3 by a dotted line. Another way to look at this game is that A and B decide simultaneously. We now show how A can use a random number generator to minimize its maximum loss by adjusting the odds for the two choices, a_1 and a_2, as y and $(1 - y)$ at points 2 and 3, respectively. If B chooses strategy b_1, then A can expect to lose

$$E_{b_1} = 7y - 4(1 - y) \tag{19.58}$$

points. If B chooses strategy b_2, then A can expect to lose

$$E_{b_2} = 3y + 5(1 - y) \tag{19.59}$$

points. If we plot E_{b_1} and E_{b_2} as shown in Figure 19.12, we see that A can minimize maximum expected loss by following strategy a_1, 9 out of 13 times, and by following strategy a_2, 4 out of 13 times. If this is a one-time decision, A can use a random number generator to pick the appropriate strategy with the odds adjusted accordingly.

The theory of games is a relatively new branch of mathematics closely related to the probability theory, information theory, and the decision theory. Since the players and payoffs could have many different forms, it has found a wide range of applications to economic phenomena such as auctions and bargaining. Other important applications are given in biology, computer science, political science, and philosophy [4–7].

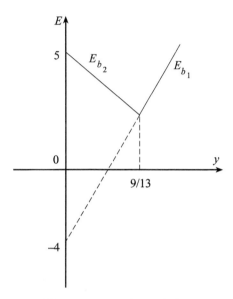

Figure 19.12 E_{b_1} and E_{b_2}.

19.2.5   Traveler's Dilemma and Nash Equilibrium

An interesting non-zero-sum game that Basu introduced in 1994, where each player tries to maximize their return with no concern to what the other player is getting, has attracted a lot of attention among game theorists. Even though the game can be played among any number of players, it is usually presented in terms of two players as the traveler's dilemma.

An airline loses two pieces of luggage, both of which contain identical antiques that belong to two separate travelers. Being afraid that the travelers will claim inflated prices, the airline manager separates the passengers and tells them that the company is liable for up to $100 per luggage and asks them to write an integer between and including 2 and 100. The manager also adds that if they both write the same number, the company will honor that as the actual price of the antique and pay both travelers that amount. In case they write different numbers, the company will take the lower number as the actual price of the antique and pay that amount to both travelers. However, in this case, the company will deduct $2 from the traveler who wrote the larger number as penalty and add $2 as bonus to the traveler who wrote the smaller number. According to these rules, we can now construct the following table, where the first column represents the choices for the first traveler, John, and the first row represents the choices for the second traveler, Mary. In the payoff matrix, the numbers in parentheses represent the reimbursements that John and Mary will receive, respectively.

Payoff matrix for the traveler's game:

↓ 1\2 →	2	3	4	⋯	98	99	100
2	(2, 2)	(4, 0)	(4, 0)	⋯	(4, 0)	(4, 0)	(4, 0)
3	(0, 4)	(3, 3)	(5, 1)	⋯	(5, 1)	(5, 1)	(5, 1)
4	(0, 4)	(1, 5)	(4, 4)	⋯	(6, 2)	(6, 2)	(6, 2)
⋮	⋮	⋮	⋮	⋱	⋮	⋮	⋮
98	(0, 4)	(1, 5)	(2, 6)	⋯	(98, 98)	(100, 96)	(100, 96)
99	(0, 4)	(1, 5)	(2, 6)	⋯	(96,100)	(99, 99)	(101, 97)
100	(0, 4)	(1, 5)	(2, 6)	⋯	(96,100)	(97,101)	(100,100)

For example, the numbers in the third column of the second row is (5,1), which means that when John chooses 3 and Mary chooses 4, John gets \$5 and Mary gets \$1. In this case, the lower number is 3; hence, they both get \$3, but since John gives the lower number 3, a \$2 bonus is added to his reimbursement, while a \$2 penalty is deducted from the Mary's, thus bringing their total reimbursements to \$5 and \$1, respectively.

The question is, What numbers or strategy should the travelers choose? Assuming that both travelers are rational, let us see what the game theory predicts. We start with John. Since his aim is to get the maximum possible amount as reimbursement, he first thinks of writing 100. However, he immediately realizes that for the same reason Mary could also write 100. Hence, by lowering his claim to 99, John expects to pick up the \$2 bonus, thus increasing his return to \$101. Then on a second thought, he realizes that Mary, being a rational person like himself, could also argue the same way and write 99, thus reducing his return to \$99. Now, John could do better by pulling his claim down to 98, which will allow him to get \$100 with the bonus. Continuing along this line of reasoning, John cascades down to the smallest number 2. Since they are both assumed to be rational, Mary also comes up with the same number. Hence, the game theory predicts that both travelers write the number 2, which is the **Nash equilibrium** for this game. However, in practice almost all participants pick the number 100 or a number very close to it. The fact that the majority of the players get such high rewards by deviating so much from the Nash equilibrium is not easy to explain mathematically. Since it is not possible to refer to the vast majority of people as being irrational, some game theorists have questioned the merits of this game, while others have proposed various modifications.

In analyzing payoff matrices, a critical concept is the Nash equilibrium. If each player has chosen a strategy and no player can improve his or her situation unilaterally, that is, when the other players keep their strategy unchanged, the set of strategies and the corresponding payoffs correspond to a Nash equilibrium. In 1950, Nash showed in his dissertation that Nash equilibrium exists for all finite games with any number of players. There is an easy way to identify

Nash equilibrium for **pure strategy** games, which is particularly helpful when there are two players with each player having more than two strategies available. In such cases, formal analysis of the payoff matrix could be quite tedious. For a given pair of numbers in a cell, we simply look for the maximum of a column and check if the second member of the pair has a maximum of the row. When these conditions are met, as in the cell with $(2, 2)$ in the traveler's dilemma, then that cell represents the Nash equilibrium. An $N \times N$ payoff matrix can have $N \times N$ pure strategy Nash equilibria. In the traveler's dilemma there is only one Nash equilibrium. A **mixed strategy** is a strategy where the players make their moves randomly according to a probability distribution that tells how frequently each move is to be made. A mixed strategy can be understood in contrast to a pure strategy, where players choose a certain strategy with the probability 1. For example, in the traveler's dilemma game, if at least one of the travelers chooses his or her number by using a random number generator, then the game is called a mixed strategy game.

The concept of stability is very important in physical systems and has been investigated for many different kinds of equilibrium. Stability of Nash equilibria in mixed strategy games can be defined with respect to infinitesimal variations of the probabilities as follows: In a given Nash equilibrium, when the probabilities for one of the players are varied infinitesimally, the Nash equilibrium is stable

(i) if the player who did not change has no better strategy in the new situation and

(ii) if the player who did change is now playing strictly with a worse strategy.

When these conditions are met, a player with infinitesimally altered probabilities will quickly return to the Nash equilibrium. An important point is that stability of the Nash equilibrium is related to but not identical with the stability of a strategy. The dilemma in the traveler's dilemma game is in our difficulty in explaining why people choose something which the game theory deems as irrational and yet get such high rewards. With the hopes of coming up with an explanation, game theorists have introduced a number of different equilibrium concepts like **strict equilibrium**, the **rationalizable solution**, **perfect equilibrium**, and more. Yet in all these cases, one reaches the prediction (2,2) for the traveler's dilemma [8].

Game theory assumes that both travelers are rational and have the time to construct and analyze the payoff matrix correctly. The fact that so many players decide diametrically opposite to what the theory predicts and do well means that there are other important factors in the decision-making process that we use. In real-life situations, aside from being rational, we can also assume that, on the average, people are honest and will not view this as an opportunity to make extra cash, hence would be glad to get out even. However, in order to be able to exercise this option, players need to have access to a critical piece of information, which the game lacks – that is, how much they have actually paid for the antique. A realistic modification of the game would be to give each player a number chosen with a certain distribution representing the actual cost of the antique for that player. Most of the players will be given numbers close to each other within a reasonable bargaining range, say 10%. Fewer players will have paid a lot more or a lot less than the mean price for various reasons. When these factors are missing from the game, players will naturally view picking any number from 2 to 100 as their declared rightful choice; hence, to maximize their return, they will not hesitate to choose a number very close to the upper limit. Since the players are not given any clue whatsoever about the actual cost of the antique and since they are not taking away the other traveler's right to write whatever he or she wants, and also considering that the airline company is the strong hand, which after all is at fault, they will have no problem in rationalizing their action.

In summary, what is rational or irrational largely depends on the circumstances under which we are forced to make a decision and how much time and information we have available. In some situations, panicking may even be a rational thing to do. In fact, in times of crisis when there is no hope or time for coming up with a strategy that will resolve our situation, to avoid freezing or decision paralysis, our brains are designed to panic. This helps us to come up with a strategy of some sort with the hopes that it will be the right one. Decisions reached through panic are not entirely arbitrary. During this time, our brain and our subconscious mind goes through a storm of ideas and potential solutions and somehow picks one. Once panic sets in, the decision that comes out is out of our control. All we can do is to hope that it turns out to be the right one or one that is close to being right. When all else fails, in line with the famous saying *A bad decision is better than no decision*, the ability to panic may actually be an invaluable advantage in evolution. Of course, unnecessary and premature activation of this mechanism, when there is still time and the means of coming up with the correct decision, is a pathology that needs to be cured.

The thought processes that lead to the experimental findings of the traveler's dilemma game still remains unknown. The game has been applied to situations like arms race and competing companies cutting prices, where the players find themselves in slowly but gradually worsening situations. While the game theory and the Nash equilibrium may not pinpoint what the *right* decision is for a given situation, they may help to lay out how far things can escalate and what some, and only some, of the options are. When one finds a stable Nash equilibrium, there is no guarantee that others, nearby or far away, with much higher rewards do not exist. In such cases, information channels called *weak links* may give players the hints or the signals of the existence of other equilibria. Once a stable Nash equilibrium is reached, whether one should stay put or decide to abandon that position and search for other equilibria with potentially higher rewards depends largely on one's insight, ability to interpret and utilize such weak signals, and courage. When these games are applied to economics, the payoff is money, and in biology, it is gene transmission, both of which are crucial in survival. All this being said, the role of leadership in critical decisions can neither be overlooked nor underestimated. It is the leadership qualities that open up new options, notice changing parameters, and create new channels of communication, which to others appear nonexistent or impossible. As evidenced in the behavior of many different complex systems, weak links play a crucial role in the processes of decision-making [9]. As for the traveler's dilemma, the two players and the airline manager are parts of the same social network; hence, they can never be considered as totally isolated. Their minds continuously gather information through weak links from their collective memory, which has important bearing on their final decisions. For example, one, or both, of the passengers may remember from the news that, once an equally reputable company has declined to pay when both passengers claimed reimbursements very close the high end. Such considerations are important even in the experimental game.

19.2.6   Classical Bit or Cbit

The classical bit is defined as the physical realization of a binary system, which can exist only in two **mutually exclusive** states and which can be read or measured by an appropriate device. If the two states are equiprobable, the amount of information gained in such a measurement is 1 bit. This is also the maximum amount of information that can be obtained from any binary device.

We now introduce **Cbit** as a device or a register, not as a unit of information. Cbit is a stable device whose state does not change by measurement. However, its state can be changed by some externally driven operation. An external device that sets or resets a Cbit is called a **gate**. In preparation to our discussion on its quantum version, Qbit, we represent the states of a Cbit by two orthonormal 2-vectors:

$$|0\rangle = \begin{pmatrix} 1 \\ 0 \end{pmatrix} \text{ and } |1\rangle = \begin{pmatrix} 0 \\ 1 \end{pmatrix}. \tag{19.60}$$

As far as the observer is concerned, H refers to the potential knowledge, that is, the knowledge that the observer does not have but expects to gain on the average after the outcome is learned. After the observation is made and the state of the Cbit is found, H collapses to zero, since subsequent observations cannot change the result for that observer. However, for a second observer, who does not know the outcome, H is still $1b$. In other words, collapse is in the mind of the observer. Cbit is always either in state $|0\rangle$ or $|1\rangle$, whether an observation has been made or not does not change this fact. Observer's state of mind, which is basically a certain pattern of neural networks, wonders between the two possibilities, $|0\rangle$ or $|1\rangle$; and after the observation is made, it collapses to a new state, from "I wonder what it is?" to "I see" or from not knowing to knowing.

This reminds us the Necker cubes (Figure 19.13). If we relax and look continuously at the center of the cube on the left, we see two surfaces oscillating

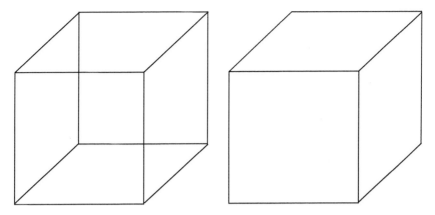

Figure 19.13   Necker cubes.

back and front. Since we are looking at the projection of a three-dimensional cube onto a plane, our brain does not have sufficient information to decide about which side is the closer one. Hence, it shows us the two possibilities by oscillating the surfaces with a definite frequency, which is probably a function of the processing speed of our brain. On the other hand, if we look at the cube on the right, which is still two-dimensional but with the missing information about the surfaces included, the two surfaces no longer oscillate. We shall not go into this any further; however, the inclusion of human mind as an information processor always makes the problem much more interesting and difficult.

For nontrivial calculations, one needs more than one Cbit. A 2-bit classical system can be constructed by combining two Cbits, $|A\rangle$ and $|B\rangle$, each of which has its own two possible states. Now the combined state, $|AB\rangle$, has four possible states given as

$$|00\rangle = |0\rangle\,|0\rangle = \begin{pmatrix} 1 \\ 0 \end{pmatrix}_A \begin{pmatrix} 1 \\ 0 \end{pmatrix}_B = \begin{pmatrix} 1 \\ 0 \\ 0 \\ 0 \end{pmatrix}_{AB}, \tag{19.61}$$

$$|01\rangle = |0\rangle\,|1\rangle = \begin{pmatrix} 1 \\ 0 \end{pmatrix}_A \begin{pmatrix} 0 \\ 1 \end{pmatrix}_B = \begin{pmatrix} 0 \\ 1 \\ 0 \\ 0 \end{pmatrix}_{AB}, \tag{19.62}$$

$$|10\rangle = |1\rangle\,|0\rangle = \begin{pmatrix} 0 \\ 1 \end{pmatrix}_A \begin{pmatrix} 1 \\ 0 \end{pmatrix}_B = \begin{pmatrix} 0 \\ 0 \\ 1 \\ 0 \end{pmatrix}_{AB}, \tag{19.63}$$

$$|11\rangle = |1\rangle\,|1\rangle = \begin{pmatrix} 0 \\ 1 \end{pmatrix}_A \begin{pmatrix} 0 \\ 1 \end{pmatrix}_B = \begin{pmatrix} 0 \\ 0 \\ 0 \\ 1 \end{pmatrix}_{AB}. \tag{19.64}$$

Any measurement will yield 2b worth of information culminating with only one of the above four states.

In general, one represents the states of n Cbits in terms of 2^n orthonormal vectors in 2^n dimensions. Since technological applications always involve multiple Cbits, it is worth getting acquainted with the nomenclature. The four states of a 2-Cbit system is usually written in terms of tensor products as

$$|0\rangle \otimes |0\rangle, |0\rangle \otimes |1\rangle, |1\rangle \otimes |0\rangle, |1\rangle \otimes |1\rangle. \tag{19.65}$$

Sometimes, we omit $\otimes$ and simply write

$$|0\rangle |0\rangle, |0\rangle |1\rangle, |1\rangle |0\rangle, |1\rangle |1\rangle. \tag{19.66}$$

Other equivalent ways of writing these are

$$|00\rangle, |01\rangle, |10\rangle, |11\rangle \tag{19.67}$$

and

$$|0\rangle_2, |1\rangle_2, |2\rangle_2, |3\rangle_2. \tag{19.68}$$

In the last case, the subscript stands for the number of Cbits in the system, which in this case is 2 and the numbers 0, 1, 2, 3 correspond to the zeroth, first, second, and third states of the 2-Cbit system. In general, for an n-Cbit system there are 2^n mutually exclusive states, hence we write

$$|x\rangle_n, \text{ where } x \text{ is integer } 0 \leq x < 2^n. \tag{19.69}$$

The analogy with vector spaces and orthonormal basis vectors gains their true meaning only in quantum mechanics, where we can construct physical states by using their linear combinations with complex coefficients in Hilbert space. In classical systems, the only meaningful reversible operations on n-Cbit systems is the $2^n!$ distinct permutations of the 2^n basis vectors. We can write the 4-Cbit state $|0\rangle |0\rangle |1\rangle |0\rangle$ in the following equivalent ways:

$$|0\rangle \otimes |0\rangle \otimes |1\rangle \otimes |0\rangle = |0\rangle |0\rangle |1\rangle |0\rangle = |0010\rangle = |2\rangle_4. \tag{19.70}$$

In general, a 4-Cbit state is written as

$$|x_3 x_2 x_1 x_0\rangle = |x\rangle_4, \tag{19.71}$$

where x is a number given by the binary expansion

$$x = 8x_3 + 4x_2 + 2x_1 + x_0. \tag{19.72}$$

The states are enumerated starting with zero on the right. For example, for the state $|0010\rangle$, we have $x_0 = 0$, $x_1 = 1$, $x_2 = 0$, $x_3 = 0$, thus x becomes

$$x = 8 \cdot 0 + 4 \cdot 0 + 2 \cdot 1 + 1 \cdot 0 = 2. \tag{19.73}$$

The tensor product is defined as

$$\begin{pmatrix} a_0 \\ a_1 \end{pmatrix} \otimes \begin{pmatrix} b_0 \\ b_1 \end{pmatrix} = \begin{pmatrix} a_0 b_0 \\ a_0 b_1 \\ a_1 b_0 \\ a_1 b_1 \end{pmatrix}. \tag{19.74}$$

For example,

$$|001\rangle = |1\rangle_3 \tag{19.75}$$

$$= \begin{pmatrix} 1 \\ 0 \end{pmatrix} \otimes \begin{pmatrix} 1 \\ 0 \end{pmatrix} \otimes \begin{pmatrix} 0 \\ 1 \end{pmatrix} = \begin{pmatrix} 1 \\ 0 \\ 0 \\ 0 \end{pmatrix} \otimes \begin{pmatrix} 0 \\ 1 \end{pmatrix} = \begin{pmatrix} 0 \\ 1 \\ 0 \\ 0 \\ 0 \\ 0 \\ 0 \\ 0 \end{pmatrix}.$$

19.2.7 Operations on Cbits

In quantum information theory, all operations on Qbits are reversible. In the classical theory, there are only two reversible operations that could be performed on a single Cbit, **do nothing** and **flip**, which can be represented as matrices:

Do nothing or the **identity** operator **I**:

$$\mathbf{I}|0\rangle = \begin{pmatrix} 1 & 0 \\ 0 & 1 \end{pmatrix} \begin{pmatrix} 1 \\ 0 \end{pmatrix} = \begin{pmatrix} 1 \\ 0 \end{pmatrix} = |0\rangle, \tag{19.76}$$

$$\mathbf{I}|1\rangle = \begin{pmatrix} 1 & 0 \\ 0 & 1 \end{pmatrix} \begin{pmatrix} 0 \\ 1 \end{pmatrix} = \begin{pmatrix} 0 \\ 1 \end{pmatrix} = |1\rangle. \tag{19.77}$$

Flip operator **X**:

$$\mathbf{X}|0\rangle = \begin{pmatrix} 0 & 1 \\ 1 & 0 \end{pmatrix} \begin{pmatrix} 1 \\ 0 \end{pmatrix} = \begin{pmatrix} 0 \\ 1 \end{pmatrix} = |1\rangle, \tag{19.78}$$

$$\mathbf{X}|1\rangle = \begin{pmatrix} 0 & 1 \\ 1 & 0 \end{pmatrix} \begin{pmatrix} 0 \\ 1 \end{pmatrix} = \begin{pmatrix} 1 \\ 0 \end{pmatrix} = |0\rangle. \tag{19.79}$$

An example for an irreversible operation is the action of the **erase** operator $\mathbf{E}_1$:

$$\mathbf{E}_1 |0\rangle = \begin{pmatrix} 1 & 0 \\ 0 & 0 \end{pmatrix} \begin{pmatrix} 1 \\ 0 \end{pmatrix} = \begin{pmatrix} 1 \\ 0 \end{pmatrix} = |0\rangle, \tag{19.80}$$

$$\mathbf{E}_1 |1\rangle = \begin{pmatrix} 1 & 0 \\ 0 & 0 \end{pmatrix} \begin{pmatrix} 0 \\ 1 \end{pmatrix} = 0. \tag{19.81}$$

$\mathbf{E}_1$ is **irreversible**, since the original states cannot be reconstructed from the output states. We also have the following operators:

$$\mathbf{Y} |0\rangle = \begin{pmatrix} 0 & -1 \\ 1 & 0 \end{pmatrix} \begin{pmatrix} 1 \\ 0 \end{pmatrix} = \begin{pmatrix} 0 \\ 1 \end{pmatrix} = |1\rangle,$$

$$\mathbf{Y} |1\rangle = \begin{pmatrix} 0 & -1 \\ 1 & 0 \end{pmatrix} \begin{pmatrix} 0 \\ 1 \end{pmatrix} = -\begin{pmatrix} 1 \\ 0 \end{pmatrix} = -|0\rangle, \tag{19.82}$$

$$\mathbf{Z} |0\rangle = \begin{pmatrix} 1 & 0 \\ 0 & -1 \end{pmatrix} \begin{pmatrix} 1 \\ 0 \end{pmatrix} = \begin{pmatrix} 1 \\ 0 \end{pmatrix} = |0\rangle,$$

$$\mathbf{Z} |1\rangle = \begin{pmatrix} 1 & 0 \\ 0 & -1 \end{pmatrix} \begin{pmatrix} 0 \\ 1 \end{pmatrix} = -\begin{pmatrix} 0 \\ 1 \end{pmatrix} = -|1\rangle. \tag{19.83}$$

Even though these operations are mathematically well defined, they are meaningless in the classical context. Only the states $|0\rangle$ and $|1\rangle$ have meaning within the context of Cbits. However, operators like $\mathbf{Z}$, which are meaningless for a single Cbit, when used in conjunction with other meaningless operators, could gain classical meaning in multi-Cbit systems. For example, the operator

$$\frac{1}{2}(\mathbf{I} + \mathbf{Z}_1 \mathbf{Z}_0) \tag{19.84}$$

acts as the **identity** operator for the 2-Cbit states $|0\rangle |0\rangle$ and $|1\rangle |1\rangle$. On the other hand, it produces **zero**, another classically meaningless result, when operated on $|0\rangle |1\rangle$ and $|1\rangle |0\rangle$. The subscript on $\mathbf{Z}$ indicates the Cbit on which it acts on. For example, on a 4-Cbit state,

$$|x\rangle_4 = |x_3 x_2 x_1 x_0\rangle = |x_3\rangle |x_2\rangle |x_1\rangle |x_0\rangle, \tag{19.85}$$

the **flip** operator $\mathbf{X}_1$, acting on the **first** Cbit is defined as

$$\boxed{\mathbf{X}_1 = \mathbf{I} \otimes \mathbf{I} \otimes \mathbf{X} \otimes \mathbf{I},} \tag{19.86}$$

where
$$\mathbf{X}_1 |x_3\rangle \, |x_2\rangle \, |x_1\rangle \, |x_0\rangle = |x_3\rangle \, |x_2\rangle [\mathbf{X}_1 |x_1\rangle] \, |x_0\rangle. \qquad (19.87)$$

We cannot emphasize enough that we start counting with the **zeroth** Cbit on the **right**. Similarly,

$$\frac{1}{2}(\mathbf{I} - \mathbf{Z}_1 \mathbf{Z}_0) \qquad (19.88)$$

is the **identity** operator for $|0\rangle \, |1\rangle$ and $|1\rangle \, |0\rangle$ and produces **zero** when operated on $|0\rangle \, |0\rangle$ and $|1\rangle \, |1\rangle$. For multiple Cbit systems, another operator that represents reversible operations is the **swap** operator $\mathbf{S}_{ij}$. It exchanges the values of the Cbits represented by the indices i and j:

$$\boxed{\mathbf{S}_{ij} |x_i\rangle \, |x_j\rangle = |x_j\rangle \, |x_i\rangle.} \qquad (19.89)$$

Another useful operation on 2-Cbit systems is the reversible **XOR** or the **controlled-NOT** gate implemented by the operator $\mathbf{C}_{10}$ as

$$\boxed{\mathbf{C}_{10} |x_1\rangle \, |x_0\rangle = \mathbf{X}_0^{x_1} |x_1\rangle \, |x_0\rangle.} \qquad (19.90)$$

The task of $\mathbf{C}_{10}$ is to flip the value of the target bit, $|x_0\rangle$, whenever the control bit, $|x_1\rangle$, has the value 1. One can easily show that $\mathbf{C}_{10}$ can be constructed from single Cbit operators as

$$\boxed{\mathbf{C}_{10} = \frac{1}{2}(\mathbf{I} + \mathbf{Z}_1 + \mathbf{X}_0 - \mathbf{X}_0 \mathbf{Z}_1).} \qquad (19.91)$$

Other examples of Cbits and their gates can be found in Mermin [10].

19.3 QUANTUM INFORMATION THEORY

In the previous sections, we have introduced the basic elements of **Shannon's theory** and discussed **Cbits**, which are binary devices with two mutually exclusive and usually equiprobable states. Even though the technological applications of Cbits involve quantum processes like photoelectric effect, semiconductivity, tunneling, etc., they are basically classical devices working with classical currents of particles.

Recently, the possibility of using single particles and quantum systems opened up the possibility of designing new information processing devices with vastly different properties and merits. Even though the classical information theory has been around for over 60 years with many interesting interdisciplinary applications, **quantum information theory** is still at its infancy with its technological applications not yet in sight. However, considering its potential in both theory and practice, quantum information theory is bound to be a center of attraction for many years to come. In what follows we start with a quick review of the basics of quantum mechanics [11].

19.3.1   Basic Quantum Theory

A quantum system is by definition a microscopic system, where the classical laws of physics break down and the laws of quantum mechanics have to be used. The most fundamental difference between classical physics and quantum mechanics is about the effect of measurement on the state of a system.

Measurement of a system always involves some kind of interaction between the measuring device and the system. For example, to measure the temperature of an object, we may bring it in contact with a mercury thermometer. During the measurement process, the thermometer and the object come to thermal equilibrium by exchanging a certain amount of heat. Finally, they reach thermal equilibrium at some common temperature and the mercury column in the thermometer settles at a new level, which allows us to read the temperature from a scale. At the end of the measurement process, neither the thermometer nor the object will be at their initial temperatures. However, the amount of mercury inside the bulb is usually so small that it reaches thermal equilibrium with the object with only a tiny amount of heat exchanged; hence, the temperature of the object does not change appreciably during this process. This shows that even in classical physics, measurement effects the state of a system. However, what separates classical physics from quantum mechanics is that in classical physics, these effects can either be minimized by a suitable choice of instrumentation or algorithms can be designed to take corrective measures. In this regard, in classical information theory, a Cbit remains in its original state no matter how many times it is measured. It changes its state only by the action of some external devices called gates.

In classical physics, there are particles and waves. These are mutually exclusive properties of matter. **Particles** are localized objects with no dimensions, while **waves** are spread out in entire space. One of the surprising features of quantum mechanics is the **duality** between the particle and the wave properties. Electrons and photons sometimes behave like particles and sometimes like waves. Furthermore, nature of the experimental setup determines whether electrons or photons behave as particles or as waves. The **de Broglie relation**:

$$\lambda = \frac{h}{p},$$ (19.92)

where $h = 6.60 \times 10^{-27} \text{cm}^2\text{g/s}$ is the famous Planck constant, establishes the relation between a wave of wavelength λ and a particle of momentum p. Similarly, the **Planck formula**

$$E = h\nu$$ (19.93)

establishes the particle property of light by giving the energy of photons in terms of the frequency ν of the electromagnetic waves.

In quantum mechanics, measurement on a system changes the state of the system irreversibly and there are limits on the accuracy with which certain pairs of observables can be measured simultaneously. **Position** and **momentum** are two such **conjugate observables**. Heisenberg's **uncertainty principle**, which could be considered as the singly most important statement of quantum mechanics, states that position and momentum cannot be determined simultaneously with greater precession than

$$\Delta x \Delta p \geq \frac{\hbar}{2}, \qquad (19.94)$$

where $\hbar = h/2\pi$. Uncertainty principle does not say that we cannot determine the position or momentum as accurately as we want. But it says that if we want to know the momentum of a particle precisely, that is, as $\Delta p \to 0$, then the price we pay is to lose all information about its position:

$$\lim_{\Delta p \to 0} \Delta x \geq \frac{\hbar}{2\Delta p} \to \infty. \qquad (19.95)$$

In other words, particles begin to act like pure waves extended throughout the entire space, hence they are *everywhere*. Similarly, if we want to know the position precisely, then we lose all information about its momentum. As Feynman et al. [12] says, "The uncertainty principle *protects* quantum mechanics. Heisenberg recognized that if it were possible to measure the momentum and the position simultaneously with greater accuracy, the quantum mechanics would collapse. So he proposed that it must be impossible." Since 1926, Heisenberg's uncertainty principle and quantum mechanics have been victorious over many experimental and conceptual challenges and still maintain their correct status.

Mathematical formulation of quantum mechanics is quite different from classical physics, and it is based on a few principles:

(**I**) The state of a system is completely described by the **state vector**, $|\Psi\rangle$, defined in the abstract **Hilbert space**, which is the linear vector space of square integrable functions. State vector is a complex valued function of real arguments. When continuous variables like position are involved, $|\Psi\rangle$ can be expressed as $\Psi(x)$. In this form, it is also called the wave function or the state function. When there is no room for confusion, we use both.

The absolute value square of the state function, $|\Psi(x)|^2$, gives the **probability density**, and $|\Psi(x)|^2 dx$ is the probability of finding a system in the interval between x and $x + dx$. Since it is certain that the system is somewhere between $-\infty$ and ∞, the state function satisfies the normalization condition:

$$\int_{-\infty}^{+\infty} |\Psi(x)|^2 dx = 1. \qquad (19.96)$$

(**II**) In quantum mechanics, the order in which certain dynamical variables are measured is important. It is for this reason that **observables** are represented by Hermitian differential operators or Hermitian matrices acting on the state vectors in Hilbert space. Due to their Hermitian property, these operators have real eigenvalues and their eigenvectors, also called the eigenstates, form a complete and orthogonal set that spans the Hilbert space. For a given operator A with the eigenvalues a_i and the eigenstates $|u_i\rangle$, or $u_i(x)$, we can express these properties as

$$A|u_i\rangle = a_i|u_i\rangle, \tag{19.97}$$

$$\textbf{Orthogonality:} \int u_i^*(x)u_j(x)\, dx = \delta_{ij}, \tag{19.98}$$

$$\textbf{Completeness:} \sum_i u_i^*(x')u_i(x) = \delta(x' - x). \tag{19.99}$$

Eigenvalues, a_i, which are real, correspond to the measurable values of the dynamical variable A. When an observable has discrete eigenstates, its observed value can only be one of the corresponding eigenvalues. Using the completeness property [Eq. (19.99)], we can express the general state vector of a quantum system as a linear combination of the eigenstates of A as

$$|\Psi\rangle = \sum_i c_i|u_i\rangle, \tag{19.100}$$

where $|u_i\rangle$ are also called the basis states. Expansion coefficients, c_i, which are in general complex numbers, can be found by using the orthogonality relation [Eq. (19.98)]:

$$c_i = \int u_i^*(x)\Psi(x)dx. \tag{19.101}$$

These complex numbers, c_i, are called the **probability amplitudes**, and from the normalization condition [Eq. (19.96)] they satisfy

$$\sum_i |c_i|^2 = 1. \tag{19.102}$$

Now, the **expectation value** $\langle A\rangle$ of a dynamical variable is found as

$$\langle A\rangle = \int \Psi^*(x)A\Psi(x)dx. \tag{19.103}$$

Using the orthogonality relation (19.98), we can write:

$$\langle A\rangle = \int \sum_i c_i^* u_i^*(x)a_i \sum_j c_j u_j(x)\, dx = \sum_i a_i \sum_j c_i^* c_j \left[\int u_i^*(x)u_j(x)\, dx\right] \tag{19.104}$$

$$= \sum_i a_i \sum_j c_i^* c_j \delta_{ij} = \sum_i a_i c_i^* c_i = \sum_i a_i |c_i|^2, \qquad (19.105)$$

hence

$$\boxed{\langle A \rangle = a_1 |c_1|^2 + a_2 |c_2|^2 + \cdots .} \qquad (19.106)$$

A quantum state prepared as

$$\boxed{|\Psi\rangle = \sum_i c_i |u_i\rangle} \qquad (19.107)$$

is called a **mixed state**.

When a measurement is performed on a mixed state of the dynamical variable A, the result is one of the eigenvalues. The probability of the result being the mth eigenvalue, a_m is given by $p_m = |c_m|^2$. The important thing here is that once a measurement is done on the system, it is left in one of the basis states, and all the information regarding the initial state, that is, all c_i in Eq. (19.107) are erased. This is called the **collapse** of the state vector or the state function. From the collapsed state vector, it is no longer possible to construct the original state vector. That is, all information about the original state contained in the coefficients c_i are lost irreversibly once a measurement is made. How exactly the collapse takes place is still debated and is beyond the conventional quantum mechanics. Expansion coefficients in the state function can only be determined by collecting statistical information on the probabilities p_i by repeating the experiment over many times on identically prepared states. It is also possible to prepare a system in the mth eigenstate of A as

$$\boxed{|\Psi\rangle = |u_m\rangle.} \qquad (19.108)$$

In this case, all c_i except c_m, which is equal to 1, are zero. Such states are called **pure states**. We have to remind the reader that for any another observable, Eq. (19.108) will not be a pure state, since $|u_m\rangle$ is now the mixed state of the new observable's eigenstates.

Applying all this to **Shannon's theory**, we see that for a pure state, that is, $p_i = 0$, $i \neq m$, and $p_m = 1$, Shannon entropy H is zero – in other words, zero new information. No matter how many times the measurement is repeated on identically prepared states, there will be no surprises. For a mixed state, **Shannon entropy** is given as

$$\boxed{H = -\sum_k^N p_k \log_2 p_k,} \qquad (19.109)$$

where $\sum_k |c_k|^2 = \sum_k p_k = 1$, and N is the dimension of the Hilbert space. From a mixed state, the maximum obtainable information is

$$H = \log_2 N, \qquad (19.110)$$

which occurs when all the probabilities are equal, that is, when $p_i = 1/N$ for all i. Note that once the state vector collapses, further measurements of A will yield the same value. In other words, with respect to that observable, the system behaves classically.

Using the definition of the expectation value, we can write the uncertainty principle for two dynamic variables, A and B, as

$$\boxed{\Delta A \Delta B \geq \frac{1}{2} |\langle [A, B] \rangle|,} \tag{19.111}$$

where $[A, B] = AB - BA$ is called the **commutator** of A and B. From the aforementioned equation, it is seen that unless A and B commute, that is, $AB = BA$, they cannot be measured precisely simultaneously. Furthermore, if two operators do not commute, $AB \neq BA$, then the order in which they are measured becomes important.

In some cases, we can ignore most of the parameters of a quantum system and concentrate on just a few of the variables. This is very useful when we are interested in some discrete set of states like the **spin** with up/down states or the **polarization** with vertical/horizontal or as in the quantum pinball machine, which we shall discuss in detail, the **path** with right/left. In such cases, we can express the state vector as a superposition of the orthonormal basis states, $|\Psi_1\rangle$ and $|\Psi_2\rangle$, corresponding to the two alternatives as

$$|\Psi\rangle = c_1 |\Psi_1\rangle + c_2 |\Psi_2\rangle, \tag{19.112}$$

where

$$|c_1|^2 + |c_2|^2 = 1. \tag{19.113}$$

When a *which-one* or *which-way* measurement is done, $p_1 = |c_1|^2$ and $p_2 = |c_2|^2$ are the respective probabilities of one of the basis states, $|\Psi_1\rangle$ or $|\Psi_2\rangle$, being seen. Statistically speaking, if we repeat the measurement N times on identically prepared states, p_1 and p_2 will be equal to the following limits:

$$p_1 = \lim_{N \to \infty} \frac{N_1}{N}, \quad p_2 = \lim_{N \to \infty} \frac{N_2}{N}, \quad N = N_1 + N_2, \tag{19.114}$$

where N_1 and N_2 are the number of occurrences of the states $|\Psi_1\rangle$ and $|\Psi_2\rangle$, respectively.

If we do not perform a *which-one* or *which-way* measurement on the system, the system will be in the **superposed state** [Eq. (19.112)] with the **probability density** given as

$$|\Psi|^2 = (c_1 \Psi_1 + c_2 \Psi_2)^* (c_1 \Psi_1 + c_2 \Psi_2) \tag{19.115}$$

$$= |c_1|^2 |\Psi_1|^2 + c_1^* c_2 \Psi_1^* \Psi_2 + c_2^* c_1 \Psi_2^* \Psi_1 + |c_2|^2 |\Psi_2|^2 \tag{19.116}$$

$$= p_1|\Psi_1|^2 + c_1^* c_2 \Psi_1^* \Psi_2 + c_2^* c_1 \Psi_2^* \Psi_1 + p_2|\Psi_2|^2. \tag{19.117}$$

The two terms in the middle are the **interference terms** responsible for the fringes seen in a quantum double slit experiment. In classical physics, the interference terms are absent and the joint probability density reduces to

$$|\Psi|^2 = p_1|\Psi_1|^2 + p_2|\Psi_2|^2. \tag{19.118}$$

In classical physics, regardless of the presence of an observer, the system is always in one of the states: $|\Psi_1\rangle$ or $|\Psi_2\rangle$. When we make an observation, we find the system in one or the other state with its respective probability, p_1 or p_2. However, in quantum mechanics, until a measurement is done and the state function has collapsed, the system is in both states simultaneously, that is in the superposed state:

$$|\Psi\rangle = c_1|\Psi_1\rangle + c_2|\Psi_2\rangle. \tag{19.119}$$

Evolution of a state vector is determined by the **Schrödinger equation**:

$$\boxed{\mathbf{H}|\Psi\rangle = i\hbar \frac{\partial}{\partial t}|\Psi\rangle,} \tag{19.120}$$

where $\mathbf{H}$ is the **Hamiltonian** operator, usually obtained from its classical expression by replacing the position and momentum with their operator counterparts. For example, for a particle of mass m moving under the influence of a conservative force field $V(\overrightarrow{r})$, the Hamiltonian operator is obtained from the classical Hamiltonian:

$$H_{\text{classical}} = \frac{\overrightarrow{p}^2}{2m} + V(\overrightarrow{r}), \tag{19.121}$$

with the replacements

$$\overrightarrow{r} \to \overrightarrow{r}, \ \overrightarrow{p} \to -i\hbar\overrightarrow{\nabla} \tag{19.122}$$

as

$$H = -\frac{\hbar^2}{2m}\overrightarrow{\nabla}^2 + V(\overrightarrow{r}). \tag{19.123}$$

In technological applications, changes in the state vector:

$$|\Psi_2\rangle = U|\Psi_1\rangle, \tag{19.124}$$

are usually managed through the actions of **reversible transformations** called **gates**, U, which are represented by **unitary transformations** satisfying the relation

$$\boxed{UU^\dagger = I,} \tag{19.125}$$

where I is the identity operator and $U^{\dagger}$ is the **Hermitian conjugate** defined as the complex conjugate of the transpose of U, that is,

$$\boxed{U^{\dagger} = \tilde{U}^{*}.}$$

(19.126)

19.3.2   Single-Particle Systems and Quantum Information

To demonstrate the profound differences between the classical and the quantum information processing devices, we start with the experimental setup shown in Figure 19.14. In this setup, we have a light source emitting coherent monochromatic light beam with intensity I. The beam impinges on a beam splitter B and then separates into two, each with intensity $I/2$. Aside from its intensity, the transmitted beam on the left goes through the beam splitter unaffected. The reflected beam on the right undergoes a phase shift of $\pi/2$ with respect to the transmitted beam. Naturally, the detectors D_0 and D_1 receive the transmitted and the reflected beams, respectively.

Next, we consider this experiment in the limit as the intensity of the beam is reduced to almost zero. Technically, it is possible to control the intensity so that we are actually sending one photon at a time to the beam splitter, which diverts these photons with equal probability to the left or the right channels. The detectors respond to individual photons with equal probability. For an experiment repeated many times, half of the time D_0 and the other half of the time D_1 will click. So far, everything looks like the pinball machine.

The difference between the two cases begin to appear when we search an answer to the question: Which path did the photon take? In the case of the

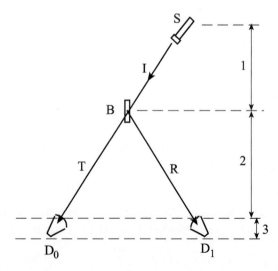

Figure 19.14   Quantum pinball machine. T for transmitted and R for reflected.

pinball machine, the ball has two possibilities; it will go to either the left or the right. The source of randomness is the pin, which diverts the ball to the left or the right with equal probabilities. Once the ball clears the pin, it has a definite trajectory and ends up in either the left or the right bins. The observer just does not know it yet. Whether the observer actually checks the bins or not has no bearing whatsoever on the result.

In the case of photons, there are two possible basis states, $|\Psi_0\rangle$ and $|\Psi_1\rangle$, corresponding to the eigenstates of the "which-way" operator. State $|\Psi_0\rangle$ corresponds to the photon following the left path, and $|\Psi_1\rangle$ corresponds to the photon following the right path. When the detectors are turned off or absent, the photon is in the superposed state

$$|\Psi\rangle = c_0|\Psi_0\rangle + c_1|\Psi_1\rangle, \tag{19.127}$$

where c_0 and c_1 are in general complex numbers satisfying the normalization condition

$$|c_0|^2 + |c_1|^2 = 1. \tag{19.128}$$

In other words, the photon is neither in the left channel, $|\Psi_0\rangle$, nor in the right channel, $|\Psi_1\rangle$, but it is in both of them, simultaneously. To find out which way the photon goes through, we turn on the detectors. One of the detectors clicks, and the state function collapses to either $|\Psi_0\rangle$ or $|\Psi_1\rangle$. Once the state function has collapsed, the photon is in a pure state, $|\Psi_0\rangle$ or $|\Psi_1\rangle$. During this process, we gain $1b$ of information about the system, but the price we pay is that we have lost (destroyed) the initial state function [Eq. (19.127)] irreversibly. It can no longer be constructed from the collapsed state:

$$|\Psi\rangle = |\Psi_0\rangle \text{ or } |\Psi\rangle = |\Psi_1\rangle. \tag{19.129}$$

By repeating the experiment many times on identically prepared setups, all we can gain is statistical information about the initial state. That is, square of the absolute values of c_0 and c_1, which are related to the probabilities of the initial state vector collapsing to either $|\Psi_0\rangle$ or $|\Psi_1\rangle$ as

$$p_0 = |c_0|^2 \text{ and } p_1 = |c_1|^2. \tag{19.130}$$

In the case of a symmetric beam splitter, the probabilities are equal, $p_0 = p_1 = 1/2$, and the state function can be given in any one of the following forms:

$$|\Psi\rangle = \frac{1}{2}[|\Psi_0\rangle + |\Psi_1\rangle] \text{ or } |\Psi\rangle = \frac{1}{2}[|\Psi_0\rangle - |\Psi_1\rangle]. \tag{19.131}$$

In the classical pinball machine, the ball is always in a "pure" state. It is following either the left or the right paths. In other words, it is either in state $|\Psi_0\rangle$ or in state $|\Psi_1\rangle$. It has nothing to do with the presence of an observer, knowing or not knowing, measuring or not measuring, peeking or not peeking through the cracks of the pinball machine. In classical physics, observation or measurement never has the same dramatic effect on the state of the system that it has in quantum systems. In the following section, we discuss how all this can be verified in laboratory through the use of the **Mach–Zehnder interferometer**.

19.3.3   Mach–Zehnder Interferometer

In a Mach–Zehnder interferometer (Figure 19.15), after the first beam splitter, B_1, the transmitted and the reflected beams are reflected at the mirrors M_L and M_R, respectively, and then allowed to go through a second beam splitter, B_2, before being picked up by the detectors D_1 and D_0. We refer to the transmitted and the reflected beams of the first beam splitter as the left, L, and the right, R, beams, respectively.

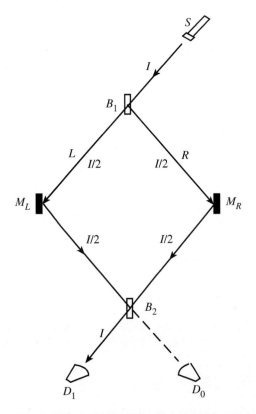

Figure 19.15   Mach–Zehnder interferometer.

We first consider the case of coherent monochromatic beam with intensity I produced by the light source S. Keep in mind that each time light gets reflected, it leads the incident wave by a phase difference of $\pi/2$, and the transmitted wave suffers no phase shift. The beam that gets transmitted at B_1 follows the left path and gets reflected at M_L, and finally splits into its reflected and transmitted parts at B_2. They are joined by the parts of the wave reflected at B_1, which follows the right path. The left beam reflected at B_2 meets the right beam transmitted at B_2. Since they both suffered two reflections, they are in phase and interfere constructively to shine on D_1 with intensity I. The part of the left beam transmitted at B_2 is joined by the part of the right beam reflected at B_2. Since the right beam has suffered three reflections, while the left beam has suffered only one, they are out of phase by π, thus interfering destructively to produce zero intensity at D_0. In summary, D_1 gets the full original beam with intensity I, while D_0 gets nothing. Note that all this is true when all the legs of the interferometer are equal in length. We now turn down the intensity so that we are sending one photon at a time to B_1. The experimental result is completely consistent with the macroscopic result; that is all photons are detected by D_1, and no photon is detected by D_0.

To understand all this in terms of the interference of electromagnetic waves is easy. However, in the case of individual photons, we find ourselves in the position of accepting the view that the photon has followed both paths to interfere with itself to produce no response at D_0 and a sure response at D_1. From the information theory point of view, we already know the answer – that is the detector D_1 – responds for sure. However, we know nothing about which path the photon has followed to get there. To learn the path that the photon follows, we remove B_2, and either D_0 or D_1 clicks. We now know which path the photon follows, and we gain 1b of information in finding that out. In the Mach–Zehnder interferometer, we know exactly which detector responds, that is, D_1, but we have no knowledge of how the photon gets there. There is a region of **irremovable uncertainty** about where the photon is in our device.

There is no classical counterpart of this. If we try the same experiment with the classical pinball machine, we see the ball half of the time in bin 1 and the other half of the time in bin 2 (Figure 19.16a). This is because the events corresponding to the ball following the left path and the right path to reach bin 1 are **mutually exclusive**, each with their respective probability of $1/4$ (Figure 19.16b). If one happens, the other one does not. Thus, their joint probability is given as their sum: $\frac{1}{4} + \frac{1}{4} = \frac{1}{2}$. A symmetric argument works for the ball reaching bin 2. In other words, no matter how many times we run the experiment, we never see a case where the ball that followed the left path colliding or interfering with itself that followed the right path. This is the strange position that we always find ourselves in when trying to understand quantum phenomenon in terms of our intuition, which is predominantly shaped by our classical experiences.

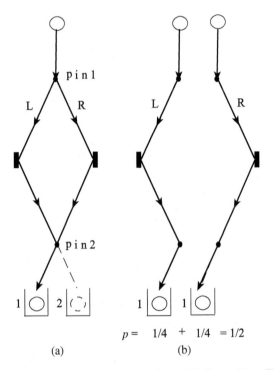

$$p = \quad 1/4 \quad + \quad 1/4 \quad = 1/2$$

(a)   (b)

Figure 19.16   Mach–Zehnder experiment with a pinball machine. The two mutually exclusive paths for the balls seen in Bin 1 are shown on the right.

19.3.4   Mathematics of the Mach–Zehnder Interferometer

In a Mach–Zehnder interferometer, let us choose the orthonormal basis states as

$$|\Psi_{left}\rangle = \begin{pmatrix} 1 \\ 0 \end{pmatrix} \text{ and } |\Psi_{right}\rangle = \begin{pmatrix} 0 \\ 1 \end{pmatrix}. \qquad (19.132)$$

These are the eigenstates of the **which-way operator**. They correspond to the undisturbed paths, that is, the paths in the absence of both beam splitters (Figure 19.15). For a photon incident from the right (Figure 19.17), the left channel is defined as $SB_1 M_L B_2 D_0$, while the second one defines the right channel for the undisturbed path for a photon incident from the left: $SB_1 M_R B_2 D_1$. Strictly speaking, these are meaningless statements. What we need is the full solution of the time-dependent Schrödinger equation, which exhibits the change to the superposition of the right and left path solutions after going through the beam splitter. However, for our purposes, it is perfectly all right to work with the reduced degrees of freedom represented by the *left* and the *right* phrases.

Between the source S and the beam splitter B_1, the photon is in the basis state (Figure 19.15)

$$|\Psi_{SB_1}\rangle = |\Psi_{left}\rangle = \begin{pmatrix} 1 \\ 0 \end{pmatrix}. \qquad (19.133)$$

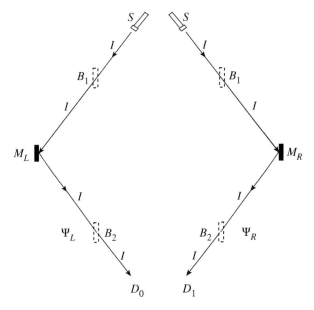

Figure 19.17 Undisturbed paths for the eigenstates of the "which way" operator: $|\Psi_{left}\rangle = \Psi_L$ and $|\Psi_{right}\rangle = \Psi_R$.

After the first beam splitter B_1 and between B_1 and B_2, the solution is transformed into the mixed, that is the superposed state of the transmitted and the reflected parts as

$$|\Psi_{B_1 B_2}\rangle = c_0 |\Psi_{left}\rangle + c_1 |\Psi_{right}\rangle. \tag{19.134}$$

We now write a mathematical expression for the action of the beam splitter B_1, which acts on state $|\Psi_{left}\rangle$ [Eq. (19.133)] to produce the superposed state [Eq. (19.134)]. Since we use a symmetric beam splitter, we have

$$|c_0|^2 = \frac{1}{2} \text{ and } |c_1|^2 = \frac{1}{2}. \tag{19.135}$$

We also know that the reflected photon in Figure 19.15 has suffered a phase shift of $\pi/2$ with respect to the state $|\Psi_{left}\rangle$; hence, without any loss of generality, we can take

$$c_0 = \frac{1}{\sqrt{2}} \text{ and } c_1 = \frac{e^{i\pi/2}}{\sqrt{2}} = \frac{i}{\sqrt{2}}. \tag{19.136}$$

Thus,

$$|\Psi_{B_1 B_2}\rangle = \mathbf{B} |\Psi_{SB_1}\rangle = \frac{1}{\sqrt{2}} \left[\begin{pmatrix} 1 \\ 0 \end{pmatrix} + i \begin{pmatrix} 0 \\ 1 \end{pmatrix} \right]. \tag{19.137}$$

We now write a matrix, $\mathbf{B}$, that acts on the initial state [Eq. (19.133)] and produces the mixed state [Eq. (19.137)] as

$$\mathbf{B} \begin{pmatrix} 1 \\ 0 \end{pmatrix} = \frac{1}{\sqrt{2}} \left[\begin{pmatrix} 1 \\ 0 \end{pmatrix} + i \begin{pmatrix} 0 \\ 1 \end{pmatrix} \right]. \tag{19.138}$$

We can easily verify that

$$\mathbf{B} = \frac{1}{\sqrt{2}} \begin{pmatrix} 1 & i \\ i & 1 \end{pmatrix} = \frac{1}{\sqrt{2}} (\mathbf{I} + i\mathbf{X}) \tag{19.139}$$

accomplishes this task, where

$$\mathbf{I} = \begin{pmatrix} 1 & 0 \\ 0 & 1 \end{pmatrix} \text{ and } \mathbf{X} = \begin{pmatrix} 0 & 1 \\ 1 & 0 \end{pmatrix}. \tag{19.140}$$

This can be understood by the fact that half of the incident wave goes unaffected into the left channel, thus explaining the factor $\frac{1}{\sqrt{2}}$ and the identity operator $\mathbf{I}$, while the other half gets reflected into the right channel with a phase shift of $\pi/2$, thus explaining the flip operator $\mathbf{X}$ and the phase shift factor $e^{i\pi/2}$:

$$\frac{1}{\sqrt{2}} e^{i\pi/2} = \frac{i}{\sqrt{2}}. \tag{19.141}$$

Note that the left channel is the path followed by the undisturbed photon incident from the right and vice versa. Some of the important properties of $\mathbf{B}$ are:

(i) It is a transformation not an observable.
(ii) Since $\mathbf{BB}^\dagger = \mathbf{I}$, where $\mathbf{B}^\dagger = \tilde{\mathbf{B}}^*$, it is a unitary transformation.
(iii) Since it is a unitary transformation, its action is reversible with its inverse given as $\mathbf{B}^{-1} = \mathbf{B}^\dagger$.
(iv) $\mathbf{B}^2 = i\mathbf{X}$.

After the second beam splitter, B_2, the final state of the photon between B_2 and the detectors D_0 or D_1 is given as

$$|\Psi_{B_2 D_1 \text{ or } 2}\rangle = \mathbf{B}|\Psi_{B_1 B_2}\rangle = \mathbf{BB}|\Psi_{SB_1}\rangle = i\mathbf{X}|\Psi_{SB_1}\rangle = i\mathbf{X} \begin{pmatrix} 1 \\ 0 \end{pmatrix} \tag{19.142}$$

$$= i \begin{pmatrix} 0 \\ 1 \end{pmatrix}. \tag{19.143}$$

Notice that the phase factor $i = e^{i\pi/2}$ is physically unimportant and has no experimental consequence. However, we shall see that the individual phases in a superposition are very important.

We now introduce the **channel blocker** represented by the operators

$$\mathbf{E}_R = \begin{pmatrix} 1 & 0 \\ 0 & 0 \end{pmatrix} \text{ and } \mathbf{E}_L = \begin{pmatrix} 0 & 0 \\ 0 & 1 \end{pmatrix}. \tag{19.144}$$

They represent the actions of blocking devices which block the right and the left channels, respectively. They eliminate one of the components in the superposition, and let the other one remain. Another useful operator is the **phase shift** operator $\mathbf{\Phi}(\phi)$,

$$\mathbf{\Phi}(\phi) = \begin{pmatrix} e^{-i\phi/2} & 0 \\ 0 & e^{i\phi/2} \end{pmatrix}, \tag{19.145}$$

which introduces a relative phase shift of ϕ between the left and the right channels. Note that we have written $\mathbf{\Phi}(\phi)$ symmetrically so that the operator delays the left channel by $\phi/2$, while the right channel is advanced by $\phi/2$. In other words, its action is equivalent to

$$\mathbf{\Phi}(\phi) = \begin{pmatrix} 1 & 0 \\ 0 & e^{i\phi} \end{pmatrix}. \tag{19.146}$$

If we insert an extra phase shift device between the two beam splitters, we can write the most general output state of the Mach–Zehnder interferometer as

$$|\Psi_{output}\rangle = \mathbf{B}\mathbf{\Phi}(\phi)\mathbf{B}|\Psi_{input}\rangle. \tag{19.147}$$

For a wave incident from the right, $|\Psi_{input}\rangle = \begin{pmatrix} 1 \\ 0 \end{pmatrix}$, this gives

$$|\Psi_{output}\rangle = \frac{1}{2}\left[(1 - e^{i\phi})\begin{pmatrix} 1 \\ 0 \end{pmatrix} + i(1 + e^{i\phi})\begin{pmatrix} 0 \\ 1 \end{pmatrix}\right]. \tag{19.148}$$

Other commonly encountered operators are

$$\mathbf{I} = \begin{pmatrix} 1 & 0 \\ 0 & 1 \end{pmatrix} : \text{ identity,} \tag{19.149}$$

$$\mathbf{X} = \begin{pmatrix} 0 & 1 \\ 1 & 0 \end{pmatrix} : \text{ flips two basis states,} \tag{19.150}$$

$$\mathbf{Z} = \begin{pmatrix} 1 & 0 \\ 0 & -1 \end{pmatrix} : \text{ shifts the phase of } \begin{pmatrix} 0 \\ 1 \end{pmatrix} \text{ by } \pi, \tag{19.151}$$

$$\mathbf{Y} = \mathbf{XZ} = \begin{pmatrix} 0 & -1 \\ 1 & 0 \end{pmatrix} : \text{ phase shift by } \pi \text{ followed by a flip,} \tag{19.152}$$

Among these, $\mathbf{I}$, $\mathbf{X}$, $\mathbf{Y}$, $\mathbf{Z}$, and $\mathbf{\Phi}$ are unitary operators, while $\mathbf{E}$ is an irreversible operator. Notice that unitary transformations produce reversible changes in the quantum system, while the changes induced by a detector are irreversible.

In addition to these, there is another important reversible operator called the **Hadamard operator**, $\mathbf{H}$, which converts a pure state into a mixed state:

$$\boxed{\mathbf{H} = \frac{1}{\sqrt{2}}\begin{pmatrix} 1 & 1 \\ 1 & -1 \end{pmatrix} = \frac{1}{\sqrt{2}}(\mathbf{X} + \mathbf{Z}).} \tag{19.153}$$

Hadamard operator is a unitary operator with the actions

$$|\Psi_{sup.}^{+}\rangle = \mathbf{H}\begin{pmatrix}1\\0\end{pmatrix} = \frac{1}{\sqrt{2}}\left[\begin{pmatrix}1\\0\end{pmatrix} + \begin{pmatrix}0\\1\end{pmatrix}\right] \tag{19.154}$$

and

$$|\Psi_{sup.}^{-}\rangle = \mathbf{H}\begin{pmatrix}0\\1\end{pmatrix} = \frac{1}{\sqrt{2}}\left[\begin{pmatrix}1\\0\end{pmatrix} - \begin{pmatrix}0\\1\end{pmatrix}\right]. \tag{19.155}$$

Hadamard operator is the workhorse of quantum computing, which converts a pure state into a superposition.

19.3.5   Quantum Bit or Qbit

We have defined a Cbit as a classical system with two mutually exclusive states. It is also possible to design binary devices working at the quantum level. In fact, the Mach–Zehnder interferometer is one example. A Qbit is basically a quantum system that can be prepared in a superposition of two states like $|0\rangle$ and $|1\rangle$. The word *superposition* on its own implies that these two states are no longer *mutually exclusive* as in Cbits. Furthermore, in general, quantum mechanics forces us to use complex coefficients in constructing these superposed states. We used the word *forces* deliberately, since complex numbers in quantum mechanics are not just a matter of convenience, as they are in classical wave problems, but a requirement imposed on us by nature. Real results that can be measured in laboratory are assured not by singling out the real or the imaginary part of the state function, which in general cannot be done, but by interpreting the square of its absolute value, $|\Psi|^2$, as the probability density and by using Hermitian operators that have real eigenvalues to represent observables.

The superposed state of a Qbit is defined as

$$|\Psi\rangle = c_0|0\rangle + c_1|1\rangle, \tag{19.156}$$

where c_0 and c_1 are complex amplitudes, which satisfy the normalization condition:

$$|c_0|^2 + |c_1|^2 = 1. \tag{19.157}$$

If we write c_0 and c_1 as

$$c_0 = a_0 e^{ia_1}, \quad c_1 = b_0 e^{ib_1}, \quad a_0 > 0, \quad b_0 > 0, \tag{19.158}$$

we obtain

$$|\Psi\rangle = a_0 e^{ia_1}|0\rangle + b_0 e^{ib_1}|1\rangle \tag{19.159}$$

$$= e^{ia_1}[a_0|0\rangle + b_0 e^{i(b_1-a_1)}|1\rangle]. \tag{19.160}$$

Since it has no observable consequence, we can ignore the overall phase factor e^{ia_1}. To guarantee normalization [Eq. (19.157)], we can also define two new real parameters, θ and ϕ, as

$$a_0 = \cos\theta, \ b_0 = \sin\theta \text{ and } (b_1 - a_1) = \phi, \tag{19.161}$$

so that $|\Psi\rangle$ is written as

$$|\Psi\rangle = \cos\theta|0\rangle + \sin\theta e^{i\phi}|1\rangle. \tag{19.162}$$

Note that the probabilities,

$$p_0 = |c_0|^2 = \cos^2\theta \text{ and } p_1 = |c_1|^2 = \sin^2\theta, \tag{19.163}$$

are not affected by the phase ϕ at all. When $\theta = \pi/4$, we have

$$p_0 = p_1 = \frac{1}{2}. \tag{19.164}$$

This is the equiprobable case with the Shannon entropy of information H of one bit, which is the maximum amount of useful average information that can be obtained from any binary system. In other words, whether we measure one Qbit or make measurements on many identically prepared Qbits, the maximum average information that can be gained from a single Qbit is one bit. This does not change the fact that we need both θ and ϕ to specify the state of a Qbit completely. What happens to this additional degree of freedom and the information carried by the phase ϕ? Unfortunately, all this wealth of information carried by the phase is lost irreversibly once a measurement is made and the state function has collapsed. Once a Qbit is measured, it behaves just like a Cbit. Given two black boxes, one containing a Cbit and the other a Qbit, there is no way to tell which one is which. If you are given a collection of 1000 Qbits prepared under identical conditions, by making measurement on each one of them, all you will obtain is the probabilities p_0 and p_1 deduced from the number of occurrences N_0 and N_1 of the states $|0\rangle$ and $|1\rangle$, respectively, as

$$p_0 = \frac{N_0}{N}, \quad p_1 = \frac{N_1}{N}, \tag{19.165}$$

where $N = N_0 + N_1$. Furthermore, there is another restriction on quantum information, which is stated in terms of a theorem first proposed by Ghirardi [13]. It is known as the **no-cloning theorem**, which says that the state of a Qbit cannot be copied. In other words, given a Qbit whose method of preparation is unknown, the no-cloning theorem says that you cannot produce its identical twin. If it were possible, then one would be able to produce as many identical copies as needed and use them to determine statistically its probability distribution as accurately as desired, while still having the original, which has not been disturbed by any measurement. In other words, there is absolute inaccessibility of the quantum information buried in a superposed state. You have to make measurement to find out, and when you do, you destroy the original state with all your expected gain as one bit.

Is there then no way to harvest this wealth of quantum information hidden in the two real numbers θ and ϕ? Well, if you do not temper with the quantum state, there is. The no-cloning theorem says that you cannot copy a Qbit, but you can manufacture many Qbits in the same state and manage them through gates representing reversible unitary operations and have them interact with other Qbits. As long as you do not meddle in the inner workings of this network of Qbits, at the end you can extract the needed information by a measurement, which itself is an irreversible operation.

If we act on a superposition:

$$|\Psi_{sup.1}\rangle = \frac{1}{\sqrt{2}} \left[\begin{pmatrix} 1 \\ 0 \end{pmatrix} + e^{i\phi} \begin{pmatrix} 0 \\ 1 \end{pmatrix} \right], \tag{19.166}$$

with the Hadamard operator $\mathbf{H}$ [Eq. (19.153)], we get

$$\begin{aligned} |\Psi_{sup.2}\rangle &= \mathbf{H}|\Psi_{sup.1}\rangle \\ &= \mathbf{H}\frac{1}{\sqrt{2}} \left[\begin{pmatrix} 1 \\ 0 \end{pmatrix} + e^{i\phi} \begin{pmatrix} 0 \\ 1 \end{pmatrix} \right] \end{aligned} \tag{19.167}$$

$$= \frac{1}{2} \left[(1 + e^{i\phi}) \begin{pmatrix} 1 \\ 0 \end{pmatrix} + (1 - e^{i\phi}) \begin{pmatrix} 0 \\ 1 \end{pmatrix} \right], \tag{19.168}$$

which is another superposition, $|\Psi_{sup.2}\rangle$, with different phases. The probabilities for the state $|\Psi_{sup.1}\rangle$,

$$p_0 = \frac{1}{2}, \ p_1 = \frac{1}{2}, \tag{19.169}$$

has now changed with the second state $|\Psi_{sup.2}\rangle$ to

$$p_0 = \frac{1}{2}(1 + \cos\phi), \ p_1 = \frac{1}{2}(1 - \cos\phi), \tag{19.170}$$

thus demonstrating how one can manipulate these phases with reversible operators and with observable consequences. Other relations between **reversible operators** that are important for designing quantum computing systems can be given as [10]

$$\boxed{\mathbf{XZ} = -\mathbf{ZX},} \tag{19.171}$$

$$\boxed{\mathbf{HX} = \frac{1}{\sqrt{2}}(\mathbf{I} + \mathbf{ZX}),} \tag{19.172}$$

$$\boxed{\mathbf{HXH} = \mathbf{Z},} \tag{19.173}$$

$$\boxed{\mathbf{HZH} = \mathbf{X} = -i\mathbf{B}^2.}$$ (19.174)

On the practical side of the problem, it is the task of quantum-computational engineering to find ways to physically realize and utilize these unitary transformations. For practical purposes, most of the existing unitary transformations are restricted to those that act on single or at most on pairs of Qbits. An important part of the challenge for the software designers is to construct the transformations that they may need as combinations of these basic elements. Any quantum system with binary states like $|0\rangle$ and $|1\rangle$ can be used as a Qbit. In practice, it is desirable to work with stable systems so that the superposed states are not lost through decoherence, that is through interactions with the background on the scale of the experiment. Photons, electrons, atoms, and quantum dots can all be used as Qbits. It is also possible to use internal states like polarization, spin, and energy levels of an atom as Qbits.

19.3.6   The No-Cloning Theorem

Using the Dirac bra–ket notation and the properties of the inner product spaces introduced in Chapter 5, we can prove the no-cloning theorem easily. Let a given Qbit to be cloned, called the **control Qbit**, be in the state $|\Psi_A\rangle$. A second quantum system in state $|\chi\rangle$, called the **target Qbit**, is supposed to be transformed into $|\Psi_A\rangle$ via a copying device. We represent the initial state of the copying device as $|\Phi\rangle$. The state of the composite system can now be written as $|\Psi_A\rangle\,|\chi\rangle\,|\Phi\rangle$. Similar to an office copying device, the whole process should be universal and could be described by a **unitary operator** U_c. The effect of U_c on the composite system can be written as

$$U_c|\Psi_A\rangle\,|\chi\rangle\,|\Phi\rangle = |\Psi_A\rangle\,|\Psi_A\rangle\,|\Phi_A\rangle,$$ (19.175)

where $|\Phi_A\rangle$ is the state of the copier after the cloning process. For a universal copier, another state, $|\Psi_B\rangle$, not orthogonal to $|\Psi_A\rangle$, is transformed as

$$U_c|\Psi_B\rangle\,|\chi\rangle\,|\Phi\rangle = |\Psi_B\rangle\,|\Psi_B\rangle\,|\Phi_B\rangle,$$ (19.176)

where all the states are normalized, that is,

$$\begin{aligned}
\langle\Psi_A\,|\Psi_A\rangle &= 1,\\
\langle\Psi_B\,|\Psi_B\rangle &= 1,\\
\langle\Phi_A\,|\Phi_A\rangle &= 1,\\
\langle\Phi_B\,|\Phi_B\rangle &= 1,\\
\langle\chi\,|\chi\rangle &= 1,\\
\langle\Phi\,|\Phi\rangle &= 1.
\end{aligned}$$ (19.177)

Note that in bra–ket notation, the inner product of two states, $\int_{-\infty}^{+\infty} \Psi^*(x)\Phi(x)dx$, is written as $\langle \Psi | \Phi \rangle$. Since $|\Psi_A\rangle$ and $|\Psi_B\rangle$ are not orthogonal, we have

$$\langle \Psi_A | \Psi_B \rangle \neq 0. \tag{19.178}$$

From the properties of the inner product, we also have the inequalities

$$|\langle \Phi_A | \Phi_B \rangle| \leq 1 \text{ and } |\langle \Psi_A | \Psi_B \rangle| \leq 1, \tag{19.179}$$

where the bars stand for the absolute value. Since a unitary operator, $U_c^{\dagger}U_c = I$, preserves inner product, the inner product of the composite states before the operation:

$$\langle \Phi | \langle \chi | \langle \Psi_A | \, U_c^{\dagger}U_c | \Psi_B \rangle \, |\chi\rangle \, |\Phi\rangle = \langle \Psi_A | \Psi_B \rangle \langle \chi | \chi \rangle \langle \Phi | \Phi \rangle = \langle \Psi_A | \Psi_B \rangle, \tag{19.180}$$

has to be equal to the inner product after the operation:

$$\langle \Psi_A | \Psi_B \rangle = \langle \Psi_A | \Psi_B \rangle \langle \Psi_A | \Psi_B \rangle \langle \Phi_A | \Phi_B \rangle \tag{19.181}$$

$$= \langle \Psi_A | \Psi_B \rangle^2 \langle \Phi_A | \Phi_B \rangle, \tag{19.182}$$

or

$$1 = \langle \Psi_A | \Psi_B \rangle \langle \Phi_A | \Phi_B \rangle. \tag{19.183}$$

Taking absolute values, this also becomes

$$|\langle \Psi_A | \Psi_B \rangle| = \frac{1}{|\langle \Phi_A | \Phi_B \rangle|}. \tag{19.184}$$

Since for nonorthogonal states the inequalities $|\langle \Psi_A | \Psi_B \rangle| \leq 1$ and $|\langle \Phi_A | \Phi_B \rangle| \leq 1$ are true, the equality in Eq. (19.184) can only be satisfied when

$$|\Psi_A\rangle = |\Psi_B\rangle. \tag{19.185}$$

Hence, the unitary operator U_c does not exist. In other words, no machine can make a perfect copy of another Qbit state $|\Psi_B\rangle$ that is not orthogonal to $|\Psi_A\rangle$. This is called the no-cloning theorem (for other types of cloning see Audretsch [14] and Bruß and Leuchs [15]).

19.3.7   Entanglement and Bell States

Let us consider two Qbits A and B with no common origin and interaction between them. We can write their states as

$$|\chi_A\rangle = \alpha_A |0\rangle_A + \beta_A |1\rangle_A \tag{19.186}$$

and

$$|\chi_B\rangle = \alpha_B|0\rangle_B + \beta_B|1\rangle_B, \tag{19.187}$$

where $|0\rangle$ and $|1\rangle$ refer to their respective basis states:

$$|0\rangle_{A \text{ or } B} = \begin{pmatrix} 1 \\ 0 \end{pmatrix}, \ |1\rangle_{A \text{ or } B} = \begin{pmatrix} 0 \\ 1 \end{pmatrix}. \tag{19.188}$$

Each pair of amplitudes satisfy the normalization condition separately as

$$|\alpha_A|^2 + |\beta_A|^2 = 1 \tag{19.189}$$

and

$$|\alpha_B|^2 + |\beta_B|^2 = 1. \tag{19.190}$$

Since there is no interaction between them, both of them preserve their identity. Their joint state $|\chi_{AB}\rangle$ is given as the tensor product $|\chi_A\rangle \otimes |\chi_B\rangle$, which is also written as $|\chi_A\rangle |\chi_B\rangle$:

$$|\chi_{AB}\rangle = |\chi_A\rangle \otimes |\chi_B\rangle = [\alpha_A|0\rangle_A + \beta_A|1\rangle_A][\alpha_B|0\rangle_B + \beta_B|1\rangle_B] \tag{19.191}$$
$$= \alpha_A\alpha_B|0\rangle_A|0\rangle_B + \alpha_A\beta_B|0\rangle_A|1\rangle_B + \beta_A\alpha_B|1\rangle_A|0\rangle_B + \beta_A\beta_B|1\rangle_A|1\rangle_B. \tag{19.192}$$

However, this is only a special two-Qbit state, which is composed of noninteracting two single Qbits. A general two-Qbit state will be a superposition of the basis vectors:

$$|0\rangle_A|0\rangle_B, \ |0\rangle_A|1\rangle_B, \ |1\rangle_A|0\rangle_B, \ |1\rangle_A|1\rangle_B, \tag{19.193}$$

which span the four-dimensional two-Qbit space as

$$|\chi_{AB}\rangle = \alpha_{00}|0\rangle_A|0\rangle_B + \alpha_{01}|0\rangle_A|1\rangle_B + \alpha_{10}|1\rangle_A|0\rangle_B + \alpha_{11}|1\rangle_A|1\rangle_B. \tag{19.194}$$

Complex amplitudes, α_{ij}, satisfy the normalization condition

$$\sum_{i,j=0}^{1} |\alpha_{ij}|^2 = 1. \tag{19.195}$$

In general, it is not possible to decompose $|\chi_{AB}\rangle$ in Eq. (19.194) as the tensor product $|\chi_A\rangle \otimes |\chi_B\rangle$ of two Qbit states: $|\chi_A\rangle$ and $|\chi_B\rangle$. Note that only under the additional assumption of

$$\alpha_{00}\alpha_{11} = \alpha_{01}\alpha_{10}, \tag{19.196}$$

Equation (19.194) reduces to Eq. (19.192). Qbit states that cannot be decomposed as the tensor product of individual single Qbit states are called

entangled. As in the single Qbit case, a measurement on a two-Qbit state causes the state function to collapse into one of the basis states in Eq. (19.193) with the probabilities given as

$$p_{ij} = |\alpha_{ij}|^2. \tag{19.197}$$

Maximum average information that can be obtained from a two-Qbit system, 2b, is when all the coefficients are equal. For entangled states, this may not be the case.

As in the case of single Qbits, where we can prepare maximally superposed states, we can also prepare maximally entangled two-Qbit states. There are four possibilities for the maximally entangled states, which are also called the **Bell states**. We write them for [3, 11]

$$|\alpha_{01}|^2 + |\alpha_{10}|^2 = 1, \quad \alpha_{00} = \alpha_{11} = 0 \tag{19.198}$$

as

$$|\Psi^-\rangle = \frac{1}{\sqrt{2}}(|0\rangle_A|1\rangle_B - |1\rangle_A|0\rangle_B), \tag{19.199}$$

$$|\Psi^+\rangle = \frac{1}{\sqrt{2}}(|0\rangle_A|1\rangle_B + |1\rangle_A|0\rangle_B) \tag{19.200}$$

and for

$$|\alpha_{00}|^2 + |\alpha_{11}|^2 = 1, \quad \alpha_{01} = \alpha_{10} = 0 \tag{19.201}$$

as

$$|\Phi^-\rangle = \frac{1}{\sqrt{2}}(|0\rangle_A|0\rangle_B - |1\rangle_A|1\rangle_B), \tag{19.202}$$

$$|\Phi^+\rangle = \frac{1}{\sqrt{2}}(|0\rangle_A|0\rangle_B + |1\rangle_A|1\rangle_B). \tag{19.203}$$

These four Bell states are orthonormal and span the four-dimensional Hilbert space. Hence, any general two-Qbit state can be expressed in terms of the Bell basis states as

$$|\chi_{AB}\rangle = C_{11}|\Psi^-\rangle + C_{12}|\Psi^+\rangle + C_{21}|\Phi^-\rangle + C_{22}|\Phi^+\rangle, \tag{19.204}$$

where C_{ij} are complex numbers. Since Bell states are constructed from the linear combinations of the original basis states in Eq. (19.193), C_{ij} are also linear combinations of α_{ij}.

If we consider the actions of the unitary operators **I**, **X**, **Y**, and **Z** on Bell states, we find

$$|\Psi^-\rangle = \mathbf{I}_A|\Psi^-\rangle, \tag{19.205}$$

$$|\Psi^+\rangle = \mathbf{Z}_A|\Psi^-\rangle, \qquad (19.206)$$

$$|\Phi^-\rangle = -\mathbf{X}_A|\Psi^-\rangle, \qquad (19.207)$$

$$|\Phi^+\rangle = \mathbf{Y}_A|\Psi^-\rangle. \qquad (19.208)$$

The subscript indicates on which Qbit the operator is acting on.

To see what all this means, consider a pair of entangled electrons produced in a common process and sent in opposite directions to observers A and B. Electrons are also produced such that if the spin of the electron going toward the observer A is up, $|0\rangle_A$, then the other one must be down, $|1\rangle_B$, and vice versa. Obviously, the pair is in one of the two Bell states given by Eqs. (19.199) and (19.200). For the sake of argument, let us assume that it is the symmetric one, that is,

$$|\Psi^+\rangle = \frac{1}{\sqrt{2}}(|0\rangle_A|1\rangle_B + |1\rangle_A|0\rangle_B). \qquad (19.209)$$

A measurement by one of the observers, say A, collapses the state function to either $|0\rangle_A|1\rangle_B$ or $|1\rangle_A|0\rangle_B$ with the equal probability of $\frac{1}{2}$. This removes all the uncertainty in any measurement that B will make on the second electron. In other words, when A makes a measurement, then any measurement of B will have zero information value, since all prior uncertainty will be gone. That is, despite the fact that there are two Qbits involved, this Bell state carries only one bit of classical information. In this experiment, separation of A and B could be as large as one desires. This immediately brings to mind action at a distance and the possibility of superluminal communication via quantum systems. As soon as A (usually called Alice) makes a measurement on her particle, spin of the electron at the location of B (usually called Bob) adjusts itself instantaneously. It would be a flagrant violation of causality if Alice could communicate with Bob by manipulating the spin of her particle. This worried none other than Einstein himself (see literature on EPR paradox). A way out of this conundrum is to notice that Bob has to make an independent measurement on his particle, and still he has to wait for Alice to send him the relevant information, which can only be done by classical means at subluminal speeds, so that he can decode whatever message was sent to him.

Let us review the problem once more. Alice and Bob share an entangled pair of electrons. Alice conducts a measurement on her electron. She has a 50/50 chance of finding its spin up or down. Let us say that she found spin up. Instantaneously, Bob's electron assumes the spin down state. However, Bob does not know this until he performs an independent measurement on his electron. He still thinks that he has 50/50 chance of seeing either spin, but Alice knows that the wave function has collapsed and that for sure he will get spin down. Bob conducts his measurement and indeed sees spin down. But to him this is normal, he has just seen one of the possibilities. Now, Alice calls Bob and tells him that he must have seen spin down. Actually, she could also call Bob before he makes his measurement and tell him that he will see spin down. In either case, it would be hard for Alice to impress Bob, since Bob will think that Alice has after all a 50/50 chance of guessing the right answer anyway.

To convince Bob, they share a collection of identically prepared entangled electrons. One by one, Alice measures her electrons and calls Bob and tells him that she observed the sequence ↑↓↓↑↓↓↑ ⋯ and that he should observe the sequence ↓↑↑↓↑↑↓ ⋯. When Bob measures his electrons, he now gets impressed by the uncanny precision of the Alice's prediction. This experiment can be repeated this time with Alice calling Bob after he conducted his measurements. Alice will still be able to predict Bob's results with 100% accuracy. In this experiment, quantum mechanics says that the wave function collapses instantaneously no matter how far apart Alice and Bob are. However, they still cannot use this to communicate superluminally. First of all, in order to communicate, they have to agree on a code. Since Alice does not know what sequence spins, she will get until she performs her measurements, they cannot do this beforehand. Once she does measure her set of particles, she is certain of what Bob will observe. Hence, she embeds the message into the sequence that she has observed by some kind of mapping. For Bob to be able to read the Alice's message, Alice has to send him that mapping, which can only be done through classical channels. Even if somebody intercepts Alice's message, it will be useless without the sequence that Bob has. Hence, Alice and Bob can establish **spy-proof** communication through entangled states. One of the main technical challenges in quantum information is **decoherence**, which is the destruction of the entangled states by interactions with the environment. It is for this reason that internal states like spin or stable energy states of atoms are preferred to construct Qbits, which are less susceptible to external influences by gravitational and electromagnetic interactions.

Example 19.7. *Quantum cryptology – The Vernam coding:* Alice wants to send Bob a message. Say the nine directions to open a safe, where the dial can be turned only one step, clockwise (CW) or counterclockwise (CCW), at a time. They agree to use binary notation, CW=1, CCW=0, to write the message as

$$101010011 \tag{19.210}$$

Afraid of the message being eavesdropped by a third party, Alice and Bob share nine ordered entangled electrons. Alice measures her particles one by one and obtains the sequence

$$010010110, \tag{19.211}$$

where 0 stands for spin up and 1 stands for spin down. Using this as a **key**, she adds the two sets of binary numbers according to the rules of modulo 2, which can be summarized as

$$0 + 0 = 0, \tag{19.212}$$

$$0 + 1 = 1 + 0 = 1, \tag{19.213}$$

$$1 + 1 = 0, \tag{19.214}$$

to obtain the coded text, that is, the **cryptograph** as

$$
\begin{array}{lccccccccc}
\text{message} & 1 & 0 & 1 & 0 & 1 & 0 & 0 & 1 & 1 \\
\text{key} & 0 & 1 & 0 & 0 & 1 & 0 & 1 & 1 & 0 \\
\text{cryptograph} & 1 & 1 & 1 & 0 & 0 & 0 & 1 & 0 & 1
\end{array} \tag{19.215}
$$

Now Alice sends the cryptograph to Bob via conventional means. Bob measures his ordered set of electrons to obtain the key and thus obtain the message by adding the key to the cryptograph with the same rules as

$$
\begin{array}{lccccccccc}
\text{cryptograph} & 1 & 1 & 1 & 0 & 0 & 0 & 1 & 0 & 1 \\
\text{key} & 0 & 1 & 0 & 0 & 1 & 0 & 1 & 1 & 0 \\
\text{message} & 1 & 0 & 1 & 0 & 1 & 0 & 0 & 1 & 1
\end{array} \tag{19.216}
$$

Since the key is a completely random sequence of zeros and ones, the cryptograph is also a completely random sequence of zeros and ones. Hence, it has no value whatsoever to anybody who intercepts it without the key that Bob has. This is called the **Vernam coding**, which cannot be broken. However, the problem that this procedure poses in practice is that for each message that Bob and Alice want to exchange they need a new key. That is, it can only be used once, which is also called a **one-time-pad** system. Another source for major concern is that during this process, the key may somehow be obtained by the eavesdropper. On top of all these, during the transmission, quantum systems are susceptible to interferences (decoherence), hence one needs algorithms to minimize and correct for errors. To attack these problems, various quantum-cryptographic methods, which are called **protocols**, have been developed (Audretsch [14, 16], Bruß and Leuchs [15], Trigg [17], and more references can be found at the back of this book).

19.3.8 Quantum Dense Coding

We have seen that entanglement does not help to communicate superluminally. However, it does play a very important role in quantum computing. Quantum dense coding is one example where we can send two bits of classical information by just using a single Qbit, thus potentially doubling the capacity of the information transfer channel. Furthermore, the communication is spy proof.

Let us say that Alice has two bits of secret information to be sent to Bob. Two bits of classical information can be coded in terms of a pair of binary digits as

$$00, \ 10, \ 01, \ 11. \tag{19.217}$$

First Alice and Bob agree to associate these digits with the following unitary transformations:

$$U_{00} = \mathbf{I}, \tag{19.218}$$

$$U_{01} = \mathbf{Z}, \tag{19.219}$$

$$U_{10} = \mathbf{X}, \tag{19.220}$$

$$U_{11} = \mathbf{Y}. \tag{19.221}$$

Then, Alice and Bob each receive one Qbit from an entangled pair prepared, say in the asymmetric Bell state $|\Psi^-\rangle$. Alice first performs a unitary transformation with the subscripts matching the pair of the digits that she is aiming to send Bob safely and then sends her Qbit to Bob as if it is a mail. Anyone who tempers with this Qbit will destroy the superposed state, hence the message. The unitary transformation that Alice has performed changes the bell state according to the corresponding formulas in Eqs. (19.205)–(19.208). When Bob receives the particle that Alice has sent, he makes a Bell state measurement on both particles to determine which one of the four states in Eqs. (19.205)–(19.208) it has assumed. The result tells him what Alice's transformation was, hence the pair of binary digits that she wanted to send him. Quantum dense coding was the first experimental demonstration of quantum communication. It was first realized by the Innsbruck group in 1996 [18]. The crucial part of these experiments is the measurement of the Bell state of the tangled pair without destroying the entanglement [3, 14, 16].

19.3.9 Quantum Teleportation

Consider that Alice has an object that she wants to send Bob. Aside from conventional means of transportation, she could somehow scan the object and send all the information contained to Bob. With a suitable technology, Bob then reconstructs the object. Unfortunately, such a technology neither exists nor can be constructed because of the no-cloning theorem of quantum mechanics. However, the next best thing, which guarantees Bob that his object will have the same properties as the original that Alice has, is possible. And most importantly, they do not have to know the properties of the original.

We start with Alice and Bob sharing a pair of entangled Qbits, A and B, which could be two electrons or two photons. We assume that the entangled pair is in the Bell state $|\Psi^-\rangle_{AB}$. A third Qbit, the **teleportee**, which is the same type of particle as A and B and is in the general superposed state

$$|\chi\rangle_T = \alpha_T |0\rangle_T + \beta_T |1\rangle_T, \tag{19.222}$$

is available to Alice. Any attempt to determine the exact state of $|\chi\rangle_T$ will destroy it. Our aim is to have Alice transport her Qbit, $|\chi\rangle_T$, to Bob without physically taking it there. In other words, Bob will have to reconstruct $|\chi\rangle_T$ at his location. In this process, due to the no-cloning theorem, we have to satisfy the following two conditions:

(i) At any time to neither Alice nor Bob, the exact state $|\chi\rangle_T$ is revealed.

(ii) At the end, the copy in Alice's hand has to be destroyed. Otherwise, there will be two copies of $|\chi\rangle_T$.

We now write the complete state vector of the three particles as

$$|\chi\rangle_{ABT} = |\Psi^-\rangle_{AB}|\chi\rangle_T \tag{19.223}$$

$$= \frac{1}{\sqrt{2}}(|0\rangle_A|1\rangle_B - |1\rangle_A|0\rangle_B)(\alpha_T|0\rangle_T + \beta_T|1\rangle_T). \tag{19.224}$$

We can express $|\chi\rangle_{ABT}$ in terms of the Bell states of the particles A and T held by Alice, that is in terms of the set [Eqs. (19.199), (19.200), (19.202), and (19.203)]

$$\{|\Psi^-\rangle_{AT}, |\Psi^+\rangle_{AT}, |\Phi^-\rangle_{AT}, |\Phi^+\rangle_{AT}\}. \tag{19.225}$$

The expansion coefficient for basis state

$$|\Psi^-\rangle_{AT} = \frac{1}{\sqrt{2}}(|0\rangle_A|1\rangle_T - |1\rangle_A|0\rangle_T) \tag{19.226}$$

is found by projecting $|\chi\rangle_{ABT}$ along $|\Psi^-\rangle_{AT}$ as

$$_{AT}\langle\Psi^-|\chi\rangle_{ABT}$$

$$= \frac{1}{2}(_T\langle1|_A\langle0| - _T\langle0|_A\langle1|)(|0\rangle_A|1\rangle_B - |1\rangle_A|0\rangle_B)(\alpha_T|0\rangle_T + \beta_T|1\rangle_T) \tag{19.227}$$

$$= \frac{1}{2}[_T\langle1|_A\langle0|0\rangle_A|1\rangle_B - _T\langle1|_A\langle0|1\rangle_A|0\rangle_B - _T\langle0|_A\langle1|0\rangle_A|1\rangle_B \tag{19.228}$$

$$+ _T\langle0|_A\langle1|1\rangle_A|0\rangle_B](\alpha_T|0\rangle_T + \beta_T|1\rangle_T)$$

$$= \frac{1}{2}[_T\langle1|\ 1\rangle_B + _T\langle0|\ 0\rangle_B](\alpha_T|0\rangle_T + \beta_T|1\rangle_T) \tag{19.229}$$

$$= \frac{1}{2}[(\alpha_T)_T\langle1|0\rangle_T|1\rangle_B + (\beta_T)_T\langle1|1\rangle_T|1\rangle_B \tag{19.230}$$

$$+ (\alpha_T)_T\langle0|0\rangle_T|0\rangle_B + (\beta_T)_T\langle0|1\rangle_T|0\rangle_B]$$

$$= \frac{1}{2}[\alpha_T|0\rangle_B + \beta_T|1\rangle_B], \tag{19.231}$$

where we have used the Dirac bra–ket notation (Chapter 5) and the orthogonality relations

$$\langle 0|0 \rangle = \langle 1|1 \rangle = 1, \tag{19.232}$$

$$\langle 0|1 \rangle = \langle 1|0 \rangle = 0, \tag{19.233}$$

for both A and T. Similarly, evaluating the other coefficients, we write the complete state vector $|\chi\rangle_{ABT}$ as

$$|\chi\rangle_{ABT} = |\Psi^-\rangle_{AB}|\chi\rangle_T \tag{19.234}$$

$$= \frac{1}{2}\{|\Psi^-\rangle_{AT}(\alpha_T|0\rangle_B + \beta_T|1\rangle_B)$$

$$+ |\Psi^+\rangle_{AT}(-\alpha_T|0\rangle_B + \beta_T|1\rangle_B)$$

$$+ |\Phi^-\rangle_{AT}(+\beta_T|0\rangle_B + \alpha_T|1\rangle_B)$$

$$+ |\Phi^+\rangle_{AT}(-\beta_T|0\rangle_B + \alpha_T|1\rangle_B)\}. \tag{19.235}$$

Now, Alice performs a Bell state measurement on her particles A and T that collapses $|\chi\rangle_{ABT}$ into one of the four Bell states in Eq. (19.225) with the equal probability of $\frac{1}{4}$. In no way this process provides Alice any information about $|\chi\rangle_T$, that is the probability amplitudes α_T and β_T, but the particle B, which Bob holds, jumps into a state connected to whatever the Bell state that Alice has observed, that is, one of

$$+\alpha_T|0\rangle_B + \beta_T|1\rangle_B = \alpha_T\begin{pmatrix}1\\0\end{pmatrix}_B + \beta_T\begin{pmatrix}0\\1\end{pmatrix}_B = \begin{pmatrix}\alpha_T\\\beta_T\end{pmatrix}_B, \tag{19.236}$$

$$-\alpha_T|0\rangle_B + \beta_T|1\rangle_B = -\alpha_T\begin{pmatrix}1\\0\end{pmatrix}_B + \beta_T\begin{pmatrix}0\\1\end{pmatrix}_B = \begin{pmatrix}-\alpha_T\\\beta_T\end{pmatrix}_B, \tag{19.237}$$

$$+\beta_T|0\rangle_B + \alpha_T|1\rangle_B = \beta_T\begin{pmatrix}1\\0\end{pmatrix}_B + \alpha_T\begin{pmatrix}0\\1\end{pmatrix}_B = \begin{pmatrix}\beta_T\\\alpha_T\end{pmatrix}_B, \tag{19.238}$$

$$-\beta_T|0\rangle_B + \alpha_T|1\rangle_B = -\beta_T\begin{pmatrix}1\\0\end{pmatrix}_B + \alpha_T\begin{pmatrix}0\\1\end{pmatrix}_B = \begin{pmatrix}-\beta_T\\\alpha_T\end{pmatrix}_B. \tag{19.239}$$

None of these states is yet the desired $|\chi\rangle_T$. However, due to the Bell state measurement that Alice has performed on $|\chi\rangle_{AT}$, they are related to the state of the transportee $|\chi\rangle_T$:

$$|\chi\rangle_T = \alpha_T|0\rangle_T + \beta_T|1\rangle_T = \begin{pmatrix}\alpha_T\\\beta_T\end{pmatrix}_T, \tag{19.240}$$

by the same unitary transformation that Alice has observed, that is one of

$$+\alpha_T|0\rangle_B + \beta_T|1\rangle_B = \mathbf{I}|\chi\rangle_T, \tag{19.241}$$

$$-\alpha_T|0\rangle_B + \beta_T|1\rangle_B = -\mathbf{Z}|\chi\rangle_T, \tag{19.242}$$

$$+\beta_T|0\rangle_B + \alpha_T|1\rangle_B = \mathbf{X}|\chi\rangle_T, \tag{19.243}$$

$$-\beta_T|0\rangle_B + \alpha_T|1\rangle_B = \mathbf{Y}|\chi\rangle_T, \tag{19.244}$$

where the operators are defined in Eqs. (19.149)–(19.152). At this point, only Alice knows which transformation to use. That is which Bell state the complete state function $|\chi\rangle_{ABT}$ has collapsed to. She calls and gives Bob the necessary two bit information, that is the two digits of the subscripts of the operator $\mathbf{U}_{ij}$, which they have agreed upon before the experiment to have the components

$$U_{00} = \mathbf{I}, \tag{19.245}$$

$$U_{01} = -\mathbf{Z}, \tag{19.246}$$

$$U_{10} = \mathbf{X}, \tag{19.247}$$

$$U_{11} = \mathbf{Y}, \tag{19.248}$$

corresponding to the four Bell states, $|\Psi^-\rangle_{AT}$, $|\Psi^+\rangle_{AT}$, $|\Phi^-\rangle_{AT}$, and $|\Phi^+\rangle_{AT}$, respectively. Now, Bob uses the inverse transformation $\mathbf{U}_{ij}^{-1}$, on particle B that he has and obtains an exact replica of the transportee $|\chi\rangle_T$. For example, if Alice observes $|\Phi^+\rangle_{AT}$ when $|\chi\rangle_{ABT}$ collapses, the two digits she gives Bob is 11, then Bob operates on the particle B that he has with $\mathbf{Y}^{-1}$, which is in state $-\beta_T|0\rangle_B + \alpha_T|1\rangle_B$, to obtain $|\chi\rangle_T$ as

$$\mathbf{Y}^{-1}(-\beta_T|0\rangle_B + \alpha_T|1\rangle_B) = \alpha_T|0\rangle_B + \beta_T|1\rangle_B = \mathbf{Y}^{-1}\mathbf{Y}|\chi\rangle_T = |\chi\rangle_T. \tag{19.249}$$

Let us summarize what has been accomplished:

(i) Since the teleportee is the same type of particle as A and B, we have obtained an exact replica of $|\chi\rangle_T$ at Bob's location who has the particle B.

(ii) The original, $|\chi\rangle_T$, that Alice had is destroyed. That is neither the particle A nor the other particle that Alice is left with, which means the transportie whose properties has been transferred to B at Bob's location, is in state $|\chi\rangle_T$.

(iii) If somebody had spied on Alice and Bob, the two bit information, that is, the subscripts of the unitary transformation, would have no practical value without the particle B that Bob holds.

Notice that Alice and Bob has communicated a wealth of quantum information hidden in the complex amplitudes of

$$|\chi\rangle_T = \alpha_T|0\rangle_T + \beta_T|1\rangle_T. \tag{19.250}$$

by just sending 2 bits. Neither Alice nor Bob has gained knowledge about the exact nature of the state $|\chi\rangle_T$. In other words, neither of them knows what

α_T and β_T are. This intriguing experiment was first realized by the Innsbruck group in 1997 [19].

REFERENCES

1. Shannon, C.E. (1948). A mathematical theory of communication. *Bell Syst. Tech. J.* 27: 379–423, 623–656.

2. Shannon, C.E. and Weaver, W. (1949). *The Mathematical Theory of Communication*. Urbana, IL: The University of Illinois Press.

3. Roederer, J.G. (2005). *Information and Its Role in Nature*. Berlin: Springer-Verlag.

4. Bather, J.A. (2000). *Decision Theory, An Introduction to Programming and Sequential Decisions*. Chichester: Wiley.

5. Miller, I. and Miller, M. (2004). *John E. Freund's Mathematical Statistics With Applications*, 7e. Upper Saddle River, NJ: Pearson Prentice Hall.

6. Osborne, M.J. (2004). *An Introduction to Game Theory*. New York: Oxford University Press.

7. Szabo, G. and Fath, G. (2007). Evolutionary games on graphs. *Phys. Rep.* 446: 97–216.

8. Basu, K. (2007). The Traveler's Dilemma. *Sci. Am.* 296 (6): 68.

9. Csermely, P. (2006). *Weak Links, Stabilizers of Complex Systems from Proteins to Social Networks*. Berlin: Springer-Verlag.

10. Mermin, N.D. (2007). *Quantum Computer Science*. Cambridge: Cambridge University Press.

11. Josza, R. (1998). *Introduction to Quantum Computation and Information* (ed. H.-K. Lo, S. Popescu, and T. Spiller). Singapore: Word Scientific.

12. Feynman, R., Leighton, R.B., and Sands, M. (1966). *The Feynman Lectures on Physics*, vol. 3. Reading, MA: Addison-Wesley.

13. Ghirardi, G. (2004). *Sneaking a Look at God's Cards*. Princeton, NJ: Princeton University Press.

14. Audretsch, J. (2007). *Entangled Systems*. Weinheim: Wiley-VCH.

15. Bruß, D. and Leuchs, G. (eds) (2007). *Lectures on Quantum Information*. Weinheim: Wiley-VCH.

16. Audretsch, J. (ed.) (2006). *Entangled World: The Fascination of Quantum Information and Computation*. Weinheim: Wiley-VCH.

17. Trigg, G.L. (ed.) (2005). *Mathematical Tools for Physicists*. Weinheim: Wiley-VCH.

18. Mattle, K., Weinfurter, H., Kwiat, P.G., and Zeilinger, A. (1966). Dense coding in experimental quantum communication. *Phys. Rev. Lett.* 76: 4656–4659.

19. Bouwmeester, D., Pan, J.W., Mattle, K. et al. (1997). Experimental quantum teleportation. *Nature*, 390: 575–579.

PROBLEMS

1. Consider two chance events, A and B, both with two possibilities, where $p(i, j)$ is the joint probability of the ith and jth possibilities occurring for A and B, respectively. We write the entropy of information for the joint event as

$$H(A, B) = -\sum_{i=1}^{2}\sum_{j=1}^{2} p(i,j)\log_2 p(i,j).$$

For the individual events, we write

$$H(A) = -\sum_{i=1}^{2}\left[\sum_{j=1}^{2} p(i,j)\log_2 \sum_{j=1}^{2} p(i,j)\right]$$

and

$$H(B) = -\sum_{j=1}^{2}\left[\sum_{i=1}^{2} p(i,j)\log_2 \sum_{i=1}^{2} p(i,j)\right].$$

Show that the following inequality holds:

$$H(A, B) \le H(A) + H(B).$$

Also show that the equality is true for independent events, where

$$p(A, B) = p(A)p(B).$$

Apply this to the case where two fair coins are tossed independent of each other.

2. Two chance events (A, B) have the following joint probability distribution:

$A \downarrow \backslash B \rightarrow$	1	2	3	4
1	$\frac{1}{8}$	$\frac{1}{16}$	$\frac{1}{16}$	$\frac{1}{4}$
2	$\frac{1}{16}$	$\frac{1}{8}$	$\frac{1}{16}$	0
3	$\frac{1}{32}$	$\frac{1}{32}$	$\frac{1}{16}$	0
4	$\frac{1}{32}$	$\frac{1}{32}$	$\frac{1}{16}$	0

Find

$$H(A/B), \ H(B/A), \ H(A), \ H(B), \ H(A, B),$$

and interpret your results.

3. Analyze the following payoff matrices [Eq. (19.53)] for zero-sum games for both players for optimum strategies:

(i)

Player A

		a_1	a_2
Player B	b_1	6	−3
	b_2	7	11

.

Note that it is foolish for the player B to choose b_1, since b_2 yields more regardless of what A decides. In such cases, we say b_2 dominates b_1 and hence discard strategy b_1. Finding **dominant strategies** may help simplifying payoff matrices.

(ii)

Player A

		a_1	a_2	a_3
	b_1	−2	5	−1
Player B	b_2	3	4	6
	b_3	−1	−5	11

.

4. Consider the following zero-sum game:

Player A

		a_1	a_2
Player B	b_1	2	−4
	b_2	−3	1

.

(i) Find the randomized strategy that A has to follow to minimize maximum expected loss.

(ii) Find the randomized strategy that B has to follow to maximize minimum expected gain.

5. The two-player competition game is defined as follows: Both players simultaneously choose a whole number from 0 to 3. Both players win the smaller of the two numbers in points. In addition, the player who chose the larger number gives up 2 points to the other player. Construct the payoff matrix and identify the Nash equilibria.

6. Identify the Nash equilibria in the following payoff matrices:

(i)

$1 \downarrow \backslash 2 \rightarrow$	1	2	3
1	$(0,0)$	$(20,35)$	$(5,10)$
2	$(35,20)$	$(0,0)$	$(5,15)$
3	$(10,5)$	$(15,5)$	$(10,10)$

,

(ii)

$1 \downarrow \backslash 2 \rightarrow$	1	2	3
1	$(40,30)$	$(20,35)$	$(50,10)$
2	$(3,5)$	$(0,0)$	$(5,30)$
3	$(10,5)$	$(15,5)$	$(20,10)$

.

7. Two drivers on a road have two strategies each, to drive either on the left or on the right with the payoff matrix

$1 \downarrow \backslash 2 \rightarrow$	Drive on the left	Drive on the right
Drive on the left	$(100,100)$	$(0,0)$
Drive on the right	$(0,0)$	$(100,100)$

,

where the payoff 100 means no crash and 0 means crash. Identify Nash equilibria for

(i) the pure strategy game,

(ii) the mixed strategy game with the probabilities (50%, 50%).

Which one of these is stable.

8. In a 2-Cbit operation, find the action of the operator

$$\frac{1}{2}(\mathbf{I} + \mathbf{Z}_1 \mathbf{Z}_0)$$

on the 2-Cbit states

$$|0\rangle |0\rangle, \ |0\rangle |1\rangle, \ |1\rangle |0\rangle, \ |1\rangle |1\rangle.$$

9. In a 2-Cbit operation, find the action of the operator

$$\frac{1}{2}(\mathbf{I} - \mathbf{Z}_1 \mathbf{Z}_0)$$

on the 2-Cbit states

$$|0\rangle |0\rangle, \ |0\rangle |1\rangle, \ |1\rangle |0\rangle, \ |1\rangle |1\rangle.$$

10. Prove the following operator representation of the operator $\mathbf{S}_{10}$, which exchanges the values of Cbits 1 and 0:

$$\mathbf{S}_{10} = \frac{1}{2}[(\mathbf{I} + \mathbf{Z}_1\mathbf{Z}_0) + \mathbf{X}_1\mathbf{X}_0(\mathbf{I} - \mathbf{Z}_1\mathbf{Z}_0)],$$

or

$$\mathbf{S}_{10} = \frac{1}{2}[\mathbf{I} + \mathbf{Z}_1\mathbf{Z}_0 + \mathbf{X}_1\mathbf{X}_0 + \mathbf{I} - \mathbf{Y}_1\mathbf{Y}_0],$$

where

$$\mathbf{Y} = \mathbf{XZ}.$$

11. Another useful operation on 2-Cbit systems is the reversible XOR or the controlled-NOT or in short c-NOT, gate executed by the operator $\mathbf{C}_{10}$ as

$$\mathbf{C}_{10}|x_1\rangle\,|x_0\rangle = \mathbf{X}_0^{x_1}|x_1\rangle\,|x_0\rangle.$$

The task of $\mathbf{C}_{10}$ is to flip the value of the target bit, $|x_0\rangle$, whenever the control bit, $|x_1\rangle$, has the value 1.

(i) Show that $\mathbf{C}_{10}$ can be constructed from single Cbit operators as

$$\mathbf{C}_{10} = \frac{1}{2}(\mathbf{I} + \mathbf{X}_0) + \frac{1}{2}\mathbf{Z}_1(\mathbf{I} - \mathbf{X}_0),$$

or as

$$\mathbf{C}_{10} = \frac{1}{2}(\mathbf{I} + \mathbf{Z}_1) + \frac{1}{2}\mathbf{X}_0(\mathbf{I} - \mathbf{Z}_1).$$

(ii) Show that the c-NOT operator can be generalized as

$$\mathbf{C}_{ij} = \frac{1}{2}(\mathbf{I} + \mathbf{X}_j) + \frac{1}{2}\mathbf{Z}_i(\mathbf{I} - \mathbf{X}_j),$$

or as

$$\mathbf{C}_{ij} = \frac{1}{2}(\mathbf{I} + \mathbf{Z}_i) + \frac{1}{2}\mathbf{X}_j(\mathbf{I} - \mathbf{Z}_i).$$

(iii) What is the effect of interchanging $\mathbf{X}$ and $\mathbf{Z}$?

12. The Hadamard operator:

$$\mathbf{H} = \frac{1}{\sqrt{2}}(\mathbf{X} + \mathbf{Z}) = \frac{1}{\sqrt{2}}\begin{pmatrix} 1 & 1 \\ 1 & -1 \end{pmatrix},$$

is classically meaningless. Show that it takes the Cbit states $|0\rangle$ and $|1\rangle$ into two classically meaningless superpositions:

$$\frac{1}{2}(|0\rangle \pm |1\rangle).$$

13. Verify the following operator relations:

$$\mathbf{X}^2 = \mathbf{I},$$

$$\mathbf{XZ} = -\mathbf{ZX},$$

$$\mathbf{HX} = \frac{1}{\sqrt{2}}(\mathbf{I} + \mathbf{ZX}),$$

$$\mathbf{HXH} = \mathbf{Z},$$

$$\mathbf{HZH} = \mathbf{X}.$$

14. Find the effect of the operator

$$\mathbf{C}_{01} = (\mathbf{H}_1\mathbf{H}_0)\mathbf{C}_{10}(\mathbf{H}_1\mathbf{H}_0),$$

on Cbits. Using the relations

$$\mathbf{HXH} = \mathbf{Z}, \ \ \mathbf{HZH} = \mathbf{X},$$

and

$$\mathbf{C}_{ij} = \frac{1}{2}(\mathbf{I} + \mathbf{X}_j) + \frac{1}{2}\mathbf{Z}_i(\mathbf{I} - \mathbf{X}_j),$$

also show that

$$\mathbf{C}_{ji} = (\mathbf{H}_i\mathbf{H}_j)\mathbf{C}_{ij}(\mathbf{H}_i\mathbf{H}_j).$$

This seemingly simple relation has remarkable uses in quantum computers [10].

15. Show that the Bell states

$$|\Psi^-\rangle = \frac{1}{\sqrt{2}}(|0\rangle_A|1\rangle_B - |1\rangle_A|0\rangle_B),$$

$$|\Psi^+\rangle = \frac{1}{\sqrt{2}}(|0\rangle_A|1\rangle_B + |1\rangle_A|0\rangle_B),$$

and

$$|\Phi^-\rangle = \frac{1}{\sqrt{2}}(|0\rangle_A|0\rangle_B - |1\rangle_A|1\rangle_B),$$

$$|\Phi^+\rangle = \frac{1}{\sqrt{2}}(|0\rangle_A|0\rangle_B + |1\rangle_A|1\rangle_B),$$

are unit vectors and that they are also orthogonal to each other.

16. Show the following Bell state transformations:

$$|\Psi^-\rangle = \mathbf{I}_A|\Psi^-\rangle,$$

$$|\Psi^+\rangle = \mathbf{Z}_A|\Psi^-\rangle,$$

$$|\Phi^-\rangle = -\mathbf{X}_A|\Psi^-\rangle,$$
$$|\Phi^+\rangle = \mathbf{Y}_A|\Psi^-\rangle.$$

The subscript A indicates the Qbit that the operator is acting on.

17. Given the states

$$|\chi\rangle_{AB} = \frac{1}{\sqrt{2}}(|0\rangle_A|1\rangle_B - |1\rangle_A|0\rangle_B)$$

and

$$|\chi\rangle_T = (\alpha_T|0\rangle_T + \beta_T|1\rangle_T),$$

verify that the state

$$|\chi\rangle_{ABT} = |\chi\rangle_{AB}|\chi\rangle_T = \frac{1}{\sqrt{2}}(|0\rangle_A|1\rangle_B - |1\rangle_A|0\rangle_B)(\alpha_T|0\rangle_T + \beta_T|1\rangle_T)$$

can be written in terms of the Bell states,

$$\{|\Psi^-\rangle_{AT}, |\Psi^+\rangle_{AT}, |\Phi^-\rangle_{AT}, |\Phi^+\rangle_{AT}\},$$

of the particles A and T as

$$
\begin{aligned}
|\chi\rangle_{ABT} = |\chi\rangle_{AB}|\chi\rangle_T \\
= \frac{1}{2}\{ & |\Psi^-\rangle_{AT}(\alpha_T|0\rangle_B + \beta_T|1\rangle_B) \\
+ & |\Psi^+\rangle_{AT}(-\alpha_T|0\rangle_B + \beta_T|1\rangle_B) \\
+ & |\Phi^-\rangle_{AT}(+\beta_T|0\rangle_B + \alpha_T|1\rangle_B) \\
+ & |\Phi^+\rangle_{AT}(-\beta_T|0\rangle_B + \alpha_T|1\rangle_B)\}.
\end{aligned}
$$

18. Verify the following transformations used in quantum teleportation and discuss what happens if Bob makes a mistake and disturbs the state of the particle B he holds?

$$+\alpha_T|0\rangle_B + \beta_T|1\rangle_B = \mathbf{I}|\chi\rangle_T,$$
$$-\alpha_T|0\rangle_B + \beta_T|1\rangle_B = -\mathbf{Z}|\chi\rangle_T,$$
$$+\beta_T|0\rangle_B + \alpha_T|1\rangle_B = \mathbf{X}|\chi\rangle_T,$$
$$-\beta_T|0\rangle_B + \alpha_T|1\rangle_B = \mathbf{Y}|\chi\rangle_T.$$

Further Reading

MATHEMATICAL METHODS TEXTBOOKS:

Appel, W. (2007). *Mathematics for Physics and Physicists*. Princeton, NJ: Princeton University Press.

Arfken, G.B. and Weber, H.J. (2005). *Mathematical Methods of Physics*, 6e. Boston, MA: Elsevier.

Bayin, S.S. (2018). *Mathematical Methods in Science and Engineering*, 2e. Hoboken, NJ: Wiley.

Butkov, E. (1968). *Mathematical Physics*. New York: Addison-Wesley.

Boas, M.L. (2006). *Mathematical Methods in the Physical Sciences*, 3e. Hoboken, NJ: Wiley.

Byron, F.W. Jr. and Fuller, R.W. (1992). *Mathematics of Classical and Quantum Physics*. New York: Dover Publications.

Dennery, P. and Krzywicki, A. (1995). *Mathematics for Physics*. New York: Dover Publications.

Hassani, S. (2002). *Mathematical Physics*, 2e. New York: Springer-Verlag.

Hassani, S. (2008). *Mathematical Methods: For Students of Physics and Related Fields*, 2e. New York: Springer-Verlag.

Hildebrand, F.B. (1992). *Methods of Applied Mathematics*, 2nd reprint edition. New York: Dover Publications.

Kreyszig, E. (2011). *Advanced Engineering Mathematics*, 10e. Wiley-VCH.

Kusse, B.R. and Westwig, E.A. (2006). *Mathematical Physics: Applied Mathematics For Scientists and Engineers*, 2e. Weinheim: Wiley-VCH.

Lebedev, N.N., Skalskaya, I.P., and Uflyand, Y.S. (1965). *Problems of Mathematical Physics*. Englewood Cliffs, NJ: Prentice-Hall.

Margenau, M. and Murphy, M. (eds.) (1964). *The Mathematics of Physics and Chemistry*. Princeton, NJ: Van Nostrand.

Mathews, J. and Walker, R.W. (1970). *Mathematical Methods of Physics*, 2e. Menlo Park, CA: Addison-Wesley.

Morse, P.M. and Feshbach, H. (1953). *Methods of Theoretical Physics*. New York: McGraw-Hill.

Rektorys, K. (1994). *Survey of Applicable Mathematics Volumes I and II*, 2nd revised edition. Berlin: Springer-Verlag.

Spiegel, M.R. (1971). *Advanced Mathematics for Engineers and Scientists: Schaum's Outline Series in Mathematics*. New York: McGraw-Hill.

Szekerez, P. (2004). *A Course in Modern Mathematical Physics: Group, Hilbert Space and Differential Geometry*. New York: Cambridge University Press.

Trigg, T. (ed.) (2005). *Mathematical Tools for Physicists*. Weinheim: Wiley-VCH.

Weber, H.J. and Arfken, G.B. (2003). *Essential Mathematical Methods for Physicists*. San Diego, CA: Academic Press.

Woolfson, M.M. and Woolfson, M.S. (2007). *Mathematics for Physics*. Oxford: Oxford University Press.

Wyld, H.W. (1976). *Mathematical Methods for Physics*. Addison Wesley.

MATHEMATICAL METHODS WITH COMPUTERS:

Kelly, J.J. (2007). *Graduate Mathematical Physics, With Mathematica Supplements+CD*. Weinheim: Wiley-VCH.

Wang, F.Y. (2006). *Physics With Maple: The Computer Algebra Resource for Mathematical Methods in Physics*. Weinheim: Wiley-VCH.

Yang, W.Y., Cao, W., Chung, T.-S., and Morris, J. (2005). *Applied Numerical Methods Using Matlab*. Wiley-Interscience.

CALCULUS/ADVANCED CALCULUS:

Apostol, T.M. (1971). *Mathematical Analysis*. Reading, MA: Addison-Wesley. 4th printing.

Buck, R.C. (1965). *Advanced Calculus*. New York: McGraw-Hill.

Franklin, P.A. (1940). *A Treatise on Advanced Calculus*. New York: Wiley.

Kaplan, W. (1984). *Advanced Calculus*, 3e. Reading, MA: Addison-Wesley.

Thomas, G.B. Jr. and Finney, R.L. (2000). *Thomas' Calculus*, alternate edition. Boston, MA: Addison-Wesley.

Titchmarsh, E.C. (1939). *The Theory of Functions*. New York: Oxford University Press.

Whittaker, E.T. and Watson, G.N. (1958). *A Course on Modern Analysis*. New York: Cambridge University Press.

LINEAR ALGEBRA AND ITS APPLICATIONS:

Anton, H. and Rorres, C. (2011). *Elementary Linear Algebra With Supplemental Applications*, 10e. Wiley.

Benjamin, A., Chartrand, G., and Zhang, P. (2015). *The Fascinating World of Graph Theory*. Princeton University Press, reprint edition.

Ficken, F.A. (2015). *The Simplex Method of Linear Programming*. Dover Publication.

Gantmacher, F.R. (1960). *The Theory of Matrices*. New York: Chelsea Publishing Company.

Gass, S.I. (2013). *An Illustrated Guide to Linear Programming*. Dover Publication.

Hoffman, K. and Kunze, R. (1971). *Linear Algebra*, 2e. Upper Saddle River, NJ: Prentice Hall.

Lang, S. (1966). *Linear Algebra*. Reading, MA: Addison-Wesley.

Lipton, R.J. and Regan, K.W. (2014). *Quantum Algorithms via Linear Algebra: A Primer*. The MIT Press.

Roman, S. (1997). *Introduction to Coding and Information Theory*. Springer.

Sultan, A. (2011). *Linear Programming: An Introduction With Applications*. CreateSpace Independent Publishing Platform.

Thie, P.R. and Keough, G.E. (2008). *An Introduction to Linear Programing and Game Theory*, 3e. Wiley-InterScience.

Trappe, W. and Washington, L.C. (2005). *Introduction to Cryptography With Coding Theory*, 2e. Pearson.

Wilson, R.A. (2012). *Introduction to Graph Theory*, 5e. Pearson.

COMPLEX CALCULUS:

Ahlfors, L.V. (1966). *Complex Analysis*. New York: McGraw-Hill.

Brown, J.W. and Churchill, R.V. (1995). *Complex Variables and Applications*. New York: McGraw-Hill.

Gamow, G. (1988). *One Two Three... Infinity: Facts and Speculations of Science*. Dover Publications.

Kyrala, A. (1972). *Applied Functions of a Complex Variable*. New York: Wiley.

Saff, E.B. and Snider, A.D. (2003). *Fundamentals of Complex Analysis With Applications to Engineering and Science*. Upper Saddle River, NJ: Prentice Hall.

DIFFERENTIAL EQUATIONS:

Ince, E.L. (1958). *Ordinary Differential Equations*. New York: Dover Publications.

Kells, L.M. (1965). *Elementary Differential Equations*, 6e. McGraw-Hill.

Murphy, G.M. (1960). *Ordinary Differential Equations and Their Solutions*. Princeton, NJ: Van Nostrand.

Nagle, R.K., Saff, E.B., and Snider, A.D. (2004). *Fundamentals of Differential Equations and Boundary Value Problems*. Boston, MA: Addison-Wesley.

Ross, S.L. (1984). *Differential Equations*, 3e. New York: Wiley.

CALCULUS OF VARIATIONS:

Akhiezer, N.I. (1962). *The Calculus of Variations*. New York: Blaisdell.

Kot, M. (2014). *A First Course in Calculus of Variations*. American Mathematical Society.

MacCluer, C.R. (2013). *The Calculus of Variations, Mechanics, Control and Other Applications*. Dover Publications, reprint edition.

Wan, F.Y.M. (1995). *Introduction to the Calculus of Variations and its Applications*. New York: Chapman and Hall.

Weinstock, R. (1974). *The Calculus of Variations With Applications to Physics and Engineering*. Dover Publications, revised edition.

FOURIER SERIES, INTEGRAL TRANSFORMS AND SIGNAL PROCESSING:

Andrews, L.C. and Shivamoggi, B.K. (1999). *Integral Transforms for Engineers*. SPIE Publications.

Candy, J.V. (2005). *Model Based Signal Processing*. New York: Wiley.

Debnath, L. and Bhatta, D. (2014). *Integral Transforms and Their Applications*, 3e. Chapman and Hall/CRC.

McClellan, J.H., Schafer, R.W., and Yoder, M.A. (2003). *Signal Processing First*. London: Pearson/Prentice Hall.

Tolstov, G.P. (2017). *Fourier Series*. New York: Dover Publications.

SERIES AND SPECIAL FUNCTIONS:

Andrews, L.C. (1985). *Special Functions for Engineers and Applied Mathematicians*. Macmillan.

Artin, E. (1964). *The Gamma Function*. New York: Holt, Rinehart and Winston.

Bell, W.W. (1968). *Special Functions for Scientists and Engineers*. Princeton, NJ: D. Van Nostrand.

Bowman, N. (1958). *Introduction to Bessel Functions*. Dover Publications.

Bromwich, T.J.I. (1991). *An Introduction to the Infinite Series*. New York: Chelsea Publishing Company.

Churchill, R.V. (1963). *Fourier Series and Boundary Value Problems*. New York: McGraw-Hill.

Erdelyi, A., Oberhettinger, M.W., and Tricomi, F.G. (1981). *Higher Transcendental Functions*, vol. I. New York: Krieger.

Lebedev, N.N. (1965). *Special Functions and Their Applications*. Englewood Cliffs, NJ: Prentice-Hall.

Watson, G.N. (1962). *A Treatise on the Theory of Bessel Functions*, 2e. London: Cambridge University Press.

MATHEMATICAL TABLES:

Dwight, H.B. (1961). *Tables of Integrals and Other Mathematical Data*, 4e. New York: Macmillan.

Erdelyi, A. (1954). *Tables of Integral Transforms*, vol. 2. McGraw-Hill.

McCollum, P.A. and Brown, B.F. (1965). *Laplace Transform Tables and Theorems*. New York: Holt, Rinehart and Winston.

Zwillinger, D. (ed.) (2018). *Standard Mathematical Tables and Formulas*. 33e. Chapman and Hall/CRC.

CLASSICAL MECHANICS:

Bradbury, T.C. (1968). *Theoretical Mechanics*. New York: Wiley.

Feynman, R., Leighton, R.B., and Sands, M. (1966). *The Feynman Lectures on Physics*, vol. 1. Reading, MA: Addison-Wesley.

Goldstein, H., Poole, C., and Safko, J. (2002). *Classical Mechanics*, 3e. San Francisco, CA: Addison-Wesley.

Hauser, W. (1966). *Introduction to Principles of Mechanics*. Reading, MA: Addison-Wesley, first printing.

Marion, J.B. (1970). *Classical Dynamics of Particles and Systems*, 2e. New York: Academic Press.

QUANTUM MECHANICS:

Bell, J.S. (1989). *Speakable and Unspeakable in Quantum Mechanics*. Cambridge: Cambridge University Press.

Dirac, P.A.M. (1982). *The Principals of Quantum Mechanics*, 4e. Oxford: Clarendon Press.

Feynman, R., Leighton, R.B., and Sands, M. (1966). *The Feynman Lectures on Physics*, vol. 3. Reading, MA: Addison-Wesley.

Gasierowicz, S. (2003). *Quantum Physics*, 3e. Hoboken, NJ: Wiley.

Merzbacher, E. (1998). *Quantum Mechanics*. New York: Wiley.

ELECTROMAGNETIC THEORY:

Feynman, R., Leighton, R.B., and Sands, M. (1966). *The Feynman Lectures on Physics*, vol. 2. Reading, MA: Addison-Wesley.

Griffiths, D.J. (1998). *Introduction to Electrodynamics*, 3e. Benjamin Cummings.

Inan, U.S. and Inan, A.S. (1998). *Engineering Electrodynamics*. Upper Saddle River, NJ: Prentice Hall.

Inan, U.S. and Inan, A.S. (1999). *Electromagnetic Waves*. Upper Saddle River, NJ: Prentice Hall.

Jackson, D.J. (1998). *Classical Electrodynamics*, 3e. Wiley.

Lorrain, P. and Corson, D.R. (1970). *Electromagnetic Fields and Waves*, 2e. Freeman.

PROBABILITY THEORY:

Andel, J. (2001). *Mathematics of Chance*. New York: Wiley.

Gnedenko, B.V. (1973). *The Theory of Probability*, 2nd printing. Moscow: MIR Publishers.

Grimmett, G.R. and Stirzaker, D.R. (2001). *Probability and Random Processes*, 3e. Oxford: Clarendon.

Harris, B. (1966). *Theory of Probability*. Reading, MA: Addison-Wesley.

Jaynes, E.T. and Bretthurst, E.L. (2003). *Probability Theory: The Logic of Science*. Cambridge University Press.

Kolmogorov, A.N. (1950). *Foundations of the Theory of Probability*. New York: Chelsea Publishing Company.

Medina, P.K. and Merino, S. (2003). *Mathematical Finance and Probability*. Basel: Birkhauser Verlag.

Pathria, R.K. (1984). *Statistical Mechanics*. Oxford: Pergamon Press.

Rozanov, Y.A. (2013). *Probability Theory: A Concise Course*. Dover Publications.

Sivia, D.S. and Skilling, J. (2006). *Data Analysis: A Bayesian Tutorial*, 2e. New York: Oxford University Press.

Todhunter, I. (1949). *A History of the Theory of Probability From the Time of Pascal to Laplace*. New York: Chelsea Publishing Company.

Venkatesh, S.S. (2012). *The Theory of Probability, Explorations and Applications*. Cambridge University Press.

Wilks, J. (1961). *The Third Law of Thermodynamics*. London: Oxford University Press.

Ziemer, R.E. (1997). *Elements of Engineering Probability and Statistics*. Upper Saddle River, NJ: Prentice Hall.

INFORMATION THEORY:

Audretsch, A. (ed.) (2006). *Entangled World: The Fascination of Quantum Information and Computation*. Weinheim: Wiley-VCH.

Audretsch, J. (2007). *Entangled Systems*. Weinheim: Wiley-VCH.

Barnett, S.M. (2009). *Quantum Information*. Oxford University Press.

Basu, K. (2007). The Traveler's Dilemma. *Sci. Am.* 296 (6): 68.

Bather, J.A. (2000). *Decision Theory, An Introduction to Programming and Sequential Decisions*. Chichester: Wiley.

Benenti, G., Casati, G., Rossini, D., and Strini, G. (2019). *Principles of Quantum Computation and Information: A Comprehensive Textbook*, 2e. World Scientific.

Berger, J.O. (1993). *Statistical Decision Theory and Bayesian Analysis*, 2e. Springer-Verlag.

Berman, P., Doolen, G.D., Mainieri, R., and Tsifrinovich, V.I. (1998). *Introduction to Quantum Computers*. World Scientific.

Bouwmeester, D., Pan, J.W., Mattle, K. et al. (1997). Experimental quantum teleportation. *Nature* 390: 575–579.

Brubeta, Brubeta and Leuchs, Leuchs (eds.) (2007). *Lectures on Quantum Information*. Weinheim: Wiley-VCH.

Cover, T.M. and Thomas, J.A. (2006). *Elements of Information Theory*, 2e. Hoboken, NJ: Wiley.

Csermely, P. (2006). *Weak Links, Stabilizers of Complex Systems from Proteins to Social Networks*. Berlin: Springer-Verlag.

Elsbach, K.D. (2006). *Organizational Perception Management*. Routledge.

Ghirardi, G. (2004). *Sneaking a Look at God's Cards*. Princeton, NJ: Princeton University Press.

Hayashi, M. (2006). *Quantum Information, An Introduction*, Springer-Verlag.

Haykin, S. (1999). *Neural Networks, A Comprehensive Foundation*. Upper Saddle River, NJ: Prentice Hall.

Jones, G.A. and Jones, J.M. (2006). *Information and Coding Theory*. London: Springer.

Josza, R. (1998). *Introduction to Quantum Computation and Information* (ed. Lo, Lo, and Spiller, Spiller). Singapore: Word Scientific.

Kantor, F.W. (1977). *Information Mechanics*. Wiley.

Lala, R.K. (2019). *Quantum Computing, A Beginner's Introduction*. McGraw-Hill Education.

Mattle, K., Weinfurter, H., Kwiat, P.G., and Zeilinger, A. (1966). Dense coding in experimental quantum communication. *Phys. Rev. Lett.* 76: 4656–4659.

McMahon, D. (2007). *Quantum Computing Explained*. Hoboken, NJ: Wiley-IEEE Computer Society Press.

Miller, I. and Miller, M. (2004). *John E. Freund's Mathematical Statistics With Applications*, 7e. Upper Saddle River, NJ: Pearson Prentice Hall.

Morsch, O. (2008). *Quantum Bits and Quantum Secrets: How Quantum Physics Is Revolutionazing Codes and Computers*. Weinheim: Wiley-VCH.

Myerson, R.B. (1991). *Game Theory, Analysis of Conflict*. Cambridge, MA: Harvard University Press.

Nielson, M.A. and Chuang, I.L. (2000). *Quantum Computation and Quantum Information*. Cambridge University Press.

Osborne, M.J. (2004). *An Introduction to Game Theory*. New York: Oxford University Press.

Parmigiani, G. and Inoue, L. (2009). *Decision Theory: Principles and Approaches*. Wiley.

Peters, E.E. (1999). *Complexity, Risk, and Financial Markets*. New York: Wiley.

Pierce, J.R. (1980). *An Introduction to Information Theory, Symbols, Signals and Noise*, 2e. Dover Publications.

Rieffel, E. and Polak, W. (2011). *Quantum Computing: A Gentle Introduction*. The MIT Press.

Roederer, J.G. (2005). *Information and Its Role in Nature*. Berlin: Springer-Verlag.

Roman, S. (1996). *Introduction to Coding and Information Theory*. Springer-Verlag.

Shannon, C.E. (1948). A mathematical theory of communication. *Bell Syst. Tech. J.* 27: 379–423, 623–656.

Shannon, C.E. and Weaver, W. (1949). *The Mathematical Theory of Communication*. Urbana, IL: The University of Illinois Press.

Stapp, H.P. (2004). *Mind Matter and Quantum Mechanics*, 2e. Berlin: Springer-Verlag.

Stolze, J. and Suter, D. (2014). *Quuantum Computing: A short Course From Theory to Experiment*. Weinheim: Wiley-VCH.

Stone, J.V. (2015). *Information Theory, A Tutorial Introduction*. Sebtel Press.

Szabo, G. and Fath, G. (2007). Evolutionary games on graphs. *Physics Reports* 446: 97–216.

Yanovski, N.S. and Mannucci, M.A. (2008). *Quantum Computing for Computer Scientists*. Cambridge University Press.

Zeilinger, A. (1997). Experimental quantum teleportation. *Nature* 390: 575–579.

Zeilinger, A. (1998). Quantum information. *Physics World* 11 (3), 35–40.

Zeilinger, A. (2000). Quantum Teleportation. *Sci. Am.* 282 (4): 32.

INDEX